Transmission and Distribution Electrical Engineering

Third edition

Dr C. R. Bayliss CEng FIET and
B. J. Hardy ACGI CEng FIET

ELSEVIER
BUTTERWORTH
HEINEMANN

AMSTERDAM · BOSTON · HEIDELBERG · LONDON · NEW YORK · OXFORD
PARIS · SAN DIEGO · SAN FRANCISCO · SINGAPORE · SYDNEY · TOKYO

Newnes is an imprint of Elsevier

Newnes

Newnes is an imprint of Elsevier
The Boulevard, Langford Lane, Kidlington, Oxford, OX5 1GB
30 Corporate Drive, Suite 400, Burlington, MA 01803, USA

First edition 1996
Second edition 1999
Third edition 2007
Reprinted 2007, 2008

Notice
No responsibility is assumed by the publisher for any injury and/or damage to persons
or property as a matter of products liability, negligence or otherwise, or from any use
or operation of any methods, products, instructions or ideas contained in the material
herein. Because of rapid advances in the medical sciences, in particular, independent
verification of diagnoses and drug dosages should be made

British Library Cataloguing in Publication Data
A catalogue record for this book is available from the British Library

Library of Congress Cataloging-in-Publication Data
A catalog record for this book is available from the Library of Congress

ISBN: 978-0-7506-6673-2

For information on all Newnes publications
visit our website at www.elsevierdirect.com

Printed and bound in *Hungary*

08 09 10   10 9 8 7 6 5 4 3

Transmission and Distribution Electrical Engineering

Contents

About the authors

Colin Bayliss gained a first class honours degree in Electrical and Electronic Engineering at Nottingham University and went on to receive a PhD in Materials Science. He has worked on major power projects both at home in the UK and throughout the world with client, contractor and consultancy organizations. He was appointed Engineering Director by the Channel Tunnel main contractors (Transmanche Link – TML) during the last 2 years of that project's construction, having been involved previously in the earlier design stages. He is currently the Major Projects and Engineering Main Board Director of the United Kingdom Atomic Energy Authority (UKAEA). Colin also holds honorary professorships with both Birmingham University and the University of the Highlands and Islands, and is a non-Executive Director of Cogent Sector Skills Council.

Brian Hardy read Electrical Engineering at Imperial College, London and received a graduate apprenticeship in the UK supply industry, where he went on to serve for 15 years, filling a series of senior engineering posts. Moving to consultancy he managed design offices in Europe and the Middle East and power system projects in a number of other countries, including the World Bank funded rehabilitation of the whole of the Tanzanian power system. After serving as joint leader of the Anglo-French design team for all the electrical aspects of the channel tunnel, he was appointed as Engineering Director of the Balfour Beatty Group's design and construct company. Brian went on to assist the Group in the establishment of a team to take advantage the newly developing private finance partnership market. The success of this team led to its evolving into Balfour Beatty Capital, a leading player in the UK PFI/PPP scene. One such project is a share in the joint venture management of the London Underground's power system. Brian acted as technical advisor to this JV before retiring to part-time consultancy.

Contributors

The preparation of a book covering such a wide range of topics would not have been possible without contributions and advice from major manufacturers, electrical supply utilities, contractors, academics and consulting engineers. Indeed, encouragement for the preparation of this book has come from the Institution of Electrical Engineers (IEE) Transmission and Distribution Professional Group, P7, under the chairmanship of David Rigden (Hawker Siddeley Switchgear) and John Lewis (Scottish Power). The names of the major contributors to the first and second editions are listed below.

D. Auckland	Professor of Electrical Engineering, University of Manchester
A. Baker	Principal Geotechnical Engineer, Balfour Beatty Projects and Engineering
R. H. Barnes	Associate Director and Principal Systems Analyst, Engineering and Power Development Consultants (EPDC)
P. Bennett	Director, Centre for Software Engineering
N. Bird	Director, Balfour Beatty Cruickshank Ltd
K. Blackmore	Senior Engineer, Interference Technology International
L. Blake	Yorkshire Electricity Group
S. A. Bleazard	Reyrolle Limited, Tyne and Wear
D. Boulu	Principal Engineer, Tractabel, Belgium
D. Brady	General Manager, Optimal Software Ltd
D. Brown	Director, BICC, Wrexham
J. Finn	Principal Engineer, Reyrolle Projects (formerly Power Systems Project Manager, TML)
H. Grant	Deputy Chief Design Engineer, Parsons Peebles Transformers, Edinburgh
G. Harris	Livingstone Hire
M. R. Hill	Marketing Director, Bowthorpe EMP Ltd

P. Hindle	Principal Engineer, GEC-Alsthom T&D Protection & Control
I. Johnston	Senior Software Engineer, Centre for Software Engineering
C. Lau	Senior Data Transmission and Control Engineer, TML
F. J. Liptrot	Technical Director, Allied Insulators
G. Little	Balfour Kilpatrick, Hackbridge, London
I. E. Massey	Senior Civil Engineer, Balfour Beatty Projects and Engineering
T. Mennel	Head of Engineering, EMMCO, Merlin Gerin
E. Meyer	Control Engineer, Technip, Paris
A. Munro	Design Engineer, Peebles Power Transformers, Edinburgh
R. Monk	Senior Applications Engineer, GEC-Alsthom T&D Protection & Control
D. Moore	Principal Engineer, National Grid Company (formerly Ewbank Preece Consulting Engineers)
P. G. Newbery	Technical Director, Cooper Bussman (formerly Hawker Fusegear)
G. Orawski	Consultant Engineer, Balfour Beatty Power
S. D. Pugh	Senior SCADA Engineer, Centre for Software Development
D. Rigden	Director, Hawker Siddeley Switchgear
A. Smith	Design Draughtsman, EPDC
M. Swinscale	Principal Technical Engineer, Furze, Nottingham
M. Tearall	Senior Building Services Engineer, Wimpey Major Projects
M. Teliani	Senior Systems Engineer, Engineering and Power Development Consultants Ltd (EPDC)
A. Thomas	Senior Communications Engineer, Ewbank Preece Consulting Engineers

In addition to the above, whose work and input remains a key element in *Transmission and Distribution Electrical Engineering*, the authors acknowledge with thanks the major additional contributions to the third edition from the following contributors:

T. E. Charlton	Managing Director, Strategy and Solutions and Managing Director, Earthing Measurements
T. M. Endersby	Systems Design Engineer, EDF Energy Powerlink
A. J. Maddox	Chief EMC Consultant, ERA Technology Ltd
L. Manning	Transmission Technology Manager PTS, ABB Ltd., Stone
U. Manmadhan	Technology Manager, Utility Automation ABB Ltd., UK
A. J. Marchbank	Technical Development Director, Haden Building Management Ltd
D. Mason	Systems Design Consultant, EDF Energy Powerlink
P. G. Newbery	Consultant (formerly Technical Director) Cooper Bussman

W. C. Sayer	Design Manager (Overhead lines), Balfour Beatty Power Networks Ltd.
A. R. Smith	Design Engineer, Cruickshanks Limited
R. A. Smith	Project Manager (Power Intakes), EDF Energy Powerlink
V. F. Temple	Technical Specialist, Balfour Beatty Capital, and Technical Advisor to EDF Energy Powerlink.
E. Tolster	Technical Manager, Communication products and systems, ABB Ltd., UK.

Preface

FIRST AND SECOND EDITIONS

This book covers the major topics likely to be encountered by the transmission and distribution power systems engineer engaged upon international project works. Each chapter is self-contained and gives a useful practical introduction to each topic covered. The book is intended for graduate or technician level engineers and bridges the gap between learned university theoretical textbooks and detailed single topic references. It therefore provides a practical grounding in a wide range of transmission and distribution subjects. The aim of the book is to assist the project engineer in correctly specifying equipment and systems for his particular application. In this way manufacturers and contractors should receive clear and unambiguous transmission and distribution equipment or project enquiries for work and enable competitive and comparative tenders to be received. Of particular interest are the chapters on project, system and software management since these subjects are of increasing importance to power systems engineers. In particular the book should help the reader to understand the reasoning behind the different specifications and methods used by different electrical supply utilities and organizations throughout the world to achieve their specific transmission and distribution power system requirements. The second edition includes updates and corrections, together with the addition of two extra major chapters covering distribution planning and power system harmonics.

C. R. Bayliss

THIRD EDITION

As this book is particularly designed to help those running projects to correctly specify, the approach has been to make frequent reference to applicable national and international standards. This is because basing specifications on such standards will ensure consistency of bids, and generally will enable bidders to offer the most economic prices for technically compliant offers. The skill of the project engineer and manager comes in applying the standards most effectively to the particular requirements of the project.

In the period between the publication of the second and third editions the work of updating electrical standards by the committees of the International Electrotechnical Commission (IEC) has proceeded apace. Moreover, within Europe the development of European Norms (ENs) has resulted in the revision and alignment of many national standards – often to result in complete consistency with the corresponding IEC standard. This has meant that every chapter in *Transmission and Distribution Electrical Engineering* has had to be carefully checked to ensure that the frequent references to standards are correct and the relevant content updated where appropriate. Developments in the recognized approach to earthing and bonding have resulted in a complete re-write of the relevant chapter, and legislation changes have necessitated updates to the chapter on electromagnetic compatibility. Recent trends in protection equipment and SCADA have needed to be mentioned and developments in the requirements of both users and public utilities in the area of power system quality have justified the expansion of the coverage of this increasingly important area of supply system engineering.

Achieving these changes, so as to continue to make this the standard reference text for practitioners in this field, would not have been possible without the valued assistance and input from the colleagues and specialists listed as contributors.

C. R. Bayliss and B. J. Hardy

1 System Studies

1.1 INTRODUCTION

This chapter describes three main areas of transmission and distribution network analysis; namely load flow, system stability and short circuit analysis. Such system studies necessitate a thorough understanding of network parameters and generating plant characteristics for the correct input of system data and interpretation of results. A background to generator characteristics is therefore included in Section 1.3.

It is now recognized that harmonic analysis is also a major system study tool. This is discussed separately in Chapter 24. Reliability studies are considered in Chapter 23.

The analysis work, for all but the simplest schemes, is carried out using tried and proven computer programs. The application of these computer methods and the specific principles involved are described by the examination of some small distribution schemes in sufficient detail to be applicable to a wide range of commercially available computer software. The more general theoretical principles involved in load flow and fault analysis data collection are explained in Chapter 26.

1.2 LOAD FLOW

1.2.1 Purpose

A load flow analysis allows identification of real and reactive power flows, voltage profiles, power factor and any overloads in the network. Once the network parameters have been entered into the computer database the analysis allows the engineer to investigate the performance of the network under a variety of outage conditions. The effect of system losses and power factor

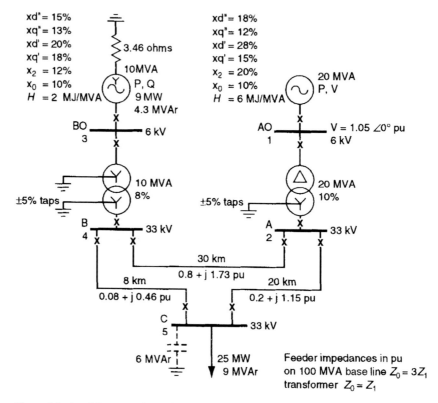

xd" = 15%
xq" = 13%
xd' = 20%
xq' = 18%
x_2 = 12%
x_0 = 10%
H = 2 MJ/MVA

3.46 ohms

10MVA
P, Q
9 MW
4.3 MVAr

xd" = 18%
xq" = 12%
xd' = 28%
xq' = 15%
x_2 = 20%
x_0 = 10%
H = 6 MJ/MVA

20 MVA
P, V

BO
3
6 kV

AO
1
V = 1.05 ∠0° pu
6 kV

±5% taps

10 MVA
8%

±5% taps

20 MVA
10%

B
4
33 kV

A
2
33 kV

30 km
0.8 + j 1.73 pu

8 km
0.08 + j 0.46 pu

20 km
0.2 + j 1.15 pu

C
5
33 kV

6 MVAr

25 MW
9 MVAr

Feeder impedances in pu
on 100 MVA base line $Z_0 = 3Z_1$
transformer $Z_0 = Z_1$

Figure 1.1   Load flow sample study single-line diagram

correction, the need for any system reinforcement and confirmation of economic transmission can then follow.

1.2.2 Sample study

1.2.2.1 Network single-line diagram

Figure 1.1 shows a simple five busbar 6 kV generation and 33 kV distribution
network for study. Table 1.1 details the busbar and branch system input data
associated with the network. Input parameters are given here in a per unit (pu)
format on a 100 MVA base. Different programs may require input data in different formats, for example per cent impedance, ohmic notation, etc. Please
refer to Chapter 26, for the derivation of system impedance data in different
formats from manufacturers' literature. The network here is kept small in
order to allow the first-time user to become rapidly familiar with the procedures for load flows. Larger networks involve a repetition of these procedures.

Table 1.1   Load flow sample study busbar and branch input data

Bus data

Bus name	Bus number	Bus type	Voltage pu	Angle	Gen MW	MVAr	Load MW	MVAr	Shunt L or C pu
Slack AO	1	1	1.05	0.0	0	0	0	0	0
A	2	2	1.0	0.0	0	0	0	0	0
BO	3	3	1.0	0.0	9	4.3	0	0	0
B	4	4	1.0	0.0	0	0	0	0	0
C	5	5	1.0	0.0	0	0	25	9	0

Branch data

Bus	Bus	Circ #	Rpu 100 MVA base	Xpu 100 MVA base	Bpu 100 MVA base	Tap ratio
1	2		–	0.5	–	1.02
2	4		0.8	1.73	–	0.0
2	5		0.2	1.15	–	0.0
3	4		–	0.8	–	1.02
4	5		0.08	0.46	–	0.0

1.2.2.2 Busbar input database

The busbars are first set up in the program by name and number and in some cases by zone. Bus parameters are then entered according to type. A 'slack bus' is a busbar where the generation values, P (real power in MW) and Q (reactive power in MVAr), are unknown; there will always be one such busbar in any system. Busbar AO in the example is entered as a slack bus with a base voltage of 6.0 kV, a generator terminal voltage of 6.3 kV (1.05 pu) and a phase angle of 0.0° (a default value). All load values on busbar AO are taken as zero (again a default value) due to unknown load distribution and system losses.

A 'P, Q generator bus' is one where P and Q are specified to have definite values. If, for example, P is made equal to zero we have defined the constant Q mode of operation for a synchronous generator. Parameters for busbar BO in the example may be specified with base voltage 6.0 kV, desired voltage 6.3 kV and default values for phase angle (0.0°), load power (0.0 MW), load reactive power (0.0 MVAr), shunt reactance (0.0 MVAr) and shunt capacitance (0.0 pu). Alternatively, most programs accept generator busbar data by specifying real generator power and voltage. The program may ask for reactive power limits to be specified instead of voltage since in a largely reactive power network you cannot 'fix' both voltage and reactive power – something has to 'give way' under heavy load conditions. Therefore busbar BO may be specified with generator power 9.0 MW, maximum and minimum reactive power as 4.3 MVAr and transient or subtransient reactance in per unit values.

  These reactance values are not used in the actual load flow but are entered in anticipation of the need for subsequent fault studies. For the calculation of oil circuit breaker breaking currents or for electromechanical protection relay operating currents it is more usual to take the generator transient reactance values. This is because the subtransient reactance effects will generally disappear within the first few cycles and before the circuit breaker or protection has operated. Theoretically, when calculating maximum circuit breaker making currents subtransient generator reactance values should be used. Likewise for modern, fast (say 2 cycle) circuit breakers, generator breakers and with solid state fast-relay protection where accuracy may be important, it is worth checking the effect of entering subtransient reactances into the database. In reality, the difference between transient and subtransient reactance values will be small compared to other system parameters (transformers, cables, etc.) for all but faults close up to the generator terminals.

  A 'load bus' has floating values for its voltage and phase angle. Busbar A in the example has a base voltage of 33 kV entered and an unknown actual value which will depend upon the load flow conditions.

1.2.2.3 Branch input database

Branch data is next added for the network plant (transformers, cables, overhead lines, etc.) between the already specified busbars. Therefore from busbar A to busbar B the 30 km, 33 kV overhead line data is entered with resistance 0.8 pu, reactance 1.73 pu and susceptance 0.0 pu (unknown in this example and 0.0 entered as a default value).

  Similarly for a transformer branch such as from busbar AO to A data is entered as resistance 0.0 pu, reactance 0.5 pu (10% on 20 MVA base rating = 50% on 100 MVA base or 0.5 pu), susceptance 0.0 pu (unknown but very small compared to inductive reactance), load limit 20 MVA, from bus AO voltage 6 kV to bus A voltage 33.66 kV (1.02 pu taking into account transformer ± 5% taps). Tap ranges and short-term overloads can be entered in more detail depending upon the exact program being used.

1.2.2.4 Saving data

When working at the computer it is always best regularly to save your files both during database compilation as well as at the end of the procedure when you are satisfied that all the data has been entered correctly. Save data onto the hard disk and make backups for safe keeping to suitable alternative media (e.g. CD, USB flash drive). Figure 1.2 gives a typical computer printout for the bus and branch data files associated with this example.

CYMFLOW   3:16 pm, Monday, December 2, 199

        NORMAL LOADFLOW 1   FILE : ZTRAIN1

Bus Data
....

	IDENT.		VOLTAGE	LOAD		SHUNT	GENERATION		CONTR. BUS	
			BASE KV	MW	NP	ACT/REAC	MW	QMAK	# NAME ZN	
NUM	NAME	ZN	INIT KV/DEG	MVAR	NQ	(PU)		QMIN	KV SPE	

....

1AO	1	6.000	0.00 0.000	0.0000	16.77	0.00		
	SW	6.300	0.00 0.000	0.0000		0.00		
		0.000						
->	2A	1 1 (R,X,B,CAP)	0.00000 0.50000	0.0000		20.0MVA FX	0.98
2A	1	33.000	0.00 0.000	0.0000				
	LOAD	33.698	0.00 0.000	0.0000				
		-4.574						
->	1AO	1 1M (R,X,B,CAP)	0.00000 0.50000	0.0000		20.0MVA FX	0.98
->	4B	1 1 (R,X,B,CAP)	0.80000 1.73000	0.0000		0.0AMP	
->	5C	1 1 (R,X,B,CAP)	0.20000 1.15000	0.0000		0.0AMP	
3BO	1	6.000	0.00 0.000	0.0000	9.00	4.30		
	GEN	5.819	0.00 0.000	0.0000		4.30		
		-3.875						
->	4B	1 1 (R,X,B,CAP)	0.00000 0.80000	0.0000		10.0MVA FX	0.98
4B	1	33.000	0.00 0.000	0.0000				
	LOAD	31.552	0.00 0.000	0.0000				
		-8.416						
->	2A	1 1M (R,X,B,CAP)	0.80000 1.73000	0.0000		0.0AMP	
->	3BO	1 1M (R,X,B,CAP)	0.00000 0.80000	0.0000		10.0MVA FX	0.98
->	5C	1 1 (R,X,B,CAP)	0.08000 0.46000	0.0000		0.0AMP	
5C	1	33.000	25.00 0.000	0.0000				
	LOAD	30.473	9.00 0.000	0.0000				
		-12.176						
->	2A	1 1M (R,X,B,CAP)	0.20000 1.15000	0.0000		0.0AMP	
->	4B	1 1M (R,X,B,CAP)	0.08000 0.46000	0.0000		0.0AMP	

CYMFLOW   3:16 pm, Monday, December 2, 199

        NORMAL LOADFLOW 1   FILE : ZTRAIN1

COMPLETE BRANCH DATA REPORT
....

FROM	TO	C#	R(p.u)	X(p.u)	B(p.u)	kV FROM	kV TO	Tap	Phase Shift

....

1	2	1	0.0	.50000	0.0	6.0000	33.660	.98039	
3	4	1	0.0	.80000	0.0	6.0000	33.660	.98039	
2	4	1	.80000	1.7300	0.0				
2	5	1	.20000	1.1500	0.0				
4	5	1	.08000	.46000	0.0				

Figure 1.2   Load flow sample study busbar and branch computer input data files

1.2.2.5 *Solutions*

Different programs use a variety of different mathematical methods to solve the load flow equations associated with the network. Some programs ask the user to specify what method they wish to use from a menu of choices (Newton–Raphson, Gauss–Seidel, Fast decoupled with adjustments, etc.). A full understanding of these numerical methods is beyond the scope of this book. It is worth noting, however, that these methods start with an initial approximation and then follow a series of iterations or steps in order to eliminate the unknowns and 'home in' on the solutions. The procedure may converge satisfactorily in which case the computer continues to iterate until the difference between successive iterations is sufficiently small. Alternatively, the procedure may not converge or may only converge extremely slowly. In these cases it is necessary to re-examine the input data or alter the iteration in some way or, if desired, stop the iteration altogether.

The accuracy of the solution and the ability to control round-off errors will depend, in part, upon the way in which the numbers are handled in the computer. In the past it was necessary to ensure that the computer used was capable of handling accurate floating-point arithmetic, where the numbers are represented with a fixed number of significant figures. Today these can be accepted as standard. It is a most important principle in numerical work that all sources of error (round off, mistakes, nature of formulae used, approximate physical input data) must be constantly borne in mind if the 'junk in equals junk out' syndrome is to be avoided. A concern that remains valid in selecting computing equipment is the need to ensure that the available memory is adequate for the size of network model under consideration.

Some customers ask their engineering consultants or contractors to prove their software by a Quality Assurance Audit which assesses the performance of one software package with another for a single trial network.

Figure 1.3 gives typical busbar and branch reports resulting from a load flow computation. It is normal to present such results by superimposing them in the correct positions on the single-line diagram as shown in Fig. 1.4. Such a pictorial representation may be achieved directly with the more sophisticated system analysis programs. The network single-line diagram is prepared using a computer graphics program (Autocad, Autosketch, GDS, etc.) and the load flow results transferred using data exchange files into data blocks on the diagram.

1.2.2.6 *Further studies*

The network already analysed may be modified as required, changing loads, generation, adding lines or branches (reinforcement) or removing lines (simulating outages).

Consider, for example, removing or switching off either of the overhead line branches running from busbars A to C or from B to C. Non-convergence

CYMFLOW   3:16 pm, Monday, December 2, 199

NORMAL LOADFLOW 1   FILE : ZTRAIN1

Complete Bus Report

#	NAME	ZN	VOLTAGE kV	PU	DEGREE	LOAD MW	MVAR	MVA	GENERATION MW	MVAR	MVA
	1AO	1	6.300	1.050	0.0	0.00	0.00	0.00	16.77	10.94	20.0
	2A	1	33.698	1.021	-4.6	0.00	0.00	0.00	0.00	0.00	0.0
	3BO	1	5.819	0.970	-3.9	0.00	0.00	0.00	9.00	4.29	9.9
	4B	1	31.552	0.956	-8.4	0.00	0.00	0.00	0.00	0.00	0.0
	5C	1	30.473	0.923	-12.2	25.00	9.00	26.57	0.00	0.00	0.0

CYMFLOW   2:59 pm, Monday, December 2, 199

LOADFLOW 1   FILE : ZTRAIN1

Branch Report

#	BRANCH IDENTIFICATION NAME	ZN	#	NAME	ZN	#C		POWER FLOW MW	MVAR	MVA	LOSSES MW	MVAR	TAP RATIO
	1AO	1	·	2A	1	1	FX	16.77	10.94	20.02	0.000	1.817	0.980
	2A	1	·	1AO	1	1M	FX	-16.77	-9.12	19.08	0.000	1.817	0.980
	2A	1	·	4B	1	1		4.63	1.83	4.97	0.190	0.410	
	2A	1	·	5C	1	1		12.12	7.29	14.14	0.383	2.205	
	3BO	1	·	4B	1	1	FX	9.00	4.29	9.97	0.000	0.845	0.980
	4B	1	·	2A	1	1M		-4.44	-1.42	4.66	0.190	0.410	
	4B	1	·	3BO	1	1M	FX	-9.00	-3.44	9.63	0.000	0.845	0.980
	4B	1	·	5C	1	1		13.43	4.87	14.29	0.179	1.027	
	5C	1	·	2A	1	1M		-11.73	-5.09	12.79	0.383	2.205	
	5C	1	·	4B	1	1	1M	-13.25	-3.84	13.80	0.179	1.027	

CYMFLOW   3:16 pm, Monday, December 2, 199

NORMAL LOADFLOW 1   FILE : ZTRAIN1

Fixed Tap Transformer Report

#	BRANCH IDENTIFICATION NAME	ZN	#	NAME	ZN	#C	LEFT TAP KV	P.U.	RIGHT TAP KV	P.U.	RATIO
	1AO	1		2A	1	1	6.000	1.0000	33.660	1.0200	0.9804
	3BO	1		4B	1	1	6.000	1.0000	33.660	1.0200	0.9804

Figure 1.3   Load flow sample study base case busbar and branch computer report

of the load flow numerical analysis occurs because of a collapse of voltage at busbar C.

If, however, some reactive compensation is added at busbar C – for example a 33 kV, 6 MVAr (0.06 pu) capacitor bank – not only is the normal load flow improved, but the outage of line BC can be sustained. An example of a computer generated single-line diagram describing this situation is given in Fig. 1.5. This is an example of the beauty of computer aided system analysis. Once

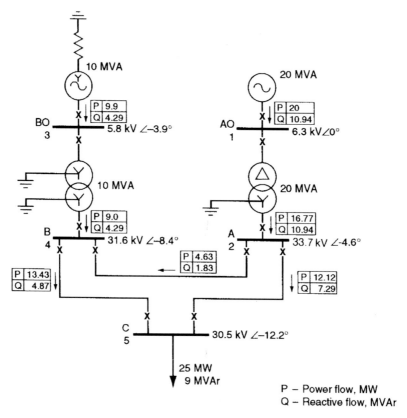

Figure 1.4   Load flow sample study. Base case load flow results superimposed upon single-line diagram

the network is set up in the database the engineer can investigate the performance of the network under a variety of conditions. Refer to Chapter 26 'Fundamentals', Section 26.8.5 regarding Reactive Compensation principles.

1.3 SYSTEM STABILITY

1.3.1 Introduction

The problem of stability in a network concerns energy balance and the ability to generate sufficient restoring forces to counter system disturbances. Minor disturbances to the system result in a mutual interchange of power between the machines in the system acting to keep them in step with each other and to maintain a single universal frequency. A state of equilibrium is retained between the total mechanical power/energy-input and the electrical power/energy-output by

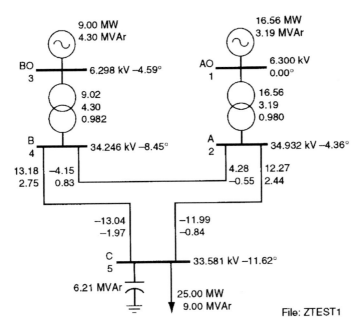

Figure 1.5   Load flow sample study. Computer generated results superimposed upon single-line diagram-reactive compensation added

natural adjustment of system voltage levels and the common system frequency. There are three regimes of stability:

(a) Steady state stability describes the ability of the system to remain in synchronism during minor disturbances or slowly developing system changes such as a gradual increase in load as the 24-hour maximum demand is approached.
(b) Transient stability is concerned with system behaviour following an abrupt change in loading conditions as could occur as a result of a fault, the sudden loss of generation or an interconnecting line, or the sudden connection of additional load. The duration of the transient period is in the order of a second. System behaviour in this interval is crucial in the design of power systems.
(c) Dynamic stability is a term used to describe the behaviour of the system in the interval between transient behaviour and the steady state region. For example, dynamic stability studies could include the behaviour of turbine governors, steam/fuel flows, load shedding and the recovery of motor loads, etc.

The response of induction motors to system disturbances and motor starting is also thought of as a stability problem. It does not relate specifically to the ability of the system to remain in synchronism.

This description is divided into two parts: the first deals with the analytical nature of synchronous machine behaviour and the different types of stability; the second deals with the more practical aspects of data collection and interpretation of transient stability study results with case studies to illustrate the main points and issues. The complexity of such analysis demands the use of computing techniques and considerable data collection.

1.3.2 Analytical aspects

1.3.2.1 *Vector diagrams and load angle*

Figure 1.6a shows the synchronous generator most simply represented on a per phase basis by an internally generated voltage (E) and an internal reactance (X). The internal voltage arises from the induction in the stator by the rotating magnetic flux of the rotor. The magnitude of this voltage is determined by the excitation of the field winding. The reactance is the synchronous reactance of the machine for steady state representation and the transient and subtransient reactance for the representation of rapid changes in operating conditions. The terminals of the generator (i.e. beyond E and X) are assumed to be connected to an 'infinite' busbar which has the properties of constant voltage and frequency with infinite inertia such that it can absorb any output supplied by the generator. In practice, such an infinite busbar is never obtained. However, in a highly interconnected system with several generators the system voltage and frequency are relatively insensitive to changes in the operating conditions of one machine. The generator is synchronised to the infinite busbar and the bus voltage (U) is unaffected by any changes in the generator parameters (E) and (X). The vector diagrams associated with this generator arrangement supplying current (I) with a lagging power factor ($\cos \phi$) are shown in Figs 1.6b to 1.6e for low electrical output, high electrical output, high excitation operation and low excitation operation, respectively. The electrical power output is $UI \cos \phi$ per phase. The angle θ between the voltage vectors E and U is the load angle of the machine. The load angle has a physical significance determined by the electrical and mechanical characteristics of the generator and its prime mover. A stroboscope tuned to the supply frequency of the infinite busbar would show the machine rotor to appear stationary. A change in electrical loading conditions such as that from Figs 1.6b to 1.6c would be seen as a shift of the rotor to a new position. For a generator the load angle corresponds to a shift in relative rotor position in the direction in which the prime mover is driving the machine. The increased electrical output of the generator from Figs 1.6b to 1.6c is more correctly seen as a consequence of an increased mechanical output of the prime mover. Initially this acts to accelerate the rotor and thus to increase the load angle. A new state of equilibrium is then reached where electrical power output matches prime mover input to the generator.

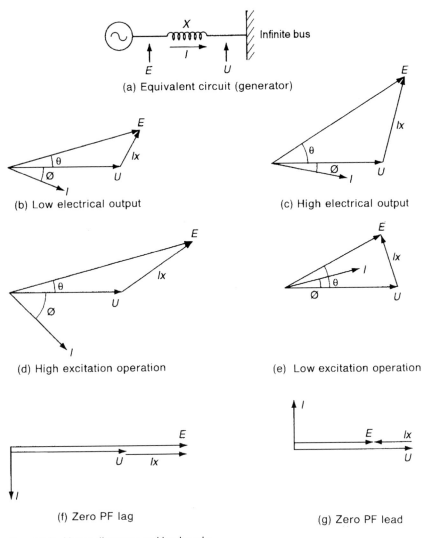

(a) Equivalent circuit (generator)

(b) Low electrical output

(c) High electrical output

(d) High excitation operation

(e) Low excitation operation

(f) Zero PF lag

(g) Zero PF lead

Figure 1.6   Vector diagrams and load angle

Figures 1.6d and 1.6e show the effect of changing the field excitation of the generator rotor at constant electrical power output and also with no change in electrical power output from the Fig. 1.6b condition – that is $UI \cos\phi$ is unchanged. An increase in (E) in Fig. 1.6d results in a larger current (I) but a more lagging power factor. Similarly, in Fig. 1.6e the reduction in (E) results in a change in power factor towards the leading quadrant. The principle effect of a variation in generator internal voltage is therefore to change the power factor of the machine with the larger values of (E) resulting in lagging power factors and the smaller values for (E) tending towards leading power factors. A secondary effect, which is important in stability studies, is also the change in load angle. The increased value of (E) shown in Fig. 1.6d (high excitation operation) has a smaller load angle compared to Fig. 1.6e (low excitation operation) for the same electrical power. Figures 1.6f and 1.6g show approximately zero lag and lead power factor operation where there is no electrical power output and the load angle is zero.

1.3.2.2 The power/load angle characteristic

Figure 1.6b represents the vector diagram for a low electrical power output:

$$P = UI \cos \phi \quad \text{(per phase)}$$

also for the vector triangles it is true that:

$$E \sin \theta = IX \cos \phi$$

substitute for I:

$$P = \frac{U \cos \phi \times E \sin \theta}{X \cos \phi} = \frac{UE \sin \theta}{X}$$

The electrical power output is therefore directly proportional to the generator internal voltage (E) and the system voltage (U) but inversely proportional to the machine reactance (X). With (U), (E) and (X) held constant the power output is only a function of the load angle θ. Figure 1.7 shows a family of curves for power output vs load angle representing this. As a prime mover power increases a load angle of 90° is eventually reached. Beyond this point further increases in mechanical input power cause the electrical power output to decrease. The surplus input power acts to further accelerate the machine and it is said to become unstable. The almost inevitable consequence is that synchronism with the remainder of the system is lost.

Fast-acting modern automatic voltage regulators (AVRs) can now actually enable a machine to operate at a load angle greater than 90°. If the AVR can increase (E) faster than the load angle (θ):

$$\frac{dE}{dt} > \frac{d\theta}{dt}$$

then stability can be maintained up to a theoretical maximum of about 130°.

This loss of synchronism is serious because the synchronous machine may enter phases of alternatively acting as a generator and then as a motor. Power surges in and out of the machine, which could be several times the machine rating, would place huge electrical and mechanical stresses on the machine.

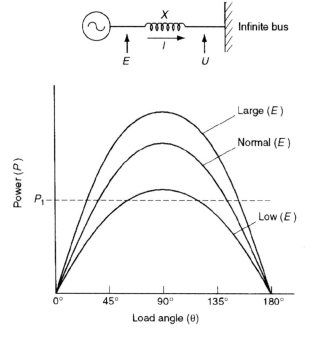

Figure 1.7   Power/load angle relationship

Generator overcurrent relay protection will eventually detect out-of-synchronism conditions and isolate the generator from the system. Before this happens other parts of the network may also trip out due to the power surging and the whole system may collapse. The object of system stability studies is therefore to ensure appropriate design and operational measures are taken in order to retain synchronism for all likely modes of system operation, disturbances and outages.

1.3.2.3 *The synchronous motor*

Operation of the synchronous motor may be envisaged in a similar way to the synchronous generator described in Section 1.3.2.2. In this case, however, the power flow is into the machine and, relative to the generator, the motor load angle is negative. An increase in load angle is in the opposite direction to shaft rotation and results in greater electrical power consumption. A leading power factor corresponds to high excitation and a lagging power factor low excitation.

1.3.2.4 *Practical machines*

In reality practical machine characteristics depart from the behaviour of the simple representations described above. However, in most cases the effects are small and they do not invalidate the main principles. The principle differences are due to saturation, saliency and stator resistance.

Saturation describes the non-linear behaviour of magnetic fluxes in iron and air paths produced by currents in the machine stator and rotor windings. Saturation effects vary with machine loading.

Saliency describes the effect of the differing sizes of air gap around the circumference of the rotor. This is important with salient pole rotors and the effect varies the apparent internal reactance of the machine depending upon the relative position of rotor and stator. Saliency tends to make the machine 'stiffer'. That is, for a given load the load angle is smaller with a salient pole machine than would be the case with a cylindrical rotor machine. Salient pole machines are in this respect inherently more stable.

The effect of stator resistance is to produce some internal power dissipation in the machine itself. Obviously the electrical power output is less than the mechanical power input and the difference is greatest at high stator currents.

1.3.3 Steady state stability

1.3.3.1 *Pull out power*

Steady state stability deals with the ability of a system to perform satisfactorily under constant load or gradual load-changing conditions. In the single machine case shown in Fig. 1.7 the maximum electrical power output from the generator occurs when the load angle is 90°.

The value of peak power or 'pull out power' is given as:

$$P_{MAX} = \frac{EU}{X} \text{ (from Section 3.2.2)}$$

With (U) fixed by the infinite bus and (X) a fixed parameter for a given machine, the pull out power is a direct function of (E). Figure 1.7 shows a family of generator power/load angle curves for different values of (E). For a generator operating at an output power P_1, the ability to accommodate an increase in loading is seen to be greater for operation at high values of (E) – increased field excitation. From Section 1.3.2 and Figs 1.6d and 1.6e, operation at high values of (E) corresponds with supplying a lagging power factor and low values of (E) with a leading power factor. A generator operating at a leading power factor is therefore generally closer to its steady state stability limit than one operating at a lagging power factor.

The value of (X), used in the expression for pull out power for an ideal machine, would be the synchronous reactance. In a practical machine the saturation of the iron paths modifies the assumption of a constant value of synchronous reactance for all loading conditions. The effect of saturation is to give a higher pull out power in practice than would be expected from a calculation using synchronous reactance. Additionally, in practical machines saliency and stator resistance, as explained in Section 1.3.2.4, would modify the expression for pull out power. Saliency tends to increase pull out power and reduces to slightly below 90° the load angle at which pull out power occurs. Stator resistance slightly reduces both the value of pull out power and the load angle at which it occurs.

1.3.3.2 Generator operating chart

An example of the effect of maximum stable power output of a generator is given in the generator operating chart of Fig. 1.8. This is basically derived as an extension of the vector diagrams of Fig. 1.6 where the value of internal voltage (E) and load angle (θ) is plotted for any loading condition of MW or MVAr. In the operating chart, the circles represent constant values of (E) and load angle is shown for an assumed operating point. The operating points for which the load angle is 90° are shown as the theoretical stability limit. Operation in the area beyond the theoretical stability limit corresponds with load angles in

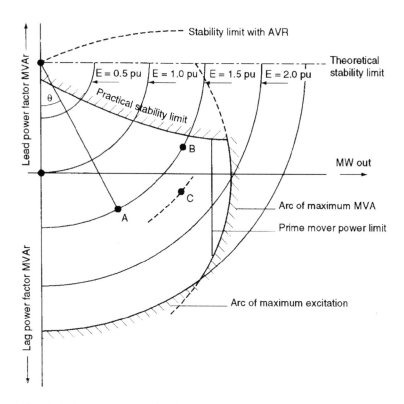

Figure 1.8   Typical generator operation chart

excess of 90° and is not permissible. The theoretical stability limit is one of the boundaries within which the operating point must lie. Other boundaries are formed by:

1. The maximum allowable stator current, shown on the chart as an arc of maximum MVA loading.
2. The maximum allowable field excitation current shown on the chart as an arc at the corresponding maximum internal voltage (E).
3. A vertical line of maximum power may exist and this represents the power limit of the prime mover.

Whichever of the above limitations applies first describes the boundaries of the different areas of operation of the generator.

In a practical situation, operation at any point along the theoretical stability limit line would be most undesirable. At a load angle of 90°, the generator cannot respond to a demand for more power output without becoming unstable. A practical stability limit is usually constructed on the operating chart such that, for operation at any point on this line, an increased power output of up to a certain percentage of rated power can always be accommodated without stability being lost. The practical stability limit in Fig. 1.8 is shown for a power increase of 10% of rated power output. The dotted line beyond the theoretical stability limit with a load angle $\theta > 90°$ shows the stabilizing effect of the AVR.

1.3.3.3 *Automatic voltage regulators (AVRs)*

The AVR generally operates to maintain a constant generator terminal voltage for all conditions of electrical output. This is achieved in practice by varying the excitation of the machine, and thus (E), in response to any terminal voltage variations. In the simple system of one generator supplying an infinite busbar, the terminal voltage is held constant by the infinite bus. In this case changes in excitation produce changes in the reactive power MVAr loading of the machine. In more practical systems the generator terminal voltage is at least to some degree affected by the output of the machine. An increase in electrical load would reduce the terminal voltage and the corrective action of the AVR would boost the internal voltage (E).

Referring to the generator operating chart of Fig. 1.8, an increase in power output from the initial point A would result in a new operating point B on the circle of constant internal voltage (E) in the absence of any manual or automatic adjustment of (E). Such an increase in power output takes the operating point nearer to the stability limit. If, at the same time as the power increase, there is a corresponding increase in (E) due to AVR action the new operating point would be at C. The operation of the AVR is therefore to hold the operating point well away from the stability limit and the AVR can be regarded as acting to preserve steady state stability.

1.3.3.4 *Steady state stability in industrial plants*

From Section 1.3.3.3 it can be seen that the steady state stability limits for generators are approached when they supply capacitive loads. Since industrial plants normally operate at lagging power factors the problem of steady state stability is unlikely to occur. Where power factor compensation is used or where synchronous motors are involved the possibility of a leading power factor condition is relevant and must be examined. Consider the Channel Tunnel 21 kV distribution scheme shown in Fig. 1.9. This consists of long 50 km lengths of 21 kV cross-linked polyethylene (XLPE) cable stretching under the Channel between England and France. Standby generation has been designed to feed essential services in the very unlikely case of simultaneous loss of both UK and French National Grid supplies. The 3 MVAr reactor shown on the single-line diagram is used to compensate for the capacitive effect of the 21 kV cable system. The failed Grid supplies are first isolated from the system. The generators are then run up and initially loaded into the reactor before switching in the cable network. The Channel Tunnel essential loads (ventilation, drainage pumping, lighting, control and communications plant) are then energized by remote control from the Channel Tunnel control centre.

1.3.4 Transient stability

1.3.4.1 *A physical explanation and analogy*

Transient stability describes the ability of all the elements in the network to remain in synchronism following an abrupt change in operating conditions. The most onerous abrupt change is usually the three phase fault, but sudden applications of electrical system load or mechanical drive power to the generator and network switching can all produce system instability.

   This instability can usually be thought of as an energy balance problem within the system. The analogy of the loaded spring is a useful aid to help visualize the situation. The general energy equation is as follows:

$$\text{Mechanical energy} = \text{Electrical energy} \pm \text{Kinetic energy (energy of motion)} + \text{Losses}$$

Under steady state conditions when changes are slow the system kinetic energy remains unchanged. However, if the disturbance to the machine is sudden (fault

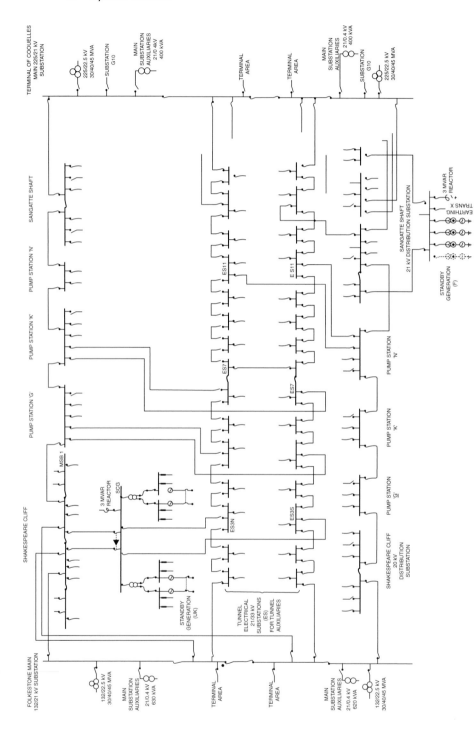

Figure 1.9   Channel Tunnel 21 kV simplified distribution network with standby generation

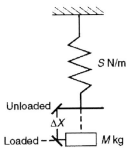

Figure 1.10   Loaded spring machine stability analogy

or load change) the machine cannot supply the energy from its prime mover or absorb energy from the electrical supply instantaneously. As explained in more detail in Section 1.3.4.6, the most common cause of instability in a generator system is a fault close to the terminals, which suddenly prevents the energy of the machine from being supplied to the system.

The excess or deficit or energy must go to or come from the machine's kinetic energy and the speed changes. As an example, if a motor is suddenly asked to supply more mechanical load it will supply it from the kinetic energy of its rotor and slow down. The slowing down process will go too far (overshoot) and will be followed by an increase in speed so that the new load condition is approached in an oscillatory manner just like the loaded spring (see Fig. 1.10).

If a spring of stiffness S is gradually loaded with a mass M it will extend by a distance Δx until the stiffness force $S\Delta x = $ Mg, the weight of the mass. The kinetic energy of the system will not be disturbed. The spring is analogous to the machine and the extension of the spring Δx is analogous to the machine load angle θ. Loading the spring beyond its elastic limit is analogous to steady state instability of a loaded machine. A machine cannot be unstable by itself, it can only be unstable with respect to some reference (another machine or infinite busbar to which it is connected) with which it can exchange a restoring force and energy. In the analogy the spring can only be loaded against a restraining mass (its attachment): an unattached spring cannot be extended.

Consider the spring analogy case with the spring being suddenly loaded by a mass M to represent the transient condition. The kinetic energy of the system is now disturbed and the weight will stretch the spring beyond its normal extension Δx to $\Delta x'$ where $\Delta x' > \Delta x$ (see Fig. 1.11).

The mass M moves past Δx to $\Delta x'$ until the initial kinetic energy of the mass is converted into strain energy in the spring according to:

$$\frac{1}{2} MV^2 = \frac{1}{2} S(\Delta x' - \Delta x)^2$$

When the weight momentarily comes to rest at $\Delta x'$ the kinetic energy of the weight has now been absorbed into the strain energy in the spring and the

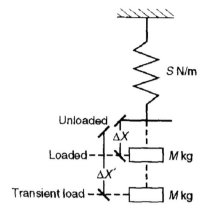

Figure 1.11   Loaded spring machine stability analogy – overshoot

spring now accelerates the mass upwards beyond Δx so there is an overshoot. The mass eventually settles down to its steady position in an oscillatory manner. It should be noted that the spring could support a weight which, if it were dropped on the spring, would cause its elastic limit to be exceeded before the downward motion of the weight was stopped. This is analogous to transient instability.

It can be seen how close the above analogy is by examining the equations of motion of the loaded spring and the synchronous machine as follows. For the spring:

$$M \frac{d^2x}{dt^2} + K \frac{dx}{dt} + Sx = \text{Force}$$

where M is the applied mass, x is the extension, d^2x/dt^2 is the acceleration or deceleration of mass M, Kdx/dt is the velocity damping and S is the restoring force.

For the synchronous machine:

$$M \frac{d^2\theta}{dt^2} + K \frac{d\theta}{dt} + Pe\sin\theta = P_m \text{ mechanical power}$$

where M is the angular momentum and $Pe\sin\theta$ is the electrical power.
For small θ, $\sin\theta$ tends to θ

$$M \frac{d^2\theta}{dt^2} + K \frac{d\theta}{dt} + Pe\theta = P_m$$

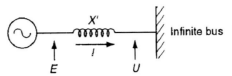

(a) - Equivalent circuit

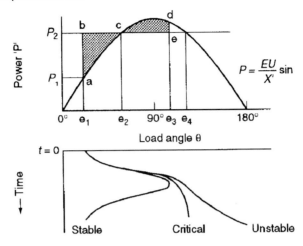

$$P = \frac{EU}{X'} \sin$$

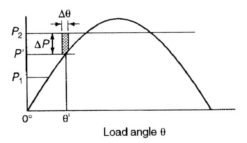

(b) - Load angle swings

(c) - Intermediate stage of load angle swing

Figure 1.12   Basic transient stability assessment

and if we change power into torque by dividing by the synchronous speed, the analogy is exact:

$$J\frac{d^2\theta}{dt^2} + K'\frac{d\theta}{dt} + Te\,\theta = T_{\mathrm{m}} \quad \text{the mechanical torque}$$

1.3.4.2 *Load angle oscillations*

The power/load angle curve shown in Fig. 1.12 can be used to show graphically the effect of a sudden change in machine load. The response shown can apply to either a synchronous generator or a synchronous motor, but the sudden loading of a motor is easier to visualize.

Suppose we consider a synchronous motor initially operating with a mechanical power P_1 and with a load angle θ_1 at a point 'a' on its characteristic operating curve. This curve is defined by the function:

$$\text{Power } P_e = \frac{EU}{x'_d} \sin \theta$$

where x'_d is the machine transient reactance. Operation at 'a' represents an equilibrium state in which the mechanical power P_1 equals the electrical power P_e, neglecting losses. The machine is operating at synchronous speed.

Suppose now that there is a sudden change in mechanical load of the synchronous motor. The mechanical power demand increases to P_2. This sudden energy demand cannot be immediately supplied from the electrical system so it must be supplied from the motor's stored rotational energy and the motor slows down. As the motor slows down its load angle increases allowing more power to be drawn from the electrical supply and the motor moves to point 'c' on its power/load angle curve where it is supplying the new power demand P_2. However, at 'c' the motor is going too slowly and therefore its load angle continues to increase. Beyond 'c' the electrical power supplied to the motor exceeds the new mechanical demand P_2 and the motor is accelerated.

The motor overswings to 'd', where the machine is again running at synchronous speed. Here, since the electrical power is still greater than the mechanical power, the motor continues to accelerate above synchronous speed and hence starts to reduce its load angle.

Back at point 'c', the electrical power is again equal to the mechanical power but the machine is operating above synchronous speed so it backswings towards 'a'. The machine will be prevented from reaching 'a' by damping; nevertheless it oscillates about 'c' until it finally stops at 'c' because of the damping effects.

1.3.4.3 The equal area criterion

The shaded areas 'a'–'b'–'c' and 'c'–'d'–'e' in Fig. 1.12, respectively represent the loss and gain of kinetic energy and for stability these two areas should balance. This is the basis of the equal area criterion of stability.

Three distinct alternative consequences occur for a sudden load change from P_1 to P_2:

1. If area 'c'–'d'–'e' can equal area 'a'–'b'–'c' at load angle θ_3 the machine is stable.
2. If the disturbance is such as to make the motor swing to θ_4, area 'c'–'d'–'e' just equals area 'a'–'b'–'c' and the motor is critically stable.

3. If area 'c'–'d'–'e' cannot equal area 'a'–'b'–'c' before angle θ_4 the motor is unstable. This is because, if the motor has not reaccelerated to synchronous speed at θ_4, where the electrical power equals the mechanical power P_2, it slows down beyond θ_4. For angles greater than θ_4 the mechanical power is greater than the electrical power and the motor continues to slow down. For angles greater than 180° the motor starts to pole slip (it becomes unstable) towards a stall. In reality, the motor protection would operate and disconnect the motor from the busbar.

From this explanation it can be seen that, unlike the steady state case, the machine can swing beyond 90° and recover. Note that a similar explanation could be applied to the generator case. Here the generator would accelerate upon a sudden fault disturbance such that the area 'a'–'b'–'c' represents a gain in kinetic energy whereas area 'c'–'d'–'e' represents a loss of kinetic energy.

1.3.4.4 Swing curves

The swing curve is generally a plot of load angle with time. The connection between the power/load angle characteristic and the dimension of time is the mechanical inertia. Actual inertias vary widely depending upon machine capacity and speed but when expressed in terms of the machine electrical rating, a narrow band of values is obtained. This gives the inertia (or stored energy) constant, H, and is defined as:

$$H = \frac{\text{Stored energy in mega Joules or kilo Joules}}{\text{MVA or kVA rating}}$$

$$= \frac{\frac{1}{2} J\omega^2 \times 10^{-3}}{\text{kVA}}$$

where $\omega = 2\pi f$ (for a two pole machine) or generally $\omega = 2\pi n$ where n is the machine speed in revolutions per second:

$$H = \frac{2\pi^2 \times 10^{-3} J n^2}{\text{kVA}}$$

J is the moment of inertia in $kg\,m^2$ and the dimensions of the stored energy constant, H are kW second/kVA or seconds. From Section 1.3.4.1 it will be remembered that the swing equation of motion takes the form:

$$M \frac{d^2\theta}{dt^2} + K \frac{d\theta}{dt} + Pe\,\theta = P_m \text{ mechanical power}$$

where M is the angular momentum of the machine. Like the inertia the angular momentum of various machines varies widely. We can replace M by H in the above equation according to:

$$H \times \text{kVA} = \frac{1}{2} M\omega = \frac{1}{2} M 2\pi n$$

so

$$M = \frac{H \times \text{kVA}}{\pi n}$$

Typical values of the stored energy constant, H, for various machines are listed below:

Machine description	H (kW second/kVA)
Full condensing steam turbine generators	4–10
Non-condensing steam turbine generators	3–5
Gas turbine generators	2–5
Diesel generators	1–3 (low speed)
	4–5 (with flywheel)
Synchronous motor with load	1–5
Induction motor with load	0.03–1.4 (100 kW–2000 kW but depends upon speed)

The load angle swing curves shown in Fig. 1.12 are obtained by solving the equation of motion of the machine or by solving the equations of motion of several machines in a group. Since the equations are non-linear numerical iterative methods computed with short time intervals (0.01 seconds or less) must be used for the solutions. Since the number of steps is enormous this is a job for computer analysis.

1.3.4.5 Transient stability during faults

In Fig. 1.13 a generator is shown feeding a load via a twin circuit transmission line. Under normal operation the load and voltage (U) are assumed to remain constant and the generator internal voltage (E) is also held constant. The power/load angle diagram for the whole system is shown in curve 1 with:

$$P = \frac{EU}{X_1'} \sin \theta \ (\text{per phase})$$

where X_1' is the total system transfer reactance with both lines in service.

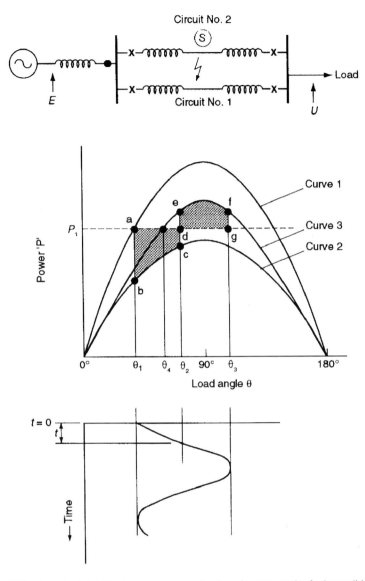

Figure 1.13   Transient stability to faults – power/load angle curve under fault conditions

A fault is now assumed at point (S). During the fault there will be a reduced possibility of power transfer from generator to load as indicated in Fig. 1.13 by curve 2 where the electrical power is given by:

$$P = \frac{EU}{X_2'}\sin\theta$$

where X_2' is the transfer reactance under the fault conditions. The fault is assumed to be cleared in time (t) by the circuit breakers isolating the faulty line. Post fault conditions are now shown by curve 3 with:

$$P = \frac{EU}{X_3'} \sin \theta \, (\text{per phase})$$

where X_3' is greater than X_1' due to the loss of the parallel overhead line section. Throughout the period under consideration the driving power to the generator is assumed constant at P_1. Prior to the fault a state of equilibrium exists with the electrical power matching the mechanical power at load angle θ_1. During the fault the driving power considerably exceeds the transmittable electrical power (shown by curve 2) and the rotor system accelerates. At the time taken to reach θ_2 the fault is assumed to be cleared and the power/load angle characteristic changes to curve 3. At θ_2 the transmitted power exceeds the driving power and the rotor decelerates. By the equal area criterion, the maximum swing angle θ_3 is determined by the area 'a'–'b'–'c'–'d' equal to area 'd'–'e'–'f'–'g'. The eventual new equilibrium angle can be seen to be θ_4 where P_1 intercepts curve 3. In this instance the swing curve shows the system to be stable. The following points should, however, be noted:

1. Had the angle θ_2 reached during the fault been larger, for example with a slower circuit breaker and protection, the system could have become unstable.
2. The value of θ_2 is determined both by the inertia constant H and the time duration of the fault. The load angle will be larger for smaller values of H and longer fault durations.
3. The decelerating power after the fault is related to the size of the post fault power/load angle curve. The larger the post fault reactance the lower the decelerating power and consequently the greater the possibility of instability.

1.3.4.6 Transient stability for close-up faults on generator terminals

The worst case fault conditions for the generator are with a three phase fault applied close to its terminals. The terminal voltage reduces to zero and the electrical power output must also reduce to zero. The whole of the prefault mechanical driving power is then expended in accelerating the generator rotor because no power can be transferred across this close-up fault. The maximum permissible fault duration to avoid instability under these conditions is a useful guide to the correct protection settings and selection of circuit breaker characteristics used in the vicinity of generators. The maximum permissible fault duration is referred to in technical literature as the critical switching time.

   The maximum (critical) fault duration is relatively insensitive to machine rating and any variation from one machine to another would largely be due to differences in inertia constant *H*. The two examples for critical fault duration given below are for identical machines with different inertia constants and give the order of fault durations for typical machines:

Drive source	Inertia constant (H)	Maximum fault duration
1. Hydro or low speed diesel	1.0 MJ/MVA	Approximately 0.14 seconds
2. Steam turbine	10 MJ/MVA	Approximately 0.50 seconds

(Figures are for generators with 25% transient reactance, no AVR action and feeding into an infinite busbar.)

1.3.4.7 *Auto-reclosing and single pole switching*

Section 1.3.4.6 shows that if the fault is of a transient nature it is advantageous (from a stability point of view) to put the system back into service rapidly by use of auto-reclosing circuit breakers once the fault has cleared. If the fault persists the generator will be subjected to a second fault impact upon reclosing the circuit breaker and a stable situation may be rendered unstable. Great care is therefore necessary when considering auto-reclosing. Applicable cases for overhead lines might be those where historical records show that the majority of faults are of a transient nature due to lightning or perhaps high impedance earth faults due to bush fires.

Over 90% of overhead line faults are single phase to earth faults. As an aid to stability auto-reclose single pole circuit breaker switching is often employed. A typical transmission system strategy is to employ single shot auto-reclose facilities only for single phase to earth faults. If the fault persists then three phase switching takes over to disconnect the circuit. Typical delay times between circuit breaker auto-reclose shots are of the order of 0.4 to 0.5 seconds allowing for a 0.3 second arc deionization time. The single pole auto-reclose technique is well established for transmission line voltages below 220 kV and stability is aided because during the fault clearance process power can be transferred across the healthy phases. It should be noted, however, that fault arc deionization takes longer with single pole switching because the fault is fed capacitively from the healthy phases. In addition the system cannot be run for more than a few seconds with one open circuit phase or serious overheating of the rotating plant may take place. Distribution systems employ three phase auto-reclose breakers and sectionalizers to isolate the fault if it persists.

1.3.4.8 *Hunting of synchronous machines*

The load angle of a stable machine oscillates about a point of equilibrium if momentarily displaced. The machine has a characteristic natural frequency associated with this period of oscillation which is influenced by its loading and inertia constant. In order to avoid large angle swings, the possibility of mechanical damage to the shaft and couplings and loss of synchronism, the natural frequency should not coincide with the frequency of pulsating loads or prime mover torque. Hunting of this type may be detected from pulsating electrical measurements seen on machine meters and excessive throbbing machine noise. Damping windings on the machine and the power system load itself assist in reduction of hunting effects. In both these cases damping arises from induced currents in the damper windings caused by rotor oscillation. The damping torques decrease with increasing resistance in the paths of the induced currents. Machines operating at the ends of long, high resistance supply lines or having high resistance damper windings can be particularly susceptible to hunting.

The possibility of hunting can be seen from the equations in Section 1.3.4.1 if the mechanical torque takes the form $T_m \sin \sigma t$. The second-order differential equation of motion has oscillatory solutions exhibiting a natural frequency ω_0. A resonance condition will arise if the mechanical driving torque frequency σ approaches the machine natural frequency ω_0.

1.3.5 Dynamic stability

Although a system may not lose synchronism in the transient interval following a disturbance, the ability to adapt in the longer term to a significantly new set of operating conditions is the subject of dynamic stability studies. In the transient period of perhaps a second or two following a disturbance, many of the slower reacting power system components can be assumed constant. Their effect on the preservation or otherwise of transient stability is negligible. In the seconds and minutes following a disturbance such slow reacting components may become dominant. Thus a thorough study of system stability from the end of the transient period to steady state must consider such effects as turbine governor response, steam flows and reserves, boiler responses and the possibility of delayed tripping of interconnectors which may have become overloaded, or load loss by frequency-sensitive load-shedding relays. In addition, during the dynamic period, motor loads shed at the start of the disturbance may be automatically restarted.

Dynamic stability studies are more normally carried out for large interconnected systems to assist with the development of strategies for system control following various types of disturbance. With smaller industrial reticulation the preservation of stability in the transient period is generally regarded as the most important case for investigation.

The adaptation of the network in the dynamic interval is left largely up to the natural properties of the system and by automatic or operator control. The control system can, for example, restore the correct frequency by adjustments to turbine governor gear and improve voltage profiles by capacitor bank switching or alteration of synchronous motor excitation.

1.3.6 Effect of induction motors

1.3.6.1 *Motor connection to the system*

The stability of an induction motor generally refers to its ability to recover to
a former operating condition following a partial or complete loss of supply.
Induction motors always run asynchronously and stability studies involve a
consideration of the load characteristics before and after a system disturbance. For a fault close to the induction motor the motor terminal voltage is
considerably reduced. Unable to supply sufficient energy and torque to the
driven load the motor slows down:

1. For a given terminal voltage the current drawn is a function only of speed. As
    the speed drops the current increases rapidly to several times normal full
    load value and the power factor drops from, say, 0.9 lag to 0.3 lag or less.
2. The torque of the motor is approximately proportional to the square of the
    terminal voltage.

Because of these characteristics substation induction motor loads are often
characterized as:

1. 'Essential' loads – Those supplying boiler feed pumps, lubricating sys-
    tems, fire pumps, etc., which must be kept running throughout a disturb-
    ance. The ability of these motors to recover and reaccelerate in the post
    disturbance period depends upon the nature of the load and system voltage
    profile. Square law loads such as centrifugal pumps will recover with
    greater ease than constant torque loads such as reciprocating compressors.
2. 'Non-essential' loads – Motors that can be shed by undervoltage relays if
    the disturbance is sufficiently severe to depress the voltage below, say,
    66%. These loads may be reconnected automatically after a delay. The sys-
    tem designer must, however, consider the possibility of voltage collapse
    upon reconnection as the starting of motors places a severe burden on gen-
    eration reactive power supply capability.

1.3.6.2 *Motor starting*

In itself motor starting constitutes a system disturbance. Induction motors
draw 5 to 6 times full load current on starting until approximately 85–90% of
full speed has been attained. The starting torque is only about 1.5 to 2 times
full load torque and does not therefore constitute a severe energy disturbance.
The motor VAr demand is, however, very large because of the poor starting
power factor. The system voltage can be severely depressed before, for example,
on-site generator AVR action comes into play. Checks should be made to

ensure that direct-on-line (DOL) starting of a large motor or group of motors does not exceed the VAr capability of local generation in industrial distribution systems. The depressed voltage should not be allowed to fall below 80% otherwise failure to start may occur and other connected motors on the system may stall.

If studies show large motor starting difficulties then DOL starting may have to be replaced with current limiting, or soft start solid state motor starting methods. The star/delta starter is not recommended without consideration of the switching surge when moving from star to delta induction motor winding connections.

1.3.7 Data requirements and interpretation of transient stability studies

1.3.7.1 *Generator representation*

The simplest generator representation for transient stability studies involving minimum data collection in the mechanical sense is by its total inertia constant H MJ/MVA. In the electrical sense by a fixed internal voltage E kV behind the transient reactance x_d' per unit or per cent. The fixed internal voltage implies no AVR action during the studies and the computer assigns a value after solving the predisturbance system load flow. This is adequate for 'first swing type' stability assessments giving pessimistic results.

Where instability or near instability is found with the simple representation, or if it is required to extend the study beyond the 'first swing' effects, a more detailed representation of the generator is necessary. AVR characteristics, saturation effects, saliency, stator resistance and machine damping are then included in the input data files. Such data collection can be time consuming and for older machines such data are not always available. A compromise is sometimes necessary whereby generators electrically remote from the disturbance, and relatively unaffected by it, can use the simple representation and those nearer can be modelled in more detail. For example, a primary substation infeed from a large grid network with high fault level to an industrial plant can usually be represented as a simple generator with large inertia constant and a transient reactance equal to the short circuit reactance. If the grid system is of a similar size to the industrial plant then a more detailed representation is necessary since the stability of the grid machines can affect plant performance.

1.3.7.2 *Load representation*

The detailed representation of all loads in the system for a transient stability study is impracticable. A compromise to limit data collection and reduce computing time costs is to represent in detail those loads most influenced by the disturbance and use a simple representation for those loads electrically remote from the disturbance. In particular where large induction motor performance is to be studied it is important to correctly represent the torque/speed characteristic of the driven load. Simple load representation to voltage variations falls into one of the following categories:

1. Constant impedance   (static loads)
2. Constant kVA   (induction motors)
3. Constant current   (controlled rectifiers).

In summary:

Induction motors (close to disturbance):
- Use detailed representation including synchronous reactance, transient reactance, stator resistance, rotor open circuit time constant, deep bar factor, inertia constant and driven load characteristics (e.g. torque varies as a function of speed).

Induction motors (remote from disturbance and represented as a static load):
- Fully loaded motors can be represented as constant kVA load. Partially loaded motors can be represented as constant current loads. Unloaded motors can be represented as constant impedance loads.

Controlled rectifiers:
- Treat as constant current loads.

Static loads:
- Generalize as constant impedance unless specific characteristics are known.

Figure 1.14 shows a flow chart indicating the stages in obtaining information for data files necessary for load flow, transient stability and dynamic stability studies.

1.3.7.3 *Interpretation of transient stability study results*

The following broad generalizations can be made in the interpretation of transient stability study results following the application and clearance of a three phase fault disturbance:

1. System faults will depress voltages and restrict power transfers. Usually, generators will speed up during the fault and the load angle will increase.
2. Generators closest to the fault will suffer the greatest reduction in load and will speed up faster than generators remote from the fault. Some generators may experience an increased load during the fault and will slow down.
3. For the same proportionate loss in load during the fault, generators with lower inertia constants will speed up more quickly. On-site generators may

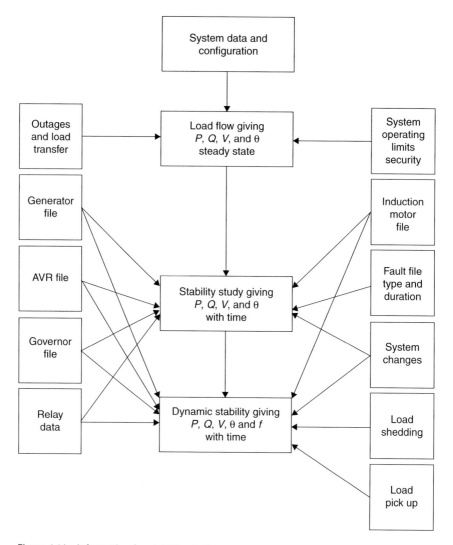

Figure 1.14   Information for stability studies

remain in step with each other but diverge from the apparently high inertia grid infeed.

4. Induction motor slips will increase during the fault.

5. After the fault, stability will be indicated by a tendency for the load angle swings to be arrested, for voltages and frequency to return to prefault values and for induction motor slips to return to normal load values.

6. If a grid infeed is lost as a result of the fault, an industrial load may be 'islanded'. If on-site generators remain in synchronism with each other but cannot match the on-site load requirements, a decline in frequency will occur. Load shedding will then be necessary to arrest the decline.

Practical examples of these principles are given in the case studies in Section 1.3.8. Faults may be classified according to their severity in terms of:

1. Type of fault (three phase, single phase to earth, etc.). A three phase fault is normally more severe than a single phase fault since the former blocks virtually all real power transfer. The single phase fault allows some power transfer over healthy phases.
2. Duration of fault. If the fault persists beyond a certain length of time the generators will inevitably swing out of synchronism. The maximum permissible fault duration therefore varies principally with the inertia constant of the generators, the type and location of the fault.

    Determination of maximum fault clearing time is often the main topic of a transient stability study. The limiting case will usually be a three phase fault close up to the generator busbars. Low inertia generators ($H = 1.0$ MJ/MVA) will require three phase fault clearance in typically 0.14 seconds to remain stable as described in Section 1.3.4.6. Note that with modern vacuum or SF_6 circuit breakers fault clearance within three cycles (0.06 seconds @50 Hz) or less is possible.
3. Location of a fault. This affects the extent of voltage depression at the generator terminals and thus the degree of electrical loading change experienced by the generator during the fault.
4. Extent of system lost by the fault. Successful system recovery, after a fault, is influenced by the extent of the system remaining in service. If a main transmission interconnector is lost, the generators may not be able to transmit total power and power imbalance can continue to accelerate rotors towards loss of synchronism. The loss of a faulted section may also lead to overloading of system parts remaining intact. A second loss of transmission due, say, to overload could have serious consequences to an already weakened system. In order to improve transient stability, fault durations should be kept as short as possible by using high speed circuit breakers and protection systems, particularly to clear faults close to the generators. The incidence of three phase faults can be reduced by the use of metal clad switchgear, isolated phase bus ducting, single core cables, etc. Impedance earthing further reduces the severity of single phase to earth faults. Appropriate system design can therefore reduce the extent of system outages by provision of more automatic sectionalizing points, segregation of generation blocks onto separate busbars, etc.

System transient reactances should be kept as low as possible in order to improve transient stability. Machines (and associated generator transformers) with low reactance values may be more expensive but may provide a practical solution in a critical case. Such a solution is in conflict with the need to reduce fault levels to within equipment capabilities and a compromise is therefore often necessary.

    A resonant link can, in principle, solve this conflicting requirement by having a low reactance under normal load conditions and a high reactance to fault

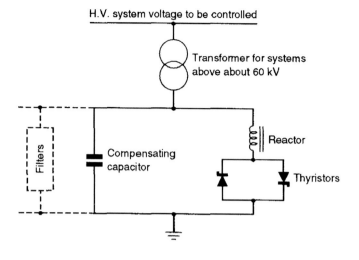

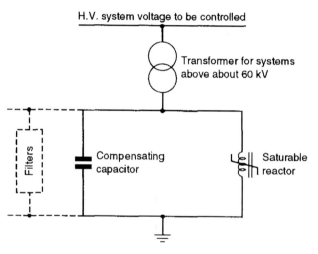

Figure 1.15   Static compensators

currents. Figure 1.15 shows the functionally different and more widely used static compensation equipment containing saturable reactors or thyristor controlled reactors. These devices can supply leading or lagging VArs to a system and thereby maintain nearly constant voltages at the point of connection in the system. For this reason they also have value in improving power quality (see Chapter 25). The characteristics of such devices are shown in Fig. 1.16. This constant voltage effect may be considered to represent a sort of inertialess infinite busbar and therefore the transfer reactance is reduced increasing the stability margin of the system. The disadvantages of such systems are their initial cost, need for maintenance, volume of equipment to be accommodated and generation of harmonics necessitating the use of filters.

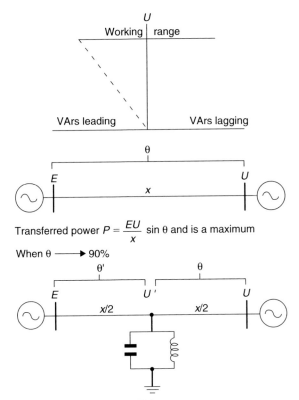

Transferred power $P = \dfrac{EU}{x} \sin \theta$ and is a maximum

When $\theta \longrightarrow 90\%$

Transferred power $P = 2\dfrac{EU'}{x} \sin \theta'$

So transferred power is doubled and may be increased further if $U' > U$

Absolute theoretical limit now occurs when $\theta + \theta' = 180°$

Figure 1.16   Characteristics of static compensators

1.3.8 Case studies

1.3.8.1 *Introduction*

Figure 1.17 shows a power transmission and distribution system feeding an industrial plant with its own on-site generation and double busbar arrangement. Normally the busbar coupler is open and grid infeed is via the non-priority busbar No. 2. On-site generation and a major 5000 hp induction motor are connected to busbar No. 1. Other smaller motor loads are connected to busbars 3, 4 and 5.

The computer data files represent the grid infeed as a generator with transient reactance equal to the short circuit reactance x_d' and a very large inertia constant of 100 MJ/MVA. The large induction motors connected to busbars 1 and 5 are represented in detail in order that slip and current variations during

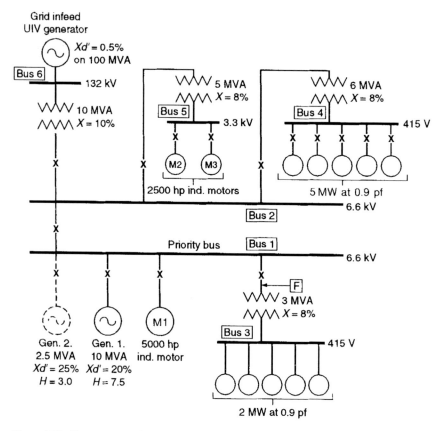

Figure 1.17   Power system for case studies

a disturbance may be studied. These motor load torque/speed characteristics are assumed to follow a square law. The two groups of smaller 415 V motors connected to busbars 3 and 4 are not to be studied in detail and are represented as constant kVA loads. On-site generator No. 1 is represented simply by its transient reactance and inertia constant and site conditions are assumed to allow full rated output during all case studies.

The results of the computer analysis associated with this system for Case studies 1–4 have been replotted in Figs 1.18 to 1.21 to allow easy comparison.

1.3.8.2 Case study 1

The system is operating as in Fig. 1.17 with industrial plant on-site generator No. 2 not connected. Generator No. 1 is delivering full power at near unity power factor. A three phase fault is imagined to occur on the 6.6 kV feeder to busbar 3 at point (F). The protection and circuit breaker are such that a total

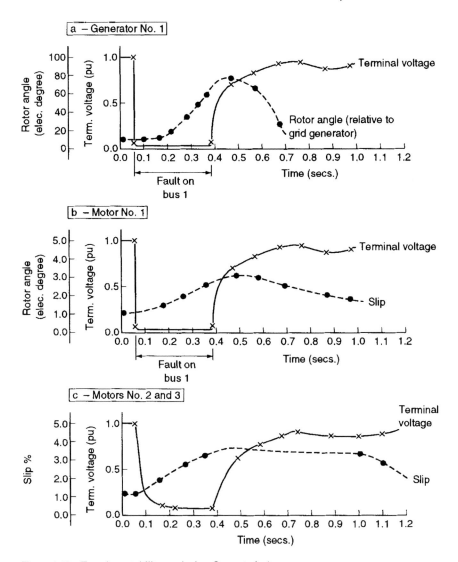

Figure 1.18   Transient stability analysis – Case study 1

fault duration of 0.35 seconds is obtained. Clearance of the fault disconnects busbar 3 and its associated stepdown transformer from busbar 1 and all other loads are assumed to remain connected.

Figure 1.18 shows the behaviour of the generator and the main motors. In Fig. 1.18a the rotor angle of generator 1 is seen to increase during the fault period. Shortly after fault clearance, a return towards the original operating load angle position is seen. The generator terminal voltage is also seen to recover towards prefault value. The on-site generator No. 1 is therefore stable to this particular fault condition.

Figure 1.18b shows the behaviour of the 5000 hp induction motor load under these fault conditions. During the fault the slip increases. However, shortly after fault clearance the terminal voltage recovers and the slip reduces towards the prefault value. Similar behaviour for motors 2 and 3 is shown in Fig. 1.18c. The main motor loads therefore seem to be able to operate under the fault condition; the smaller motor loads have not been studied.

The situation in this configuration is therefore stable and only one busbar is lost as a result of the fault.

1.3.8.3 *Case study 2*

In this study it is assumed that a decision has been made to use surplus industrial plant gas to generate more electrical power and thus reduce grid infeed tariffs. A 2.5 MVA generator No. 2 is added to busbar 1. This machine has a relatively low inertia constant compared to the existing on-site generator No. 1. No changes are proposed to the existing protection or circuit breaker arrangements. Both site generators are supplying full load.

Figure 1.19 shows the consequences of an identical fault at (F) under these new system conditions. Figure 1.19a shows generator 1 to continue to be stable. Figure 1.19b shows generator 2 has become unstable. The duration of the fault has caused generator 2 to lose synchronism with generator 1 and the grid infeed. The ensuing power surging is not shown in Fig. 1.19 but can be assumed to jeopardize the operation of the whole of the power system.

Acting as a consultant engineer to the industrial plant owner what action do you recommend after having carried out this analysis?

1. Do you have anything to say about protection operating times for busbar 5 feeder or generator 2 breaker?
2. The client, not wishing to spend more money than absolutely necessary, queries the accuracy of your analysis. Generator 2 is a new machine and good manufacturer's data is available including AVR characteristics, saliency, saturation, damping and stator resistance. Would you consider a further study under these conditions with more accurate generator modelling?

This study demonstrates the need to review plant transient stability whenever major extensions or changes are contemplated. In this example a solution could be found by decreasing protection and circuit breaker operating times. Alternatively, if generator 2 has not already been purchased a unit with a similar inertia constant to generator 1 (if practicable) could be chosen.

1.3.8.4 *Case study 3*

The system is as for study 1 – i.e. generator 2 is not connected. Generator 1 is supplying full load at unity power factor and the grid infeed the balance of

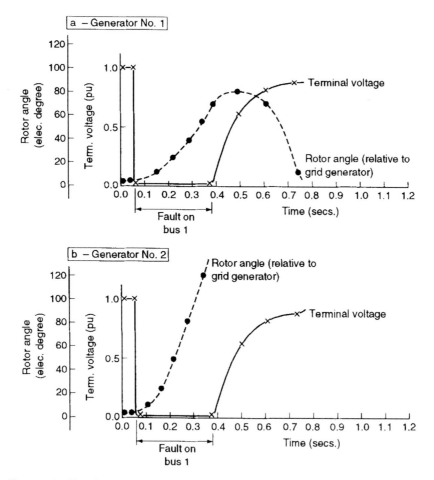

Figure 1.19   Transient stability analysis – Case study 2

site demand. It is now imagined that the grid infeed is lost due to protection operation.

The site electrical load now considerably exceeds the on-site generation capacity and a decline in frequency is expected. The mechanical driving power to generator 1 is assumed to remain constant. Stability in the sense of loss of synchronism is not relevant here since the two power sources are isolated by the 132 kV transmission line and 132/6.6 kV transformer disconnection.

Figure 1.20a shows the predicted decline in plant system frequency. As the grid supply engineer in charge of this connection you have been called by the plant manager to explain what precautions could be taken to prevent plant shut down under similar outage conditions in the future. You have some knowledge of protection systems, although you are not an expert in this field. You propose an underfrequency relay associated with the bus-coupler circuit breaker

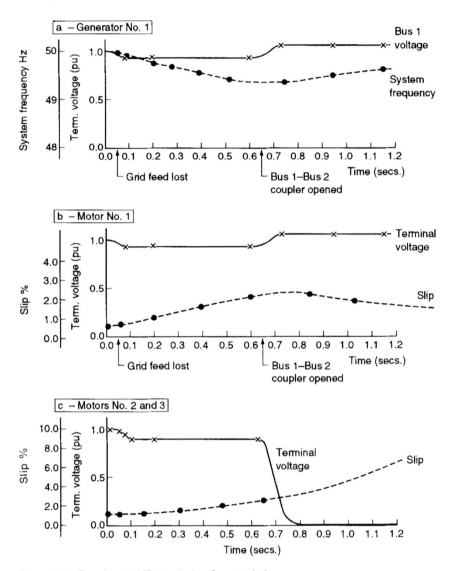

Figure 1.20   Transient stability analysis – Case study 3

separating busbars 1 and 2. From the transient stability studies shown in Fig. 1.20a you recommend an underfrequency relay setting of approximately 49.4 Hz. The hoped for effect of bus-coupler opening is for recovery in system frequency.

The plant manager considers that too much load will be shed by utilizing the bus coupler in this way although he is thinking more about plant down-times than system stability. Again as grid engineer you acknowledge the point and indeed you are worried that such a large load shed could leave generator 1

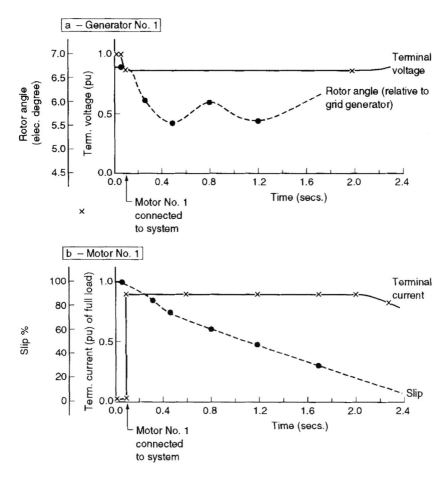

Figure 1.21   Transient stability analysis – Case study 4

underloaded. Unless some adjustment is made to the generator 1 driving power an overfrequency situation could arise. With more thought what similar action to plant protection could be taken?

In this example the crude bus coupler protection allows motor 1 to recover successfully. Motors 2 and 3 connected to busbar 5 will decelerate to a standstill due to loss of supply as will all motors connected to busbar 4.

1.3.8.5. Case study 4

The system of Fig. 1.17 is originally operating without the 5000 hp motor 1 or the second on-site generator 2 connected.

The result of direct-on-line (DOL) starting of motor 1 is shown in Fig. 1.21. Since the fault level is relatively high (the system is said to be 'stiff' or 'strong') the induction motor starts with only slight disturbance to operating conditions. Fig. 1.21a shows only minor changes to generator 1 load angle (note sensitivity of the scale). The deflection is in the direction of decreasing rotor angle and indicates that the motor starting has initially acted to slow down generator 1 relative to the grid generation. There is, however, no instability since the rotor angle is seen to recover towards its original position.

As consultant to the plant manager are you able to confirm successful DOL starting and run up of motor 1? Would you wish to place any provisos on your answer? With fast electronic protection, together with vacuum or SF_6 circuit breakers, would you consider fault durations reduced from some 0.35 to 0.175 seconds to be more representative of modern practice?

1.4 SHORT CIRCUIT ANALYSIS

1.4.1 Purpose

A short circuit analysis allows the engineer to determine the make and break fault levels in the system for both symmetrical and asymmetrical, low or high impedance faults. This in turn allows the correct determination of system component ratings; for example the fault rating capability of circuit breakers. A full analysis will allow investigation of protection requirements and any changes to the system that might be necessary in order to reduce fault levels.

1.4.2 Sample study

1.4.2.1 *Network single-line diagram*

The system described in Section 1.2 for the load flow case is now analysed under fault conditions. Figures 1.1 and 1.2 detail the system single-line diagram, busbar and branch data.

1.4.2.2 *Input data*

The main input data file created for the load flow case using positive sequence impedances is again required for the short circuit analysis. A second data file containing generator parameters is also now needed if not already available from the load flow case. Induction and synchronous motor contributions to the faults may also be considered in most commercially available computer system analysis programs and the creation of a motor data file is necessary for this purpose. In this example a 5 MVA, 0.85 power factor induction motor load is assumed to form part of the total 25 MW load at busbar C.

Zero sequence data is required for the simulation of faults involving ground or earth. The zero sequence data file is not necessary if only three phase symmetrical faults are being investigated. Guidance concerning the derivation of zero sequence impedances is given in Chapter 26. Sample zero sequence, generator and motor files for the network are given in Fig. 1.22. Line zero sequence impedances are assumed to be three times the positive sequence impedance values. Transformer zero sequence impedances are taken as equal to positive sequence values in this example for the vector groups used. The generator earthing resistance appears in positive, negative and zero sequence impedance circuits for earth faults and is therefore represented as 3×3.46 ohms or 9.61 pu (100 MVA base, 6 kV).

1.4.2.3 Solutions

A summary short circuit report from a software program covering this example is given in Fig. 1.23. Three phase (3-PH), single phase to earth or line

```
02   380    1       9.61   1.
0                        Ro  Xo                        Ro   Xo
03   1A0    1    2A    1   1                            0.   .5
03   380    1    4B    1   1   0.   0.8
03   2A     1    4B    1   1   2.4  5.19
03   2A     1    5C    1   1   .6   3.45
03   4B     1    5C    1   1   .24  1.38
0
```

(a) : ZERO-SEQUENCE DATA FILE (p.u., 100 MVA base)

Generator			x'_d		x''_d	kV	MVA		
G	1	0.	.28	0.	.18	6.	20.	4	.14
G	3	0.	.2	0.	.15	6.	10.	4	.10

(b) : MACHINE (GENERATOR) DATA FILE

```
                     kV    MVA  pf
M   5 IM  1    0.    .4    33.   5.  .85   0.   .20   0   0 4.   0
```

(c): MOTOR DATA FILE

Figure 1.22 Fault analysis sample study. Zero sequence, generator and motor files (System single-line diagram as per Fig. 1.1)

CYMFAULT   11:38 am, Wednesday, December 4, 199

FAULT STUDY NO.1   FILE:ZTRAIN1

NETWORK SUMMARY REPORT

. . . .

TITLE: FAULT STUDY NO.1   FILE:ZTRAIN1

	BUS			3-PH		L-G		L-L		L-L-G	
#	NAME	ZN	KV	KA	MVA	KA	MVA	KA	MVA	KA	MVA
1	AO	1	6.30	10.7	117	0.0	0	9.3	101	0.0	0
2	A	1	33.70	1.6	94	2.0	116	1.4	81	2.0	114
3	BO	1	5.82	8.3	83	5.1	51	7.2	72	8.0	80
4	B	1	31.55	1.5	80	1.0	55	1.3	69	1.4	76
5	C	1	30.47	1.4	74	1.0	52	1.2	64	1.3	68

Figure 1.23   Fault analysis sample study. Summary short circuit report

to earth (L-G), phase to phase (L-L) and two phase to earth (L-L-G) fault currents at each busbar are given together with busbar voltage and fault MVA. More detailed short circuit busbar reports are also available from most programs and an example of such a report is given in Fig. 1.24 for busbar 5c. The fault infeed contributions from the different branches, including the induction motor contribution, into the busbar are shown.

As in the load flow case the results can also be drawn up in a pictorial manner by placing the fault level results against each busbar on the associated single-line diagram. The effect of changes to the network can be seen simply by altering the input data. This is particularly useful when carrying out relay protection grading for the more complex networks. A variety of operational and outage conditions can make backup IDMT grading particularly difficult. The computer takes the drudgery out of the analysis. An example of computer aided protection grading is given in Chapter 10.

1.4.2.4 Asymmetrical fault levels

An interesting aspect of fault level analysis is that the three phase solid symmetrical type of fault does not always lead to the highest fault level currents. For highly interconnected transmission systems the ratio of the zero phase sequence impedance (Z_0) and positive phase sequence impedance (Z_1) may be less than unity (i.e. (Z_0)/(Z_1) < 1.)

The Zambian Copperbelt Power Company 66 kV transmission system stretches for about 150 km close to the border between Zambia and Zaire. The major power generation infeed is from the hydroelectric power station at Kariba Dam some 450 km to the south via 330 kV overhead lines and

CYMFAULT   11:38 am, Wednesday, December 4, 199

FAULT STUDY NO.1   FILE:ZTRAIN1

Short circuit bus report
....

TITLE: FAULT STUDY NO.1   FILE:ZTRAIN1
BUS IDENT.   INIT.VOLTAGE (kv) ANGLE(degree)
    TYPE ·············· CURRENT ·············· ····· VOLTAGE P·G ·····
            MODULE ANGLE   MODULE POW. BREA MODULE ANGLE   MODULE
            (p.u.) (degree) (amp) (mva)   (p.u.) (degree) (kv)
....

FAULT AT
    5C   1   30.5   -12.2
    LLL·A   0.81   ·83.99   1416   75   0%   0.00   0.00   0.00
    LG·A   0.57   ·83.05   1003   53   0%   0.00   0.00   0.00
    LL·B   0.70   ·173.99   1226   65   0%   0.46   167.82   8.80
    LL·C   0.70   6.01   1226   65   0%   0.46   167.82   8.80
    LLG·B   0.74   168.57   1295   68   0%   0.00   0.00   0.00
    LLG·C   0.73   23.71   1277   67   0%   0.00   0.00   0.00
    Z00·1 = 0.3561+j 1.0842 (p.u.)   X/R =   3.045
    Z00·0 = 0.8706+j 2.3968 (p.u.)
CURRENT TO CONSTANT IMPEDANCE LOAD DURING FAULT
    LLL·A   0.00   0.00   0   0
    LG·A   0.12   149.93   217   11
    LL·B   0.12   ·29.23   206   11
    LL·C   0.12   ·29.23   206   11
    LLG·B   0.10   ·29.56   168   9
    LLG·C   0.10   ·29.56   168   9
FIRST RING CONTRIBUTIONS
  ·> 2A   1 1   33.7   -4.6
    LLL·A   0.33   ·78.73   583   31   0.39   1.40   7.41
    LG·A   0.32   ·75.32   565   30   0.64   ·3.42   12.13
    LL·B   0.34   ·177.29   594   31   0.58   ·149.40 11.08
    LL·C   0.25   23.09   431   23   0.64   143.87 12.20
    LLG·B   0.34   163.78   599   32   0.49   ·140.92   9.27
    LLG·C   0.32   41.23   553   29   0.54   137.02 10.26
    Z0P·1 = 0.1931+j 0.7612 (p.u.)
    Z0P·0 = 0.0815+j 0.4354 (p.u.)
  ·> 4B   1 1   31.6   -8.4
    LLL·A   0.41   ·78.34   718   38   0.19   1.79   3.65
    LG·A   0.34   ·64.46   586   31   0.23   15.19   4.44
    LL·B   0.42   ·174.99   727   38   0.48   ·168.42   9.10
    LL·C   0.30   20.81   528   28   0.53   153.75 10.16
    LLG·B   0.44   173.24   769   41   0.24   ·118.72   4.60
    LLG·C   0.32   36.96   568   30   0.17   136.48   3.32
    Z0P·1 = 0.2769+j 0.9083 (p.u.)
    Z0P·0 = 0.9462+j 1.8013 (p.u.)
CONTRIBUTION FROM 1 IND. MOTOR(S) EACH RATED   5.00 MVA
    LLL·A   0.10   ·130.08   172   9
    LG·A   0.05   ·162.24   91   5
    LL·B   0.06   141.72   102   5
    LL·C   0.11   ·41.02   196   10
    LLG·B   0.06   133.28   112   6
    LLG·C   0.11   ·36.14   188   10

Figure 1.24   Fault analysis sample study. Detailed report for busbar 5c

330/220 kV stepdown autotransformers located at 'Central' and 'Luano' sub-stations. Consider the case of reinforcement works at the 66 kV 'Depot Road' which requires the use of additional 66 kV circuit breakers. Bulk oil breakers from the early 1950s were found in the stores with a fault rating of approximately 500 MVA. A fault analysis on the system showed that the three phase fault level on the 66 kV busbars at 'Depot Road' substation to be some 460 MVA while the two phase to earth fault level could be as high as 620 MVA. For a single phase to earth fault the fault current is given by the equation:

$$I_F = \frac{3E}{Z_1 + Z_2 + Z_0}$$

where E is the source phase to neutral e.m.f. and Z_1, Z_2 and Z_0 are the positive, negative and zero sequence impedances from source to fault. This indicates that the sequence networks for this type of fault are connected in series. In this example Z_0 is small because (i) the 66 kV overhead lines in the copperbelt area are very short; (ii) the 66/11 kV transformers are star–delta connected with the high voltage star point solidly earthed and (iii) the 330/220 kV and 220/66 kV transformers have a low zero sequence impedance. The parallel effect of these low zero sequence impedances swamps the zero sequence impedance of the long overhead lines from the power source at Kariba making Z_0 tend to a very small value.

Because of this effect the old spare oil circuit breakers could not be used without further consideration of the financial aspects of purchasing new switchgear or fault limiting components. Figure. 1.25 shows a plot of fault current against the ratio of Z_0/Z_1 for the different types of symmetrical and asymmetrical fault conditions and shows how the phase to earth and two phase to earth fault current levels maybe higher than the three phase symmetrical fault level if the zero sequence impedance is very low in relation to the positive sequence impedance.

1.4.2.5 *Estimations for further studies*

Possibly the biggest single obstacle in fault calculations is obtaining reliable information on system constants. Equipment nameplate data and equipment test certificates are the best starting point followed by contacting the original manufacturers. However, checking the authenticity of information, particularly where old machines are concerned, can be quite fruitless. Some approximate constants are given in this book as a guide and they may be used in the absence of specific information. Refer to Chapter 26, Section 26.6.2.

Longhand working of fault calculations is tedious. The principle employed is that of transforming the individual overhead line, generator, cable, transformer,

CONDITIONS

Z_1 = Positive phase sequence impedance

Z_2 = Negative phase sequence impedance

Z_0 = Zero phase sequence impedance

Z_2 = Taken as equal to Z_1

Resistance taken as zero

r = Ratio $\dfrac{Z_0}{Z_1}$

CURVE FORMULAE

L-L Fault line current $= -j\,0.866 \times 3\emptyset$ Fault line current

L-E Fault line current $= \dfrac{3}{2+r} \times 3\emptyset$ Fault line current

L-L-E Fault line current $= \dfrac{-1.5 - j(\sqrt{3}/2 + \sqrt{3}r)}{1 + 2r} \times 3\emptyset$ Fault line current

L-L-E Earth fault current $= \dfrac{3}{1 + 2r} \times 3\emptyset$ Fault line current

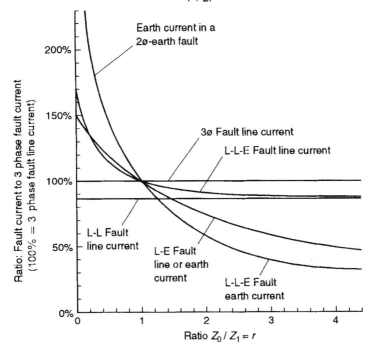

Figure 1.25   Effect of network zero to positive sequence impedances on system fault levels

etc., system impedances to a per unit or percentage impedance on a suitable MVA base. These impedances, irrespective of network voltage, may then be added arithmetically in order to calculate the total impedance per phase from source to fault. Once this has been determined it is only necessary to divide the

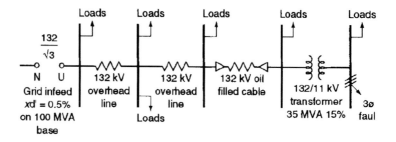

a – Simple radial network (load contribution to fault ignored)
   3ø fault as shown on 11 kV busbar

	Source	OHL	OHL	OFC	Transformer
Z% 100 MVA base	j 0.5	6.19 + j 8.26	2.9 + j 3.9	6.89 + j 1.89	+ j 42.8
Z ohms 132 kV basis	j 0.87	10.8 + j 14.4	5.1 + j 6.8	0.12 + j 0.33	j 74.6
I Z I ohms scaler	0.87	18	8.5	0.35	74.6

b – % Impedance and ohmic impedance network values

Source to fault resistance $\quad\quad\quad\quad R \;= 16.02$

Total source to fault scalar impedance $\quad$ I Z I = 102.3
Total source to fault reactance $\quad\quad\quad X \;= 97$
Total source to fault vector impedance $\quad Z \;= 98.3$

c – Source to fault impedance values

Scaler vs vector impedance simplification gives –4% to resulting fault level
Reactance vs vector impedance simplification gives +1.3% to resulting fault level

Figure 1.26   Simple radial network and system resistance, reactance and scalar
impedance values

value by the phase to neutral voltage to obtain the total three phase fault current.
Consideration of even a small section of the system usually involves at least one
delta–star conversion. Obviously hand calculation of earth fault currents involving sequence impedance networks is even more time consuming and hence the
computer is the best option for all but the simplest system.

Sometimes it is a knowledge of phase angle (e.g. in directional relay protection studies) that may be important. More usually, as is the case of circuit
breaker ratings or stability of balance protection schemes, it is the magnitude
of the fault level which is of prime interest. Some assumptions may be made

for hand calculations in order to simplify the work and avoid vector algebra with errors of approximately ±10%. For example:

1. Treat all impedances as scalar quantities and manipulate arithmetically. This will lead to an overestimation of source to fault impedance and hence an underestimation of fault current. This may be adequate when checking the minimum fault current available with suitable factors of safety for protection operation.
2. Ignore resistance and only take inductive reactance into account. This will lead to an underestimation of source to fault impedance and hence an overestimation of fault current. This may be satisfactory for circuit breaker rating and protection stability assessment where the results are pessimistic rather than optimistic.
3. Ignore the source impedance and assume a source with infinite MVA capability and zero impedance. This may be satisfactory for calculating fault levels well away from the source after several transformations. In this case the transformer reactance between source and fault will swamp the relatively low source impedance. Such an assumption would not be valid for an assessment of fault level near the source busbars.
4. Ignore small network impedances such as short lengths of cable between transformers and switchgear. This may be valid at high voltages but at low voltages of less than 1 kV the resistance of cables compared to the inductive reactance will be significant.
5. If exact earth resistivity measurements are unavailable use known data for the type of soils involved. Remember that the zero sequence impedance is proportional to the log of the square root of the resistivity and therefore wide variations in resistivity values will not cause such large variations in zero sequence impedance approximations. For the calculation of new substation earth grids a soil resistivity survey should always be carried out.

Figure 1.26 shows a simple radial network and system resistance, reactance and scalar impedance values. The simplifications described above are taken into account to derive the total vector impedance, the total reactance and the total scalar impedance of the system components from source to fault. The errors resulting from the use of reactance only or scalar impedance compared to the more correct vector impedance are less than 10% in this example.

2 Drawings and Diagrams

2.1 INTRODUCTION

This chapter describes different types of electrical diagrams. It explains how the diagrams are developed from original concepts into drawings which describe the full operation of the system and how further drawings, schedules and diagrams are produced in order to enable the system to be constructed at the factories, installed, tested and commissioned on site. Examples are given of a variety of different styles of presentation based upon manufacturers and National Standards. This chapter concentrates on substation control and protection schemes but the principles apply equally to most electrical plant.

2.2 BLOCK DIAGRAMS

The starting point for new substation work is the block diagram or single line diagram (SLD). A typical example is given in Fig. 2.1. The various elements (current transformers (CTs), metres, control and relay equipment, etc.) are shown symbolically superimposed upon the substation single line diagram. Figure 2.1 uses symbols based upon international practice.

The advantage of this type of diagram is that the complete system can be seen as a whole in semi-pictorial form. Although not meant to be a detailed guide for the layout of the controls and instruments on the control panels, it is sufficiently concise to enable the designer to check that all the facilities required by the operator are present. Similarly for the relay cubicle, the block diagram only illustrates the general requirements for the siting of the relays. A single item on the block diagram could, for example, represent a complex relay scheme which in itself could occupy several racks on the protection panel.

The correct location of CTs for different functions, the summation of CT windings, overlapping of protection zones, selection of voltage transformers (VTs), etc., can all be easily checked on the block diagram. Location of such items as auto-reclose on/off switches (which could be mounted on the control or relay panel or elsewhere) can be seen. Details such as CT ratios, current ratings of switchgear, etc., are also included on these drawings.

The block diagram is usually included in the contract document as part of a tender specification for new works. A fundamental requirement in any such documentation is for the engineer specifying the equipment to leave no doubt as to the exact requirements. The block diagram is therefore usually completed before the scheduling of equipment since its pictorial representation makes it easy to visualize the completed equipment and therefore no major item is missed or wrongly placed. Lack of definition at the tender stage will only result in claims for variations to the contract and extra costs later by the contractor. In particular the need to define local and remote alarms, metering and control may be further clarified using the form of protection block diagram detailed in Fig. 2.2.

Three line diagrams are similar to single line diagrams but show all three phases. This added detail gives further information and is sometimes useful to assist in the location of VTs and especially to help describe three phase to single phase traction substation schemes.

2.3 SCHEMATIC DIAGRAMS

2.3.1 Method of representation

Schematic diagrams describe the main and auxiliary circuits for control, signalling, monitoring and protection systems. They are drawn in sufficient detail to explain to the user the circuitry and its mode of operation. They allow circuits to be 'followed through' when tracing faults.

With the increasing complexity of electronic circuitry discrete functional blocks, such as relays, are often represented as 'black boxes' on the overall schematic diagrams with only the input and output terminals to these units clearly identified. The characteristics of the 'black boxes' are identified by standard symbols and further reference to the complex circuitry inside the 'black boxes' may be obtained from separate manufacturers' drawings. Circuit arrangements are usually shown on schematic diagrams according to their functional aspects. They seldom follow the actual physical layouts of the different component parts.

The key to understanding schematic diagrams is that the switchgear units and relay contacts on the diagrams are shown in their non-energized or standard reference condition. For example, normally open (NO) and normally closed (NC) contacts are shown in their open or closed states respectively. Any exceptions to this rule must be clearly marked on the diagram.

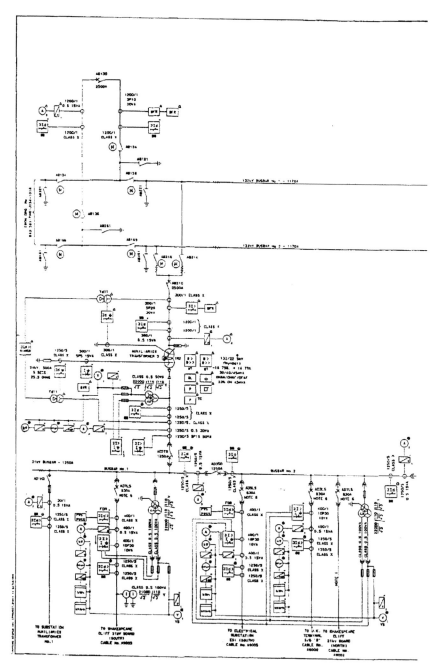

Figure 2.1   Substation block diagram

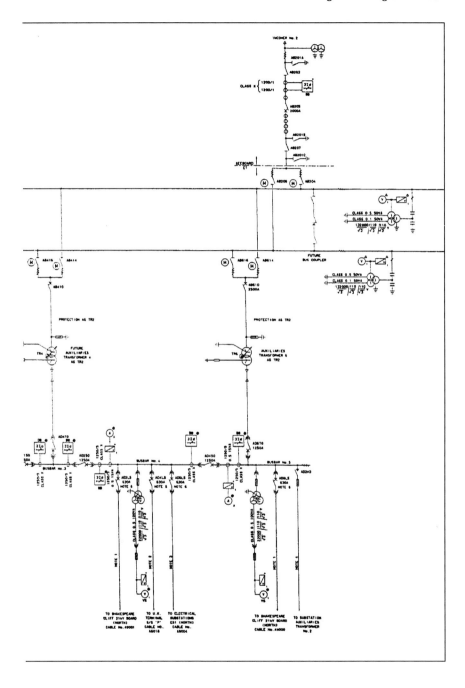

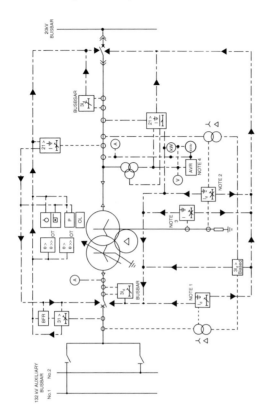

Notes :-

1. Balanced earth fault relay
2. Restricted earth fault relay
3. Stand by earth fault realy
4. Tap changer blocking relay included in AVR scheme
5. Master trip relays to be incorporated on 132 kV and 20 kV CB's.

Figure 2.2   Protection block diagram

2.3.2 Main circuits

Main circuits are shown in their full three phase representation as indicated in the motor starter circuit of Fig. 2.3. The diagrams are drawn from power source (top) to load (bottom) showing the main power connections, switching and protection arrangements.

2.3.3 Control, signalling and monitoring circuits

European practice is to draw circuits between horizontal potential lines (positive at top and negative at bottom for DC control). The signal and information flow in open-loop control systems is also generally from the top of the drawing down and for closed-loop systems from left to right. Figure 2.4 shows a typical control schematic arising from a motor starter system. Relay auxiliary contacts (in their standard reference condition) are shown beneath each relay coil. Exact codification varies from manufacturer to manufacturer as explained in Section 2.4. Some examples of European practice are given in Fig. 2.5a–e.

2.4 MANUFACTURERS' DRAWINGS

Manufacturers use different styles and presentations for their drawings. This section compares some of the advantages and disadvantages of different manufacturers' styles by comparing a simple control and protection scheme for a feeder circuit. The concept block diagram for this feeder is shown in Fig. 2.6. For simplicity in all these examples the CT connections have not been shown.

2.4.1 Combined wiring/cabling diagrams

Figure 2.7 illustrates this type of drawing. Usually drawn on a large sheet of A0 or A1 size paper it shows the rear view of all equipments in the control and relay cubicles, the equipment in the circuit breaker and the multicore cable interconnections. The internal arrangements of the contacts in switches and relays are shown. In practice, the drawing would also show CT connections, relay coils, etc., which would make the circuits even more difficult to follow.

This type of drawing dates back to the days when control switches were mounted on flat slabs of stone or slate supported on a steel frame. (This is where the term 'panel' originates and it is still used by many engineers to refer to a relay or control cubicle.) It is found on circuits for older, relatively simple equipment.

The advantage of this type of drawing is that it is possible to trace the complete circuit of part of a control scheme from only one diagram. It is a pictorial

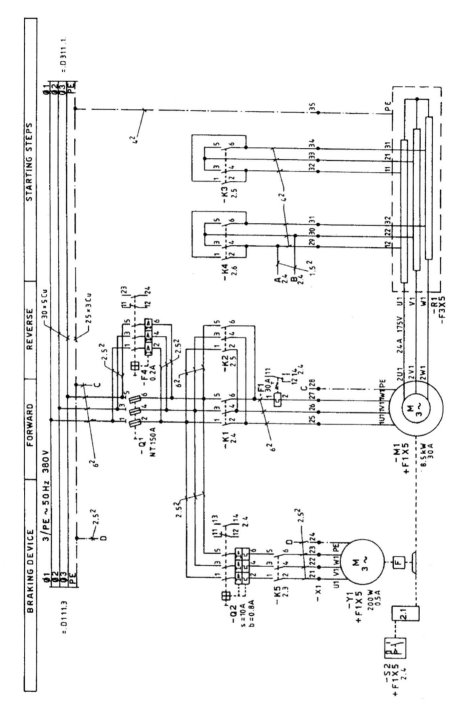

Figure 2.3   Typical main circuit diagram motor starter circuit

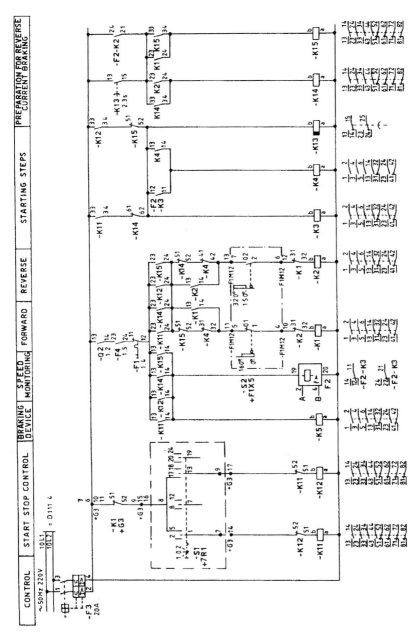

Figure 2.4 Typical control schematic (associated with Fig. 2.3 motor circuit)

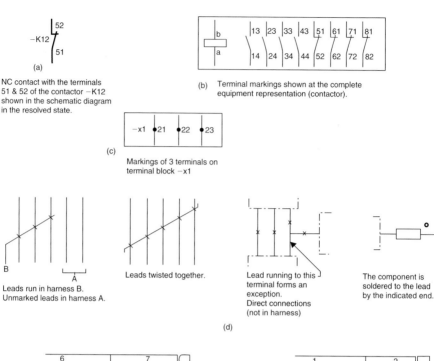

(a)

NC contact with the terminals
51 & 52 of the contactor −K12
shown in the schematic diagram
in the resolved state.

(b)   Terminal markings shown at the complete
equipment representation (contactor).

(c)

Markings of 3 terminals on
terminal block −x1

B

A

Leads run in harness B.
Unmarked leads in harness A.

Leads twisted together.

Lead running to this
terminal forms an
exception.
Direct connections
(not in harness)

The component is
soldered to the lead
by the indicated end.

(d)

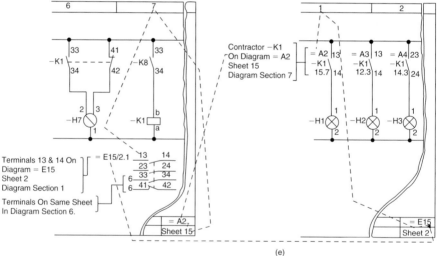

(e)

Figure 2.5   Equipment markings – European practice; (a) NC contact with the terminals
51 & 52 of the contactor −K12 shown in the schematic diagram in the resolved state; (b) terminal markings shown at the complete equipment representation (contactor); (c) markings of
3 terminals on terminal block −X1; (d) method for representing wiring; (e) partial representation of an item of equipment in different diagrams

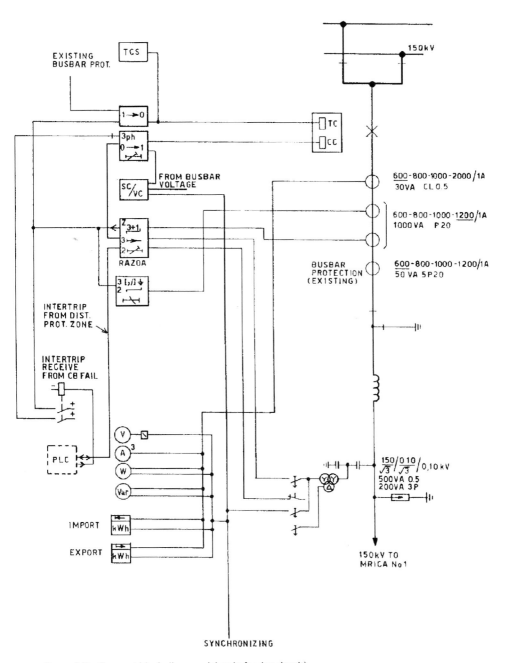

Figure 2.6 Concept block diagram (simple feeder circuit)

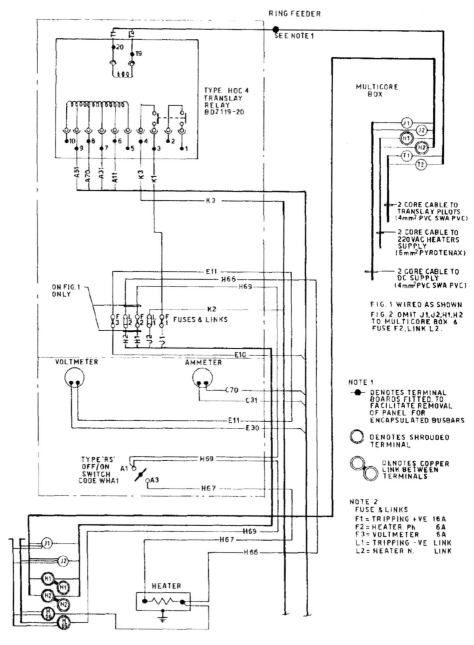

Figure 2.7   Combined schematic wiring cabling diagram for 11 kV circuit breaker feeder

representation of the physical wiring layout and may be used by the electrician or wireman at the manufacturer's works to install the wires in the cubicles, used by the site staff to install and terminate multicore cables, by commissioning staff to check operation and testing and also later by operating staff for fault tracing or future modifications. For site use the wires are often coloured by operations staff to assist in circuit tracing.

A major disadvantage of this type of drawing is that it is difficult to trace all branches and sub-branches of complex control and relay circuits. When the large format drawing has been folded and unfolded many times it becomes difficult to follow the individual wires as they cross or follow the drawing creases. The advent of more complex protection schemes and multicore electronic data lines has meant that drawings of this type are rarely produced today.

2.4.2 British practice

Although traditional, older British practice is described here, the same basic system has also been used in US, Japan and other countries.

Working straight from the block diagram, two separate drawings are produced for each HV circuit showing VT/CT circuits in one and DC/auxiliary AC circuits in the other. Large A0 or A1 sheets of paper are employed. VT and CT circuits are fully drawn out showing schematically every CT wire, tapping, relay coil, terminals, etc. The second drawing has the complete DC control system drawn out schematically from +ve to −ve, left to right. The DC circuit diagram also usually includes any auxiliary AC circuits (for heaters, motors, etc.). Figure 2.8 shows the DC schematic for a simple distribution voltage level feeder circuit.

Each circuit (closing circuit, trip circuit, etc.) is clearly shown, each branch can be traced and isolation points such as fuses or links are indicated. Some manufacturers put on terminal markings for different types of equipment (as shown in Fig. 2.8), others draw dotted lines around the limits of control cubicles, protection cubicles, etc.

The principal difference between the traditional British practice and most other systems is that wires in the British system are identified by a wire number and not by a terminal number. The wire number has an alphabetic prefix which identifies the type of circuit. For example, trip circuit (K), alarm circuit (L), CT circuit (A, B or C, depending on the type of protection), etc. The number identifies the relative position in the circuit. The number changes at each contact, coil, indicator lamp, etc., but the same number is used for all branches of a circuit which are directly connected together. Most British manufacturers write a full description of the protection relay or an abbreviation (e.g. $2 \times$ O/C, $1 \times$ E/F IDMT for two pole overcurrent and one pole earth fault inverse definite minimum time-lag relay) next to the unit together with the manufacturer's type number. There is also a numerical code system for relay identification widely used by US and Japanese manufacturers. A list is given in Appendix A.

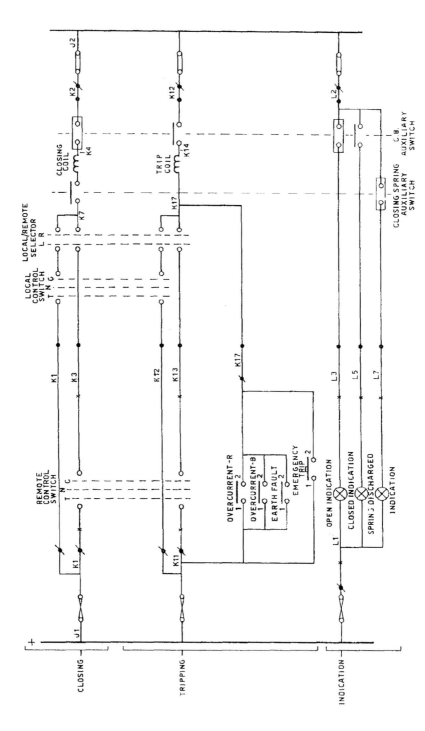

Figure 2.8   Traditional British style DC schematic diagram

Figure 2.9 shows the type of drawing produced by the manufacturer to enable the wiring to be carried out in the works. Sometimes the multicore cable connections are also shown on the same diagram. The equipment and terminals blocks are drawn as seen by the wireman from the back of the cubicle and is very useful to assist the site staff during fault tracing, service and for future

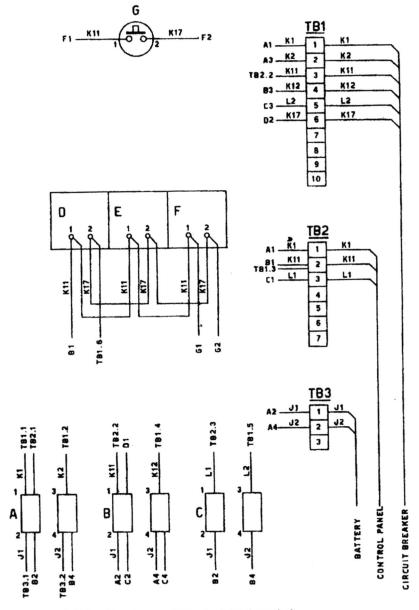

Figure 2.9 Cubicle wiring diagram (British traditional practice)

modifications. Similar drawings are supplied for all relay and control cubicles, marshalling kiosks, circuit breakers, etc.

Multicore cable schedules are also supplied by the manufacturers. These schedules give the number of cores required, the wire numbers to be given to the cores, types of cable, lengths, sizes, etc.

2.4.3 European practice

Many European manufacturers used to produce schematic diagrams on paper 30 cm high (A4 height) but in fold-out format to be as long as necessary (often several metres long) to show all the circuits. This type of drawing produced on rolls of tracing paper could be reproduced easily in long lengths using dyeline printers. With the advent of computer aided drafting (CAD) and cheap electrostatic photocopying machines most European manufacturers now produce a series of A3 format schematic drawings. Some manufacturers break down their large drawings into several A3 sheets based upon 'functional' aspects of the circuitry and some produce 'unit'-orientated drawings.

In the 'function' drawings, one sheet (or group of sheets) will cover a complete circuit such as a circuit breaker and isolator interlocking scheme, indication and alarm scheme, CT circuit, etc.

In the 'unit'-orientated drawings, one or more drawings will show an item of plant and all parts of all circuits wired through that plant. For example, the drawings showing a circuit breaker will include part of the trip, close, synchronizing, alarms, interlocking supervisory functions, etc.

Manufacturers using these types of diagrams have developed their own codes for identification of equipment function and location. They are all generally based upon the German DIN standard. Their starting point for producing drawings is to take the concept block diagram and produce further block diagrams for each HV circuit using their own symbols and terminology. Schedules are also provided to explain the uninitiated meaning of the various symbols and codes used.

Figures 2.10a–2.10d illustrate the type of diagrams produced by ABB (previously known as ASEA Brown Boveri) from the block diagram of Fig. 2.6. These are basically 'unit' rather than 'function' orientated.

Figure 2.10a shows the DC supply sources for the various trip, close and indications. On a more complex scheme these may be on separate sheets. Incoming supplies are shown on the lower side of the dashed line indicating the boundary of the relay cubicle, RP. The symbols R+, R−, etc., shown on the wires do not physically exist on the wires themselves. They are added on the schematics to enable the same wire to be identified on the continuation sheets, Figs 2.10b–2.10d. Looped connections for the +ve and −ve lines within the cubicles are not shown to avoid cluttering the diagram. It is therefore not possible to tell from the diagram if the +ve wire from terminal B50.2 loops first to switch F1 terminal 3, switch F2 terminal 3 or switch F3 terminal 3. For this information you would have to consult the separate wiring schedule. The D29

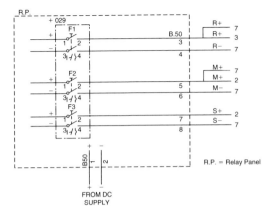

Figure 2.10a   DC supplies

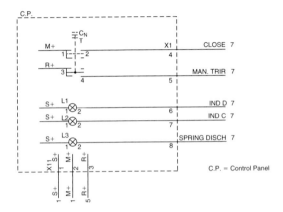

Figure 2.10b   Control and indication

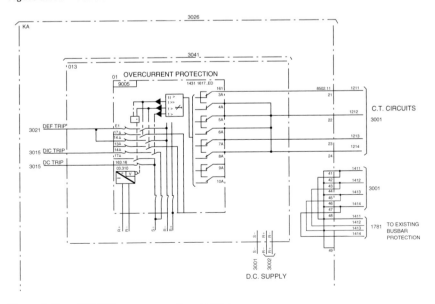

Figure 2.10c   150 kV Line overcurrent protection

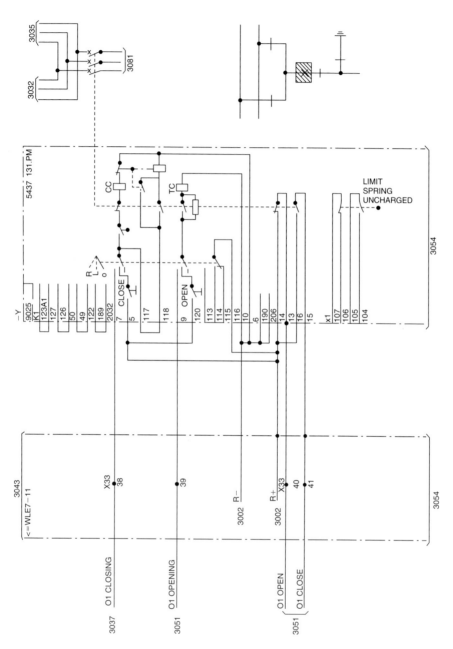

Figure 2.10d   Circuit breaker

indicates that all the items in the small dotted rectangle (F1, F2 and F3) are mounted in location D29. One firm's practice is to designate the terminal block for power supplies, etc., in any cubicle as B50, even if the miniature circuit breakers are mounted at location +D29.

Figure 2.10b shows some of the circuits on the control panel. Again incoming supplies are at the lower edge of the dotted line representing the boundary of the control panel. The 'close', 'main trip', etc., descriptions on the wires from terminals X1.4, X1.5, etc., are to enable the continuation of the same circuits to be identified on Fig. 2.10d.

Figure 2.10c shows in block form the protective relays and connections. The 74310027-EB refers to a separate drawing which shows the full internal wiring of the relay. In some cases the manufacturers might not be willing to reveal the full details of their electronic 'black boxes'. The +D13 refers to the location of the relay within the relay cubicle +RP. For simplicity, the CT circuits have not been included in the drawing but in practice they would continue on one or more other sheets.

Figure 2.10d shows the circuit breaker. As can be seen, the drawing includes trip, close and indication circuits.

The main disadvantage of this type of schematic diagram is that it is difficult to follow a complete circuit because it runs from page to page. Even with the simple example shown, to trace the working of the 'spring discharged' indication circuit requires reference to Figs 2.10a, 10b, 10d and back to 10a. On a double busbar substation, to trace out the workings of the interlocking scheme might need reference to say 20 drawings. The drawings are also difficult to modify since a single wire change may affect many diagrams and schedules. With this type of schematic diagram the manufacturer also produces a wiring table or schedule for all the wires between the terminal blocks, relays and other equipment in the cubicle, the inter-cubicle wiring and to switchgear. When checking and commissioning equipment it is possible to miss a parallel circuit and it is sometimes useful to colour the diagrams as each branch of a circuit is tested and checked off.

The advantage of this type of drawing is that all the drawings are of a manageable size, they can be easily stored and handled at the work place using A3 files. Standard schemes can be worked up by the manufacturer and the diagrams assist the moves to European standardization.

2.4.4 Other systems

Three basic systems have been described: the older all-in-one drawings which are still encountered on relatively simple schemes and may still be encountered in existing substations, the older British traditional system which is similar to the Japanese, US and other countries where large drawings are used and the more modern systems with numerous small (A3 and A4) drawings and schedules. It is not practical to compare drawings from every main manufacturer but variations and combinations of each of these types exist. The important thing is that whenever a system of drawings is encountered for the first time, the engineer concerned should ensure that the manufacturer supplies full details of the schemes employed together with lists of standard symbols and codes. This is necessary to ensure that the operating staff have sufficient information to operate and maintain the plant and update the drawings if future modifications are carried out. Appendix B gives a comparison between some German, British, US/Canadian and International symbols.

2.5 COMPUTER AIDED DESIGN (CAD)

Most manufacturers now use computer aided design (CAD) to assist in the preparation of diagrams. The designer's basic sketch is used by the computer operator and standard symbols for relays, switches, etc., are called up from the computer memory files and placed in the appropriate positions on the visual display unit (VDU) representation of the drawing. The wiring interconnections are added, location details entered and when satisfactory a hard copy printout of the drawing is produced. At the same time by correct programming the computer can produce all the interconnection and multicore cable schedules directly from the drawing input. CAD is not only limited to A3 drawings but can be used to produce drawings in A0 sheet format using suitable printers or plotters. Figure 2.2 is a typical example of a block diagram prepared using CAD with the relay symbols called up from a memory bank.

The main disadvantage with CAD is that the finished drawings may be more useful from the manufacturer's point of view since they will be difficult to update or modify on site without access to the original drawing data files. It is therefore usual to specify that final drawings are provided in electronic form as well as in hard copy.

Any disadvantage is outweighed by the clarity of the diagrams and circuit standardization which the control or protection engineer quickly recognizes. Modifications to schematics, for example during commissioning, can now be carried out on site using a personal computer and issued back to head office electronically for updating the master copies. In addition the ability to 'layer' CAD drawings is especially useful. The 'background layer' file can contain the manufacturer's standard scheme so that customization of standard designs can be quickly achieved. Substation cable routing diagrams can be based upon the civil substation trench layout contained in one file layer and the cable routing design superimposed upon it by the designer in another layer. Strict drawing office procedures are necessary to ensure layers are not muddled. Another word of caution here is that when exporting your data exchange files (DXF) to another manufacturer or customer they will then have access to your symbol data bank which may have taken many hours to construct.

2.6 CASE STUDY

The block diagram shown in Fig. 2.1 has formed part of the tender documentation for the Greater New Town Transmission and Distribution Phase V Project. The project is being sponsored by a large international aid agency which has appointed a firm of consulting engineers to prepare the contract documentation and administer the contract. At the present time assume that the contract has already been let on a 'turnkey basis' after competitive tender with one of five well-known design and construct contractors, namely S. U. B. Betabuilder Plc. The client and substation operator's representative, Mr Ali of Greater New Town Ministry of Electricity and Water, has called the consultant and contractor to a meeting at short notice because of a dispute regarding the substation works. The client is refusing to accept the 'completed' works. The contractor is refusing to carry out further modifications to the metering alarms and indications associated with the protection and control scheme of the feeder circuits, although the threat of S. U. B. Betabuilder's performance bond not being released by the client is hanging over them. S. U. B. Betabuilder's line is that this is a 'turnkey contract' and that they have built the substation in accordance with the tender drawings. In addition all the substation detailed design drawings have been approved during the design process by the consultant. Further, the aid agency (which pays the consultant's fees) has got to hear of the dispute and is applying pressure on the consultant to resolve rapidly the issue 'in accordance with the terms and conditions of the contract'.

The meeting is strained and Mr Ali thumps the table and asks the consultant, 'Well, what did you exactly specify for the monitoring and control of the feeder circuits in this substation? My engineers tell me that additional work is necessary for interfacing with the Greater New Town Stage VI system control and data acquisition (SCADA) scheme, which is already out to tender. Are you expecting me to start issuing variation orders even before the SCADA contract is let?'

The consultant lays out on the table the substation block diagram (Fig. 2.1) and the schedule of requirements from the tender document as shown below in Table 2.1. 'It is quite clear what the requirements are. In any case any competent contractor would have enquired exactly what was required before entering the definitive detailed design stage'.

Do you agree?

How could the design definition, if required, have been improved in the tender documentation? Illustrate with a diagram for inclusion in a future tender document by the consultant if they get another job on the Greater New Town Project.

Do you think a variation order issued to S. U. B. Betabuilder would resolve the problem?

2.7 GRAPHICAL SYMBOLS

IEC 60617 is a very important reference for graphical symbols used in drawings and diagrams. It is divided into several parts, but is only now accessible

Table 2.1 Schedule of requirements – overhead line protection circuits

Item	Description	Quantity
1.	Main protection – distance relay	1
2.	Direction earth – fault relay (short lines only ~20 km)	1
3.	Back-up protection – IDMTL O/C and E/F elements	1
4.	Inter-tripping	Set
5.	Auto-reclosing – high-speed single pole (delayed auto-reclosing with check synchronizing where shown on the single line diagram)	Set
6.	All items necessary whether fully described or not are deemed to be included in the scope of the contract and will be checked during the course of the works	Set

through an online data base IEC 60617-DB-12M. CENELEC has therefore withdrawn the identical EN60617 as a set of publications, on the basis that the symbols are more conveniently available from the IEC web site. The UK equivalent BSEN60617 (which superseded BS 3939) has been similarly treated. The useful BS 7845, a guide to 60617, is currently available but has been declared 'obsolescent'.

Part 7 of 60617 covering switchgear, control gear and protective devices is especially useful. Protection relays are drawn as a box (see Fig. 2.1) with a symbol describing its function. For example:

$\boxed{I>}$ Delayed overcurrent relay with inverse time-lag characteristic

$\boxed{Z<}$ Under-impedance relay

Recommendations for printing symbols and numbers to represent electromagnetic quantities (volt, V, etc.) are described in IEC60027, 'Letter symbols to be used in electrical technology'.

Appendix A
RELAY IDENTIFICATION –
NUMERICAL CODES

In addition to the use of international symbols for indicating protection relay functions (see EN 60617-7,1996) American ANSI Standard C37-2 numbering is also used. Where more detailed relay functionality is required this is achieved by supplementing the numbering system with letters. An underfrequency relay then has the ANSI coding of 81 U and an IEC 60617 symbol: $\boxed{f<}$. This appendix describes this numbering system and suffix letters.

Device number, definition and function

1. **master element** is the initiating device such as a control switch voltage relay float switch, etc., which serves either directly or through such permissive devices as protective and time delay relays to place an equipment in or out of operation.
2. **time-delay starting, or closing, relay** is a device which functions to give a desired amount of time delay before or after any point of operation in a switching sequence of protective relay system except as specifically provided by device functions 63 and 79 described later.
3. **checking or interlocking relay** is a device which operates in response to the position of a number of other devices or to a number of predetermined conditions in an equipment to allow an operating sequence to proceed to stop or to provide a check of the position of these devices or of these conditions for any purpose.
4. **master contactor** is a device generally controlled by device … or equivalent and the necessary permissive and protective devices which serves to make and break the necessary control circuits to place an equipment into operation under the desired conditions and to take it out of operation under other or abnormal conditions.
5. **stopping device** functions to place and hold an equipment out of operation.

6. **starting circuit breaker** is a device whose principal function is to connect a machine to its source or starting voltage.

7. **anode circuit breaker** is one used in the anode circuits of a power rectifier for the primary purpose of interrupting the rectifier circuit of an arc back should occur.

8. **control power disconnecting device** is a disconnecting device – such as a switch, circuit breaker or pullout fuse block – used for the purpose of connecting and disconnecting, respectively the source of control power to and from the control bus or equipment.
   Note: Control power is considered to include auxiliary power which supplies such apparatus as small motors and …

9. **reversing device** is used for the purpose of reversing a machine held or for performing any other reversing functions.

10. **unit sequence switch** is used to change the sequence in which units may be placed in and out of service in multiple-unit equipments.

11. Reserved for future application.

12. **over-speed device** is usually a direct-connected speed switch which functions on machine overspeed.

13. **synchronous-speed device**, such as a centrifugal-speed switch, a slip-frequency relay, a voltage relay, an undercurrent relay or any type of device which operates at approximately synchronous speed of a machine.

14. **under-speed device** functions when the speed of a machine falls below a predetermined value.

15. **speed or frequency, matching device** functions to match and hold the speed or the frequency of a machine or of a system equal to or approximately equal to that of another machine source or system.

16. Reserved for future application.

17. **shunting, or discharge, switch** serves to open or to close a shunting circuit around any piece of apparatus (except a resistor) such as a machine held a machine armature a capacitor or a rectifier.
   Note: This excludes devices which perform such … operations as may be necessary to the process of starting a machine by devices 6 or 42 or their equivalent . . . excludes device 73 function which serves for the switching on resistors.

18. **accelerating or decelerating device** as used to close or to cause the closing of circuits which are used to increase or to decrease the speed of a machine.

19. **starting-to-running transition contactor** is a device which operates to initiate or cause the automatic transfer of a machine from the starting to the running power connection.

20. **electrically operated valve** is a solenoid or motor-operated valve which is used in a vacuum, air, gas, oil, water or similar lines.
   Note: The function of the valve may be indicated by the insertion of descriptive words such as Brake in Pressure Reducing in the function name, such as Electrically Operated Brake Valve.

21. **distance relay** is a device which functions when the circuit admittance impedance or reactance increases or decreases beyond predetermined limits.

22. **equalizer circuit breaker** is a breaker which serves to control or to make and break the equalizer or the current-balancing connections for a machine held, or for requesting equipment, in a multiple-unit installation.

23. **temperature control device** functions to raise or to lower the temperature of a machine or other apparatus, or of any medium when its temperature falls below, or rises above, a predetermined value.
    Note: An example is a thermostat which switches on a space heater in a switchgear … when the temperature falls to a desired value as distinguished from a device which is used to provide automatic temperature registering between close limits and would be designated as 90T.

24. Reserved for future application.

25. **synchronizing, or synchronism-check, device** operates when two AC circuits are within the desired limits of frequency, phase angle or voltage to permit or to cause the paralleling of these two circuits.

26. **apparatus thermal device** functions when the temperature of the shunt held or the armortisseur winding of a machine, or that of a load limiting or load shunting resistor or of a liquid or other medium exceeds a predetermined value or if the temperature of the protected apparatus such as a power rectifier or of any medium decreases below a predetermined value.

27. **undervoltage relay** is a device which functions on a given value of undervoltage.

28. Reserved for future application.

29. **isolating contactor** is used expressly for disconnecting one circuit from another for the purposes of emergency operating, maintenance or test.

30. **annunciator relay** is a non-automatically reset device which gives a number of separate visual indications upon the functioning of protective devices, and which may also be arranged to perform a lock-out function.

31. **separate excitation device** connects a circuit such as the shunt held of a synchronous converter to a source of separate excitation during the starting sequence or one which energizes the excitation and ignition circuits of a power rectifier.

32. **directional power relay** is one which functions on a desired value of power flow in a given direction or upon reverse power resulting from arc-back in the anode or cathode circuits of a power rectifier.

33. **position switch** makes or breaks contact when the main device or piece of apparatus which has no device function number reaches a given position.

34. **motor-operated sequence switch** is a multi-contact switch which fixes the operating sequence of the major devices during starting and stopping, or during other sequential switching operations.

35. **brush-operating, or slip-ring short-circuiting, device** is used for raising, lowering or shifting the brushes of a machine, or for short-circuiting its slip rings, or for engaging or disengaging the contacts of a mechanical rectifier.

36. **polarity device** operates or permits the operation of another device on a predetermined polarity only.

37. **undercurrent or underpower relay** is a device which functions when the current or power flow decreases below a predetermined value.

38. **bearing protective device** is one which functions on excessive bearing temperature or on other abnormal mechanical conditions, such as undue wear which may eventually result in excessive bearing temperature.
39. Reserved for future application.
40. **field relay** is a device that functions on a given or abnormally low value or failure of machine-held current or on an excessive value of the reactive component or armature current in an arc machine indicating abnormally low-held excitation.
41. **field circuit breaker** is a device which functions to apply or to remove the held excitation of a machine.
42. **running circuit breaker** is a device whose principal function is to connect a machine to its source of running voltage after having been brought up to the desired speed on the starting connection.
43. **manual transfer** or selector device transfers the control circuits so as to modify the plan of operation of the switching equipment or of some of the devices.
44. **unit sequence starting relay** is a device which functions to start the next available unit in a multiple unit equipment on the failure or on the non-availability of the normally preceding unit.
45. Reserved for future application.
46. **reverse-phase, or phase-balance, current relay** is a device which functions when the polyphase currents are of reverse-phase sequence or when the polyphase currents are unbalanced or contain negative phase-sequence components above a given amount.
47. **phase-sequence voltage relay** is a device which functions upon a predetermined value of polyphase voltage in the desired phase sequence.
48. **incomplete sequence relay** is a device which returns the equipment to the normal, or off, position and locks it out if the normal starting operating or stopping sequence is not properly completed within a predetermined time.
49. **machine, or transformer, thermal relay** is a device which functions when the temperature of an AC machine armature, or of the armature or other load carrying winding or element of a DC machine or converter or power rectifier or power transformer (including a power rectifier transformer) exceeds a predetermined value.
50. **instantaneous overcurrent, or rate-of-rise relay** is a device which functions instantaneously on an excessive value of current, or on an excessive rate of current rise, thus indicating a fault in the apparatus or circuit being protected.
51. **AC time overcurrent relay** is a device with either a definite or inverse time characteristic which functions when the current in an AC circuit exceeds a predetermined value.
52. **AC circuit breaker** is a device which is used to close and interrupt an AC power circuit under normal conditions or to interrupt this circuit under fault or emergency conditions.
53. **exciter or DC generator relay** is a device which forces the DC machine-held excitation to build up during starting or which functions when the machine voltage has built up to a given value.

54. **high-speed DC circuit breaker** is a circuit breaker which starts to reduce the current in the main circuit in 0.01 s or less, after the occurrence of the DC overcurrent or the excessive rate of current rise.

55. **power factor relay** is a device which operates when the power factor in an AC circuit becomes above or below a predetermined value.

56. **field application relay** is a device which automatically controls the application of the field excitation to an AC motor at some predetermined point in the slip cycle.

57. **short-circuiting or grounding device** is a power or stored energy-operated device which functions to short-circuit or to ground a circuit in response to automatic or manual means.

58. **power rectifier misfire relay** is a device which functions if one or more of the power rectifier anodes fails to fire.

59. **overvoltage relay** is a device which functions on a given value of overvoltage.

60. **voltage-balance relay** is a device which operates on a given difference in voltage between two circuits.

61. **current-balance relay** is a device which operates on a given difference in current input or output of two circuits.

62. **time-delay stopping, or opening, relay** is a time-delay device which serves in conjunction with the device which initiated the shutdown stopping, or opening operation in an automatic sequence.

63. **liquid or gas pressure, level, or flow relay** is a device which operates on given values of liquid or gas pressure, flow or level or on a given rate of change of these values.

64. **ground protective relay** is a device which functions on failure of the insulation of a machine transformer or of the other apparatus to ground or on flashover of a DC machine to ground.
Note: This function is designed only to a relay which detects the flow of current from the frame of a machine or enclosing case or structure or a piece of apparatus to ground, or detects a ground on a normally ungrounded winding or circuit. It is not applied to a device connected in the secondary circuit or secondary neutral of a current transformer or current transformers, connected in the power circuit or a normally grounded system.

65. **governer** is the equipment which controls the gate or valve opening of a prime mover.

66. **notching, or jogging, device** functions to allow only a specified number of operations of a given device, or equipment, or a specified number of successive operations within a given time of each other. It also functions to energize a circuit periodically, or which is used to permit intermittent acceleration or jogging of a machine at low speeds for mechanical positioning.

67. **AC directional overcurrent relay** is a device which functions on a desired value or AC overcurrent flowing in a predetermined direction.

68. **blocking relay** is a device which initiates a pilot signal for blocking of tripping on external faults in a transmission line or in other apparatus under predetermined conditions, or co-operated with other devices to

block tripping or to block reclosing on an out-of-step condition or on power swings.

69. **permissive control device** is generally a two-position, manually operated switch which in one position permits the closing of a circuit breaker, or the placing of an equipment into operation, and in the other position prevents the circuit breaker or the equipment from being operated.

70. **electrically operated rheostat** is a rheostat which is used to vary the resistance of a circuit in response to some means of electrical control.

71. Reserved for future application.

72. **DC circuit breaker** is used to close and interrupt a DC power circuit under normal conditions or to interrupt this circuit under fault or emergency conditions.

73. **load-resistor contactor** is used to shunt or insert a step of load limiting, shifting or indicating resistance in a power circuit, or to switch a space heater in circuit or to switch a light or regenerative, load resistor or a power rectifier or other machine in and out of circuit.

74. **alarm relay** is a device other than an annunciator as covered under device No. 30 which is used to operate, or to operate in connection with a visual or audible alarm.

75. **position changing mechanism** is the mechanism which is used for moving a removable circuit breaker unit to and from the connected, disconnected and test positions.

76. **DC overcurrent relay** is a device which functions when the current in a DC circuit exceeds a given value.

77. **pulse transmitter** is used to generate and transmit pulses over a telemetering or pilot-wire circuit to the remote indicating or receiving device.

78. **phase angle measuring, or out-of-step protective relay** is a device which functions at a predetermined phase angle between two voltages or between two currents or between voltage and current.

79. **AC reclosing relay** is a device which controls the automatic reclosing and locking out of an AC circuit interruptor.

80. Reserved for future application.

81. **frequency relay** is a device which functions on a predetermined value of frequency either under or over or on normal system frequency or rate or change of frequency.

82. **DC reclosing relay** is a device which controls the automatic closing and reclosing of a DC circuit interruptor generally in response to load circuit conditions.

83. **automatic selective control, or transfer, relay** is a device which operates to select automatically between certain sources or conditions in an equipment or performs a transfer operation automatically.

84. **operating mechanism** is the complete electrical mechanism, or servo mechanism including the operating motor, solenoids position switches, etc., for a tap changer induction regulator of any piece of apparatus which has no device function number.

85. **carrier, or pilot-wire, receiver relay** is a device which is operated or restrained by a signal used in connection with carrier-current or DC pilot-wire fault directionally relaying.

86. **locking-out relay** is an electrically operated band or electrically reset device which functions to shut down and … and equipment out of service on the occurrence of abnormal conditions.

87. **differential protective relay** is a protective device which functions on a percentage or phase angle or other quantitative difference of two currents or of some other electrical quantities.

88. **auxiliary motor, or motor generator** is one used for operating auxiliary equipment such as pumps, blowers, exciters, rotating magnetic amplifiers, etc.

89. **line switch** is used as a disconnecting or isolating switch in an AC or DC power circuit when this device is electrically operated or has electrical accessories such as an auxiliary switch, magnetic lock, etc.

90. **regulating device** functions to regulate a quantity or quantities, such as voltage current, power, speed frequency temperature and load, at a certain value or between certain limits for machines, the lines or other apparatus.

91. **voltage directional relay** is a device which operates when the voltage across an open circuit breaker or contactor exceeds a given value in a given direction.

92. **voltage and power directional relay** is a device which permits or causes the connection of two circuits when the voltage difference between them exceeds a given value in a predetermined direction and causes these two circuits to be disconnected from each other when the power flowing between them exceeds a given value in the opposite direction.

93. **field changing contactor** functions to increase or decrease in one step the value of field excitation on a machine.

94. **tripping, or trip-free, relay** is a device which functions to trip a circuit breaker contactor or equipment or to permit immediate tripping by other devices, or to prevent immediate reclosure of a circuit interruptor, in case it should open automatically even though its closing circuit is maintained closed.

95. to 99. Used only for specific applications on individual installations where none of the assigned numbered functions from 1 to 94 is suitable.

SUFFIX LETTERS

Suffix letters are used with device function numbers for various purposes. In order to prevent possible conflict, any suffix used singly, or any combination of letters, denotes only one word or meaning in an individual equipment. All other words should be written out in full each time they are used or use a clearly distinctive abbreviation. Furthermore, the meaning of each single suffix letter, or combination of letters should be clearly designated in the legend on the drawings or publications applying to the equipment.

The following suffix letters generally form part of the device function designation and thus are written directly behind the device number, such as 23X, 90V or 52RT.

These letters denote separate auxiliary devices, such as:

X	auxiliary relay
Y	auxiliary relay
Z	auxiliary relay
R	raising relay
L	lowering relay
O	opening relay
C	closing relay
CS	control switch
CL	'a' auxiliary switch relay
OP	'b' auxiliary switch relay
U	'up' position switch relay
D	'down' position switch relay
PB	push button

Note: In the control of a circuit breaker with so called X-Y relay control scheme, the X relay is the device whose main contacts are used to energize the closing coil and the contacts of the Y relay provide the anti-pump feature for the circuit breaker.

These letters indicate the **condition** or **electrical quantity** to which the device responds, or the medium in which it is located, such as:

A	air or amperes
C	current
E	electrolyte
F	frequency or flow
L	level or liquid
P	power or pressure
PF	power factor
O	oil
S	speed
T	temperature
V	voltage, volts or vacuum
VAR	reactive power
W	water or watts

These letters denote the **location of the main device in the circuit**, or the type of circuit in which the device is used or the type of circuit or apparatus with which it is associated when this is necessary, such as:

A	alarm or auxiliary power
AC	alternating current
AN	anode

B	battery or blower or bus
BK	brake
BP	bypass
BT	bus tie
C	c … or condenser, compensator or carrier current
CA	cathode
DC	direct current
E	exciter
F	feeder or field or filament
G	generator or ground**
H	heater or housing
L	line
M	motor or metering
N	network or neutral**
P	pump
R	reactor or rectifier
S	synchronizing
T	transformer or test or thyratron
TH	transformer (high-voltage side)
TL	transformer (low-voltage side)
TM	telemeter
U	unit

** Suffix 'N' is generally in preference to 'G' for devices connected in the secondary neutral or current transformers, or in the secondary or a current transformer whose primary winding is located in the neutral or a measure or power transformer, except in the case of transmission line receiving, where the suffix 'G' is more commonly used for those relays which operate on ground faults.

These letters denote parts of the main device, divided in the two following categories: all parts, except auxiliary contacts and limit switches as covered later.

Many of these do not form part of the device number and should be written directly below the device number, such as $\frac{43}{A}$.

BB	bucking bar (for high-speed DC circuit breaker)
BK	brake
C	coil or condenser or capacitor
CC	closing coil
HC	holding coil
OS	inductive shunt
L	lower operating coil
M	operating motor
MF	fly-ball motor
ML	load-limit motor
MS	speed adjusting or synchronizing motor
S	solenoid

TC	trip coil
U	upper operating coil
V	valve

All auxiliary contacts and limit switches for such devices and equipments as circuit breakers, contactors, valves and rheostats. These are designated as follows:

a	Auxiliary switch, open when the main device is in the de-energized or non-operated position.
b	Auxiliary switch, closed when the main device is in the de-energized or non-operated position.
aa	Auxiliary switch, open when the operating mechanism of the main device is in the de-energized or non-operated position.
bb	Auxiliary switch, closed when the operating mechanism of the main device is in the de-energized or non-operated position.

e, f, h, etc., ab, ad, etc., or ba, bc, bd, etc., are special auxiliary switches other than a, b, aa and bb. Lower-case (small) letters are to be used for the above auxiliary switches.

Note: If several similar auxiliary switches are present on the same device they should be designated numerically, 1, 2, 3, etc., when necessary.

LC	Latch-checking switch, closed when the circuit breaker mechanism linkage is relatched after an opening operation of the circuit breaker.
LS	Limit switch.

These letters cover **all other distinguishing features or characteristics or conditions**, not previously described which serve to describe the use of the device or its contacts in the equipment such as:

A	accelerating or automatic
B	blocking or backup
C	close or cold
D	decelerating, detonate or down
E	emergency
F	failure or forward
H	hot or high
HR	hand reset
HS	high speed
IT	inverse time
L	left or local or low or lower or leading
M	manual
OFF	off
ON	on

O	open
P	polarizing
R	right, or raise, or reclosing, or receiving, or remote or reverse
s	sending or swing
T	test or trip, or trailing
TDC	time-delay closing
TDO	time-delay opening
U	up

Suffix numbers

If two or more devices with the same function number and suffix letter (if used) are present in the same equipment they may be distinguished by numbered suffixes as for example 52X-1, 52X-2 and 52X-3, when necessary.

Devices performing more than one function

If one device performs two relatively important functions in an equipment so that it is desirable to identify both of these functions this may be done by using a double function number and name such as:

27–59 undervoltage and overvoltage relay.

Appendix B
COMPARISON BETWEEN GERMAN, BRITISH, US/CANADIAN AND INTERNATIONAL SYMBOLS

B1 General circuit elements

Description	German symbols	British symbols	US/Canadian symbols	International symbols
Resistor		≃ or  -�-	= or -�- 	= or -�- 
with tappings	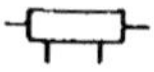	≃	=	=
Winding, inductor		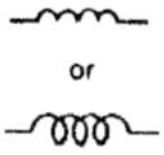 or		= or or 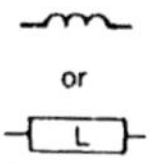
with tappings			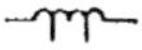	 or =
Capacitor		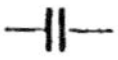	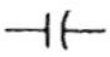	= or
with tapping	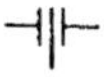			=
Polarized capacitor	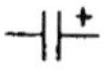 +	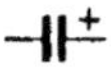 +	=	=
Polarized electrolytic capacitor	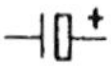 +	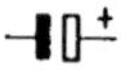 +	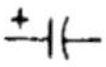 +	= or +
Permanent magnet	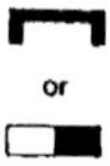 or	=	= or PM	=
Accumulator cell, battery (long line = positive pole)	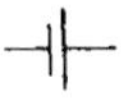	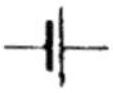	=	=
Earth (ground) connection		=	=	=
Frame or chassis connection		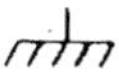		
Electrical driven fan or blower	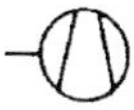	 ∞	—	 ∞

Description	German symbols	British symbols	US/Canadian symbols	International symbols
Variable in operation / – continuously ⌐ – stepwise				
Variable for testing (pre-set adjustment)				
Variable under the influence of a physical quantity	linearly non-linearly			
Spark gap				
Surge diverter, general			or	
Thermocouple				or
Clock			(π)	
Converter, transmitter			–	
Amplifier, general symbol	or	or		or
Single-phase bridge-connected rectifier			– +	

Description	German symbols	British symbols	US/Canadian symbols	International symbols
Isolating fuse		—	—	—
Fuse	supply side	== or	= or	= or =
Isolating link				
Plug and socket device	or			or
Filament lamp	X			=
Discharge lamp				

B2 Operating mechanisms

Description	German symbols	British symbols	US/Canadian symbols	International symbols
Hand operated mechanism		=	=	=
Foot operated mechanism		see page 28	see page 28	
Cam operated mechanism	3 -- 2 -- 1 --		=	or =
Pneumatic operating mechanism			PNEU --	=
Power operating mechanism (stored energy type)		=	—	=
Motor operated mechanism	M	=	= or MOT	=
Valve, general symbol		. =	—	=
Unidirectional latching device		=	—	=
Bidirectional latching device		latched unlatched	—	latched unlatched
Notch			with annotation	=
Device for time delayed operation, following actuating force to right		—		=
Device for cyclic actuation		—	with annotation	—
Latching mechanism		—	SW MECH	—

Description	German symbols	British symbols	US/Canadian symbols	International symbols
Operating element with automatic return on discontinuation of actuating force for contactors, relays, releases		or = or	= or	=
Operating coil energized (the arrow denotes the operating state, if this deviates from the standard representation)		—	—	—
Relay with two coils acting unidirectionally	or or or			= or
Undervoltage relay	U<	UV	U< or	=
Time-delay for electro-mechanic operating elements Coil of slow-releasing relay		(slow releasing) (very slow releasing)	= or SR or SR	= (slow releasing) (very slow releasing)
Coil of slow-operating relay		(slow operating) (very slow operating)	= or SO	= (slow operating) (very slow operating)
Coil of a slow-operating and slow-releasing relay			= or SA	=

Description	German symbols	British symbols	US/Canadian symbols	International symbols
Coil of a polarized relay with permanent magnet		P	+ or P	P or =
Coil of a remanent relay		∫	—	∫ or =
Coil of a mechanically-resonant relay	—	=	—	=

B3 Switchgear

Description	German symbols	British symbols	US/Canadian symbols	International symbols
Make contact (NO)				
Break contact (NC)				
Change-over contact				
Change-over contact make-before-break				
Time-delayed contacts				
Make contact, delayed make			TC or TDC	
Break contact, delayed break			TO or TDO	
Make contact, delayed break			TO or TDO	
Break contact, delayed make			TC or TDC	

Description	German symbols	British symbols	US/Canadian symbols	International symbols
Contactor with thermal overload releases				
Triple-pole circuit breaker with latching mechanism, electro-magnetic release and 3 thermal overload relays				
Isolating circuit breaker			—	
Circuit breaker				
Triple-pole load-break switch			—	
Triple-pole fused isolator				
Triple-pole isolator				
Isolating link, change-over type				

Description	German symbols	British symbols	US/Canadian symbols	International symbols
Single-throw switch, manually operated				—
Spring-return switches, manually operated				=
with 1 NO contact				
with 1 NC contact				=
foot operated				
cam operated			— or CO	
flow speed actuated				=
pressure actuated				=
temperature actuated				= or
liquid level actuated				
over/under normal flow speed	$v > / v <$	—	—	
over/underpressure	$p > / p <$	—	P / P	= / =
over/undertemperature	$\vartheta > / \vartheta <$	—	T / T	= / =
over/under normal liquid level	$V > / V <$	—	L / L	
over/underspeed	$n > / n <$	—	SP / SP	$v >	v <$
Examples: Spring-return switch opens at overspeed		—	SP	
Spring-return switch closes at undertemperature		—	T	=

3 Substation Layouts

3.1 INTRODUCTION

Substations are the points in the power network where transmission lines and distribution feeders are connected together through circuit breakers or switches via busbars and transformers. This allows for the control of power flows in the network and general switching operations for maintenance purposes. This chapter describes the principal substation layouts, the effects of advancements in substation equipment, modular design, compact substations and the moves towards design and construction 'turnkey' contract work. The descriptions concentrate on air insulated switchgear (AIS) outdoor open terminal designs at rated voltages of 72 kV and higher. The design of distribution voltage switchgear and gas insulated switchgear (GIS) is described in Chapter 13, in which terminology is also defined.

3.2 SUBSTATION DESIGN CONSIDERATIONS

3.2.1 Security of supply

In an ideal situation all circuits and substation equipment would be duplicated such that following a fault or during maintenance a connection remains available. This would involve very high cost. Methods have therefore been adopted to achieve a compromise between complete security of supply and capital investment. A measure of circuit duplication is adopted whilst recognizing that duplication may itself reduce the security of supply by, for example, providing additional leakage paths to earth.

Security of supply may therefore be considered in terms of the effect of this loss of plant arising from fault conditions or from outages due to maintenance. The British Code of Practice for the Design of High Voltage Open Terminal

Substations BS 7354 categorizes substation service continuity; recognizing that line or transformer faults destroy service continuity on the affected circuits:

Category 1　No outage necessary within the substation for either maintenance or fault; e.g. the 1½ breaker scheme under maintenance conditions in the circuit breaker area.

Category 2　Short outage necessary to transfer the load to an alternative circuit for maintenance or fault conditions; e.g. the double busbar scheme with bypass disconnector and bus-coupler switch under fault or maintenance conditions in the circuit breaker or busbar area.

Category 3　Loss of a circuit or section; for example the single busbar with bus section circuit breaker scheme for a fault in the circuit breaker or busbar area. The transformer feeder scheme also comes under category 3 service continuity and for this arrangement the addition of incoming circuit breakers, busbar and transformer circuit breakers does not improve the classification.

Category 4　Loss of substation; for example the single busbar scheme without bus sectionalization for a fault in the busbar area.

3.2.2 Extendibility

The design should allow for future extendibility. Adding bays of switchgear to a substation is normally possible and care must be taken to minimize the outages and outage durations for construction and commissioning. Where future extension is likely to involve major changes (such as from a single to double busbar arrangement) then it is best to install the final arrangement at the outset because of the disruption involved. When minor changes such as the addition of overhead line or cable feeder bays are required then busbar disconnectors may be installed at the outset (known as 'skeleton bays') thereby minimizing outage disruption. The use of gas insulated switchgear (GIS) tends to lock the user into the use of a particular manufacturer's switchgear for any future extension work. In comparison an open terminal switchyard arrangement allows the user a choice of switchgear for future extension work.

3.2.3 Maintainability

The design must take into account the electricity supply company system planning and operations procedures together with a knowledge of reliability and maintenance requirements for the proposed substation equipment. The need for circuit breaker disconnector bypass facilities may therefore be obviated by an understanding of the relative short maintenance periods for modern switchgear. Portable earthing points and earthing switch/interlock requirements will also need careful consideration. In a similar way the layout must allow easy access for winching gear, mobile cranes or other lifting devices if maintenance downtimes are to be kept to a minimum. Similarly standard minimum clearances (see

Section 3.4.2) must be maintained for safe working access to equipment adjacent to operational live switchgear circuits or switchgear bays, bearing in mind that some safety authorities now resist the use of ladder working and require access from mobile elevated working platforms or scaffolding.

3.2.4 Operational flexibility

The physical layout of individual circuits and groups of circuits must permit the required power flow control. In a two transformer substation operation of either or both transformers on one infeed together with the facility to take out of service and restore to service either transformer without loss of supply would be a normal design consideration. In general a multiple busbar arrangement will provide greater flexibility than a ring busbar.

3.2.5 Protection arrangements

The design must allow for the protection of each system element by provision of suitable CT locations to ensure overlapping of protection zones. The number of circuit breakers that require to be tripped following a fault, the auto-reclose arrangements, the type of protection and extent and type of mechanical or electrical interlocking must be considered.

For example a 1½ breaker substation layout produces a good utilization of switchgear per circuit but also involves complex protection and interlocking design which all needs to be engineered and thus increases the capital cost.

See Section 3.2.8 regarding the use of circuit breakers with CTs in the bushings.

3.2.6 Short circuit limitations

In order to keep fault levels down parallel connections (transformers or power sources feeding the substation) should be avoided. Multi-busbar arrangements with sectioning facilities allow the system to be split or connected through a fault limiting reactor. It is also possible to split a system using circuit breakers in a mesh or ring type substation layout although this requires careful planning and operational procedures.

3.2.7 Land area

The cost of purchasing a plot of land in a densely populated area is considerable. Therefore there is a trend towards compact substation design. This is made possible by the use of indoor gas insulated switchgear (GIS) substation designs or by using such configurations as the transformer-feeder substation layout. In addition compact design reduces civil work activities (site preparation, building costs, requirements for concrete cable trenches, surfacing and access roads). Long multicore control cable runs and switchyard earth grid requirements are also reduced. The reduction in site work by using compact layouts and in particular by using modular elements results in an overall shorter substation project design and construction duration to the advantage of the client. Figure 3.1 dramatically shows the reduction in land area required for an indoor GIS substation as a direct replacement for the previous conventional outdoor open terminal switchyard arrangement.

3.2.8 Cost

A satisfactory cost comparison between different substation layout designs is extremely difficult because of the differences in performance and maintainability. It is preferable to base a decision for a particular layout on technical grounds and then to determine the most economical means of achieving these technical requirements.

Busbar span lengths of about 50 m tend to give an economical design. However, the gantry structures involved have a high environmental impact and the current trend is for low profile substations. Tubular busbars tend to offer cost advantages over tensioned conductor for busbar currents in excess of 3000 A. Taking into account some of the factors mentioned and the savings in cost of land (see Section 3.2.7) resulting from a reduced 'footprint' manufacturers now consider that a 400 kV GIS substation may produce overall savings when compared to a conventional open terminal arrangement, although this varies greatly dependent on site and territory and the reduced bay centres can result in clearance difficulties where there are incoming overhead lines.

The use of circuit breakers with CTs in the appropriate bushings, available up to 275 kV, saves the use of separate post-CTs, with their associated plant, structural, civil and space costs, and may result in overall economy compared to the use of initially cheaper breakers without this facility.

3.3 ALTERNATIVE LAYOUTS

3.3.1 Single busbar

The single busbar arrangement is simple to operate, places minimum reliance on signalling for satisfactory operation of protection and facilitates the economical addition of future feeder bays.

Figure 3.2 illustrates a five circuit breaker single busbar arrangement with four feeder circuits, one bus section and ten disconnectors. Earth switches (not shown) will also be required.

1. Each circuit is protected by its own circuit breaker and hence plant outage does not necessarily result in loss of supply.

Figure 3.1   GIS substation replacement for conventional open terminal outdoor arrangement. A striking comparison between land area requirements for a conventional open terminal 132 kV double switchyard arrangement and replacement indoor GIS housing to the top right-hand corner of the picture (Yorkshire Electricity Group ptc)

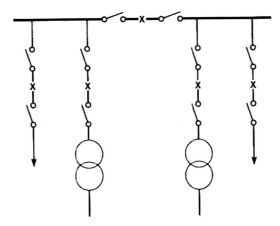

Figure 3.2   Five circuit breaker single busbar arrangement

2. A fault on a feeder or transformer circuit breaker causes loss of the trans-
    former and feeder circuit one of which may be restored after isolating the
    faulty circuit breaker.
3. A fault on a bus section circuit breaker causes complete shutdown of the sub-
    station. All circuits may be restored after isolating the faulty circuit breaker
    and the substation will be 'split' under these conditions.
4. A busbar fault causes loss of one transformer and one feeder. Maintenance
    of one busbar section or disconnector will cause the temporary outage of
    two circuits.
5. Maintenance of a feeder or transformer circuit breaker involves loss of that
    circuit.
6. The introduction of bypass isolators between the busbar and circuit isol-
    ator (Fig. 3.3a) allows circuit breaker maintenance facilities without loss
    of the circuit. Under these conditions full circuit protection is not available.
    Bypass facilities may also be obtained by using a disconnector on the out-
    going ways between two adjacent switchgear bays (Fig. 3.3b). The circuits
    are paralleled onto one circuit breaker during maintenance of the other. It
    is possible to maintain protection (although some adjustment to settings
    may be necessary) during maintenance but if a fault occurs then both
    circuits are lost. With the high reliability and short maintenance times
    involved with modern circuit breakers such bypasses are not nowadays so
    common.

3.3.2 Transformer feeder

The transformer-feeder substation arrangement offers savings in land area
together with less switchgear, small DC battery requirements, less control and
relay equipment, less initial civil works together with reduced maintenance
and spares holding in comparison with the single busbar arrangement.

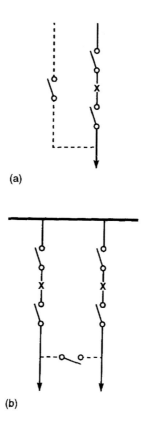

(a)

(b)

Figure 3.3   (a) Bypass isolator for circuit breaker maintenance. (b) Bypass isolator facilities between two adjacent line bays

Figure 3.4 shows the single line diagram for a typical transformer feeder, two transformer substation arrangement. A comparison of land area requirements between a conventional single busbar fully switched outdoor 33/11 kV distribution substation (2150 m²), a fully switched one-storey indoor substation (627 m²) and for the transformer-feeder arrangement (420 m²) is shown in Fig. 3.5.

The major practical service continuity risk for the transformer-feeder substation is when the substation supply cables are both laid in the same trench and suffer from simultaneous damage. Much of the substation cost savings would be lost if the supply cables were laid in separate trenches since the civil trench work, laying and reinstatement costs are typically between 33% and 40% of the total supply and erection contract costs for 132 kV oil filled and 33 kV XLPE, respectively. In congested inner city areas planning permission for separate trenches in road ways or along verges is, in any case, seldom granted. The civil works trenching and backfill costs for two separate trenches (one cable installation contract without special remobilization) are typically 1.6 times the cost of a single trench for double circuit laying. The choice depends upon the degree of risk involved and the level of mechanical protection, route markers and warnings utilized. The

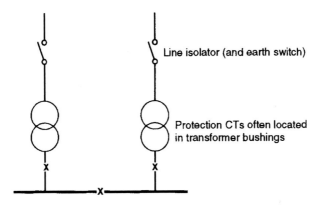

Figure 3.4 Transformer-feeder arrangement

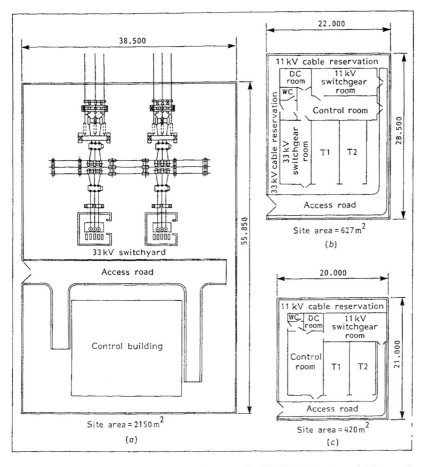

Figure 3.5 Comparison of land area requirements for 33/11 kV substations. (a) Conventional outdoor fully switched single busbar. (b) Fully switched indoor. (c) Transformer feeder

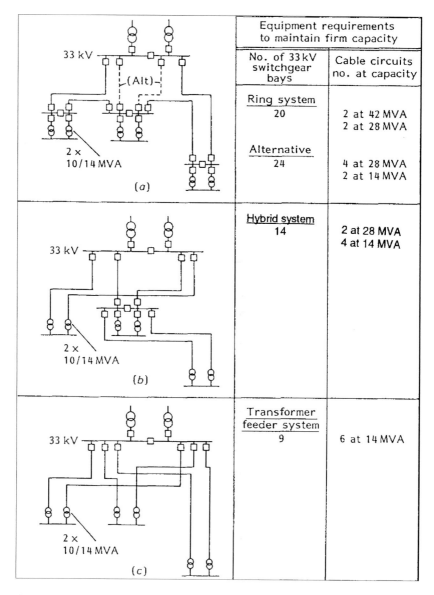

	Equipment requirements to maintain firm capacity	
	No. of 33 kV switchgear bays	Cable circuits no. at capacity
	Ring system 20	2 at 42 MVA 2 at 28 MVA
	Alternative 24	4 at 28 MVA 2 at 14 MVA
	Hybrid system 14	2 at 28 MVA 4 at 14 MVA
	Transformer feeder system 9	6 at 14 MVA

Figure 3.6 Comparison of equipment requirements: (a) ring system; (b) hybrid system; (c) transformer feeder

cable routes for ring systems do not normally present such problems since the feeder cables usually run in different directions and only come in close proximity adjacent to the substation.

A comparison of equipment requirements between a ring, hybrid and transformer-feeder arrangement is given in Fig. 3.6.

The usual practice for the cable supplied transformer-feeder substation is to terminate the supply cables on outdoor sealing ends with bare busbar connections to the transformer HV bushings. On first examination it might appear more sensible to terminate the HV cables directly into a transformer cable box. This would reduce the length of exposed live conductor and hence reduce the likelihood of insulation failure due to pollution, debris, animals or birds, etc. However, difficulties arise with this solution when, say, after cable damage, isolation and earthing, repair and DC pressure testing is required. On lower voltage systems (11 kV) disconnection chambers may be specified on transformers but this is not practical at the higher voltage (36 kV and above) levels. With outdoor bushings and busbar it is easy to apply portable earths and isolate the transformer or cable for maintenance, repair or test.

An isolator and earth switch may be added at the transformer HV connections depending upon the electrical supply company's operational procedures. With the development of metal-clad SF_6 insulated equipment the possibility exists for provision of an HV isolator and earth switch all within an SF_6 insulated environment connected directly to the transformer windings without the need for additional land space. With an overhead line fed transformer-feeder substation a line disconnector/earth switch is desirable since the probability of a fault (insulator failure, development of hot spots on connections, etc.) is greater than with a cable circuit.

3.3.3 Mesh

An arrangement known as a three switch mesh substation is shown in Fig. 3.7a. It utilizes only three circuit breakers to control four circuits. The scheme offers better features and facilities than the single busbar without a bus section switch:

1. Any circuit breaker may be maintained at any time without disconnecting that circuit. Full protection discrimination will be lost during such maintenance operations. In order to allow for all operating and maintenance conditions all busbars, circuit breakers and disconnectors must be capable of carrying the combined loads of both transformers and line circuit power transfers.
2. Normal operation is with the bypass disconnectors or optional circuit breaker open so that both transformers are not disconnected for a single transformer fault.
3. A fault on one transformer circuit disconnects that transformer circuit without affecting the healthy transformer circuit.
4. A fault on the bus section circuit breaker causes complete substation shutdown until isolated and power restored.

A development of the three switch arrangement for multiple circuit substations is the full mesh layout as shown in Fig. 3.7b. Each section of the mesh is included in a line or transformer protection zone so no specific separate busbar protection is required. Operation of two circuit breakers is required to

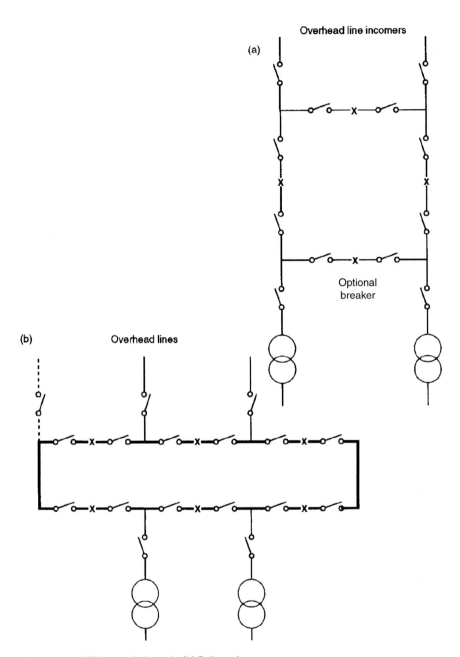

Figure 3.7   (a) Three switch mesh. (b) Full mesh

connect or disconnect a circuit and disconnection involves opening the mesh. Line or transformer circuit disconnectors may then be used to isolate the particular circuit and the mesh reclosed:

1. Circuit breakers may be maintained without loss of supply or protection and no additional bypass facilities are required. The particular circuit may be fed from an alternative route around the mesh.
2. Busbar faults will only cause the loss of one circuit. Circuit breaker faults will involve the loss of a maximum of two circuits.
3. Generally not more than twice as many outgoing circuits as infeeds are used in order to rationalize circuit equipment load capabilities and ratings. Maximum security is obtained with equal numbers of alternatively arranged infeeds and load circuits. Sometimes banked pairs of feeders are arranged at mesh corners.

3.3.4 Ring

The ring busbar offers increased security compared to the single busbar arrangement since alternative power flow routes around the ring busbar are available. A typical scheme which would occupy more space than the single busbar arrangement is shown in Fig. 3.8. The ring is not so secure as the mesh arrangement since

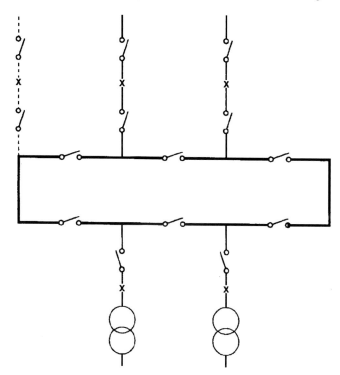

Figure 3.8   Ring

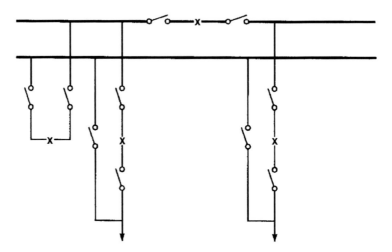

Figure 3.9   Transfer busbar

a busbar fault causes all circuits to be lost until the fault has been isolated using the ring busbar isolators. Unless busbar disconnectors are duplicated maintenance on a disconnector requires an outage of both adjacent circuits. The inability of disconnectors to break load current is also an operational disadvantage.

3.3.5 Double busbar

3.3.5.1 *Transfer bus*

The double busbar arrangement is probably the most popular open terminal outdoor substation arrangement throughout the world. It has the flexibility to allow the grouping of circuits onto separate busbars with facilities for transfer from one busbar to another for maintenance or operational reasons. A typical transfer busbar arrangement is shown in Fig. 3.9:

1. This is essentially a single busbar arrangement with bypass disconnector facilities. When circuit breakers are under maintenance the protection is arranged to trip the bus-coupler breaker.
2. The system is considered to offer less flexibility than the full duplicate double busbar arrangement shown in Fig. 3.10.

3.3.5.2 *Duplicate bus*

1. Each circuit may be connected to either busbar using the busbar selector disconnectors. On-load busbar selection may be made using the bus-coupler circuit breaker.

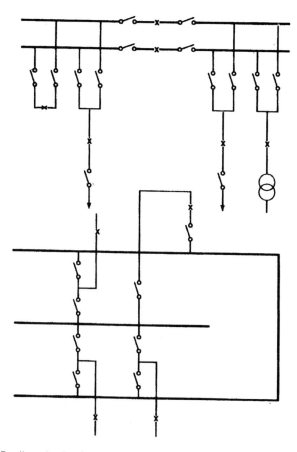

Figure 3.10   Duplicate busbar (and wrap around arrangement)

2. Motorized busbar selector disconnectors may be used to reduce the time to reconfigure the circuit arrangements.
3. Busbar and busbar disconnector maintenance may be carried out without loss of supply to more than one circuit.
4. The use of circuit breaker bypass isolator facilities is not considered to offer substantial benefits since modern circuit breaker maintenance times are short and in highly interconnected systems alternative feeder arrangements are normally possible.
5. A variant on the scheme uses a 'wrap around' busbar layout arrangement as shown in Fig. 3.10 in order to reduce the length of the substation.

3.3.6 1½ Circuit breaker

The arrangement is shown in Fig. 3.11. It offers the circuit breaker bypass facilities and security of the mesh arrangement coupled with some of the flexibility

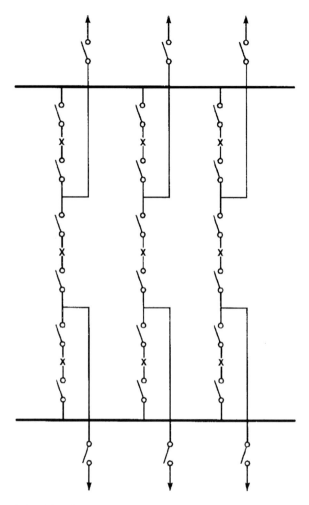

Figure 3.11   1½ circuit breaker

of the double busbar scheme. The layout is used at important high voltage sub-stations and large generating substations in the USA, Asia and parts of Europe where the cost can be offset against high reliability requirements. Essentially the scheme requires 1½ circuit breakers per connected transmission line or trans-former circuit and hence the name of this configuration:

1. Additional costs of circuit breakers are involved together with complex pro-tection arrangements.
2. It is possible to operate with any one pair of circuits, or group of pairs of circuits separated from the remaining circuits. The circuit breakers and

other system components must be rated for the sum of the load currents of two circuits.

3. High security against loss of supply.

3.4 SPACE REQUIREMENTS

3.4.1 Introduction

Having selected the required substation single line diagram arrangement it is then necessary to convert this into a practical physical layout. It is essential to allow sufficient separation or clearances between substation equipment to withstand voltage stresses and to allow safe operation and maintenance of the equipment. The designer will have to consider the following.

Actual site selection. The substation configuration and number of circuits involved (including any allowance for future expansion) will largely determine the land area requirements. The ideal site will have the following characteristics:

1. Reasonably level and well drained so minimum surface dressing and civil ground works are required.
2. Low lying and not in a prominent position so that planning permission will be relatively easy to obtain. For an open terminal air insulated switchgear (AIS) switching substation a location as far from a built-up area as possible. However, for a primary distribution substation this will conflict with the technical and economic requirement to have the substation as close to the load centre as possible. Consider the use of indoor gas insulated switchgear (GIS), the cost of which in some locations will be largely offset by reduced land area costs.
3. Good access from public highways for easy transportation of materials and especially heavy items such as transformers to the site.
4. Good overhead line wayleave substation entry routes.
5. Pollution-free environment. If the substation is to be sited in a highly polluted industrial area (next to a quarry, cement works, etc.) or close to a coastal salty atmosphere then a meteorological study will be required to determine the prevailing wind direction. The substation should then be sited upwind of the pollution source. Again an indoor GIS arrangement should also be considered.

High or low level, catenary or solid, busbar arrangements. A high busbar is exposed and must span complete switchgear bays. Low busbars are more shielded, may be more suitable for connection of portable earths but may need frequent supports. They may be considered to be more visually or environmentally acceptable. Space savings are also possible from the use of different

Table 3.1   Safety clearances to enable operation, inspection, cleaning, repairs, painting and normal maintenance work to be carried out (BS7354)

1	2	3	4	5	6
Nominal system voltage/BIL/SIL (kV)	Basic electrical clearance (phase–earth) (m)	Safety working clearance (vertical)[a] (m)	Safety working clearance (horizontal)[a] and see note 1 (m)	Insulation height (pedestrian access) and see note 2 (m)	Phase–phase clearance (m)
6.6/7.5	0.5[b]	2.9	2.3	2.1	0.25
11/95	0.5[b]	2.9	2.3	2.1	0.25
33/170	0.5[b]	2.9	2.3	2.1	0.43
66/325	0.7	3.1	2.5	2.1	0.78
132/550/650	1.1	3.5	2.9	2.1	1.4
275/1050/850	2.1	4.8	3.9	2.4	2.4
400/1425/1050	2.8	5.5	4.6	2.4	3.6

(a) Increased allowance for effects of hand tools.
(b) Increased value to 500 mm for 170 kV BIL systems and below.
Note 1: The safety working clearances (horizontal) in column 4 are regarded as minimum clearances. For systems where predominant voltages are higher (275 kV and 400 kV), it is practice in the UK that the horizontal and vertical clearances are equal, conforming to the minimum figures in column 3.
Note 2: The insulation height (pedestrian access) figures in column 5 are shown in respect of the appropriate system voltage as follows:
(i) For systems predominantly involving distribution voltages, i.e. up to and including 132 kV, 2.1 m is the recommended minimum requirement.
(ii) For systems where the predominant voltages are higher, i.e. 275 kV and 400 kV, 2.4 m is the recommended minimum.

The above figures are regarded as minimum clearances. Users and operators may wish to allow further clearance greater than that shown for the voltage level in column 5, particularly for equipment at the higher distribution voltages situated in a single compound with higher voltage equipment present in a compound subject to greater clearance. In these circumstances, it might be convenient to design and operate to the clearance for the highest voltage equipment.

types of switchgear, for example by using pantograph instead of horizontal swivel isolators.

3.4.2 Safety clearances

The safety distance means the minimum distance to be maintained in air between the live part of the equipment or conductor on the one hand and the earth or another piece of equipment or conductor on which it is necessary to carry out work on the other. A basic value relates to the voltage impulse withstand for the substation. To this must be added a value for movements for *all* methods necessary to maintain and operate the equipment so that a safety zone may be determined. Section clearances and ground clearances based on British practice (BS7354) are given in Table 3.1. Figures 3.12 and 3.13 illustrate diagrammatically the clearances required between the different items of substation equipment for

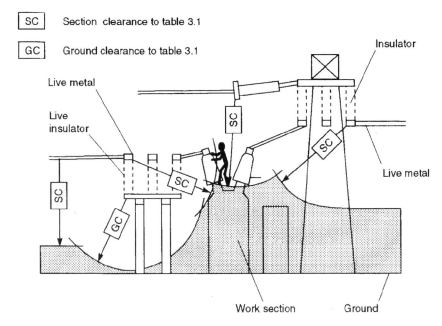

Figure 3.12   Substation work section boundaries, section and ground clearances

maintenance and safe working limits (see also Tables 3.2 and 3.3). Note that safety clearance must also be allowed to any necessary working platforms.

CIGRE is an organization of electricity authorities which meets to discuss and exchange information on matters of electricity generation, transmission and distribution. Working groups study various problems and report back to various committees. Their work is published in *Electra* and excellent reports have been issued which form guides for the selection of substation clearances. CIGRE recommendations are technically coherent and essentially the same as BS7354 but slightly more difficult to apply. A basic curve is drawn on the layout drawings first and separate horizontal and vertical clearances added.

3.4.3 Phase–phase and phase–earth clearances

IEC 60071 deals with insulation co-ordination and proposes standard insulation levels and minimum air distances. BS7354 also specifies phase–phase and phase–earth clearances. Extracts from BS covering International practice are enclosed in Table 3.3 and from IEC in Table 3.4. Phase–phase clearances and isolating distances are usually specified as 10–15% greater than phase–earth clearances. The justification is that phase–phase faults or faults between equipment terminals usually have more serious consequences than phase–earth faults. It should be noted that the configuration of conductors and adjacent earthed structures and equipment also affects these clearances. Therefore

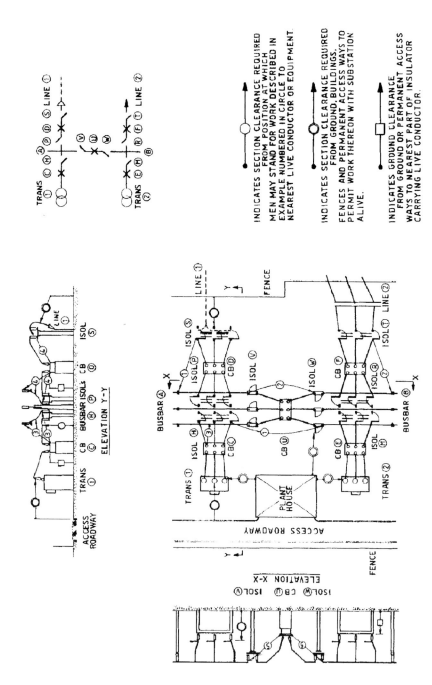

Figure 3.13   Work section clearance, example

Table 3.2   Necessary operations for maintenance work on different items of open terminal outdoor substation plant as shown in the substation layout, Fig. 3.13

Example No.	For work upon	Isolate at	Access via	Equipment remaining alive	Remarks
1	Busbar isolate H. Insulators supporting busbar A	Isolators and W. L.V. side of transformer 1	Ground level and temporary means only.	Line 1 equipment up to isolator S. Busbar B up to isolator W. Line 2 equipment. Transformer 2.	Access also to: Transformer 1. Circuit breakers C, D and U. Isolators P and V.
2	Bus section isolator W. Insulators supporting busbar B.	Isolators T. and V. L.V. side of transformer 2.	Ground level and temporary means only.	Busbar A up to isolator V. Line 1 equipment. Transformer 1 equipment. Line 2 equipment up to isolator T.	Access also to: Transformer 2. Circuit breakers E, F and U. Isolators M and R.
3	Transformer 1 and circuit breaker C.	Isolator H and L.V. side of transformer 1.	Ground level and temporary means only.	Busbar A up to isolator H. Busbar B. Line 1 equipment. Line 2 equipment. Transformer 2 equipment.	
4	Circuit breaker D.	Isolators P and S.	Ground level and temporary means only.	Busbar A up to isolator P. Busbar B. Line 1 equipment up to isolator S. Line 2 equipment. Transformers 1 and 2 equipments.	
5	Circuit breaker U.	Isolators V and W.	Ground level and temporary means only.	Busbar A up to isolator V. Busbar B up to isolator W. Lines 1 and 2 equipments. Transformers 1 and 2 equipments.	
6	Line 2 anchorage structure	Isolator T and remote end of line 2.	Ground level and temporary means only.	Busbars A and B. Transformers 1 and 2 equipments. Line 1 equipment. Line 2 equipment up to isolator T.	
7	Line isolator T.	Isolator R and remote end of line 2.	Ground level and temporary means only.	Busbar A. Busbar up to isolator R. Transformers 1 and 2 equipments. Line 1 equipment.	Access also to: Circuit breaker F. Line 2 and line anchorage structure.

Table 3.3   International practice – electrical clearances for open terminal outdoor switchgear (BS7354)

1	2	3		4		5		6	7
BIL (kV)	SIL (kV)	Basic electrical clearance (phase–earth) (m) See Note 1 A	B	Safety working clearance (vertical) (m) A	B	Safety working clearance (horizontal) (m) A	B	Insulation height (pedestrian access) (m) See Note 2	Phase–phase clearance (m) See Note 3
20		0.15		2.6		1.6		2.4	
40		0.15		2.6		1.6		2.4	
60		0.15		2.6		1.6		2.4	
75		0.15		2.6		1.6		2.4	
95		0.16		2.6		1.6		2.4	
125		0.22		2.7		1.7		2.4	
145		0.27		2.7		.7		2.4	
170		0.32		2.8		1.8		2.4	
250		0.48		2.9		1.9		2.4	
325		0.63		3.1		2.1		2.4	
450		0.9		3.3		2.3		2.4	
550		1.1		3.5		2.5		2.4	
650		1.3		3.7		2.7		2.4	
750		1.5		3.9		2.9		2.4	
850	750	1.6	1.9	4.0	4.3	3.0	3.3	2.6	
950	750	1.7	1.9	4.1	4.3	3.1	3.3	2.6	
950	850	1.8	2.4	4.2	5.2	3.2	4.2	2.7	
1050	850	1.9	2.4	4.3	5.2	3.3	4.2	2.7	
	950	2.2	2.9	4.9	5.8	3.9	4.8	2.7	
	1050	2.6	3.4	5.3	6.4	4.3	5.4	2.7	
	1175	3.1	4.1	6.0	7.3	5.0	6.3	2.8	
	1300	3.6	4.8	6.7	8.2	5.7	7.2	2.9	
	1425	4.2	5.6	7.6	9.2	6.6	8.2	3.0	
	1550	4.9	6.4	8.5	10.3	7.5	9.3	3.1	

Note 1: Clearances under columns marked 'A' are appropriate to 'conductor-structure' electrode configuration.
Clearances under columns marked 'B' are appropriate to 'rod-structure' electrode configuration.
Note 2: The 'rod-structure' configuration is the worst electrode configuration normally encountered in service.
The 'conductor-structure' configuration covers a large range of normally used configurations. The heights in column 6 related to SIL (switching impulse levels) are based upon 'conductor-structure' electrode configurations. Higher values should be agreed if the more onerous electrode configuration applies.
Note 3: Phase–phase clearances are under consideration

care must be taken when applying these criteria. For example, the clearance required from an open contact on a disconnector to an adjacent structure will be greater than that from a continuous busbar to ground level in order to achieve the same insulation level.

Once the various minimum allowable phase–phase and phase–earth clearances have been chosen it is necessary to ensure that the design maintains these

Table 3.4   Phase–phase and phase–earth clearances (IEC 60071)

Highest voltage for equipment, Um (kV rms)	Rated SIL (kV peak)	Rated SIL (kV peak)	Phase–earth air clearance (mm)	Phase–phase air clearance	Safety clearance (mm)	Horizontal safety clearance (mm)	Height (mm)
3.6		20	60	60	3000	2000	2500
7.2		40	60	60	3000	2000	2500
12		60	90	90	3000	2000	2500
17.5		75	120	120	3000	2000	2500
		95	160	160	3000	2000	2500
24		125	220	220	3000	2000	2500
		145	270	270	3000	2000	2500
36		170	320	320	3000	2000	2500
52		250	480	480	3000	2000	2500
72.5		325	630	630	3100	2100	2500
123		450	900	900	3400	2400	2500
145		550	1100	1100	3600	2600	2500
170		650	1300	1300	3800	2800	2500
245		750	1500	1500	4000	3000	2500
		850	1600	1700	4100	3100	2500
		950	1700	1900	4200	3200	2500
		1050	1900	2100	4400	3400	2500
300	750	850	1600	2400	4100	3100	2500
		950	1700	2400	4200	3200	2500
362	850	950	1800	2700	4400	3400	2600
		1050	1900	3100	4400	3400	2600
420	950	1050	2200	3100	4800	3800	2600
		1175	2200	3100	4800	3800	2600
	1050	1175	2600	3500/3900*	5300	4300	2700
		1300	2600	3500/3900	5300	4300	2700
		1425	2600	3500/3900	5300	4300	2700
525	1175	1300	3100	4300	5800	4800	2700
		1425	3100	4300	5800	4800	2700
		1550	3100	4300	5800	4800	2700
	1300	1425	3600	6300	6400	5400	2800
		1530	3600	6300	6400	5400	2800
		1800	3600	6300	6400	5400	2800
765	1425	1550	4200	7100	7100	6100	2900
		1800	4200	7100	7100	6100	2900
		2100	4200	7100	7100	6100	2900
	1550	1800	4900	7900	7900	6900	3000
		1950	4900	7900	7900	6900	3000
		2400	4900	7900	7900	6900	3000

*3500/3900 mm for 420/525 kV, respectively.

at all times. Allowance must be made for movement of conductors in the wind and temperature sag effects. Under short circuit conditions flexible phase conductors may first repel each other (reducing clearances to adjacent equipment) and then swing together (reducing phase–phase clearances). The coincidence of an overvoltage on one phase with an overvoltage or peak value of system voltage of opposite polarity on an adjacent phase can produce an increase in voltage between phases. The 10–15% margin in phase–phase clearances allows for a degree of protection against this occurrence.

At high altitudes the reduced air density lowers the flashover voltage and clearances should be increased by approximately 3% for each 305 m (1000 ft) in excess of 1006 m (3300 ft) above sea level.

Allowances must also be made for variations in the level of the substation site and the positioning of foundations, structures and buildings. At lower voltages an additional margin may be added to avoid flashovers from birds or vermin. A common mistake is not to take into account the substation perimeter fence and thereby infringe phase–earth clearances.

4 Substation Auxiliary Power Supplies

4.1 INTRODUCTION

All but the smallest substations include auxiliary power supplies. AC power is required for substation building small power, lighting, heating and ventilation, some communications equipment, switchgear operating mechanisms, anti-condensation heaters and motors. DC power is used to feed essential services such as circuit breaker trip coils and associated relays, supervisory control and data acquisition (SCADA) and communications equipment. This chapter describes how these auxiliary supplies are derived and explains how to specify such equipment.

4.2 DC SUPPLIES

4.2.1 Battery and charger configurations

Capital cost and reliability objectives must first be considered before defining the battery and battery charger combination to be used for a specific installation. The comparison given in Table 4.1 describes the advantages and disadvantages of three such combinations.

Figure 4.1 details the main electrical features associated with these battery and charger combinations. Charger units are used to supply either just a battery to provide an autonomous DC supply or a battery/inverter combination to provide an autonomous AC supply. The level of 'autonomy' is usually defined in terms of the number of hours or minutes the equipment will enable a specified load to function correctly after loss of input mains AC supply. The capacity of the charger must also be such that after a severe discharge it has the capacity to supply the full DC system load current and the full charging current

Table 4.1   Capital cost and reliability objectives must first be considered before defining the battery/battery charger combination to be used for a specific installation. The comparison given describes the advantages and disadvantages of three such combinations

Type	Advantages	Disadvantages
1. Single 100% battery and 100% charger	Low capital cost	No standby DC System outage for maintenance Need to isolate battery/charger combination from load under boost charge conditions[1] in order to prevent high boost voltages appearing on DC distribution system[2]
2. Semi-duplicate 2 × 50% batteries and 2 × 100% chargers	Medium capital cost Standby DC provided which is 100% capacity on loss of one charger Each battery or charger can be maintained in turn. Each battery can be isolated and boost charged in turn without affecting DC output voltage	50% capacity on loss of one battery during AC source failure
3. Fully duplicate 2 × 100% batteries and 2 × 100% chargers	Full 100% standby DC capacity provided under all AC source conditions and single component (charger or battery) failure	High capital cost Greater space requirement Increased maintenance cost

Notes: [1]Not all batteries have a boost facility.
[2]It is possible to specify the voltage operating range of auxiliary devices to match any possible imposed voltages.

simultaneously. The technique used for battery charging is called 'float' charging and involves the battery being permanently connected to the load (possibly via an inverter) in parallel with a charger. Therefore the charger must satisfy the requirements of both the battery and the load. The exact charger functional requirements will depend upon the type of battery (lead acid, nickel cadmium – NiCad, sealed recombination, etc.) being used and this is discussed in Section 4.3. In general the charger must provide a combination of constant voltage and constant current charging profiles within close tolerances. For some battery types it must also be able to be switched to a 'boost' charge function that will apply a larger voltage to the battery in order that the charging period may be reduced. The control unit is relatively complicated but may be seen as an analogue feedback loop which samples the output voltage and current and uses these signals to control a single or three phase thyristor bridge rectifier. Switched mode power supplies are also employed in the smaller units and by

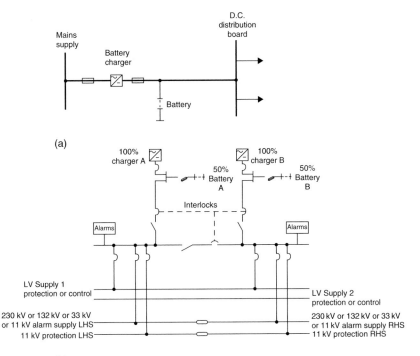

(a)

(b)

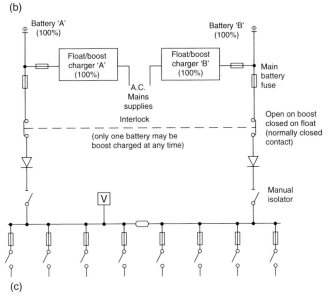

(c)

Figure 4.1   Battery/battery charger combination: (a) Single 100% battery and 100% charger; (b) semi-duplicate 2 × 50% batteries and 2 × 100% chargers; (c) fully duplicate 2 × 100% batteries and 2 × 100% chargers (courtesy of Balfour Beatty Projects and Engineering Ltd)

using an oscillator frequency of around 20 kHz small wound components help to reduce charger size and weight.

The simple single battery/single charger combination is suitable for the small distribution substation where, with perhaps only a few metres between the switchgear and the DC distribution board; 30 V DC was often specified to operate trip coils and relays in the past – 50 V is more common now. A useful low-cost addition to such a simple system would be a facility to connect an emergency (or 'hospital') battery via the DC distribution.

The option of using 2 × 50% batteries and 2 × 100% chargers may be used for primary substation applications where this is the practice of the supply authority or where costs are to be kept to a minimum in keeping with high reliability. It is very important to specify clearly the operating regime for such a system before going out to tender as manufacturers will need to understand fully the interlock requirements involved. A DC supply float of 125 V is a typical IEC standard voltage for such applications with 110 V nominal system voltage.

For the larger substations the cost of the DC supply will be small in comparison with the total substation and the full 2 × 100% battery and 2 × 100% charger combination is usually chosen. Separate systems are often used for substation switchgear control and communications equipment.

As an alternative to two separate 50% batteries, a single battery composed of two strings of batteries in parallel can be employed. This has the advantage of enabling limited service to be maintained in the event of one cell failing in open-circuit mode. Further, as modern chargers can often be quickly returned to service after failure by simply replacing an electronic 'card', a single alarmed charger in conjunction with a two string battery is often considered satisfactory, depending on the risk assessment for the installation in question. In making such assessments, which must consider the overall reliability of the installation, it is essential to consider also the frequency of inspection and the extent of remote supervision and alarms. Even a fully duplicated system is at risk of collapse if a single, initially non-critical, failure is left untended for a considerable time (see also Sections 4.2.2.5 and 4.2.2.8).

4.2.2 Battery charger components

The function of the different components shown on the block diagrams in Fig. 4.1 is as follows.

4.2.2.1 *Interlocks and cross connecting batteries and chargers*

The interlocks between the battery/battery charger combination and the DC distribution board are necessary to prevent boost charging voltages appearing on the DC distribution system which could exceed the ratings of trip coils and other equipment. For NiCad batteries approximate voltages would be:

Float	116%
Boost/commissioning	135%
Minimum	84%

In the semi-duplicate system the interlocks must ensure:

(a) Only one battery/charger can be selected to boost charging at any one time.
(b) Busbars have to be interconnected prior to boost charging commencing.
(c) Boost voltage is not to be applied to the DC distribution busbar and system.

A busbar-section switch is used to achieve this.

End cell tapping is a low cost method used to prevent boost voltages appearing on the DC system. However, it has the disadvantage of reduced reliability owing to additional switching components and series cells with differing states of charge. Alternatively for low power chargers (<1 kW) DC series regulation may be used with low output impedance common collector transistor/zener diode combinations. The disadvantages here are the costs involved for high current systems, heat losses and again reduced reliability.

The fully and semi-duplicated systems may also be specified such that the batteries and chargers may be either manually or automatically cross connected so either battery may be charged from either charger. This improves the availability of the DC supply but does so at the expense of increasing complexity. Failure of the cross connecting switches at a point of common connection could reduce reliability.

4.2.2.2 Anti-paralleling diodes

These are intended to prevent high circulating currents in the duplicated and semi-duplicated systems. Should one battery be faulty, the fully charged battery should not be allowed to discharge into it. Such diodes have very high reliability with low forward voltage drop. They are only likely to fail to short circuit and therefore will maintain a connection between the battery and the DC distribution system.

4.2.2.3 Battery fuses

These are positioned in both the positive and negative battery leads so as to minimize unprotected cable or equipment and should be accessible so as to provide an easy method of battery isolation for maintenance. The fuses are intended to protect against fire and to limit fault durations. The fuse rating for

normal lead acid or NiCad cells may need to be at least three times the maximum battery demand current at the highest boost charge voltage. In assessing this maximum demand, take account of any short term requirements (e.g.
motor starting currents). It is important for the designer to ensure the positioning or type of fuse presents no danger of gas ignition upon fuse operation.

4.2.2.4 *Radio frequency interference suppression*

The steep wave fronts associated with fast thyristor switching are rich in harmonics. The system design engineer must therefore satisfy Electro Magnetic
Compatibility (EMC) requirements (typically to EN 5022 or BS6527 Class B
conducted and Class A radiated levels). Simply specifying DC output ripple (to
be typically 5–10%) and noise levels is insufficient if sensitive electronic equipment is involved in the substation installation. Adequate filtering will involve
radio frequency chokes (RFCs) in the supply source and load connections
together with bypass capacitors to short RF to earth and adequate screening.
Refer also to Chapter 24. In the UK, Engineering Recommendation G5/4 sets
limits on the connected pulsed rectifiers on the public supply network.

4.2.2.5 *Protection and alarms*

Typically some of the following may be specified:

AC fail	Loss of AC supply detection
Battery fault	Voltage per cell or string of cells monitored*
Charger fail	Output ripple, firing pulse fault or output tolerance
DC voltage	Voltage high/low detection and tripping
AC earth leakage	Earth leakage module
Overtemperature	Shut down and auto reset as temperature reduces
Overload	Overcurrent limiting
Reverse polarity	Tripping

4.2.2.6 *Metering and controls*

Typically some of the following may be specified with remote monitoring
connections as required:

AC supply present	Lamp or AC voltmeter with or without phase selection
Protection operation	Local or remote combined or individual indication
DC voltage	Battery voltage and/or DC system voltage*

| DC current | Battery charging current and/or system load current* |
| Isolation Float/boost | AC source and DC supply |

*Note: Monitoring of battery condition through impedance measurement is sometimes used on critical installations. If recorded on a historic basis the technique can give warning of imminent risk of failure, rather than just alarm after failure. For further guidance on monitoring lead acid batteries see IEC/TR62060 – user guide on monitoring lead acid stationary batteries.

4.2.2.7 *DC switchboard*

The DC switchboard should comply with the requirements of IEC 60439. Double pole switches and fuses, switch fuses or MCBs (miniature circuit breakers) may be used for incomers and outgoing ways to the DC distribution system. Links or switches may be used to sectionalize the busbars as necessary.

The complete charger, battery and DC distribution board may be housed in a single cabinet for the smaller units. The danger of vented gas causing corrosion problems or gas ignition is small if sealed recombination cells are correctly used but in very critical locations the probability of a number of co-incident failures (e.g. cell sealing plus charge rate control plus inspection oversight) must be assessed and the risk mitigated to an acceptable level.

Larger installations require separate battery racks with combined or separate charger/DC distribution board combinations.

4.2.2.8 *DC distribution supply monitoring*

A healthy DC supply is essential for the correct operation of the substation controls, relays and circuit breakers. A regime of DC distribution supply monitoring must therefore be defined so that immediate remedial action may be taken should the DC supply fail. In addition to the alarms on the battery/battery charger combination itself alarms may be derived from failure within the DC distribution. A typical scheme is shown in Fig. 4.2. In this case the DC supply is duplicated to each control and relay panel by sectionalizing the DC distribution board and having separate feeders to each panel. Each relay and/or control panel DC circuit associated with each power substation circuit is also monitored for loss of DC supply. Since DC failure could in itself prevent alarms from operating small DC/DC converters may be specified to drive the annunciator modules.

4.2.3 Installation requirements

4.2.3.1 *General*

Since acid or alkaline liquids and vapours are toxic, a separate battery room is traditionally provided in the substation control building to house the battery

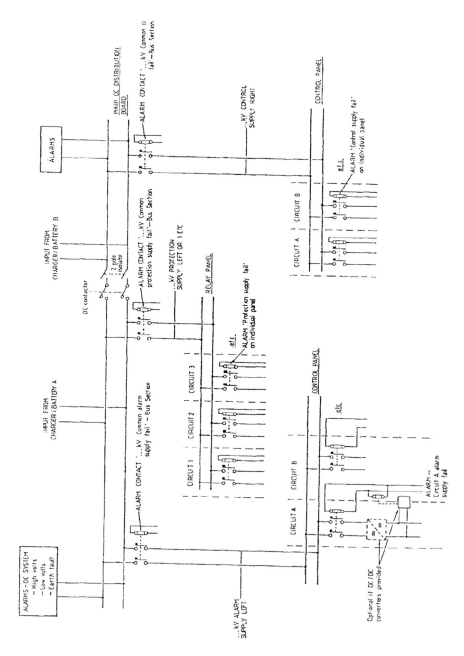

Figure 4.2 DC Distribution supply monitoring (courtesy of Balfour Beatty Projects and Engineering Ltd)

banks. The room has to have adequate ventilation (possibly forced), an acid resistant concrete or tiled floor and sink unit with running water and eye wash facilities. Division II explosion-proof lighting and ventilation fan installations are required for large vented battery installations. In addition notices must be displayed about the corrosive materials and to prohibit smoking. Most lead acid and NiCad batteries are now manufactured in enclosed containers with special plugs to permit ventilation without excessive loss of electrolyte. A typical battery room as built for the Channel Tunnel Main 132 kV/25 kV/21 kV Intake Substation at Folkestone, UK, is shown in Fig. 4.3.

4.2.3.2 Ventilation

The ventilation requirements for other than the sealed recombination type cells is determined from manufacturers' literature. It can be shown that in the case of a lead acid battery 1 gram of hydrogen and 8 grams of oxygen will be evolved with an input of 26.7 ampere hours to a fully charged cell. One gram of hydrogen will occupy 11.2 litres, or $0.0112 \, m^3$. The volume of hydrogen produced by a battery will therefore be equal to:

$$\text{no. of cells} \; \times \text{charge current} \times \frac{0.0112}{26.7}$$

or

$$\text{no. of cells} \; \times \text{charge current} \times 0.000\,419\,4 \, m^3$$

This value may be expressed as a percentage of the total volume of the battery room and assuming that a mixture of 2% hydrogen and air is a safe limit (based on 50% of the 'Lower Explosive Limit' of 4% for hydrogen), the number of air changes per hour to keep the concentration of hydrogen within this limit can be calculated. A typical small battery requiring a charging current of 17 amps will require about three changes of air per hour if installed in a $4 \times 2 \times 2.5$ m room.

As an *indication* of the amount of air to be replaced in order to consider the battery room to be adequately ventilated the following practical formula is used:

$$Q = 55 \times N \times I \; \text{litres/hour}$$

where Q = volume rate of air replacement (litres per hour)
55 = factor for allowable air and hydrogen volume plus a safety factor (per Ah)
N = number of battery cells
I = charging current causing formation of hydrogen gas (A) (Note: 7 amps per 100 Ah battery capacity typical)

Figure 4.3 Primary substation battery room DC distribution and charger (courtesy of Balfour Beatty Projects and Engineering Ltd)

Therefore a 110 V lead acid, 400 Ah capacity substation battery will consist of approximately 54 cells and $Q = 83\,160$ litres/hour ($83\,\text{m}^3$/hour). An equivalent NiCad system would have more cells and a slightly greater ventilation requirement.

The amount of hydrogen quoted above that is released during charging is appropriate only to the period near the end of a boost charge. Therefore full forced ventilation will strictly only be necessary for a few hours every 1 or 2 years and it is important not to get this problem out of perspective (but remember that a charger fault or abnormal ambient temperatures can also affect gas generation). In installations with vented lead acid batteries of the order of or greater than 20 kVAh capacity the hydrogen production and temperature rise during boost charging makes the provision of a separate ventilated room mandatory.

Since temperature affects battery performance, temperature effects must also be considered in designing the ventilation system.

4.2.3.3 *The installation process*

Designs must take account of the need to safely install, maintain and dismantle a battery installation. Some battery units come in multi-cell blocks, and can be very heavy. Racks – either open or in-cabinet – must enable safe handling, taking account of human weight handling limits. If handling trolleys are used, access routes must be designed accordingly.

4.2.4 Typical enquiry data – DC switchboard

1. Maximum physical dimensions – width $\times$ depth $\times$ height (mm)
2. Enclosure IP rating (IEC 60529)
3. Single line diagram drawing number
4. Unequipped spare ways
5. Equipped spare ways[g]
6. Relevant standards

DC distribution boards	IEC 60439
Moulded case circuit breakers[a]	IEC 60157
Fuses	IEC 60269
Contactors	IEC 60158
Isolators and earth switches	EN 60129

7. Busbar maximum current rating (A)
8. Switchgear type[b]
9. Manufacturer[c]
10. Manufacturer's drawings[c]
11. Metering, alarms and protection[d]
12. Boost charge contactors[e]
13. Anti-paralleling diodes[f]

Notes: (a) Recommend P2 category for repeated short circuit capability.
　　　 (b) Metal-clad, metal enclosed, etc.

(c) To be completed by the manufacturer.

(d) To be clearly indicated upon enquiry drawing or detailed circuit by circuit here. Quiescent and operated power consumption should be noted.

(e) Maximum current rating, coil rating and method of interlocking if applicable.

(f) I^2t and reverse blocking voltage diode details.

(g) Including possible provision of 'hospital battery' connection.

4.3 BATTERIES

4.3.1 Introduction

Batteries consisting of a series of individual cells are used to store electricity and are relied upon to provide the required power for a specified period within specified voltage limits. Different battery types have different characteristics best suited to different applications. The choice for substation auxiliary supplies lies between lead acid and nickel cadmium cells and variants within these categories. At the time of writing, alternatives such as fuel cells do not provide economic or sufficiently proven options for this use.

4.3.2 Battery capacity

The capacity of the battery is determined by the capacity of the individual series connected cells. Parallel connection of cells can be made to increase capacity. In the past this practice was discouraged because a weak or defective cell in one of the batteries means that this battery on discharge does not carry its share of the total load. Also, on charge the battery with a defective cell tends to accept a greater share of the available charging current to the detriment of healthy cells in parallel with it. However, provision of parallel strings provides continuity of service, albeit with reduced capacity, in the event of one cell failing in open-circuit mode – something not detected by the simpler battery voltage monitoring alarms. Selection of approach must depend on the manufacturer's data as to failure mode probability, and the required installation reliability.

   Capacity is expressed in ampere-hours (Ah) and is a measure of the electricity that the battery is able to deliver. The following factors affect its capacity:

1. The rate of discharge. If a lead acid battery has a capacity of 100 Ah at a 10 hour discharge rate it can deliver 10 A for 10 hours while maintaining the load voltage above a certain value. Rapid discharge over a 1 hour period will reduce its capacity to typically 50 Ah, i.e. a constant current of 50 A for 1 hour. This effect is not so severe with NiCad batteries.

2. The output voltage reduces as the battery is discharged. It is therefore necessary to specify required current delivery over a given period within voltage limits. In particular the required 'end voltage' at the end of the discharge period must be detailed when specifying battery capacity.
3. Battery capacity varies with temperature. The maximum and minimum temperature range at which the battery will be expected to supply the required capacity must be specified. A battery with 100 Ah capacity at 15°C might have a capacity of 95 Ah at 10°C. Typically the variation in capacity with temperature is as follows:

NiCad batteries	0.6% increase per °C from 0°C to +30°C
	1.5% decrease per °C from 0°C to −20°C
Lead acid batteries	1% increase per °C from 0°C to +60°C
	1.5% decrease per °C from 0°C to −10°C

4.3.3 Characteristics of batteries

The characteristics of different battery types and their relative advantages and disadvantages for different applications are given in Table 4.2. Essentially NiCad battery banks maintain their capacity better at lower temperatures. NiCad life expectancy is good (typically 15 years), better than the standard pasted or tubular lead acid battery (typically 12 years) but not quite as long as the rugged lead acid Planté cell (typically 20 years). NiCad batteries lose their capacity over time under float charge conditions more so than lead acid types. NiCad battery chargers can therefore be programmed automatically to boost charge the NiCad battery bank at regular intervals. Sealed gas recombination batteries have lower life (typically 10 years) and require a strict charging regime. They may be of either lead acid or NiCad type and have the advantage of not requiring special battery room provisions. There is a huge range of international codes governing batteries; it includes IEC 60623 – Specification for vented nickel cadmium rechargeable single cells (see also BSEN 2570) and IEC 60896 – Specification for stationary lead acid batteries. Valve-regulated lead acid batteries perform similarly to sealed batteries but precautions must be taken to deal with the limited level of gas generation under emergency conditions (e.g. if overcharged). Note that ambient temperatures above the design specification (typically 20°C) will not only affect capacity (see Section 4.2.2) but will significantly reduce design life. If this is likely to be a problem, careful consultation with the manufacturer is advised, and engineers should ensure that specifications clearly define conditions to be encountered. Increased life can be achieved by overspecifying the Ah requirement, which has a cost implication, and so, as with other aspects, the cost saving due to increased life has to be weighed against increased initial capital cost.

The discharge period of the battery is the time required before a full capacity battery becomes discharged to a specified end voltage which will still

Table 4.2   Characteristics of different battery types

Description	Lead acid–Planté	Lead acid–Pasted	Nickel Cadmium	Sealed gas recombination (lead acid in this case – NiCad also available)
Life expectancy	20 years	12 years	15 years	10 years
Relative cost	100	70	250	80
Average Watthour per Kg	10 (=5.4 Ah/kg) (100%)	20 (=10 Ah/kg) (50%)	16 (=13 Ah/kg) (55%)	23 (=15 Ah/kg) (35%)
Relative volume indicator				
Open circuit voltage per cell	2.15–2.27 V per cell	2.15–2.27 V per cell	1.35–1.46 V per cell	2.15–2.35 V per cell
Nominal voltage	2.0 V per cell	2.0 V per cell	1.2 V per cell	2.0 V per cell
Gassing voltage	2.4 V per cell	2.4 V per cell	1.55 V per cell	2.4 V per cell (not recommended)
Final discharge voltage	1.8–1.72 V per cell	1.8–1.72 V per cell	1.1–0.8 V per cell	1.60 V per cell
Electrolyte	Dilute sulphuric acid	Dilute sulphuric acid	Dilute potassium hydroxide	Dilute sulphuric acid
Minimum time to full charge	10.5 hours	10.5 hours	7 hours	10.5 hours
Cycle duty	Mainly shallow discharges	Good	Mainly shallow discharges	Mainly shallow discharges
Temperature tolerance	−10°C to +50°C	−10°C to +40°C	−30°C to +45°C	−10°C to +40°C
Precautions	– Boost charge voltage – Gassing	– Boost charge voltage – Gassing	– Boost charge voltage – Gassing–Regular boost required say every 3–6 months	– Should not boost charge – Close voltage tolerance charging
Maintenance	Specific gravity varies with charge. SG should be between 1.200 and 1.225 and typically 1.210 at 15°C.	Specific gravity varies with charge	Specific gravity does not vary significantly with charge condition. SG of about 1.210 is normal. Change of electrolyte is desirable if SG drops to 1.190 and mandatory at about 1.170	Check charger performance carefully every 6 months
	Top up every 9 months. Do not discharge below recommended values, say 1.65 V per cell.	Top up every 6 months. Do not discharge below recommended values, say 1.65 V per cell	Top up every 12 months	Minimal maintenance
	Clean & grease terminals, check connections. Do not leave in discharged state	Clean & grease terminals, check connections. Do not leave in discharged state	Clean & grease terminals, check connections. May be left for long periods in any state of charge	Clean & grease terminals, check connections. Do not leave in discharged state

Charger characteristics	Float charging: Float voltage 2.25 V per cell. If constant current, limit charge current to specified value at end of charge period – typically 7% of rated Ah capacity, i.e. 7A per 100 Ah capacity If constant voltage, EMF of battery will rise during the charging period and current drawn will reduce. Use current limitation at start of charge, i.e. a combination of constant current initially then constant voltage charging to completion Boost charging: Every 2 years (min 2.4 V per cell) or when SG falls below 1.200 at 15°C.	Float charging: As for Planté cells.	Float charging: Float voltage 1.4 V per cell. If constant current, limitation at end of charge period is less stringent than for lead acid batteries and may be 15% of rated capacity, i.e. 15A per 100 Ah capacity Use current limitation at start of charge Boost charging: Every 6 months (min 1.55 V per cell) to help prolong life.	Float charging: Float voltage 2.27 V per cell. Constant voltage preferred method for maximum life Apply temperature compensation to charging, to optimize life Boost charging: Do not boost charge except in emergency at approx. 2.4V per cell with current limiting.
Appropriate usage	Long life float duty applications: Telecommunications Uninterruptible power supplies Power generation and transmission Switch tripping and closing Emergency lighting	Medium life low capital cost applications: Telecommunications Uninterruptible power supplies Power generation and transmission Switch tripping and closing Emergency lighting Engine starting	Long life with wide operating temperature range: Switch tripping and closing Telecommunications	Sealed for life, no gas evolution, minimum maintenance, low space requirement, low weight applications: Telecommunications Uninterruptible power supplies Switch tripping and closing Emergency lighting

ensure correct equipment operation. A comparison of discharge characteristics for different types of lead acid cells together with the characteristics for a 110 V DC substation battery system using NiCad cells is given in Fig. 4.4. Superimposed upon the substation NiCad characteristic are the maximum and minimum circuit breaker closing coil voltage tolerance limits ($\pm 15\%$), the minimum relay operating voltage limit (-20%) and the minimum trip coil operating voltage limit (-30%) around the 110 V nominal 110 V DC level.

One aspect of battery comparison not covered in Table 4.2 is the environmental impact. This is a rapidly changing situation, and specifiers should check current regulations in the region of installation. For example, lead, cadmium and mercury may all be used in one or another cell design, and may all be affected by environmental legislation.

4.3.4 Battery sizing calculations

4.3.4.1 *Capacity and loads*

The required battery capacity is calculated by determining the load which the battery will be expected to supply, the period for which the supply is required and the system voltage limits. Reference is then made to manufacturers' tables of capacity, discharge current capability and final voltages. This should take account of the expected temperature range – see Section 4.3.2 item 3.

The load on the battery is calculated from the power consumption characteristics of the loads taking into account their nature:

- Continuous – (indicating lamps, relays, alarm systems or other items that draw steady current over the whole battery discharge period).
- Time limited – (motors, emergency lighting or other systems which consume power for longer than 1 minute but shorter than the battery discharge period).
- Momentary (particularly the power needed to close or trip switchgear).

Good design practice is to adopt common nominal voltages for substation loads in order to avoid additional batteries or voltage tappings on the battery bank. Standard voltages used are 24, 30, 48 and 110 V. A 48 V DC supply to control and communications equipment is often used and is physically separated from other 110 V DC substation switchgear, control, relay and services load supplies. The control and communications equipment is more locally confined, more suited to a lower operating voltage, voltage drop is not such a problem and different maintenance personnel are involved.

Some typical substation loads are listed below:

Trip coils   load requirements approximately 150 W for less than 1 second. Note that in complex protection schemes

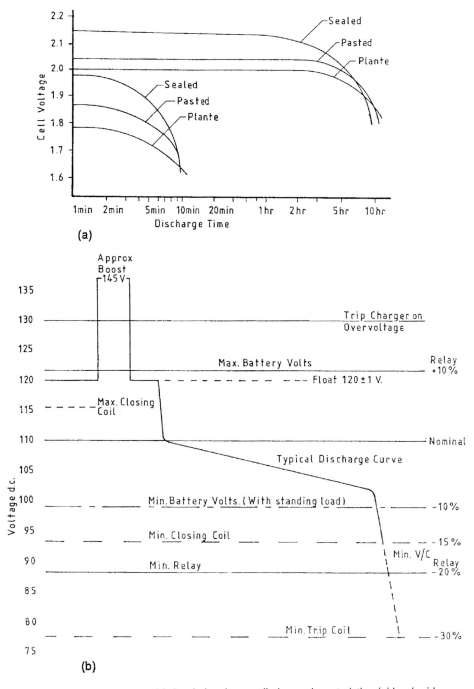

Figure 4.4   Lead acid and nickel cadmium battery discharge characteristics: (a) Lead acid cell typical discharge characteristics; (b) substation 110 V DC. NiCad battery system discharge characteristics

(e.g. busbar protection) several trip coils may be simultaneously energized and the sum of the individual loads must therefore be used in the battery sizing calculations.

Controls/relays   Continuous loads such as indicator lamps will contribute to battery discharge on loss of mains supply.

Closing coils   Older oil circuit breaker coils may take 10–30 kW depending upon design for less than 1 second at 110 V. More modern vacuum or SF$_6$ circuit breaker motor wound spring charged mechanisms and solenoid closing coils have 300–600 W ratings.

DC motors   Diesel generator 'black start' pump and cranking, isolator or switchgear drives, air blast circuit breaker air compressor motor drives.

The standby period or autonomy varies according to the particular power supply authority standards. For industrial consumers 30 minutes is typical, power utilities 60 minutes and 120 minutes minimum on major installations. Where standby generation is also available the battery standby period may be reduced to say 15 minutes after which it is assumed that the local diesel generator will have successfully started automatically.

4.3.4.2 *Practical example*

A distribution substation having 17 No. 13.8 kV oil circuit breakers is to be refurbished with a new battery/battery charger configuration comprising 100% 110 V NiCad battery and 100% charger unit for 3 hour autonomy with the following duties:

1. Momentary loads
   Switchgear closing   13.8 kV breakers, 15 kW each – consecutive load switchgear closing current = 15 kW/110 V = 136 A 20 No. 380 V breakers, manual close.
   Switchgear tripping   17 13.8 kV breakers, 150 W each – simultaneous or 20 380 V breakers, 100 W each – simultaneous. Take maximum switchgear tripping current from either the 13.8 kV or 380 V breakers. 17 No. × 150 W/110 V = 23 A approx.
2. Time limited/continuous loads
   Control and switchgear
   building emergency lighting   15 No., 40 W fittings = 600 watts for 3 hours
   Indicator lamps   37 No., 15 W units = 555 watts for 3 hours

Trip circuit healthy   4 No., 15 W units   = 60 watts for 3 hours
Control panel transducers   = 230 watts for 3 hours
Relay panel components   = 270 watts for 3 hours
Total time limited/continuous load   = 1715 watts
Capacity of time limited/continuous load =

$$= \frac{\text{watts} \times \text{period of autonomy (hours)}}{\text{voltage}}$$

$$= \frac{1715 \times 3}{110} = 47\text{Ah approx.}$$

Average continuous load   = 16 Amps
Allowance for future expansion   = 25%
Maximum momentary load   = 136 Amps

(In this case occurs on switchgear closing. Switchgear tripping only presents a small load in comparison and may be ignored.)

Allowance for future expansion   = 5%

From manufacturers' tables a suitable battery may be selected with the most onerous of the calculated capacity, maximum current or continuous load current taking precedence, after taking account of any ambient condition limitations.

4.3.5 Typical enquiry data

4.3.5.1 *General*

It is normal practice for both the batteries and the charger units to be purchased from the same supplier in order to ensure correct compatibility. The following enquiry forms may be used to assist the vendor to understand fully the requirements for the particular installation. If battery or charger life, or reliability levels (mean time between failures = mtbf) are critical these must also be scheduled.

4.3.5.2 *Battery*

Type of battery and relevant IEC standard
Electrolyte
Nominal system voltage   (V)
Ambient temperature maximum, minimum and average   (°C)
Number of cells
Float voltage per cell   (V)
Normal system float voltage required   (V)
Normal float charging current required   (A)
Minimum recommended battery voltage   (V)
Recommended boost charging voltage per cell   (V)

Recommended boost charging current   (A)
Dimensions of cells – width $\times$ depth $\times$ height   (mm)
Overall dimensions of battery bank – width $\times$ depth $\times$ height   (mm)
Overall weight of battery bank   (kg)
Weight of individual battery cells or blocks   (kg)
Material of battery cases
Battery capacity at _____ hour discharge rate   (Ah)
Duty cycle requirements[a]
Battery voltage at end of duty cycle   (V)
Normal standing load   (A)
Maximum DC current capability (short circuit)   (A)
Battery mounting[b]
Connections[c]
Volume of hydrogen produced during boost charging   (l)
Manufacturer, type reference and manufacturer's drawings

Notes: (a) To be clearly specified in the tender documents.
       (b) Wood or metal stands or racks, internal batteries to charger, access
           for topping up, etc.
       (c) Markings, connecting links and cabling, etc.

4.3.5.3 *Battery charger*

Maximum physical dimensions – width $\times$ depth $\times$ height[a]   (mm)
Enclosure IP rating[b] (IEC 60529)
Ambient temperature maximum, minimum and average   (°C)
Charger to suit following type of battery
   (Cell type)
   (Separate rack or internal to charger)
AC input supply for which specified output must be
maintained
   (No. phases)
   (1/3 ph voltage)   (V)
   (Voltage tolerance) $\pm$ (%)
   (Frequency)   (Hz)
   (Frequency tolerance) $\pm$ (%)
   (AC input)[h]   (kVA)
DC output[c]
   (Float voltage)   (V)
   (Boost voltage)   (V)
   (Float current)   (A)
   (Boost current)   (A)
   (Ripple)[g]   (%)
   (DC output)   (kW)
Psophometric output noise level (for loads between 0%

and 100% to CCITT Regulations)   (mV @__Hz)
Noise level limit   (mV rms hum
                                                  @__Hz)
Current limitation range ±   (A)
Voltage limitation range ±   (% V)
Time to recharge battery to 90% capacity from fully
discharged state   (hours)
Charger efficiency   (%)
Overload protection
Controls, indications and alarms[d]
Applicable standards[e]
Manufacturer and type reference[f]

Notes: (a) Add details of gland plate, top or bottom cable entry, etc. as
            required.
       (b) Often best left to manufacturer unless specific housing conditions
            are required. For example a high IP rating could necessitate forced
            air cooling which in turn could reduce overall reliability. Alter-
            natively, specify ambient conditions which might affect electronic
            charger components (e.g. humidity level, high dust content).
       (c) Output voltage range as per IEC 62271-100 and IEC 60694 for
            nominal switchgear DC voltage and shunt trip coil voltage ranges.
       (d) See Sections 4.2.2.5 and 4.2.2.6.
       (e) For example IEC 60146 for semiconductor rectifier equipment.
       (f) To be completed by manufacturer unless nominated supplier
            sought.
       (g) Manufacturer also to confirm compliance with relevant limits on
            harmonics generated back into the AC supply.
       (h) Subject to supply kVA limits as specified, manufacturer to indi-
            cate maximum AC current required (in amps).

4.4 AC SUPPLIES

4.4.1 Power sources

Substation auxiliary AC supplies may be derived from dedicated sources or
from additional circuits on low voltage distribution switchgear forming part
of the substation's outgoing distribution system. Three examples are given in
Fig. 4.5:

• Simple 380–415 V three phase circuit allocations fed by the distribution
  substation transformer(s).

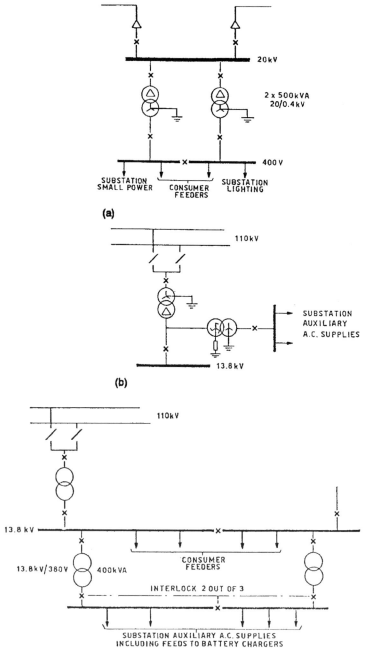

Figure 4.5   Derivation of substation auxiliary low voltage (LVAC) supplies: (a) Distribution substation; (b) primary substation with AC auxiliary supplies derived from earthing transformer(s). (c) dedicated and duplicated auxiliary transformers (courtesy of Balfour Beatty Projects and Engineering Ltd)

- Tertiary windings on substation main transformer(s) or from earthing transformer (zigzag star-star) windings.
- Dedicated substation auxiliary transformers and switchgear.

The essential factors to be considered are the level of security of supply required (duplicated transformers, independent source of supply, LVAC sectionalized switchboard, key interlocks, etc.), the fault level of the LVAC switchgear (possible high fault levels at primary substation sites) and allowances for future substation extensions (additional future switchgear bays, future use of presently equipped or unequipped spare ways).

4.4.2 LVAC switchboard fault level

The substation auxiliary LVAC switchboard will typically be fed by auxiliary transformers in the range 100–630 kVA. Transformers in this range normally have impedance values of the order of 4–5% and will therefore act as the main fault limiting element in the system between generation source and substation LVAC switchboard. Neglecting source impedance this implies auxiliary transformer secondary fault levels of some 12 MVA without having transformers in parallel. Key-interlock systems are usually employed to prevent paralleling of substation auxiliary transformers and thereby avoid exceeding the fault rating of the switchgear. Air circuit breakers are often employed as incomers and bus-section switches on the LVAC switchboard. They can be specified to cater for high fault levels and load currents over a wide temperature range in withdrawable format and as an integral part of a larger switchboard.

4.4.3 Auxiliary transformer LV connections

A single auxiliary transformer is normally connected to the LVAC distribution switchboard earth and neutral busbars via links as shown in Fig. 4.6a–c. A current transformer (CT) associated with secondary unrestricted earth fault protection is located on the earth side of the neutral to earth link. In this way the CT is in the path of the earth fault current. At the same time unbalanced or harmonic currents involving the neutral (3rd harmonics and multiples) will not be 'seen' by the CT in this position. This is the 'classic' standby earth fault (SBEF) protection CT location and transformer connection to the LVAC switchboard for a *single* transformer source of supply. This arrangement is unsatisfactory when applied to a *multi-source* supply system.

Consider the case of two auxiliary transformers, A and B, used to derive the substation auxiliary LVAC supply. Fig. 4.6b shows such a system with a key interlocked normally open LV busbar-section circuit breaker and the same 'classic' SBEF CT location. For an earth fault, $I_F = I_{FA} + I_{FB}$, on the left-hand side LV switchgear busbar fed by transformer A a proportion of the fault current,

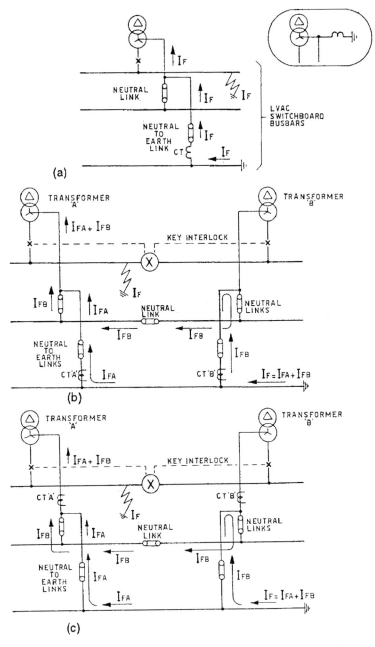

Figure 4.6   Auxiliary transformer LV connections: (a) Single transformer LVAC auxiliary supply source and simplistic single line diagram representation (inset); (b) substation auxiliary supply derived from duplicate transformer source with incorrect SBEF CT location; (c) substation auxiliary supply derived from duplicate transformer source with correct SBEF CT location. I_F = busbar earth fault current. I_{FA} = component of earth fault current returning via transformer 'A' neutral to earth link. I_{FB} = component of earth fault current returning via transformer 'B' neutral to earth link. $\otimes$ = open bus-section circuit breaker (courtesy Balfour Beatty Projects and Engineering Ltd)

I_{FB}, may return to the neutral of transformer A via the earthing path of transformer B. This fault current could therefore cause the relay associated with transformer B to operate. If the earth fault current path is particularly unfavourable it is possible for the LVAC switchboard incomer from transformer B to be tripped, thereby losing the healthy side of the switchboard. In addition it should be noted that the I_{FB} proportion of the fault current will bypass, and not be summated by, the neutral CT associated with the transformer supplying the fault.

The auxiliary transformer connections to the LVAC switchboard shown in Fig. 4.6c involve relocating the SBEF CTs close to the transformer neutrals. Even with both earth fault components I_{FA} and I_{FB} present, the SBEF CT associated with transformer A summates the currents to operate the appropriate relay and in turn correctly disconnects transformer A from the faulty busbar. At the same time maloperation of transformer B protection is avoided and transformer B continues to supply the healthy busbar and associated substation LVAC loads. The disadvantage of this connection arrangement is that the SBEF CTs will now register 3rd harmonic or out of balance load currents. Relays with harmonic restraint filters can be employed in cases where the harmonic component of the load (such as with discharge lighting) is high.

4.4.4 Allowance for future extension

It is good engineering practice to formulate a policy for spare capacity on auxiliary LVAC transformers and associated switchgear in keeping with capital cost constraints.

This is especially true in developing countries and a typical policy guide might be to allow an overall 25% spare switchboard capacity with 10% equipped spare ways and 15% unequipped spare ways within the switchboard physical dimensions.

4.4.5 Typical enquiry data

The table given below describes the essential characteristics for a substation auxiliary LVAC distribution board. This type of enquiry data sheet should be used in conjunction with a full enquiry specification of requirements which details all general and specific requirements (LVAC supply characteristics, etc.):

Maximum physical dimensions – width $\times$ depth $\times$ height (mm)	
Enclosure IP rating (IEC 60529)	
Single line diagram drawing number[a]	
Unequipped spare ways	
Equipped spare ways	
Operating voltage (max.)	(V)
One minute power frequency voltage	(kV rms)
System frequency	(Hz)

Phases
Short time current (3 seconds or 1 second as appropriate) (kA)
Floor mounting/free standing, etc.
Front access/rear access
Busbars and switchboard allowable for future expansion   (Yes/No)
Painting finish
Earth bar (internal, full size, etc.)
Panel indicators
Panel anti-condensation heaters   (Yes/No)
Wiring
− Standard
− Control wiring size   (mm^2)
− CT wiring size   (mm^2)
− Ferrule/cable core identification standard
Relevant standards

AC distribution boards	IEC 60439
Moulded case circuit breakers[b]	IEC 60157
Fuses	IEC 60269
Contactors	IEC 60158
Isolators and switches	IEC 62271-102 & IEC 60265

Busbar maximum current rating   (A)
Terminal details
Switchgear type[c]
Manufacturer[d]
Manufacturer's drawings[d]
Metering, alarms, remote indication outputs and protection[e]
Remote control facilities
Auxiliary power supplies

Notes: (a) Include method of interlocking (mechanical key interlocks) if applicable for incoming supply with switchgear bus-section circuit breaker.
   (b) Recommend P2 category for repeated short circuit capability.
   (c) Metal-clad, metal enclosed, withdrawable fuse carriers, circuit padlock arrangements, gland plate details, labelling, ACB incomer details, MCCBs, MCBs, fuses, etc.
   (d) To be completed by the manufacturer unless nominated supplier required.
   (e) To be clearly indicated upon enquiry drawing or detailed circuit by circuit here. Quiescent and operated power consumption should be noted.

4.4.6 Earthing transformer selection

It is often necessary to derive the substation LVAC supply from the main power transformers. The lowest primary substation distribution voltage level

(10 kV, 20 kV, etc.) is also often provided by a delta secondary. Provision of a medium voltage earthing point is necessary in order to limit and better control the medium voltage earth fault level. This earthing point and derivation of a useful LVAC substation auxiliary power source may be provided by using an earthing transformer. Refer to Chapter 14, Section 14.5.6.

The options available are:

- interconnected star/star
- star/interconnected star
- star/delta/interconnected star

The zero sequence impedance on the MV side must limit the earth fault current to a specific value of typically 1000 A. The earthing transformer must exhibit low positive and zero sequence impedance on the LV side in order to permit unbalanced loads and minimize voltage regulation difficulties. The relative merits of these different earthing transformer connections are described when fed from the delta connected secondary of a primary substation power transformer.

(a) Interconnected star/star

An interconnected star winding on its own has sufficiently low reactance to provide an MV earthing point in conjunction with a main delta connected power transformer secondary winding.

Figures 4.7a and 4.7b show the winding connection/flux diagram for an 11/0.415 kV interconnected star/star earthing transformer under MV earth faults and under LV unbalanced load or earth fault conditions respectively. An ampere-turn balance is achieved for the external 11 kV earth fault condition and so the earthing transformer presents a low reactance to such faults. An unbalanced LVAC load or a phase-to-neutral earth fault on the secondary side of the earthing transformer produces no corresponding ampere-turn balance with this vector grouping. Therefore the magnetic circuit to secondary zero sequence currents must pass out of the core, returning via the air/oil interface to the tank sides. In practice, for the usual 3-limb core arrangement, the resulting zero sequence impedance is sufficiently low to allow limited unbalanced loading. However, a transformer design that does not rely on external flux paths for certain loading or fault conditions can be more precisely designed. If a 5-limb or shell type arrangement is used the resulting magnetizing current would be very low and unbalanced loading impossible. Interconnected star/star earthing transformers for substation auxiliary LVAC loads in the range 250–500 kVA are perfectly feasible. However, as the transformer rating increases so does the percentage reactance and load regulation becomes difficult.

(b) Star/interconnected star

Figures 4.8a and 4.8b show the winding connection/flux diagram for this vector grouping again under MV earth fault and LV unbalanced loading or earth

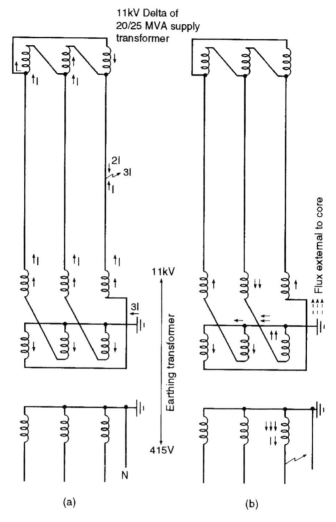

Figure 4.7   Interconnected star/star winding: (a) 11 kV earth fault; (b) LV earth fault single phase loading

fault conditions, respectively. For the 11 kV earth fault case the ampere-turns in the earthing transformer star winding are not balanced against the delta connected primary substation transformer secondary. The interconnected star earthing transformer secondary winding has no effect in providing balancing ampere-turns for this fault condition. Therefore the earthing transformer presents a high reactance to 11 kV earth faults and is not particularly useful for this substation application. Under LVAC unbalanced load conditions an ampere-turns balance is achieved and the earthing transformer presents a low reactance to out-of-balance secondary loads.

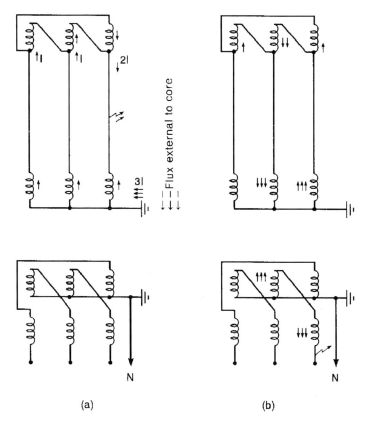

Figure 4.8   Star/interconnected star winding: flux diagrams

(c) Star/delta/interconnected star
Refer to Fig. 4.9a and 4.9b for the winding connection and flux diagrams for this vector grouping again under MV earth fault and LV unbalanced loading or earth fault conditions, respectively. Under 11 kV earth fault conditions balancing ampere-turns are provided by the circulating current in the earthing transformer delta winding. The earthing transformer therefore provides a low reactance to 11 kV earth faults. An ampere-turns balance is also achieved for the LVAC out-of-balance or earth fault conditions such that the earthing transformer with this vector grouping presents a low reactance. The cost of the additional third delta winding makes this earthing transformer connection less economic than the simple and more common interconnected star/star type. However, the connection offers greater flexibility in the design of satisfactory impedances and as the earthing transformer LVAC load requirement increases this connection offers better regulation than the interconnected star/star arrangement.

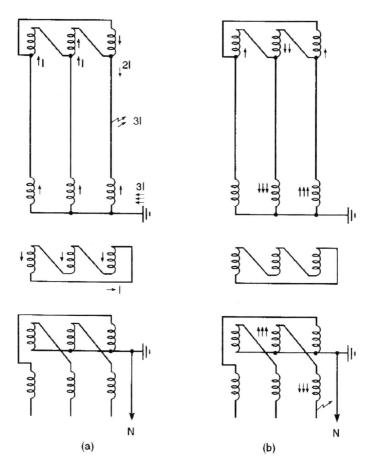

Figure 4.9   Star/delta/interconnected star winding flux diagrams

4.4.7 Uninterruptible power supplies

4.4.7.1 *Introduction*

Static uninterruptible power supply (UPS) units producing a secure AC (or DC) output usually consist of an AC to DC rectifier, battery unit and (for an AC output) a DC to AC inverter as shown in Fig. 4.10. As well as being used to provide supply security, it may also be installed to provide power of con-trolled quality to sensitive electronic equipment.

4.4.7.2 *Operation*

The rectifier float or boost charges the battery bank. The battery is sized for a given autonomy of supply under mains power failure conditions in the same

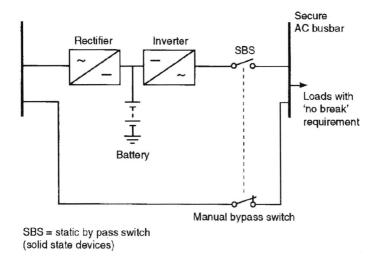

SBS = static by pass switch
(solid state devices)

Figure 4.10   Uninterruptible power supply (UPS)

manner as described in Section 4.3 above. The autonomy may be specified as typically between 15 minutes to 3 hours under full load conditions. The inverter produces, from the stored energy contained within the battery, an independent AC supply with very close tolerances. It is usually an isolated supply, and the UPS may be considered and treated as a generator in many respects. For example, it is vital to examine the earthing arrangements, and also safety isolation when working on the system.

The unit may be continuously connected in circuit. This configuration is particularly applicable where different input to output voltages and/or frequencies are required. Alternatively a very fast acting solid state transfer switch (SBS) may be used in conjunction with a voltage-sensing electronic control circuit to connect the unit upon brief mains supply voltage dips, spikes or longer-term interruptions. This ensures that the load supply is maintained with an 'uninterrupted' changeover in the event of a fault or an overload condition. Such systems are usually specified for computer power supplies.

4.4.7.3 Particular requirements

Apart from the autonomy required special consideration must be given in the UPS specifications to the speed of changeover (fraction of a cycle) achievable by the static bypass transfer switch. The various parts of IEC 60146 detail methods for specifying the UPS. The tolerance of the connected load to voltage disturbances must also be matched with those likely to be caused with the UPS in service. In particular, the specifications must cover limitations to harmonic disturbances caused by the solid state rectifier and inverter units, both as regards the load and also in the supply network. A typical technical data sheet for use at the enquiry stage for UPS systems is detailed in Table 4.3.

Table 4.3 Uninterruptible power supply technical particulars and guarantees

Description			Type and requirement or manufacturers guarantee
Manufacturer			
Standards			
Source supply:	Voltage (rms)	(V ±)	
	Frequency	(Hz ±)	
	No. of phases		
	Minimum power factor		
UPS output:	Voltage (rms)	(V ±)	
	Frequency	(Hz ±)	
	No. of phases		
	Minimum load power factor		
	Maximum load current (A)		
Types of output switching devices			
Indications and controls			
Remote control or indication requirements			
Rectifier output current range:			
Float	(A)		
Boost (if applicable)	(A)		
End boost	(A)		
Battery charging time from fully discharged (end) condition to 90% fully charged capacity (hours)			
Battery capacity (rating) (Ah)			
Cell type			
Cell range/operating			
voltage: Float		(V to V)	
Boost		(V to V)	
Commission		(V to V)	
Cell voltage when fully discharged		(V)	
Equipment function			Statements about changeover times, transient behaviour under changeover and supply side interruptions, undervoltage, overvoltage and voltage spike conditions
Load duty			
Operation mode:			
Continuously connected without static bypass switch			Continuously conditioned power to load even in event of mains failure. Especially suitable when different input and output voltages and frequencies are required
Continuously connected with mechanical bypass switch			Load is normally supplied from the UPS with an electro-mechanical contactor for short break changeover to the alternative supply when required. Useful if supply is poor quality but not a true no-break system
Continuously connected with static bypass switch			Uninterrupted changeover in the event of a fault or overload condition. Often specified for computer power supplies
Active standby mode with no break static transfer switch			Useful configuration if mains supply variations are acceptable to the load. Rectifier maintains the battery in charged condition and UPS used immediately upon mains failure
UPS space requirement (L × W × D) (mm)			
UPS weight (kg)			
Battery space requirement (If separate) (L × W × D) (mm)			
Weight of one cell and total battery (filled) bank (kg and kg)			
Ventilation requirements:			
Environmental conditions:			
– Temperature (min., average and max.)	(°C)		
– Relative humidity range	(% ±)		

Figure 4.11   Conventional vented lead acid battery room showing battery bank, sink, tiles and ventilation arrangements

Bearing in mind that a major reason for installing a UPS facility is reliability of supply, it is worth mentioning again that the reliability and availability of the UPS unit itself should be carefully investigated. Availability depends, among other things, on time to restore after failure, and that in turn depends upon speed of detection of failure. Choice of test facilities – local or remote, automatic or manual, and frequency of test are significant. For AC output units a manual bypass facility (preferably external), to allow supply continuity during maintenance or repair, is strongly recommended. For further treatment of the subject of reliability, see Section 23.5.

Figure 4.12   Combined sealed lead acid battery and charger unit (courtesy Telepower)

5 Current and Voltage Transformers

5.1 INTRODUCTION

Current and voltage transformers are required to transform high currents and voltages into more manageable quantities for measurement, protection and control. This chapter describes the properties of current transformers (CTs) and voltage transformers (VTs), and how to specify them for particular applications.

5.2 CURRENT TRANSFORMERS

5.2.1 Introduction

A current transformer is used to transform a primary current quantity in terms of its magnitude and phase to a secondary value such that in normal conditions the secondary value is substantially proportional to the primary value. IEC 60044 (instrument transformers) covers CTs and VTs, superseding IEC 60185 (current transformers for measuring and protective applications) and its predecessor IEC 185.

5.2.2 Protection CT classifications

Protection CTs, unlike measuring CTs, may be required to operate at many times full load current. Linearity under these conditions is not of great importance. The essential point is that saturation must be high enough to drive the magnetizing current and the secondary current under fault conditions.

5.2.2.1 *5P or 10P classification*

Several terms are used in connection with CTs and these are described below:

Rated primary (or secondary) current	This value, marked on the rating plate of the CT, is the primary or secondary current upon which the performance of the transformer is based.
Rated transformation ratio	The rated transformation ratio is the ratio of rated primary current to rated secondary current and is not necessarily exactly equal to the turns ratio.

The magnetizing current depends upon the magnitude of the primary voltage which in turn depends upon the magnitude and power factor of the burden. It is possible partially to compensate for the magnetizing current ratio error in CT designs by slightly reducing the number of turns on the secondary. However, no similar compensation is available for small phase errors. The standards to which the CTs are specified may not detail a continuous overload rating. It is therefore prudent to choose a primary current rating at least equal to the circuit rating. An accuracy class of 5P (P stands for protection) is usually specified for large systems where accurate grading of several stages of IDMTL overcurrent relay protection is required. An accuracy class of 10P is also often acceptable and certainly satisfactory for thermal overload relays on motor circuits. These accuracy classes correspond to 5% or 10% composite error with rated secondary burden connected at all currents up to the primary current corresponding to the rated accuracy limit factor.

Composite error	Under steady-state conditions the r.m.s. value of the difference between the instantaneous values of the primary current and the actual secondary current multiplied by the rated transformation ratio.
Rated output at rated secondary current	The value, marked on the rating plate, of the apparent power in VA that the transformer is intended to supply to the secondary circuit at the rated secondary current.

The rated VA should be specified to correspond to the relay and connecting lead burden at rated CT secondary current. If relays are mounted on the switchgear adjacent to the CTs then the lead burden can often be neglected. It is best to allow a margin for greater than anticipated burden but this should be included in the specification for the rated accuracy limit factor.

Rated accuracy limit factor (RALF)	The primary current up to which the CT is required to maintain its specified accuracy with rated secondary burden connected, expressed as a multiple of rated primary current.

Ideally the RALF current should not be less than the maximum fault current of the circuit up to which IDMTL relay grading is required, and should be based upon transient reactance fault calculations. If a switchboard is likely to have future additional fault-in-feeds then it is sensible to specify an RALF corresponding to the switchgear fault-breaking capacity. Rated outputs higher than 15 VA and rated accuracy limit factors higher than 10 are not recommended for general purposes. It is possible to make a trade-off between RALF and rated output but when the product exceeds 150 the CT becomes uneconomic with large physical dimensions. An RALF of 25 is an economic maximum figure. A reduction in RALF is not always possible and therefore the following measures should be considered:

- Use the highest possible CT ratio.
- Investigate relays with a lower burden. Solid-state relays have burdens of 0.5 VA or less and do not change with tap setting.
- At lower system voltage levels (15 kV and below) consider the use of fuses on circuits of low rating but high fault level.

Typical electromagnetic protection relays have a burden of about 3 VA at the setting current. The burden increases on the minimum plug setting (50% for a typical overcurrent relay). Precautions are therefore taken in protection relay designs to ensure that the increase in burden does not exceed half the nominal value as the tap setting is changed. In addition to the relay burden the CT leads and connecting cables must be taken into account. A 100 m length of typical 2.5 mm^2 cable would have a burden of about 0.74 ohms per core or 0.74 VA for a 1-amp secondary rating and 18.5 VA for a 5-amp rating. Hence the advantage of using 1-amp secondary CTs for substations with long distances between relays and CTs.

A typical marking on a protection CT would be 15 VA Class 5P 10, where 15 VA is the VA output at rated secondary current, Class 5P indicates that this is a protection (P) CT with a composite error of <5% at rated accuracy limit primary current and 10 is the rated accuracy limit factor (RALF) for the CT, i.e. overcurrent = 10 × rated normal current.

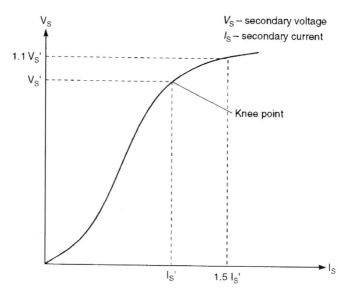

Figure 5.1   Typical magnetizing characteristic

5.2.2.2 *Knee point*

For protective purposes current transformer specifications may be defined in terms of the 'knee point'. This is the voltage applied to the secondary terminals of the CT with all other windings being open circuited, which, when increased by 10%, causes the exciting current to be increased by 50%. A typical CT magnetizing characteristic is shown in Fig. 5.1. Older standards (BS3938) catered for the specification of such 'Class X' CTs in terms of:

- Rated primary current.
- Turns ratio.
- Rated knee point e.m.f.
- Maximum exciting current at a stated percentage of rated knee point e.m.f.
- Maximum resistance of secondary winding.

In addition the CT must be of the low reactance type and the turns ratio error must not exceed 0.25%. Bar-type CTs with jointless ring cores and evenly distributed secondary windings will provide negligible secondary winding leakage reactance and will usually satisfy this reactance requirement. For Class X CTs turns compensation is not permitted and a 400/1 Class X CT should have exactly 400 turns. Such carefully controlled CTs are used in pilot wire and balanced differential protection schemes and the manufacturer usually provides an excitation curve at the design stage which may be later confirmed by routine testing and site tests. Such CTs could be specified for use with IDMTL relays but this is not usual.

5.2.2.3 *Other standards*

American Standards designate CTs with negligible secondary leakage react-ance as Class C and the performance may be calculated in a similar manner to the now obsolete BS3938 Class X CTs. Class T CTs have some leakage and tests are called for in the ANSI Standards to establish relay performance. In addition to the leakage reactance classification, the CTs are specified with a permissible burden in ohms equivalent to 25, 50, 100 or 200 VA for 5-A-rated CTs. The secondary terminal voltage rating is the voltage the transformer will deliver to a standard burden at 20 times rated secondary current without exceed-ing 10% ratio correction. This is not exactly equivalent to the Class X CT knee point voltage since the terminal voltage will be of a lower value due to losses in the secondary winding resistance.

BS3938, referred to above, remained current until 1999 because it referred to class X CTs. IEC 60044-1 and its European equivalent EN 60044-1 (and the UK national BSEN60044-1) deal with class PX and completely replace class X. Hence the eventual withdrawal of 3938.

5.2.3 Metering CTs

For non-protection purposes metering CTs need perform very accurately but only over the normal range of load up to, say, 120% full load current. Metering CTs are specified in terms of:

- ratio,
- rated-VA secondary burden,
- accuracy class.

Accuracy classes recognized by IEC 60044 are 0.1, 0.2, 0.5 and 1. Accuracy classes 3 and 5 are also available from manufacturers. For each class the ratio and phase angle error must be within specified limits at 5%, 20%, 100% and 120% of rated current. A class 0.2 metering CT means that at 100–120% of the rated current, the percentage ratio error will be ± 0.2; i.e. for a class 0.2 CT with a rated secondary current of 5 A the actual secondary current would be 5 ± 0.01 A. Phase displacement error is also specified in the IEC standard. For special applications an extended current range up to 200% may be specified. Above these ranges accuracy is considered to be unimportant since these condi-tions will only occur under abnormal fault conditions. There is an advantage in the CT being designed to saturate under fault conditions so that the connected metering equipment will have a lower short-time thermal withstand requirement. It is preferable not to use common CTs to supply both protection and metering equipment. If, for example, only one set of protection CTs is available then it is good practice to separate the measuring instrumentation from the protection relays by means of saturable interposing CT or by adding saturable shunt reactors. This has the advantage of protecting the instrumentation and reducing the overall burden under fault conditions. A typical marking on a metering CT would be 15 VA Class 0.2 120%.

- The VA output at rated secondary current is 15 VA.
- The percentage error is ±0.2 at rated current.
- The extended current rating is 120% of rated secondary current.

5.2.4 Design and construction considerations

The power system design engineer should appreciate the following points with regard to CT design.

- Core materials.
- Non-oriented silicon steel usually least expensive.
- Grain-oriented cold-rolled silicon steel gives a higher knee point voltage and lower magnetizing current.
- Mumetal may be used for high accuracy metering CTs having a very low magnetizing current and low knee point voltage.
- Special cores with air gaps may be used for linear output.

Knee point:
The knee point of a CT is directly proportional to the cross-sectional area of the core. The magnetization current of a CT at a particular voltage is directly proportional to the length of the magnetic core around its mean circumference.

Secondary winding:
The knee point voltage is directly proportional to the number of secondary turns which are usually determined by the turns ratio. High voltages can appear across the open circuit secondary terminals of CTs. Therefore switching contact arrangements must be added to protection schemes such that when relays are withdrawn from service (e.g. for maintenance) their associated secondary CT terminals are automatically short circuited.

Space considerations:
The design of a CT is based upon the best compromise between choosing maximum core cross-section for the highest knee point voltage and choosing maximum cross-section of copper for the secondary winding to achieve the lowest winding resistance.

Transient behaviour:
The transition from steady-state current to fault current conditions is accompanied by a direct current component. The magnitude of the DC component depends upon the point on the wave at which the fault occurs. The DC component will then decay with an exponential time constant proportional to the ratio of resistance to inductance in the circuit. While the DC component is changing a unidirectional flux is built up in the CT core in addition to the AC working flux. If the protection scheme requires a constant transformation ratio without significant saturation under all possible fault conditions then the DC time constant must be allowed for in the knee point derivation formula.

Some high impedance relay protection schemes are designed to operate correctly under saturated CT conditions. Distance relays would tend to operate more slowly if the CTs are not designed to avoid transient saturation. Low impedance-biased differential protection, pilot wire protection and phase comparison schemes would tend to be unstable and operate under out-of-zone fault conditions if the CTs are allowed to saturate.

Some typical CT knee point requirements, all based on 5-A secondary CTs, for different types of protection are detailed below:

Distance impedance measuring schemes
$$V_{kp} > I_f (1 + X/R) \cdot (0.2 + R_{ct} + 2R_t)$$

Phase comparison scheme
$$V_{kp} > 1.5 \cdot X/R \cdot I_f (0.2 + R_{ct} + 2R_t)$$

Pilot wire differential scheme
$$V_{kp} > 50/I + I_f (R_{ct} + 2R_t)$$

Electromagnetic overcurrent relay scheme
15 VA 5P 20. (Note that the rated accuracy limit factor (RALF) is dependent on the maximum fault level, CT ratio and type of relay.)

Solid-state overcurrent relay scheme
5 VA 5P 20

High impedance relay scheme
$$V_{kp} > 2I_f (R_{ct} + 2R_t)$$

where V_{kp} = CT knee point voltage
R_{ct} = CT secondary wiring resistance (75°)
I = CT, and relay, secondary rating (5 A assumed)
I_f = maximum symmetrical fault level divided by the CT ratio (for distance protection relays use I_f at the end of zone 1, otherwise use the maximum through fault level)
R_t = Resistance per phase of CT connections and leads.

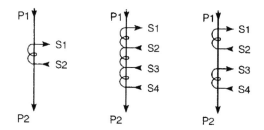

Figure 5.2   CT terminal markings

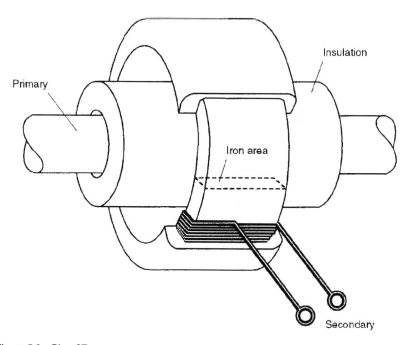

Figure 5.3   Ring CT

5.2.5 Terminal markings

The terminals of a CT should be marked as indicated in the diagrams shown in Fig. 5.2. The primary current flows from P1 to P2 and it is conventional to put the P1 terminal nearest the circuit breaker. The secondary current flows from S1 to S2 through the connected leads and relay burden. A typical ring CT is shown in Fig. 5.3. Checking the correct polarity of CTs is essential for differential protection schemes and a simple method is explained in Chapter 19.

Table 5.1 Current transformers (to IEC60044-1)

Location	Circuit	Type	Ratio	Rated output (VA)	Accuracy class	Rated short time thermal current (kA) 3 or 1 second	Rated accuracy limit factor (RALF)	Knee point voltage	DC resistance

5.2.6 Specifications

Table 5.1 gives a typical format for setting out CT requirements on a substation circuit-by-circuit basis. Open terminal substation CTs will also require insulator details (creepage, arcing horns, impulse withstand, etc.) to be specified (see Chapter 6).

5.3 VOLTAGE TRANSFORMERS

5.3.1 Introduction

IEC 60044 applies to both electromagnetic (inductive) and capacitor type voltage transformers, superseding IEC 60186 and its predecessor IEC186. For protection purposes VTs are required to maintain specified accuracy limits down to 2% of rated voltage:

- Class 3P may have 3% voltage error at 5% rated voltage and 6% voltage error at 2% rated voltage.
- Class 6P may have 6% voltage error at 5% rated voltage and 12% voltage error at 2% rated voltage.

5.3.2 Electromagnetic VTs

These are also referred to as inductive voltage transformers and are fundamentally similar in principle to power transformers but with rated outputs in VA rather than kVA or MVA. It is usual to use this type of voltage transformer up to system rated voltages of 36 kV. Above this voltage level capacitor VTs become cost effective and are more frequently used. The accuracy depends upon the control of leakage reactance and winding resistance, which determines how the phase and voltage errors vary with burden. Permeability and core losses affect the magnetizing current and the errors at low burdens. Therefore electromagnetic measurement VTs normally operate at lower flux densities than power transformers. The derivation of residual voltages for earth fault protection using open delta tertiary windings and five limb or three single-phase VTs is explained in Chapter 10.

It is usual to provide fuse protection on the HV side of electromagnetic VTs up to 36 kV, although some utilities prefer to dispense with these at voltages below 15 kV, on the basis that fuse failure is much more likely than VT failure in modern equipment at these voltages. In addition fuses or MCBs are used on the secondary side to grade with the HV protection and to prevent damage from secondary wiring faults.

5.3.3 Capacitor VTs

Capacitor voltage transformers (CVTs) use a series string of capacitors to provide a voltage divider network. They are the most common form of voltage transformer at rated voltages of 72 kV and higher. A compensating device is connected between the divider tap point and the secondary burden in order to minimize phase and voltage errors. In addition a small conventional voltage transformer is used to isolate the burden from the capacitor chain. Tapping connections are added to this wound isolating transformer in order to compensate for manufacturing tolerances in the capacitor chain and to improve the overall accuracy of the finished CVT unit. Coupling transformers may also be added to allow power line carrier signalling frequencies to be superimposed upon the power network. A typical arrangement is shown in Fig. 5.4. In addition to the

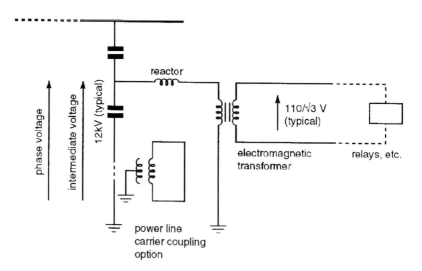

Figure 5.4   Capacitor voltage transformer arrangement

Table 5.2   Coupling capacitors

Manufacturer	
Type	
Intermediate phase-to-earth voltage (kV)	
Total capacitance at 100 kHz (pF)	
1 minute power frequency withstand (kV)	
Impulse withstand 1.2/50 μs (kV)	
Insulating medium	
Dielectric power factor @ kHz	Choose frequency to suit power line carrier system
Weight (kg)	
CVTs	
Rated burden per Class (VA)	
Temperature coefficient of ratio per °C	
Maximum errors with 5% primary voltage ratio (%)	
Phase angle (minutes)	
Intermediate voltage (kV)	
Secondary output voltage and electromagnetic transformer tapping range (V, ±V)	

accuracy class limits described for electromagnetic transformers CVTs must be specified to avoid the production of over voltages due to ferroresonant effects during transient system disturbances.

5.3.4 Specifications

Capacitor voltage transformers and coupling capacitors may be specified in the format shown in Table 5.2 for open terminal 145 kV-rated voltage equipment.

5.4 FUTURE TRENDS

In conjunction with the development of complete substation automation systems (see Chapter 10), future trends include optical data communication to 'optical' CTs and VTs. IEC standard 61858 covers this optical communication from the process side.

An optical CT HV installation with optic communication to the relay is pictured in Fig. 5.7.

Figure 5.5   LVAC switchboard busbars and ring CTs. Note the clearly displayed CT terminal markings P1, P2 and S1, S2

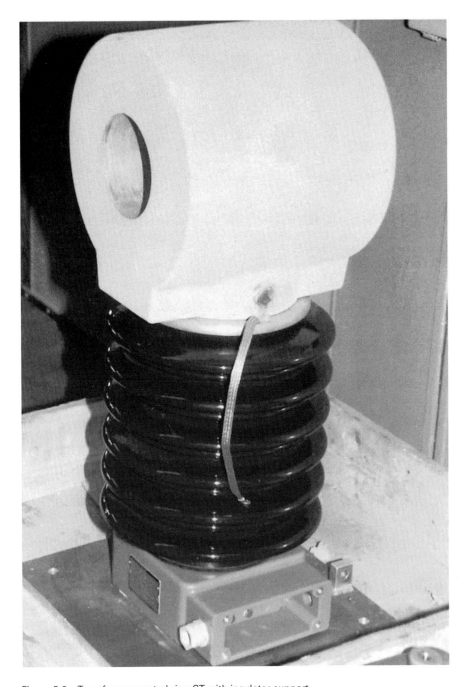

Figure 5.6   Transformer neutral ring CT with insulator support

Figure 5.7   Optical CT HV installation (courtesy of ABB)

6 Insulators

6.1 INTRODUCTION

This chapter describes the different types of overhead line and substation insulators, their design characteristics and their application. Conductors are attached to their support by means of an insulator unit. For overhead lines up to 33 kV and for outdoor substation equipment, the insulator is typically of the post insulator type. For overhead lines above 33 kV and substation aerial conductor busbars, suspension or tension cap and pin or long rod insulator units are employed. Insulators must be capable of supporting the conductor under the most onerous loading conditions. In addition, voltage flashover must be prevented under the worst weather and pollution situations with leakage currents kept to negligible proportions.

6.2 INSULATOR MATERIALS

Three basic materials are available: polymeric composite, glass and porcelain types.

6.2.1 Polymeric and resin materials

Overhead line polymeric insulators (sometimes called 'silicone insulators') are a development dating from the 1960s. They have the advantage of reduced weight, high creepage offset and resistance to the effects of vandalism since the sheds do not shatter on impact. Epoxy resin cast insulators are extensively used in indoor substation equipment up to 66 kV and metal enclosed switchgear. Epoxy resins have been used to a limited extent on medium voltage current transformers installed outdoors and in particular on neutral connections where the insulation is not subject to the same dielectric stress as the phase conductor supports.

Cap and pin insulators used on low voltage distribution lines and long rods at transmission voltage levels may employ composite insulators based on a high tensile strength core of glass fibre and resin. The insulator sheds are bonded to this core and made from silicone and ethylene propylene flexible elastomers. Considerable satisfactory experience under various climatic conditions has now been collected over the last 20 years using these materials. However, there is still some reluctance to specify polymeric insulators for overhead line work because of the conservative nature of the electricity supply industry and doubts about their long-term resistance to ultraviolet exposure and weathering. They have yet to be generally applied to substation installations but have shown good pollution withstand. The advantage of lightweight has an overall cost reduction effect on new substation steelwork support structures.

6.2.2 Glass and porcelain

Both glass and porcelain (see Section 6.5.1 regarding the use of these terms) are commonly used materials for insulators and have given excellent service history backed by years of manufacturing experience from reputable firms. There is little difference in the cost or performance between glass and porcelain. Toughened glass has the advantage for overhead lines that broken insulators tend to shatter completely upon impact and are therefore more easily spotted during maintenance inspections. In practice, the type to be used on overhead lines will depend partly upon the existing spares holdings and spares rationalization practices employed by the particular electricity supply company.

On the other hand, glass insulators are rarely used in substation practice since on shattering they leave only some 15 mm between the top metal cap and the pin. Porcelain insulators, which may be chipped or cracked but not shattered, are therefore preferred for substation use since access for replacement may require a busbar outage.

6.3 INSULATOR TYPES

6.3.1 Post insulators

Post insulators comprise of pedestal posts and solid core cylindrical types. Figure 6.1 illustrates the general construction.

6.3.1.1 Pedestal post insulators

Pedestal post insulator stacks used in substations are available as single units with a range of lightning impulse withstand ratings (LIW) from 60 to 250 kV

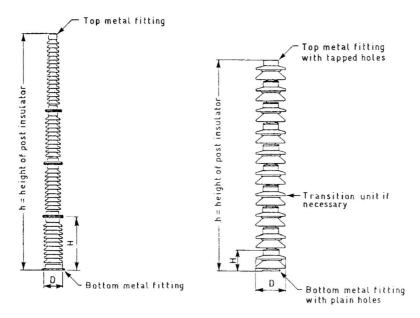

Figure 6.1   Solid core cylindrical and pedestal post insulators. (a) Example of an outdoor cylindrical post insulator with external metal fittings. (The example shown is composed of four units, but cylindrical post insulators may consist of one or more units.) (b) Example of an outdoor pedestal post insulator. (The example shown is composed of ten units, but pedestal post insulators may consist of one or more units.) H = height of one unit; D = insulating part diameter

per unit. An example of a unit is shown in Fig. 6.2. They have a high bending strength of up to typically 310 kN and in this regard are superior to the solid core types. The units have standardized top and bottom fixing arrangements such that insulator stacks may be built up with bending strengths varying from the maximum required at the base to the minimum at the top. As many as 12 such units may be required to form a post insulator for a 550 kV-rated voltage system. For a given insulator height the total creepage distance is comparable to that of cylindrical posts but a greater protected creepage distance (the distance measured along the underside of the insulator sheds) is feasible and this can be an advantage in certain environments.

6.3.1.2 Solid core cylindrical posts

Cylindrical post type insulators are shown in Fig. 6.3. Shed shapes are usually simplified for ease of production since the units are cast in cylindrical form and machined on vertical milling machines before firing.

   Cylindrical post insulators may also be made up from individual sheds cemented together. This allows more complicated shed profiles to be achieved at the expense of cost and use of special cements to overcome any degradation

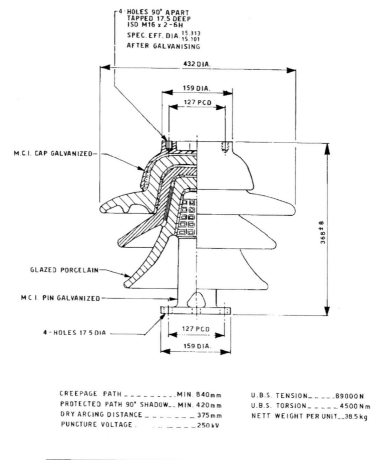

CREEPAGE PATH _ _ _ _ _ _ _ _ _ MIN. 840mm
PROTECTED PATH 90° SHADOW _ _ MIN. 420mm
DRY ARCING DISTANCE _ _ _ _ _ _ _ _ 375mm
PUNCTURE VOLTAGE .   · · _ _ _ _ _ 250kV

U.B.S. TENSION_ _ _ _ _89000N
U.B.S. TORSION_ _ _ _ _4500Nm
NETT WEIGHT PER UNIT _ _38.5kg

UNITS PER STACK	O/A HEIGHT	ELECTRICAL CHARACTERISTICS (kV)							U.B.S. CANTILEVER 'N'	
		FLASHOVER		WITHSTAND		50% FLASHOVER		IMPULSE		
								WITHSTAND	UPRIGHT	INVERTED
		DRY	WET	DRY	WET	POS	NEG			
1	368.3	150	83	130	80	220	240	200	40000	31200
2	737	255	175	240	150	370	500	350	15500	12500

I.E.C. REF. No. E56

Figure 6.2   Pedestal post insulator detail

problems in service. An alternative method of increasing the creepage distance without increasing the overall insulator stack height is shown in Fig. 6.3 using alternate long short (ALS) insulator shed profiles. These have alternate sheds of a lesser diameter with the distance between the sheds being of the order of 15 mm. Again standardized top and bottom fixing arrangements allow stacks to be formed from an assembly of single units. Typically one unit may be used for a 72 kV-rated voltage system and up to four units for a 550 kV system. In practice, the number employed is usually determined in conjunction with the insulator manufacturer for a specified bending strength.

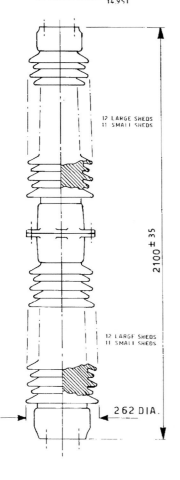

4-HOLES EQUI-SPACED ON 127 P.C.D.
TAPPED ISO M16×2-6H
SPEC. EFF. DIA. 15.163
              14.951

12 LARGE SHEDS
11 SMALL SHEDS

2100 ± 35

12 LARGE SHEDS
11 SMALL SHEDS

262 DIA.

DRY ARCING DISTANCE = 1850
MIN. TOTAL CREEPAGE = 6125

Solid Core Alternate Long Short Profile

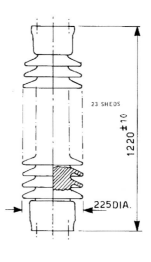

4-HOLES EQUI-SPACED ON 127 P.C.D.
TAPPED ISO M16×2-6H
SPEC. EFF. DIA. 15.163
              14.951

23 SHEDS

1220 ± 10

225 DIA.

4-HOLES EQUI-SPACED ON 127 P.C.D.
TAPPED ISO M16 2-6H
SPEC. EFF. DIA. 15.163
              14.951

DRY ARCING DISTANCE = 1080
MIN. TOTAL CREEPAGE = 3000

Solid Core Conventional Profile

Figure 6.3   Cylindrical post insulator detail

6.3.1.3 *Hollow insulators*

Hollow insulators are employed by substation equipment manufacturers to house post type current transformers (CTs), voltage transformers (VTs and CVTs), cable bushings, circuit breaker supports with central operating rods and interrupting chamber assemblies, isolator supports, etc. The specifications are determined between the substation equipment and insulator manufacturers. In particular, the mechanical strength of hollow porcelains must be determined in this way since insulators used in such applications may be subject to sudden pressures such as in circuit breakers or surge arresters. Torsion failing loads are also important where insulators form part of a circuit breaker or isolator drive mechanism. The main issue from the substation designer's point of view is to adequately specify the required creepage distance (to suit the environmental conditions) and shed shape (to conform with the electricity supply company's spares holdings or standards).

6.3.2 Cap and pin insulators

Cap and pin insulators of porcelain or glass predominate as overhead line suspension or tension sets above 33 kV and they are also used for substation busbar high level strained connections. Almost any creepage distance may be achieved by arranging the required number of individual units in a string. Upper surface shed shapes are similar with the top surface having a smooth hard surface to prevent the accumulation of dirt and moisture and a slope greater than 5° to assist self-cleaning. The undersides have a considerable variation in shape which depends upon aerodynamic and creepage distance requirements. Figure 6.4 illustrates standard, anti-fog and aerofoil disc profiles. Suspension insulator sets are rarely used in substation designs and substation busbar tension sets avoid the use of the anti-fog profiles because the deep ribs may not be naturally cleaned by rainfall when mounted nearly horizontally with short spans. Such substation short span applications do not require high strength cap and pin units and the insulators are often specified with 80 kN minimum failing electromechanical failing test load to meet a 3 × safety factor requirement. Overhead line cap and pin insulators are specified with correspondingly higher failing loads of 125 and 190 kN.

6.3.3 Long rod

Long rod insulators are similar to porcelain solid core cylindrical post insulators except that the top and bottom fittings are of the cap and pin type. Long rods are an alternative to the conventional cap and pin insulator sets with the possibility of providing longer creepage paths per unit length. Long rod insulators have not, however, exhibited for overhead line work any marked improvements in performance under heavy pollution conditions. In addition, the mechanical performance of porcelain under tension is such that brittle fracture could easily cause a complete failure of the whole unit leading to an outage condition. In contrast, cap and pin insulators using toughened glass or porcelain are designed so that they do not exhibit a brittle fracture characteristic. Cap and pin insulators are able to support the full tensile working load with the glass shed shattered or all the porcelain shed broken away. For these

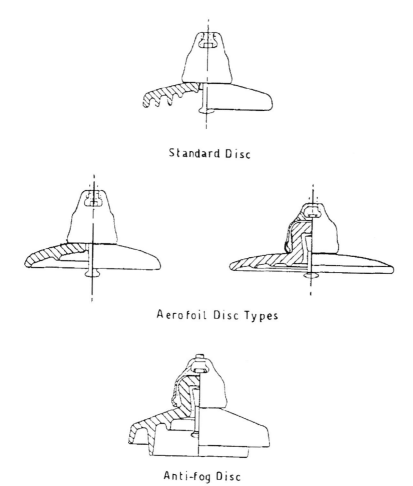

Standard Disc

Aerofoil Disc Types

Anti-fog Disc

Figure 6.4   Cap and pin insulators

reasons cap and pin insulators are more often specified for overhead line work. A selection of long rod porcelain units are shown in Fig. 6.5 and a composite insulator in Fig. 6.6.

6.4 POLLUTION CONTROL

6.4.1 Environment/creepage distances

Specific insulator creepage distance is determined for the particular environment by experience from earlier installations. If no such information is available then for major projects the establishment of an energized test station

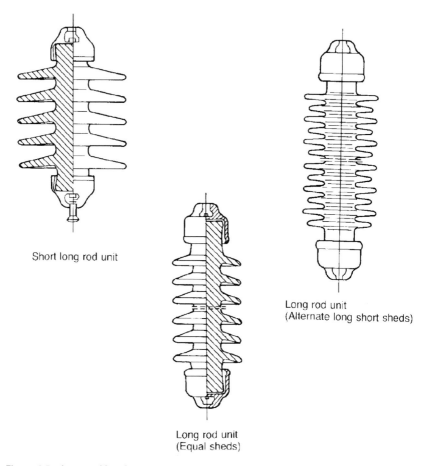

Short long rod unit

Long rod unit
(Alternate long short sheds)

Long rod unit
(Equal sheds)

Figure 6.5　Long rod insulators

or overhead line test section 1 or 2 years in advance of the insulator material procurement project programme should be established. Various types of insulators
and shed profiles may be tested and those with the best performance selected. If
insufficient time is available for meaningful testing then the alternative is simply to overinsulate on the basis of results or performance records of insulators
used in similar environments elsewhere. A further alternative is to refer to IEC
60815 for pollution level descriptions and associated suggested insulator creepage distances. Pollution levels are categorized as shown in Table 6.2.

The corresponding minimum nominal specific insulator creepage distances
(allowing for manufacturing tolerances) are then selected as shown in Table
6.3. The actual insulator shed profile is also an important factor. For example,
the aerofoil profile which does not have 'skirts' on the underside (see Fig. 6.4)
discourages the adherence of sand and salt spray and has been shown to be
successful in desert environments.

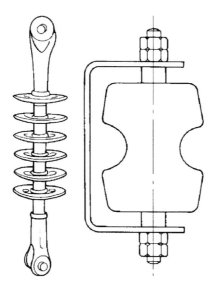

Figure 6.6   Polymeric material insulators. Left: composite insulator; Right: epoxy resin distribution line insulator

Table 6.1   Relative merits of insulator materials

Type	Comments
Porcelain	Good service history. High impact and self-cleaning support. Good compressive strength.
Glass	Good service history. Obvious failures may be spotted with glass shattering.
Polymeric	Early problems now overcome. Advantages for special applications. Some composite polymer insulators have excellent pollution withstand performance.

Some terminology associated with creepage distances and puncture resistance is explained below:

Total creepage distance	The surface distance over the total upper and lower portions of the insulator surface.
90° protected creepage	The surface distance over the undersurface of the insulator. This is a useful measure for insulators with deep ribs and grooves as used for anti-fog profiles.
Puncture-proof insulators	Those insulators for which all credible surges will flashover the surface of the insulator rather than puncture through the insulator shed material. Ceramic long rod and composite insulators are therefore classed as puncture-proof. Because of their construction cap and pin insulators are classified as non-puncture-proof and such faults, if not accompanied by shed breakage, may be difficult to observe by visual inspection.

Table 6.2   Description of polluted conditions for insulator selection

Pollution level	Examples of typical environments
I–Light	– Areas without industries and with low density of houses equipped with heating plants. – Areas with low density of industries or houses but subjected to frequent winds and/or rainfalls. – Agricultural areas.[a] – Mountainous areas. All these areas must be situated at least 10–20 km from the sea and must not be exposed to winds directly from the sea.[b]
II–Medium	– Areas with industries not producing particularly polluting smoke and/or with average density of houses equipped with heating plants. – Areas with high density of houses and/or industries but subjected to frequent clean winds and/or rainfalls. – Areas exposed to wind from the sea but not too close to the coast (at least several kilometres).[b]
III–Heavy	– Areas with high density of industries and suburbs of large cities with high density of heating plants producing pollution. – Areas close to the sea or in any case exposed to relatively strong wind from the sea.[b]
IV–Very heavy	– Areas generally of moderate extent, subjected to conductive dusts and to industrial smoke producing particularly thick conductive deposits. – Areas generally of moderate extent, very close to the coast and exposed to sea spray or to very strong and polluting winds from the sea. – Desert areas, characterized by no rain for long periods, exposed to strong winds carrying sand and salt, and subjected to regular condensation.

Notes: [a]Use of fertilizers by spraying, or when crop burning has taken place, can lead to higher pollution level due to dispersal by wind.
[b]Distances from sea coast depend on the topography of the coastal area and on the extreme wind conditions.

6.4.2 Remedial measures

In practice, it may be found that insulators in certain existing substations or overhead line sections are causing trouble due to pollution. Remedial measures include:

- Installation of creepage extenders and booster sheds. Increasing the number of insulators in a string has to be carefully considered since if not done at the initial design stage clearances to ground or the supporting structure will be reduced and possibly infringed.
- Use of anti-fog insulator disc profiles (see Fig. 6.4).
- Regular overhead line insulator live line washing using carefully monitored low conductivity water droplet sprays or washing under outage conditions.
- Silicone grease coating of the insulator sheds. The action of the grease is to encapsulate the particles of dirt thus preventing the formation of an

Table 6.3   Guide for selection of insulator creepage distances to suit polluted conditions

Pollution level	Minimum nominal specific creepage distance[a] (mm/kV)[b]
I–Light	16
II–Medium	20
III–Heavy	25
IV–Very heavy	31

Notes: [a]For actual creepage distances the specified manufacturing tolerances are applicable (IEC 60273, 60305, 60433 and 60720).
[b]rms value corresponding to the highest voltage for equipment (phase to phase).
[c]In very light polluted areas creepage distances less than 16 mm/kV can be used depending upon service experience. 12 mm/kV seems to be a lower limit.
[d]In cases of exceptional pollution a specific nominal creepage distance of 31 mm/kV may not be adequate. Depending upon service experience and/or laboratory test results a higher value of specific creepage distance can be used.

electrolyte with moisture from rain or condensation. When the grease becomes saturated with dirt it must be removed and a new coat applied. Greasing is not recommended for use with anti-fog insulator shed profiles. The grease tends to get trapped in the grooves on the underside of the shed and bridged by conductive dirt forming on the grease surface thereby reducing the overall insulator creepage distance.

6.4.3 Calculation of specific creepage path

Typical creepage values for either cap and pin insulator strings or post insulators at 132 kV under very heavy pollution, classification IV, would be determined as follows:

Nominal voltage	132 kV rms
Rated voltage	145 kV rms
Maximum rated voltage	phase to phase
Creepage (from Table 6.3)	31 mm/kV
Test creepage distance	145 kV × 31 mm/kV = 4495 mm

From manufacturer's details:
Typical substation insulator	80 kN minimum failing load
Creepage	330 mm
Spacing distance	127 mm

Number of insulators required to provide 4495 mm creepage using cap and pin porcelain string insulators = 4495/330 = 13.6. Therefore the minimum total number of insulators per string would be 14. In this case failure of one unit would reduce the creepage distance to 4165 mm, equivalent to an effective value of 28.7 mm/kV @ 145 kV-rated voltage. However, at the nominal system voltage of 132 kV the design value is maintained at 31.5 mm/kV. Minimum length of suspension insulator string = 14 × 127 = 1778 mm.

Alternatively, 15 insulators per string could be specified in order to maintain the recommended creepage distance at 145 kV-rated voltage under a one insulator shed failure condition. This allows full overhead line operation between normal line maintenance outages at the expense of increased overall string length of 1905 mm.

6.5 INSULATOR SPECIFICATION

6.5.1 Standards

Standards are under regular review. Some of the important International Electrotechnical Commission (IEC) insulator standards are listed in Table 6.4, and generally national standards are in line with these. There are however some differences with ANSI standards worth mentioning:

- IEC defines sockets, clevis and tongues up to 530 kN. ANSI standards and US practice generally has a limit of 50 000 lbs (=444 kN).
- The ANSI practice is to define the electrical values of a suspension or tension insulator according to its flashover values. IEC defines the insulator according to its withstand strength. Therefore, even with slightly different test proced_ures, the IEC withstand is always lower than the ANSI flashover definition.

Some comparison of terms may also be useful:

- The ANSI term 'design tests' equates to the IEC term 'type tests'.
- The ANSI term 'prototype test' equates to the IEC term 'design test'.
- The ANSI term 'leakage distance' equates to the IEC term 'creepage distance'.
- For ANSI the term 'ceramic' includes both porcelain and glass; the IEC term 'ceramic' means porcelain. Glass is separately defined.
- In ANSI terms 'cap and pin' is a pedestal post insulator. In IEC terms it is a suspension insulator.

6.5.2 Design characteristics

6.5.2.1 *Electrical characteristics*

The principal electrical characteristics are:

- Power frequency withstand.
- Lightning impulse withstand voltage (LIWV).
- Switching impulse withstand voltage (SIWV).

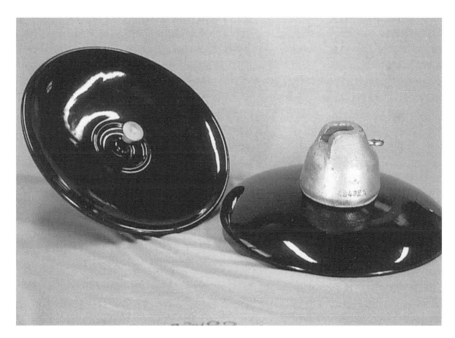

Figure 6.7   Some typical porcelain insulators (courtesy Allied Doulton)

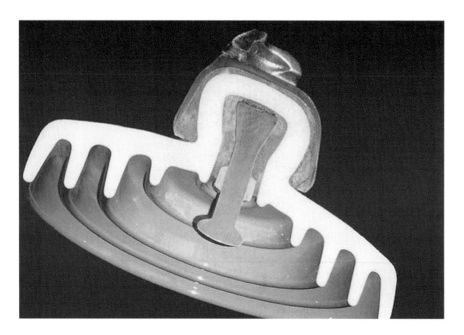

Figure 6.7   *Continued*

Figure 6.7   *Continued*

Table 6.4   Some IEC insulator standards

IEC standard	Short title
60120*	Dimensions of ball and socket couplings of string insulators
60137	Bushings for AC above 1000 V
60168 and 60273	High voltage post insulators – tests and dimensions
60233	Hollow insulators – tests
60305* and 60383	Cap and pin insulators – characteristics and tests
60372*	Locking devices for ball and socket couplings
60433	Characteristics of long rod insulators above 1000 V
60437	Radio interference on HV insulators (report)
60438	Tests and dimensions of HV DC insulators
60471	Dimensions of clevis and tongue couplings
60507	Artificial pollution tests (conductivity and withstand levels vs pollution)
60575	Thermal–mechanical tests on string insulators
60660	Indoor post insulators of organic material 1–300 kV
60720	Characteristics of line post insulators
60815	Guide for selection of insulators vs polluted conditions
61109	Composite insulators greater than 1000 V
61211	Puncture testing ceramic or glass insulators greater than 1000 V
61325	Ceramic or glass insulators for DC above 1000 V
61462	Composite hollow insulators for electrical equipment
61466-1	Composite string insulators for overhead lines greater than 1000 V – standard strength classes and end fittings
61467	Insulators for overhead lines greater than 1000 V – AC power arc tests
61952	Composite line post insulators for overhead lines greater than 1000 V
62073	Technical specification – guidance on measurement of wettability of insulator surfaces
62155	Hollow pressurized and unpressurized ceramic and glass insulators for electrical equipment greater than 1000 V

*Note: In UK BS3288 Part 2 remains in use to cover certain fittings outside the IEC range.

- Pollution withstand (typically expressed as equivalent salt solution density for the system voltage concerned).
- Radio interference.

Note that it is the lightning impulse withstand voltage that appears in manufacturers' and IEC insulator selection tables to categorize all insulators and not the system-rated voltage U_n. The switching impulse withstand voltage has more influence on designs above about 300 kV.

6.5.2.2 Mechanical characteristics

The principal mechanical characteristics are:

- Tensile strength (string insulators).
- Cantilever strength (rigid insulators).

Table 6.5   Cap and pin overhead line insulator technical particulars and guarantees

Description	Type and requirement or manufacturers guarantee
Manufacturer	
Maximum working load per string (kN)	
Spacing of units (mm)	
Maximum elastic limit of fittings (kN)	
Electromechanical minimum failing load per unit (kN)	
Electrostatic capacity per unit (pF)	
Weight per unit (kg)	
No. units per string	
Overall creepage path distance per unit (mm)	
90° creepage path distance per unit (mm)	
Length of string (mm)	
Weight of insulator set complete with all clamps and fittings (kg)	
Lift of arcing horn over line end unit (mm)	
Reach of arcing horn from centre line of string (mm)	
(note these dimensions may require adjustment to suit impulse test requirements)	
50 or 60 Hz withstand voltage per string (wet) (kV) (dry) (kV)	
Minimum 50–60 Hz puncture voltage per string (kV)	
Impulse flashover voltage per string (50%) (+kV) (−kV)	
Impulse withstand voltage per string (+kV) (−kV)	

For overhead lines these mechanical ratings are determined by the factor of safety (usually three or greater), the wind span, the weight span, the uplift and the broken wire requirements.

6.5.2.3 *Insulator selection*

Table 6.5 gives characteristics for outdoor pedestal post insulators taken from IEC 60273 and allows identification of an insulator for the required service conditions. The creepage distance is derived as shown in Section 6.4.3. The required mechanical strength for a substation post insulator is derived from a calculation of wind force expressed as a cantilever load at the top of the insulator. A factor of safety of 2.5 or 3 times is applied and the insulator selected from the tables with the appropriate minimum bending failing load.

6.6 TESTS

6.6.1 Sample and routine tests

Routine and type tests for substation and overhead line insulators are clearly set out in the relevant IEC standards listed in Table 6.4. Note that it is unusual to call for tests on hollow insulators to be witnessed at the manufacturer's works in substation construction contract technical specifications. This is because the substation equipment manufacturers and insulator manufacturer have agreed the requirements between themselves; for example, the appropriate torsion failing load where the insulator is to be used in an isolator drive mechanism. Overall visual inspections on final assembly are, however, worthwhile before allowing release of material for shipment. These checks should include dimensional correctness within the permitted tolerances, alignment and any glaze defects.

For cap and pin overhead line insulators one suspension and one tension string of each type should be shown to have been tested in accordance with the provisions of IEC 60383 and 60575 covering:

dimensional checks
temperature cycle tests
radio interference
dry withstand impulse tests
wet power frequency withstand including any string fittings as may be incorporated in service – for example arc horns and dampers short circuit tests on complete strings with fittings mechanical test to demonstrate failing load
puncture test
porosity test galvanizing test.

6.6.2 Technical particulars

Technical particulars associated with overhead line cap and pin insulators and fittings will typically be as shown in Table 6.5.

7 Substation Building Services

7.1 INTRODUCTION

This chapter introduces some of the main principles involved with substation control building and switchyard services. Such work is often left to specialist building services, engineering groups and architects. Therefore this chapter covers basic internal and external lighting design, heating, ventilation and air-conditioning (HVAC) practice together with the different types of low voltage distribution systems in sufficient detail as likely to be encountered by the power engineer.

7.2 LIGHTING

7.2.1 Terminology

7.2.1.1 General

Light consists of electromagnetic radiation with wavelengths between 760 and 380 nm to which the eye is sensitive. This energy spectrum (red, orange, yellow, yellow-green, green, green-blue, blue and violet) lies between the infrared (700–2000 nm) and the ultraviolet (200–400 nm) wavelength ranges. Lighting schemes may be necessary for the following substation areas:

- Indoor and outdoor schemes for control buildings and indoor switchrooms under both normal and emergency (loss of alternative current (AC) supply) conditions.
- Floodlighting and emergency schemes for outdoor switchyards and door or gate access.
- Security/access road lighting together with any supplementary lighting for video camera surveillance.
- Enclosed transformer pen lighting.

Emergency lighting involves battery backup derived either from cells within the fittings or from the main substation DC supply. The units are designed to have a given 'autonomy' such that upon AC failure the lighting continues to operate for a specified number of hours from the battery source.

7.2.1.2 Types of luminaires

The type of light source chosen for an unmanned substation control building or switchyard will be less influenced by aesthetic than technical characteristics such as colour rendering, glare, efficacy, life and cost. Table 7.2 gives the chief technical characteristics of common types of luminaires.

An overcast sky in temperate regions gives an illuminance of some 5000 lux. Typical substation lighting levels are shown below:

Substation area	Standard illuminance (lux)	Limiting glare index
(a) Indoor switchroom	200	–
(b) Control rooms	400	25
(c) Telecommunication rooms	300	25
(d) Battery rooms	100	–
(e) Cable tunnels and basements	50	–
(f) Offices	500	19
(g) Entrance halls, lobbies, etc.	200	19
(h) Corridors, passageways, stairs	100	22
(i) Messrooms	200	22
(j) Lavatories and storerooms	100	–
(k) Outdoor switchyard (floodlighting)	20	–
(l) Perimeter lighting	10	–
(m) Exterior lighting (control buildings, etc.)	15	–

Note that Division II (flameproof) lighting fittings may be necessary in the battery room because of fumes given off from unsealed batteries.

7.2.1.3 Harmonics

Discharge lamps generate harmonics. Red, yellow and blue phase components take the form:

$$I_R = i_1 \sin \omega t \qquad\qquad + i_2 \sin 2\omega t \qquad\qquad + i_3 \sin 3\omega t \qquad\qquad + \cdots$$
$$I_Y = i_1 \sin (\omega t - 2\pi/3) \quad + i_2 \sin 2(\omega t - 2\pi/3) + i_3 \sin 3(\omega t - 2\pi/3) + \cdots$$
$$I_B = i_1 \sin (\omega t - 4\pi/3\pi) + i_2 \sin 2(\omega t - 4\pi/3) + i_3 \sin 3(\omega t - 4\pi/3) + \cdots$$

Table 7.1   Details some standards and useful references covering indoor and outdoor lighting schemes

Standard or Reference	Description
IEC 60598	Luminaires – Covers general requirements for the classification and marking of luminaires, their mechanical and electrical construction, together with related tests. Applicable to tungsten filament, tubular fluorescent and other discharge lamps on supply voltages not exceeding 1000 V.
IEC 60972	Classification and interpretation of new lighting products.
CISPR 15 (International Special Committee on Radio Interference)	Limits and methods of measurement of radio disturbance characteristics of electrical lighting and similar equipment.
Illumination Engineering Society (IES) Code for interior lighting	Complete guide. Also Lighting Division of the Chartered Institute of Building Services Engineers (CIBSE)
Philips Lighting Manual	Handbook of lighting installation design prepared by members of the staff of the N.V. Philips' Gloeilampenfabrieken Lighting Design and Engineering Centre, Eindhoven, The Netherlands.

Fundamental
Normal 120° ($2\pi/3$) phase relation in sequence RYB. (This is also true for 4th, 7th, 10th … harmonics).

2nd harmonic
120° phase relation but in sequence RBY – phase reversal. (This is also true for 5th, 8th, 11th … harmonics)

3rd harmonic
All in phase and will produce zero phase sequence (ZPS) system voltages and currents. (This is also true for 6th, 9th, 12th … harmonics)

If $I_R = I_Y = I_B$ then:

$$I_R + I_Y + I_B = I_R (1 + h^2 + h) = 0 \text{ for a balanced three phase system.}$$

where $h = -\frac{1}{2} + j\sqrt{\frac{3}{2}}$ and $h^2 = -\frac{1}{2} - j\sqrt{\frac{3}{2}}$. Since ZPS components are present with 3rd harmonics then the neutral line will be involved. On large discharge lighting installations this can result in a substantial neutral current and cables must be sized accordingly. In addition third harmonic restraint may be necessary on standby earth fault (SBEF) protection relays located in the supply transformer neutral.

In general for the nth harmonic the resultant harmonic line voltage in two successive phases $2\pi/3$ out of phase is:

$U_n \sin n\,\omega t$ (red phase, n^{th} harmonic) $- U_n \sin n\,(\omega t - 2\pi/3)$ (yellow phase, n^{th} harmonic) $= 2\,U_n \cos n\,(\omega t - \pi/3) \times \sin (n\pi/3)$

The harmonic rms line voltage $= 2U_n \sin (n\pi/3)$

For 3rd harmonics (6th, 9th, etc.) rms line voltage $= 2U_3 \sin (3\pi/3) = 0$.

Table 7.2   Characteristics of different light sources

Type of luminaire	Approximate life (hours)	Luminous efficacy (lm/W)	Colour rendering	Illuminance (lm)	Notes
(a) Tungsten filament (GLS)	1000	10–20	Excellent, warm white	40 W–430 lm 60 W–730 lm 100 W–1380 lm	Cheap, easy to replace, no control gear required. Zero restrike time.
(b) Tungsten halide (TH)	2500	60–90	Excellent	75 W–5000 lm 150 W–11 250 lm 250 W–20 000 lm	Relatively low initial cost. Moderate efficacy. Zero restrike time.
(c) Fluorescent (MC)	7500	30–90	Good, various types available	10 W–630 lm 20 W–1100 lm 40 W–3000 lm 65 W–5000 lm	Cheap tubes and relatively cheap control gear as integral part of fitting.
(d) Low pressure sodium (SOX)	10 000	70–150	Very poor, colour recognition impossible, monochromatic yellow/orange	35 W–4800 lm 135 W–22 500 lm 180 W–33 000 lm	Long life and very high efficacy. Control gear required. Historically used for road lighting. 7–12 minute restrike time.
(e) High pressure sodium (SON)	9000	50–120	Fair, warm yellow/white	50 W–3500 lm 100 W–10 000 lm 150 W–17 000 lm 400 W–48 000 lm	Generally favoured exterior lighting source. Control gear required. 4–5 minute restrike time.
(f) High pressure mercury vapour (MBF)	6000–9000	30–90	Good, neutral white	80 W–3700 lm 125 W–6300 lm 250 W–13 000 lm 400 W–22 000 lm	White light but less efficient than high pressure sodium type. Control gear required. 4–5 minute restrike time.

7.2.1.4 *Definitions*

Candela	The illuminating power of a light source in a given direction. The unit of luminous intensity.
Colour rendering	General expression for the effect of an illuminant on the colour appearance of objects in conscious or subconscious comparison with their colour appearance under a reference illuminant.
Efficacy	Luminous efficiency of lamps measured in lumens per watt (lm/W).
Flicker	Impression of fluctuating luminance or colour, occurring when the frequency of the variation of the light stimulus lies within a few hertz of the fusion frequency of the retinal images.
Glare	Condition of vision in which there is discomfort or a reduction in the ability to see significant objects, or both, due to an unsuitable distribution or range of luminance or to extreme contrasts in space or time.
Glare index	Introduced by the British Illumination Engineering Society (IES) to specify and evaluate the degree of discomfort glare for most types of working interior, for a range of luminaires with standardized light distributions.
Illuminance, E (lux, lx)	The measure of light falling on a surface. The illumination produced by one lumen over an area of one square metre measured in lux. $E = \Phi/A$ where A = area in m^2
Intensity, I (candela, cd)	A measure of the illuminating power of a light source in a particular direction, independent of the distance from the source.
Luminous flux, Φ (lumen, lm)	Unit of light flux. The amount of light contained in one steradian from a source with an intensity of one candela in all directions. The amount of light falling on unit area of the surface of a sphere of unit diameter from a unit source.
Luminance, l (cd/m^2)	Measure of light reflected from a surface or in some cases emitted by it. A measure of brightness of a surface. The units of measured brightness are candela per square metre and the apostilb being the lumens emitted per square metre. Luminance and illuminance are not to be confused. For example, if a sheet of paper has a reflectance of 80% and an illuminance of 100 lux its luminance will be 80 apostilbs. 1 apostilb $= 1/\pi$ cd/m^2

Luminance = $0.318 \times$ illuminance $\times$ reflectance
$L = 0.318 \times E \times \rho$

Luminosity — Attribute of visual sensation according to which an area appears to emit more or less light. Luminance and luminosity are not to be confused. Substation switchyard floodlighting seen by day and night will appear different to the observer. The luminance is the same but the luminosity in daylight will appear low while at night may appear perfectly satisfactory.

Maintenance factor — Ratio of the average illuminance on the working plane after a specified period of use of a lighting installation to the average illuminance obtained under the same conditions for a new installation.

Room index, RI — Code number representative of the geometry of the room used in calculation of the utilization factor. Unless otherwise indicated: RI = $(l \times b)/(H_m [l + b])$ where l = length of room, b = width of room and H_m is the distance of the luminaires above the working plane.

Stroboscopic effect — Apparent change of motion or immobilization of an object when the object is illuminated by a periodically varying light of appropriate frequency. May be eliminated by supplying adjacent luminaires on different phases, or using twin-lamp fittings on lag-lead circuits, or providing local lighting from tungsten filament lamps rather than discharge types.

Utilization factor — Ratio of the utilized flux to the luminous flux leaving the luminaires. Table 7.3 details utilization factors for various indoor lighting fittings.

7.2.2 Internal lighting

The following design procedure is suggested:

1. Decide upon the level of illumination required. The illuminance required can be obtained from various guides or the values given in Section 7.2.1.2 above.
2. Determine the mounting height of the fittings above the working plane. Note that a desktop height is usually taken as 0.85 m above floor level. Note that the fittings should be mounted as high as possible in order to permit wider spacing between fittings and reduced glare.
3. Ascertain the minimum number of fittings to be employed from the spacing factor (normally taken as 1.5 for batten type fluorescent fittings) and the mounting height.

For $H_m = 2.25$ and spacing factor 1.5 then minimum number of fittings for a 20 m × 10 m room = {20/(2.25 × 1.5)} × {10/(2.25 × 1.5)} = 20/3.37 × 10/3.37 = 18.

4. Calculate the Room index. RI = $(l × b)/(H_m[l + b])$ where l = room length, b = room width, and H_m = mounting height above the working plane. For the 20 m × 10 m room mentioned above RI = 20 × 10/2.25(30) = 2.96 or 3 approximately.

5. Having decided upon the general type of fitting to be used, ascertain the utilization factor, UF, from manufacturers' tables taking into account the reflectance factors of the room (see Table 7.3). Since accurate information is rarely available it is common to take reflectance factors of 20%, 70% and 50% for the working plane, ceiling and walls respectively, assuming finishes normally found is a substation control room.

6. Decide upon the maintenance factor, MF, to be used; typically taken as 0.8. This allows for a reduced light output from the fittings due to ageing and formation of dust on the luminaires.

7. Calculate the total light input to the room necessary, Φ lumens, to give the illumination level, E lux, required. Total luminous flux $\Phi = (E × l × b)/(UF × MF)$.

8. Calculate the number and size of fittings required. Note that there may be preconditions regarding the physical size and wattage of the luminaires to meet the electrical supply company standards. If the size of the fittings has already been decided then the number of luminaires required is obtained by dividing the total calculated lumens figure by the output per fitting. If the result obtained is greater than the minimum determined by the spacing requirement the higher number of fittings must be used even though this may not be the most economical arrangement. The actual number of fittings used may require adjustment in order to give a symmetrical arrangement.

A typical calculation sheet for such work is given in Fig. 7.1 and an example using this based on a 30 m × 15 m control and relay room, 3.1 m ceiling height, spacing factor 1.5 and required 400 lux illuminance is given in Fig. 7.2. Reflectance and utilization factors are taken from design guides.

Following the determination of the number of luminaires required it is essential to take into account the position of the equipment and air-conditioning ductwork in the room in order to produce a satisfactory lighting fitting layout.

7.2.3 External lighting

The illuminance E_p lux at a location d metres from a point source of intensity I candela is given by the cosine law of illumination, $E_p = I \cos \theta/d^2$. The point calculation method of determining the illuminance at a point then consists of taking the sum of the contributions of partial illuminances produced by all the individual lighting fittings involved in the design (see Figs 7.3 and 7.4).

Table 7.3 Indoor Lighting (utilization factors)

Description and Typical Downward Light Output Ratio %	Typical outline	Basic Downward LOR %	Ceiling Walls Room Index	Reflectance %								
				Light			Medium			Dark		
				50	30	10	50	30	10	50	30	10
(M) Aluminium Industrial Reflector (72.76) (T) High Bay Reflector Aluminium (72) or Enamel (66)		70	0.6	0.39	0.36	0.33	0.39	0.36	0.33	0.39	0.35	0.33
			0.8	0.48	0.43	0.40	0.46	0.43	0.40	0.46	0.43	0.40
			1.0	0.52	0.49	0.45	0.52	0.48	0.45	0.52	0.48	0.45
			1.25	0.56	0.53	0.50	0.56	0.53	0.49	0.56	0.52	0.49
			1.5	0.60	0.57	0.54	0.59	0.57	0.53	0.59	0.55	0.53
			2.0	0.65	0.62	0.59	0.63	0.60	0.59	0.63	0.59	0.57
			2.5	0.67	0.64	0.62	0.65	0.62	0.61	0.65	0.62	0.60
			3.0	0.69	0.66	0.64	0.67	0.64	0.63	0.67	0.64	0.62
			4.0	0.71	0.68	0.67	0.69	0.67	0.65	0.69	0.66	0.64
			5.0	0.72	0.70	0.09	0.71	0.69	0.66	0.71	0.67	0.66
(M) Reflectorized Colour-Corrected Mercury Lamp MBFR/U (80–90)		95	0.6	0.40	0.34	0.30	0.39	0.33	0.29	0.37	0.32	0.29
			0.8	0.53	0.46	0.41	0.51	0.45	0.40	0.49	0.43	0.40
			1.0	0.62	0.55	0.49	0.58	0.52	0.48	0.56	0.51	0.46
			1.25	0.68	0.60	0.55	0.64	0.58	0.53	0.61	0.56	0.51
			1.5	0.72	0.65	0.59	0.68	0.62	0.57	0.65	0.59	0.54
			2.0	0.81	0.73	0.67	0.75	0.69	0.64	0.69	0.65	0.61
			2.5	0.85	0.78	0.72	0.79	0.73	0.69	0.73	0.68	0.65
			3.0	0.90	0.83	0.78	0.83	0.78	0.75	0.77	0.73	0.70
			4.0	0.94	0.89	0.84	0.87	0.83	0.80	0.80	0.77	0.75
			5.0	0.97	0.92	0.89	0.90	0.87	0.84	0.83	0.79	0.77
(F) Enamel Slotted trough Louvered (45–55) (f) Louvered recessed (module) fitting (40–50)		50	0.6	0.27	0.24	0.22	0.26	0.24	0.22	0.26	0.23	0.22
			0.8	0.32	0.30	0.27	0.32	0.29	0.27	0.31	0.29	0.27
			1.0	0.35	0.32	0.30	0.35	0.32	0.30	0.34	0.31	0.30
			1.25	0.38	0.35	0.32	0.38	0.35	0.33	0.38	0.34	0.33
			1.5	0.41	0.38	0.36	0.40	0.38	0.35	0.40	0.37	0.35
			2.0	0.45	0.42	0.40	0.43	0.41	0.39	0.43	0.40	0.39
			2.5	0.47	0.44	0.42	0.45	0.43	0.41	0.45	0.42	0.41
			3.0	0.48	0.45	0.44	0.46	0.45	0.43	0.46	0.44	0.42
			4.0	0.49	0.47	0.46	0.48	0.47	0.45	0.47	0.45	0.44
			5.0	0.50	0.49	0.48	0.49	0.48	0.47	0.48	0.47	0.46

(F) Closed-end enamel trough (65–83)

(T) Standard dispersive industrial reflector (77)

(T) Enamel deep bowl reflector (73)

75

Room index									
0.6	0.36	0.31	0.35	0.28	0.31	0.35	0.28	0.31	0.35	0.28
0.8	0.45	0.40	0.44	0.37	0.40	0.44	0.37	0.40	0.44	0.37
1.0	0.49	0.45	0.49	0.40	0.44	0.49	0.40	0.45	0.48	0.40
1.25	0.55	0.49	0.53	0.46	0.49	0.53	0.45	0.49	0.52	0.45
1.5	0.58	0.54	0.57	0.49	0.53	0.57	0.49	0.54	0.55	0.49
2.0	0.64	0.59	0.67	0.55	0.58	0.67	0.55	0.59	0.60	0.54
2.5	0.68	0.63	0.65	0.60	0.62	0.65	0.59	0.63	0.61	0.58
3.0	0.70	0.65	0.67	0.62	0.64	0.67	0.61	0.65	0.63	0.61
4.0	0.73	0.70	0.70	0.57	0.67	0.70	0.65	0.67	0.66	0.64
5.0	0.75	0.72	0.73	0.69	0.70	0.73	0.67	0.70	0.68	0.67

F–Fluorescent lamps
M–Mercury colour corrected lamp
T–Tungsten filament lamp

(F) plastic trough louvered (45–55)

50

Room index									
0.6	0.26	0.22	0.25	0.19	0.21	0.25	0.19	0.24	0.20	0.18
0.8	0.34	0.29	0.32	0.26	0.28	0.32	0.25	0.31	0.27	0.24
1.0	0.39	0.34	0.36	0.30	0.32	0.36	0.29	0.34	0.31	0.28
1.25	0.43	0.38	0.39	0.34	0.36	0.39	0.33	0.37	0.34	0.31
1.5	0.46	0.41	0.42	0.37	0.39	0.42	0.36	0.39	0.36	0.33
2.0	0.50	0.46	0.43	0.43	0.42	0.43	0.40	0.43	0.39	0.37
2.5	0.53	0.49	0.49	0.46	0.46	0.49	0.43	0.45	0.42	0.40
3.0	0.55	0.51	0.51	0.49	0.48	0.51	0.46	0.47	0.45	0.43
4.0	0.58	0.54	0.53	0.52	0.51	0.53	0.49	0.48	0.47	0.45
5.0	0.60	0.57	0.55	0.55	0.53	0.55	0.51	0.50	0.48	0.47

(F) Plastic trough unlouvered (60–70)

70

Room index									
0.6	0.33	0.28	0.32	0.25	0.28	0.32	0.25	0.31	0.27	0.25
0.8	0.42	0.37	0.41	0.33	0.36	0.41	0.33	0.40	0.36	0.33
1.0	0.48	0.43	0.46	0.38	0.42	0.46	0.38	0.45	0.42	0.38
1.25	0.52	0.47	0.50	0.43	0.46	0.50	0.43	0.49	0.45	0.42
1.5	0.56	0.51	0.54	0.47	0.50	0.54	0.46	0.52	0.48	0.45
2.0	0.62	0.56	0.58	0.53	0.55	0.58	0.51	0.56	0.52	0.50
2.5	0.65	0.60	0.61	0.57	0.58	0.61	0.55	0.59	0.56	0.53
3.0	0.67	0.63	0.64	0.60	0.61	0.64	0.58	0.62	0.59	0.56
4.0	0.70	0.66	0.67	0.64	0.64	0.67	0.61	0.64	0.62	0.59
5.0	0.73	0.69	0.69	0.67	0.67	0.69	0.64	0.66	0.64	0.62

Table 7.3 Continued

(T) Near spherical diffuser open beneath (50) — 50

RI									
0.6	0.28	0.22	0.18	0.25	0.20	0.17	0.22	0.18	0.16
0.8	0.39	0.30	0.26	0.33	0.28	0.23	0.27	0.25	0.22
1.0	0.43	0.36	0.32	0.38	0.34	0.29	0.31	0.29	0.26
1.25	0.48	0.41	0.37	0.42	0.38	0.33	0.34	0.32	0.29
1.5	0.52	0.46	0.41	0.46	0.41	0.37	0.37	0.35	0.32
2.0	0.58	0.52	0.47	0.50	0.46	0.43	0.42	0.39	0.36
2.5	0.62	0.56	0.52	0.54	0.50	0.47	0.45	0.42	0.40
3.0	0.65	0.60	0.56	0.57	0.53	0.50	0.48	0.45	0.43
4.0	0.68	0.64	0.61	0.60	0.56	0.54	0.51	0.48	0.46
5.0	0.71	0.68	0.65	0.62	0.59	0.57	0.53	0.50	0.48

(F) Bare lamp on ceiling / (F) Batten fitting (60–70) — 65

RI									
0.6	0.29	0.24	0.19	0.27	0.22	0.19	0.24	0.21	0.19
0.8	0.37	0.31	0.27	0.35	0.30	0.25	0.31	0.28	0.24
1.0	0.44	0.37	0.33	0.40	0.35	0.31	0.35	0.32	0.29
1.25	0.49	0.42	0.38	0.45	0.40	0.36	0.39	0.36	0.33
1.5	0.54	0.47	0.42	0.50	0.44	0.40	0.43	0.40	0.37
2.0	0.60	0.52	0.49	0.54	0.49	0.45	0.48	0.44	0.41
2.5	0.64	0.57	0.53	0.57	0.53	0.49	0.52	0.48	0.45
3.0	0.67	0.61	0.57	0.60	0.57	0.53	0.56	0.52	0.49
4.0	0.71	0.66	0.62	0.64	0.61	0.57	0.59	0.55	0.52
5.0	0.74	0.70	0.66	0.68	0.64	0.61	0.62	0.58	0.54

(F) Injection moulded prismatic wrap around enclosure (55–65) — 55

RI									
0.6	0.32	0.28	0.25	0.30	0.27	0.25	0.27	0.24	0.22
0.8	0.40	0.36	0.32	0.39	0.34	0.31	0.36	0.32	0.30
1.0	0.45	0.41	0.38	0.43	0.38	0.36	0.39	0.35	0.33
1.25	0.50	0.45	0.42	0.47	0.44	0.41	0.43	0.39	0.37
1.5	0.53	0.48	0.45	0.50	0.46	0.41	0.46	0.43	0.41
2.0	0.58	0.53	0.49	0.54	0.50	0.47	0.50	0.47	0.45
2.5	0.61	0.57	0.53	0.57	0.53	0.51	0.52	0.50	0.47
3.0	0.64	0.59	0.56	0.58	0.55	0.53	0.53	0.51	0.49
4.0	0.66	0.63	0.60	0.61	0.58	0.55	0.55	0.53	0.51
5.0	0.68	0.65	0.62	0.62	0.60	0.58	0.56	0.56	0.54

	Room index									
	0.6	0.8	1.0	1.25	1.5	2.0	2.5	3.0	4.0	5.0
50 — (F) Enclosed plastic diffuser (45–55)										
	0.27	0.34	0.40	0.44	0.47	0.52	0.55	0.58	0.61	0.63
	0.21	0.29	0.35	0.39	0.42	0.47	0.51	0.54	0.57	0.59
	0.18	0.26	0.31	0.35	0.38	0.44	0.48	0.51	0.54	0.57
	0.24	0.32	0.37	0.40	0.43	0.47	0.50	0.52	0.55	0.57
	0.20	0.28	0.33	0.36	0.39	0.44	0.47	0.49	0.52	0.55
	0.18	0.25	0.30	0.33	0.36	0.41	0.44	0.47	0.50	0.53
	0.22	0.29	0.33	0.36	0.38	0.41	0.44	0.47	0.49	0.51
	0.19	0.26	0.30	0.33	0.35	0.39	0.42	0.45	0.47	0.49
	0.17	0.24	0.28	0.31	0.33	0.37	0.40	0.43	0.45	0.47
50 — (F) Shallow fitting with diffusing sides optically designed downward reflecting surfaces (55); (T) Industrial reflector with diffusing globe (50)										
	0.24	0.30	0.33	0.36	0.39	0.43	0.45	0.46	0.48	0.50
	0.21	0.27	0.30	0.33	0.36	0.39	0.41	0.43	0.45	0.47
	0.19	0.24	0.27	0.30	0.33	0.37	0.39	0.41	0.44	0.46
	0.23	0.29	0.32	0.35	0.38	0.40	0.42	0.43	0.45	0.47
	0.21	0.26	0.29	0.32	0.34	0.37	0.40	0.42	0.44	0.45
	0.19	0.24	0.27	0.30	0.32	0.36	0.38	0.40	0.42	0.44
	0.23	0.29	0.32	0.35	0.38	0.40	0.42	0.43	0.45	0.47
	0.20	0.26	0.29	0.32	0.34	0.37	0.40	0.42	0.44	0.45
	0.19	0.24	0.27	0.31	0.32	0.36	0.38	0.40	0.42	0.44
45 — (T) Opal sphere (45) and other enclosed diffusing fittings of near spherical shape										
	0.23	0.30	0.36	0.41	0.45	0.50	0.54	0.57	0.60	0.63
	0.18	0.24	0.29	0.34	0.39	0.45	0.49	0.52	0.56	0.60
	0.14	0.20	0.25	0.29	0.33	0.40	0.44	0.48	0.52	0.56
	0.20	0.27	0.31	0.35	0.39	0.43	0.46	0.49	0.52	0.54
	0.16	0.22	0.26	0.30	0.34	0.38	0.42	0.45	0.48	0.51
	0.12	0.18	0.22	0.26	0.30	0.34	0.38	0.42	0.46	0.49
	0.17	0.23	0.26	0.29	0.31	0.34	0.37	0.40	0.43	0.45
	0.14	0.19	0.23	0.26	0.28	0.32	0.35	0.38	0.41	0.43
	0.11	0.16	0.19	0.22	0.25	0.29	0.32	0.34	0.37	0.40

	Parameters	Formulae	Notes
1	Room length, m	l	
2	Room width, m	b	
3	Room height, m	$H1$	
4	Working plane above floor level, m	$H2$	
5	Fitting suspension height	$H3$	
6	Height above working plane, H_m	$H_m = H1 - (H3 + H2)$	
7	Room Index, RI	$RI = (l \times b)/(H_m[l+b])$	
8	Spacing factor, SF and minimum number of fittings	Minimum No. fittings = $l/(SF . H_m) \times b/(SF . H_m)$	
9	Reflection factor - ceiling	RF_C	
10	Reflection factor - walls	RF_W	
11	Reflection factor - floor	RF_F	
12	Average illumination required, E lux	E_{AV}	
13	Utilization factor	UF	
14	Maintenance factor (assume 0.8)	MF	
15	Lamp flux required, lumens	$(E_{AV} \times l \times b)/(UF \times MF)$	
16	Power and type of fittings		
17	Selected lamp luminous flux, Φ lumens		
18	Number of fittings required, N	$N = (E_{AV} \times l \times b)/(\Phi \times UF \times MF)$	
19	Actual number of fittings, N'		
20	Projected actual illuminance, lux	$(N' \times \Phi \times UF \times MF)/(l \times b)$	

Layout plan / drawing   **Project Ref:-**

Drawn by:-   **Checked by:-**   **Approved by:-**

Figure 7.1   Interior lighting calculation sheet

For the more general case the illuminance, E_γ lux, at point p on any plane the normal of which makes an angle γ with the direction of incidence of the light is given by:

$$E_\gamma = I_\theta \cos^2\theta \cos\gamma/h^2 \text{ (see Figs 7.4 and 7.5)}$$

For substation access road lighting calculations the reflective properties of the road surface must really be known. The surface luminance may therefore also be calculated by the point calculation method from a knowledge of the luminous intensity of the lighting fittings involved and the addition of the contributions of their partial luminances. Such hand computations would be time

	Parameters	Formulae	Notes
1	Room length, m	l	30
2	Room width, m	b	15
3	Room height, m	$H1$	3.25
4	Working plane above floor level, m	$H2$	0.85
5	Fitting suspension height, m	$H3$	0.1
6	Height above working plane, H_m	$H_m = H1 - (H3 + H2)$	2.3
7	Room index, RI	$RI = (l \times b)/(H_m [l + b])$	30	15 / 2.3(30+ 15) = 4.35
8	Spacing factor, SF and minimum number of fittings	Minimum No. fittings = $l/(SF . H_m)$ x $b/(SF . H_m)$	{30×(1.5×2.3)} × {15/(1.5×2.3)} = 8.7 × 4.35 = 38
9	Reflection factor - ceiling	RF_C	0.7
10	Reflection factor - walls	RF_W	0.5
11	Reflection factor - floor	RF_F	0.2
12	Average illumination required, E lux	E_{AV}	400 lux
13	Utilization factor	UF	0.67 (see Table 7.3 for a prismatic wrap around enclosure and light reflectance)
14	Maintenance factor (assume 0.8)	MF	0.8
15	Lamp flux required, lumens	$(E_{AV} \times l \times b)/(UF \times MF)$	$\frac{400 \times 30 \times 15}{0.67 \times 0.8}$ = 335821 lumens
16	Power and type of fittings	Electricity supply utility Standard = 1800 mm, 75 W twin fluorescent in prismatic wrap around enclosure	
17	Selected lamp luminous flux, Φ lumens	5750 lumens per tube. (after 2000 hours) Use 1.8 × individual tube output for a twin fitting	1.8 × 5750 = 10 350 lumens per fitting
18	Number of fittings required, N	$N = (E_{AV} \times l \times b)/(\Phi \times UF \times MF)$	33 5821 / 10 350 = 32.4 , say 33
19	Actual number of fittings, N'	For a symmetrical layout	34
20	Projected actual illuminance, lux	$(N' \times \Phi \times UF \times MF)/(l \times b)$	419 lux

Layout plan/drawing		Project Ref:-	
Drawn by:	Checked by:-		Approved by:-

Figure 7.2   Interior lighting calculation sheet (as completed for control room example p. 187)

consuming and assistance may be obtained from the major lighting fitting manufacturers who have computer programs to determine the correct type, spacing and mounting height of floodlighting fittings to obtain the required average illuminance over a substation switchyard or luminance on a road surface. Manufacturers also produce isolux and isocandela diagrams for their different fittings in order to reduce the calculation work involved. For example the illuminance at a point is derived from the contour values given in an isolux diagram using formulae of the form:

Illuminance (lux)

$$= \frac{\text{contour value (lux per lumen)} \times \text{lamp flux (lumens)} \times \text{Maintenance Factor}}{\text{mounting height (metres)}, h^2}$$

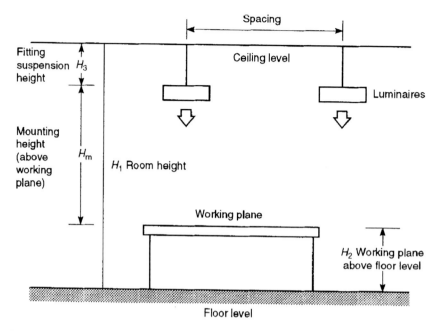

Figure 7.3   Relationship between floor level, working plane, fitting mounting height and spacing between each fitting

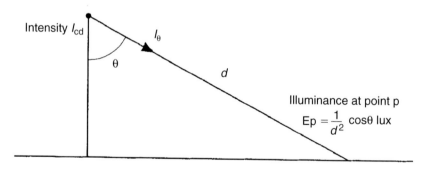

Figure 7.4   Illuminance from a point source

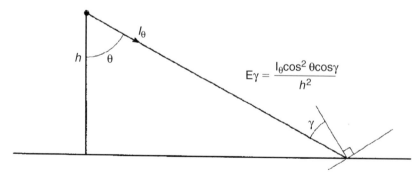

Figure 7.5   Illuminance on any plane

Consider the following example:

A Client requires boundary or security wall lighting around his substation. The arrangement of the high pressure mercury vapour (MBF/U) 125 W lighting fittings which each have a luminous flux of 6300 lumens is given in Fig. 7.6a.

1. To find the illuminance at point P
(a) Determine the distance from the row of lighting fittings to the point P. Distance from the line of fittings to the point $P = 2.5\,m = 0.6\,h$ where h is the mounting height of the fittings.

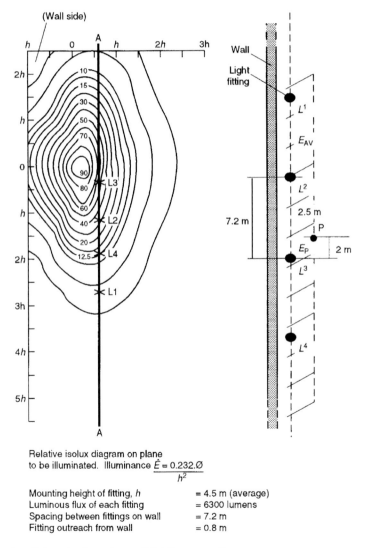

Relative isolux diagram on plane
to be illuminated. Illuminance $\hat{E} = \dfrac{0.232.\varnothing}{h^2}$

Mounting height of fitting, h	= 4.5 m (average)
Luminous flux of each fitting	= 6300 lumens
Spacing between fittings on wall	= 7.2 m
Fitting outreach from wall	= 0.8 m

Figure 7.6a   Substation boundary wall lighting example

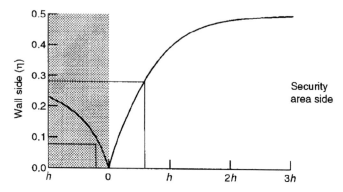

Figure 7.6b   Boundary wall lighting utilization factor curve

Table 7.4   Spotlighting calculation results

E_p lux	d (m)
370	2
250	4
160	6
110	8

(b) Draw the line A–A on the isolux diagram this same distance from, and parallel to, the longitudinal axis of the lighting fitting.

(c) Determine the distance from the point P to the transverse axis of each fitting and represent these points on the relative isolux diagram as points L_1, L_2, L_3 and L_4 on the axis A–A. Ignore the contributions from more distant lighting fittings in this example.

$$L_1 \text{ to } P = 12.2 \,\text{m} = 2.7\,h$$
$$L_2 \text{ to } P = 5.2 \,\text{m} = 1.2\,h$$
$$L_3 \text{ to } P = 2 \,\text{m} = 0.4\,h$$
$$L_4 \text{ to } P = 9.2 \,\text{m} = 2.0\,h$$

(d) From the isolux diagram read off the relative illuminance at each of these four points and calculate the total illuminance at point P:

$$EL_1 = 4\% \text{ of } E_{max}$$
$$EL_2 = 35\% \text{ of } E_{max}$$
$$EL_3 = 55\% \text{ of } E_{max}$$
$$EL_4 = 11\% \text{ of } E_{max}$$

Total $= 105\%$ of E_{max}

Using the isolux diagram formula for this particular fitting $E_{max} = 0.232 \times$ luminous flux/$h^2 = 0.232 \times 6300/4.5^2 = 72 \,\text{lux}$.

(e) The illuminance at point P, $E_p = 72 \times 105/100 = 76 \,\text{lux}$.

2. To find the average illuminance from the substation boundary or security wall to a distance 2.5 m from the line of the lighting fittings.

The simplest way of calculating the average illuminance from a long straight line of fittings is by using the manufacturer's Utilisation Factor (UF) curves and the formula:

$E_{av} = UF \times \phi/W \times S$

where W = width of area under consideration (m)
S = distance between each fitting (m)
ϕ = luminous flux of each fitting (lumens)

(a) The Utilisation Factor of the area is taken from the manufacturers UF curve as shown in Fig. 7.6b.
UF_1 at base of substation boundary or security wall = 0.08
UF_2 at 2.5 m ($0.6 h$) from substation boundary or security wall = 0.27
$UF = UF_1 + UF_2 = 0.35$
(b) Average illuminance, $E_{av} = UF \times \phi/W \times S$
$$= 0.35 \times 6300/3.3 \times 7.2$$
$$= 92\,\text{lux}$$
(c) An overseas client requires security spotlighting from an observation post along the boundary wall of a substation site. Calculate the illuminance, E_p lux d metres away from the observation post if the spotlight has a 300 W halogen lamp with a main lobe luminous intensity of 9000 cd (see Table 7.4 for calculation results). The arrangement is shown in Fig. 7.7. Why would sodium or mercury vapour lamps be unsuitable in this application?

The illuminance at point P, $E_p = I/(x^2) = I/(h^2 + d^2)$ lux
so $E_p = 9000/(4.5^2 + d^2)$ (approximately) lux

7.2.4 Control

Adjustment of the brightness may be achieved by varying current, voltage or delay angle.

Given sufficient voltage current control is applicable to fluorescent and other discharge lamps using variable inductance circuits. Such control is not normally applicable to AC tungsten filament lamp operation.

Voltage control is not particularly applicable to fluorescent and other discharge lamps because of the need to maintain a threshold voltage (typically 50% of rated voltage) below which the lamp is not extinguished and therefore to avoid erratic operation. A control range in terms of rated luminous flux from 30% to 100% over an operating voltage range of 70–100% of rated voltage is achievable for discharge lamps. Voltage control will dim incandescent tungsten filament lamps to zero light output at approximately 10% of rated voltage when some 30% of rated current will flow (see Fig. 7.8).

Delay angle control is suitable for both incandescent and discharge lamps. Control units consist of a pair of thyristors connected in inverse parallel (or a triac) so that each thyristor may conduct a proportion of each half cycle

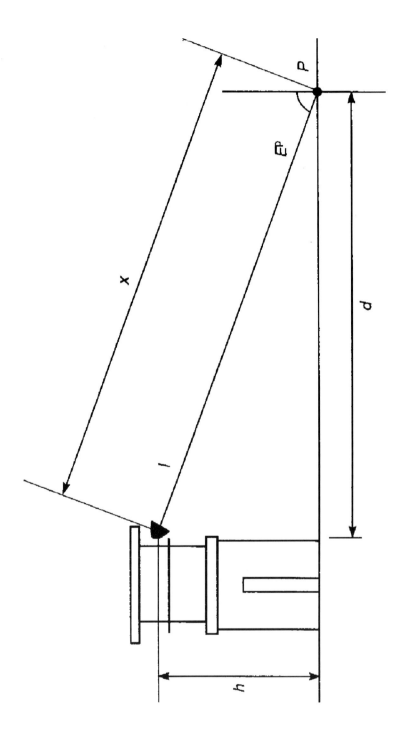

Figure 7.7   Observation post spotlighting example. Illuminance at point P = E_p lux

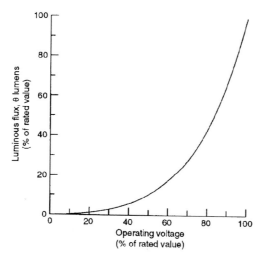

Figure 7.8   Incandescent lamp light output variation with operating voltage

of the AC supply. The thyristor or triac trigger pulses are arranged to initiate conduction at any point in the half cycle. Particular attention has to be made to interference suppression because of the fast thyristor switching rise times involved with this type of device.

7.3 DISTRIBUTION CHARACTERIZATION

The characterization of LV building services AC distribution systems may be described by the type of earthing arrangements used. Effective earthing is essential:

1. To prevent the outer casing of the apparatus and conductors rising to a potential which is dangerously different from that of the surroundings. Where there is an explosive risk there may be a danger from very small voltage differences causing sparking.
2. To allow sufficient current to pass safely in order to operate the protective devices without danger. This requirement may conflict with the necessity to keep potentials at a low level and restrict the available methods of protection. Regulations therefore require the 'earth loop impedance' of the completed system to be measured in order to ensure protection operation is not compromised.
3. To suppress dangerous earth potential gradients.

Earthing methods are defined by a three letter coding:

First letter – defines the state of the supply system in relation to earth.
T = Directly earthed system at one point.
I = Either all live parts are insulated from earth or one point connected to earth through an impedance.

Second letter – defines the state of the exposed conductive parts of the installation in relation to earth.

T = Exposed conductive parts connected directly to earth, independent of any
    earthing of a point on the supply system.

N = Exposed conductive parts connected directly to the earthed point of the
    supply system, normally the supply transformer neutral point.

Third letter – defines the earthing arrangement of the system conductors.

C = Combined neutral and earth conductors.

S = Separate neutral and earth conductors.

Five common arrangements are shown in Fig. 7.9. The TN-S system, using
separate neutral and protective conductors throughout the network, is the recommended method for installations in hazardous areas such as substations
feeding chemical plants. The neutral and protective conductors may also be
combined into a single conductor on part of the system (TN-C/S) or combined
into a single conductor throughout (TN-C). In the TT system, with exposed
metal bonded directly to earth, quick acting sensitive earth leakage protection
is required. The IT system employs a neutral conductor isolated from earth or
earthed through an impedance. It may be utilized on high security lighting circuits where the first fault is monitored by a current detector in the transformer
neutral and initiates an alarm. A second fault causes further current to flow
and is arranged to initiate a trip.

Irrespective of what earthing system is used it will not be effective unless it is
frequently checked to ensure that all earth bonds are mechanically strong and
free from corrosion. Furthermore, earth impedance should be monitored and
recorded so that any change can be detected and the appropriate action taken.

7.4 HEATING, VENTILATION AND AIR-CONDITIONING

7.4.1 Air circulation

The correct air circulation or number of air changes per hour is essential to
ensure comfort of substation operations and maintenance personnel. The
number of air changes depends on the number of personnel and size of the
room but a minimum of four fresh air changes per hour is recommended. In
addition, it is necessary to prevent the build up of dangerous gases such as
may occur in a battery room using vented cells. Typical air changes per hour
for different substation building areas are listed below:

Substation area	Air changes per hour
MV and/or HV switchrooms	4 to 8 (30 to 60 for smoke removal)
LVAC and/or DC switchrooms	4 to 8
Control and relay rooms	4 to 8
Battery rooms	6 to 10

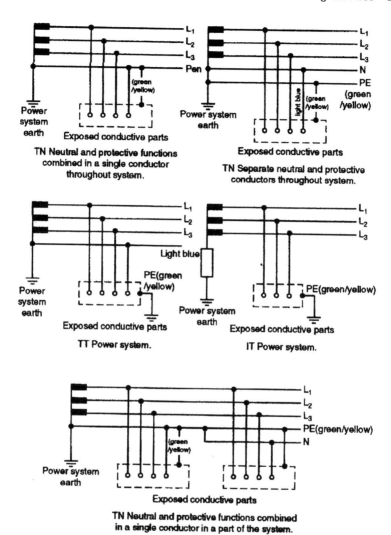

Figure 7.9   Methods of earthing

Control and communication rooms	4 to 8 (overpressures via a filter may be specified to prevent ingress of dust into sensitive equipment partly depending upon the equipment enclosure protection (IP) rating and need for adequate heat dissipation)
Offices	4 to 8
Toilet and wash rooms	10 to 12
Mess room	10 to 12
Corridors	3 to 6

Consider a 21/3.3 kV transformer housed in a transformer pen (7.5 m × 6 m × 11.7 m). The maximum ambient conditions are 28°C and the design engineer has specified a maximum pen air temperature of 34°C. The transformer enclosure has louvres to the atmosphere along one side (6.0 m × 10.0 m) with 30 m² open space. Air circulation is assisted by a ducted ventilation system. What air flow is necessary to maintain the transformer pen within the specified upper temperature limit?

The necessary ventilation calculations may be based upon the 'Sol-Air' temperature concept as described in Section A2 pages 69 and 70 of the CIBSE Guide together with the derivations given in Appendix 2 of Section A5 of the Guide. The formulae are incorporated into a spreadsheet programme to give the number of air changes required. As an alternative the total summer maximum heat gain may be determined by using W/m² figures as described in Section 7.4.2.3. The necessary ventilation air flow heat transfer Q_V is then calculated with a simple thermal heat balance equation:

$$Q_V = c\rho NV (t_i - t_o)/3600$$

where: Q_V = ventilation heat transfer, W
$\quad\quad c$ = specific heat of air, J/kg°K
$\quad\quad \rho$ = density of air, kg/m³
$\quad\quad N$ = number of air changes per hour, h^{-1}
$\quad\quad V$ = volume of room, m³
$\quad\quad t_i$ = inside temperature, °C
$\quad\quad t_o$ = outside temperature, °C

For practical purposes $c\rho/3600 = 0.33$. It should also be noted that the major element of the heat gain in this example will arise from the transformer losses rather than the solar heat gain through the building fabric.

All ductwork must be designed in conjunction with the fire safety engineers in order to ensure that the fire zoning is reflected in the ductwork. Zonal control is usually achieved by using automatic fire dampers where ductwork passes through floors or across fire zone walls. Dampers should have 1- or 2-hour ratings and be released from a fusible link or by a solenoid interconnected into the fire detection system. This is especially important where migration of smoke from one zone to another could be a substantial hazard. The need for zonal control also applies to cable trenches running through different fire zones within the substation building.

7.4.2 Air-conditioning

7.4.2.1 Introduction

The internationally recognized standard used for the design of Heating, Ventilation and Air-Conditioning (HVAC) is that produced by the American

Society of Heating, Refrigeration and Air-Conditioning Engineers (ASHRAE). However, in the UK the standards usually acknowledged are those produced by the Chartered Institute of Building Services Engineers (CIBSE). This is a specialist subject and therefore only a brief introduction is given in this section such that the power engineer can correctly specify requirements and understand the terminology used by HVAC engineers.

Areas of substation buildings should be air-conditioned in the following ways:

Substation area	*HVAC requirements*
Switchgear rooms	Ventilation with air tempering to prevent the formation of condensation on cooled surfaces is the ideal approach. The most controllable solution is to temper the supply air using electrical or low pressure water coils. Alternatively, the switchgear should be specified to operate at ambient temperatures up to 40°C and 90% relative humidity with ratings calculated accordingly and panel heaters should be installed to prevent condensation and freezing.
LVAC and/or DC switchrooms	As for MV and HV switchgear rooms.
Control and relay rooms	HVAC. If because of economies switchgear and control and relay panels and operators' desks are all provided together in a single room then air-conditioning will be required.
Battery room	Extract ventilation with high quality acid fume resistant fans.
Control and communication rooms	Manufacturer's standard light current equipment often demands a stringent environment. Such sensitive equipment should be maintained typically at 20°C and 50% relative humidity.
Offices	HVAC to between 20°C and 25°C with relative humidity maintained between 40% and 60% for comfort.
Toilet and wash rooms	Extract ventilation fans with supply air drawn in through transfer grilles to ensure smells do not enter the working area.
Mess room	Extract ventilation discharging to the outside with air transfer grilles as described for the toilet and wash rooms. Possible HVAC depending on budget.
Corridors	Air-conditioning if in a manned substation where people are frequently moving between rooms.

Note: Depending on the volumes of the rooms and therefore the volume of air to be moved to achieve the desired air changes per hour when applying air-conditioning, noise may be an issue. As well as continually occupied spaces, account must be taken of operatives working in substations, switchrooms, etc. during maintenance periods. In the UK the CDM Regulations require that a noise assessment is made to ensure that the Noise at Work regulations are not breached during periods of occupancy.

7.4.2.2 *Calculation methodology*

Calculating the cooling load involves estimating and compensating for the heat gains associated with radiation from equipment (usually taken under full load conditions and including lighting) and people in the room plus solar radiation effects. The cooling load consists of the 'sensible' load associated with the cooling required to maintain the desired temperature and the 'latent' load which is necessary for dehumidification and the release of latent heat of vaporization. Detailed calculations therefore involve the orientation of the room walls relative to the sun, wall, window, ceiling and floor materials and thickness, coefficients of heat transmission (K factors in W/m^2 K), temperature difference, air infiltration, the number of people (that produce heat and moisture) and the heat produced from equipment.

Because of the traditional dominance of North American manufacturers in this field, air-conditioning plant capacity is often described in imperial units of BThU/hour, being the measure of heat removal; 1 BThU is the amount of heat required to raise the temperature of 1 pound of water by 1°F. (12 000 BThU/hour is equivalent to 3024 kcal/hour).

Note that 12 000 BThU/hour is called 1 ton of refrigeration (to freeze 1 ton of ice/24hours).

1 kW = 3413 BThU
1 kW = 860 kcal
1 kcal = 3.968 BThU

7.4.2.3 *Simplified cooling load estimate for window mounted air-conditioners*

It is sometimes necessary for the substation engineer to have a rough estimate of the air-conditioning requirements before detailed calculations are made. The room dimensions are recorded. Doors and windows are assumed closed to reduce unnecessary infiltration and exhaust only when necessary or under smoky conditions. All windows and glass areas exposed to direct sunlight are assumed to have blinds, awnings or shades and an additional heat gain is added where windows face summer afternoon sun. Roofs exposed to direct sunlight are assumed to have at least 50 mm insulation material and all false ceiling space is assumed to be well ventilated with outside air. A further heat gain is added to compensate if this is not the case. The cooling load estimations given in Table 7.5 are then made.

Table 7.5   Simplified cooling load estimations for window air-conditioner selection

Type of use	W/hour/m² floor area	BThU/hour/ft² floor area	Notes
Control and relay rooms (load depends on manning level and heat emissions from equipment)	250–380	75–120	Minimum fresh air requirement per person depends upon outside temperatures and level of pollution (smoking, etc.). Use 20 m³/hour per person for outside temperatures 0–25°C and assumption that smoking is not allowed.
Offices	200	60	
Mess room	160–200	50–60	Allow 300 W per person
Corridors	100	30	

For example, consider a control and relay room in a normally unmanned Middle East distribution substation 10.8 m length × 7.3 m wide × 4.1 m high. Air-conditioning is a specified requirement and the initial rough estimated cooling load would be:

Floor area = $78.85\,\text{m}^2$
Control room volume = $323.25\,\text{m}^2$
Temperature to be maintained in range 25°C ± 3°C

New energy codes in the UK specify maximum coefficients of heat transmission (K values in W/m²°K) for different constructions and materials together with typical heat requirements (q in W/m²) for various types of building. As an initial estimate assume the cooling load requirement from the table giving simplified estimations above is 320 W/m² giving a cooling requirement of $320 \times 78.85 = 25\,\text{kW}$. This may be checked against a slightly more rigorous approach taking into account individual solar and equipment heat gains:

	m^2	W/m^2	Total, W
(a) Windows and doors under direct sunlight:			
(i) South* direction	25.55	230	5877
(ii) East direction (double door)	5.37	260	1397
(b) Outer walls under direct sunlight	38.91	50	1946
(c) Other outer walls	74.21	25	1855
(d) Flat roof	78.85	60	4731
(e) Floor	78.85	9	710
(f) Lighting			720
(g) Equipment			6300
			23536 W

*NB – This assumes a location North of the tropics.

Window mounted 24 000 BThU air-conditioning units are standard Electrical Supply Utility store items. 23 536 W = 23.536 × 3413 BThU = 80 316 BThU.

Therefore number of units required = 80 316/24 000 = 3.3. Therefore four window mounted units would be sufficient. Following such a calculation the building civil engineers or architects would then become involved for more accurate estimations and to detail the aesthetic appearance of the system. Such estimations assist in the sizing of the substation auxiliary transformers at the early planning stages of a project.

7.4.2.4 Air-conditioning plant

'All-refrigerant' air-conditioning systems are normally sufficient for substation control buildings. The advantages of such systems are that they are compact, cheap and simple to install. However, they require a relatively high level of maintenance, have a relatively short working life and single package units may be noisy. The classification is confined to 'single package' room air-conditioners and 'split' systems in which the cooling effect is produced by refrigerant gas being compressed and passed through an evaporator whereby it absorbs heat. Single package units are typically sized up to 24 000 BThU and split units up to 64 000 BThU with air volume flows of 200–1000 m^3/hour. The units come complete with controls and protective devices. In the single package unit air is drawn through a filter into the unit by a centrifugal fan. The air is cooled as it passes over the evaporator cooling coil and discharged into the room. The most common form of such units is window mounted. A room thermostat actuates the system. Such units may also contain an electric heating element (ratings between 1 and 4 kW) for use as a fan heater in winter conditions.

In a 'split' air-conditioning system the condenser and evaporator sections are supplied separately, usually for interconnection by means of flexible precharged refrigerant piping with quick connection fittings. The condenser is mounted outside the building (to reduce noise and allow heat loss to the outside air) while the evaporator section with a fan may take the form of a free-standing console mounted inside the room being air-conditioned. A suitable water piping system is necessary to carry condensate to an external drain or soakaway.

Local 'air-handling' units supplied with cooling water or direct refrigerant expansion as the cooling medium coupled with electric or low pressure hot water (LPHW) heater elements using a fan fresh air supply are used for the larger substation buildings. Supply and return ductwork provides a uniform air distribution within the different areas being served. Such units may be provided with humidifiers, efficient air filtration and can generally control the internal environment within close limits of temperature and humidity.

'Fully ducted' air systems connected to a central substation air-handling unit are used where a number of similar rooms are to be air-conditioned or where space prohibits the installation of air-handling units adjacent to the air-conditioned area. It is essential to adequately filter all input or make-up air in order to avoid dust build-up and possible premature equipment failure. Energy efficiency is increased by recirculating a proportion of already cooled or heated substation building air back into the air-conditioning system. The disadvantage of all air-ducted systems is the space requirement for the ducts.

Larger 'all-water' plants are unlikely to be necessary for substation buildings. Piped, heated or chilled water derived from a central refrigerant or boiler plant is circulated to different parts of the building. Fan coil heat exchanger units dissipate the heating or cooling effect via air grilles in ductwork into the rooms.

7.4.3 Heating

Heating is not normally required in climates where the minimum ambient temperature does not fall below 15°C although units may be installed where close control of temperature and humidity is necessary. The following methods are available:

'Air heater' batteries using electrically heated elements operated in stages, or low pressure hot water coils, located in air-conditioning ductwork.
'Electric space heating' from room heaters.
'Low pressure hot water' (LPHW) space heating using radiators fed by circulating hot water.

Normally the air heater option would be used in conjunction with an air-conditioning scheme. The heat gains from personnel, switchgear and control equipment are not normally taken into account when sizing heating plant so as to cater for conditions when the substation is not operational. Anti-condensation heaters with ratings of a few tens of Watts are often specified for installation within switchgear and control panels.

7.5   FIRE DETECTION AND SUPPRESSION

7.5.1 Introduction

Heat, fuel and oxygen are required together for a fire to exist. Removal of any of these components will extinguish the fire. The fire safety philosophy is to:

1. Safeguard personnel.
2. Maintain the functional state of the substation.

Substations have a comparatively low fire risk because the designs are such as to introduce little chance of an internal fire spreading to an adjoining property or causing danger by smoke contamination. However, certain equipment (traditional bulk oil or low oil volume indoor switchgear installations) and materials (transformer oil, cable joint compounds, solid forms of insulation,

Table 7.6   Commonly available fire extinguishers and their usage

Type	Colour code bands	Application	Extinguishing action
Water	Red	Fires involving wood and other solid, organic or carbon based material.	Cools fuel to below the temperature at which sustained flaming occurs.
CO_2	Black	Mainly electrical equipment fires.	Cools and inert the atmosphere.
Dry powder	Blue	Flammable liquid fires/electrical fires.	Inhibits the chemical reactions in the flames.
Halon (BCF)[1]	Green	Flammable liquid fires, electrical fires and fires in carbon based solids.	Same as dry powder but also has a cooling effect.

(1) Note: Halon has now been effectively banned for future installations because of its ozone depletion potential. The EU regulation 2037/2000 required mandatory decommissioning of existing halon fire extinguishing systems by 31st December 2003. There are a number of inert gas products that can be used instead. Use of fluorinated gases, some of which are also currently available as an alternative, will be controlled by an EU directive by 2006.

etc.) may ignite and therefore fire detection and suppression systems should be incorporated. Further, special precautions have to be taken concerning the spread of fire caused by oil leakage from transformers or the oil-filled types of switchgear. This chapter describes general fire detection and suppression schemes for substation installations covering fires originating from transformers, switchgear, control and protection equipment and cables.

7.5.2 Fire extinguishers

Hand or trolley mounted fire extinguishers are the cheapest form of manual fire extinguishing. Such extinguishers should be mounted in the substation control or switchgear building as well as in the switchyard. The types of commonly available fire extinguishers and their usage under different conditions are as shown in Table 7.6. See BS EN3, BS 7863:1996 and BS 5306.

7.5.3 Access, first aid and safety

7.5.3.1 *Access*

The exits from substation control and switch rooms must always be kept clear. Panic release bars should be fitted to the doors such that the doors will quickly open outwards from the inside by pressure against the release bar. Doors should be sized greater than 750 mm × 2000 mm and areas at the rear of switchboards should not be less than 12 m from an exit. Doors between equipment containing bulk oil should have a 2-hour fire resistance rating. Two exits should always be included in the layout design (especially in switchgear rooms containing switchboards greater than 5 m long) such that escape is possible by at least two different routes from any area in the substation building. In this way the chance of personnel being cut off from an exit by the fire is reduced. Emergency lighting fittings should be installed over each exit.

7.5.3.2 *First aid and signage*

Fire and emergency signage should be installed in accordance with the appropriate national standard (In the UK the Safety Signs and Signals Regulations 1966 apply). A sample of signs in accordance with BS5378 and BS5499 is shown in Fig. 7.10. It should be noted that in accordance with the European Regulations and the UK Health and Safety Executive rulings it is necessary to examine the risks associated with any engineering design. The Electricity at Work Regulations require precautions to be taken against the risk of death or personal injury from electricity in work activities. Basic first aid kits (first aid dressings, medication, eyewash, blankets, etc.) and safety barriers to screen off work areas should therefore be available on the substation site. Signage describing the actions to be taken in the event of electrical shock should also be clearly on display.

It should be noted that modern legislation requires a risk assessment to be made of the activities to be performed in the substation building or compound and the standards, directives and regulations given in Table 7.7 are important.

7.5.3.3 *Good housekeeping*

Manual or automatic fire detection and suppression systems must be regularly maintained to ensure their effectiveness. A regime of maintenance checks must be included in the substation operations procedures. In addition, regular disposal of waste materials (e.g. oil cleaning rags, etc.) must be enforced together with careful removal of smoker's materials or a ban on smoking (especially important in cable basements).

7.5.4 Fire detection

7.5.4.1 *Manual call points*

The fire may be detected by personnel manning or working on the substation site at the time of the fire. The alarm may therefore be raised by the breaking of glass at a manual call point. These should be mounted at approximately 1.4 m

ELECTRICITY AT WORK SAFETY SIGNS ●Signs comply with Electricity at Work Regulations 1989

ELECTRICAL MAINTENANCE SIGNS

Figure 7.10   Fire and emergency signs

Table 7.7   Relevant European standards, directives or regulations

Standard, directive or regulation	Description	Notes
UK Fire Precautions Act 1971	Covers emergency lighting	UK Local authorities or fire officers will look for emergency lighting compliance with BS 5266 Part 1: 1988.
89/106/EEC	The Construction Products Directive	Concerned with products used in construction. Introduces six essential requirements: – mechanical resistance and stability – safety in case of fire – hygiene, health and environment – safety in use – protection against noise – energy, economy and heat retention
89/659/EEC 89/654/EEC	The Framework Directive The Workplace Directive	General enabling directive. Covers a broader range of premises than the UK Fire Protections Act, imposes legal obligations for compliance upon the employer or Electrical Supply Utility rather than the Fire Authority and applies retrospectively.
92/58/EEC	The Safety Signs Directive	Legalised in UK by the safety signs and signals regulations 1996. Introduces new symbols for safety exit signs replacing those previously defined in BS 5499 Part 1: 1990. For example pictograms rather than wording are introduced onto signs and an exit sign shows a 'running man' with an arrow pointing towards a symbol for a door.

above floor level and located on exit routes both inside and outside the substation building so that no person need travel more than approximately 30 m from any position in the building in order to raise the alarm.

7.5.4.2 *Sensors*

Two types of detector or sensor are found in substation applications:

1. Heat detectors
   Depending upon the type these sense changes in the thermal environment very locally or in the immediate vicinity of the unit. Bimetallic strips and thermistors are commonly used devices in such sensors. Maximum mounting heights depend upon the grade of the detector but lie in the range 6–13.5 m. In order to differentiate between the normal temperature changes

and fire conditions the sensors detect the temperature above a preselected limit and the rate of rise of temperature (which will be rapid in the case of a fire) in order to initiate an alarm or automatic fire suppression system.

2. Smoke detectors

   These sense small particles of matter or smoke in the air which are the result of a fire. Ionization detectors work on the principle that the current flowing through an ionization chamber reduces when smoke particles enter the chamber. Electronic alarm circuitry detects this change and initiates the alarm.

   Optical detectors note the scattering or absorption of light due to the smoke particles in a light beam. Maximum mounting heights are typically 10.5–15 m. It is considered that optical types are best suited to detect slow smouldering fires where large smoke particles are formed and that ionization types are best suited for fast burning fires where small smoke particles are formed. Smoke detectors tend to give a faster response than heat detectors but may also be liable to give more false alarms. Because of this and since prediction of the fire type may not be possible both types of smoke detector are often found in a single installation together with heat detectors.

3. Radiation (flame) detectors.

   These detect ultraviolet or infrared radiation and are mainly suitable for supplementing heat and smoke detectors or as a general surveillance of a large switchyard area.

7.5.5 Fire suppression

It is important to avoid anomalous operation of a fire detection and suppression system. Therefore a 'double knock' system is usually employed where two sensors have to detect the alarm before the suppression system is activated. Radial circuits are used with the detectors effectively cabled in parallel together with an 'end-of-line' resistor. The circuit is monitored for both short circuit (typically less than 1000 ohms) and open circuit conditions and a maintenance alarm raised if the circuit is out of tolerance. Sensor circuits are arranged on a 'zonal' basis in order to isolate the fire into certain areas. For example, the substation switchgear room may be split into two, both physically (with a fire wall) and electrically (with a bus-section switch). Each half of the switchboard would then be covered by a separate zone of the fire detection and suppression system. In a similar manner the control and relay room might also be covered by a separate zone of protection. The zone where the fire has occurred is indicated on the fire detection control panel. The panel sends signals to alarm sounders to alert personnel and to send signals to the automatic fire extinguishing systems or to shut down the HVAC plants which could spread the fire. Both inert gas and CO_2 gas systems require the rooms to be enclosed. 'Fire stopping' is the term used to describe the sealing of small openings in fire barriers. Ventilation louvres should be fitted with temperature sensing or remote controlled closing devices.

It is not considered essential for modern gas insulated switchgear to be housed in rooms with fire suppression systems.

Halon 1301 gas, previously used as a fire extinguishing medium in substations is an ozone layer depleting gas and most electrical supply utilities now ban its use. There are available a number of inert gas mixtures of argon, nitrogen and carbon dioxide, and also a heavily fluorinated compound with the generic term HFC227ea, any of which are used as a replacement. (HFC227ea breaks down to give hydrofluoric acid under severe fire conditions, and as a fluorinated gas is likely to be regulated by the EU directive on fluorinated gases to be made statute in member states during 2006 – it is not therefore the recommended option).

Alternatively the older CO_2 flooding systems (which require a larger concentration of gas for the same extinguishing effect) may be used.

Gas bottles are suspended from the ceiling in the room being protected, or a central set of cylinders with a piping system to ceiling mounted nozzles in the different rooms may be employed. The required concentration of gas to extinguish the fire, while small, is considered dangerous to personnel. It is therefore considered necessary to avoid personnel being in the zone during gas discharge. This is an absolutely essential requirement for CO_2 flooding since the necessary 28% CO_2 gas concentrations will be lethal. A door/gas suppression system interlock is therefore necessary such that the suppression system is deactivated whilst maintenance staff are working in the room. In addition a delay between alarm and suppression activation is built into CO_2 flooding systems.

Water sprinkler systems may be employed in cable basements. The normal sprinkler has a liquid filled glass bulb valve which is activated by the expansion of the liquid and shattering of the glass. This is not sufficiently fast for cable fire protection. Therefore the glass ampoule is fitted with a 'percussion' hammer which is activated electronically from the smoke or heat detectors.

The fire resistant properties of cables are described in Chapter 12. Cables may be coated with protective paints and mastics to reduce fire risk without affecting cable current carrying thermal capacity. In tumescent coatings swell up over an elevated temperature range to form an insulating foam layer.

Automatic water spray systems may be used for transformer protection in outdoor areas where the gas would leak away. The oil is contained in a bund as part of the transformer civil works installation design. The water spray cools the oil to a temperature below its fire point at which sustained flaming can occur. Other techniques involve oil temperature sensors within the transformer. Upon activation the transformer is electrically isolated and a small proportion of oil drained from the core. Dry nitrogen gas is then injected at the base of the transformer which bubbles through the oil causing mixing, heat transfer within the oil and lowering of oil temperature.

7.5.6 Cables, control panels and power supplies

It makes sense to ensure that the cabling associated with the substation fire detectors is both flame retardant and flame resistant if it is to operate successfully and reliably initiate an alarm. Mineral insulated copper sheathed cable is therefore favoured. An oversheath is not essential but if used should preferably be LSF and coloured red to differentiate from other services. Other types of $0.5\,mm^2$ or $1.0\,mm^2$ minimum cross-section (depending upon load) copper conductor PVC or EPR insulated cables should only be used in conjunction with a suitable conduit, trunking or ducted system.

Control panels with key access should be located in an area of low fire risk close to the entrance of the substation building or guard house. The panels should display the status of each protected zone, detector faults (insulation or loop resistance, detector removal, rupture of any fuse or operation of protection devices associated with the system) and power supply. A fire alarm must take precedence over any other indication that the control panel may be giving. Key switch isolation facilities must be available for maintenance and test functions. The audible alarm generated within the control panel may be both locally and remotely sounded.

The fire alarm system should have its own battery backed power supply with sufficient autonomy (say, 1 or 2 hours) for correct operation upon loss of AC supply and to provide an evacuation alarm for at least 30 minutes upon fire detection. The connection to the AC power source must be clearly labelled 'Fire alarm: do not switch off' and arranged such that continuity of supply is ensured. On larger substation sites it may be possible to integrate the fire alarm system with any local standby generation.

8 Earthing and Bonding

8.1 INTRODUCTION

The purpose of a substation earthing system is to ensure safe conditions for personnel and plant in and around the site during normal and earth fault conditions. To achieve this, it needs to perform the following functions:

- Provide equi-potential grading by electrodes or similar, to control touch and step potentials.
- Enable connection (bonding) of necessary exposed extraneous conductive parts to earth.
- Provide a route for the passage of fault current which does not result in any thermal or mechanical damage to connected plant and allows protective equipment to operate.
- Provide an earth connection for transformer neutrals, sometimes via an impedance to restrict the fault current magnitude.
- Minimise electromagnetic interference between power and other systems such as control or communication cables or pipelines.

This Chapter is an over-view of the subject, dealing mainly with the design limits and the general procedure followed to design an electrode system, including a computer analysis.

Small power and lighting installation earthing is covered in Chapter 7.

8.2 DESIGN CRITERIA

8.2.1 Touch and step voltages

Substation 'earthed' metal work (consisting of switchgear enclosures, supports, fencing, etc.) and overhead line steel supports all have an impedance to true earth. When fault current flows through them to earth, a voltage rise will occur.

This 'earth potential rise' (EPR) is the maximum voltage that the earthing system of an installation may attain relative to a remote point assumed to be at true (zero) earth potential. The EPR is the product of the current that returns to its remote sources via the soil and the earthing system impedance. The overall (gross) fault current calculated will normally include amounts that may return via metallic routes such as the other (un-faulted) phase conductors, the metal screens/sheaths of buried cables and the earthwires (if fitted) of an overhead line. These current flows are subtracted from gross value to leave the amount that will return via the soil. This current will flow through the local electrode system and any other electrodes connected in parallel to it. These include sheet steel foundations, the steel legs of transmission towers, large pipes and the lead sheath of cables that have a conductive outer covering (such as Hessian).

Whilst the EPR exists, voltages will occur in and around the installation. A number of voltage definitions are used to characterise the situation at any point. They include:

- The 'touch voltage' which is the potential difference between the EPR on a structure and the surface potential at a point where a person is standing (normally 1 m away), whilst at the same time having one or both hands in contact with the structure.
- The 'step voltage', which is the difference in surface potential experienced by a person bridging a distance of 1 m with their feet.
- The 'transfer potential' is that between steelwork (physically distant from the installation, but bonded to it) and the remote local earth. This could occur at any point along a cable, pipe or steel fence. The design normally seeks to prevent dangerous transfer potentials occurring, by limiting the EPR or removing the electrical connection between the steelwork and installation earth.
- The 'mesh voltage' is a quantity used in the American Standard, IEEE 80, and is the touch voltage seen at the centre of a mesh of the substation earthing grid.

Calculations seek to find the worst case value of touch and step voltage for the design and compare it against tolerable voltage limits. These different scenarios are illustrated in Fig. 8.1, extracted from the North American Standard IEEE 80.

8.2.2 Touch and step voltage limits

Electro-pathological effects on the human body are produced when current passes through it and to avoid death we are concerned about the proportion which flows in the region of the heart. IEC 60479-1 (2004) 'The effects of current on Human Beings and Livestock' provides the internationally accepted best available guidance on the subject and is used as the basis of touch and step voltage limits in many standards – particularly those in Europe.

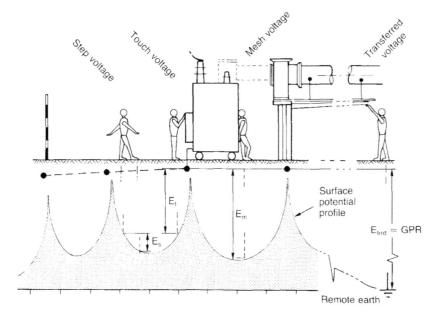

Figure 8.1   Illustration of the voltages that occur during an earth fault (from IEEE 80)
E_S = step voltage; E_t = touch voltage; E_m = mesh voltage; E_{trrd} = transfer potential

The first edition (called IEC 479) was published in 1974 and was influenced by the early research work of Dalziel. This data was subsequently improved in the 1984 and 1994 editions, which are still used in many standards. A new and extended version of IEC 60479-1 has recently been completed. Original work in the area included that by Biegelmeier, who confirmed some observations by subjecting himself to electric shocks! The most often referenced part of IEC 60479-1 is illustrated in Fig. 8.2. This uses current flowing in the region of the heart against its duration to establish the boundaries between different threshold levels.

The document shows that humans are particularly vulnerable to alternating currents of between 50 and 60 Hz with a threshold of perception of about 1 mA. With increasing current, the effects move from muscular contraction, unconsciousness, fibrillation of the heart, respiratory nerve blockage to burning. Humans can generally withstand higher currents at DC and higher frequencies. The current limits are time dependant, because the threshold of fibrillation rapidly decreases if the alternating current flow persists for more than one cardiac cycle. For shock durations below one cardiac cycle, the ventricular fibrillation threshold current is nearly constant down to very short times. The risk of ventricular fibrillation is much lower at short intervals because there is less chance of current flow during the vulnerable period in the 'T' phase of the cardiac cycle. This is illustrated in Fig. 8.3 and only occupies approximately 10% to 20% of the total cardiac cycle. For this reason there is a pronounced 'kink' in threshold values in the standards. Earthing standards

Zones	Physiological effects
1.	Usually no reaction effects.
2.	Usually no harmful physiological effects.
3.	Usually no organic damage to be expected. Likelihood of muscular contractions and difficulty in breathing, reversible disturbances of formation and conduction of impulses in the heart, including atrial fibrillation and transient cardiac arrest without ventricular fibrillation increasing with current magnitude and time.
4.	In addition to the effects of zone 3, probability of ventricular fibrillation increasing up to about 5% (curve C_2), up to about 50% (curve C_3) and above 50% beyond curve C_3. Increasing with magnitude and time, pathophysiological effects such as cardiac arrest, breathing arrest and heavy burns may occur.

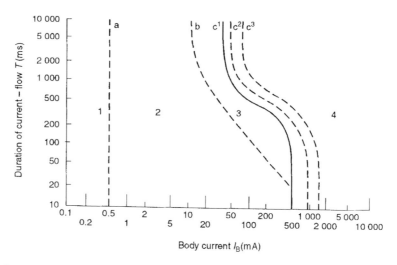

Figure 8.2   Time/current zones of effects of AC currents (15–100 Hz) on persons (IEC 60479-1)

convert the tolerable currents into touch and step potential limits because it is easier to make voltage calculations and measurements at electrical installations. A typical set of touch and step potential limits are shown in Fig. 8.5.

As can be seen, the tolerable voltages are surprisingly high at short clearance times, although it should be noted that a person would still experience a considerable shock. The aim is to avoid the shock being sufficient to cause ventricular fibrillation.

The earthing coverage in the standards is very varied and even taken together, the guidance offered to designers until quite recently was sparse and often conflicting. The most recent versions of the standards now include reasonably detailed design guidance. Of further concern is the fact that National and International Standards still quote different touch and step potential limits and different ways of calculating them. For example, within Europe they are

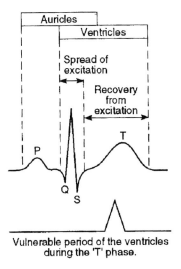

Figure 8.3   Typical human heart cycle

based on the IEC 60469 curves, whilst in America they are based upon equations such as those illustrated in Section 8.3.4 later.

Partly to address the above problems, a chapter in the new IEC standard (IEC 61936-1) was developed. This sets out the methodology to be used to establish the touch and step voltage limits and provides a design flow chart, together with supporting guidance to show how the earthing system needs to be designed. Work has now started on a new European earthing standard to bring together the earthing related guidance that is present in many different standards.

Most standards (including IEEE 80 and BS 7354) take into account the resistivity of the substation surface material and whether a small thickness high resistivity surface layer such as crushed rock is used. The results provide slightly higher admissible touch and step potential voltages whilst stressing the advantages of rapid fault clearance times.

8.3 SUBSTATION EARTHING CALCULATIONS

8.3.1 Environmental conditions

8.3.1.1 *Introduction*

In order to calculate the required earthing parameters at a new site, a measurement of the soil resistivity (ohm metres), site dimensions and site specific earth fault levels (kA) are necessary. An initial grid layout is designed, based upon the site dimensions and influenced by the soil structure. Its resistance,

EPR, touch and step voltages are then calculated. These are compared to the tolerable voltage limits and the grid adjusted, as necessary, to enable these requirements to be met. In some cases the calculated resistance is needed to revise the fault current and the voltages it creates. The process continues until the calculated safety voltages are all lower than the limits.

8.3.1.2 Relevant Fault Conditions

The fault conditions are calculated using short circuit analysis techniques as described in Chapters 1 and 26. The substation may be a switching centre and/or voltage transformation point. The substation earthing is common to the different voltage levels, so the calculations need to cater for a fault within the site at the highest voltage level and external to the site at the lower voltage levels. Only in exceptional cases will faults within the substation at the lower voltage level create a significant EPR, because the majority of the fault current in this case should normally circulate within the earth grid rather than flow into the soil.

Depending upon the voltage level and applicable standard, a fault duration of 1 to 3 seconds is used for the conductor sizing calculations. The conductor cross section required can be calculated using the following formula:

$$S = \frac{\sqrt{I^2 t}}{K}$$

where S = required cross sectional area of conductor in mm²
   I = value (AC rms.) of fault current in amperes
   t = standard substation design time or operating time of backup disconnecting device, in seconds
   K = factor dependent on the material of the conductor, the insulation
   and other parts and the initial and final temperatures. Some K factors that are mainly used for building services for different conductors are listed in Table 8.1 (selected from BS 7671, but the
   same figures may be found in IEC 60364). Different values are used
   within substations for the buried grid and structure connections –
   based for example on higher final temperatures.

For bare copper:

$$K = 226 \sqrt{\ln \frac{1 + [\theta_f - \theta_i]}{[234.5 + \theta_i]}}$$

$$= 159$$

For example a 13 kA, 1 second design fault level will require a bare copper conductor size (based upon a K factor of 159) of 82 mm² or its nearest equivalent standard size.

Table 8.1   K factors used for calculating minimum earth conductor sizes

(i)	Bare copper conductors		
	Initial temperature of conductor	= 30°C	
	Maximum temperature of conductor	= 150°C or 200°C	(fire risk conditions)
	K factor (BS 7671)	= 138	(normal conditions)
		or 159	(fire risk)
			(normal)
(ii)	XLPE insulated copper/aluminium cable	Single core	Multicore
	Initial temperature of conductor	= 30°C	90°C
	Maximum temperature of conductor	= 250°C	250°C
	K factor (BS 7671) copper	= 176	143
	K factor (BS 7671) aluminium	= 116	94
(iii)	Bare steel electrode		
	Initial temperature of conductor	= 30°C	(normal conditions)
	Maximum temperature of conductor	= 200°C	
	K factor (BS 7671)	= 58	

Table 8.2   Typical ground resistivity values

Ground type	Resistivity (ohm metres)
Loams, garden soils, etc.	5–50
Clays	10–100
Chalk	30–100
Clay, sand and gravel mixture	40–250
Marsh, peat	150–300
Sand	250–500
Slates and shales	300–3000
Rock	1000–10 000

The conductor size needs to include an amount for corrosion over the lifetime of the site and can take into account the effect of duplicate paths.

The exposure time for touch and step voltages depends upon the fault clearance time and this is provided from the fault level study mentioned earlier. As a general guide, 132 kV and above circuits have a clearance time of less than 0.2 seconds. At lower voltages, clearance times in the range 0.3 to 1 second are the most common.

8.3.1.3 *Earth resistivity*

Measurements are normally carried out using the Wenner method and the data is used to arrive at a representative soil model for the site. Whilst the measurements would best be carried out in representative weather conditions, this is clearly not always possible, so allowance for seasonal effects may need to be made in the model. This would normally be done by modifying the resistivity and/or depth of the surface layer. Some typical soil resistivity values are shown in Table 8.2.

Measurements are taken for a range of probe separations, each of which is a general indicator of the depth to which the value applies. Measurements in a number of directions would be taken and averaged values (excluding obvious errors) for each separation distance would be used to derive the initial soil model.

A number of computer programmes are commercially available and used to translate the data into a representative soil model. It is useful to have both the average model and the data spread, so that the error band is known, as this will influence the subsequent calculations or suggest that the derived soil model be modified to improve its accuracy.

It is possible to use formulae or graphical methods to derive a two layer model. The formula below compares the resistivity, σ_1 of the upper layer of depth h_1 with the lower layer of resistivity, σ_2:

$$\frac{\sigma_s}{\sigma_1} = 1 + 4 \sum_{n=1}^{\infty} \frac{k^n}{[1 + (2n \cdot h_1/a)^2]^{1/2}} - \frac{k^n}{[4 + (2n \cdot h_1/a)^2]^{1/2}}$$

The value σ_s is the resistivity measured at a depth a. IEEE 80 includes a number of graphs to achieve the same result, based on the work of Sunde.

It is unusual to use formulae now, because the interactive computer programmes available can quickly provide a model which may have a number of vertical or horizontal interfaces. Often a three layer model is necessary to provide sufficient accuracy.

The soil model values are used in formula or a computer programme to calculate the earth resistance and hazard voltages (see Section 8.2).

8.3.2 Earthing materials

8.3.2.1 Conductors

The conductors used must be capable of carrying the anticipated fault current and cope with corrosion over the lifetime of the installation. Bare copper is normally used for a substation earthing grid, being buried at depths between 0.6 and 1.0 m in rectangles of between 3 and 7 m side length. Equipment connections are generally laid at a shallower depth of about 0.2 m. Because of mechanical and thermal criteria, it is unusual for copper of less than 70 mm^2 to be used.

Aluminium is often used for above ground connections and could be used below ground if it is certain that the soil will not cause corrosion problems, but most standards prohibit this. Some protection, such as painting with bitumastic paint, is recommended in the area where the conductor emerges from the ground, as corrosion may occur here and just below. Where not connected directly to the electrode, all metallic substation plant is bonded via above ground conductors.

8.3.2.2 Connections

The connection methods used for below ground application are welded, brazed, compression or exothermic. For above ground use, bolted connections are used in addition to these.

Soldering is not permitted, because the heat generated during fault conditions could cause failure.

Bolted joints should normally have their contact faces tinned and particular care is necessary for connections between dissimilar metals, such as copper and aluminium. The standards (such as IEEE 837) and national codes of practice offer much advice on connection methods.

8.3.2.3 *Earth rods*

The earth grid's horizontal electrodes may be supplemented by vertical rods to assist the dissipation of earth fault current, further reduce the overall substation earthing resistance and provide some stability against seasonal changes. This is especially useful for small area substation sites (such as GIS substations) or where the rod is of sufficient length to enter the water table. Rods may be of solid copper or copper clad steel and are usually of 1.2 m length with screw threads and joints for connecting together in order to obtain the required length for installation in the soil. The formula for the effective resistance, R_{ROD} ohms of a single earth rod is given by:

$$R_{ROD} = \frac{\sigma}{2\pi l} \cdot [\ln(8l/d) - 1]$$

where R_{ROD} = bare vertical earthing rod effective resistance (ohms)
σ = resistivity of soil (ohm metres)
l = length of earthing rod (metres)
d = diameter of earthing rod (metres)

Long rods are often fitted with test facilities so that their resistance can be measured during regular maintenance checks. Individual rods are also specified for line traps, current and voltage transformers (CVTs) and surge arresters. Care must be taken with CT and VT earthing to ensure that current loops do not cause protection maloperation.

Some examples of standard manufacturers earthing fittings and connections are shown in Fig. 8.4.

8.3.2.4 *Substation fence earthing*

Two substation fence earthing practices are generally used, namely:

- Extending the substation earth grid 0.5–1.5 m beyond the fence perimeter and bonding it to the grid at regular intervals.
- Placing the fence beyond the perimeter of the switchyard earthing grid and providing it with its own earth rod system which is independent of the main substation earth grid.

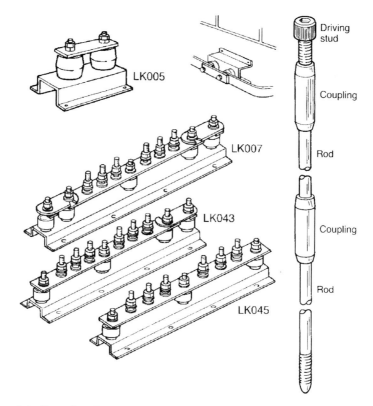

Figure 8.4 Examples of manufacturers' earthing fittings

Including the fence in the earth grid system can reduce the substation earth resistance and allows the earthing system design to be simplified, but requires an increased land area, puts electrode into publicly accessible areas, requires protection against theft or damage and can produce higher touch voltages external to the site than otherwise necessary.

Isolation of the fence from the main grid system is a safer option, but does involve an ongoing maintenance responsibility to ensure that isolation is maintained. Inadvertent connections could give rise to dangerous potentials under fault conditions.

Special fence earthing arrangements are necessary near single phase reactors or other substation plant generating high electromagnetic fields. It is necessary to electrically separate the fence into short, individually earthed sections to avoid large induced circulating currents. Experience shows that this must not be overlooked, to avoid cases such as substation gates being literally welded together due to induced circulating current associated with large static VAr compensation equipment.

Table 8.3   Typical substation earthing grid resistances (IEEE 80)

Parameter	Sub 1 Sand and gravel	Sub 2 Sandy loam	Sub 3 Sand and clay	Sub 4 Sand and gravel	Sub 5 Soil and clay
Resistivity (Ω m)	2000	800	200	1300	28
Grid area (m²)	1408.3	5661.4	1751.1	1464	5711.6
Buried length (m)	951	2895.6	541	1164.3	914.4
Calculated resistance (Ω)	25.7	4.97	2.55	16.15	0.19
Measured resistance (Ω)	39.0	4.10	3.65	18.2	0.21

8.3.3 Earth resistance and earth potential rise

A simplified formula for approximating the resistance, R ohms, of a sub-station earth grid of horizontal electrode is:

$$R = \sigma \left[\frac{1}{4r} + \frac{1}{L} \right]$$

This needs more terms to be added if rods are present, to account for their combined effect. An example from BS 7354 is:

$$R = \sigma \left[\frac{1 + (r/r + 2.5h)}{8rK_R} + \frac{1}{L} \right]$$

where σ = soil resistivity (ohm metres)
   r = equivalent circular plate radius (m)
   h = depth of buried grid (m)
   L = total length of buried conductors (m)
   K_R = constant concerned with the number of vertical earthing rods used in the overall substation earthing grid design, which is normally supplied as a table or graph in the relevant standard.

Equations from IEEE 80 and BS 7354 are normally introduced into mathematical packages or a spreadsheet to enable the resistance values to be calculated quite easily. It is important to be able to compare calculated and measured values to see how accurate the formula are in different situations.

Table 8.3 (from IEEE 80) compares calculated and measured earth grid resistances for five different substation sites and configurations, the calculations being based upon the simplified equation. Note that some other soil resistivity figures are provided in Table 8.2.

Because of the approximations introduced by the formulae and especially the fact that the soil more often needs to be represented as a three or more layer model, it is more common now for designers to use purpose-designed computer programmes to carry out the necessary calculations. These account

Table 8.4   Data needed and outputs from substation earthing calculations

---

Input parameters
Site length (m)
Site width (m)
Fault current (A)
Fault maximum duration (s) (typically 1 second)
Normal fault duration (s) (typically 0.2 second)
Soil resistivity (ohm metres)
Crushed rock resistivity (ohm metres) – (typically 3000 ohm metres)
Minimum depth of crushed rock (m)
Grid depth (m)
Grid spacing (m)
Required resistance to earth (ohms), if known

Calculated parameters
Conductor diameter (m)
Number of parallel conductors
Number of cross conductors
Reduction factor, C
Coefficient, Kim
Coefficient, Km
Coefficient, Kis
Coefficient, Ks
Theoretical conductor length (m)

Actual grid conductor length (m)
Grid corner conductors (m)
Total rod length (m)

Overall resistance to earth (ohms)

Tolerable step and touch voltage on crushed rock (volts)
Tolerable step and touch voltage on natural soil (volts)
Generated maximum step voltage (volts)
Generated maximum mesh voltage (volts)

Data used for budget and material ordering
Total length of tape or stranded conductor required
Total length of earth rods

---

for the actual electrode/grid shape and the soil structure and some can also account for the longitudinal impedance of the earth conductors at a range of different frequencies. Use of these programmes enables a closer match between calculated (predicted) and measured values than suggested in Table 8.3. Most designers will continually examine difference between these values and amend procedures to improve the outcome of their design studies.

The input data and examples of the output parameters provided when an earthing design study is undertaken using formulae or a computer programme are listed in Table 8.4.

For small substation sites, where access may not be available for the installation of an earth grid of sufficient size, a satisfactory earthing arrangement may often be achieved by installing copper tape in the ground around the periphery of the substation buildings. This may have additional earth rods connected and the building floor reinforcement may also be used to supplement

the design. When necessary to account for the effect of the steel reinforcements, a typical value for the resistivity of damp concrete is 90 ohm metres.
More often a sufficiently accurate result is obtained by assuming the same
resistivity as the surrounding soil.

The substation EPR is then the product of the total substation earth impedance, Z, and the amount of fault current that flows through it into the soil. Use
of reduction factors or separate calculations enables the current flowing into the
earth grid to be determined. Simplified equations from the standards should be
used with caution. The value of soil resistivity used in them must be selected
carefully and large errors can result if the soil has several layers of markedly different resistivity, because this will react in a completely different way to a uniform soil. The substation EPR, U_E, is then used to calculate the touch and step
potentials that are compared against the applicable limits.

8.3.4 Hazard voltage tolerable limits

An example of some tolerable touch and step potential limits, as derived from
IEC 60479-1 and IEEE 80 current/maximum disconnection time curves, are
shown in Fig. 8.5. From this figure the large difference between the limit values calculated using different standards is evident.

The touch and step voltage limits may be individually calculated for the
specific substation soil conditions if required, using formulae from standards.
For example, the IEEE 80 formulae are:

$$U_{t(tol)} = \frac{k_w + 0.17 \cdot \sigma_s}{\sqrt{t_E}}$$

$$U_{s(tol)} = \frac{k_w + 0.7 \cdot \sigma_s}{\sqrt{t_E}}$$

where $U_{t(tol)}$ = tolerable touch voltage (volts)
  $U_{s(tol)}$ = tolerable step voltage (volts)
  σ_s = resistivity of earth surface layer (say 3000 Ω m for crushed
    rock)
  t_E = maximum exposure time to be taken into account for touch
    and step voltages (say 0.5 seconds)
  k_w = body weight factor (116 and 157 for 50- and 70-kg body
    weight, respectively)

For a 50-kg person, 3000 Ω m crushed rock surface layer and 0.5 second maximum exposure time, values for allowable touch and step voltages using the
IEEE 80 formula are 885 and 3134 V, respectively. Note that this differs from
the value obtained using IEC 60479-1.

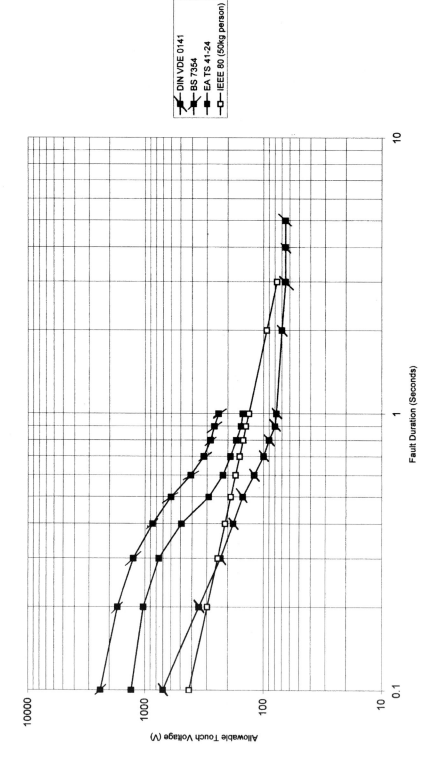

Figure 8.5   Comparison of allowable touch voltage curves (no crushed rock) (courtesy of Strategy and Solutions)

8.4 COMPUTER SIMULATION

Most major engineering project companies and designers now have in-house computer programmes to evaluate different substation earthing arrangements. A summary of some of the more important stages in the design process is shown in Fig. 8.6.

FIELD DATA
Substation and equipment layout.
Soil resistivity survey data (if required).
Details of pile and building reinforcing bars, if to be included in earth grid
Maximum earth fault current magnitude and ground return component.
Fault current durations for both conductor sizing & hazard voltage calculation.

CONDUCTOR
Conductor material.
Jointing methods
Conductor size.

HAZARD VOLTAGES TOLERABLE LIMIT VALUES
Standard Applicable (E.g. IEEE 80, BS 7430).
Calculation or look up values from tables.

INITIAL GRID DESIGN
Surround area available
Run cross conductors near structure connection points
Basic grid dimensions.
Number of parallel conductors in x and y directions.
Total length of grid conductors.
Supplementary earth rods, fencing, cable sheaths,
substation control building floor reinforcements, etc.
Calculate earth impedance, EPR and hazard voltages

MODIFY DESIGN
Change grid spacing or add electrodes.
Change grid conductor length.
Add supplementary earth rods.
Change area occupied by grid
Account for soil structure

CHECK VOLTAGES AGAINST TOLERABLE OR OTHER LIMIT VALUES
Transferred Potential
Touch voltage.
Step voltage.
Mesh voltage.
External voltage contours

DEFINITIVE DESIGN
Earthing layout drawing updated.
Grid and earth rod connection details.
Finalize calculation notes, QA procedures/checks.
Materials take-off.
Order materials.

Figure 8.6   Some calculation stages and data used for substation earth grid design

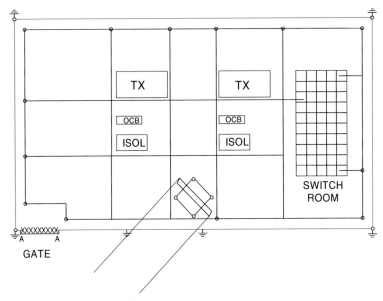

Figure 8.7   Basic electrode/grid design for a 132 kV/11 kV substation (courtesy of Strategy and Solutions)

A simplified layout for a 132/11 kV open terminal, two transformer sub-station is shown in Fig. 8.7. The substation has a separately earthed fence. The two incoming circuits are carried in on a steel tower line that has a terminal tower within the substation. Overhead connections are provided from this to switchgear and each transformer. The busbar supports, surge arresters and other equipment have been excluded for clarity. The earthing design is quite straightforward, with an electrode loop (or perimeter) about 2 m inside the fence, electrode rectangles surrounding the equipment and an electrode quite close to the plant items to enable short connections to them. The steel mesh rebar of the switchroom is assumed to be connected to the electrode system.

The procedure followed in modelling the electrode is generally:

1. Site soil survey measurements are first taken and analysed.
2. The resistance of the earth grid (grounding resistance of the primary elec-trode) is calculated.
3. The total impedance of the earthing installation is calculated. In this case there is a parallel earth contribution via the tower line and its overhead earth wire to include.
4. The fault current data is analysed and the amount flowing into the grid and tower line calculated. This is multiplied by the total earth impedance to provide the EPR.
5. Contour plots of the required potentials are produced by the computer soft-ware. For example, in Fig. 8.8 the touch voltages across the site are shown, whilst the external potential contours are shown in Fig. 8.9. The plots can be provided to show actual potentials, but in the figures they are shown as

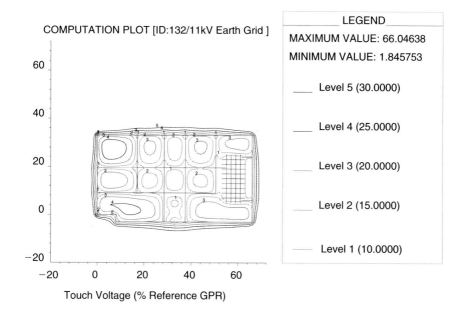

COMPUTATION PLOT [ID:132/11kV Earth Grid]

Touch Voltage (% Reference GPR)

LEGEND

MAXIMUM VALUE: 66.04638

MINIMUM VALUE: 1.845753

___ Level 5 (30.0000)

___ Level 4 (25.0000)

___ Level 3 (20.0000)

___ Level 2 (15.0000)

······· Level 1 (10.0000)

Figure 8.8   Touch voltages across a 132 kV/11 kV substation (courtesy of Strategy and Solutions)

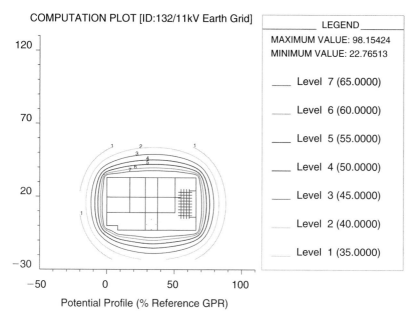

COMPUTATION PLOT [ID:132/11kV Earth Grid]

Potential Profile (% Reference GPR)

LEGEND

MAXIMUM VALUE: 98.15424

MINIMUM VALUE: 22.76513

___ Level 7 (65.0000)

___ Level 6 (60.0000)

___ Level 5 (55.0000)

___ Level 4 (50.0000)

___ Level 3 (45.0000)

___ Level 2 (40.0000)

___ Level 1 (35.0000)

Figure 8.9   Surface potentials external to a 132 kV/11 kV substation (courtesy of Strategy and Solutions)

percentages of the EPR. If this was calculated as 2000 V, then the 30% touch voltage contours in Fig. 8.8 correspond to a voltage of 600 V, whilst in Fig. 8.9 the 'level 2' contour is 40% of the EPR, or a value of 800 V.

6. Once the final studies have been completed, the total quantities of earthing materials necessary to achieve the design can be produced from the model and used for budgeting purposes.

The studies were carried out using the CDEGS computer package. Note that this package actually provides additional clarification of the output diagrams with the use of colour.

REFERENCES

1. IEC 60479-1, *The Effects of Current on human beings and livestock* 2004.
2. IEEE Standard 80, *Guide for Safety in AC Substation Grounding* 2000.
3. G. F. Tagg, *Resistances*, George Newnes Ltd, London, 1964.
4. C. R. Bayliss & H. Turner, *Electrical Review*, 18–33 May 1990, Shock voltage design criteria.
5. IEC 61936-1, *System Engineering and Erection of Power Installations in Systems with Nominal Voltages above 1 kV A.C. and 1.5 kV D.C, particularly concerning safety aspects.*
7. BS 7430: 1998 (formerly CP. 1013:1965),Code of Practice for Earthing.
8. IEEE Std 837-1989 (Revision of ANSI/IEEE Std 837-1984), *IEEE Standard for Qualifying Permanent Connections Used in Substation Grounding.*
9. ANSI/IEEE Std 81-1983 (Revision of IEEE Std 81-1962), *IEEE Guide for Measuring Earth Resistivity, Ground Impedance, and Earth Surface Potentials of a Ground System.*
10. Engineering Recommendation G12/3, National code of practice on the application of protective multiple earthing to low voltage networks. [Note: A new version (G12/4) is presently prepared, to include guidance on locating earth electrodes and ensuring compliance with the latest UK electricity supply, quality etc regulations.]

9 Insulation Co-ordination

9.1 INTRODUCTION

Insulation co-ordination is the technique used to ensure that the electrical strengths of the various items of plant making up the transmission and distribution system and their associated protective devices are correlated to match the system characteristics and expected range of voltages. The objective of the analysis and application of its conclusions are to reduce the probability of plant failure and supply interruptions caused by insulation breakdown to an operationally and economically acceptable level.

IEC 60071 covers the subject of insulation co-ordination as indicated in Table 9.1. The standard recognizes that insulation may occasionally fail since it is not economically feasible to eliminate failure completely. A proposed order of priorities for an insulation co-ordination policy is to:

- Ensure safety to public and operating personnel.
- Avoid permanent damage to plant.
- Minimize interruption of supplies to consumers.
- Minimize circuit interruption.

9.2 SYSTEM VOLTAGES

9.2.1 Power frequency voltage

It should be noted that insulation levels are dependent upon the highest system operating voltage and not the nominal voltage. IEC 60038 gives details of standard transmission and distribution voltage levels. Thus for a 132 kV system, the highest voltage is 145 kV. Plant may be subjected to the normal power frequency voltages which do not exceed the highest rated voltage for which the

Table 9.1   IEC 60071 insulation co-ordination

IEC 60071 insulation co-ordination
60071–1 Part 1 Terms, definitions, principles and rules Specifies the insulation for the various items of plant used in a given installation. Applies to plant for AC systems having a higher voltage for plant above 1 kV, and covers phase-to-earth insulation.
60071–2 Part 2 Application guide Provides guidance on the selection of the electric strength of plant, of surge arresters or protective spark gaps, and on the extent for which it will be useful to control switching overvoltages. Indicates the lines to be followed to obtain rational and economic solutions. Deals with phase-to-phase and phase-to earth insulation co-ordination, completing the principles and rules laid down in IEC 60071-1 Having specified the general principles, gives the standard insulation levels for the ranges I (1 kV to 245 kV) and II (above 245 kV).
TR 60071-4 Part 4 Computational guide to insulation co-ordination and modelling of electrical networks Gives guidance on conducting insulation co-ordination studies. Gives information in terms of methods, modelling and examples, allowing for the application of the approaches presented in IEC 60071-2, and for the selection of insulation levels of equipment or installations, as defined in IEC 60071-1
60071-5 Part 5 Procedures for high-voltage direct current (HVDC) converter stations

equipment has been designed. Obviously the insulation must be able to withstand these steady state power frequency voltages and plant must be specified accordingly. Breakdown does, however, occur due to pollution, heavy rain, etc. Chapter 6 describes how insulators should be specified to minimize this risk and how adequate insulator creepage distances may be determined to match the environmental conditions.

9.2.2 Overvoltages

9.2.2.1 *Internal overvoltages*

As well as steady state power frequency overvoltages it is also necessary to ensure that plant is able to withstand short duration power frequency overvoltages or other types of weakly damped oscillatory voltages with harmonic content which may last in the worst cases for tens of seconds. Such phenomena can occur during transformer saturation. Also, distribution systems with lightly loaded large cable networks involving high capacitance when fed from a source rich in harmonics can greatly magnify the voltage distortion during switching operations. In general, the principal causes of temporary power frequency overvoltages are:

- Phase-to-earth faults – on normal systems it may be assumed that the temporary overvoltages will not exceed:
  1.4 per unit for solidly earthed networks
  1.7 per unit for resistance earthed networks
  2.0 per unit for reactance earthed networks.

- Load rejection (supplying capacitive current through a large inductive reactance, e.g. a small generator connected to a long cable or overhead line).
- Ferro resonance (interchange of stored energy for series or parallel combinations of inductive and capacitive reactance).
- Ferranti effect (receiving end voltage greater than sending end voltage under no load or for lightly loaded lines).

Sustained overvoltages involving resonance and arcing ground faults are normally eliminated by careful system design and correct neutral earthing. At distribution voltage levels (below 145 kV) the method of earthing will normally determine the level of temporary overvoltage.

9.2.2.2 Switching surges

Switching surges are of short duration, of irregular or impulse form and highly damped. A typical switching impulse standard form is the 250/2500 microsecond, time-to-crest/time-to-half value wave. Overvoltages due to switching phenomena become important at the higher transmission voltage levels (above 245 kV). Chapter 13 describes the effect of various types of switching surges which are well understood. The magnitude of internally generated switching surges is related to the system operating voltage. On a system where the circuit breakers are not subject to multiple restriking the switching surges will rarely exceed 3 per unit and 2.5 per unit would be a typical maximum upon which the discharge duty for surge arresters may be assessed. On range II installations (above 245 kV) it may be necessary to suppress maximum switching surges to 2 per unit or less by the installation of shunt reactors and/or closing resistors on the circuit breakers.

At voltage levels below 245 kV some practical aspects of switching surges found on networks are listed below:

- Resonance effects when switching transformer feeders or combinations of cable and overhead line. Resonance can occur between the lumped reactive and capacitive elements and the overhead line. If the frequency of the reflections of the travelling waves along the line approximates to the natural frequency of the lumped elements, high voltages can be generated.
- Ferro resonance encountered on transformer feeder circuits greater than 5 to 10 km in length when one feeder/transformer on a double circuit is switched out but the parallel feeder remains energized. The dead circuit draws energy by capacitive coupling from the parallel live circuit which resonates with the transformer impedance at a subharmonic frequency. Operational procedures such as opening the line isolator at the transformer end on the disconnected circuit will eliminate the problem.

- In addition to the transformer feeder energization cases listed above, line energization can also create large switching surges particularly at the remote end of the line being energized. Such circumstances include:
  - Very long lines particularly if there is no shunt reactor compensation.
  - Lines already energized with a standing charge such as might occur from auto-reclose conditions.
  - Current chopping during shunt reactor, transformer and motor switching. Nowadays modern circuit breakers should be restrike free, or virtually so. This was a particular problem with early vacuum circuit breaker designs and air blast circuit breakers where the current may be broken before the natural cyclic current zero. Overvoltages due to these sudden interruptions may be of the order of 2.5 to 3 times the normal voltage. When a circuit breaker interrupts reactive current any magnetic energy in the reactor is exchanged with electrical energy according to the relationship:

$$\tfrac{1}{2}LI_c^2 = \tfrac{1}{2}CV^2$$

where L = inductance
        I_c = chopped current level
        C = shunt capacitance
        V = voltage created by current chopping

  - Existing reactor switching installations may have this phenomenon resolved by installation of suitable surge arresters.
  - The possibility of circuit breaker arc restriking when switching large capacitive currents. It is therefore very important to specify the correct capacitive current which the circuit breaker may have normally to switch and to match the circuit breaker manufacturer guarantees for restrike-free operations with the network application. Early low oil volume (LOV) circuit breaker designs were vulnerable to this phenomena when low surge impedances (cables or capacitor banks) were connected to both sets of switchgear terminals.

9.2.2.3 *External overvoltages/lightning surges*

On power systems operating at 145 kV and below overvoltages due to lightning will predominate rather than overvoltages generated by internal phenomena (fault conditions, resonance, etc.) or switching operations. Such overvoltages arise from lightning discharges which are usually of very short duration, unidirectional and of a form similar to the standard impulse wave shape 1.2/50 microsecond, front time/time-to-half value wave.

The point of insulation flashover in the system depends upon a number of independent variables:

- The geographical position of the stroke.
- The magnitude of the stroke.

- The rise time of the voltage wave.
- The system insulation levels.
- The system electrical characteristics.
- The local atmospheric or ambient conditions.

The damaging part of the lightning flash is the 'return stroke' where a charged cell in a thunder cloud is discharged to earth. The current in the return stroke varies from about 2 kA to 200 kA in accordance with a log-normal distribution:

$$1\% > 200\,\text{kA}$$
$$10\% > 80\,\text{kA}$$
$$50\% > 28\,\text{kA}$$
$$90\% > 8\,\text{kA}$$
$$99\% > 3\,\text{kA}$$

Impulse rise times are of the order of 10 microseconds for the more common negative flow from cloud to ground (and considerably longer for strikes from a positive part of the cloud) together with a relatively slow decay time of approximately 100 microseconds or less. For design purposes the most severe peak lightning current and rate of rise of 200 kA and 200 kA/microsecond may be considered.

The cloud potential is of the order of 100 MV and therefore high enough to ensure that the potential of the object struck is controlled by the current flow and impedance to ground. When a lightning strike takes place on an overhead line support structure the potentials along the current path will rise to very high values due to even the smallest inductive and resistive impedance to true earth. If the effective impedance to true earth is high enough to break down the insulation then a flashover will take place either from the earth wire or tower to the phase conductor(s), usually across the insulator strings. This type of lightning fault is known as a 'back flashover'. A reduction in lightning outages requires adequate overhead line shielding angles and low tower footing resistances of less than 10 to 20 ohms. An unearthed woodpole structure offers superior lightning performance and hence higher reliability through the reduced risk of back flashover because of the inherent insulating properties of wood.

The short duration of a lightning strike is usually insufficient to present temperature rise problems to the earthing and shielding conductors. A minimum cross-sectional area of 50 mm^2 is recommended in order to reduce surge impedance and temperature rise. In contrast, the conductivity of an arc path through air is high and with the large currents involved the air adjacent to the flash will experience a rapid temperature rise with a resulting explosive expansion. Large mechanical forces will also be present for parallel conductors or conductors with sharp bends.

The lightning flash density N_g is the number of flashes to ground per year per km^2 and maps are available with this or the number of thunderstorm days per year data. The relationship between such data is given in Table 9.2. The effective

Table 9.2 Relationship between thunderstorm days per year and lightning flashes per km² per year. ('Lightning parameters for engineering application,' *Electra*, 1980, 69, 65–102)

Thunderstorm days per year	Flashes per km² per year (mean)	Flashes per km² per year (limits)
5	0.2	0.1 to 0.5
10	0.5	0.15 to 1
20	1.1	0.3 to 3
30	1.9	0.6 to 5
40	2.8	0.8 to 8
50	3.7	1.2 to 10
60	4.7	1.8 to 12
80	6.9	3 to 17
100	9.2	4 to 20

collection area A_c is a function of a structure's dimensions. The probability, P, of the number of strikes to a structure per year is given by $P = A_c \cdot N_g \cdot 10^{-6}$ to which weighting factors based on experience are applied to cover different types of structure, construction, contents, degree of local isolation and profile of the surrounding country. For buildings risks less than 10^{-5} do not generally require lightning protection.

9.2.2.4 *Substation lightning shield protection*

Outdoor substations may be shielded by overhead earthwire screens strung across the substation site or by the use of shielding towers. The zone of protection provided by an earthed structure is the volume within which it gives protection against a direct and/or attracted lightning strike. British and German Standards differ as to the extent of the coverage offered (see Figs. 9.1a and 9.1b). The function of the overhead earthwire shield or shielding towers is to divert to itself a lightning discharge which might otherwise strike the phase conductors or substation plant. The use of shielding towers alone tends to require high structures in order to give adequate coverage. The shielding wire system allows lower height structures for a given coverage and the lightning current will be attenuated by increasing the number of paths to earth and thereby reducing the risk of back flashover. Often substation overhead line termination towers act as suitable support points for the shielding wire earth screen. Some electricity supply companies in areas with low lightning activity believe that the risk of an overhead earth wire screen falling onto the substation and causing a major outage is greater than an outage due to a lightning strike.

Electrogeometric lightning theory considers that the lightning arc stroke distance, r_{sc}, is a function of the lightning stroke leader current:

$$r_{sc} = 8.5 I_c^{2/3}$$

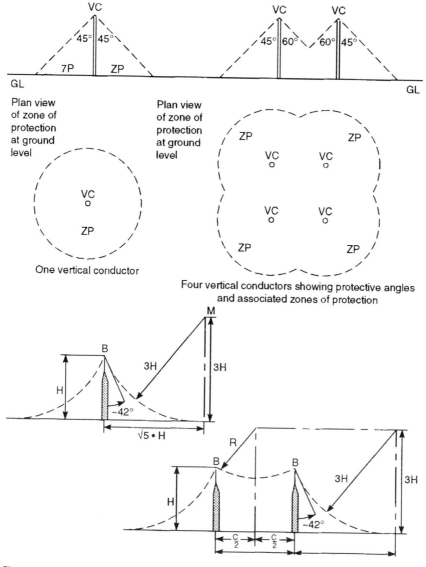

Figure 9.1a   Lightning protection using shielding towers. Top: Zones of protection from vertical conductor (VC) shielding towers according to British Standards. Bottom: German research association for high voltage and current technology (FGH) equivalent

where I_c is the critical stroke current which is the peak value of impulse current which will cause failure of the insulation. Then:

$$I_c = \frac{V_i}{0.5Z}$$

where V_i = impulse voltage withstand for the insulation
Z = surge impedance of the conductor

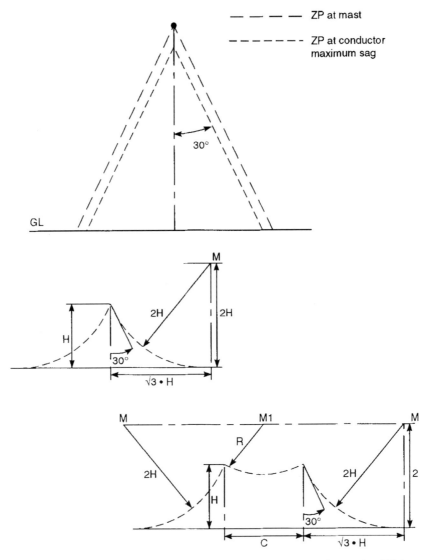

Figure 9.1b Lightning protection using aerial earth wires. Top: Zone of protection (ZP) from aerial earth wire according to British Standards. Bottom: German (FGH) equivalent

By knowing V_i and Z then I_c may be determined and hence the strike distance, r_{sc}. A series of arcs is drawn around the substation phase conductors with radius r_{sc} and around the earth wire screen with radius r_{se}. Similarly a line is drawn at a height r_{sg} parallel to the ground with:

$$r_{sc} = r_{se} \approx r_{sg}$$

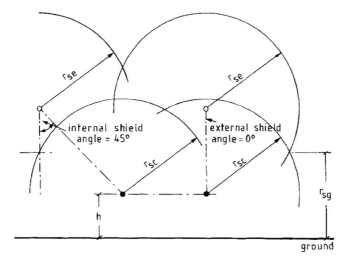

r_{sg}	strike distance to ground
r_{sc}	strike distance to phase conductor
r_{se}	strike distance to earth shield wire
h	maximum height of phase conductor
o	overhead earth shield wire
●	overhead phase conductor

Figure 9.2   Electrogeomagnetic model for lightning screen

If the lightning arc stroke distance cuts either the line above the earth or one of the earth wire radius arcs before it cuts an arc whose centre is the phase conductor perfect shielding will be obtained. Examples are shown in Fig. 9.2. In practice electricity supply companies and engineering consultants tend to adopt specific shield designs similar to those shown in Fig. 9.1.

9.2.2.5 Surges in transformers

The winding of a transformer can be represented as a distributed capacitance to steep fronted waves as shown in Fig. 9.3.

As the steep fronted surge U_p travels down the winding it can be shown that the voltage U at any point in the winding is given by:

$$ U = \frac{U_p \sinh \sqrt{\frac{Co}{C}} x}{\sinh \sqrt{\frac{Co}{C}} D} \qquad \text{(1) for an earthed winding} $$

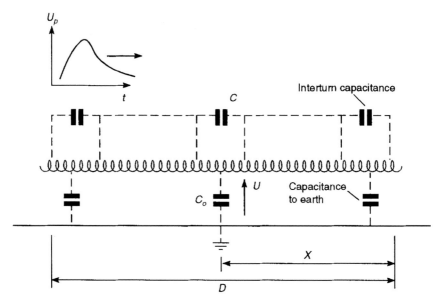

Figure 9.3   Representation of a transformer winding with distributed capacitance undergoing a voltage surge

$$U = \frac{U_p \cosh\sqrt{\dfrac{Co}{C}}x}{\cosh\sqrt{\dfrac{Co}{C}}D} \qquad \text{(2) for an open circuited winding}$$

where Co = capacitance to earth
           C = interturn capacitance

The presence of capacitance to earth causes a non-uniform distribution of voltage in the winding and the greater the value of $\sqrt{Co/C}$ $(=\alpha)$ the greater will be the concentration of voltage at the line end of the winding and the larger the interturn insulation stress on the first few turns of the transformer winding. Such a phenomenon has been responsible for many unprotected distribution transformer insulation failures.

After the surge has travelled down the winding the picture becomes complicated by multiple reflections and natural frequency oscillations in the winding (see Fig. 9.4). In high voltage transformer design the value of the interturn capacitance can be artificially increased by screening and by winding interconnections. These measures improve the transformer surge response and reduce the stressing of the line end turns.

Another factor to bear in mind is the near voltage doubling effect that occurs when a surge travelling down a line encounters the high surge

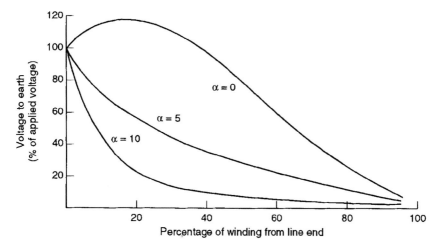

Figure 9.4   Peaks of natural frequency oscillations

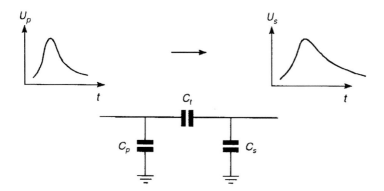

Figure 9.5   Representation of the appearance of a transformer to a steep fronted wave

impedance of a transformer. This effect can be virtually eliminated by the presence of a short length of cable of low surge impedance between the transformer and overhead line. However, because of improvements in transformer insulation and the high cost of such cable and fixings this practice is diminishing, certainly in the UK.

9.2.2.6 *Transferred surges*

Waves in one part of a circuit can be transferred to other circuits by inductive and capacitive coupling. As indicated above a transformer appears to a steep fronted wave as a distributed capacitance which can be crudely represented by a simple pi network (Fig. 9.5). From the figure, C_p and C_s are the lumped capacitances to earth and C_t is the lumped interturn capacitance so that the transferred wave U_s is given by:

$$U_s = U_p \cdot \frac{C_t}{C_t + C_s}$$

The values of these capacitances are not easily obtainable so IEC 60071-2 and the identical EN 60071-2 give various formulae for transferred waves. It is considered that the initial voltage on the secondary side of the transformer is given by:

$$U_s = spU_p$$

where s can range from 0 to 0.4 and is typically about 0.2 and p for a star/delta or delta/star is about 1.15 and for a star/star or delta/delta transformer is typically about 1.05.

Consider an 800 kV steep fronted lightning surge impinging upon the high voltage side of a 295/11 kV star/delta transformer. The transferred surge is:

$$U_s = 800 \times 0.2 \times 1.15 = 184\,\text{kV} \text{ which will appear on the 11 kV side.}$$

Alternatively if capacitance values are available:

$C_p = 0.0029\,\mu\text{F})$ (These are actual values measured by Hawker
$C_s = 0.0102\,\mu\text{F})$ Siddeley power transformers for transformers used
$C_t = 0.0032\,\mu\text{F})$ on a large, privately funded power station project in
                     the UK)

Transferred surge $U_s = 800 \times 0.0032/(0.0032 + 0.0102) = 191\,\text{kV}.$

In reality the transferred wave is complicated by multiple reflections inside the transformer and is attenuated by the transformer and any connected load. Nevertheless, the problem of wave transference should be recognized and low voltage equipment should be protected by surge arresters if such an event is likely to occur. The presence of external cables and loads further modifies the voltage wave appearing at the transformer terminals.

Slow surges such as switching surges that have rise times of the order of a few tens of microseconds or with an effective frequency of the order of 5–10 kHz will transfer through transformers electromagnetically. IEC 60071-2 gives an equation:

$$U_s = p \cdot q \cdot r \cdot \frac{U_p}{N}$$

where again p depends upon the winding configuration and is 1.15 for a star/delta transformer, q is a response factor for the lower voltage circuit with a value of 0.9 to 1.3, r is a correction factor and N is the transformer phase-to-phase voltage ratio.

Consider a star/delta 295/11 kV transformer with a 500 kV incident wave on the HV side:

$$U_s = 1.15 \times 0.9 \times 0.866 \times 500 \times \frac{11}{295} = 16.7\,\text{kV}$$

The magnitude of this surge would be modified, more or less, by whatever is connected to the 11 kV side of the transformer.

9.3 CLEARANCES

9.3.1 Air

Recommendations for insulation clearances are given in IEC 60071-2. For system nominal voltages up to 245 kV it implies use of the same insulation levels and electrical clearances for phase-to-phase as phase-to-earth cases although it warns against use of the lowest insulation levels without great caution and very careful study. Before publication of this standard, electricity supply companies developed their own policies regarding insulation levels and clearances. In the UK it was assumed that the phase-to-phase insulation should be able to withstand a full lightning impulse on one phase simultaneously with a peak power frequency voltage of opposite polarity on the adjacent phase. This policy has resulted in a satisfactory, reliable and possibly conservative design with phase-to-phase insulation levels 15% to 25% higher than the phase-to-earth level.

For the higher voltages, including 500 kV, when air clearances are determined by the level of switching surges, IEC 60071- 2 recommends withstand voltages between 1.5 and 1.8 per unit greater than the phase-to-earth level. The recommendations in the standard give a choice of two clearances depending on the conductor-to-conductor symmetrical or unsymmetrical configuration. Thus for a 525 kV system with a rated switching surge withstand level of 1175 kV, the IEC document recommends the adoption of a phase-to-phase switching surge withstand of 1800 kV, and clearances of either 4.2 metres or 5.0 metres depending on the gap configuration. It should be possible to avoid the use of unsymmetrical gaps between phases and therefore permit the use of the reduced clearances. In the UK, where the main transmission nominal voltage level is 400 kV, the reduced phase-to-phase clearance of 3.56 metres has been used without any reliability problems. The IEC document recommends for such a system clearances of either 3.6 metres or 4.2 metres. Table 9.2a shows the choice of impulse insulation strengths for systems operating at some typical rated voltages in accordance with the recommendations of IEC 60071.

Table 9.2a   IEC insulation levels for some standard system rated voltages

| System highest voltage (kV) | Range | Standard short-duration power frequency withstand voltage (kV rms) | Standard lightning impulse withstand voltage[a] (kV peak) | Standard switching impulse withstand voltage | | | Minimum clearance (mm) based on standard lightning impulse withstand[b] | | Minimum clearance phase-to-earth/phase-to-phase (mm) based on standard switching impulse withstand | |
				Longitudinal insulation[d] (kV peak)	Phase-to-earth (kV peak)	Phase-to-phase (kV peak)	Rod-structure	Conductor-structure	Rod-structure/ rod-conductor	Conductor-structure/ conductor-conductor[c]
12	I	28	60				90			
			75				120			
			95				160			
24	I	50	95				160			
			125				220			
			145				270			
36	I	70	145				270			
			170				320			
72.5	I	140	325				630			
145	I	230	550				1100			
		275	650				1300			
245	I	360	850				1700	1600		
		395	950				1900	1700		
		460	1050				2100	1900		
300	II		850	750	750	1125	1700	1600	1900/3100	1600/2600
			950		850	1275	1900	1700	2400/3400	1800/2900
			950				1900	1700		
			1050				2100	1900		

420	II	1050	850	850	1360	2100	1900	2400/3400	1800/2900
		1175	950	950	1425	2350	2200	2900/3600	2200/3100
		1175		1050	1575	2350	2200	3400/4200	2600/3600
		1300				2600	2400		
		1300				2850	2400		
		1425					2600		
525	II	1175	950	950	1615	2350	2200	2900/4300	2200/3700
		1300		1150	1840	2600	2400	3750/4650	2850/3950
		1300		1175	1763	2600	2400	4100/5000	3100/4200
		1425				2850	2600		
		1425				3100	2600		
		1550					2900		

Notes:
a The standard lightning impulse is applicable phase-to-phase and phase-to-earth

b For phase-to-earth clearances based on lightning impulse withstand, rod structure and conductor structure figures are applicable. For phase-to-phase clearances based on lightning impulse withstand, rod structure structure figures are applicable.

c The phase–phase conductor–conductor clearances based on switching impulse withstand voltage assume parallel conductors.

d Longitudinal overvoltages between the terminals during energization are composed of the continuous operating voltage at one terminal and the switching overvoltage at the other.

This table shows a selection of figures derived from IEC standards, but does not cover all options; see Tables 2 and 3 of IEC 60071-1 and Tables A21, A2 and A3 of IEC 60071-2.

9.3.2 SF_6

The use of SF_6 as an insulating medium requires special insulation co-ordination attention. Insulation failure in gas insulated switchgear (GIS) is not self-restoring and long repair times are likely to be involved. The withstand level of SF_6 for various impulse wave fronts and polarity varies significantly from air. As with air for very fast wave fronts the negative breakdown voltages are higher than the positive. For wave fronts slower than 1 microsecond the SF_6 positive voltage withstand level is greater than the negative. Also the SF_6 voltage withstand level does not reduce so markedly for the longer switching surge voltage wave fronts as does air insulation.

GIS disconnectors often have to break small magnitude capacitive charging currents. This can cause high frequency discharges across the contacts as the disconnector commences opening. The resulting overvoltages of 3 to 4 per unit must not be allowed to cause flashovers from the phase contacts to earth. Considerable design effort has been involved in reducing this problem since the mid-1970s. Surge arresters must be located very close to any open ended busbar if they are to be effective in attenuating such high frequency surges.

Where possible GIS switchgear should be transported to site in pre-assembled and pre-impulse tested sections. Guidance on selection on a test procedure to be adopted for a particular equipment is provided in IEC 60060.

9.4 PROCEDURES FOR CO-ORDINATION

9.4.1 The IEC standard approach

The detailed procedure for insulation co-ordination set out in IEC 60071-1 (European standard EN60071-1 is identical) consists of the selection of a set of standard withstand voltages which characterize the insulation of the equipment of the system. This set of withstands correspond to each of the different stresses to which the system may be subject:

- Continuous power frequency voltage (the highest voltage of the system for the life of the system).
- Slow-front overvoltage (a standard switching impulse).
- Fast-front overvoltage (a standard lightning impulse).
- Very-fast-front overvoltage (depends on the characteristics of the connected apparatus).
- Longitudinal overvoltage (a voltage between terminals combining a power frequency voltage at one end with a switching (or lightning) impulse at the other).

These voltages and overvoltages need to be determined in amplitude, shape and duration by system study. For each class of overvoltage, the analysis then determines a 'representative overvoltage', taking account of the characteristics of the insulation. The representative overvoltage may be characterized by one of:

- an assumed maximum,
- a set of peak values,
- a complete statistical distribution of peak values.

The next step is the determination of 'co-ordination withstand' voltages – the lowest values of the withstand voltages of the insulation in use which meet the system or equipment performance criteria when subjected to the 'representative overvoltages' under service conditions. Factors are then applied to compensate for:

- the differences in equipment assembly,
- the dispersion of the quality of the products within the system,
- the quality of installation,
- the ageing of installation during its lifetime,
- atmospheric conditions,
- contingency for other factors.

This results in so-called 'required withstand voltages' – test voltages that must be withstood in a standard withstand test. In specifying equipment the next step is then to specify a standard test withstand voltage (a set of specific test voltages is provided in IEC 60071-1) which is the next above the required withstand voltage, assuming the same shape of test voltage. A test conversion factor must be applied to the required withstand voltage if the test voltage is of a different shape to the class of overvoltage in question.

Figure 9.6 sets this procedure out in diagrammatic form, and full details of what is involved with each step is provided in IEC 60071.

9.4.2 Statistical approach

The statistical approach is especially valuable where there is an economic incentive for reducing insulation levels and where switching overvoltages are a problem. The method is therefore particularly applicable at the higher voltage range II installations above 245 kV.

The risk of insulation failure, R, may be expressed by the formula:

$$R = f_0(U) \cdot P_\mathsf{T}(U) \cdot \mathrm{d}U$$

where $f_0(U)$ = the overvoltage probability density

$P_\mathsf{T}(U)$ = the probability of insulation failure in service at voltage U

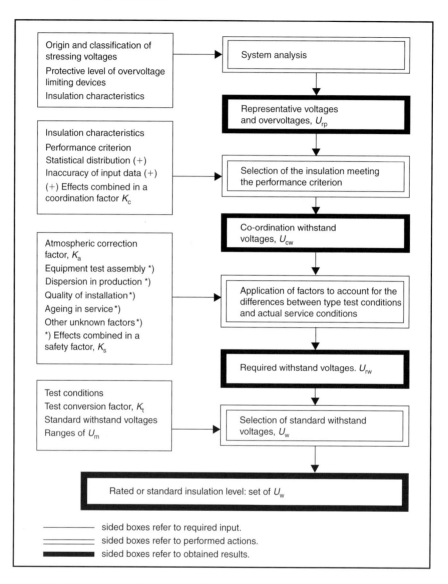

Figure 9.6   Flow chart for the determination of rated or standard insulation level (from IEC 60071-1)

Since it is difficult to determine $f_0(U)$ and $P_\tau(U)$ in practice IEC 60071 recommends a simplified method of assessment taking a 90% withstand level for a given insulation system equated with a 2% probability of an overvoltage being exceeded. From this different safety factors, γ, may be applied and the risk of failure determined. Modelling of the network on a computer may also be used to determine possible overvoltage conditions although obviously the accuracy of such simulations is only as good as the input data.

Laboratory tests on insulation will give an assessment of withstand capability. If the insulation is to be installed in outdoor conditions then the effects of rain and pollution on insulation strength must also be simulated. For a given state of the insulation there is a statistical spread in the breakdown voltage coupled with time effects and variations in environmental conditions. This may be expressed as:

$$\sigma_T = \sqrt{\sigma_t^2 + \sigma_n^2}$$

where σ_t = standard deviation at a given instance in time
σ_n = standard deviation due to environmental conditions

Some standards suggest that σ_t may be assumed to be equal to 0.06 for switching surges and 0.03 for lightning impulses. The 50% breakdown voltage U_{T50} is related to the required withstand voltage U_{RW} by the relationship $U_{T50} = k \cdot U_{RW}/(1 - 1.3\sigma_t)$. Constant k is dependent upon weather but typically may be made equal to 1 and σ_n is associated with pollution levels and may be made equal to 0.6.

9.4.3 Non-statistical approach

The conventional procedure is based on adopting an adequate margin between produced overvoltages and the withstand strength of the plant. The margin determines the safety factor, and is to some extent provided by the various factors mentioned in Section 9.4.1. However, it should never be less than the value found to be adequate from experience. This method is generally applied to the lower transmission (upper end of range I: 52 kV to 245 kV) installations because of the practical difficulty of determining $f_o(U)$ and $P_T(U)$ with any degree of accuracy. Computer simulations are recommended for range II system voltage levels above 245 kV.

Transient overvoltages are limited to a protective level established by the use of surge arresters and/or co-ordinating spark gaps. The insulation requirements of the various items of plant are selected to be above this protective level by a safe margin of 15% to 25%. Overhead lines are generally regarded as the main collectors of lightning surges on a system and transformers, cables and switchgear associated with overhead lines will require protection.

9.5 SURGE PROTECTION

9.5.1 Rod or spark gaps

Rod or spark gaps are easy and cheap to install. They are usually installed in parallel with insulators between the live equipment terminal and earth. The gap distance setting is arranged such that the sparkover occurs at overvoltages well below the breakdown insulation level of the plant the gaps are protecting.

Gaps have the following disadvantages:

- When they operate a short circuit fault is created which will cause protection to operate and isolate the circuit. However, the alternative of insulation failure of the plant being protected is much more serious.
- Sudden reduction in voltage during gap operation places high stress on transformer interturn insulation.
- The breakdown of plant insulation varies with the duration of the overvoltage. A gap has a relatively slow response to fast rise time overvoltage surges and performance is influenced by polarity and atmospheric conditions.
- Short distance gaps applicable at the lower distribution voltages are vulnerable to maloperation due to wind-borne debris, birds, etc.

Notwithstanding these disadvantages the rod gap is widely used for the protection of small distribution transformers and as a backup protection for transformers protected by surge arresters. In the UK the present National Grid Company practice is to use rod gaps in preference to surge arresters at all voltage levels. However, internationally, because of the disadvantages, surge arresters are used as the principal form of substation plant overvoltage protection. Air gaps are used across insulators on overhead lines up to several kilometres from substations in order to protect the substation plant from surges emanating from the overhead lines. The gap settings are reduced as the overhead line approaches the substation. Gaps may also be used as back-up protection to surge arresters at cable sealing ends and transformer bushings. The gaps are arranged so that the distance can be easily adjusted. The rods are angled such that the power arc is directed away from the associated insulator sheds in order to avoid possible damage during flashover.

Typical backup transformer spark gap settings are given in Table 9.3. Normally the rod gap characteristic should lie just above the surge arrester characteristic by, say, 20% so that the rod gap will protect the transformer or other plant against all but the steepest surges (rise times less than 1 or 2 μs), if the surge arrester fails. This philosophy also applies in the absence of surge arresters when the minimum gap setting for flashover should be at least 20% above the highest possible power frequency system voltage. For example, on a 132 kV system with a highest phase-to-earth voltage under transient fault conditions of $132 \times 110\% = 145$ kV the rod gaps should be set to operate at $145 \times 120\% = 174$ kV.

Under impulse conditions the breakdown characteristics of the equipment to be protected are normally not known and only a BIL figure will be available. In such cases the rod gaps may be set to give a flashover on impulse, with a 1.2/50 μs wave, of 80% of the BIL of the protected equipment with a 50% probability. Thus a 132 kV system designed to a BIL of 550 kV might be given a rod gap setting on surge impulse of 440 kV. The gap setting may be taken from graphs giving both positive and negative surge impulse and power frequency

Table 9.3   Typical spark gap settings

Transformer basic impulse Insulation level (BIL, kV peak)	Spark gap setting (mm)
75	2 × 32
170–200	2 × 95
325	400
550	650
650	775
850	1000
1050	1200
1300	1200–1410
1425	1500
1550	1575

Note: The voltage withstand/time characteristic of spark or rod gaps rises steeply below about 2 μs so that they may not protect transformers against very steep fronted surges.

gaps. In this particular case a minimum gap setting of 560 mm (22 inches) using ½ inch (1.27 cm) square rod gaps would be suitable (see Fig. 9.7).

At the higher transmission voltage levels rod gaps are not used because of the corona discharge effect and radio frequency interference associated with high electric fields around pointed objects. Loops are therefore used instead with a radius sufficient to reduce these effects.

9.5.2 Surge arresters

9.5.2.1 Zinc oxide types

Modern surge arresters (also and perhaps more accurately known as surge diverters) are of the gapless zinc oxide (ZnO) type. Under nominal system operating voltages the leakage current is of the order of a few milliamperes. When a surge reaches the arrester only that current necessary to limit the overvoltage needs to be conducted to earth. ZnO has a more non-linear resistance characteristic than the previously used silicon carbide (SiC) surge arrester material. It is therefore possible to eliminate the series of gaps between the individual ZnO blocks making up the arrester. A change in current by a factor of some 10^5 will result in a change of voltage across the ZnO arrester of only about 56% thus yielding a high but finite energy discharge capability. IEC 60099 details the standards applicable to both gapped SiC and ZnO non-spark-gapped surge arresters. Typical ZnO surge arrester characteristics are shown in Fig. 9.8. The devices have a particularly good response to fast rise time overvoltage impulses.

The construction of ZnO surge arresters is relatively simple. It is essential that good quality control is employed when manufacturing the non-linear resistor blocks since the characteristics are very dependent upon the temperature firing range. Good electrical contact must be maintained between the non-linear resistor blocks by well-proven clamping techniques. SiC arresters employing series spark gaps must ensure equal voltage division between the gaps under all operating and

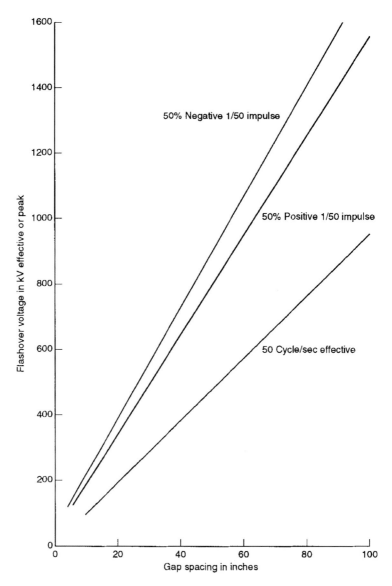

Figure 9.7   Flashover voltage of ½ inch square rod gaps

environmental conditions. The power frequency sparkover of such arresters should be greater than 1.5 times the rated arrester voltage. Figure 9.9 shows a selection of typical surge arresters together with the individual ZnO elements. Pressure relief diaphragms are fitted to the porcelain housings in order to prevent shattering of the units should the arrester fail.

Ratio of minimum power frequency withstand voltage/power frequency
rated voltage V's time curve for surge arrester type MB MC

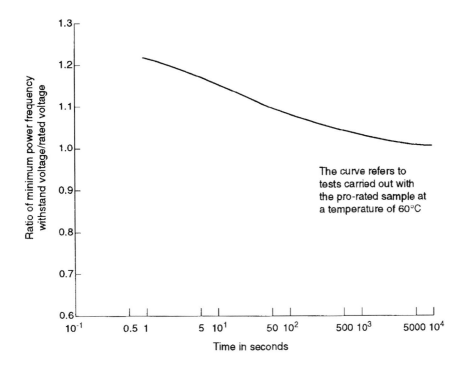

MB Series: Voltage ratings from 3 kV to 150 kV
IEC Arrester classification: 37 (co) 38
10 000 amp line discharge class 1 and 2
ANSI Arrester classification
10 000 amp station class

MC Series: Voltage ratings from 3 kV to 288 kV
IEC Arrester classification: 37 (co) 38
10 000 amp and 20 000 amp line discharge class 3
ANSI Arrester classification
10 000 amp station class

Figure 9.8   Typical ZnO surge arrester characteristics

9.5.2.2 Selection procedure

The principles for the application of surge arresters to allow a sufficient margin between the plant breakdown insulation level and surge arrester protection capability are shown in Fig. 9.10. Withstand voltages as a function of the operating

Figure 9.9   Individual ZnO elements

voltage within the two phase-to-phase insulation level ranges are shown in Fig. 9.11.

The application process is described below:

1. Determine the continuous operating arrester voltage – normally the system rated voltage.
2. Select a rated voltage for the arrester (IEC 60099).
3. Determine the nominal lightning discharge current. At distribution voltage levels below 36 kV when it is necessary to keep costs to a minimum 5 kA ratings are often specified. In most circumstances 10 kA surge arresters should be considered. For insulation > 420 kV 20k A rating may be appropriate.
4. Determine the required long duration discharge capability. At system-rated voltages of 36 kV and below light duty surge arresters may be specified unless the duty is particularly onerous (e.g. surge arresters connected adjacent to large capacitor banks). At rated voltage levels between 36 kV and 245 kV and where there is a risk of high switching, long duration fault currents (discharge of long lines or cable circuits) heavy duty surge arresters are normally specified. If any doubt exists the network parameters should be discussed with the surge arrester manufacturer. At rated voltages above 245 kV (IEC range II insulation level) long duration discharge capabilities may be important.
5. Determine the maximum prospective fault current and the protection tripping times at the location of the surge arresters and match with the surge arrester duty (including pressure relief class per IEC 60099).

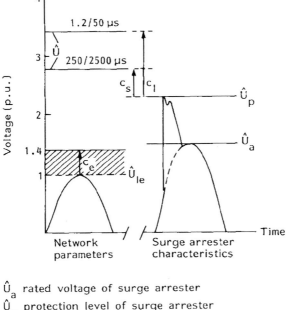

Figure 9.10   Plant breakdown insulation level and surge arrester protection capability

$\hat{U}_a$ rated voltage of surge arrester

$\hat{U}_p$ protection level of surge arrester

$\hat{U}_{le}$ rated voltage peak value
   conductor–earth $= \hat{U}_m \frac{\sqrt{2}}{\sqrt{3}} \; \hat{=} \; 1 \; p.u.$

c_e earthing factor $\hat{U}_a = c_e \cdot \hat{U}_{le}$

$\hat{U}$ withstand voltages

c_l, c_s safety margin: withstand voltage/protection level

6. Select the surge arrester housing porcelain creepage distance in accordance with the environmental conditions and state to the manufacturer if live line washing is electricity supply company practice.
7. Determine the surge arrester protective level and match with standard IEC 60099 recommendations. Typical protective levels are given in Table 9.4.

In order to assist specifying surge arresters details of typical technical particulars and guarantees are given in Table 9.5.

Rated voltage
The power frequency voltage across an arrester must never exceed its rated voltage otherwise the arrester may not reseal and may catastrophically fail after absorbing the energy of a surge. As a rule of thumb if the system is effectively earthed the maximum phase-to-earth voltage is 80% of the maximum

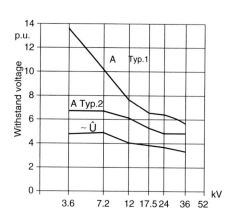

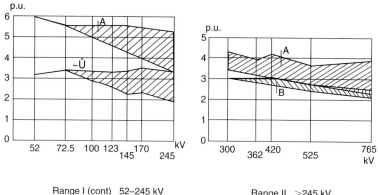

Range I (cont)   52–245 kV   Range II   >245 kV

A = 1.2/50 μs lightning surge voltage
B = 250/2500 μs switching surge voltage
$1\text{p.u.} = U_m\dfrac{\sqrt{2}}{\sqrt{3}}$
U_m = operating voltage
$\check{U}$ = withstand voltage

Figure 9.11   Withstand voltages as a function of operating voltage for insulation ranges I and II

Table 9.4   Surge arrester ratings and protective characteristics

Arrester rating, U_r, kV	Front of wave (kV)	Lightning/discharge voltage (kV)
12	46	40
36	145	125
138	400	350
240	645	550
426	1135	970

Note: The voltage withstand/time characteristic of spark or rod gaps rises steeply below about 2 μs so that they may not protect transformers against very steep fronted surges.

Table 9.5 Surge arrester technical particulars and guarantees

Characteristic	Requirement or manufacturer's guarantee
System highest voltage (kV)	
Insulation levels of protected systems:	
– transformers (kV)	
– switchgear (kV)	
Manufacturer	
Type No.	
Class of surge arrester (IEC 60099):	
– duty	
– long duration discharge class	
– pressure relief class	
Rated voltage (kV rms)	
Nominal discharge current (kA)	
Number of separate units per arrester	
Discharge residual voltage based on 8/20 wave at:	
(a) 5 kA (kV peak)	
(b) 10 kA (kV peak)	
(c) 20 kA (kV peak)	
Power frequency voltage capability for:	
(a) 1 sec (kV rms)	
(b) 3 sec (kV rms)	
(c) 10 sec (kV rms)	
(d) continuous (kV rms)	
Switching impulse residual voltage for .... wave shape (kV)	
Total height of arrester (mm)	
Total weight of arrester (kg)	
Minimum creepage distance per unit:	
– specified (mm)	
– guaranteed (mm)	
Porcelain housing cantilever strength (kN)	
Surge monitor required (yes/no)	
Surge monitor type reference	

line voltage. For a non-effectively earthed system the maximum phase-to-earth voltage is equal to the maximum line voltage (see Chapter 1). Consider a 132 kV system with a maximum line or phase-to-phase voltage 110% of the nominal system voltage 1. effectively earthed and 2. not effectively earthed

1. Arrester voltage rating $> 0.8 \times 132 \times 1.1 = 116$ kV and 120 kV arresters are usually selected.
2. Arrester voltage rating $> 132 \times 1.1 = 145$ kV.

Rated current
Arresters are tested with 8/20 μs discharge current waves of varying magnitude: 1.5 kA, 2.5 kA, 5 kA, 10 kA and 20 kA yielding increasing values of residual discharge voltage. Maximum residual discharge voltages are detailed in IEC 60099-1 and this parameter is usually taken care of in the manufacturer's design specification. For areas with high isokeraunic levels (e.g. the tropics) or at locations near to generators or for unshielded lines 10 kA arresters should be

specified. Lower-rated arresters can be selected for well-screened systems if it can be demonstrated that the surge discharge current is less than 10 kA. However, the cost of arresters is small compared to the overall system cost and therefore if some doubt exists regarding the discharge current it is safer to specify the higher-rated heavy duty type of arrester.

Although lightning strikes have impressive voltage and current values (typically hundreds to thousands of kV and 10–100 kA) the energy content of the discharge is relatively low and most of the damage to power plant is caused by the 'power follow-through current'. The lightning simply provides a suitable ionized discharge path. The likelihood of power follow-through current after a lightning discharge is statistical in nature and depends in a complicated way on the point on the wave of lightning discharge relative to the faulted phase voltage.

9.5.2.3 *Location*

Surge arrester and spark gap devices are installed in parallel with the plant to be protected between phase and earth. They should always be located as close as possible to the items of plant they are protecting consistent with maintenance requirements. This is to avoid back flashovers caused by any surge impedance between the surge arrester and the plant. The earth terminals should be connected directly and separately to earth as well as to the tank or frame of the plant being protected. Dedicated earth rods will provide the necessary low inductive path together with additional connections to the substation earth grid.

Note that generator windings have a low impulse strength, typically 50 kV for the 1.2/50 μs wave. Arresters for generators should therefore be heavy duty (10 kA station type) which may be shunted by 0.1–0.25 μF capacitors which absorb very fast surges with rise times less than 1 μs. Surge protection of generators becomes particularly important when they feed directly onto an overhead line without the benefit of an interposing generator transformer. Shunting capacitors may be essential in such applications.

Probably the best way to understand insulation co-ordination is by way of worked examples.

Insulation co-ordination Example 1
The example is for co-ordination for a typical 132 kV substation on an effectively earthed system having transformers of 550 kV impulse withstand level and the other apparatus having an impulse level of 650 kV. It is assumed that the altitude is below 1000 m and that the pollution level is not unduly heavy. The positive voltage/time breakdown curves for the various devices have been plotted as shown in Fig. 9.12 to demonstrate the co-ordination which can be obtained. Normally it is not necessary to plot curves in this way since a simple tabulation of figures is usually adequate. The curves could also include breakdown characteristics for substation post type insulators and overhead line cap and pin insulator strings with different numbers of units for completeness.

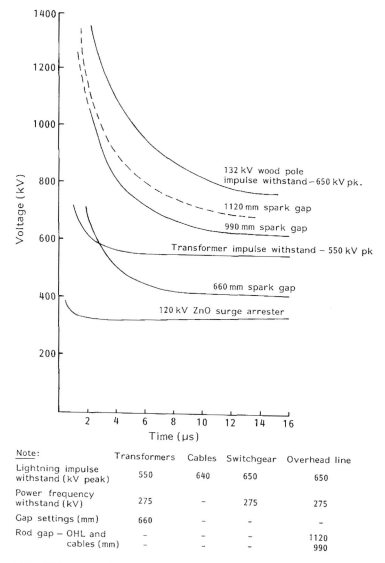

Figure 9.12   132 kV system insulation co-ordination

For the protection of the transformers and other equipment either a surge arrester or a rod gap system may be used. Since this is an effectively earthed system an '80% arrester' would be used; that is one rated at 120 kV (see Section 9.5.2.2 above). If the particular arrester chosen has a maximum residual impulse discharge voltage of 350 kV when discharging a 10 kA surge then using the 20% safety margin the capability is 350 × 120% = 420 kV. This is well below the transformer impulse withstand level of 550 kV, assuming the arrester is located within about 20 m of the transformer terminals.

GENERATOR

Nominal voltage	11 kV continuous
Maximum nominal voltage	12.1 kV continuous
Maximum transient 50 Hz overvoltage on loss of load	14.4 kV for a few seconds (131%)
Generator BIL	44.9 kV
Generator winding capacitance	0.46 μF
The generator is resistance earthed	Not effectively earthed

11 kV SYSTEM

Nominal voltage	11 kV
Maximum nominal voltage	12.1 kV (10%)
Maximum transient 50 Hz overvoltage	14.4 kV rms
System BIL	60 kV

GENERATOR TRANSFORMER

Ratio		295/11 kV
Winding		Star/delta (Ynd11)
Winding capacitances: Primary/earth	$Cp =$	0.0029 μF
Secondary/earth	$Cs =$	0.0102 μF
Primary/secondary	$Ct =$	0.0032 μF

275 kV GRID SYSTEM

Nominal voltage range	Max 302.5 kV (10%)
	Min 247.5 kV (−10%)
Temporary 50 Hz overvoltage	385 kV for a few seconds
Design BIL	1050 kV
Typical lighting overvoltage	850 kV
Typical switching overvoltage	550 kV
Surge protection policy	Gaps co-ordinated for probability at 835 kV
Chopped wave	1200 kV chopped on wave front rising at 100 kV/μs

Figure 9.13   System data

If a rod gap is to be used for protection, then from Fig. 9.7 a value of 560 mm (22 inches) could initially be thought of as suitable. This gap also gives protection to the transformer even for waves with rise times as short as 1 μs. However, on longer duration surges (possible switching surges) that are below the impulse strength of the transformer such a setting could give the occasional flashover. The gap setting could therefore be increased to 660 mm (26 inches) in order to reduce such a possibility. This larger setting does not, however, give adequate transformer protection against very fast rise time waves. A degree of judgement and experience is therefore required in order to determine the final rod gap setting. Such experience should also take into account whether or not the substation or incoming overhead lines have overhead earth wire screens.

Insulation co-ordination Example 2
This is an example of insulation co-ordination carried out for a large gas turbine power station in the UK feeding directly into the National Grid at 275 kV.

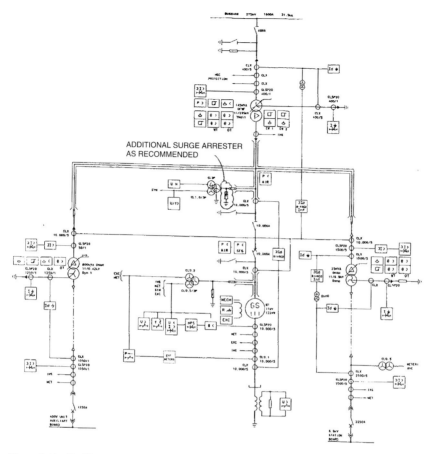

Figure 9.14   Barking Reach power station. (MECH = turbine generator mechanical protection; EXC = excitation control system; INS = instruction circuits; MET = metering circuits; SYUN = synchronizing scheme; AVC = Trans. tap change control)

The Grid is insulated to the highest (BIL) level whereas, for economic reasons, the power station equipment has been specified to lower levels and is protected by surge arresters. The system for one generator is shown in Fig. 9.14. This example stresses the need for engineers to question the reliability or meaning of system data presented to them. It also shows how a technical understanding of a subject can lead to innovative and cost effective design solutions.

HV arrester selection

The Grid is insulated to the highest BIL for 300 kV, namely 1050 kV. Below this level two other standard BIL ratings are possible in accordance with IEC 60071-1 at 950 kV and 850 kV. The 950 kV level has been specified in the transformer enquiry documents as a compromise between cost and closeness to the typical 275 kV nominal voltage Grid system overvoltage level. In accordance with IEC 60071 a surge arrester is required that will intercept surges 20% below the rated equipment BIL. That is, with a maximum 'impulse protective level' (IPL) voltage:

$$IPL \leq \frac{950}{1.2} = 792 \, kV$$

We now encounter a conflict. If we require the surge arrester to intercept the switching surge, then for the arrester:

$$IPL \leq \frac{550}{1.2} = 458 \, kV$$

However, the peak temporary 50 Hz overvoltage is:

$$385 \times 0.8 \times \sqrt{2} = 435 \, kV$$

These figures are too close. With an adequate transformer BIL of 950 kV the solution in this case could be to accept that the switching surges may not be intercepted by the HV transformer surge arresters. Such surges will be transferred through the transformer without damage if correctly intercepted by arresters on the 11 kV side. From manufacturers' catalogue data available at the time a 275 kV nominal system voltage 10 kA heavy duty (Bowthorpe 2 MC 240) arrester could be chosen with the following characteristics:

Rated voltage	240 kV rms
Maximum continuous operating voltage	180 kV rms
One second temporary overvoltage	288 kV rms
Maximum residual voltage (MRV @ 10 kA, 8/20 μs wave)	718 kV
Steep current residual voltage (1.2/50 μs wave)	790 kV

The maximum continuous 275 kV system voltage is $302.5/\sqrt{3} = 175 \, kV$ rms and we note that the maximum residual arrester voltage of 718 kV > 790/1.15 kV so we choose the former maximum residual voltage for the arrester impulse protective level.

The 50 Hz temporary overvoltage under earth fault conditions for an effectively earthed 275 kV nominal voltage system would be expected to be $302.5 \times 0.8 = 242 \, kV$. The Grid Company data, however, includes a figure for temporary overvoltage of 385 kV and under earth fault conditions this implies that the phase-to-earth voltage on the healthy phases could reach $385 \times 0.8 = 308 \, kV$. Should a higher rated arrester (such as a Bowthorpe 3 MC 264 unit) therefore be chosen?

Let us look at some other figures. According to IEC 60099-1 the BIL/IPL ⩾ 1.2. From manufacturer's data:

$$2 \, MC \, 240 \, arrester \, BIL/IPL = 950/718 = 1.32$$
$$3 \, MC \, 264 \, arrester \, BIL/IPL = 950/790 = 1.20$$

In addition, it should be noted that the 2 MC 240 arrester may give some protection against the switching surge whereas the 3 MC 264 unit would not. The
final selection must therefore be made on the basis of the foregoing figures
and engineering judgement. The possibility of the 275 kV Grid actually reaching the quoted 385 kV rms level for several seconds should be technically
researched with Grid company systems engineers. Consider, for example, if
the Grid voltage approached 385 kV (140%) all Grid transformers would be
driven hard into saturation. A huge reactive power demand would result making it difficult, if not impossible, to actually achieve such a voltage level sustained for the several seconds specified. Such a figure should therefore be
questioned as possibly incorrect data.

LV arrester selection
Transferred surges will pass through the transformer.
   For *fast transients* using the transformer capacitance figures and assessing
an equivalent pi circuit the surge

$$U_s = 790 \times \frac{0.0032}{0.0032 + 0.0102}$$
$$= 190\text{kV}$$

Alternatively using the IEC 60071-2 formula

$$U_s = s \times p \times U_p$$

where U_p = incident surge
            s = 0 to 0.4 (assume a value of 0.2)
            p = 1.15 for a star/delta transformer

$$U_s = 790 \times 0.2 \times 1.15 = 181\,\text{kV}$$

which is of the same order of magnitude. In addition, it should be noted that
these figures do not allow for a load or generator connected to the 11 kV side
of the transformer which would otherwise reduce the value of the surge.
   For *electromagnetically transferred surges* IEC 60071-2 gives the equation:

$$U_s = pqr\,\frac{U_p}{N}$$

For the 295/11 kV star/delta transformer

$p = 1.15$
$q = 0.9$
$r = 0.866$
$N = 295/11 = 26.8$
$U_p = 550\,\text{kV}$ (the switching surge)

Thus

$$U_s = 1.15 \times 0.9 \times 0.866 \times \frac{550}{26.8}$$
$$= 18.4\,\text{kV (again assuming no connected load)}$$

The 11 kV system is not effectively earthed and the maximum nominal voltage will be 12.1 kV. The temporary overvoltage could be 14.4 kV from the generator and $385 \times 11/295 = 14.35$ kV from the Grid (assuming the unrealistic case of no transformer saturation). The 11 kV system BIL is 60 kV. However, the generator BIL is only 44.9 kV and this figure will be used for protection purposes. From manufacturers' data a heavy duty (station-type) arrester rated at 12 kV is chosen (Bowthorpe type 1 MC 12). IEC 60099-1 recommends BIL/IPL, $\geqslant 1.2$. Therefore arrester IPL $\leqslant$ BIL/1.2 = 44.9/1.2 = 37.4 kV.

The 12 kV 1 MC 12 arrester has an MRV = 35.9 kV at 10 kA for a 8/20 μs wave and 39.5/1.15 = 34.3 kV at 10 kA for the steep fronted 1.2/50 μs wave.

Again in accordance with IEC 60099-1 BIL/IPL = 1.25 > 1.2 which is judged to be satisfactory for the generator since the rise time of the transferred surge would be much greater than 1.2 μs and is certainly satisfactory for the general 11 kV system. Such an arrester could therefore be applied to both the 11 kV terminals of the 295/11 kV transformer and to the generator terminals for further added protection.

The system diagram shown in Fig. 9.14 also shows an 11/6.9 kV Dzn0 station transformer connected to the 11 kV busbar. This transformer will also be subjected to electromagnetically transferred surges but the 11 kV incident surge will be limited to the residual voltage of the 11 kV surge arrester. Again, a check can be made as to the transferred surge value appearing on the 6.9 kV side of this transformer in a similar way to that indicated above:

$$U_s = pqrU_p/N$$

where $p = 1.15$
            $q = 0.9$
            $r = 1/\sqrt{3}$
            $N = 11/6.9 = 1.59$
            $U_p = 35.9\,\text{kV (the limited switching surge)}$

Thus

$$U_s = 1.15 \times 0.9 \times \frac{1}{\sqrt{3}} \times \frac{35.9}{1.59}$$
$$= 13.5\,\text{kV (again assuming no connected load)}$$

This is only twice the normal 50 Hz line voltage and therefore constitutes a temporary overvoltage lasting for a few milliseconds at most. Furthermore, the actual voltage appearing on the 6.9 kV side would be significantly reduced by cables and loads. Since the 6.9 kV system should have a BIL of 40 kV no additional surge protection is required.

9.5.2.4 *Monitoring*

Surge counters are often specified for plant rated voltages of 145 kV and above. In such cases the base of the surge arrester is supported on small insulators and the surge counter fitted at the earthy end of the surge arrester in the lead to earth. The counters should be located so that they may be easily read from ground level.

9.5.2.5 *Testing*

As for all substation or overhead line plant type test certificates should be obtained from the manufacturer. ZnO surge arrester type tests include:

- residual voltage test,
- current impulse withstand test,
- operating duty tests,
- power frequency voltage vs time curve,
- pressure relief tests,
- tests on arrester disconnectors (if applicable).

Routine tests include:

1. On all arrester sections:
   - radio interference tests
   - test to check sealing or gas leakage from completed housing
2. On SiC gapped arrester sections:
   - power frequency sparkover test
3. On a sample number of surge arresters to be supplied
   - lightning voltage impulse sparkover on the complete arrester (SiC types) or time voltage characteristic (ZnO types)
   - residual voltage at nominal discharge current on complete arrester or section
   - leakage current with 40% to 100% of rated voltage applied
4. On all gapless arrester sections
   - measurement of grading current when energized at maximum continuous operating voltage
   - measurement of power frequency voltage at a resistive current level to be determined between manufacturer and purchaser (1–10 mA peak)
   - residual voltage at a discharge current level to be determined between manufacturer and purchaser

REFERENCES

1. IEC 60038 IEC standard voltages
2. IEC 60060 High voltage test techniques
   Part 1 – General definitions and test requirements
   Part 2 – Measuring systems
   Part 3 – Definitions and requirements for on-site testing
3. IEC 60071 Insulation Co-ordination
   Part 1 – Terms, definitions principles and rules
   Part 2 – Application guide
   Part 4 – Computational Guide to insulation co-ordination and modelling of electrical networks
4. IEC 60099 Surge Arresters
   Part 1 – Non-linear resistor type gapped surge arresters for a.c. systems
   Part 3 – Artificial pollution testing of surge arresters
   Part 4 – Metal oxide surge arresters without gaps for a.c. systems
   Part 5 – Selection and application principles
   Part 6 – Surge arresters containing both series and parallel gapped structures rated 52 kV and less
5. IEC 61643 Low voltage surge protective devices
6. IEC 62272-203 – Gas insulated metal enclosed switchgear for rated voltages of 72.5 kV and above
7. L. J. H. White, D. H. A. Tufnell and G. G. Gosling, A Review of Insulation Co-ordination Practice on A.C. Systems, CEPSI 1980
8. D. H. A. Tufnell, Insulation Co-ordination: A Review of Present Practices and Problems, IEC Symposium, Indonesian Institute of Sciences, Jakarta, 1983
9. U. Berger, Insulation Co-ordination and Selection of Surge Arresters, Brown Boveri Review No. 4, April 1979, Volume 66

10 Relay Protection

10.1 INTRODUCTION

Switchgear, cables, transformers, overhead lines and other electrical equipment require protection devices in order to safeguard them during fault conditions. In addition, the rapid clearance of faults prevents touch and step potentials on equipment from reaching levels which could endanger life. The function of protection is not to prevent the fault itself but to take immediate action upon fault recognition. Protection devices detect, locate and initiate the removal of the faulted equipment from the power network in the minimum desirable time. It is necessary for all protection relays, except those directly associated with the fault clearance, to remain inoperative during transient phenomena which may arise during faults, switching surges or other disturbances to the network. Protection schemes are designed on the basis of:

- safety,
- reliability,
- selectivity.

The requirements for CTs and VTs associated with relay protection are described in Chapter 5 and fuse and MCB protection devices in Chapter 11. Standard reference texts are provided in the references section at the end of this chapter. They very adequately cover protection theory and particular relays in UK[1], US[2] and general[3] practice. Graphical symbols for switchgear, control gear and protective devices are given in IEC 60617-7. This chapter therefore concentrates on the principal relay protection schemes and typical applications with practical calculation and computer assisted examples.

   While the earliest relays were electromechanical in construction, technological developments led to the introduction of solid state or static relays using discrete devices such as transistors, resistors, capacitors, etc. Advent of microprocessors led to the development of microprocessor-based relays and this culminated with today's state of the art system of numerical relaying where the measurement principles themselves changed from analogue to numerical. Other recent advances are discussed in the following sections.

10.2 SYSTEM CONFIGURATIONS

10.2.1 Faults

All power system components are liable to faults involving anomalous current flow and insulation breakdown between conductors or between conductors and earth. The insulation material may vary from air, in the case of a transmission line, to oil, SF_6 or a vacuum, in the case of switchgear. The transmission and distribution engineer is concerned with symmetrical faults involving all three phases with or without earth, and asymmetrical faults involving phase-to-phase and one or two phase-to-earth faults. In addition, interturn winding faults also occur in transformers and electrical machines. Chapter 1 describes computer assisted methods of deriving fault levels in power system networks and Chapter 26 describes the basic fundamentals involved.

10.2.2 Unearthed systems

Such arrangements are only found in small isolated networks. At first sight the earth fault current would seem to be negligible with this connection. In practice, for all but the smallest networks the capacitive current becomes significant and dangerous transient overvoltages can occur due to low power factor arcing faults to earth. Unearthed systems therefore require high insulation levels and are limited to low voltage distribution where insulation costs are less significant. The main application is for very critical systems where continuity of supply is of paramount importance; two separate faults are required before an outage occurs and the first earth fault simply causes alarms which enables damage to be located and repaired before the critical supply is lost.

10.2.3 Impedance earthed systems

In this configuration a resistance or reactance is placed between the transformer neutral and earth. The earth fault current may be limited by the sizing of the impedance. This has the advantage of limiting:

- possible damage to equipment from the fault current;
- interference to control and communication circuits from the resulting induced currents.

High insulation still has to be incorporated in the impedance earthed system since voltage to earth levels on the unfaulted phases during a phase-to-earth fault will exceed 80% of the normal system phase-to-phase voltage. This is not normally a problem at system-rated voltage levels of 145 kV and below. The impedance earthed system is known as a 'non-effectively' earthed system. Normally a resistor rather than a reactor is used and the value is chosen still to ensure sufficient fault current to operate reliably the protection under all fault conditions. For a single supply point infeed the protection sensitivity may be set at, say, approximately 10% of the associated transformer full load rating and the earthing resistor to give a fault current equal to the transformer rating for a solid earth fault close to the transformer terminals.

10.2.4 Solidly earthed systems

Solidly earthed systems have the transformer neutral connected directly to earth. This has the advantage that it limits the likely overvoltages during fault conditions and is applied by most electricity supply companies for rated voltages above 145 kV. The voltage-to-earth levels on the unfaulted phases should not exceed 80% of the normal system phase voltage with the solidly earthed arrangement. The system is then known as 'effectively' earthed and is considered to be satisfied for ratios of $X_0/X_1 < 3$ and $R_0/X_1 < 1$ throughout the system under all conditions. In practice, these ratios will vary according to the network switching conditions and connected generation. The disadvantage is that the earth fault current can exceed the three phase fault current depending upon the ratio of zero-to-positive sequence impedance (see Fig. 10.1). Substation equipment must be rated accordingly. Sufficient current to operate the protection relay equipment is, however, not normally a problem. In addition, it should be noted that a high earth fault current will lead to high touch and step potentials during the fault conditions. This must be limited to safe levels by adequate substation earthing. Further, control and communication circuits must be protected against induced currents and possible interference resulting from the earth fault.

10.2.5 Network arrangements

10.2.5.1 *Radial*

A simple radial feeder is shown in Fig. 10.2. The fault level is highest closest to the source and limited by the impedances from source to fault location. Clearance of a fault near the source will result in loss of supply to downstream loads. Protection selectivity must be such that a fault on busbar A must be isolated by only tripping the circuit breaker X via relay R_1 and maintaining supply to load busbar B.

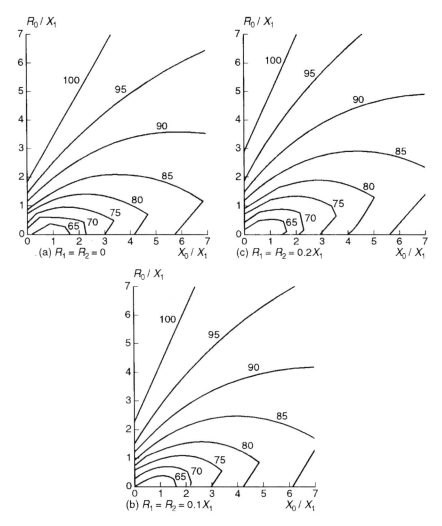

Figure 10.1   Maximum line-to-earth voltage at the fault for earthed neutral systems under any fault conditions

10.2.5.2 *Parallel*

A parallel feeder arrangement is shown in Fig. 10.3. A fault on one parallel feeder should be cleared by suitable protection such that it is quickly isolated from the supply. There should be no loss of supply via the remaining healthy feeder to the load.

10.2.5.3 *Ring*

A ring feeder arrangement is shown in Fig. 10.4. Two routes exist for the power inflow to a faulted feeder in a closed ring system. It is therefore

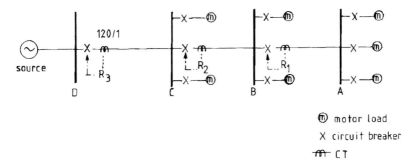

Figure 10.2   Typical radial distribution system

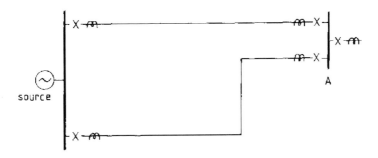

Figure 10.3   Typical parallel feeder arrangement

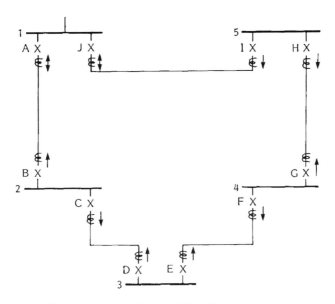

Figure 10.4   Typical ring system showing use of directional relays
Notes: Arrows represent current flow direction upon which relays will act. A, B, C, etc. are
       circuit breakers operated by associated relay. 1, 2, 3, etc.: busbar identification

necessary for the protection devices only to isolate the faulted section and not disconnect the whole system from the source. Often such ring systems use directional relay protection which requires both VT and CT connections. Alternatively, they may be operated with a mid-point feeder circuit breaker open in order to simplify the protection arrangements.

10.2.5.4 *Interconnected*

This is a more complex arrangement of interconnected parallel and radial feeders, often with multiple power source infeeds. More sophisticated protection schemes are necessary in order selectively to disconnect only the faulted part of the system.

10.2.5.5 *Substations*

Busbars, transformers, cables and other important plants are all involved in the different substation layouts described in Chapter 3. The switchgear arrangements will help to dictate the types of relay protection devices used throughout the particular substation.

10.3 POWER SYSTEM PROTECTION PRINCIPLES

10.3.1 Discrimination by time

For simple radial circuits discrimination is achieved by giving the minimum tripping time setting to the relay furthest away from the power source. A small time delay is then added to each relay in turn, moving nearer to the source each time. This ensures that the relay closest to the fault trips first, and as a result leaves the rest of the system between the source and the faulty section in service.

It is necessary to allow a minimum grading interval or delay between successive relay settings in order to take account of:

1. Circuit breaker tripping times – typically from 150 msec for an older oil circuit breaker to 50 msec for the latest vacuum or SF_6 switchgear.
2. Relay time delay errors – variation from the characteristic time delay curve for the relay as allowed by the appropriate specification standard, say, 150 msec.
3. Relay reset time – the relay must definitely fully reset when the current is 70% of pick-up value. Electromechanical relays reset at 90–95% of setting and a figure of 85% is taken for calculation purposes. Solid state relays have an even better improved characteristic in this regard.
4. Relay overshoot – an electromechanical relay must stop all forward movement or overshoot of the induction disc within 100 msec of the removal of

current. Again solid state or numerical relays have an advantage over electro-mechanical types in this regard.

If all these items are additive then for discrimination to be achieved typical time grading intervals of 0.4–0.5 s are used for electromechanical relays with oil circuit breakers and 0.25 s for modern solid state or numerical relays which are tripping vacuum or SF_6 switchgear. The effect of current transformer errors on relay operating times is not expected to be additive and such errors (say ±5%) are normally neglected when establishing a discrimination margin.

The disadvantage of discrimination by time delay alone is that the longest tripping times are those nearest the source. The fault current will be the highest and most damaging at this point in the circuit. Therefore shorter tripping times near the source would be an advantage.

10.3.2 Discrimination by current magnitude

The impedance of the power circuit between source and fault limits the fault current flowing at any point. Therefore by suitably selecting the current setting at which the particular relay operates discrimination can be achieved. In practice this is quite difficult for transmission and distribution feeder circuits because various interconnection arrangements significantly alter the fault level at any point in the network. The method works well for power transformer protection where instantaneous high set overcurrent relays can be used to protect the HV windings. Similarly, but for different reasons, instantaneous earth fault relays can be applied to the delta winding of a delta star (Dy) power transformer. Here the zero sequence currents generated in the secondary star winding during earth faults in that winding or system do not appear in the primary delta winding. In this case an instantaneous earth fault relay on the transformer primary delta will not respond to LV earth faults.

10.3.3 Discrimination by time and fault direction

It is possible to add directional sensing elements to the relay protection system such that the relay responds to both the magnitude and one particular direction of the current flow. Typical applications are for closed ring feeder systems, parallel feeders and parallel transformers. It is vitally important during commissioning of such protection schemes that the polarity of operation is properly checked or maloperation and lack of discrimination may result.

10.3.4 Unit protection

In these schemes the CTs located at either end of a feeder, transformer or 'unit' of plant to be protected (the protected zone) are interconnected. A comparison of magnitude and phase angle of the current entering the protected zone with that leaving is made. Two requirements are checked:

1. If the currents entering and leaving the protected zone are equal, operation of the protection must be prevented – this is known as the through fault stability requirement.
2. If the currents entering and leaving the protected zone are unequal the protection must operate – this is known as the sensitivity to internal faults requirement.

A number of unit protection schemes rely on a current balance principle. Obviously for power transformer differential unit protection the primary and secondary CT connections have to be arranged to take into account the different phase relationships associated with different power transformer connections (Dy1, Dy11, etc.) as explained in more detail in Section 10.5.1. Because the correct current balance performance relies upon CT characteristics it is essential that the associated CTs are matched and dimensioned correctly. Stability under fault conditions outside the zone of protection is vital. Therefore it is necessary to ensure that the spill current is minimized by limiting the degree of CT saturation or to use high impedance relays which are designed to remain stable even under saturated CT conditions. Conventionally, this is achieved by ensuring that the voltage which appears across the relay circuit with one CT fully saturated is insufficient to operate the relay at a given current rating. Modern numerical relays are also available in a low or medium impedance version with differing operating principles to combat problems related to CT saturation, remanance, ratio differences, etc.

The advantage of unit protection is that it provides a very fast (typically 200 msec or less) disconnection only of the plant being protected. The disadvantage is that the interconnection between the relays requires communication systems which make the overall schemes more expensive than simple time graded schemes for long feeder lengths.

10.3.5 Signalling channel assistance

Rapid protection operation may be necessary for system stability reasons as explained in Section 1.3, Chapter 1. The speed of response of a protection system may be enhanced by the use of interconnecting signalling channels between relays. For example, this enhancement can be applied to a distance protection scheme for improving the fault clearance time over the last 15–20% of the feeder length, as explained in more detail in Section 10.6. Such signalling channels may be by the use of hard wire circuits (dedicated pilot wires, rented telephone cables, etc.) using on/off or low frequency signals. Alternatively, signal information superimposed upon carrier frequencies of several hundred kHz may be used over the power circuits (power line carrier (PLC)) to convey the information. A more modern development is to use fibre optic cables which may, for example, form an integral part of an overhead line earth wire. The transmission times are essentially instantaneous but delays associated with interposing relays and electronics must all be considered when checking for correct selective grading.

10.4 CURRENT RELAYS

10.4.1 Introduction

The following types of current relays are considered in this section:

- Plain overcurrent and/or earth fault relays with inverse definite minimum time lag (IDMTL) or definite time delay (DT) characteristics.
- Overcurrent and/or earth fault relays as above but including directional elements. Note that a directional overcurrent relay requires a voltage connection and is not therefore operated by current alone.
- Instantaneous overcurrent and/or earth fault relays. For example, a high set overcurrent (HSOC) relay.
- Sensitive earth fault (SEF) relays.

Differential and unit protection schemes that require connections from more than one set of current transformers are described in Section 10.5 of this chapter. Current operated relays are applied almost universally as the 'main' or only protection on power distribution systems up to 36 kV. For very important feeders and at transmission voltage levels above about 36 kV current operated relays tend to be used as 'back-up' protection to more sophisticated and faster acting relay systems. Current relays may also form part of special schemes such as circuit breaker failure protection.

10.4.2 Inverse definite minimum time lag (IDMTL) relays

Historically this type of relay characteristic has been produced using electromagnetic relays, and many such units still exist in power systems. A metal disc is pivoted so as to be free to rotate between the poles of two electromagnets each energized by the current being monitored. The torque produced by the interaction of fluxes and eddy currents induced in the disc is a function of the current. The disc speed is proportional to the torque. As operating time is inversely proportional to speed, operating time is inversely proportional to a function of current. The disc is free to rotate against the restraining or resetting torque of a control spring. Contacts are attached to the disc spindle and under preset current levels operate to trip, via the appropriate circuitry, the required circuit breaker. The theoretical characteristic as defined in IEC 60255-3 is based on the formula:

$$t = \frac{K}{\left(\dfrac{G}{G_b}\right)^a - 1}$$

where
t = theoretical operating time
G = value of applied current
G_b = basic value of current setting
K and a = constants

With $K = 0.14$ and $a = 0.02$ the 'normal' inverse curve is obtained as shown in Fig. 10.5. This characteristic is held in the memory of modern microprocessor controlled solid state relays. Electronic comparator circuits are used to measure the source current and initiate tripping depending upon the relay settings. In comparison with grading by time settings alone the IDMTL relay characteristic is such that it still allows grading to be achieved with reduced operating times for relays located close to the power source.

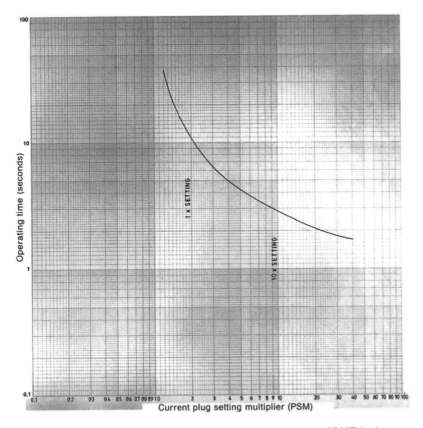

Figure 10.5   Normal characteristic inverse definite minimum time lag (IDMTL) relay curve

This type of relay has two possible adjustments:

1. The current setting by means of tap 'plugs' on electromagnetic relays or 'DIP' switches on solid state relays for values between 50% and 200% in 25% steps (the plug setting multiplier or PSM) for overcurrent relays and between 10% and 40% or 20–80% in 10% or 20% steps for earth fault relays. The 100% PSM corresponds to the normal current rating of the relay which may be 5, 1 or 0.5 amps to suit the CTs employed. Thus on a 100% tap a 5 A relay is stable under power circuit full load conditions with up to 5 A flowing in the CT secondary and relay input circuit. From Fig. 10.5 it can be seen that the relay will operate in approximately 30 s for overloads in the primary circuit of 1.3 × full load or with 5 × 1.3 = 6.5 A in the relay input circuit.
2. The operating time at a given current PSM. This is achieved by a continuously adjustable time multiplier torsion head wheel on an electromagnetic relay and potentiometer or DIP switches on solid state relays. The time setting may be varied between 0.05 and 1.0 s (the time multiplier setting or TMS).

Consider an overcurrent relay set at 175% working with a 300/5 ratio CT, the equivalent primary current sensitivity is, then:

$$300 \times \frac{175}{100} = 525A$$

Suppose the power circuit fault current for which the relay is required to trip is 5000 A then the overall PSM is given by:

$$PSM = \frac{\text{fault current}}{\text{maximum circuit primary current}}$$

$$= \frac{5000}{525}$$

$$= 9.5$$

From the normal characteristic curve given in Fig. 10.5 with a TMS = 1 the relay operating time under the 5000 A fault condition is 3.2 s. With a TMS = 0.5 the operating time would be half this at 1.6 s.

The 'normal' curve gives:

1. 10.03 (say 10) second theoretical relay operating time at a current equal to 2 × PSM.
2. 4.28 s theoretical relay operating time at a current equal to 5 × PSM.
3. 2.97 (say 3) second theoretical operating time at a current equal to 10 × PSM.
4. 2.27 s theoretical operating time at a current equal to 20 × PSM.

The actual pick-up level is best obtained on site by secondary current injection. Operation of IDMTL relays at currents greater than 20 × PSM is not covered by the standards and ideally the protection engineer tries to use CT ratios and relay

Table 10.1   Typical IDMTL electromagnetic relay operating times at high current levels (normal characteristic)

Operating time (s)	Operating current multiple of PSM
1.99	25
1.84	30
1.72	35
1.62	40
1.47	50
1.36	60
1.28	70
1.21	80
1.15	90
1.1	100
1.06	115

settings which avoid operation in this region. This is because the capability and characteristic of the CT used to drive the relay under heavy fault conditions may be far from linear. In addition, at the larger values of current the thermal rating of the relay must be considered. Some solid state relays operate to the normal IEC 60255-3 characteristic up to $20 \times$ PSM and then follow a definite-time characteristic above this current level. Some typical, but not guaranteed, values of operating time at TMS $= 1$ above $20 \times$ PSM are given in Table 10.1.

10.4.3 Alternative characteristic curves

Alternatives to the 'normal' IDMTL characteristic are available. A 'very' long time inverse curve is obtained with the constants $K = 13.5$ and $a = 1.0$ in the theoretical operating time equation. This very inverse characteristic is useful as a last stage of back-up earth fault protection (for example when used in conjunction with a CT associated with a transformer neutral earthing resistor). The 'extremely' inverse curve characteristic ($K = 80$, $a = 2.0$) is useful for ensuring the fastest possible operation whilst still discriminating with a fuse. The extremely inverse characteristic does not exhibit such a useful definite minimum time and it is difficult to accommodate more than one or two such relay stages in an overall graded protection scheme. A variety of commonly used characteristic curves are illustrated in Fig. 10.6.

10.4.4 Plotting relay curves on log/log graph paper

The characteristic curves shown in Figs 10.5 and 10.6 are plotted on log/log graph paper with time on the vertical scale and current on the horizontal scale. Three or 4 cycle log/log paper is the most useful in practice for manual relay grading exercises. If a template for the normal characteristic is used based on the same cycle log paper then the 10 or 3 s operating time at 2 or $10 \times$ PSM may be used as guide points. The actual circuit current being monitored is transformed

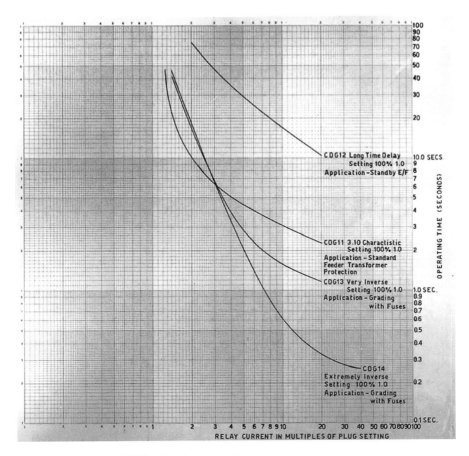

Figure 10.6   Typical IDMTL relay characteristics

by the actual CT ratio and relay PSM being used. The operating curve for other than TMS = 1 relay settings may be approximately drawn by moving the template vertically up the log paper so that the 10 × PSM mark coincides with an operating time in seconds equivalent to 3 multiplied by the actual TMS in use.

Relay characteristics are now available on disc for use with microcomputer relay grading programs. An example of such a computer assisted relay grading exercise is given in Section 10.8.

10.4.5 Current relay application examples

10.4.5.1 *IDMTL main protection*

(a) Radial network
Consider the industrial radial feeder system shown in Fig. 10.2. All motor loads are considered identical and have a normal full load current (FLC) of 20 A

and a starting current of 6 × FLC. Only one motor may be started at a time and the starting run up time is 5 s. The motors are protected by thermal relays which follow the characteristics of the motor windings and they incorporate high set overcurrent (HSOC) elements set to 160 A.

All CT ratios – 120/1 A
Maximum 3 phase fault level at busbar A – 1000 A
Maximum 3 phase fault level at busbar B – 1400 A
Maximum 3 phase fault level at busbar C – 1800 A
Full load current to be carried by relay R_1 due to the loads at busbar A = 3 × 20 = 60 A
Maximum current to be carried by relay R_1 under one motor starting and two running conditions for which no tripping must occur = 6 × 20 + 2 × 20
$$= 160 \text{ A}$$
Full load current to be carried by relay R_2 due to loads at busbars A and B = 60 + 2 × 20 = 100 A
Maximum current to be carried by relay R_2 under one motor starting and four running conditions for which no tripping must occur = 6 × 20 + 4 × 20
$$= 200 \text{ A}$$
Full load current to be carried by relay R_3 due to loads at busbars A, B and C = 60 + 40 + 40 = 140 A
Maximum current to be carried by relay R_3 under one motor starting and six running conditions for which no tripping must occur = 6 × 20 + 6 × 20
$$= 240 \text{ A}$$

Consider relay R_1

$$\text{Current setting} = \frac{60}{0.85 \times 120} = 0.59 \text{ (motor running condition)}$$

$$\text{or} = \frac{160}{1.15 \times 120} = 1.16 \text{ (motor starting condition)}$$

The factor 0.85 is used with electromechanical relays to give assurance that the fault is cleared by the downstream relay taking resetting factors into account. For solid state relays a factor of 0.9 would be applicable.

The factor 1.15 is a conservative relay operating figure to ensure relay pick-up with motor starting conditions.

Therefore set relay R_1 to PSM = 125% to ensure no anomalous tripping under motor starting conditions.

At a fault level of 1000 A at busbar A the relay high set elements will operate in approximately 0.1 s. The IDMTL relay must discriminate and operate in 0.5 s.

Relay R_1 nominal setting with PSM of 125% = 120 A × 1.25 = 150 A

$$\text{Multiple of setting} = \frac{\text{fault current}}{\text{maximum circuit primary current}} = \frac{1000}{150} = 6.7$$

From the 'normal' IDMTL curve at multiple of 6.7 the theoretical relay tripping time = 3.6 s at TMS = 1.

For tripping time of 0.5 s set the time multiplier setting to 0.5/3.6 = 0.139, set to 0.14.

At a fault level of 1400 A, multiple of setting = 1400/150 = 9.3.

From the 'normal' IDMTL curve at multiple of 9.3 the theoretical relay tripping time = 3.1 s at TMS = 1.

With TMS = 0.14 the theoretical relay tripping time = 0.14 × 3.1 = 0.43 s.

Consider relay R_2

$$\text{Current setting} = \frac{100}{0.85 \times 120} = 0.98 \text{ (motor running condition)}$$

$$\text{or} = \frac{200}{1.15 \times 120} = 1.45 \text{ (motor starting condition)}$$

Therefore set relay R_2 to PSM = 150% to ensure no anomalous tripping under motor starting conditions.

At a fault level of 1400 A at busbar B and assuming an adequate discriminating time delay between the two relays R_1 and R_2 of 0.4 s then relay R_2 must operate in 0.4 + 0.43 = 0.83 s.

Relay R_2 nominal setting with PSM = 150% = 120 × 1.5 = 180 A.

$$\text{Multiple of setting} = \frac{\text{fault current}}{\text{maximum circuit primary current}} = \frac{1400}{180} = 7.8$$

From the 'normal' IDMTL curve at multiple of 7.8 × the theoretical relay tripping time = 3.4 s at TMS = 1

For tripping time of 0.83 s set the Time Multiplier Setting to 0.83/3.4 = 0.244, set to 0.25.

At a fault level of 1800A, Multiple of setting = 1800/180 = 10.0

From the 'normal' IDMTL curve at multiple of 7.8 the theoretical relay tripping time = 3.4 s at TMS = 1.

With TMS = 0.25 the theoretical relay tripping time = 0.25 × 3.0 = 0.75 s.

Consider relay R_3

$$\text{Current setting} = \frac{140}{0.85 \times 120} = 1.38 \text{ (motor running condition)}$$

$$\text{or} = \frac{240}{1.15 \times 120} = 1.74 \text{ (motor starting condition)}$$

Therefore set relay R_3 to PSM = 175% to ensure no anomalous tripping under motor starting conditions.

At a fault level of 1800 A at busbar C and assuming an adequate discriminating time delay between the two relays R_2 and R_3 of 0.4 s then relay R_3 must operate in $0.4 + 0.75 = 1.15$ s.

Relay R_3 nominal setting with PSM $= 175\% = 120 \times 1.75 = 210$ A

$$\text{Multiple of setting} = \frac{\text{fault current}}{\text{maximum circuit primary current}}$$

$$= \frac{1800}{210} = 8.6$$

From the 'normal' IDMTL curve at multiple of 8.6 the theoretical relay tripping time $= 3.2$ s at TMS $= 1$.

For tripping time of 1.15 s set the time multiplier setting to $1.15/3.2 = 0.36$, set to 0.36.

The relays would therefore have the following settings:

Relay	Plug setting	Time multiplier
R_1	125%	1.14
R_2	150%	0.25
R_3	175%	0.36

Now consider the system if one of the motors at busbar C is subsequently increased in size to have a normal full load current (FLC) of 50 A and a starting current again of $6 \times$ FLC $= 300$ A.

Full load current to be carried by relay R_3 due to loads at busbars A, B and C $= 100 + 20 + 50 = 170$ A.

Maximum current to be carried by relay, R_3 under the new larger motor starting and six running conditions for which no tripping must occur $= 6 \times 50 + 6 \times 20 = 420$ A

Consider again relay R_3:

$$\text{Current setting} = \frac{170}{0.85 \times 120} = 1.67 \text{ (motor running condition)}$$

$$\text{or} = \frac{420}{1.15 \text{ x } 120} = 3.06 \text{ (motor starting condition)}$$

A setting of 306% is not possible and so the highest relay tap of 200% must be used and anomalous relay operation under motor starting conditions avoided by using the time multiplier.

$$\text{Current setting} = \frac{420}{2.0 \times 120} = 1.75 \text{ (motor running condition)}$$

From the 'normal' IDMTL curve at multiple of 1.75 the theoretical relay tripping time $=$ 12.0 s at TMS $=$ 1. The motor starting surge disappears after a run up time of approximately 5 s. Therefore no relay operation will result if the TMS is set to a value greater than $5/12 = 0.416$ (set to 0.42).

(b) Ring network

The use of inverse time overcurrent relays on a ring system necessitates the use of directional relays at all points where fault current could flow in both directions. Figure 10.4 shows a typical ring feeder arrangement together with the location of directional and non-directional relays. A directional relay is merely a combination of the inverse time overcurrent relay and a directional sensing unit.

For an electromagnetic overcurrent directional relay the voltage supply to the directional element may be supplied from a conventional star/star VT. Connections are made for the voltage to lag the current by typically 30° or 90/45° in order to ensure the maximum disc operating torque. Voltage supplies to directional earth fault elements must be such as to ensure that the voltage is less than 90° out of phase with the current supplied to the relay under all fault conditions. This is achieved by supplying the relay with a residual voltage derived from the vector sum of all the line voltages using an open delta VT winding. This is usually provided as a tertiary winding on a five limb magnetic circuit as shown in Fig. 10.9a. The VT must be solidly earthed so that the HV phase winding receives line/earth voltage under both healthy and fault conditions. If a three-limb VT were used with a star/star winding the fluxes in the three limbs would sum to zero and there would be no residual flux.

On a ring system grading is usually carried out by initially considering the ring fed at one end only. With circuit breaker A open relays B, D, F, H and J are then graded in a similar way to the radial system described in Section 10.4.5.1(a) above. The method is repeated with circuit breaker J open and relays I, G, E, C and A graded. This method takes into account the maximum fault current at which the discrimination is required.

With circuit breaker A open the initial step is made by considering relay B. Relay B can only carry current when there is a fault on the feeder between busbars 1 and 2. Under normal circumstances with breaker A open load current will not flow between busbars 2 and 1. Therefore a low current setting and time multiplier for relay B can be used.

Relay D is then set to discriminate with relay B and also any relays on the other feeders at busbar 2. Relay D must have a current setting high enough to avoid anomalous tripping under a full load and motor starting or other surge current condition.

Relay F is arranged to discriminate with relay D and so on around the ring to the slowest operating non-directional relay J at busbar 1. The calculations are then repeated for the relays in the opposite direction with circuit breaker J open.

Having fixed the current and time settings for all the relays in this manner the time of operation is then checked for the fully closed ring system under the associated redistributed fault levels. Consider a fault between busbars 2 and

3 and let the fault infeed via busbar 3 be 1000 A and via busbar 2 be 1500 A. Suppose under these fault conditions relay D operates in 0.7 s at 1000 A and relay C in 1.3 s at 1500 A. Suppose after relay D and the associated oil circuit breaker operating time (say $0.7 + 0.15 = 0.85$ s) the fault current level on the feeder between busbars 2 and 3, now fed only from busbar 2, becomes 2000 A. Relay C will now trip at this fault level in 1.0 s. The total time for relay C to trip under these fault conditions will be:

$$0.85 + 1.0 \times \left(1 - \frac{0.85}{1.30}\right) = 1.2 \text{ s}$$

If a high number of grading steps are required in the network it may be necessary to take this 'sequential' tripping into account when calculating relay multiplier settings.

10.4.5.2 IDMTL back-up protection

In more important networks and at higher voltage levels current relays are used as a 'back-up' to more sophisticated and faster acting 'main' protection systems. The back-up protection should be graded to achieve selective tripping if the main protection fails to operate. However, it must be noted that this is not always possible on highly interconnected networks involving widespread generation sources. If discrimination is not possible throughout the network then it must at least be ensured that the back-up protection ensures overall circuit breaker tripping times within the thermal capability of the plant being protected.

If all the IDMTL back-up relays use the same characteristic curve and time multiplier then they will all tend to grade naturally. The faulty circuit will normally carry much more current than the many other interconnected circuits supplying the fault and the relays associated with the faulty feeder will therefore tend to trip faster. An example is shown in Fig. 10.7a. In this arrangement the IDMTL relay gives a greater measure of selectivity than if definite minimum time relays were being used.

10.4.5.3 Instantaneous high set overcurrent relays

The tripping times at high fault levels associated with the IDMTL characteristic may be shortened by the addition of instantaneous high set overcurrent (HSOC) elements. These may be an integral part, for optional use depending upon front panel settings, of a modern solid state relay. Alternatively, the elements may be specified as an extra requirement when ordering electromagnetic type relays and included in the same relay case. The high set overcurrent or earth fault characteristic is such that at a predetermined current level the relay initiates essentially instantaneous tripping. This allows relays 'up stream'

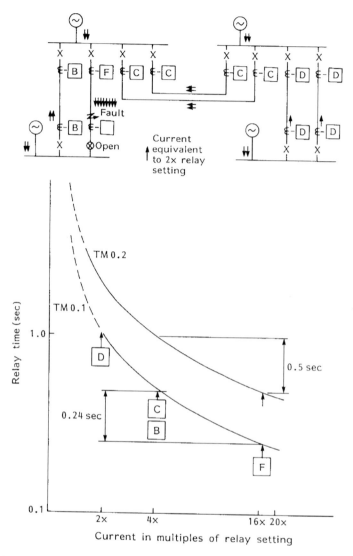

Figure 10.7a   IDMTL overcurrent relay discrimination with common relay setting on all feeders

towards the power source to be graded with the instantaneous HSOC 'down stream' relay setting and not the maximum fault level that would be required to trip using the normal IDMTL curve.

The relays are often specified for use with transformer protection. The primary side HSOC setting is chosen to be high enough so as to ensure no operation for secondary side faults. This is possible because of the fault limiting effect of the transformer reactance which allows adequate grading with transformer secondary protection. Some manufacturers produce 'low transient

Figure 10.7b   Traditional electromechanical IDMTL relay. The illustration shows an earth fault relay (type CDG) with a 'very inverse' characteristic. The plug settings are at the front of the relay and time settings at the top (courtesy GEC Alsthom T & D Protection & Controls)

over-reach' HSOC elements. These include a means of rejecting the DC component of fault current and improve the grading margin between HSOC setting and maximum transformer through fault current.

HSOC protection may be applied to cable feeders, overhead lines or series reactors where sufficient reactance is available to avoid anomalous tripping for remote faults in the network. Care must also be taken into account to ensure any initial energization or switching inrush currents do not cause the relays to trip anomalously.

Figure 10.7c   Modern solid state IDMTL relay – type KCGG (courtesy of GEC Alsthom T & D Protection & Controls)

Consider the network shown in Fig. 10.8. Three 33/11 kV transformers are fed via overhead lines through circuit breakers A, B and C. What setting should be applied to the HSOC relays operating the circuit breakers A, B and C?

Fault current (minimum generation, transformer primary 33 kV side terminals)

$$= \frac{100}{(0.15 + 0.15)} \times 17.5 = 5827 \text{ A}$$

CT secondary current under these fault conditions $= \dfrac{5827}{400} = 14.6 \text{ A}$

In order to ensure relay operation set to 7.3 A.

The relay must be stable under a secondary side 11 kV fault. Consider an 11 kV busbar fault with maximum generation and only one transformer in service. This is a worst case condition because it gives the most fault current flowing through a particular transformer feeder CT. With three transformers

Figure 10.7d   Modern microprocessor controlled solid state IDMTL relay – type MCGG. A wide variety of characteristics is available. PSM and TSM settings are configured using the small switches accessible from the front of the relay (courtesy GEC Alsthom T & D Protection & Controls)

in parallel the 11 kV fault level would be higher but the current would divide via the three paths.

Fault current (maximum generation, 11 kV side, one transformer in circuit)

$$= \frac{100}{(0.1 + 0.15 + 0.95)} \times 52.5 = \frac{5250}{1.2}$$

Fault current (same conditions, 33 kV side) $= \dfrac{5250}{1.2} \times \dfrac{11}{33} = 1{,}453 \text{ A}$

CT secondary current under these fault conditions $= \dfrac{1453}{400} = 3.63 \text{ A}$

The relay setting must ensure stability under these conditions and not operate at 2×3.63 A $= 7.26$ A. A relay setting of 7.3 A is therefore appropriate in this example.

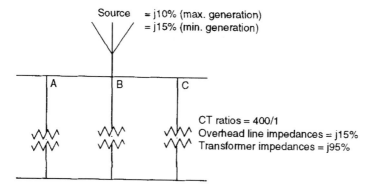

Figure 10.8   High set protection exercise network (all impedances to 100 MVA base)

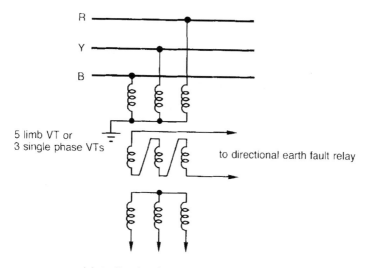

(a) to directional overcurrent relay

Figure 10.9a   VT connections for directional relays

10.4.5.4 *Sensitive earth fault relays*

Under certain high resistance conditions ordinary distance protection and some other sophisticated relays may fail to operate. Such conditions have been experienced with flashovers from overhead transmission lines above bush fires, distribution lines adjacent to overgrowing vegetation and rubber tyred gantry crane vehicles touching overhead lines. The fault is said to be outside the 'reach' of the relay.

Historically, earth fault relays with very sensitive, low setting ranges, typically 1–5% of circuit rating, have been used to detect such faults. They may be connected in the residual circuit of the overcurrent back-up protection overcurrent CTs as shown in Fig. 10.9b. When used in conjunction with main protection they

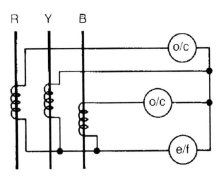

Figure 10.9b   Earth fault relay connected in residual circuit of protection CTs (2 × o/c, 1× e/f protection)

usually incorporate a sufficient time delay to allow the main protection and possible associated auto-reclose scheme to operate first. In addition, when used in conjunction with parallel feeder arrangements the settings must not cause tripping due to unequal load sharing between the same phases of the two line circuits.

However, relays with high impedance fault detection feature (HIF) have been available for some time now, based on advanced wavelet & statistical algorithms.

10.5 DIFFERENTIAL PROTECTION SCHEMES

10.5.1 Biased differential protection

10.5.1.1 *Introduction*

Basically, unit protection schemes compare the current entering and leaving the protected zone. Any difference will indicate the presence of a fault within the zone. By operation of the appropriate relays the associated circuit breakers can be made to trip thus isolating the faulty equipment from the power network.

With perfect CTs, relays and symmetry of connections, stability should be theoretically possible under steady state conditions and no anomalous tripping due to faults outside the zone of protection occur. In practice, there will be differences in the magnitude and phase of the currents entering and leaving the zone of protection. CT characteristics will not be perfectly matched and DC components will be involved under fault conditions. There is an upper level of through fault current at which both steady state and transient unbalance rapidly increases as shown in Fig. 10.10. Restraint features, known as 'bias', are therefore added to the relays to desensitize under through fault conditions. The discrimination quality of a differential system can be defined in terms of a factor given by the ratio of the degree of correct energization of the relay under internal fault conditions to that which occurs (and is unwanted) under external fault conditions at the same time as the specified CT primary current. If the system were perfect the factor would be infinite.

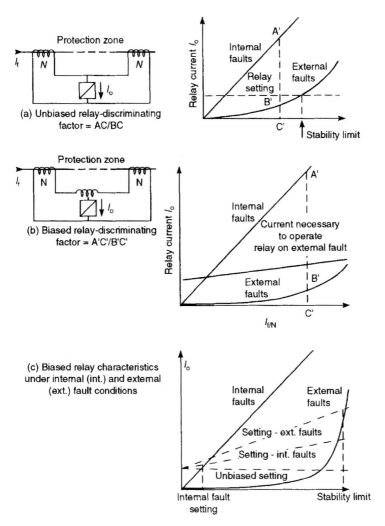

Figure 10.10   Biased differential protection-discriminating factor and stability: (a) unbiased relay-discriminating factor = AC/BC; (b) biased relay-discriminating factor A'C'/B'C' (c) biased relay characteristics under internal (int.) and external (ext.) conditions

10.5.1.2 *Biased differential transformer protection*

Figure 10.11 shows a typical arrangement. It is necessary to take into account the following:

1. Transformer ratio – the magnitudes of the currents on the primary and secondary sides of the power transformer will be inversely proportional to the turns ratio. The CTs located on the primary and secondary sides of the power transformer should therefore be selected to match this difference so that CT secondary currents are as equal as possible and spill currents

through the relay operating circuit are kept to a minimum. CT ratios are standardized and often do not match the power transformer ratio. In such cases interposing correction CTs are employed.

2. Transformer vector grouping – the primary and secondary power transformer currents will have differing phase relationships due to the vector grouping. The CT connections must be made in such a way as to compensate for this change and bring the CT secondary currents back into phase. This phase correction can be achieved either with the main or interposing CTs.

3. Transformer tap changer – if the transformer has a tap changer then the ratio between the magnitude of the power transformer primary and secondary currents will vary depending upon the tap position. The mean tap position should be taken for calculations and any spill current compensated by the relay bias circuit.

4. Magnetizing inrush current – when a transformer is energized a magnetizing inrush current up to 10 × full load current can occur. This current is present only on the primary or source side of the transformer and if not

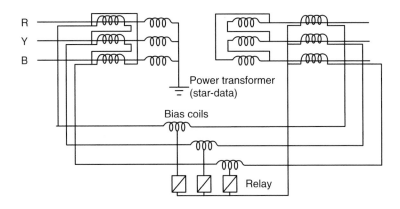

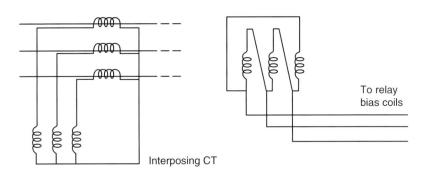

Figure 10.11   Transformer biased differential protection. CT connections arranged to compensate for vector grouping and turns ratio

compensated will introduce an unbalance and cause the differential protection system to trip.

In fact magnetizing inrush current has a high second harmonic component which does not specifically appear under fault conditions. A compensation circuit is therefore introduced into the relay to detect second harmonic components. These are recognized as switching transients and the relay is restrained from operating under this condition.

An alternative method of inrush current recognition and relay operation restraint used by some manufacturers is the detection of the zero period in the magnetizing inrush current wave form and blocking relay initialization.

Restraint and/or blocking for other harmonics such as the 5th are also catered to in modern transformer differential schemes.

10.5.2 High impedance protection

Typical applications for high impedance relays include busbar and restricted earth fault protection schemes. Under through fault conditions the differential relay unit protection scheme must remain stable. During such an *external fault* one CT may become saturated producing no output while a parallel connected CT might remain unsaturated and continue to produce full output. This condition is shown in Fig. 10.12. The current from the remaining operating CT will divide between the relay and the other arm of the network comprising of connecting cable and lead impedance and the effectively short circuited saturated CT. By using a relay of sufficiently high impedance (not a problem with

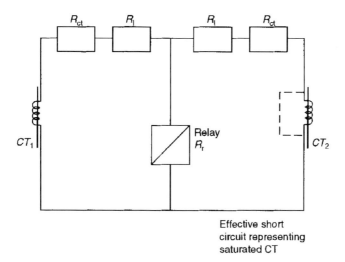

Effective short
circuit representing
saturated CT

Figure 10.12   High impedance protection under external fault conditions. R_{ct}: CT resistance; R_1: cable and lead resistance between CT and relay; R_r: relay resistance (CT magnetizing circuit neglected)

modern solid state electronic relays) the proportion of current flowing through the relay is reduced below the operating level of the relay making for stable operation under these heavy through fault conditions. (It must be noted that modern low or medium impedance relays are designed so that they are largely unaffected by some of these issues and cater for larger amounts of CT saturation, remanance, etc.)

Because the relay has a high impedance the voltage developed across it during an *internal fault* may be sufficiently high so as to force the CTs into saturation. It is therefore necessary when applying high impedance relay protection to calculate the relay circuit resistance and CT 'knee point' voltage such that the CT is still able to produce sufficient output under saturated conditions. The relay circuit resistance, R_1, is calculated for the highest combination of $R_r + R_{ct}$ in accordance with the formula:

$$R_r = I_{SC} (R_1 + R_{ct})/I_r$$

where R_r = required relay circuit resistance
   I_{SC} = CT secondary current equivalent to design through fault current
   R_1 = cable and lead resistance between saturated CT and the relay connection point
   R_{ct} = resistance of saturated CT
   I_r = relay setting current

Having selected a ratio for R_r the minimum knee point voltage is then determined in order to meet the relay manufacturers' requirements. Most relays require a CT knee point voltage equal to twice the voltage developed across the relay circuit when passing a current equal to the relay setting. Assuming the knee point voltage, V_{kn}, to be at least this value then:

$$V_{kn} = 2 \times I_r \times R_r$$

10.5.3 Transformer protection application examples

10.5.3.1 *Balanced earth fault protection*

Consider the arrangement shown in Fig. 10.13a of two 33/11 kV cable fed delta/star transformers. The HV balanced earth fault protection at the substation must be set to detect faults on the transformer primary and feeder cable.

- The CT ratio on the transformer HV side is 400/1.
- The lead resistance of the CT wiring to a common point is 2.1 ohms.
- The charging current of the transformer feeder cable is 2.1 A/phase/km.
- The system impedances are shown to a 100 MVA base.

It is necessary to calculate the relay current and voltage setting and the required CT knee point requirement.

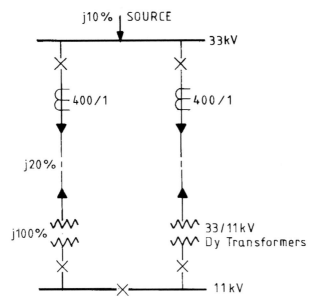

Figure 10.13a   Balanced earth fault and restricted earth fault protection examples – network configuration (all impedances on 100 MVA base)

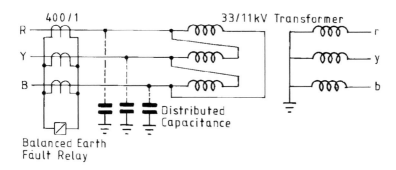

Figure 10.13b   BEF protection CT arrangement

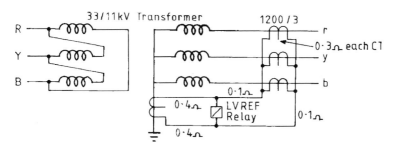

Figure 10.13c   LVREF protection CT arrangement and lead-CT resistances

In order to calculate the relay setting and ensure stability under through fault conditions it is necessary to consider the following.

1. *The capacitance currents due to an earth fault on the 33 kV HV side.* The effect of distributed cable capacitance associated with the transformer feeder is to introduce a capacitive current under phase-to-earth conditions. This is equivalent to $\sqrt{3}$ × normal cable charging current per phase and results in a total current through the balanced earth fault (BEF) relay of 3 × normal phase charging current.

   Therefore capacitive current, I_c = 3 × feeder length × charging current/phase/km = 3 × 5 × 2.1 = 31.5 A.

$$\text{CT secondary capacitive current} = \frac{31.5}{400} = 79 \, \text{mA}$$

   Therefore apply a relay setting of twice this value = 160 mA. Such a setting is achieved by suitable shunting resistors and plug adjustments.

2. *The situation with one of the 33 kV CTs saturated during an 11 kV LV fault.* The balanced earth fault protection is arranged with the relay connected in the residual circuit of the transformer feeder CTs. It is necessary to ensure that the relay does *not* operate for faults on the LV side of the power transformers. Stability is checked by considering the worst case through fault current condition. The highest secondary side fault current will be with the two power transformers in parallel. However, the maximum CT current will flow when the fault current is not shared and this occurs with only one transformer in circuit. Therefore primary fault current for secondary side fault = 100/1.3 × Amps/MVA @ 33 kV = 100/1.3 × 17.5 = 1347 A.

$$\text{Therefore CT secondary fault current} = \frac{1347}{400} = 3.37 \, \text{A}$$

   Voltage appearing across the relay = CT secondary current × lead resistance = 3.37 × 2.1 = 7.1 V.

   Therefore set the relay to 2 × voltage value, say, 15 V. Such a setting is achieved by insertion of a suitable resistor in series with the relay. The knee point of the CTs must be 2 × voltage setting = 2 × 15 V = 30 V.

10.5.3.2 *Restricted earth fault protection*

Consider the same network as shown in Fig. 10.13a. Low voltage restricted earth fault (LVREF) protection is applied to the LV 11 kV transformer star winding and neutral connections as shown in Figs 10.13b and 10.13c. Under fault free or external fault conditions the CT currents will sum to zero and the relay will not operate. Under internal fault conditions within the transformer

LV winding or connections the CT currents will become unbalanced and the LVREF protection will operate. It is necessary to verify the relay voltage setting, a suitable CT knee point, and, given the relay voltage – current relationship, the effective relay primary CT setting.

- CT ratio on transformer LV side = 1200/3.
- Assume worst case earth fault with solid earthing equals 3 phase fault level with one transformer feeding maximum CT current.
- As an alternative consider a resistance-earthed neutral with earth fault current limited to 1000 A.
- Lead and CT resistances to common LVREF relay connection point = 0.4 ohm neutral to relay cable resistance per core
    = 0.1 ohm phase-to-relay cable resistances per core
    = 0.3 ohm for each CT resistance.
- CT knee point = 100 V.
- CT and relay characteristics:

Voltage (V)	10	20	30	40	60	80	90	100	110	200
CT current (mA)	20	30	40	45	50	70	80	100	200	500
Relay current (mA)	5	10	15	20	25	30	35	40	42	43

Consider an out-of-zone fault with one CT saturated. Maximum 11 kV fault current $I_f = 1347 \times 33/11 = 4042$ A (see Section 10.5.3.1).

$$\text{Therefore CT secondary fault current} = 4042 \times \frac{3}{1200} = 10.1 \text{ A}$$

Maximum voltage across relay = current $\times$ lead and CT resistances = $10.1 \times (0.3 + 2 \times 0.4) = 11.1$ V.

Therefore set the relay to 2 $\times$ voltage value = 22 V.

Therefore CTs must have a minimum knee point value of 44 V.

From the CT characteristics given in the table above the knee point is approximately 100 V and therefore suitable for this application.

At 22 V the relay current is approximately 11 mA and the CT magnetizing current 32 mA. Under worst case conditions for an internal fault with three CTs receiving magnetizing current from the CT in the faulted phase, then CT output required to operate the relay = $11 + (3 \times 32) = 107$ mA. Equivalent CT primary current = $1200/3 \times 107 \times 10^{-3} = 43$ A approximately.

With a resistance-earthed neutral the maximum earth fault current is 1000 A. This will result in desensitizing the LVREF protection and approximately

$43/1000 = 4.3\%$ of the transformer windings at the neutral end will not be protected.

10.5.4 Pilot wire unit protection

10.5.4.1 *Pilot cables*

Since currents at each end of the protected zone must be compared a signalling channel is necessary using:

- pilot cables – typically two core, 2.5 mm^2 cross-section;
- private or rented telephone circuits;
- radio links;
- fibre optic links.

The selection is based on economic and reliability factors. With increasing length, or difficult installation conditions, buried hard wire or fibre optic cable links become expensive.

Telephone lines have typical limitations on the peak applied voltage and current as follows:

- Maximum applied voltage – 130 V DC.
- Maximum current – 0.6 A.

All connected circuits must be insulated to 5 or 15 kV.

Telephone pilots may be subjected intermittently and without warning to ringing tones, open or short-circuit conditions. Therefore their use requires more complicated terminal equipment and if rented may have a lower level of reliability.

Most systems use armoured and well-screened twisted pair two core copper cable installed in the same trench as the power feeder cable or under the overhead transmission line in the associated wayleave. This type of pilot cable is known as 'high grade'. The twisted pairs and minimum power-to-pilot cable spacing of approximately 300 mm help to minimize pick-up noise and induced currents. At present the typical economic distance for such a unit protection relay and pilot cable scheme is 25 km. After this distance the cost of cable and trenching begins to exceed the cost of a comparable distance relay scheme. The pilot cable may be represented as a complex pi network of uniformly distributed series resistance and inductance with shunt capacitance and conductance. If the cable is in good condition the leakage conductance will be very small. Unlike a communications circuit (which is terminated with the cable characteristic impedance in order to achieve a resistive load and maximum power transfer) a protection pilot cable is terminated in an impedance which varies almost between open and short circuit (see Fig. 10.14). On open circuit the pilot cable has a predominantly capacitive impedance

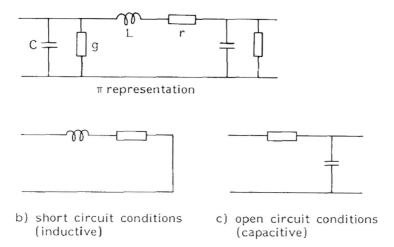

π representation

b) short circuit conditions   c) open circuit conditions
   (inductive)   (capacitive)

Figure 10.14   Pilot cable equivalent circuits: (a) π representation; (b) short-circuit conditions (inductive); (c) open circuit conditions (capacitive)

while on short circuit the impedance is inductive. The impedance of the pilot cable is an important factor in the design of the protection relay system and it must match the relay manufacturer's requirements.

10.5.4.2 Summation transformers

In order to minimize pilot cable costs the three phase primary currents are converted via the matched CTs and summation transformers at each end of the link to a single phase secondary current. This is justified since as explained in Chapter 25, the primary currents bear definite relationships for different types of fault condition. The output from the summation transformer may be set, by altering the proportion of n turns as shown in Fig. 10.15, to give greater sensitivity to earth faults. This is a useful feature when using a unit protection scheme on a resistance-earthed system which produces low earth fault currents.

10.5.4.3 Basic schemes

The fundamental difference between biased differential protection and pilot wire differential protection is that relays are required at each end of the pilot wire scheme. Only one relay is required for the biased differential schemes described in Section 10.5.1.

The long distances involved between the two ends of a feeder cable or an overhead line circuit necessitates the use of a relay at each end of the protection zone. The relays control the associated circuit breakers and minimize the

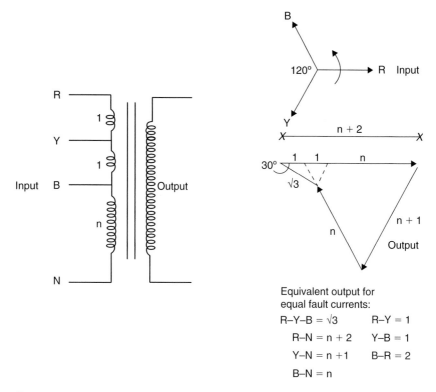

Equivalent output for
equal fault currents:

R–Y–B = √3	R–Y = 1
R–N = n + 2	Y–B = 1
Y–N = n +1	B–R = 2
B–N = n	

Figure 10.15   Summation transformer arrangement (three phase to one phase)

effects of pilot cable characteristics on relay performance. In addition, a single relay scheme is not used because the CTs would have to be impossibly large in order to avoid saturation on through fault current when used in conjunction with a pilot cable burden of approximately 1000 ohms. The relays are arranged to operate simultaneously with intertripping to provide very rapid fault clearance irrespective of whether the fault is fed from one or both ends of the protected zone.

Practical schemes are based on circulating current or balanced voltage principles as shown in Figs 10.16a and 10.16b. Since the pilot cables may run in parallel with power cables or overhead lines for long distances isolation transformers and non-linear resistors are used in order to prevent unacceptable induced voltages appearing at each end on the relays. The condition of the pilot cables can also be monitored by incorporating pilot wire supervision modules into the relay. Such supervision will not prevent anomalous operation but can be arranged to raise an alarm due to open circuit, short circuit or cross-connected faulty pilots. When there is sufficient fault current anomalous operation due to faulty pilots may be positively prevented by the addition of overcurrent check features into the relay scheme. Figure 10.17 shows a modern pilot wire protection relay based on the voltage balance principle which can incorporate all these features.

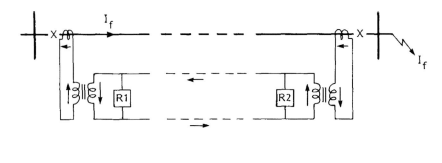

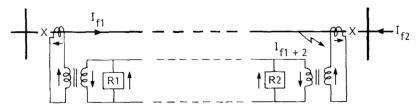

Figure 10.16a   Pilot wire differential protection through and in-zone fault conditions (circulating current)

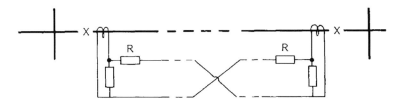

Figure 10.16b   Balanced voltage scheme

10.5.5 Busbar protection

10.5.5.1 *Introduction*

Busbar reliability is of paramount importance since failure will result in the loss of many circuits. In practice, busbar faults are rare and usually involve phase-to-earth faults. Such faults may be due to:

- insulation failure resulting from deterioration over time;
- flashover due to prolonged or excessive overvoltages;
- circuit breaker failure to operate under through fault conditions;
- operator/maintenance error (especially leaving earths on busbars after a maintenance operation);
- foreign objects falling on outdoor catenary busbars.

Some electricity supply companies prefer not to employ busbar protection because they do not wish to incur the occasional outage due to protection

Figure 10.17   Modern pilot wire protection relay – type MBCI (courtesy GEC Alsthom T & D Protection & Control)

maloperation which could be more likely than a true busbar fault which might only occur once in 20 years. Such a risk may be reduced by employing a separate 'check' feature in the busbar protection relay scheme which must also recognize the fault before tripping is initiated. In a similar way, in areas of low thunderstorm activity, an electricity supply company may decide not to employ outdoor substation overhead lightning screens because the likelihood that they may fall and cause a busbar fault is considered to be more probable than an outage due to a lightning strike.

10.5.5.2 *Frame leakage detection*

This is the cheapest form of busbar protection for use with indoor metal-clad or metal-enclosed switchgear installations. Since the probability of a busbar fault on such modern equipment is very small, busbar protection on such equipment is only considered for the most important installations. The switchboard is lightly insulated from earth (above, say, 10 ohms) and currents in a single connection to earth measured via a CT and frame leakage relay. This arrangement requires care to ensure all main and multicore cable glands are insulated and that bus sections are not shorted by bolted connections through the concrete floor rebar or switchgear steel floor fixing channel arrangements (Fig. 10.18a).

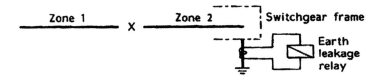

Figure 10.18   (a) Frame leakage detection

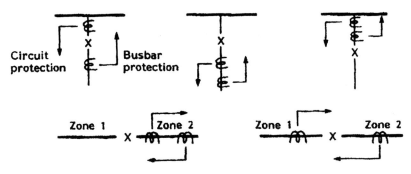

Figure 10.18   (b) CT location arrangements and overlapping zones of protection

To avoid anomalous tripping an additional check feature should be incorporated when possible by using a CT on the star point neutral of the supplying transformers. The installation must be in a dry substation and in practice is not really suitable for retrofitting on existing switchboards. Equally, care must be taken in extending switchboards that do already have frame leakage detection, since some new switchgear types do not permit the necessary insulating features.

The frame insulation resistance is effectively in parallel with the substation main earthing system with typical resistance values to true earth of less than 1 ohm. The earth leakage relay will therefore only 'see' approximately 9–10% of the earth fault current when a 10 ohm switchgear-to-earth insulation level is employed. To ensure stability under external out-of-zone fault conditions the frame leakage relay may be set at 30% of the minimum earth fault current.

10.5.5.3 *Bus zone*

A comparison is made between the currents entering and leaving the busbar or busbar zone. CTs are therefore required on all circuits and the CT locations are arranged to maximize the required zone of protection coverage as shown in Fig. 10.18b. Conventionally, main and check high impedance relays are used in conjunction with these CTs to measure the sum of all the currents. Very fast operating times (40 msec) are feasible with such schemes. An example of the practical application of traditional busbar protection principles is given in Section 10.9.

The check feature makes tripping dependent upon two completely separate measurements of fault current using separate CTs and different routing for

CT wiring to the protection relays. In a double busbar arrangement a separate protective relay is applied to each bus section (zones 1 and 2) and an overall check system arranged to cover all sections of both main and reserve busbars.

In current practice, schemes based on multiple operating principles and sometimes based on a distributed concept, are applied for EHV substations. These schemes are typically low impedance types and handle differing CT ratios, a high amount of CT saturation, diverse issues such as evolving faults and CT remanance. Continuous self-supervision and the ability to detect abnormalities on CT's, CT wiring and the auxiliary contacts used to provide a bus image make the scheme complete and highly reliable, thus obviating needs for a separate set of check zone CT's and relays.

10.5.5.4 *CT selection*

In order to ensure stability under load, switching transient and external fault conditions the CTs must be all carefully matched up to the maximum fault level with the same ratio and characteristics. As explained in Section 10.5.2 it is the voltage required to operate the relay rather than its current setting which determines the stability level of the scheme. The CT 'knee point' voltage must be kept as high as possible and at least three times the relay voltage setting. The testing of busbar protection schemes will therefore necessitate particular care over CT polarities, correct operation of busbar selector auxiliary contacts and primary operating current at the selected relay settings.

An interruption in CT wiring will cause an unbalance and anomalous busbar protection operation. Wiring supervision is therefore a feature of most schemes in order to raise an alarm with typical settings of unbalance at 10% of minimum circuit rating.

See also Section 10.10 regarding optical communication links.

10.6 DISTANCE RELAYS

10.6.1 Introduction

The operation of distance relays is not governed by the current or power in the protected circuit but by the ratio of applied voltage to current. The relays therefore effectively measure the impedance of the circuit they are protecting. Under fault conditions the voltage is depressed and the current flow greatly increases. The impedance therefore falls during the fault; this is sensed by the distance relay and a trip initiated. Such relays are used on power network overhead lines and feeder cables when the required operating times cannot be achieved by conventional current operated relays. Since the impedance of an overhead line or cable is proportional to its length such impedance measuring relays are known as distance relays or distance impedance (DI) protection.

10.6.2 Basic principles

When a fault occurs on a feeder the fault current, I_f, is dependent upon the sum of the fault impedance, Z_f, from the 'relay point' to the fault and the source impedance, Z_s

$$I_f = E/(Z_f + Z_s)$$

where E is the line to earth voltage.

The voltage at the relaying point is proportional to the ratio of the fault impedance to total source-to-fault impedance.

$$V_f = E \times Z_f/(Z_f + Z_s)$$

If the relay is designed to measure and compare both voltage and current then $V_f/I_f = Z_f$ and the relay measurement of fault impedance is effectively independent of the source impedance. If a fault occurs on the feeder a long way from the relaying point then there may be insufficient fault current to operate the relay and the relay will not trip. The point at which a fault occurs which just fails to cause relay operation is known as the 'balance point' or relay 'reach'. This impedance measuring type of relay is the simplest form of distance relay and the relay is set to have a 'reach' sufficient to cover the highest impedance setting likely to be required. A radial feeder arrangement is shown in Fig. 10.19a. Ideally the relay, R_1, would be set to cover faults arising on any section of the feeder from busbar A to busbar B. In a practical situation the accuracy of such a relay setting is insufficient and the reach impedance is usually set to cover 80–85% of the overhead line or cable feeder length in order to ensure that faults outside the zone of protection do not cause an anomalous trip. To protect the rest of the feeder and to preserve discrimination a time delay is introduced into the relay electronics such that a fault on the last 15–20% of the line being protected will still initiate a trip but after a preset time. To give complete back-up protection the relay is adjusted after a further time delay to cover faults on all the following feeders between busbars B and C. This is known as a three-step characteristic covering protection zones 1, 2 and 3. The relay will trip for faults in 'zone 1' in essentially instantaneous zone 1 time.

10.6.3 Relay characteristics

The simple, plain impedance measuring relay has a circular non-directional characteristic which may be plotted on a resistance–reactance (R–X) diagram with the relaying point at the centre as shown in Fig. 10.20. The reach of the relay is represented by the radius of the circle and the impedance of the feeder being protected is shown as a straight line passing through the origin. A trip will be initiated for any value of impedance falling within the relay trip setting

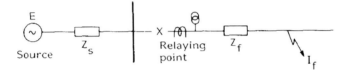

Z_s = source impedance
Z_f = line impedance from relaying point to fault

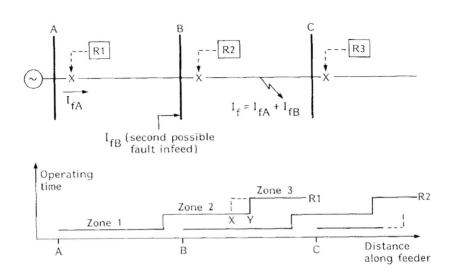

Figure 10.19   (a) Plain impedance relay 3 zone distance protection

Figure 10.19   (b) Modern solid state microprocessor controlled distance impedance relay: type optimo (Courtesy GEC Alsthom T & D Protection and Control)

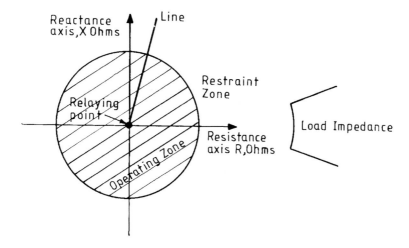

Figure 10.20   Plain impedance relay characteristic (R–X diagram)

radius. If this relay was used at relaying point B in Fig. 10.19a a trip would be initiated due to its non-directional nature for faults lying on the feeders between both busbars AB and BC – that is both 'behind' and in 'front' of the relaying point. This plain impedance relay characteristic would not therefore, be used in practice.

In order to improve on this situation a directional element may be added to the relay such that both impedance and directional measurements have to operate before a trip command is given. The resulting directional characteristic, Fig. 10.21a, is still not ideal as there is the problem of a possible 'race' between the directional and impedance measuring elements. In addition, when applied to long lines, with relatively high reactance to resistance ratios, the circular operating area to the right of the reactance line can be so large as to cause operation from load currents or under power swing conditions.

Further improvements with directional properties are possible using the admittance or 'mho' relay characteristic as shown in Fig. 10.21b. The standard mho relay uses the voltage from the faulty phase to derive the directional feature. The operating value of impedance passes through the origin but with this characteristic the reach point setting varies with fault angle.

One problem faced with distance relays is that of fault resistance which may be represented on the R–X diagram as a straight line parallel to the resistance axis for a single end fed line. The arc fault resistance bears no relation to the neutral earthing resistance used in resistance-earthed systems as regards the relay setting value. The neutral earthing resistor value is used to determine the minimum reach of the relay. Arc resistance is empirically proportional to the arc length and inversely proportional to a power of the fault current and as such may well fall outside the zone 1 reach of the relay such that the majority of earth faults are only detected and cleared in zone 2 times. A number of distance relay characteristics such as reactance, cross-polarized and quadrilateral have been developed to improve the amount of fault resistance coverage.

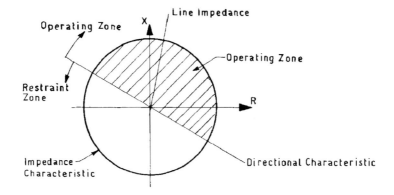

Figure 10.21   (a) Directional impedance relay characteristic

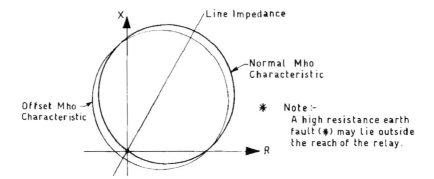

Figure 10.21   (b) Mho relay characteristic

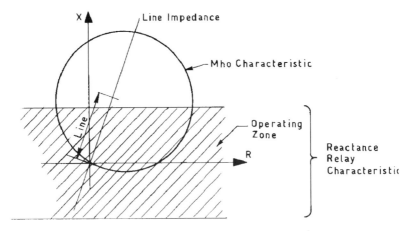

Figure 10.21   (c) Combined mho-reactance relay characteristic

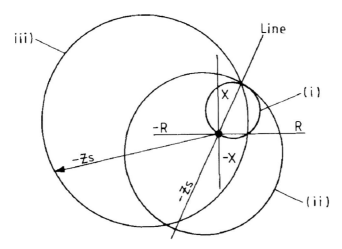

Figure 10.22   Cross-polarized mho relay characteristics under different conditions: (i) standard mho characteristic; (ii) cross-polarized mho characteristic, solidly earthed system; (iii) cross-polarized mho characteristic, resistance-earthed system. Z_s = source impedance

The reactance relay characteristic is represented by parallel straight lines above and below the resistance axis on the R–X diagram. A combined mho and reactance relay characteristic is shown in Fig. 10.21c. The mho element of the relay introduces a directional feature and limits the reach of the relay such that tripping on load current is avoided. A practical mho setting would be approximately twice the line length as indicated.

The 'cross-polarized mho' relay uses a voltage for the directional or polarizing feature derived from phases other than those involved in the fault. For example, phases R and Y will be used for a B phase fault condition. Therefore the relay has different characteristics for different types of fault condition as shown in Fig. 10.22. For a three phase fault all phases are affected in the same manner and the relay follows the standard mho relay characteristic. Unlike the plain impedance relay the characteristic is a function of the source impedance as well as the relay setting. The larger the source impedance the larger the operating R–X diagram circle becomes and the greater the fault resistance capacity. For a single phase-to-earth fault the relay source impedance has positive, negative and zero phase sequence components. For a resistance-earthed system the zero sequence impedance will include a value of earthing resistance equal to three times its nominal value. This will swamp the other sequence impedance components with the effect of moving the centre of the operating circle away from the −X reactance axis to the −R resistance axis compared to the solidly earthed system case. The relay may in fact be directional even though the characteristic shown envelopes the R–X diagram origin. For a fault 'behind' the relay the characteristic is different. The cross-polarized mho relay has the essential stability of the standard mho relay with regard to power swing and load impedance performance for symmetrical faults. It also provides an improved fault resistance coverage.

Solid state relays have the advantage of being able to produce quadrilateral characteristics with independently adjustable R and X settings and some modern relays have complex polygonal characteristics or even multiple characteristics for differing faults such as mho for multiphase faults and polygonal-shaped ones for single phase to earth faults. Typical resistance settings may be 4 to 8 times reactance settings allowing for improved arc resistance characteristics and therefore better short-line protection. It is, however, necessary to be careful to avoid anomalous over-reach tripping. On a double end fed system the fault resistance measured by the relay at one end is also affected by the current infeed from the remote end. These two fault current infeeds will make the fault resistance appear to the measuring relay to be larger than it really is. In addition, the fault currents are likely to be out of phase. This makes the measured resistance have a reactive component which can appear either positive or negative on the R–X diagram. If negative a fault beyond the reach of the relay can appear within the relay tripping zone and cause anomalous operation. To help overcome this problem solid state electronic relay manufacturers produce quadrilateral impedance characteristics with an inclined reactance line as shown in Fig. 10.23.

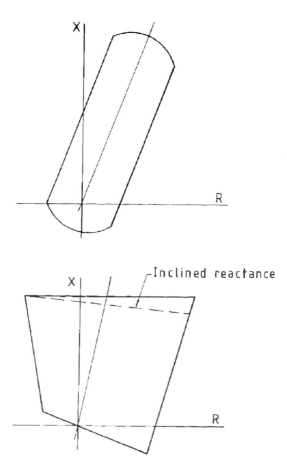

Figure 10.23   Quadrilateral impedance relay characteristics

10.6.4 Zones of protection

The settings normally applied to a three-stage distance impedance relay must take into account:

- the accuracy and performance of the associated CTs and VTs;
- the accuracy of the relay calibration;
- the accuracy of the transmission line or cable feeder impedance data.

Consider the radial feeder arrangement shown in Fig. 10.19.

Zone 1 setting is based on actual line impedance. Relay R_1 at relaying point A is set to 85% of the protected line:

$$0.85 \times Z_{AB} \times \frac{\text{CT ratio}}{\text{VT ratio}}$$

where Z_{AB} is the line impedance from busbar A to busbar B. The setting should be reduced to suit the accuracy of the relay being used. For a reactance relay line reactance X would be used instead of impedance Z. For a teed feeder 85% of the appropriate distance between the relay and the nearest adjacent relay should be chosen.

The zone 1 time is essentially instantaneous ; in most cases it is not actually set but is provided as a basic operating time for zone one at approximately 20–40 ms; older relays may require to be set at 50–100 msec.

Zone 2 settings are arranged to cover the remaining 10–20% of the line AB and should not normally reach further than 75% of the next section of line BC. The minimum setting should be 120% of the protected line and on teed feeders not less than 1.2 times the distance of the further of the remote ends. As explained in Section 10.6.3 above if there is an infeed from both ends of a line or at an intermediate busbar zone 2 reach will be reduced because the relay will not 'see' all the fault current. In the example in Fig. 10.19 the relay R_1 zone 2 reach is reduced by the additional infeed at busbar B in the ratio BX/BY $= I_{fA}/(I_{fA} + I_{fB})$.

The zone 2 time setting should lag behind the longest zone 1 operating time of the protection and circuit breakers on adjacent feeders. Typical time settings are up to 0.4 s where older switchgear is being used. In the Fig. 10.19 example, since zone 2 overlaps zone 1 it is necessary to delay zone 2 time by approximately 0.4 s. In addition, the zone 2 time setting must allow an adequate margin over adjacent relay R_2 zone 2 time delay settings.

Zone 3 is set to cover the whole of the adjacent line BC. The relay should be set to cover 120% of the protected line and the longest adjacent feeder, provided there is no overlap with the zone 3 protection setting on a shorter adjacent feeder.

In the case of the dual source infeed shown in Fig. 10.19, line BC apparent impedance is:

$$Z_{BC} \times (I_{fA} + I_{fB})/I_{fB}$$

the zone 3 setting should be:

$$1.2 \times [Z_{AB} + Z_{BC} \times (I_{fA} + I_{fB})/I_{fB}]$$

It is also necessary to ensure that this setting is not high enough to cause zone 3 operation on load and the adjustments available on quadrilateral characteristics are useful in this respect.

The zone 3 time setting is delayed beyond all other protection within its reach and should be determined for maximum fault infeed. The time delay should be at least equal to the maximum operating time of the protection on any adjacent feeder or transformer circuit plus a discrimination time of, say, 0.2 to 0.4 s depending upon the relays and switchgear being used.

Modern relays have even more set table zones, up to say 5. One or more of these zones can be reversed to provide back-up protection for the busbar itself.

10.6.5 Switched relays

As noted above, some relays have more than three zones. The different zone settings may be achieved by using individual relays for each zone or by having one relay and extending its reach after each time interval has elapsed. Six sets of electronics may be built into the overall distance relay: three for phase faults and three for earth faults. The directional elements detect the direction of the fault such that the relay is only 'started' for faults in the required direction. The use of separate relays is known as a full or non-switched scheme and has traditionally been used at the highest transmission voltage levels above 300 kV. However, advances in technology and reduction of costs mean that only such full schemes are available these days, whatever the voltage level.

Projects incorporating existing networks will come across 'switched' systems. In a switched system, one measuring relay is used to cover all types of faults. Starting elements are used to detect the type of fault, which phases are involved and to apply the appropriate voltage and current to the measuring relay and to initiate the appropriate zone time delay. The starting elements may be undervoltage, underfrequency or underimpedance types. When setting starting elements it is necessary to take into account the load impedance and the current in the sound phases during single phase-to-earth faults. With modern solid state relays the reliability differences between full and switched schemes have become blurred. The time differences for starter switching operations have also been reduced and selective fault detection using quadrilateral characteristics is easier. Considerable flexibility is available and basic distance protection schemes have become cheaper such that their use is justified at distribution volt-age levels as well as the more traditional use at transmission voltages.

10.6.6 Typical overhead transmission line protection schemes

10.6.6.1 *Permissive under-reach*

For a zone 1 fault the relay operates and trips its associated circuit breaker. At the same time the relay sends a signal via pilot wires or using power line carrier to the remote end of the line. This signal, in conjunction with the already initi-ated starting elements or some other form of double checking security which avoids false tripping, initiates tripping at the remote end thus completely isolat-ing the faulted section of line. Typical time delays due to transmission and oper-ation of auxiliary relays are of the order of 40 msec. This is a tenfold improvement over the normal zone 2 operating time (typically 400 msec) of the remote end relay (see Fig. 10.24).

10.6.6.2 *Zone extension*

In a zone extension scheme the remote end receives a signal which extends the reach of the remote end relay to zone 2 without waiting for the zone 2 timer delay. The zone 2 element confirms a fault and initiates tripping of the remote end breaker. The scheme is slower than permissive under-reach but consid-ered to be slightly more secure (see Fig. 10.25).

10.6.6.3 *Permissive over-reach*

The zone 1 coverage is set to beyond the length of the feeder being protected at, say, 120% line length. For an in-zone fault on the protected feeder both local and remote end relays detect the fault and send a signal to a receiver at the opposite end of the line. The relay tripping contacts are in series with receiver recognition contacts and tripping is allowed (see Fig. 10.26). For an

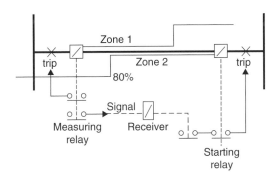

Figure 10.24   Permissive under-reaching scheme

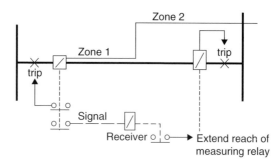

Figure 10.25   Zone extension scheme

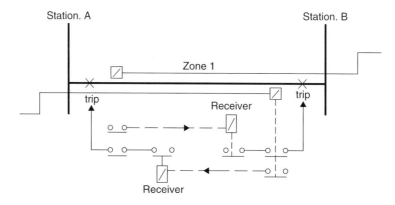

Figure 10.26   Permissive over-reach scheme

out-of-zone fault lying beyond substation B but within the 20% over-reach, the relay at substation A detects the fault but not the relay at substation B. Relay B will receive a signal from relay A but will not operate. Relay B will not therefore send a signal to relay A which in turn will be inhibited. Neither relay will therefore cause circuit breaker tripping to occur. Should the transmission path fail the scheme will still operate in zone 2 time with an over ride of the receiver contact.

10.6.6.4 Blocking

Forward and reverse 'looking' distance measuring relays are required at each end of the line and set to approximately 120% line length coverage. For an in-zone fault both relays 'see' the fault, no signal transmission is required to take place, and the circuit breakers at each end of the line operate (see Fig. 10.27). For an out-of-zone fault lying beyond substation B but within the 20% over-reach, the relay at substation A will detect the fault. Its tripping operation will be inhibited by receipt of a signal from the reverse looking relay at substation B. The scheme introduces a slight time delay due to transmission times and auxiliary relay operating times. The advantage of the scheme is that

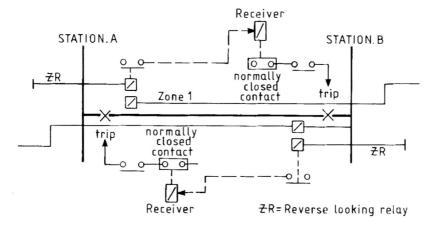

Figure 10.27 Blocking scheme

when power line carrier signalling is used the 'blocking' is achieved by transmission of the inhibit signal over a healthy rather than a faulty line.

10.6.6.5 *Phase comparison*

High frequency signals are transmitted between substations A and B as shown in Fig. 10.28. These signals are arranged to be sent in bursts only during the positive half cycles of the power frequency current. The continuity of the combined signals generated is monitored. Under fault conditions attenuation of signals over the associated faulted lines or those lines carrying fault current is likely. This scheme does not rely upon magnitude of received signal but its continuity.

Consider the case of an internal fault on line AB between substations A and B. The power frequency fault currents entering the zone of protection at A and B will only be slightly out of phase. Upon detection of the fault both ends will therefore transmit an intermittent signal simultaneously which, when combined, will still consist of bursts. The relays at substations A and B are arranged to initiate tripping under these conditions.

Under external out-of-zone through fault conditions the current is entering at one end of the line and leaving at the other. The power frequency fault currents at substations A and B will therefore be displaced by approximately 180°. The high frequency signals generated at each end of the line will now, when combined, form a continuous signal. The detection circuits are arranged such that under reception of such a continuous signal tripping is inhibited.

10.6.6.6 *Auto-reclosing*

Records indicate that 80–90% of overhead line faults are developed from transient causes such as lightning strikes or objects coming into contact with the

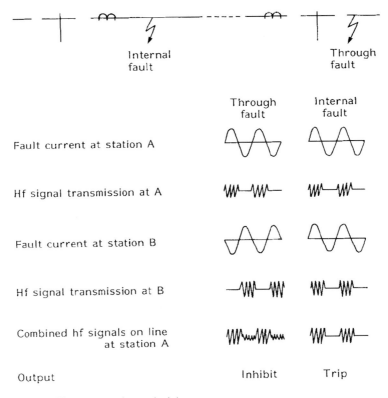

Figure 10.28 Phase comparison principles

lines. Service may be preserved and system stability maintained by rapid fault clearance and re-energization of the line. The theoretical minimum dead time between reclosing attempts are only of the order of a few cycles. To this must be added:

- The actual circuit breaker operating time. When 'multi-shot' schemes (repetitive reclosing usually followed by a 'lock-out' command to inhibit further attempts) are designed the dead time is increased above the single shot requirement in order to take into account circuit breaker recovery times and ratings.
- A de-ionization period to allow for the dispersal of ionized air around the fault path is of the order of 10 to 20 cycles depending upon voltage level.
- When electromagnetic IDMTL protection relays are used in conjunction with auto-reclose schemes it is essential to allow full resetting during the dead time before the next circuit breaker closure attempt. This is of the order of 10 s or more and sets a limitation for auto-reclose dead times on distribution systems with simple current operated protection relays.
- The reclaim time must be sufficient to allow the protection to operate correctly on a persistent fault. Again for IDMTL relays this could be up to 30 s

under low fault level conditions. For definite-time protection 3 s reclaim times or less are common practice.

• The auto-reclosing timer and trip relay must provide a circuit breaker trip signal sufficiently long enough to allow the circuit breaker motor wound spring charge, hydraulic or pneumatic mechanism and solenoid closing to be correctly initiated. This is of the order of 0.1–1 s. Once initiated a practical spring charge mechanism may take 30–60 s to complete spring charging.

10.7 AUXILIARY RELAYS

10.7.1 Tripping and auxiliary

In order to trip circuit breakers and operate alarms more than one relay contact is usually required. In addition, the relay contacts may have to be rated to carry heavy and highly inductive currents. This requires large contacts and high contact pressures. Tripping relays are therefore used to multiply the number of contacts available, provide isolation between the source and system operating element and to meet the required duty. Examples of basic circuit breaker tripping schemes are given in Fig. 10.29. The disadvantage of intermediate tripping relays is that they increase the overall fault clearance time by adding another stage in the fault clearance sequence. Therefore alternatives such as opto-isolators, reed relays and solid state switches may be considered to provide good isolation and faster responses.

Changeover 'all or nothing' relays may be divided into auxiliary and tripping classes. Tripping relays must have the following characteristics:

• Fast, with less than 10 msec operating times.
• High operating contact ratings to operate circuit breaker trip coils.
• Mechanically stable and shock resistant.
• Must not be prone to maloperation.
• Have provision for a visual 'flag' or operation indicator in the relay case.

Relay contacts are referred to as being 'normally open' (n.o.) or 'normally closed' (n.c.) in the normal unenergized state. In order to understand schematic diagrams it must be emphasized that this n.o. or n.c. state has nothing to do with the usual state of the relay when included in its particular circuit. For throwover or bistable relay contacts the convention is for illustration in the state which will apply when the initiating device is in the normally open state. For circuit breakers the auxiliary switches are shown with the circuit breaker main contacts in the open position. Standard symbols for these contacts and different types of relays are illustrated in Chapter 2.

IEC 60255 'Electrical Relays' gives the following nominal DC relay coil voltages: 6 V, 12 V, 24 V, 48 V, 60 V, 110 V, 125 V, 220 V, 250 V and 440 V. The IEC preferred working voltage for all or nothing relays is 110–80% of nominal. The systems designer should be careful when matching battery system

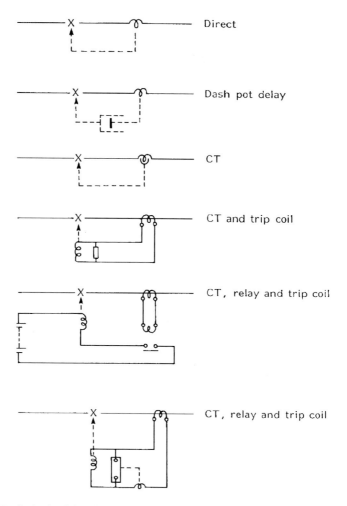

Figure 10.29   Basic circuit breaker trip initiation schemes

and relay coil voltages. A 110 V nominal DC system when using Planté lead acid cells will have a float charge of some 124 V DC. Therefore 125 V relay coils should be used and *not* 110 V coils. In addition, the substation battery voltage at the end of the discharge period could well be below the 80% minimum. British Electricity Supply Standards have recognized this and an assured trip operation at <53% of rated voltage is specified. This is a difficult or at least unusual requirement for many manufacturers to meet and is probably over stringent. For general application it is prudent to stick to the IEC duty range of 70–110% to avoid the risk of excessive tender prices. Addition assurance can be achieved by slightly oversizing the substation battery.

It is essential that the trip circuits do not maloperate owing to earth fault currents.

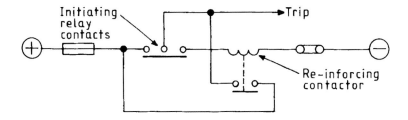

a) Typical shunt reinforcing contactor circuit

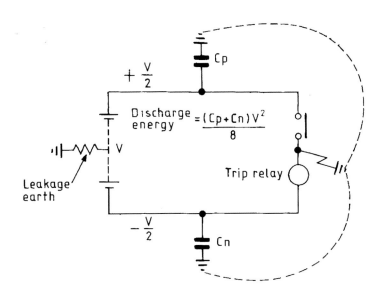

b) Earth leakage paths

Figure 10.30   Effect of earth faults on trip relay operation: (a) typical shunt reinforcing contactor system; (b) earth leakage path

In the circuit shown in Fig. 10.30 an earth fault on the wire between the protection relay contact and the trip relay could cause current to flow through the relay as the wiring capacitance on the negative side of the battery discharges. In a transmission voltage substation there is a large amount of secondary wiring permanently connected to the substation auxiliary supply battery. Therefore a standard test for a tripping relay is that it should not operate when a $10 \,\mu\mathrm{F}$ capacitor charged to $150\,\mathrm{V}$ is connected across the relay coil. If the substation battery centre point is resistance earthed then there is also another path for current to circulate. It is usual to size this resistor to limit the maximum earth fault

Figure 10.31   Attracted armature relay type VAA (courtesy GEC Alsthom T & D Protection & Controls)

current to about 10 mA. As very light duty auxiliary relays will certainly operate at this current level they should not be used in protection trip schemes if spurious signals are to be avoided.

A typical trip relay is illustrated in Fig. 10.31. The attracted armature relay is used for high speed operation (20–40 ms) at high current levels. In order to avoid damage to the protection relay contact if it has to break the trip relay current or sustain it for a long length of time a self-cut-off contact is incorporated which breaks the coil current when operation is complete. For self-reset relays an 'economy' resistor is placed in series with the coil to limit the current once the relay has operated.

Trip relays may be divided into two classes:

1. Lock-out trip relays. Used to ensure that once a circuit breaker has been tripped by the protection scheme it cannot be reclosed either manually or automatically until the trip relay has been reset. The reset function may be either manual or electrical.
2. Self-reset trip relays. Useful for auto-reclose schemes.

Table 10.2   Performance of some trip and auxiliary relays

Manufacturer ABB	Heavy (button contacts)	Medium (finger contact)	Light (reed relay)
Current carrying capacity (1 s)	30–75 A	50 A	2.4 A
Continuous rating	6–10 A	5 A	2 A
Making current L/R > 10 ms	30 A	30 A	2 A
Breaking capacity			
DC L/R <ms @ 24 V	10–20 A	4 A	1 A
@ 125 V	0.7–5 A	0.3 A	0.16 A
Operating times	35–50 ms	20–40 ms	1–5 ms
Operating burdens	2.7–4 W	1.5 W	0.8 W

Manufacturer GEC Alsthom	Tripping	Standard auxiliary	Heavy duty auxiliary
Make and carry capacity (0.5 s)	30 A	30 A	30 A
Continuous rating	5 A	5 A	5 A
Breaking capacity @ 24 V	2 A	2 A	
@ 125 V	0.4 A	0.4 A	2 A
Operating times	10 ms	>10 ms	>10 ms
Operating burdens	25 W	3 W	up to 150 W

Manufacturer ABB	High speed tripping	Auxiliary (button contacts)	
Continuous rating	5 A	3 A	
Making current	25 A	10 A	
Breaking capacity @ 24 V	5 A	–	
(Inductance load) @ 125 V	4 A	0.35 A	
Operating times	20–50 ms	1.8 ms	
Operating burdens	6 W	up to 11 W	

Some typical relay characteristics are detailed in Table 10.2.

10.7.2 AC auxiliary relays

Since small and reliable silicon diodes are available it is nowadays common practice to rectify the AC and use ordinary highly reliable DC relays. If AC relays are used the design must take into account the fact that the holding flux in the armature will fall to zero every half cycle. Small copper shading rings are sometimes employed to cause a phase shift sufficient to damp any flux reversal vibrations or relay chatter.

10.7.3 Timers

Highly accurate solid state timers are available by using digital electronics to divide down and count pulses from a known standard. The time delay may be derived from 50 Hz or 60 Hz mains, a simple free running RC oscillator, an internal piezo electric quartz crystal oscillator, reception of a calibrated radio beacon or satellite transmission. Setting ranges of 0.02 to many hours may be obtained with an accuracy down to less than 0.3% of the set value. Clockwork, wound spring mechanisms and synchronous motors have also been used in the past with 10 to 1 timer ranges and 5–10% accuracy levels. The delayed reset relay is arranged to change the state of the output contacts immediately upon energization. It will also allow a preset delay period of time to elapse before change of state on loss of supply. The delayed operate timer relay allows change of state of its output contacts after some adjustable preset period following energization of the coil.

Simple diodes or RC networks may be applied to relays in order to slow down operation. Some circuits are shown in Fig. 10.32. The application of these delays helps prevent 'races' between circuits operating at slightly different speeds. A mechanical dash pot damper device is also used on some 400 V distribution air circuit breaker operating mechanisms.

D.C.-supplied auxiliary relays drop-out
delayed by a diode

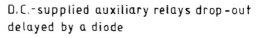

Drop-out delay

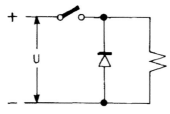

Pick-up delay

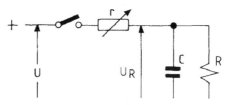

Figure 10.32   Diode and RC circuits used to provide short-time delays (~200 ms)

10.7.4 Undervoltage

DC undervoltage relays are used to provide continuous trip circuit supervision or main substation DC distribution voltages. AC undervoltage relays are used to protect large motors from damage that could occur from running at low voltage, in mains failure monitors and in VT secondary circuits. Again, solid state relays offer advantages over older electromechanical types in terms of accuracy and have typical drop-off and pick-up voltage settings within 1% of each other without cost penalty. Specifications should cover the following points:

- Single or adjustable setting.
- Setting or setting range required.
- Instantaneous- or time-delayed operation.
- Single-or-three phase supervision.
- Resetting ratio or adjustment range required.

10.7.5 Underfrequency

Underfrequency relays are applied to generation plant to avoid fatigue or excess vibration when running outside synchronous speed. The relays operate in conjunction with load shedding schemes in order quickly to bring the generation back into synchronism and up to speed. Digital electronic underfrequency counter relays divide down and compare the supply frequency with an internally derived stable clock reference. A delay is incorporated in order to prevent anomalous operation under transient conditions (DC offsets under fault conditions, harmonics, voltage dips on large motor starting, etc.). The advantage of the underfrequency relay is that it may be installed almost anywhere throughout the network and wired to initiate tripping of particular loads on a customer's premises. In contrast, telecontrol methods tend to operate out to primary substations rather than customer's premises and therefore are less selective in the particular loads shed. The characteristics required are explained in Chapter 1. A useful formula for designing a load shedding scheme is:

$$f_t = f_o\sqrt{\{[e^{-tL/H} \times (L - G) + G]/L\}}$$

where H = Total system inertia – MJ
   L = System load – MW
   G = System generator output – MW
   f_t = Frequency at time t following load or generation change
   f_o = Initial frequency
   t = Time in seconds following disturbance

Figure 10.33 shows a typical set of time – frequency curves for various degrees of system overload with underfrequency relay trip setting points at 49 and 48 Hz. Several stages of load shedding are established in order to

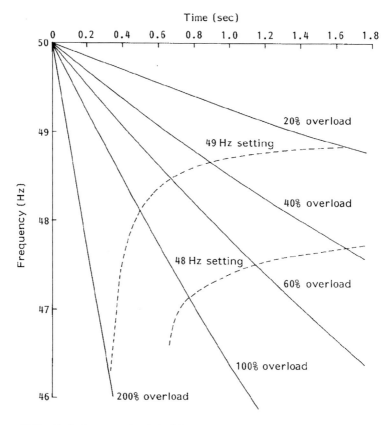

Figure 10.33   Underfrequency load shedding curves

minimize outages whilst maintaining system stability. The load shed should always slightly exceed the theoretical minimum required in order to avoid further, more drastic stages of shedding. A time delay between stages is necessary such that the effect of the first load shedding (which will take several hundred milliseconds depending upon circuit breaker operating times, generator inertia and generator governor response) can be determined before the second stage is started. On a large steam generation system up to five stages of load shedding may be applicable, whereas on a single gas turbine or light diesel installation only one or two stages with 300 ms delays might be applicable.

10.8 COMPUTER ASSISTED GRADING EXERCISE

10.8.1 Basic input data

Consider the simple 11/3.3/0.415 kV distribution network shown in Fig. 10.34. It is necessary to derive suitable protection settings and grading margins for the

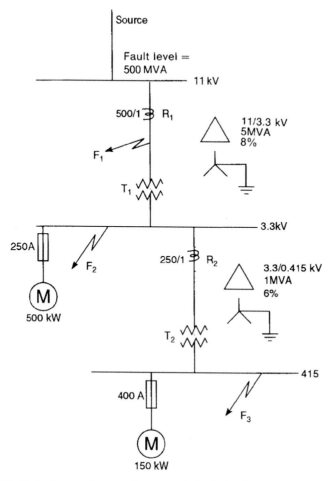

Source

Fault level =
500 MVA

11 kV

500/1 R₁

11/3.3 kV
5MVA
8%

F₁

T₁

3.3kV

250A

F₂

250/1 R₂

3.3/0.415 kV
1MVA
6%

M

500 kW

T₂

415

400 A

F₃

M

150 kW

Figure 10.34   Protection grading exercise: network single the diagram

network. The fundamental formulae for determining network impedance parameters on a similar base is given in Chapter 26, Section 26.7.2. The basic input data and protection requirements are as follows:

- 11 kV source fault level = 500 MVA.
- The fault contribution from large motor loads is ignored in this example. See Chapter 1, Section 1.3.6 for the effects of simplifying assumptions concerning induction motor loads and Chapter 14, Section 14.3.9 for a sample calculation taking induction motor fault contribution into account.
- The protection must only disconnect from the power source the minimum of faulted plant such that services to other users or loads are maintained.

- On a 1 MVA base the network impedances are:
  - (a) Source impedance, $X_{source} = 1/500\,MVA = 0.002\,pu$.
  - (b) 11/3.3kV transformer impedance, $X_{T1} = $ Base MVA $\cdot X$ pu $_{@\,BMVA}/B$
    MVA $= 1 \cdot 0.08/5 = 0.0162\,pu$.
  - (c) 3.3/0.415 kV transformer impedance, $X_{T2} = $ Base MVA $\cdot X$ pu $_{@\,BMVA}/B$
    MVA $= 1 \cdot 0.06/1 = 0.06\,pu$.
  - (d) FLC, $I_{flc\,T1}$ of transformer T1 $= 5000\,kVA/\sqrt{3} \cdot 11\,kV = 262$ A at 11 kV
  - (e) FLC, $I_{flc\,T2}$ of transformer T2 $= 1000\,kVA/\sqrt{3} \cdot 3.3\,kV = 175$ Amps
    at 3.3 kV.

10.8.2 Network fault levels

(a) Fault level on the 415 V busbar, $F_3 = 1/(0.06 + 0.0162 + 0.002) = $
12.8 MVA or in terms of fault current $= 12.8 \cdot 10^6/\sqrt{3} \cdot 415 = 17.8\,kA$.
(b) Fault level on the 3.3 kV busbar, $F_2 = 1/(0.0162 + 0.002)$ 55 MVA or in
terms of fault current $= 55 \cdot 10^6/\sqrt{3} \cdot 3300 = 9.6\,kA$.
(c) Source fault level on the 11 kV busbar, $F_1 = 500$ MVA or in terms of fault
current $= 500 \cdot 10^6/\sqrt{3} \cdot 11000 = 26.2\,kA$.

10.8.3 CT ratios and protection devices

(a) 11 kV primary side of 11/3.3 kV transformer CT ratio taken as 500/1.
(b) 3.3 kV primary side of 3.3/0.415 kV transformer CT ratio taken as 250/1.
(c) A 400 A fuse in conjunction with a motor starter is used to protect the
150 kW, 415 V motor.
(d) A 250 A fuse in conjunction with a motor starter is used to protect the
500 kW, 3.3 kV motor.
(e) IDMTL (extremely inverse characteristic) relays are used to grade with
the fuses.

10.8.4 Relay settings

The characteristics of the relays and fuses are held on a data file and loaded into
the computer. The chosen CT ratios are then entered into the computer. A reference voltage is selected and all protection device characteristics are plotted on a
log–log time–current scale at the reference voltage. With a simple computer
assisted grading program the operator can 'move' the relay characteristics on the
screen to give the required discrimination margin between successive protection
devices. The required plug setting multiplier, (PSM) and time setting multiplier
(TSM) to give this margin can then be read off from the computer screen.

For the relay R_1 located on the primary side of the 11/3.3 kV transformer a
PSM of 100% (i.e. 500 amps) is chosen with a TSM of 0.1 s to discriminate
with the 500 kW, 3.3 kV motor fuse.

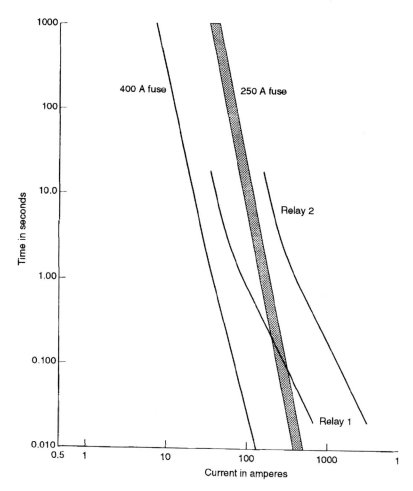

Figure 10.35   Computer generated grading curves

For the relay R_2 located on the primary side of the 3.3/0.415 kV transformer a PSM of 150% (i.e. 375 amps) is chosen with a TSM of 0.1 s to discriminate with the 150 kW, 0.415 kV motor fuse.

Figure 10.35 is an example of computer generated grading curves associated with this example.

10.9   PRACTICAL DISTRIBUTION NETWORK CASE STUDY

10.9.1   Introduction

This section shows how the principles and the different types of protection are used on an actual distribution network. The network chosen for this purpose is the Channel Tunnel power supply network which embodies a traction supply system and a reconfigurable distribution network. The traction supply system feeds a 25 kV single phase-to-earth 2500 A catenary to provide the power to the electric locomotives which haul the international Eurostar trains as well as the special shuttle trains operated by Eurotunnel.

The distribution system is required to support the requirements of the passenger terminals at Folkestone and Coquelles as well as the auxiliary services of the tunnels, such as ventilation and cooling.

To ensure the highest level of security for the power infeed to the tunnel system, the power demand is shared approximately equally between the French and British power systems. The connections to these power systems are at 400 kV to minimize the effects of unbalance and harmonics and then transformed to 225 kV for the French infeed and 132 kV for the UK infeed. These grid infeeds supply main high voltage substations on each side of the Channel. It is in these substations that the segregation of traction and auxiliary systems occurs.

10.9.2 Main substation protection

The main substations on both sides of the Channel are fed from the main grid system by underground cable circuits. On the French side the cables operate at 225 kV and are approximately 2.5 km in length. These cable circuits are protected by a pilot wire scheme using one pair of pilots to transmit a signal proportional to the three phases. On the UK side the cable circuits operate at 132 kV but the feeder length is approximately 14 km. Because of the high capacitance of the charging currents these circuits are protected by three independent pilot wire protection schemes on a phase-by-phase basis.

The 225 kV busbars at the French main substation are protected by a modern electronic type of busbar protection with monitoring systems. The 225 kV system is also fitted with circuit breaker fail protection initiated by all main protections which causes back tripping of the busbars if the current has not ceased within a preset time of the tripping request.

The 132 kV busbars are protected by a high impedance circulating current busbar protection scheme. This scheme utilizes two sets of current transformers on each circuit for check and discriminate relays as described in Section 10.5.5.3. The UK system is also fitted with circuit breaker fail protection.

The auxiliary transformers (225/21 kV) in the French main substation are protected by Buchholz and frame leakage earth fault protection. The frame leakage protection on the transformers works on the same principle as the busbar protection system described in Section 10.5.5.2. The transformer tank is insulated from earth and only one connection via a frame leakage current transformer is provided. On the UK side these transformers (132/21 kV) are protected by biased differential protection as described in Section 10.5.1.2. The transformers are basically the same, however; the difference in the protection used reflects the different philosophies applied in France and the UK.

The 21 kV busbars in both the French and UK substations are protected by high impedance busbar protections using check and discriminating zones as used on the 132 kV busbars.

The protection on the other circuits in the main substations are described under the traction system or 21 kV distribution system in the following sections.

10.9.3 Traction system protection

On the French side, there are three single phase transformers, one dedicated to each phase with a fourth transformer suitable to replace any phase. These transformers are also protected by Buchholz relays and frame leakage earth fault protection with a high set overcurrent protection.

On the UK side because of the interaction with a specially designed load balancer, it is necessary to feed the traction system by three 132/43.3 kV star zigzag transformers. These transformers are protected by biased differential protection as described in Section 10.5.1.2 with IDMTL overcurrent and earth fault backup.

The catenaries have to be capable of delivering up to 2500 A of load current whilst the maximum fault current is limited to 12 kA. This means that the minimum fault current when feeding the full length of the tunnel from one side in downgraded conditions can be lower than the load current. The catenaries are protected by a special distance protection which develops a parallelogram-shaped characteristic with adjustable angle by the use of phase comparators (see Fig. 10.36). This ensures correct detection of the low fault currents compared to the high load current. The protection also incorporates an overcurrent element and a thermal overload alarm.

However, as the overcurrent relay cannot provide a full back-up for low levels of fault current, second distance elements have been added. On the French side

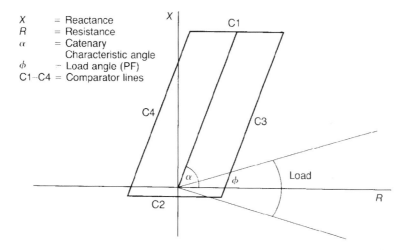

Figure 10.36   Traction protection characteristic

these are fitted on the low voltage side of the traction transformers and act to trip the appropriate phase traction transformer. On the UK side, the second elements are fitted on the catenary circuits but with a breaker fail system operating from the catenary circuit breakers to back trip the traction transformer infeeds.

10.9.4 21 kV distribution system and protection philosophy

The auxiliary distribution system operates at a nominal voltage of 21 kV. The terminal systems being ring main on the UK side and radial with interconnection at the end of the French side. These systems are provided with overcurrent and earth fault protection applied as described in Section 10.4.5.1.

For the tunnel distribution system, the security of the system has been of considerable concern. The basic system is illustrated in Fig. 10.37. Within the tunnel 21 kV distribution system, the system can function satisfactorily with the loss of two circuits. The distribution system consists basically of four circuits, two of which are used to feed the pumping stations (PS) and two which are used to feed the electrical substations (ES) in the tunnel. The electrical substation feeds are split at substation 7 which is approximately the centre of the tunnel, whilst the pumping station feeds are split at Coquelles on the north feeder and Folkestone on the south feeder. The cooling plants which are considerable non-priority loads are connected at Shakespeare Cliff North and Sangatte Shaft South, and two additional feeders are connected from the main substations to these busbars. The system is normally run split as indicated but, on loss of total infeed from either France or the UK, the normally open circuit breakers are closed and the system run solid from the healthy infeed side.

On this system, it is important that faults are cleared quickly and that uncleared faults do not persist. This has led to the use of main protection typically of the unit type with separate back-up protection usually in the form of overcurrent and earth fault protection. The 21 kV busbars at all switching locations are provided with high impedance busbar protection, basically as described in Section 10.5.5.3 but without a check zone. The larger transformers are fitted with differential and restricted earth fault protection with overcurrent and earth fault backup.

The transformers used to feed the cooling plant 3.3 kV busbars, however, do not use restricted earth fault protection. This is because the 3.3 kV earth fault current is limited by an earthing resistor to 30 A, and with a transformer rated at 12.5 MVA with line CTs of 2500/1 A it is impossible to achieve satisfactory sensitivity under these conditions. In this case frame leakage earth fault protection has been used as described for the French auxiliary and traction transformers.

The cable circuits are equipped with pilot wire protection as described in Section 10.5.4 as the main protection. On the pumping station feeders and interconnectors, these are straightforward two-ended pilot schemes. On the electrical substation feeders, the pilot wire protection covers the sections between the major bussing points (ES3, 5, 7, 11, 13) and has to make allowance for the tee-offs in between. The special aspects of the pilot wire protection are discussed in

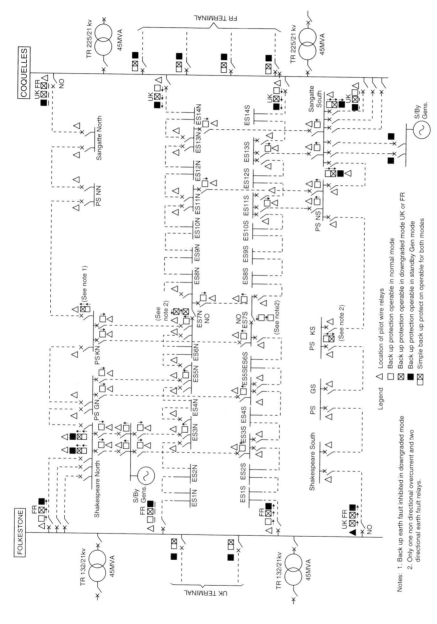

Figure 10.37   Tunnel distribution system showing main and back-up protection

the next section. Fault clearance within 300 msec was the target for the main protection to ensure back-up fault clearance times able to provide full thermal protection of the equipment could be achieved.

Backup overcurrent and earth fault protection (refer to Section 10.4.5.2) are graded to minimize the amount of system lost in the event of the main protection failing to clear the fault. This requirement is complicated by the different configurations required for normal split operation, solid downgraded operation fed from either France or the UK or the modified split arrangement used when operating with the much lower fault infeeds available from the standby generators. This aspect is discussed in detail later.

10.9.5 21 kV pilot wire unit protection

A number of constraints were encountered in applying pilot wire protection to some of the 21 kV feeder sections, which arose from the following system aspects:

1. High in-zone charging current (longest section is 13.6 km).
2. Limited earth fault current (resistance earthing) and low overall fault level with emergency generators.
3. In-zone, fuse-protected, transformer tee-offs on the electrical substation (ES) feeders.

Considering the way in which the pilot wire protection sums the healthy phase in-zone charging currents for an external earth fault on a resistance-earthed system, the overall sensitivity of the pilot wire protection was adjusted to give a 20% stability margin with respect to the operating quantity generated for the most onerous external PH-E fault. This margin allows for system operation at above nominal voltage and for measurement errors without imposing practical sensitivity limitations for all types of internal fault.

Multi-terminal pilot wire protection schemes were ruled out in terms of performance and complexity because of security/reliability concerns. It was decided to apply pilot wire protection with an operating time characteristic that would co-ordinate with fuse operation for a fault on any in-zone transformer. The protection also has to be stable for aggregate magnetizing inrush current in the case of multiple transformers. To meet the latter requirement, the manufacturer's recommendation that the aggregate teed transformer load current should not exceed 50% of the protection three-phase fault sensitivity was adhered to.

The pilot wire relays could have been applied with external definite-time delays to co-ordinate with the transformer fuses, but the allowance for an unpredictable pilot wire relay reset time, following fuse clearance of a transformer fault, would have created unacceptably high back-up protection operating times. Interlocking the pilot wire protection with suitable IDMT overcurrent relays at each end of a feeder section was also ruled out since breaker tripping would be prevented at both ends of a faulted feeder section in the normal case of single end feedings. Double end tripping is necessary to ensure that a fault cannot be reapplied when a supply is rerouted following a fault. The best compromise was

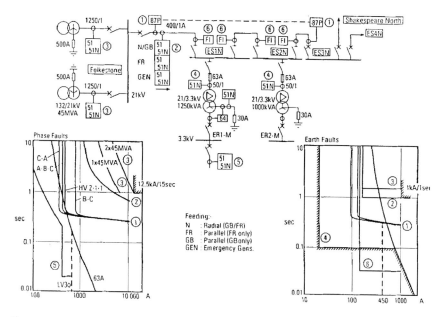

Figure 10.38   Example of 21 kV protection grading

to utilize a pilot wire relay design where the measuring element itself produces an adjustable operating time characteristic. The way in which this grades with the transformer HV fuse is illustrated in Fig. 10.38.

Under normal operation, the earth fault level (1 kA) is high enough to give P/W relay co-ordination with the transformer fuse protection. However, it can be seen from Fig. 10.38 that for low level earth faults full co-ordination cannot be achieved because of the constraints on overall pilot wire sensitivity dictated by co-ordination between transformer fuse protection and back-up protection for HV and LV phase faults. For this reason earth fault relays have been fitted to the transformers to give remote indication if a transformer earth fault causes the pilot wire protection to operate, thus allowing rapid reconfiguration.

10.9.6 21 kV system back-up protection

Sets of microprocessor-based, directional, time-delayed, overcurrent and earth fault protection were strategically located at the positions indicated in Fig. 10.37.

The use of modern, multi-characteristic, microprocessor-based relays provided great flexibility and significant advantages when the relay setting study was carried out. Project lead times were also reduced since contract work could commence before the protection study had been completed in detail. In selecting the type of back-up protection, significant service experience was an important consideration, in addition to economic and technological factors.

In setting phase fault protection, with supplies from one or both countries, three forms of dependent-time (IDMT) relay characteristic were utilized on the multi-characteristic relays. The reduction in phase fault levels away from the intake substations meant that discrimination between feeder relays in the tunnel could be maintained without infringing a 12.5 kA/1 s damage constraint at the main intake substations.

In the cases where one intake substation (S/S) feeds the whole tunnel, with the remote ends of tunnel feeders tied together, natural discrimination is exhibited by IDMT relays on contributing feeders with those on a faulted feeder, even if some have similar settings.

As the system is resistance earthed, earth fault levels do not vary significantly with fault location, and the earth fault current is restricted to 500 A per supply transformer. This meant that it was possible to set up relays with definite-time E/F characteristics to provide discrimination without infringing a cable-screen constraint of 1 kA/1 s. IDMT characteristic grading with the limited earth fault current was not feasible since the relay currents would include a significant, variable, neutral charging current component.

The high accuracy, low overshoot time and consistency of electronic protection, coupled with high speed SF_6 breakers, were factors fully exploited in determining the minimum grading margins. It was necessary to adopt:

dependent-time relay margin $= 0.2t(d) + 0.135$ s

definite-time relay margin $= 0.06t(d) + 0.135$ s

where $t(d)$ is the downstream relay time.

To ensure full discrimination of phase and earth fault back-up protection, in the case of one national supply only, it is necessary for protection settings to be modified at the remote intake substation to give a faster response. It is also necessary effectively to adjust the definite-time delay of some sets of earth fault protection within the tunnel. Figure 10.37 indicates where and under what circumstances protection settings are effectively altered.

For operation with supplies from emergency generators only, additional back-up protection setting changes are required at the intake substations and also at Shakespeare Cliff and Sangatte Shaft. Under such circumstances it is more appropriate for the O/C relays at these locations to have definite-time characteristics and more sensitive settings, to ensure that the protection will respond to the limited generator fault current contribution with its decrement. Earth fault protection also needs to be made more sensitive at these locations.

Where setting modifications are required for O/C and E/F protection, mainly at the intake substations, the setting changes are brought about by switching between two or three sets of protective relays; relay selection being under the control of electrically operated/reset latching relays with contacts acting on control terminals of the protective relays. Where it is necessary only to alter the definite-time delay of earth fault protection, for example within the tunnel, this is accomplished by the selection and insertion of external time delay relays; again under the control of latching relays. The latching relays are switched by a common command from the control centre which trips circuit breakers, load sheds and sets the protection in readiness for the appropriate configuration.

10.9.7 Use of earth fault indicators

On the electrical substation (ES) feeders, because circuit breakers are not fitted at each substation, it is necessary to identify the location of a fault quickly so that the system can be reconfigured, by opening disconnect switches, and re-energized. This is achieved by installing strategically located earth fault passage indicators. The same technique has also been used on the 3.3 kV distribution system which also uses ring main units.

The indicators, and their associated core balance CTs, are custom built to ensure an operating current of less than half the minimum system E/F level, but in excess of three times the normal charging current for the longest length of cable that could be fed downstream of an indicator (to prevent sympathetic operation on a healthy feeder). The indicator design operating currents are 150 A for 21 kV and 15 A for 3.3 kV. A contact and a DC reset coil have been provided on each indicator to allow remote indication and resetting, via telecontrol, in addition to the local facilities.

10.9.8 Summary

It can be seen how, in a complex distribution network such as that of the Channel Tunnel, each of the different protection principles described in the earlier sections of this chapter has a part to play in providing high speed selective and reliable protection:

- Current operated relays with differing characteristics to provide main protection in the terminal and back-up protection in the tunnel.
- Differential protection schemes in the form of busbar protection, restricted earth fault, biased differential transformer protection and plain and biased pilot wire schemes for cable protection.
- Distance protection to protect the traction system catenary with a specially shaped parallelogram characteristic to give the discrimination between low fault current and high load current.
- Tripping and auxiliary relays used for protection switching.

10.10 RECENT ADVANCES IN CONTROL, PROTECTION AND MONITORING

10.10.1 Background

Substations contain amongst other systems, subsystems specific to control & protection. They are:

(a) Control panels with:
- control switches for manual control;
- annunciators and lamps for alarms;

- selector switches and meters for metering of current, voltage, power, energy, etc.
- recorders for trend monitoring and historical logging;

(b) Relay panels with:
  - protective and auxiliary relays, suitably arranged;
  - fault & event recorders for fault analysis & logs;

(c) Communication panels that provide interfaces to the PLC communication system

(d) RTU panels which:
  - digitise binary and analogue information from the substation for transfer via the communication system to the SCADA control centre;
  - execute control commands from the control centre.

(e) Interface panels to provide level of interface and isolation between engineers from the SCADA communication, protection and operations department of a utility.

Conventionally all this equipment gets interconnected with miles of cabling, which results in lengthy engineering and testing processes during the substation installation stage. In modern substations new technology has been implemented to increase the reliability of the installation as well as reduce its size and cost. To this end, a large amount of integration has taken place in the above systems and products, shrinking the footprint and reducing the overall complexity.

10.10.2 Developments

Integration has resulted in the terms such as relays, control panels, etc. being replaced by the term Intelligent Electronic Device (IED) and Substation Automation (SA) systems (see Figure 10.39).

A substation engineered on this basis would have one or a number of such IEDs per HV bay connected to the process (CT, VT, CB, isolator...) on one side, galvanically, and communicating via an optic digital Ethernet bus to a computerized control & monitoring (SA) system on the other side. The control system also communicates upwards to the SCADA system at the control centre via fibre optic or other channels. In some installations, security and operational reasons dictate the segregation of control from protection.

An IED today is a compact cost effective product that could cover protection, local control, recording, monitoring and communication all in one box. Communication standards such as IEC 61850 ensure that the communication protocol and format is standardized across various vendors paving the way towards IED inter-operability and, it is to be hoped, IED inter-changeability in the future. All of this reduces the number of panels and the wiring. Furthermore, the IED is 'multi-function' and it is not uncommon to have a large number (20–30 or more) of protection functions in one device due to its high processing capacity. Self-supervision and watchdog functions ensure high availability for such devices (see Figure 10.40).

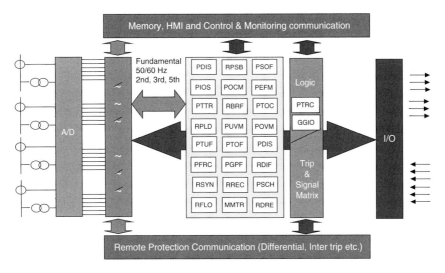

Figure 10.39   IED with a modular concept (courtesy of ABB Ltd UK)

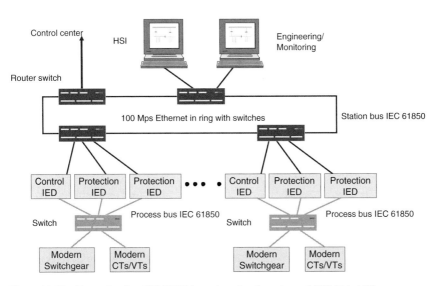

Figure 10.40   Example of an IEC 61850 based station (courtesy of ABB Ltd., UK)

Some of these IEDs, with special functions such as line differential, are even provided with built-in geographical positioning system (GPS) cards to achieve highly reliable microsecond accuracy time stamps of internal signals at source. This is a prerequisite for accurate and reliable line differential protection and advanced monitoring applications.

Such systems will in future include optical data communication to 'optical' CT's and VT's (see also Chapter 5), intelligent breakers, etc. with the ultimate result perhaps of moving towards a substation with 'copper less' controls and protection.

REFERENCES

1. Electricity Training Association, *Power System Protection*, Vols 1–4, IEE, Electricity Association Services Ltd., London, 1995.
2. Anderson, P. M., *Power System Protection,* Wiley-IEEE Press, December 1998.
3. *Protective Relays, Application Guide*, GEC ALSTHOM Protection & Control Ltd.,` Stafford, 1995.

11 Fuses and Miniature Circuit Breakers

11.1 INTRODUCTION

Fuses act as a weak link in a circuit. They reliably rupture and isolate the faulty circuit under overload and short circuit fault conditions so that equipment and personnel are protected. Following fault clearance they must be manually replaced before that circuit may be put back into operation. Striker pins are available on some designs such that remote alarms may be initiated on fuse operation.

Miniature circuit breakers (MCBs) or moulded case circuit breakers (MCCBs) are also overcurrent protection devices often with thermal and magnetic elements for overload and short circuit fault protection. Earth leakage protection, shunt trip coils and undervoltage releases may also be incorporated in the designs. As a switch they allow isolation of the supply from the load. Normally the MCB requires manual resetting after a trip situation but solenoid or motor driven closing is also possible for remote control.

This chapter describes the various types of fuse and MCB together with their different uses and methods of specification. Examples and calculations for correct selection of different applications are also given.

11.2 FUSES

11.2.1 Types and standards

11.2.1.1 *General*

Table 11.1 gives a summary of different fuse types, their uses, advantages and disadvantages. Table 11.2 summarizes some current relevant standards

Table 11.1   Summary of fuse types

Category	Types	Use	Advantages and disadvantages
1. High voltage fuses above 1000 V AC	Expulsion types IEC 60282-2	Outdoor and indoor network protection	Cheap replaceable element. Arc extinguished by expulsion effect of gases and therefore needs good clearances. Not current limiting.
	Capacitor protection IEC 60549, 60871, 60143	Protection of shunt and series power capacitors.	Unit fuses. Clearance of faults within capacitor unit. Permits continued operation of residue of shunt capacitor bank. Line fuses to isolate faulted bank from the system. Refillable types available. Resist a stated level of repetitive discharge I^2t. System requires mechanical switching devices.
	Transformer protection IEC 60787	Transformer circuit protection and co-ordination.	Good selection guide.
	Motor circuit applications IEC 60644	For use with direct on-line (DOL) AC motors.	Withstands motor starting currents. Slow operation in the 10 s region (high K) combined with fast operation below 0.1 s to retain good short circuit limitation. Usually back-up types.
	Current limiting types IEC 60282-1	Networks and high power industrial uses with strikers. Switchgear tripping. Prevents single phasing. Gives indication of operation.	Limits short circuit energy. Cheaper than circuit breakers. Special types may be oil immersed. Takes time to replace fuses when restoring supply but exact restoration of characteristics ensured, whereas circuit breakers may need maintenance. Isolating switch required.
	For correct striker-pin operation see IEC 62271-105.		
2. Low voltage fuses below 1000 V AC – IEC 60269-1 – BS88	High rupturing capacity (HRC) types IEC 60269-2	Supply networks. Industrial protection with ratings up to 1250 A, and breaking capacity 80 kA. For use by authorized persons.	Interchangeable ratings within range of fuse carrier size. Comparatively inexpensive limitation of short circuits. Accurate time/current characteristics for a variety of applications. Quick and easy replacement with cartridge of the correct

Table 11.1 Continued

Category	Types	Use	Advantages and disadvantages
			type but longer than reclosing circuit breakers. Very fast acting on short circuit I^2t and overvoltage very carefully controlled.
	Protection of consumer units Domestic types IEC 60269-3, BS1361	Special semi-conductor protection IEC 60269-4 Breaking capacity to 33 kA	Ratings may not be interchangeable for safety reasons when replaceable by a domestic consumer. Special types for electricity supply utility replacement ensure discrimination giving negligible chance of anomalous rupture of Supply Utility fuse.
		Fuse links in plugs BS1362	Ratings interchangeable– 13 A for power to 3 A for lighting with 5 A, 2 A and 1 A ratings for other applications. High breaking capacity for small size. Cheap and easy to replace. Remain stable when carrying current for long periods.
	Semi-enclosed rewirable types BS 3036	Protection of subcircuits, breaking capacity 1 to 4 kA.	Economical where frequent short circuits occur. High fusing factor, low breaking capacity. Less efficient limitation of short circuits. Variability of characteristics after rewiring. Deterioration in use.
3. Miniature fuses. Breaking capacity below 2 kA – IEC 60127	Higher breaking capacity sand-filled cartridge types IEC 60127-2 Low breaking capacity air-filled cartridge types IEC 60127-2 Low breaking capacity subminiature types IEC 60127-3 Fuse holders IEC 60127-6	Protection of electronic and similar apparatus	Cheap, large range of characteristics from quick acting to long time delay. Interchangeable. Assist rapid mainten-ance by isolating parts of electronic circuitry. Avoid use on high prospective fault current circuits and replacement by incorrect types.

Table 11.2   Summary of IEC, BS and North American fuse standards

Description	IEC	BS	USA
Definitions	60050-441	EN60269	UL 248.1
		2692	ANSI/IEEE C37.40
Low voltage	60269	88	UL248
Semi-enclosed	–	3036	–
LV contactors	60947-4	5424	
Industrial	60269-1 and 2	88-2	
Fuse switchboard	–	5486	
High voltage	60282	EN60282-1	IEEE C 37.1
Motor circuits (HV)	60644	EN60644	IEEE C 37.46
HV contactors	60470	EN60470	
HV starters	60470	EN60470	
Distribution type	60282-2	2692-2	ANSI/IEEEC37.40,41,42,47
Semiconductors	60269-4	88-4	UL248-13
Capacitors (HV)	60549	5564	
Isolators and switches	60129	EN60129	
	60265-2	EN60947-3	ANSI/IEEE C37 series
		EN60265	
Oil immersed type			ANSI/IEEE C37.44
Design tests			ANSI/IEEE C37.42

covering fuses. There are various categories ranging from subminiature electronic and solid state device protection fuses, power types (expulsion and high rupturing capacity (HRC)) to 72 kV.

A trend to harmonize fuse types (National Standards – BS, etc.; European – CENELEC; and International – IEC) is currently in progress speeded up by the pan-European mergers between large manufacturers. For example, revisions to BS88 Parts 1 to 5 were introduced in 1988 to coincide with IEC Standards and an additional Part 6 introduced. General purpose fuses are given the classification 'gG' where 'g' indicates full range breaking capacity and 'G' indicates general application. Fuses for application in motor circuits are given the classification 'gM' and are characterized by having essentially two current ratings, I_n and I_{ch}. I_n denotes the rated current of the associated fuse holder and the second value, I_{ch} gives the operational characteristics. For example, a 32M63 fuse link has the operational characteristics of a 63 A fuse link but its continuous rating and size is restricted to that of a 32 A fuse holder.

11.2.1.2 Standard conditions of operation

As the behaviour of fuses is affected by environmental conditions, it is important to check this aspect before determining ratings. The following are usually included in a specification:

- Ambient temperature – the IEC standards call for low voltage fuses to be suitable for ambient temperatures between -5 and $+40°C$, while high voltage fuses must operate satisfactorily from -25 to $+40°C$.

- Humidity – a typical requirement is that satisfactory operation should be obtained in relative humidities up to 50% at +40°C (and higher levels at lower temperatures).
- Altitude – sometimes overlooked – LV fuses meeting IEC specifications must be suitable for operation up to 2000 m, but the IEC specification for HV is only 1000 m.
- Pollution – it is usual for standards to contain statements to the effect that the ambient air 'should not be excessively polluted' by dust, smoke, corrosive or flammable gases, vapour or smoke. Specifiers should therefore make special note of coastal or industrially polluted atmosphere.

11.2.2 Definitions and terminology

The major terms and definitions associated with fuses are described in Table 11.3. A fuller range is provided in Ref. (5), which in turn derives its list from IEC standards 60127, 60269 and 60282.

11.2.3 HRC fuses

The high rupturing capacity (HRC) fuse has excellent current and energy limiting characteristics and is capable of reliable operation at high prospective rms symmetrical current fault levels (typically 80 kA at 400 V and 40 kA at 11 kV). Fuses are available in ratings up to 1250 A at low voltages and, say, 100 A at 11 kV, and normally packaged in cartridge format. The fuse operates very rapidly under short circuit fault conditions to disconnect the fault within the first half cycle and therefore limit the prospective peak current.

The fuse element traditionally consists of a silver element. Recent research and development by some manufacturers has allowed copper to be used when problems of increased pre-arcing I^2t, less pronounced eutectic alloying ('M') effect and surface oxidation are overcome. In some cases the performance of the copper element fuses actually surpasses that of the silver types.

The silver or copper strip element is perforated or waisted at intervals to reduce power consumption and improve the tolerance to overloads as shown in Fig. 11.3. The fuse operation consists of a melting and an arcing process. Under high fault currents the narrow sections heat up and melt. Arcing occurs across the gaps until the arc voltage is so high that the current is forced to zero and the fuse link ruptures. The operation of a typical 100 A HRC-rated fuse under short circuit conditions is shown in Fig. 11.4.

Under low fault current or overload conditions the whole centre part of the fuse element heats up uniformly as heat is conducted from the narrow sections to the wider parts. The centre section then eventually melts. Low melting point alloys deposited at points on silver or copper fuse elements (Fig. 11.3) are used to delay the fuse operation. Alloys with melting points of approximately 180°C

Table 11.3   Useful terms and definitions

Item	Description
Fuse	The complete device including the fuse holder and fuse link. Fig. 11.1 shows a semi-enclosed rewirable fuse and a filled cartridge type with bolted end connection arrangements.
Fuse link	The replaceable part, normally in cartridge form containing a fuse element that melts under overload or short circuit conditions.
Fuse holder	The combination of fuse base and fuse carrier.
Ambient air temperature	The temperature of the air outside the fuse enclosure. Note that the performance of fuses, and to an even greater extent MCBs, is affected by the ambient temperature and the type of thermal characteristics of the enclosure. Cartridge fuses have different characteristics when mounted in a fuse holder compared to the standard (IEC 60269) test rig.
Switch fuse	A switch in series with a fixed fuse.
Fuse switch	A switch where the fuse link (or carrier) forms the moving contact of the switch.
Fusing current	The rms current which will melt the fuse element in any specified time from the commencement of current flow.
Fuse breaking capacity rating	The maximum prospective current that can be broken by a fuse at its voltage rating under prescribed conditions.
Prospective current	The rms value of the alternating component of current that would flow in the circuit if the fuse were replaced by a solid link.
Minimum fusing current	The minimum current capable of causing the fuse to operate in a specified time.
Current rating	The current that the fuse link will carry continuously without deterioration.
Cut off	If the melting of a fuse element prevents the current reaching the prospective current then the fuse link is said to 'cut off'. The instantaneous minimum current obtained is then the 'cut-off current'.
Pre-arcing time	The time between the commencement of a current large enough to cause a break in the fuse element and the instant that the arc is initiated.
Arcing time	The time between the instant when the arc is initiated and the instant when the circuit is broken and the current becomes permanently zero.
Total operating time	The sum of the pre-arcing and arcing times.
Let through I^2t (Joule integral)	The integral of the square of the current over a given time.
Fusing factor (At present fuses tend to have different characteristics depending not only upon type but also on the standard to which they are manufactured. IEC 60269 sets time 'gates' for maximum and minimum fusing currents at set times. see Fig. 11.2).	A fuse must reliably carry full load current and small overloads such as transformer magnetizing inrush currents, capacitor charging currents and motor starting currents for a short time. The ratio between the rated current and the minimum fusing current is the fusing factor and is normally 1.45 or as low as 1.25. At such overloads the fuse will melt in about 1 hour and at higher currents more quickly.

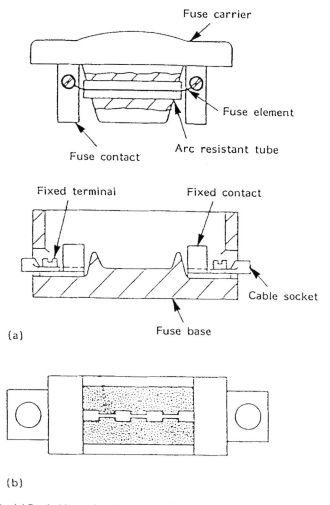

Figure 11.1   (a) Rewirable semi-enclosed fuse; (b) quartz sand-filled cartridge fuse

and 230°C are used for silver- and copper- (melting points approximately 1000°C) based fuse elements. When the alloy reaches its melting point it combines, after a delay, with the main fuse element material to produce a eutectic with a slightly higher melting point than the alloy alone but a much reduced overall fuse material melting point. This allows the main fuse element to melt at low overcurrents.

Especially fast acting, low I^2t let through (low Joule constant, see Section 11.3.2.1) and high current HRC fuses are required to protect power semiconductor devices because of the low thermal mass and very short times for the semiconductor devices to achieve thermal runaway to destruction.

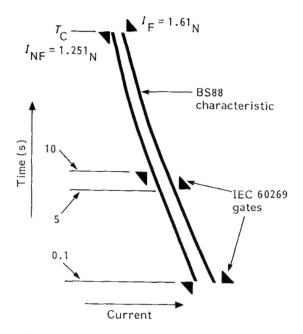

Figure 11.2   IEC 60269 time/current gates for type gG fuses

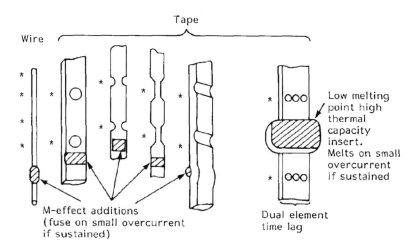

Figure 11.3   Techniques of time delay on a selection of types of fuse element

11.2.4 High voltage fuses

11.2.4.1 *HRC types*

The construction is similar to the low voltage type except that the element must be longer, with more constrictions because of the higher arc voltage that

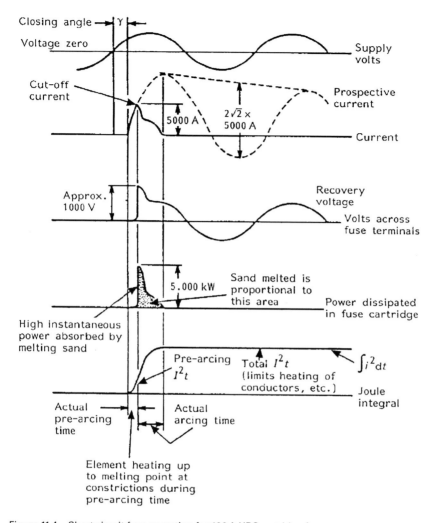

Closing angle → γ ←

Voltage zero
⎯⎯⎯⎯⎯⎯⎯⎯ Supply volts

Cut-off current
Prospective current

5000 A
$2\sqrt{2} \times$ 5000 A
Current

Approx. 1000 V
Recovery voltage
Volts across fuse terminals

5.000 kW
Sand melted is proportional to this area
Power dissipated in fuse cartridge

High instantaneous power absorbed by melting sand

Pre-arcing I^2t
Total I^2t (limits heating of conductors, etc.)
$\int i^2 dt$
Joule integral

Actual pre-arcing time
Actual arcing time

Element heating up to melting point at constrictions during pre-arcing time

Figure 11.4   Short circuit fuse operation for 100 A HRC cartridge fuse

must be developed to interrupt the current. Such designs must have safe low-overcurrent operation, time/current characteristics to suit the application (for example, distribution transformer HV protection), fully adequate current and energy limitation characteristics under short circuit conditions, be of robust mechanical construction and available in standard fuse dimension packages. The element is commonly helically wound on a ceramic former.

Such an arrangement is not particularly suitable for motor protection fuse applications at, say, 3.3 kV to 11 kV because of the thermal stresses imposed upon the element under frequent starting conditions. Fuses required for such applications (for example, in series with vacuum contactors of insufficient fault rating) have straight corrugated self-supporting elements to accommodate the stresses involved (Fig. 11.5).

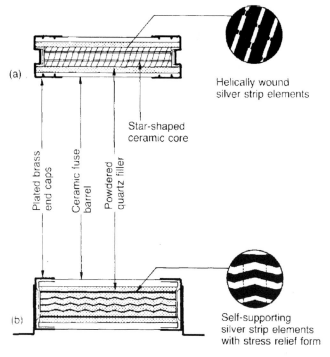

Helically wound
silver strip elements

Star-shaped
ceramic core

Plated brass
end caps

Ceramic fuse
barrel

Powdered
quartz filler

Self-supporting
silver strip elements
with stress relief form

Figure 11.5   Main constructional features of GEC high voltage fuses: (a) typical distribution
fuse; (b) typical motor circuit fuse

11.2.4.2 *Expulsion types*

Unlike the HRC type the fuse element is contained within a narrow bore tube
surrounded by air. Under fault conditions the fuse element melts and an arc is
struck across the break. The heat of the arc vaporizes a material such as resin-
impregnated fibre lining the inner tube wall and this, added to the arc vapour,
rushes out of the tube ends at high velocity. The gas movement assisted by the
cooling and de-ionizing effect of the vaporized tube wall products extinguishes
the arc. Suitable dimensioning allows reliable fault clearance down to the minimum fuse element melting current. The spring tension arrangement shown in
Fig. 11.6 allows the fuse break to enlarge when arcing commences thus increasing the arc voltage and aiding extinction. Such fuse types are usually employed
on outdoor distribution equipment and overhead line poles. The mechanism
allows the top contact to disengage on fuse operation so that the fuse carrier
tube falls outwards about the lower hinge. This allows isolation and avoids leakage along the tube due to build up of arc deposits. It also makes it easy to spot
fuse operation by the overhead line inspection/repair team. Expulsion fuses are
not silent in operation and additional clearances are required to avoid ionized
gases causing flashover.

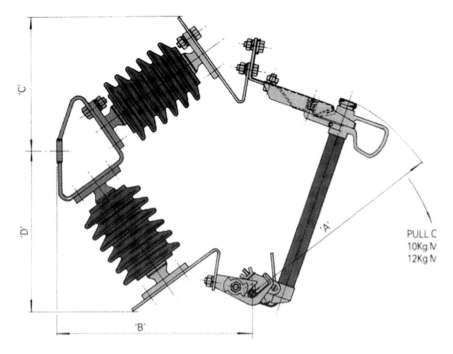

Figure 11.6   Expulsion fuse arrangement

The advantage of the expulsion fuse element is that it has characteristics well suited to small distribution transformer protection. Slow and fast blowing characteristics are available (Fig. 11.7). Its small surface area and air surround produces rapid operation under moderate fault conditions. Lack of current restrictions gives a much slower high fault current operation time. The device is not current limiting and therefore has a rather low breaking capacity limit.

11.2.4.3 *Maximum instantaneous short circuit current, I_S, limiter*

A practical difficulty exists in producing high voltage HRC fuses at the higher current ratings. Following the installation of additional generation onto a system or perhaps the reinforcement of a system by the introduction of various inter-connections the fault levels will inevitably increase. Sometimes this increase is beyond the capability of the existing switchgear. A choice then has to be made on whether to replace the switchgear or introduce fault limiting devices such as series reactors or the I_S limiter.

The I_S limiter acts like an HRC fuse and may be placed in series with the equipment to be protected. It is available for rated voltages in the range 0.75 to 36 kV. It limits the mechanical stresses on equipment by limiting the maximum instantaneous short circuit current. The cut off is very rapid such that the

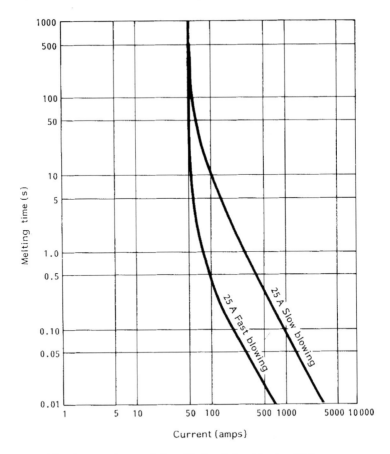

Figure 11.7   Time/current characteristics of fast and slow blowing HV fuses

short circuit current reaches only about 20% of the unrestrained prospective current peak and is completely cut off in typically 5 ms with a low resulting overvoltage. The AC component of the fault, which stresses the equipment thermally by the heat generated, is also minimized. An oscillogram of the interruption of a single phase with an I_S limiter is shown in Fig. 11.8.

The I_S limiter essentially consists of three components as shown in Fig. 11.9:

- An adjustable electronic sensing circuit and integral current transformer which are set to interrupt the fault depending on the rate of rise of fault current and a minimum fault current value.
- A main current conductor which contains a small explosive charge. When the tripping signal is received from the sensing circuit the main conductor is interrupted by this charge.
- A quenching circuit consisting of a lower current-carrying capacity HRC fuse which is connected in parallel with the main current conductor.

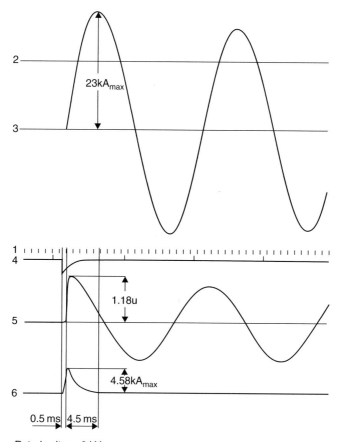

Rated voltage 6 kV

1 Time base
2 Voltage across I_S-limiter insert
   bridged by copper bar
3 Short-circuit current without I_S-limiter
4 Tripping pulse
5 Voltage across I_S-limiter under
   operation
6 Short circuit current with I_S-limiter

Figure 11.8   Oscillogram of a single-phase interruption with an I_S-limiter

Following breaking of the main current conductor the fuse circuit rapidly
(0.5 ms) completes the fault clearance.

The advantages of the I_S limiter are:

● Considerable cost savings compared to alternatives such as replacement of
  existing switchgear with higher fault-rated equipment or the introduction
  of fault limiting reactors into a system.

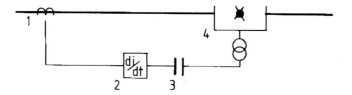

1) A CT measures the short circuit current

2) An electronic circuit controls the tripping

3) A pulse condenser provides the power via a pulse transformer for firing the detonator

4) Is-Limiter

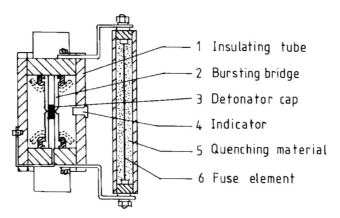

1 Insulating tube

2 Bursting bridge

3 Detonator cap

4 Indicator

5 Quenching material

6 Fuse element

Figure 11.9 I_S-limiter components

- Operating costs for the I_S limiter are nil. Reactors would introduce losses into the system.
- Further increases in current rating may be obtained by operating individual units in parallel per phase.

The disadvantages of the I_S limiter are:

- Control circuitry is somewhat complex and a risk of maloperation does exist.
- Replacement of inserts in the event of operation involves expenditure and is necessary before supply can be restored. A spares holding is therefore necessary.

- Compliance with the Health and Safety Executive Electricity at Work Regulations in the UK and similar regulations abroad requires periodic testing of the units and a record of test results to be maintained. A test kit is available from the manufacturers for this purpose.

11.2.4.4 *Automatic sectionalizing links (ASLs)*

These devices, sometimes referred to as 'smart fuses', are not fuses in the sense of operating through the fusion of metallic elements. However, they are designed as retrofit replacements for conventional expulsion fuse carriers and are installed and maintained using fuse operating poles, so they are included in this chapter.

The ASL fulfills the same function in a distribution network as an automatic sectionalizing switch or sectionalizer. It is installed in an overhead system, downstream of an auto-reclose circuit breaker, and minimizes the extent of the system outage in the event of a permanent network fault.

Making use of advances in solid-state microelectronics, a miniaturized logic circuit is fitted inside the dimensions of the carrier tube of a conventional expulsion fuse, and installed accordingly. The tube is a conductor instead of an insulator, so the current flows through the tube and encircling current transformers feed information regarding the state of this current into the logic circuit within.

On the occurrence of a transient overcurrent, the upstream auto-recloser or circuit breaker will trip and then reclose. The ASL records this event and retains it in its memory for several seconds. If on reclosure the load current has returned to normal, the ASL erases that memory and the circuit returns to normal. If however the fault is permanent, the fault will be present when the recloser re-energizes the circuit. The ASL logs this second event and awaits the second trip of the recloser. When the ASL detects that the line current has fallen to zero, it fires a chemical actuator (similar to a fuse striker) which de-latches the ASL carrier tube so that it swings down and provides safe isolation of the downstream fault. The remainder of the network reverts to normal operation. In some versions of the ASL, a resettable magnetic latch is in place of the chemical actuator. This obviates the need to replace the actuator, but is more expensive.

11.2.5 Cartridge fuse construction

The cartridge fuse consists of a fuse element surrounded by a pure quartz granular filler contained in a tough ceramic enclosure. The filler allows the fuse element arc vapours rapidly to condense and avoid pressure build-up within the enclosure. It also aids heat dissipation from the fuse element thereby allowing a smaller quantity of fuse element material to be used, again reducing the pressure in the cartridge. Having a number of fuse elements within the cartridge in parallel increases the surface area in contact with the filler, assists heat dissipation and helps arc extinction. Good filler material quality control is essential for repeatable minimum fusing current characteristics.

The striker-pin fuse is a variation of the standard cartridge fuse link. A high resistance wire in parallel with the fuse element melts when the fuse operates and detonates an explosive charge. The charge fires out the striker pin from the fuse end cap (see Fig. 11.10). The striker-pin operation from any one phase is normally arranged in the associated three-phase switchgear to trip out all three phases virtually simultaneously.

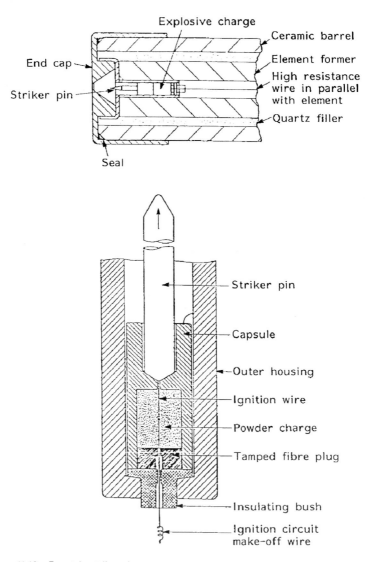

Figure 11.10   Fuse trip striker-pin arrangements

11.3 FUSE OPERATION

11.3.1 High speed operation

Normally the short circuit current will reach a very high value limited by the system source impedance to the fault and the fault impedance itself. Where the fault impedance is approximately zero the fault current will equal the 'maximum prospective short circuit current' (see Fig. 11.4). Most fuses are designed to interrupt the fault so quickly that the current never reaches its maximum value and therefore acts as a current limiting device. On the other hand, an expulsion fuse acts quite slowly and does not limit the current.

On AC circuits the rise of current depends on the circuit parameters and the point in the cycle when the short circuit occurs. At high fault currents the very short clearance times will vary according to the phase angle, and below 100 ms a range of clearing times is possible for each type of fuse. Fuse performance is therefore generally tested for two onerous conditions:

- Arcing (after the fuse element melts) must commence at a point on the voltage wave between 40° and 65° to test thermal stresses.
- Between 65° and 90° to test electromagnetic stresses.

11.3.2 Discrimination

11.3.2.1 *Joule integral*

The energy required to melt the fuse element varies only slightly with the prospective fault current and is almost constant for a particular type of fuse. This is measured by a constant called the 'Joule integral' or 'I^2t' value. Therefore for short operating times below 100 ms the 'I^2t' value is used for grading series fuses. Discrimination is achieved when the total I^2t of the minor (downstream – load side) fuse link does not exceed the pre-arcing I^2t of the major (upstream – power source side) fuse. I^2t characteristics and tabulated values for low voltage fuses in the range 125–400 A are given in Table 11.4.

At longer operating times above 100 ms cooling occurs and more energy has to be input into the fuse element so that current/time discrimination curves must be used. A variety of fuse characteristics are available. An example of typical time/current characteristics for general purpose low voltage HRC fuses in the range 125–400 A are given in Fig. 11.11.

Satisfactory discrimination time/prospective current curves between fuses and between an IDMTL relay and fuses in two alternative 11 kV/415 V radial connected circuits are shown in Fig. 11.12. It should be noted that an 'extremely inverse' IDMTL relay characteristic is available for grading between relay operated circuit breakers and fuse protected circuits.

Table 11.4 Tabulated I^2t characteristics for HRC fuses, 125–400 A (courtesy of Cooper–Bussmann)

| Fuse link reference | I^2t (amps² secs) | | | | Nominal watts loss at full load |
	Pre-arcing	Total at 415 V	Total at 550 V	Total at 660 V	
125N	30 000	52 000	75 000	150 000	12
160N	67 000	120 000	170 000	335 000	13
200N	120 000	210 000	300 000	590 000	15
250N	220 000	390 000	550 000	1 100 000	19
315N	340 000	600 000	870 000	1 700 000	25
355P	490 000	870 000	1 250 000	2 450 000	28
400P	670 000	1 200 000	1 700 000	3 350 000	32

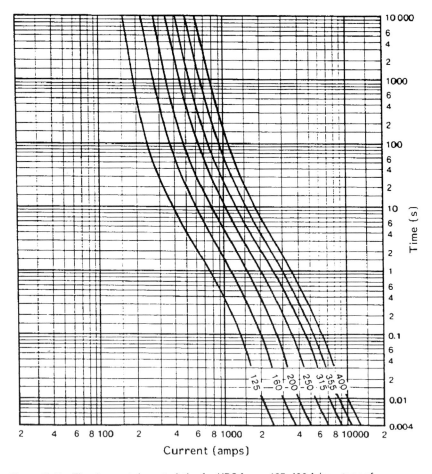

Figure 11.11 Time/current characteristics for HRC fuses, 125–400 A (courtesy of Cooper–Bussmann)

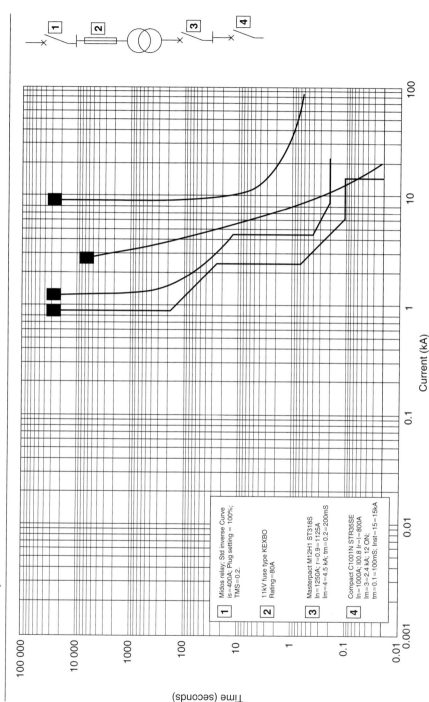

Discrimination study

1 Midos relay; Std inverse Curve
is = 400A; Plug setting = 100%;
TMS = 0.2.

2 11kV fuse type KEXBO
Rating = 80A

3 Masterpact M12H1 ST318S
In = 1250A; r = 0.9 = 1125A
Im = 4 = 4.5 kA; tm = 0.2 = 200mS

4 Compact C1001N STR35SE
In = 1000A; I00.8 Ir = 800A
Im = 3 = 2.4 kA; 12 ON;
tm = 0.1 = 100mS; Inst = 15 = 15kA

Current (kA)

Time (seconds)

Figure 11.12   Discrimination on a radial feeder

Cut-off characteristics for 4–1250 A, Cooper–Bussman HRC fuses are given in Fig. 11.19 for comparison with Merlin Gerin Compact series MCCB current limitation curves.

11.3.2.2 *Earth loop impedance*

IEC 60364-1 gives guidance for safe installations and maximum permissible disconnection time/touch voltages in building services applications. The earth loop impedance must be such that under earth fault conditions sufficient fault current flows to trip the circuit breaker or operate the fuse and isolate the fault in time. Where this is not possible residual current circuit breakers (RCCBs) must be used to ensure rapid isolation of the fault. RCCBs designed for domestic applications only have a low fault breaking capability (typically 1 kA at 0.8 pf). Therefore it is very important to check this parameter before applying such devices to high fault level industrial applications.

11.3.3 Cable protection

11.3.3.1 *Overload*

The IEC requirements for low voltage cable overload protection (IEC 60364-4-43) are that the characteristics of a device protecting a cable against overload (e.g. a fuse) shall satisfy the following conditions:

$$I_B \leqslant I_n \leqslant I_z$$
$$I_2 \leqslant 1.45 \times I_z$$

where I_B is the design current of that circuit;
I_n is the continuous current-carrying capacity of the cable under specified conditions;
I_z is the nominal current of the protective device;
I_2 is the current ensuring effective operation in the conventional time of the protective device.

Type gG fuse links to IEC 60269 are tested to a conventional cable overload test at $1.45 \times I_z$. Therefore compliance with the first equation inherently satisfies the requirement of the second equation. Typical fuse protection of 3 core copper conductor PVC/SWA/PVC LV cables is shown in Table 11.5.

11.3.3.2 *Short circuit*

Cable manufacturers give short circuit current/time curves which must not be exceeded for different cable constructions and insulating materials. An

Table 11.5   Conventional cable overload fuse protection

I_n – Fuse rating (amps)[a]	Conductor size (mm²)	I_z – Cable rating (amps)[b]
16	1.5	18
32	4	33
63	10	68
125	35	135
250	120	290
400	240	445

[a]Extract from IEC 60269-1.
[b]For 3 or 4 core PVC/SWA/PVC 600/1000 V cables in air, Bungay & McAllister, *BICC Electric Cables Handbook*, BSP Professional Books, 2nd Edition, 1992.

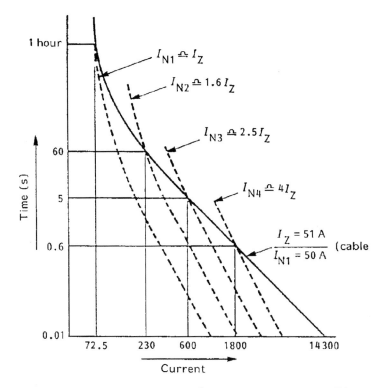

Figure 11.13   Short circuit protection of 10 mm² PVC insulated cable by type T fuses

example of a 10 mm² PVC insulated copper conductor cable is shown by the solid curve in Fig. 11.13, together with four fuse current/time characteristics (shown by the dotted curves) to be selected for short circuit protection of this cable.

Table 11.6   Short circuit fuse protection of PVC/SWA/PVC copper cables in motor circuits

PVC insulated, copper conductor, 3 core armoured cable size (mm)^2	Maximum sustained cable rating in air (A)[a]	Maximum associated 9G fuse rating (A)
1.5	18	32
4	33	63
10	58	125
25	110	250
50	163	400
70	207	630
120	290	800

[a]Bungay & McAllister, *BICC Electrical Handbook*, BSP Professional Books, 2nd Edition, 1992.

11.3.4 Motor protection

It is normal to use contactor motor starters to control motor circuits. In some instances these do not have adequate fault rating characteristics to withstand short circuit conditions or break high fault currents. A series fuse capable of withstanding the repeated motor starting current stresses (see Section 11.2.4.1) is therefore added with the contactor in the motor control centre (MCC). Such fuses are designed to operate quickly at high overcurrents and dissipate low power. Therefore motor starting fuses may be made physically smaller than those designed for protection over a wide range of fault currents. In motor protection applications fuse ratings from two to three times the motor feeder cable rating are necessary, as shown in Table 11.6. Guidance on fuse application taking into account the recent amendments to motor control gear IEC 60947-4-1 standard is available from leading fuse manufacturers.

A 3.3 kV, 950 kW (196 A full load current) motor running and starting current/time characteristic is shown in Fig. 11.14. In order to check adequate motor and cable protection and discrimination the following characteristics are shown superimposed on this diagram:

- A typical contactor maximum fault interrupting capability of 7 kA.
- The hot and cold motor thermal overload relay characteristics.
- 95 mm^2 PVC insulated copper conductor motor feeder cable short circuit current/time capability.
- 250 A motor fuse protection characteristic.

On large motor drives the fuses will incorporate striker pins which trip all three phases of the contactor to prevent single phasing.

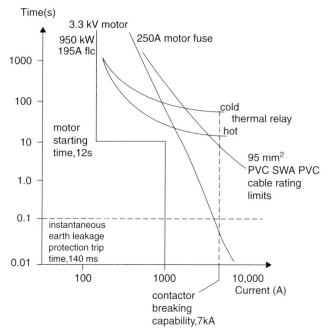

Figure 11.14   Motor starting characteristic showing motor thermal overload and graded fuse short circuit protection

11.3.5 Semiconductor protection

Special care is necessary because of the low tolerance of semiconductor devices to high-overcurrent conditions. Semiconductor fuses are therefore designed to be faster acting than conventional HRC industrial fuses (Fig. 11.15). In addition, it should be noted that semiconductors often operate in a switched mode fashion with high current fluctuations but relatively low rms current values. The fuse must be selected such that the I^2t value is not exceeded in order to avoid anomalous fuse operation.

11.4 MINIATURE CIRCUIT BREAKERS

11.4.1 Operation

The miniature circuit breaker (MCB) and moulded case circuit breaker (MCCB) offer the overload protection characteristics of the fuse, good short circuit current limiting protection together with the advantage of a switching function. If correctly specified the MCB also has the added advantage of not requiring replacement after breaking a short circuit within its rated capability.

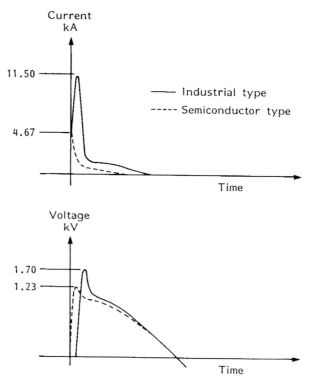

Figure 11.15   Comparison between semiconductor and normal industrial HRC 63 A fuse characteristics (operating on 80 kA, 750 V system, 0.5 pF fault)

To achieve good fault current limitation the current-carrying contacts are arranged such that a magnetic repulsion effect proportional to the square of the fault current rapidly separates the contacts. An arc is then developed and extended across arc chutes (Fig. 11.16). Typical contact opening times are of the order of 0.5 ms and total fault clearance time 8 ms with a 50% reduction in prospective current peak for a modern 15 kA MCB. Such devices do not, therefore, meet the very fast fuse fault clearance times and prospective short circuit current limitation. Enhanced current limiting characteristics are, however, available from some manufacturers. Improved contact layouts and gas producing polyamide arc chutes which smother the arc give 0.2 ms opening times and total clearance times of only 2.5 ms. For reliable repeated operation up to at least 10 times at a 150 kA prospective short circuit current, the installation protected by such an enhanced modern breaker would see less than 9% of the peak prospective current and less than 1.3% of the calculated thermal stresses. Careful MCB selection may therefore offer short circuit protection characteristics nearly as good as a fuse.

Overload protection is achieved by the thermal distortion effects produced by a bimetallic element. After a preset, and often adjustable, amount of

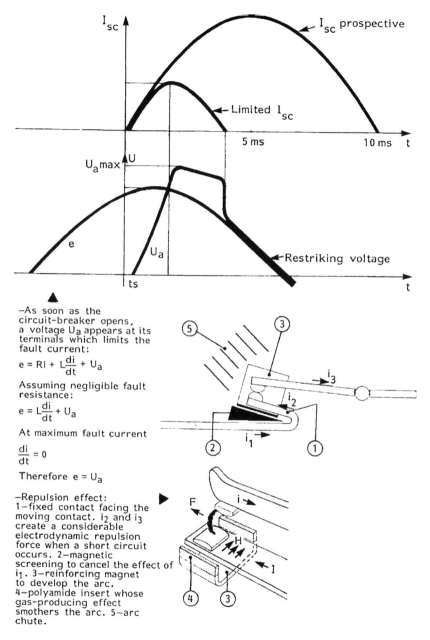

-As soon as the circuit-breaker opens, a voltage U_a appears at its terminals which limits the fault current:

$$e = Ri + L\frac{di}{dt} + U_a$$

Assuming negligible fault resistance:

$$e = L\frac{di}{dt} + U_a$$

At maximum fault current

$$\frac{di}{dt} = 0$$

Therefore $e = U_a$

-Repulsion effect: 1-fixed contact facing the moving contact. i_2 and i_3 create a considerable electrodynamic repulsion force when a short circuit occurs. 2-magnetic screening to cancel the effect of i_1. 3-reinforcing magnet to develop the arc. 4-polyamide insert whose gas-producing effect smothers the arc. 5-arc chute.

Figure 11.16   Current limiting effect of the miniature circuit breaker (MCB) (courtesy of EMMCO/Merlin Gerin)

thermal overload the main current-carrying contacts are arranged to open rapidly. Manual or motor driven reset is then necessary to restore supply. Single, double and three pole MCBs are shown in Fig. 11.20. Figure 11.17 shows typical time/current characteristics of a 160 A HRC fuse and high speed current

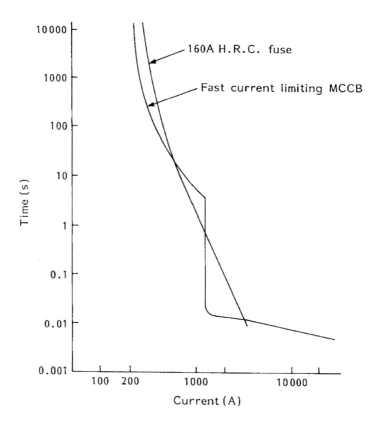

Figure 11.17   Time/current characteristics of an HRC fuse and MCCB both rated at 160 A.
I^2t and temperature characteristics must also be investigated

limiting 160 A MCCB. At high fault levels the MCB or MCCB has a definite minimum time characteristic. Therefore special care must be paid to achieve adequate discrimination between breakers on radial circuits at these higher fault levels (see Section 11.4.3.1).

A modern 250 A MCCB with enhanced discrimination characteristics (courtesy of Merlin Gerin) is shown in Fig. 11.21. These devices are designed to ensure protection discrimination for short circuit currents greater than the rated breaking capacity of the circuit breaker. The devices are characterized as follows and may be fitted with electronic tripping units to allow a wide degree of adjustment for the tripping current thresholds and times:

I_n – circuit breaker current rating
I_r – overload protection current tripping threshold
I_m – short circuit current tripping threshold
t_r – overload tripping time adjustment
t_m – short circuit tripping time adjustment

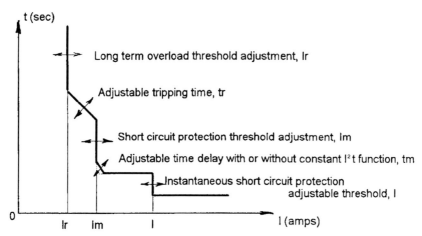

Figure 11.18   Adjustable characteristics of a modern MCCB fitted with an electronic tripping unit

11.4.2 Standards

At low voltage domestic and distribution levels there is a trend away from the use of fuses in favour of MCBs in order to avoid the inconvenience and cost of replacing cartridge or rewirable fuses. At the higher voltages and at high fault currents the fuse remains a highly cost effective solution to the protection of equipment. IEC 60947-2 has now replaced the older IEC 157 as the present standard covering MCBs up to 1000 V AC. It should be very carefully noted that MCBs are given a short circuit category (P1, P2, etc.) depending upon their capability to operate repeatably under short circuit conditions. Not all categories represent MCBs that are usable after clearing a fault and some, like a fuse, have to be replaced. Two utilization categories are defined:

Category A – Breakers without an intentional short time delay provided for selectivity under short circuit conditions.
Category B – Breakers with an intentional short time delay provided for selectivity under short circuit conditions.

The term moulded case circuit breaker (MCCB) normally applies to the higher current-carrying capacity three pole units typically in the range 100–1250 A at up to 1000 V. Miniature circuit breakers (MCBs) are applied at the final distribution feeder level in single, two, three and four pole varieties up to 100 A. In comparison the traditional air circuit breaker (ACB) has low voltage (<1000 V) current-carrying capacity ratings at least up to 6300 A. Advice must be sought from the manufacturers for operation in other than temperate climates. The current/disconnection time characteristics are sensitive to wide temperature variations and current-carrying capacity derating factors should be checked for operation

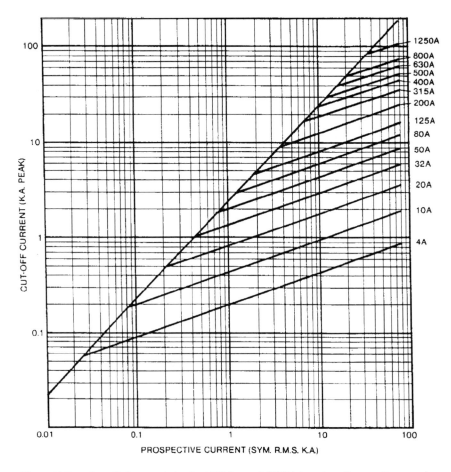

Figure 11.19a   Cut-off characteristics for HRC fuses, 4–1250 A (courtesy Cooper–Bussman)

above 40°C. Application of several MCBs in close proximity may also require grouping derating factors to be applied.

11.4.3 Application

11.4.3.1 *Cascading and prospective fault current limitation*

Application principles, especially for the modern enhanced current limiting MCB types, are similar to those mentioned for fuses in Section 11.3.

In a radial network, upstream circuit breakers, installed near to the source, must be selected to have an adequate fault breaking capacity greater than the prospective short circuit current at the point of installation. Two or more breakers may be cascaded in a network in this way such that the current limiting

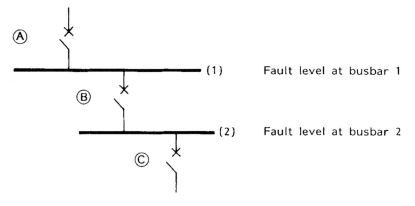

<table>
<tr><td></td><td>(1)</td><td>Fault level at busbar 1</td></tr>
<tr><td></td><td>(2)</td><td>Fault level at busbar 2</td></tr>
</table>

Example 1	Example 2
The upstream breaker A is a C250 L (breaking capacity 150 kA) for a prospective Isc of 80 kA across its output terminals. A C125 N (breaking capacity 22 kA) can be used for circuit breaker B for a prospective Isc of 50 kA across its output terminals, since the 'reinforced' breaking capacity provided by cascading with the upstream C250 L is 150 kA. A C45N (breaking capacity 6 kA) can be used for circuit breaker C for a prospective Isc of 24 kA across its output terminals since the "reinforced" breaking capacity provided by cascading with the upstream C250 L is 25 kA.	The upstream breaker A is a C400 H (breaking capacity 50 kA) for a prospective Isc of 48 kA across its output terminals. A C125 N (breaking capacity 22 kA) can be used for circuit breaker B for a prospective Isc of 40 kA across its output terminals, since the 'reinforced' breaking capacity provided by cascading with the upstream C400 H is 50 kA. A C45N (breaking capacity 6 kA) can be used for circuit breaker C for a prospective Isc of 14 kA across its output terminals since the "reinforced" breaking capacity provided by cascading with the upstream C125 N is 15 kA.

The upstream breaker A is a C250 L (breaking capacity 150 kA) for a prospective Isc of 80 kA across its output terminals. A C125 N (breaking capacity 22 kA) can be used for circuit breaker B for a prospective Isc of 50 kA across its output terminals, since the 'reinforced' breaking capacity provided by cascading with the upstream C250 L is 150 kA. A C45N (breaking capacity 6 kA) can be used for circuit breaker C for a prospective Isc of 24 kA across its output terminals since the "reinforced" breaking capacity provided by cascading with the upstream C250 L is 25 kA.
Note that the 'reinforced' breaking capacity of the C45N with the C125N upstream is only 15 kA, but:
A + B = 150 kA
A + C = 25 kA

The upstream breaker A is a C400 H (breaking capacity 50 kA) for a prospective Isc of 48 kA across its output terminals. A C125 N (breaking capacity 22 kA) can be used for circuit breaker B for a prospective Isc of 40 kA across its output terminals, since the 'reinforced' breaking capacity provided by cascading with the upstream C400 H is 50 kA. A C45N (breaking capacity 6 kA) can be used for circuit breaker C for a prospective Isc of 14 kA across its output terminals since the "reinforced" breaking capacity provided by cascading with the upstream C125 N is 15 kA.
Note that the breaking capacity of the C45N is not 'reinforced' by cascading with the upstream C400H, but:
A + B = 50 kA
B + C = 15 kA

Figure 11.19b   415 V network with three MCCBs in series and installation examples for correct cascading (courtesy EMMCO/Merlin Gerin based upon current limiting capacity)

capacity of the upstream devices allows installation of lower rated and therefore lower cost downstream (away from the source) circuit breakers. This is a recognized and permitted technique under IEC 60364-1 and related national LV installation standards: 'A lower breaking capacity is permitted if another protective device having the necessary breaking capacity is installed on the supply side. In that case the characteristics of the device shall be co-ordinated so that the energy let through of these two devices does not exceed that which can be withstood without damage by the device on the load side and the conductors protected by these devices.'

Correct cascading characteristics may only be obtained after laboratory testing and details of possible combinations for a particular application are detailed in manufacturers' literature. Consider the 415 V network and associated MCCB current limitation curves shown in Fig. 11.19b.

Table 11.7   Reinforced cascaded MCCB breaking capacities in a radial network

Breaker	Busbar short circuit level	MCCB short circuit breaking capacity	Reinforced cascaded breaking capacity at breaker position[a]
Example 1			
A – Type C250L, 200 A	80 kA	150 kA	150 kA @ $I_{SC} \sim 80$ kA
B – Type C125N, 63 A	50 kA	22 kA @ $I_{SC} = 50$ kA	A = B = 150 kA
C – Type C45N, 25 A	25 kA	6 kA	A + C = 25 kA
			(note B + C = 15 kA)
Example 2			
A – Type C400H, 400 A	48 kA	50 kA	50 kA @ $I_{SC} = 48$ kA
B – Type C125N, 100 A	40 kA	22 kA @ $I_{SC} = 40$ kA	A + B = 50 kA
C – Type C45N, 25 A	14 kA	6 kA	B + C = 15 kA

Note: [a]From manufacturers' literature (Merlin Gerin Compact C series MCCBs).

For correct cascading the following two criteria must be met:

1. The upstream device, A, is co-ordinated for cascading with both devices B and C even if the cascading criteria is not achieved between B and C. It is simply necessary to check that the fault breaking capacity combinations A + B and A + C meet the downstream requirements.
2. Each pair of successive devices is co-ordinated i.e. A with B and B with C, even if the cascading criteria between A and C is not fulfilled. It is simply necessary to check that the combinations A + B and B + C have the required breaking capacity.

The results of this approach to the examples shown in Fig. 11.19b are shown in Table 11.7.

In the same way as shown in Section 11.3.4 using fuses, a contactor or switch disconnector with limited breaking capacity and short circuit withstand or a cable with a thermal stress limitation may be short circuit and overload protected by a suitably rated series MCCB or MCB.

Using the Merlin Gerin Compact series MCCB 415 V system voltage current limitation curves shown in Fig. 11.21, the following typical design questions may be answered:

1. To what value is a prospective short circuit current, $I_{SC} = 100$ kA rms, limited when upstream protection is provided by using a C400L type MCCB (660 V-rated voltage, 400 A-rated current @ 20°C, 150 kA rms breaking capacity, IEC Class P1)?
   • From the current limitation curves at 415 V approximately 42 kA peak.
2. A 25 mm², aluminium conductor PVC insulated cable has a maximum permissible thermal stress limit of $3.61 \times 10^6 \text{A}^2\text{s}$. Will the cable be adequately short circuit protected by a C250H type MCCB (660 V-rated

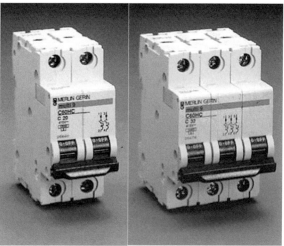
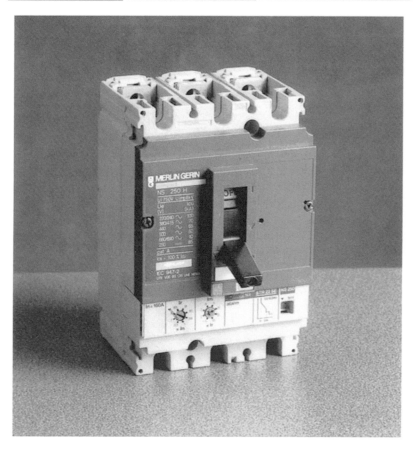

Figure 11.20   Single, double and three pole MCBs for low voltage applications (courtesy of Merlin Gerin)

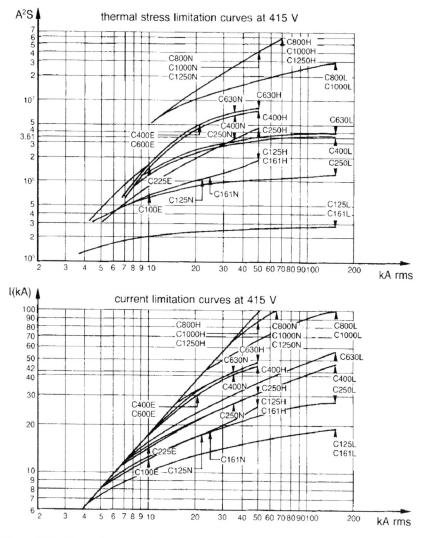

Figure 11.21   Thermal stress and current limitation MCCB curves

voltage, 250 A-rated current @ 40°C, 85 kA rms breaking capacity, IEC Class P1)?
- From the thermal stress limitation curves at 415 V the protection is limited to a short circuit current of approximately 38 kA.

3. A C161L type MCCB (660 V-rated voltage, 150 A-rated current @ 40°C, 150 kA rms breaking capacity, IEC Class P1) feeds via a long length of cable, a distribution board with one 120 A and one 30 A outgoing way. The prospective short circuit level at the C161L MCCB is $I_{SC} = 40$ kA and this is reduced by the cable impedance to $I_{SC} = 24$ kA at the distribution board busbars. Check if it is possible to install a C45N MCB as the 30 A breaker

and a C125N MCCB as the 120 A breaker. The maximum operating temperature inside the cubicle and at the busbar connections is 40°C.

- From Fig. 11.21 and the data given above the C161L MCCB will limit the 40 kA rms prospective short circuit current to approximately 15 kA peak. A C125N MCCB has a current rating of 125 A @ 40°C and from Fig. 11.21 a short circuit breaking capacity of approximately 22 kA. It is therefore suitable for this application.

A C45N MCB has a current rating of 60 A @ 40°C and from manufacturers' literature a breaking capacity of only 6 kA rms. However from Fig. 11.21 the C161L MCCB will limit the 24 kA prospective short circuit level to 13.5 kA peak and from manufacturers' literature (not included here) it is recommended by reinforced cascaded fault limitation as adequately rated for this application.

In both cases a further check is necessary to ensure adequate discrimination with both up- and downstream protective devices.

11.4.3.2 *Discrimination*

Discrimination or selectivity is the co-ordination of automatic devices in such a manner that a fault appearing at a given point in the network is cleared by the protective device, and by that device alone, installed immediately upstream of the fault. A full explanation is given in Chapter 10. If the loading conditions of two circuit breakers connected in series in a radial circuit is similar, then a comparison of the cold MCCB time/current characteristic curves will provide a reasonable assessment of overload discrimination. In order to determine short circuit discrimination the relevant I^2t let through and operating thresholds must be considered. In the example given in Fig. 11.19b for a fault on busbar 2 only breaker B must operate such that other supplies fed from busbar 1 are maintained. This discrimination must be satisfied up to the full short circuit levels and may be difficult to achieve owing to the similar definite minimum time characteristics of the cascaded MCBs at high fault levels.

REFERENCES

1. P. Rosen & P. G. Newbery, 'Recent Advances in H.R.C. Fuse Technology', *Electronics & Power*, June 1983.
2. P. G. Newbery (Cooper Bussman–Hawker Fusegear Ltd.) & T. Mennell (Merlin Gerin), 'Update on Fuse and MCB Developments', IEE Discussion Meeting, Savoy Place, London, 1 October 1993.
3. H. W. Turner & C. Turner, *Power System Protection*, Volume 1, Chapter 5, Edited by The Electricity Council, Peter Peregrinus Ltd, 2nd Edition, 1981.
4. A. Wright, *IEE Power Engineering Journal*, Vol. 4, No. 6, November 1990.
5. A. Wright & P. G. Newbery, Electric fuses, 3rd Edition, *IEE power and Energy Series No. 49*, 2004.

12 Cables

12.1 INTRODUCTION

The selection of cables for particular applications is best done with reference to the latest and specific manufacturer's cable data and application guides. It is not therefore appropriate to include here comprehensive tables giving cable dimensional, weight and current rating information. This chapter concentrates on the properties of different types of LV, MV and HV power cables, their merits for different applications, cable sizing and loss calculations, useful installation practices and cable management systems. A section is also included on control and communication, including fibre optic, cables. Technical specification details are included such that competitive quotations from leading manufacturers may be obtained. Consideration is given to the safety implications associated with gases and smoke emitted from cables under fire conditions, especially where installed in public places.

12.2 CODES AND STANDARDS

Table 12.1 details some useful IEC and National cable Standards.

Standard cable nomenclature based upon IEC 60183 used to designate appropriate cable voltage ratings is as follows:

U_0 = rated rms power frequency voltage, core to screen or sheath.
U = rated rms power frequency voltage, core to core.
U_m = maximum rms power frequency voltage, core to core (highest core to core voltage under normal operating conditions).
U_p = peak lightning impulse withstand voltage, core to screen or sheath.

A cable voltage classification may therefore be designated as $U_0/U(U_m)$. The selection of cables with the appropriate voltage rating for the particular

Table 12.1   Useful IEC standards

IEC standard	Brief description and comment
60055	Paper-insulated metal sheathed cables for rated voltages up to 18/30 kV (with copper or aluminium conductors and excluding gas-pressure and oil-filled cables). Covers tests and general construction requirements. In the UK BS 6480 is used.
60096	Radio frequency cables various parts cover general requirements, measuring methods, specifications and tests.
60141	Tests on oil-filled and gas-pressure cables and their accessories. Includes oil-filled (normal and high pressure) cables up to 400 kV and gas-pressure cables to 275 kV.
60183	Guidance for the selection of HV cables, the conductors size, insulation level and cable construction to be used on three phase AC systems operating at voltages exceeding 1 kV.
60227	PVC-insulated cables of rated voltages up to and including $U_0/U = 450/750$ V. Covers small power and lighting cables mainly applicable to building services applications. Gives details of core identification, test methods, non-sheathed single core cables, light PVC $U_0/U = 300/500$ V cables, flexible cables for lift applications, etc. See also BS 6004.
60228	Conductors of insulated cables. Standardized nominal cross-sectional areas from 0.5 mm^2 to 2000 mm^2, numbers and diameters of wires and resistance values. Solid and stranded copper and aluminium conductor classes.
60229	Tests on cable oversheaths which have a special protective function and are applied by extrusion. Appropriate to particular conditions in addition to corrosion protection, such as reduced sheath losses. See also EN60811 and BS6469-99-2.
60230	Impulse tests on cables and their accessories. Guide for rationalization between different laboratories.
60287	Calculation of the continuous current rating of cables (100% load factor). Deals with steady-state operation at AC voltages up to 5 kV for cables direct buried, ducts, troughs, steel pipe and cables in air installations. Appendices include details of ambient temperatures and soil thermal resistivities in various countries, information required from the purchaser for the selection of the appropriate type of cable plus digital calculation of quantities given graphically.
60331	Fire resisting characteristics of electric cables. See Table 12.11 for more comments and related standards.
60332	Tests on electric cables under fire conditions. Test methods and flame propagation of power and control/communication cables.
60364	Electrical installations in buildings. Part 5, Chapter 52 covers wiring systems and current carrying capacities for cables not exceeding 0.6/1 kV. Provides a series of tables containing the relationship between cross-sectional area of conductors and the load depending upon type of conductor material, type of insulation and method of installation.
60502	Extruded and dielectric insulated power cables for rated voltages from 1 kV to 30 kV. Specifies construction, dimensions and test requirements for PVC ($U_0/U = 1.8/3$ kV), PE-, EPR- and XLPE-insulated cables.
60702	Mineral insulated cables and their terminations with a rated voltage not exceeding 750 V. See also BS6387.
60724	Guide to the short circuit temperature limits of electric cables with a rated voltage not exceeding 0.6/1.0 kV. Concerns insulating materials and gives guidance on calculation of permissible short circuit currents.
60754	Test of gases evolved during combustion of electric cables. Covers emissions of halogen acid gas (as might be expected from PVC-sheathed and -insulated cables) and degree of acidity.

(Continued)

Table 12.1 *Continued*

IEC standard	Brief description and comment
60811	Common test methods for insulating and sheathing materials of electric cables. Concerns dimensional tolerances, elongation, water absorption, thermal stability, etc. More applicable to manufacturers but specific features may be quoted by purchasers in technical enquiry specification.
60840	Power cables with extruded insulation and their accessories for rated voltages above 30 kV (U_m = 36 kV) up to 150 kV (U_m = 170 kV) – Test methods and requirements (see also 62067).
60853	Calculation of the cyclic and emergency current rating of (a) cables up to 18/30 (36) kV and (b) cables >18/30 (36) kV. Supplements the 100% loading calculations given in IEC 60287.
60859	Cable connections for gas-insulated metal-enclosed switchgear for rated voltages of 72.5 kV and above – Fluid-filled and extruded insulation cables Fluid-filled and dry type cable terminations.
60885	Electrical test methods for electric cables up to and including 450/750 V including partial discharge tests.
61034	Measurements of smoke density of electric cables burning under defined conditions.
61042	A method for calculating reduction factors for groups of cables in free air, protected from solar radiation. Applicable to cables of equal diameter emitting equal losses.
61084	Cable trunking and ducting systems for electrical installations. Gives some guidance on cable segregation. See also CP1022 concerning power and control/communication cable segregation.
61443	Short-circuit temperature limits of electric cables with rated voltages above 30 kV (U_m = 36 kV).
62067	Power cables with extruded insulation and their accessories for rated voltages above 150 kV (U_m = 170 kV) up to 500 kV (U_m = 550 kV) – Test methods and requirements (see also 60840).

Note: See also Section 12.7.3.5 for standards relating to optic fibres.

application is dependent upon the system voltage and earthing category. These categories are defined as follows:

- Category A – A system in which, if any phase conductor comes in contact with earth or an earth conductor, it is automatically disconnected from the system.
- Category B – A system which, under fault conditions, is operated for a short time with one phase earthed. These conditions must not exceed 8 hours on any occasion with a total duration, during any 12 month period, not exceeding 125 hours.
- Category C – A system which does not fall into categories A and B.

Examples of cable voltage ratings are given in Table12.2. The maximum sustained voltage should exclude transient overvoltages due to switching surges, lightning surges, fault conditions, etc. For system voltages at intermediate levels from those given in Table 12.2 the cable should be selected with the next higher rating. For example, in Saudi Arabia with an MV category A or B and system voltage of 13.8 kV, an 8700/15 000 V cable voltage rating could be selected.

Table 12.2   Standard power cable ratings

U_0/U (kV) (Minimum cable voltage rating)	U_m (kV) (Maximum sustained voltage between phases)
1.8/3 and 3/3	3.6
1.9/3.3 and 3.3/3.3	3.6
3.6/6 and 6/6	7.2
3.8/6.6[a] and 6.35/11[b]	7.2
6.35/11[a] and 8.7/15[b]	12
8.7/15[a] and 12.7/22[b]	17.5
12.7/22[a] and 19/33[b]	24
19/33[a]	36

Notes: [a]Category A or B.
       [b]Category C.

12.3 TYPES OF CABLES AND MATERIALS

12.3.1 General design criteria

The following factors govern the design of power cables:

1. The cross-sectional area of the conductors chosen should be of the optimum size to carry the specified load current or short circuit short term current without overheating and should be within the required limits for voltage drop.
2. The insulation applied to the cable must be adequate for continuous operation at the specified working voltage with a high degree of thermal stability, safety and reliability.
3. All materials used in the construction must be carefully selected in order to ensure a high level of chemical and physical stability throughout the life of the cable in the selected environment.
4. The cable must be mechanically strong, and sufficiently flexible to withstand the re-drumming operations in the manufacturer's works, handling during transport or when the cable is installed by direct burial, in trenches, pulled into ducts or laid on cable racks.
5. Adequate external mechanical and/or chemical protection must be applied to the insulation and metal or outer sheathing to enable it to withstand the required environmental service conditions.

Types of cables are detailed in Table 12.3.

After voltage selection cables tend to be specified by describing the materials and their properties from the phase conductors to the outer covering. Manufacturers will provide a drawing showing a cross-section through the cable and the relevant technical parameters and guarantees associated with the design. A typical physical technical specification sheet for, say, 19 000/33 000 XLPE power cable is shown in Table 12.4.

Table 12.3   Types of cables

Voltage level	Usage	Voltage range	Insulation
Low voltage (LV)	Telephone	50 V	PVC or PE
	Control	600/1000 V	PVC
	Solid dielectric	600/1000 V	PVC, XLPE, EPR
	MI or MIND	600/1000 V	Paper
	Fire resistant/	600 & 1000 V	Mineral, Silicone Rubber,
	retardant[a]	or 600/1000 V	LSF, LSOH
Medium voltage (MV)	Solid dielectric	3 kV to 7.2 kV	PVC, PE, XLPE, EPR
	MI or MIND	3 kV to 7.2 kV	Paper
High voltage (HV)	Solid dielectric	10 kV to 150 kV	XLPE, EPR
	MIND	10 kV to 36 kV	Paper
	Oi- filled, gas pressure	80 kV to 150 kV	Paper
	Gas insulated ducts	10 kV to 150 kV	SF_6
Very high voltage (VHV)	Solid dielectric	150 kV to 300 kV	XLPE
	Oil-filled	150 kV to 300 kV	Paper, PPL
	Gas insulated ducts	150 kV to 300 kV	SF_6
Extra high voltage (EHV)	Solid dielectric	above 300 kV	XLPE
	Oil-filled	above 300 kV	Paper, PPL
	Gas insulated ducts	above 300 kV	SF_6

Notes: [a]Refer to Section 6 of this chapter.
XLPE – Cross-linked polyethylene
PVC – Polyvinylchloride
PE – Polyethylene
EPR – Ethylene propylene rubber
LSF – Low smoke and fume
LSOH – Low smoke zero halogen
MI – Mass impregnated
MIND – Mass impregnated non-draining
PPL – Polypropylene paper laminate

12.3.2 Cable construction

12.3.2.1 *Conductor materials*

Copper is still the predominant conductor material in stranded, shaped, segmental sectorial and milliken formats. Solid or stranded, shaped or segmental aluminium is also often specified on the basis of cost in the manufacturer's country at the time of tender. Aluminium is also lighter and assists with ease of handling large cables. Additional care has to be taken when jointing aluminium cables. It is necessary to ensure that the contact surfaces are free from oxide and that when connecting to copper or brass terminals no corrosion cell is formed.

12.3.2.2 *Insulation*

Paper insulation
Oil-impregnated, paper-insulated cables have a history of satisfactory use at all voltage levels. They are nowadays rarely specified for new installations

Table 12.4   Cable physical parameters

Description	Requirements or guarantees
VOLTAGE	
Rated rms power frequency core to earth voltage	(kV)
Rated rms power frequency core to core voltage	(kV)
CORES	
Number of cores	(No.)
CONDUCTOR	
Cross-sectional area	(mm^2)
Material	
Design (stranded, sectoral, etc.)	
Overall dimensions	(mm)
CONDUCTOR SCREEN	
Material	
Thickness (10 kV and above)	(mm)
INSULATION	
Type of curing	
Thickness – nominal	(mm)
– minimum	(mm)
Diameter of insulation	(mm)
CORE SCREEN	
Material	(mm)
Thickness	
Diameter over screen (10 kV and above)	(mm)
FILLERS	
Material	
BINDER OVER LAID-UP CORES	
Material	
Nominal thickness	(mm)
Diameter over binder (3 core cables only)	(mm)
METALLIC LAYER/SCREEN	
Material	
No. wires and size	(No./mm)
No. tapes and size	(No./mm)
Cross-section	(mm^2)
ARMOUR BEDDING	
Type	
Nominal thickness	(mm)
ARMOUR	
Type of wire or tape	
No. wires or tapes	(No.)
Diameter or thickness of tapes (generally 3 core cables only)	(mm^2 or mm)
OUTER COVERING	
Material	
Insect or worm attack repellants	
Minimum average thickness	(mm)
COMPLETED CABLE	
Overall diameter	(mm)
Weight per metre	(kg/m)
Maximum drum length	(m)
Drum overall diameter/width/loaded weight	(m/m/kg)

except at voltage levels of 66 kV and above or for reinforcement of existing installations where standard cable types are required throughout the network. Until the development of XLPE or EPR cables paper tape insulation was the most stable form at high temperatures and better able to withstand the stresses occurring under short circuit conditions. However, paper insulation deteriorates rapidly because of its hygroscopic nature if exposed to moisture. In order to prevent this, the paper layers are protected against ingress of water, usually by a lead/lead alloy or corrugated aluminium alloy metal sheath. Furthermore, during installation special attention has to be paid to the quality of joints and terminations which often require special materials and highly skilled jointers.

For cables up to 36 kV, mass impregnated non-draining (MIND) cables are employed. IEC 60055-2 covers the general and construction requirements of these cables but in the UK British standard BS6480 Part 1 is generally used as a specification for these cables. Routine, sample and site tests should be carried out to IEC 60055-1. The conductor screen consists of carbon paper and/or metallized paper tapes applied over the conductors which are of stranded copper or aluminium wires to IEC 60228. The insulation is made up of 6 to 12 layers of dry paper tapes, each layer wound in the opposite direction to the previous layer. The core screen is made up of carbon paper and/or non-ferrous metallized paper tapes applied over the insulation to a constant thickness in order to obtain a maximum electrical stress level at the conductor screen of 5 to 6 kV/mm. After construction of the insulated conductor cores the paper is impregnated with an oil resin which has a consistency of a soft wax at 20°C. After impregnation the metal sheath is extruded onto the cable. The 'belted' type cable, which does not have individually screened cores, is used for 3 core cables up to 12 kV. Instead of individual screening, an overall belt of paper is applied round all 3 cores and this type of cable may be seen designated as 11 000/11 000 V rating, whereas screened cable of the same rating would be designated 6350/11 000 V.

Cable types have 1, 2, 3, 3½ and 4 cores with conductors to IEC 60228 for stranded copper or aluminium wires. The maximum conductor temperature is 60°C to 70°C and the power factor (dielectric loss angle) should not exceed 0.006 at 60°C or 0.013 at 70°C. Three or four core cables are generally armoured for direct burial in ground with cable tests to IEC 60055-1.

Oil-filled (OF) cable is used up to 525 kV. Single core cables have a hole approximately 12 mm in diameter in the centre of the conductor through which the oil may flow during expansion and contraction of the cable as it heats up and cools down. Three core oil-filled cables up to 630 mm² have ducts between the cores to allow for the necessary oil movement at typical working pressures of between 80 kPa and 350 kPa. Reinforcement tapes, generally made of stainless steel or phosphor bronze, are applied over the lead sheath of oil-filled cables to assist withstanding abnormal oil pressures up to some 600 kPa. Maximum conductor temperatures are 85°C to 90°C with power factors between 0.0028 and 0.0035 for cable core-to-earth nominal voltage ratings between 66 kV and 400 kV. A typical oil-filled 132/150 kV cable has a power factor of some 0.0033. The pressurized oil in the dielectric reduces the chance of partial discharge under normal conditions; however, impulse tests are still important in order to verify the performance of the cable under lightning strike and switching surge conditions. Stress levels at the conductor vary between 8 MV/m at 33 kV and 15 MV/m at 400 kV.

Gas pressurized cables have a similar construction to mass impregnated (MI) types but because of the gas pressure the insulation thickness is less. Dry or impregnated paper tape is used for the insulation and nitrogen gas at typical normal working pressures of up to 1400 kPa is used. Again reinforcement of the lead sheath is necessary at these pressure ratings.

PPL Insulation

Oil-filled cables above 200 kV are now increasingly manufactured using polypropylene paper laminate (PPL), rather than paper. This is because at higher voltages paper suffers from high dielectric losses which reduce the cable rating. PPL consists of a film of polypropylene coated on each side with a thin layer of paper. The material consists typically of 50% polypropylene and 50% paper. The physical properties of the material means that it can be readily substituted for paper in the cable manufacturing process. The material has a much lower dielectric loss factor than paper (0.0021 at 90°C compared to 0.0078 for paper) and hence the heat generated within the insulation at high voltages is significantly reduced, allowing higher current ratings for a given conductor size. PPL cables have a higher impulse strength compared to paper and can operate at higher stress levels reducing the amount of insulation required. They also have a lower permittivity reducing the capacitive charging current. PPL material is more expensive than paper, but this is offset against the benefits listed above.

PVC insulation

PVC insulation is nowadays being rapidly superseded by XLPE cables with LSF (see Section 12.6) or MDPE oversheathing. However, it is still specified and is suitable for cables rated up to 7.2 kV. PVC has the advantage over paper insulation in that it is non-hygroscopic and does not therefore require a metallic sheath. The absence of such a sheath simplifies jointing by the elimination of plumbing operations on the lead sheath. Moreover, it is both lighter and tougher and inherently more flexible than paper. Therefore, PVC-insulated cables may be bent through smaller radii than paper-insulated cables thus easing installation problems. PVC is resistant to most chemicals although care must be taken with installations in petrochemical environments. It is a thermoplastic material which softens at high temperatures and therefore cannot withstand the thermal effects of short circuit currents as well as paper insulation. The maximum conductor temperature is 65°C to 70°C. Multicore cables are generally armoured when laid direct in ground. At low temperatures PVC hardens and becomes brittle and installations should not be carried out at temperatures below 0°C.

XLPE insulation

XLPE is a thermo setting material achieved by a process akin to the vulcanization of rubber. The resulting material combines the advantages of PVC insulation (high dielectric strength, good mechanical strength and non-hygroscopic nature) with thermal stability over a wide temperature range. XLPE has no true melting point and remains elastic at high temperatures therefore permitting greater current carrying capacity, overload and short circuit performance in comparison with PVC- and paper-insulated cables. IEC 60502-2 covers the design and testing of these single and 3 core cables up to 36 kV. Cables with voltages between 36 kV and 150 kV are also manufactured generally in accordance with IEC 60502-2 with testing carried out to IEC 60840. Above 150 kV where single core cables are normally employed, IEC 62067 provides test methods and other requirements.

The type of curing affects the electrical strength of the insulation against partial discharges. Originally XLPE cables were steam cured for voltages in the range 24 kV to 145 kV in USA, Scandinavia and Japan. Faults in service due to partial discharges in small voids caused carbon deposits to form. Further breakdown led to the formation of water and dust 'trees' and 'bow ties' eventually leading to full insulation breakdown. Since the early 1970s cable breakdown caused by voids and/or contaminating particles such as dust or humidity within the dielectric has been avoided by improvements in the cable insulation materials and manufacturing techniques, in particular by the introduction of the 'dry curing' process. This has also led to higher impulse withstand test results. Because of the importance of reliability and long service for XLPE cables partial discharge values are extremely important. Improvements in manufacturing techniques and good service history records allow XLPE insulation thickness to be reduced such that cable stress levels are increasing. Stress levels at the conductor of 3 to 3.5 MV/m at 36 kV- and 7 to 8 MV/m at 150 kV-rated cable voltages are typical. XLPE cables up to 36 kV are generally manufactured with water tree retardant material and do not require a metal sheath. It is recommended that cables with stress levels above 6.5 MV/m are protected by a metal sheath.

XLPE cables have greater insulation thickness than their equivalent paper-insulated cables. This results in XLPE cables having larger overall diameters and for a given cable drum size slightly less overall cable length can be transported.

The power factor of XLPE cables is very low compared to paper-insulated cables; 0.001 at the nominal system voltage to earth. Cable capacitance affects voltage regulation and protection settings. The 'star capacitance' is normally quoted on manufacturers' data sheets for 3 core screened cables operating at 6.6 kV and above, i.e. the capacitance between the conductor and screen (Fig. 12.1a). Unscreened cables are only normally used at voltages less than 6.6 kV.

EPR insulation
Ethylene propylene rubber cables have a cross-linked molecular structure like XLPE and are produced by a similar process. Both EPR and XLPE have the same durable and thermal characteristics but EPR has a higher degree of elasticity which is maintained over a wide temperature range. This EPR flexibility characteristic is somewhat mitigated when such cables are used in conjunction with steel armouring. Between 6 and 12 ingredients are used in the production of EPR which necessitates great care to maintain purity and avoid contamination during the production process. EPR insulation is water resistant and cables

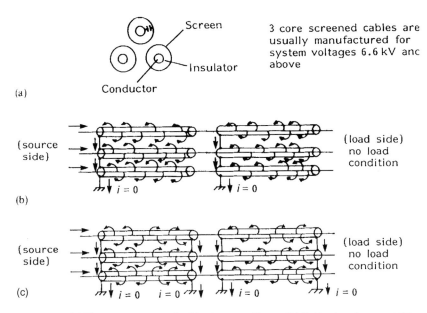

(a) Conductor

Screen

Insulator

3 core screened cables are usually manufactured for system voltages 6.6 kV anc above

(b) (source side) ... (load side) no load condition

(c) (source side) ... (load side) no load condition

Figure 12.1 (a) Three core screened cable star capacitance. (b) Cable charging currents with 3 core screened cable and screen earthed at one only. Earth current $i = 0$ because charging currents are balanced. (c) Cable charging currents with 3 core screened cable and screen earthed at both ends. Earth current $i = 0$ because charging currents are balanced. (d) R-phase-to-earth fault, 3 core screened cable earthed at one end to earth fault relay. (e) To maintain earth fault relay stability with balanced ring type CT for a single phase-to-earth fault the setting has to be higher than that shown in (d)

are able to operate without a metallic sheath in waterlogged areas. This initially gave EPR insulation an advantage over XLPE insulation. However, this advantage has largely gone with the increasing use of water tree retardant XLPE. EPR insulation tends to be more expensive than XLPE insulation and its use has reduced in recent years with only a relative few manufacturers now offering EPR-insulated cables.

Mineral insulation
Mineral insulated copper conductor (MICC) cables are manufactured for 600 V (light grade) and 1000 V (heavy duty) installations which could involve high temperatures, rough mechanical handling, surface knocks or contact with oils. The cables consist of copper or aluminium conductors insulated with highly compressed magnesium oxide compound surrounded by a copper or stainless steel tube. They have a small overall diameter for a given current rating and will continue to operate continuously under fire conditions at sheath temperatures up to 250°C. Such cables are specified for high security applications and in particular for use with fire alarm systems. Particular care has to be taken during the installation and storage of such cables in order to ensure that moisture does not penetrate into the magnesium oxide material. In addition the impulse withstand of such insulation is not as good as more conventional cable insulation.

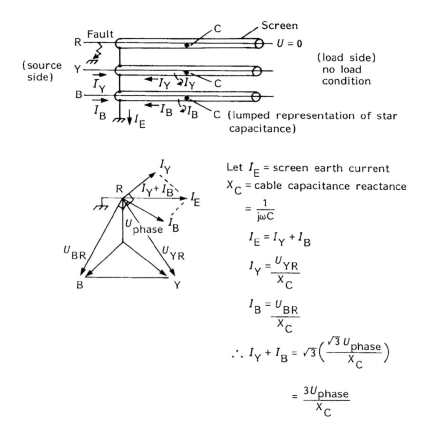

Let I_E = screen earth current

X_C = cable capacitance reactance

$$= \frac{1}{j\omega C}$$

$$I_E = I_Y + I_B$$

$$I_Y = \frac{U_{YR}}{X_C}$$

$$I_B = \frac{U_{BR}}{X_C}$$

$$\therefore I_Y + I_B = \sqrt{3}\left(\frac{\sqrt{3}\,U_{phase}}{X_C}\right)$$

$$= \frac{3U_{phase}}{X_C}$$

= 3 × steady state per phase charging current

(d)

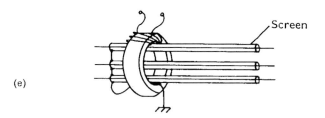

(e)

Figure 12.1   *Continued*

12.3.2.3 *Sheaths*

Very little lead sheathing is now specified except for special HV cables. Lead and lead alloy sheaths have been traditionally used to prevent the ingress of moisture into paper-insulated cables or other cables installed in particularly marshy conditions. Lead corrosion and fatigue resistance properties are important and improvements are obtained by the addition of other elements.

Table 12.5   Lead and lead alloys for cable sheaths

Description	Nominal composition	Application
Lead	Lead (99.8% minimum purity)	Forms a pliable sheath, but because of the somewhat low fatigue resistance is used only for armoured cables.
Lead alloy E	0.4% tin 0.2% antimony, remainder lead, including impurities	For general use and where some vibration may be encountered. The most normally specified type of lead sheathing.
Lead alloy B	0.85% antimony, remainder lead, including impurities	Provides increased resistance to vibration at the expense of ductility. Used for cables on bridges, in roadway crossings or near railway tracks.

Alloy sheaths are used with unarmoured cables where vibration problems might be encountered. Table 12.5 summarizes the materials normally used.

As a cheaper and nowadays far more popular alternative to lead an aluminium alloy sheath is specified. The composition is an important factor in reducing the possibility of corrosion in service. A corrugated aluminium sheath construction helps to improve overall cable flexibility.

For XLPE cables above 60 kV, foil laminate sheaths have become increasingly popular. The foils are made from either copper or aluminium typically 1 mm–2 mm thick, with a polyethylene coating. If necessary additional copper wires can be added to the design to provide for a higher sheath fault rating. The foil is applied longitudinally and folded around the cable and overlapped onto itself. Heating the polyethylene at the overlap bonds the foil to itself creating a water tight design. Although not as robust as a lead or aluminium sheath, a foil laminate cable can offer a cheaper alterative as it avoids need for expensive metal extrusion presses.

12.3.2.4 *Insulation levels and screening*

The correct selection of appropriate cable voltage designation depends upon the type of network and network earthing arrangements as described in Section 12.2. Generally, if the network is solidly earthed the voltage will not rise above the maximum system phase-to-neutral voltage under fault conditions. However, if under fault conditions the earthing arrangement is such as to allow the voltage to neutral to rise to the line voltage then the cable insulation must be specified accordingly.

To minimize the possibility of discharges at the inner surfaces of cable core dielectric a grading screen is introduced. This screen comprises of one or two layers of semiconducting tapes or compounds over the core insulation. Such measures are introduced at the following voltage levels:

PILC-insulated cables – 6350/11 000 V
PVC-insulated cables – 7200/12 500 V
XLPE-insulated cables – 3300/6000 V

Typical three phase screened cable systems with screens earthed at one or both ends are shown in Figs 12.1b and 12.1c. During a single phase earth fault the earth fault current will be three times the steady state per phase charging current (Fig. 12.1d). Earth fault relay settings have to be sufficiently high to ensure stability for upstream single phase-to-earth faults when high cable capacitance effects due to the cable design or long lengths of cable are involved.

12.3.2.5 *Armouring*

In order to protect cables from mechanical damage such as pick or spade blows, ground subsidence or excessive vibrations cable armouring is employed. For 3 core cables this consists of one or two layers of galvanized steel tapes, galvanized steel wire braid or galvanized steel wires helically wound over the cable. Galvanized steel wire armour (SWA) is preferred since it gives a more flexible construction, is easy to gland and gives better performance where the cable may be subjected to longitudinal stresses in service. In addition, the overall cross-sectional area of steel wire armour tends to be greater than that for the equivalent steel tape armour mechanical protection and therefore SWA presents a lower impedance if the armour is used as the earth return conductor. If armouring is required on single core cables aluminium should be used instead of steel wire in order to avoid losses. Armour protection for lead-sheathed cables was traditionally laid up on suitably impregnated fibrous bedding material. For PVC- and XLPE-insulated cables PVC-, LSF- or MDPE-extruded bedding is now normally specified rather than the older PVC tapes.

12.3.2.6 *Finish*

One of the most important factors that can affect cable life is the degree of protection afforded by the cable finish against the harmful effects of chemical corrosion, electrolytic action, insect or rodent attack and mechanical damage. Compounded fibrous materials were originally employed but these have now been replaced by extruded MDPE or LSF outer sheaths which may be impregnated with chemicals to deter insects such as termites. The integrity of the outer sheath may be tested after installation. A graphite outer coating on the cable may be specified to allow for an electrical connection to the outside of the cable sheath.

Typical electrical XLPE cable properties could be specified in the tabular format as shown in Table 12.6 as part of an overall cable specification. Similar formats may be used for other types of cable.

12.3.3 Submarine cables

Submarine cables require additional tensile strength to permit laying on or under the sea or river bed under high tension conditions. Paper, PVC or XLPE

Table 12.6   Typical MV cable electrical parameters

Description	Requirements or guarantees
CONTINUOUS CURRENT CARRYING CAPACITY BASED ON FOLLOWING CONDITIONS: Laid in ground:	
One circuit	A
Two circuits	A
Three circuits	A
Drawn into ducts: One circuit	A
Two circuits	A
Three circuits	A
In air: One circuit	A
PERMISSIBLE OVERLOAD IN SERVICE CONDITIONS	
% ...... for a duration of ...... hours	% hours
% ...... for a duration of ...... months	% months
MAXIMUM CONDUCTOR TEMPERATURE	
Laid direct in ground	°C
Drawn into ducts	°C
Erected in air	°C
CONDUCTOR SHORT CIRCUIT CURRENT Carrying capacity for 1 second cable loaded as above prior to short circuit and final conductor temperature at 250°C	kA
SCREEN EARTH FAULT CURRENT Carrying capacity for 1 second cable loaded as above prior to earth fault	kA
Final screen temperature	°C
MAXIMUM DIELECTRIC STRESS at the conductor screen (assumed smooth)	MV/m
MAXIMUM BENDING RADIUS around which cable may be laid	(Note: often expressed as a multiplier of the cable overall outer diameter)
Laid direct	m
In ducts	m
In air	m
During installation without damage	m
At terminations	m
DUCTS Nominal internal diameter of pipes of ducts through which cable may be pulled	mm
Maximum recommended distance between cable draw pits	m
MAXIMUM DC RESISTANCE per km of cable at 20°C	
Of conductor	ohm
Of metallic layer	ohm
MAXIMUM AC RESISTANCE of conductor per km of cable at maximum normal conductor temperature	ohm
MAXIMUM INDUCED VOLTAGE on screen under fault conditions (single core cables above 10 kV)	V

Table 12.6   Continued

Description	Requirements or guarantees
INSULATION RESISTANCE	
per km of cable per core	
At 20°C	Megohm
At maximum rated temperature	Megohm
EQUIVALENT STAR REACTANCE	
per metre of three phase circuit at nominal frequency	Micro ohm
EQUIVALENT CAPACITANCE	
per metre of cable	pF
MAXIMUM CHARGING CURRENT	
per core of cable per metre at nominal voltage	mA
MAXIMUM DIELECTRIC LOSS	
of cable per metre of 3 phase circuit when laid direct	
in ground at nominal voltage and frequency at	
maximum conductor temperature	W/m
MAXIMUM DIELECTRIC LOSS ANGLE	
of charging VA of cable when laid direct in ground at	
nominal voltage and frequency at:	
conductor temperature of 20°C	tan
maximum conductor temperature	tan
MAXIMUM DIELECTRIC LOSS ANGLE	
of charging VA of cable at nominal frequency and	
conductor temperature of 20°C at:	
50% rated voltage	tan
200% rated voltage	tan
CONDITIONS UPON WHICH CURRENT CARRYING	
CAPACITIES ARE BASED	
Soil thermal resistivity at burial depth of ...... m	°Km/W
Ground temperature at burial depth of ...... m	°C
Air temperature	°C
Axial spacing between circuits	mm
Axial spacing and installation arrangement of single	
phase cables	mm (trefoil, etc.)
Type of earth bonding over ...... m cable route length	

insulation is used together with additional protection measures against water ingress and mechanical damage and with special sheath compositions to repel worm attack. Such cables are manufactured in the longest possible lengths in order to minimize the number of underwater cable joints. When preparing the design for submarine cables an accurate knowledge of the prevailing currents and tidal variations is essential to assist in deciding the best cable route and most favourable times for the cable laying work.

12.3.4 Joints and terminations

Techniques for jointing and terminating paper-insulated cables are well established. With the trend away from paper-insulated cables the traditional practices of highly skilled jointers for soldering and plumbing and application of paper rolls and tapes have been revolutionized. Key factors in the design of cable joints and terminations include:

- safe separation between phases and between phase and earth;
- capability to avoid dielectric breakdown at the interface and around rein-stated jointing insulation under normal load and impulse surge conditions;
- adequate stress control measures to avoid high fields around screen dis-continuities and cable/joint interfaces.

In all cases great care should be taken to ensure dry clean conditions during the jointing process on site. Tents may be erected over the jointing area to pre-vent ingress of dust or moisture.

Cable conductor connections are normally achieved using compression lugs and ferrules. These provide good mechanical grip and electrical contact, are designed to avoid any oxide layer build-up in cases using aluminium cores and provide a more repeatable solution than soldered connections. Specially designed hand-operated or hydraulic tools and dies are used.

Soldered connections with operating temperature limits of some 160°C are not compatible with the 250°C short circuit temperature rating of XLPE cables. In addition, such soldered connections require a well-trained workforce if high resistance connections are to be avoided. Mechanical clamps are also used to connect cable cores together. Metal inert gas (MIG) welding is favoured for alu-minium conductor connections.

LV and MV XLPE cable joints up to about 24 kV employ two pack resin systems. The resin components are mixed just prior to pouring into the joint shell where they harden to provide good mechanical and waterproof protec-tion. It is important to follow the manufacturer's temperature and humidity storage recommendations and to monitor the useful shelf-life of such resins. Figure 12.16 shows a resin filled through joint with LSF properties for 24 kV 3 core XLPE cable.

Modern, fully moulded type plug-in connectors, pre-moulded push-on sleeves and heat shrink sleeve terminations allow for repeatable and rapid ter-minations to be prepared. Single and three phase cable connections to SF$_6$ GIS are described in IEC 60859 for rated voltages of 72.5 kV and above. When connecting oil-filled cables to SF$_6$ switchgear special barriers are intro-duced to prevent problems of gas and cable oil pressure differentials.

12.4 CABLE SIZING

12.4.1 Introduction

After correct cable voltage classification the following considerations apply:

- Current carrying capacity.
- Short circuit rating.

- Voltage drop.
- Earth loop impedance.
- Loss evaluation.

It should be noted that very valuable research has been carried out by the Electrical Research Association (ERA) in the UK with regard to cable current carrying capacities.

Typical calculations for a 20 kV transformer feeder cable, 3.3 kV motor feeder and a 400 V distribution cable are enclosed.

At voltages of 36 kV and above, current rating calculations are normally undertaken for individual installations in accordance with IEC 60287. This standard provides detailed algorithms for calculation of current ratings taking into account details such as: cable design, installation conditions (direct buried, in air, in ducts), ambient temperatures, phase spacing and arrangement, type of sheath bonding and proximity of other cables.

12.4.2 Cables laid in air

Current rating tables are generally based on an ambient air temperature of 25°C (Europe) or 40°C (Japan). Separate manufacturer's tables state the factors to be applied to obtain current ratings for the particular site conditions.

A 36 kV, 3 core, 300 mm², Cu conductor cable is to be laid in an ambient air temperature of 35°C. The rating is given in manufacturers tables as 630 A at 25°C and a derating factor of 0.9 is applicable for 35°C operation. Therefore cable rating at 35°C = 630 × 0.9 = 567 A.

In the case of cables laid in a concrete trench, the ambient temperature in the trench would be higher than the outside ambient air temperature. In addition, the proximity to other power cables laid in the same trench will have an effect on the cable current carrying capacity. Derating factors are included in manufacturers' literature. It should also be noted that cables laid outdoors should be protected from direct sunrays with appropriate sunshields. Metallic shields should certainly not fully surround single core cables because of their effect as a closed loop magnetic circuit to stray induced currents from the cable.

12.4.3 Cables laid direct in ground

Current rating tables are generally based upon thermal aspects and the following environmental data:

Ground thermal resistivity $G = 1.0°C$ m/W (Japan and Scandinavia)
$$= 1.2°C \text{ m/W (UK)}$$

More accurate data for the particular application may be collected from site measurements. Typically values range from 0.8 to 2.5°C m/W and occasionally

Figure 12.2   24 kV hear shrink sleeve termination awaiting final connections

to 3.0°C m/W in desert areas. Derating factors in comparison with the 1.2°C m/W reference may be obtained from ERA Report 69-30. For a G value of 2.5°C m/W a derating of approximately 75% would result.

Ground temperature $t = 25$°C (Japan)

$= 15$°C (Europe)

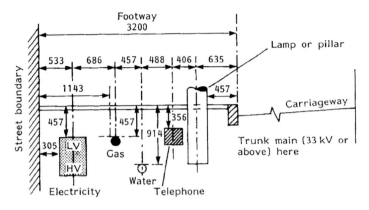

Figure 12.3   Standard installation details at a roadway verge

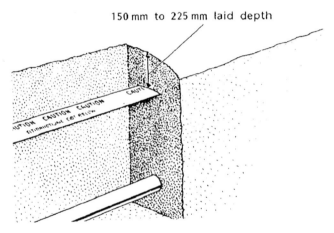

Figure 12.4   Warning, location and identification tape (sometimes used to supplement protective tiles, and allowed or even preferred in some countries instead of tiles at low or medium voltage)

Installations at variance from the standard 15°C ground temperature are taken into account by suitable derating factors with values deviating from unity by approximately 1% per °C.

Cable laying depth d = typically 1 m – but see typical laying arrangements in Figs 12.3–12.8. Actual depths will vary according to voltage and to regulations in the territory concerned.

Where cables are laid together in one trench the proximity will necessitate derating factors to be applied to obtain the correct current carrying capacity for the site conditions. In some cases the use of special trench backfill materials may improve the situation by improving heat transfer.

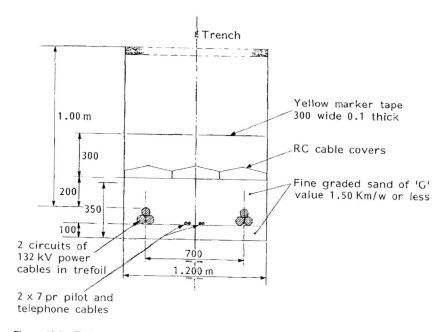

Figure 12.5   Typical trench cross-section for 132 kV cables

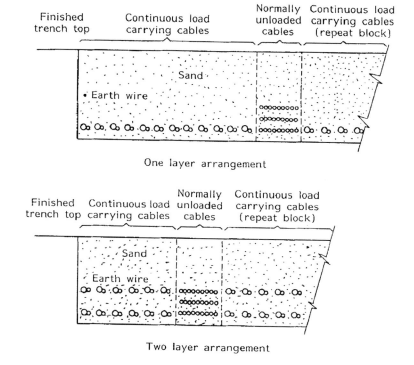

Figure 12.6   Cables laid in sand-filled trenches

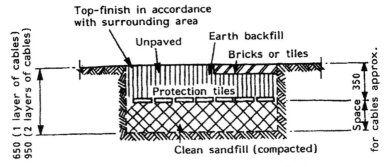

Electrical cables in unpaved, brick-paved or tiled areas

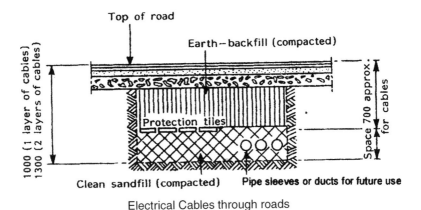

Electrical Cables through roads

Figure 12.7   Electrical cables in unpaved, brick-paved or tiled areas and through roads

Example
Two 12 kV three phase circuits comprising of single core, 500 mm², XLPE insulated, Al conductor cables are each laid in ground in trefoil formation in parallel at a nominal 0.7 m depth and 0.25 m apart and with their 35 mm~ copper screens bonded at both ends. The 90°C XLPE cable rating is 655 A. What is the rating of each circuit in this configuration?

- depth
- temperature
- ground thermal resistivity
- proximity of parallel circuits (grouping)

0.7 m derating factor   = 1.0
25°C derating factor f_1 = 0.93
1.5°C m/W derating factor f_2 = 0.91
0.25 m apart derating factor f_3 = 0.86

Therefore the maximum current rating at 90°C per circuit

$$= 655\,A \times 1.0 \times 0.93 \times 0.91 \times 0.86 = 477\,A.$$

Some typical arrangements for cable installations are given in Figs 12.3–12.9. For installations in roadway verges and other public areas it is very important to

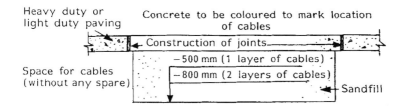

(a) Permanently covered cable trench

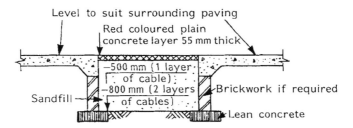

(b) Cable trench not wider than 1 m in light duty paving

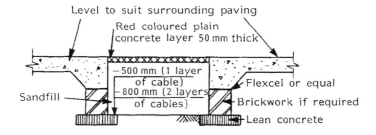

(c) Cable trench not wider than 1 m in heavy duty paving

Figure 12.8   Cable trenches in concrete-paved area

install to agreed standards and to maintain accurate records of the cable location (depth, lateral distance from known reference points, location of cable joints, oil tanks, etc.). Suitable cable management systems are described in Section 12.8.

12.4.4   Cables laid in ducts

Cables may be installed in ducts buried in the ground with an earth, sand or concrete surround. Generally, it is good practice to install only one power cable per duct and the internal diameter of the duct should be at least 35 mm greater than the cable diameter. Cable ratings in ducts correspond to typically

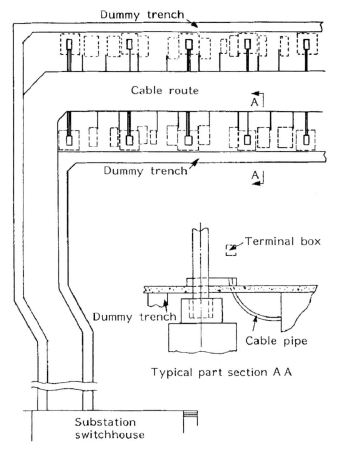

Figure 12.9   Typical plot plan of cable routes

80% of the direct in-ground burial rating. In order to improve the thermal conduction from the cable to the surrounding ground and improve this derating factor the cable ducts may be filled with a bentonite slurry after cable pulling. Bentonite is a clay containing minerals of the montmorillonite (smectite) group which give it its characteristic swelling upon wetting and malleable nature for sealing around cables at cable duct ends. Filling the ducts has the added benefit of preventing cable thermomechanical movement which can lead to fatigue failure of the cable sheath and mechanical damage to joints.

12.4.5 Earthing and bonding

12.4.5.1 *General*

Sheaths and/or armouring on successive lengths of cable are bonded together and earthed to prevent stray voltages in uninsulated or lightly insulated metal in the event of a phase-to-earth fault occurring, or due to the transformer action of the conductor and sheath. A mechanically sound and strong connection is essential.

When cable sheaths are bonded together, the induced voltages are short circuited but a current flows in the closed loop and this gives rise to heat loss. In addition to this loss in the sheath, a circulating eddy current due to the asymmetrical flux distribution in the sheath is also present whether the cables are bonded or not. Therefore two types of heating loss occur in the sheath:

- sheath circuit loss (bonded sheaths only);
- sheath eddy loss (normally small compared to loss in the bonded sheath circuit).

12.4.5.2 *Three core cables*

Three core cable circuits are normally solidly bonded such that the cable sheath, screen and/or armour are connected together to a grounding point at both ends. Each joint along the route is also bonded to earth (see Section 12.2.3.4).

12.4.5.3 *Single core cables*

Single core cable circuits require special consideration because of the voltages, which are proportional to the conductor current and frequency, being induced in the metal sheath and the introduction of circulating sheath currents. Single core cables may be solidly bonded (bonded at both ends) and this is the normal practice up to 36 kV with trefoil configurations. With larger conductor sizes and higher voltages specially bonded systems are more economic.

Single point bonding over short 500 m lengths is used to keep induced voltages between the cable screen free ends within permissible limits. The sheath or screen is insulated from ground at one end and often fitted with sheath voltage limiters. The method is sometimes known as 'end point' earthing. Single point bonding is also often employed for communications cables to prevent ground current loops.

On route lengths too long to employ end point earthing *mid-point earthing* may be used. In this system the cable is earthed at the mid-point of the route generally at a joint and is insulated from ground and provided with sheath voltage limiters at each termination. The maximum length of mid-point-bonded circuits is about 1 km. A separate earth continuity conductor should be provided for fault currents that would normally be carried by the sheath for both single point and mid-point-bonded circuits.

Cross-bonding or *cyclic transposition* is also employed to minimize the effect of induced voltages. In the cross-bonding system the cable route is split into groups of three drum lengths and all joints are fitted with insulating flanges. The cables laid in flat formation are normally transposed at each joint position. At each third joint position, the sheaths are connected together and grounded. At the other joint positions the sheaths occupying the same position in the cable trench are connected in series and connected to earth via sheath voltage limiters (see Fig. 12.10).

12.4.6 Short circuit ratings

Each conductor in a three phase circuit must be capable of carrying the highest symmetrical three phase short circuit through fault current at that point in the network. Ratings are normally taken over a 1 second fault duration period for a conductor temperature not exceeding 250°C for XLPE-insulated cables and 160°C for paper-insulated cables. This temperature must not adversely affect the conductor or the lead sheath and the armour wires (if present and being used as an earth return path). Mechanical strength to restrain the bursting forces and joint damage due to through fault currents is also a major design factor.

The earth fault condition affects both the phase conductors, screen wires/ metallic sheath and the armour wires On smaller cables the short circuit rating of the phase conductor is the limiting feature but on larger sizes the effect of the fault current on the metallic sheath, screen wires and/or armouring is an

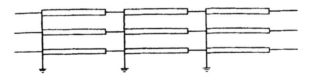

Splitting sheath into sections and earthing at one end

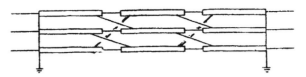

Cross-bonding of sheaths

Cyclic transposition of cables

Figure 12.10   Methods of bonding and earthing cable sheaths to minimize the effects of induced voltages

overriding consideration. In the unlikely event of an internal 3 core cable fault, the intensity of the arc between the conductor and screen will normally cause rupture of the core bedding. This results in the fault current taking the least resistance along the steel tape or wire armour and will involve earth. The sheath or screen and armour must be able to carry the full specified earth fault current. On single core unarmoured cables care must be taken to specify correctly the screen fault current carrying capability such that a fuse type failure of thin copper screen wires or tapes does not occur. The sheath or screen should be able to carry at least one-half of the specified earth fault current as sharing between cores is not always even. The general formula for calculating the allowable short circuit current I_{SC} is:

$$I_{SC} = \frac{KA}{\sqrt{t}} \text{ (amps)}$$

where K = a constant depending upon the conductor material and on the initial and final temperatures associated with the short circuit conditions

t = duration of short circuit in seconds

A = cross-sectional area of the conductor in square mm (i.e. the number of wires $\times$ cross-sectional area of each wire)

Typical values of K for paper-, PVC- and XLPE-insulated cables are given in Tables 12.7a–c.

Table 12.7a   Paper-insulated cable K values

| | | K | |
| | Conductor | | |
Voltage	temperature rise °C	Copper conductor	Aluminium conductor
600/1000 V			
1.9/3.3 kV	80–160	108.2	69.6
3.8/6.6 kV			
6.35/11 kV			
Single core	70–160	115.6	74.4
Three core belted	65–160	119.3	76.7
Three core screened	70–160	115.6	74.4
8.7/15 kV	70–160	115.6	74.4
12.7/22 kV	65–160	119.3	76.7
19/33 kV	65–160	119.3	76.7

Notes: [a] The upper limit of 160°C is fixed by the melting point of solder in the plumbed joints and terminations.
[b] The short circuit current rating of the lead sheath based upon initial and final temperatures of 60°C and 250°C can be calculated by the formula $I_{SC} = 29.6\ A_S/t$ where A_S is the cross-sectional area of the sheath in mm^2.

Table 12.7b   PVC-insulated cable _K_ values

	K	
Conductor temperature rise °C	Copper conductor	Aluminium conductor
70–130	96.4	62
70–150	109.8	70.6

Table 12.7c   XLPE-insulated cable _K_ values

	K	
Conductor temperature rise °C	Copper conductor	Aluminium conductor
90–250	136	87

12.4.7 Calculation examples

12.4.7.1 20 kV transformer feeder

Consider a 20/3.3 kV, 12.5 MVA transformer to be fed by direct buried, 3 core XLPE, SWA, PVC, copper conductor cable.

Cable current carrying capacity

$$\text{Transformer full load current} = \frac{12.5 \times 10^6}{1.73 \times 20 \times 10^3} = 361\text{A}$$

Derating factors

Manufacturers provide data sheets for cables including appropriate derating factors based upon IEC 60287 (Table 12.8).

For a ground temperature at depth of laying of 20°C, the derating factor is 0.97.

The group derating factor based upon 3 cables laid in trench at 0.45 m centres is 0.84.

Ground thermal resistivity taken as the normal of 1.2°C m/W for a UK installation and 1.00 rating factor.

Cable installation depth to be 0.8 m and 1.00 rating factor.

$$\text{Therefore subsequent current rating of cable to be } \frac{361}{0.97 \times 0.84} = 443\text{A}$$

From manufacturers tables selected cable size = 240 mm².

Table 12.8   Derating factors based on IEC 60287

(a)(i)   Variation in ground temperature, direct burial in ground

Ground temperature, °C	15	20	25	30	35	40	45	50
Derating factor	1.00	0.97	0.93	0.89	0.86	0.82	0.76	0.72

(a)(ii)   Variation in ground temperature, cables installed in single way ducts

Ground temperature, °C	15	20	25	30	35	40	45	50
Derating factor	1.00	0.97	0.93	0.89	0.86	0.82	0.76	0.72

(b)(i)   Variation in soil thermal resistivity, single core cables, direct burial in ground

	Thermal resistivity, G °C m/W						
Depth of laying (m)	0.8	0.9	1.0	1.5	2.0	2.5	3.0
Conductor area, mm^2							
50	1.16	1.11	1.07	0.91	0.81	0.73	0.68
70	1.16	1.12	1.07	0.91	0.81	0.73	0.68
95	1.16	1.12	1.07	0.91	0.81	0.73	0.68
120	1.16	1.12	1.07	0.91	0.81	0.73	0.68
150	1.16	1.12	1.07	0.91	0.81	0.73	0.68
185	1.17	1.12	1.07	0.91	0.81	0.73	0.68
240	1.17	1.12	1.07	0.91	0.80	0.73	0.68
300	1.17	1.12	1.07	0.91	0.80	0.73	0.68
400	1.17	1.12	1.07	0.91	0.80	0.73	0.67
500	1.17	1.12	1.07	0.91	0.80	0.73	0.67
630	1.17	1.12	1.07	0.91	0.80	0.73	0.67
800	1.17	1.12	1.07	0.91	0.80	0.72	0.66
1000	1.18	1.12	1.07	0.91	0.80	0.72	0.66

(b)(ii)   Variation in soil thermal resistivity, multicore cables, direct burial in ground

	Thermal resistivity, G °C m/W						
Depth of laying (m)	0.8	0.9	1.0	1.5	2.0	2.5	3.0
Conductor area, mm^2							
25	1.13	1.09	1.05	0.93	0.83	0.77	0.71
35	1.13	1.09	1.06	0.92	0.83	0.76	0.71
50	1.13	1.09	1.06	0.92	0.83	0.76	0.71
70	1.14	1.09	1.06	0.92	0.83	0.75	0.70
95	1.14	1.09	1.06	0.92	0.83	0.75	0.70
120	1.14	1.10	1.06	0.92	0.82	0.75	0.69
150	1.14	1.10	1.06	0.92	0.82	0.75	0.69
185	1.14	1.10	1.06	0.92	0.82	0.74	0.69
240	1.15	1.10	1.07	0.92	0.81	0.74	0.69
300	1.15	1.10	1.07	0.92	0.81	0.74	0.69
400	1.15	1.10	1.07	0.92	0.81	0.74	0.69

Table 12.8   Continued

(b)(iii)   Variation in soil thermal resistivity, single core cables installed in single way ducts

Depth of laying (m)	Thermal resistivity, G °C m/W						
	0.8	0.9	1.0	1.5	2.0	2.5	3.0
Conductor area, mm²							
50	1.08	1.06	1.04	0.94	0.87	0.82	0.77
70	1.09	1.06	1.04	0.94	0.87	0.81	0.76
95	1.09	1.06	1.04	0.94	0.87	0.81	0.76
120	1.10	1.07	1.04	0.94	0.86	0.80	0.75
150	1.10	1.07	1.04	0.94	0.86	0.80	0.75
185	1.10	1.07	1.04	0.93	0.86	0.79	0.75
240	1.11	1.07	1.04	0.93	0.86	0.79	0.74
300	1.11	1.08	1.05	0.93	0.85	0.79	0.74
400	1.11	1.08	1.05	0.93	0.85	0.78	0.73
500	1.11	1.08	1.05	0.93	0.85	0.78	0.73
630	1.12	1.08	1.05	0.93	0.84	0.78	0.72
800	1.12	1.09	1.05	0.93	0.84	0.77	0.72
1000	1.13	1.09	1.05	0.92	0.84	0.77	0.71

(b)(iv)   Variation in soil thermal resistivity, multicore cables installed in single way ducts

Depth of laying (m)	Thermal resistivity, G °C m/W						
	0.8	0.9	1.0	1.5	2.0	2.5	3.0
Conductor area, mm²							
25	1.05	1.03	1.02	0.96	0.92	0.88	0.84
35	1.05	1.03	1.02	0.96	0.92	0.87	0.83
50	1.05	1.03	1.02	0.96	0.91	0.87	0.83
70	1.05	1.04	1.02	0.96	0.91	0.86	0.82
95	1.06	1.04	1.02	0.96	0.91	0.86	0.82
120	1.06	1.04	1.03	0.95	0.90	0.85	0.81
150	1.06	1.04	1.03	0.95	0.90	0.85	0.80
185	1.07	1.05	1.03	0.95	0.89	0.84	0.80
240	1.07	1.05	1.03	0.95	0.89	0.84	0.79
300	1.07	1.05	1.03	0.95	0.88	0.83	0.78
400	1.07	1.05	1.03	0.95	0.88	0.83	0.78

Table 12.8   Continued

(c)   Variations in depth of laying measured from ground surface to the centre of a cable or to the centre of a trefoil group, direct burial in ground

Depth of laying (m)	Cables up to 300 mm²	Cables over 300 mm²
0.80	1.00	1.00
1.00	0.98	0.97
1.25	0.96	0.95
1.50	0.95	0.93
1.75	0.94	0.91
2.00	0.92	0.89
2.50	0.91	0.88
3.0 or more	0.90	0.86

(d)(i)   Variations due to grouping, single core cables (in close trefoil formation), direct burial in ground

No. of groups	2	3	4	5	6
Groups touching	0.78	0.66	0.59	0.55	0.52
Groups at 0.15 m between centres*	0.82	0.71	0.65	0.61	0.58
Groups at 0.3 m between centres	0.86	0.77	0.72	0.68	0.66
Groups at 0.45 m between centres	0.89	0.80	0.77	0.74	0.72

* This spacing not possible for some of the larger diameter cables.

(d)(ii)   Variations due to grouping, single core cables (in flat spaced formation), direct burial in ground

No. of groups	2	3	4	5	6
Groups at 0.15 m between centres*	0.80	0.69	0.63	0.59	0.56
Groups at 0.3 m between centres	0.84	0.75	0.70	0.66	0.64
Groups at 0.45 m between centres	0.87	0.79	0.75	0.72	0.70

* This spacing not possible for some of the larger diameter cables.

(d)(iii)   Variations due to grouping, multicore cables, direct burial in ground

No. of cables in group	2	3	4	5	6
Cables touching	0.80	0.68	0.62	0.57	0.54
Cables at 0.15 m between centres	0.85	0.76	0.71	0.66	0.64
Cables at 0.3 m between centres	0.89	0.81	0.77	0.73	0.71
Cables at 0.45 m between centres	0.91	0.84	0.81	0.78	0.77

Short circuit rating
The maximum system fault level in this application is 8.41 kA. From Section 4.6 of this chapter and IEC 60364-5-54 (Electrical Installations In Buildings – Earthing arrangements, protective conductors and protective bonding conductors):

$$I_{SC} = \frac{KA}{\sqrt{t}}$$

where K = constant, 143 for XLPE cable
A = cable cross-section, 240 mm² based on current carrying capacity
t = short circuit duration, for MV cables use 1 second

$$I_{SC} = 240 \times 143/\sqrt{1}$$
$$= 34.3 \text{kA}$$

From manufacturers tables and/or Figs 12.11a–c for working voltages up to and including 19 000/33 000 XLPE based insulated cable the selected 240 mm² cable is just capable of this 1 second short circuit rating. Note tables are conservative and assume a fully loaded cable. At the initiation of the fault conductor temperature = 90°C and at the end of the fault conductor temperature = 250°C.

Voltage drop (V_d)
Consider a 100 m route length of cable with resistance, R = 0.0982 Ω/km and inductive reactance, X_L = 0.097 Ω/km.
   At full load current, I_{fl} = 361 A @ 0.85 pf the cable voltage drop over a 100 m cable length,

$$V_d = I_{fl} \times X_L \times \sin \phi + I_{fl} \times R \times \cos \phi \text{ volts}$$
$$= (361 \times 0.097 \times 0.53 + 361 \times 0.0982 \times 0.85)100/1000$$
$$= (18.56 + 30.13)100/1000$$
$$= 4.87 \text{ V}$$
$$= 0.042\%$$

Notes: (a) At 20 kV the voltage drop is negligible over such a short length of cable.
(b) IEE Wiring Regulations require a voltage drop for any particular cable run to be such that the total voltage drop in the circuit of which the cable forms part does not exceed 2½% of the nominal supply voltage, i.e.10.4 volts for a three phase 415 V supply and 6 volts for a single phase 240 V supply.
(c) Industrial plant users may use different specifications and apply ±5% (or even ±10%) under no load to full load conditions and perhaps −20% at motor terminals under motor starting conditions.
(d) Manufacturers data for building services installations is often expressed in terms of voltage drop (volts) for a current of 1 ampere for a 1 metre run of a particular cable size.

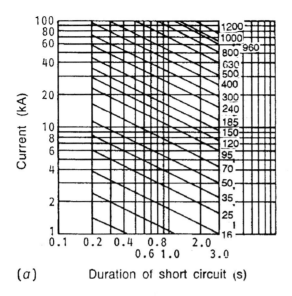

(a)   Duration of short circuit (s)

(cable sizes in mm²)

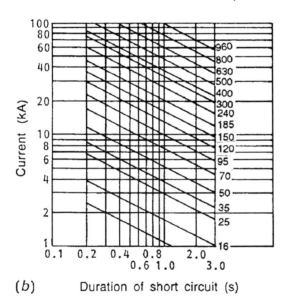

(b)   Duration of short circuit (s)

Figure 12.11   (a) Paper-, (b) PVC- and (c) XLPE-insulated copper conductor cable short circuit ratings

Earth loop impedance

For building services work it is important with small cross-section wiring and low fault levels to ensure that sufficient earth fault current flows to trip the MCB or fuse protection. For distribution power networks with more sophisticated protection the check is still necessary and allows the calculation of the

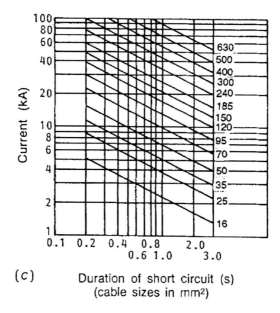

(c)   Duration of short circuit (s)
(cable sizes in mm²)

Figure 12.11   *Continued*

likely touch voltages arising from the earth fault. This in turn can then be
checked against the allowable fault duration to avoid danger. See Chapter 8 for
a consideration of the design criteria associated with touch and step potentials.

- Consider the earthing resistance at the source substation $= 0.5\,\Omega$.
- The source substation 20 kV neutral is approximately 10 km from the 100 m
  cable under consideration. In addition parallel copper conductor earth cable
  is run to supplement and improve power cable armour resistance values from
  equipment back to the primary substation infeed neutral. For this example
  assume power and supplementary earth copper cables and armour over the
  10 km distance have a combined effective resistance of 0.143 Ω.
- The combined resistance of the 100 m, 240 mm², cable armour (0.028 Ω/
  100 m) and in parallel 2 × 95 mm² copper supplementary earth cables
  (0.00965 Ω/100 m) $= 7.18 \times 10^{-3}\,\Omega$.
- Consider the earthing resistance at the cable fault to be 0.5 Ω.
- The effective earth circuit is shown in Fig. 12.12. The effective primary
  substation neutral-to-fault cable resistance $= 0.15\,\Omega$.
- The maximum earth fault current at 20 kV has to be determined. Sometimes this is limited by a neutral earthing resistor and the maximum limited
  current may be taken for calculation. Maximum earth fault current for this
  calculation is 1000 A. For a fault to earth at the end of the 100 m cable,
  10 km from the primary power infeed the fault current, $I_f = (1000 \times 0.15)/$
  $(1 + 0.15) = 131$ A. Therefore touch voltage to earth at the cable fault $=$
  $131 \times 0.5 = 65.3$ V.

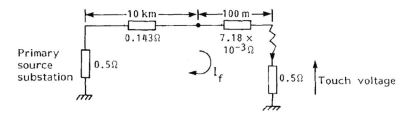

Figure 12.12   Calculation example – earth loop impedance

12.4.7.2 *3.3 kV motor feeder*

Cable current carrying capacity
Current input to a motor is given by

$$I = P/\sqrt{3} \times U \times \eta \times \cos\phi \text{ (3 phase)}$$
$$I = P/U \times \eta \times \cos\phi \text{ (1 phase)}$$

where P = motor shaft power output
            U = phase voltage
            η = motor efficiency
            ϕ = phase angle

Consider a 3.3 kV, 340 kW fan motor. Full load current

$$= \frac{340 \times 10^3}{1.73 \times 3.3 \times 10^3 \times 0.9}$$
$$= 66\text{A}$$

Cable derating factors
Apply a group derating factor of 0.78 based upon cables touching on trays.

$$\text{Necessary cable rating} = \frac{66}{0.78} = 85\,\text{A}.$$

From manufacturers' data a 3 core, 16 mm², XLPE/SWA/PVC copper conductor cable is suitable.

Short circuit rating
System fault level is 3.5 kA for 1 second.

$$16\,\text{mm}^2 \text{ cable fault capability } I_{SC(sqmm)} = \frac{KA}{\sqrt{t}} = \frac{16 \times 143}{1} = 2.23\,\text{kA}.$$

Try next larger, 25 mm² standard size cable

$$I_{SC(25sqmm)} = \frac{KA}{\sqrt{t}} = \frac{25 \times 143}{1} = 3.6\,\text{kA}$$

and therefore complies.

Voltage drop (V_d)
Consider a 75 m route length of cable at full load running current with resistance, $R = 0.927\,\Omega/\text{km}$ and inductive reactance, $X_L = 0.094\,\Omega/\text{km}$.

$$
\begin{aligned}
V_{d(flc)} &= I_{fl} \times X_L \times \sin\phi + I_{fl} \times R \times \cos\phi \text{ volts} \\
&= (66 \times 0.094 \times 0.53 + 66 \times 0.927 \times 0.85)75/1000 \\
&= (3.29 + 52)\ 75/1000 \\
&= 4.14\,\text{V}
\end{aligned}
$$

Consider a starting current $= 4 \times$ full load current ($4 \times$ flc) and 0.2 pf.

$$
\begin{aligned}
V_{d(starting)} &= I_{fl} \times X_L \times \sin\phi + I_{fl} \times R \times \cos\phi \text{ volts} \\
&= (4 \times 66 \times 0.094 \times 0.98 + 4 \times 66 \times 0.927 \times 0.2)75/1000 \\
&= (24.32 + 48.95)75/1000 \\
&= 5.5\,\text{V}
\end{aligned}
$$

Earth loop impedance
$25\,\text{mm}^2$ cable armour resistance (from manufacturers' literature) $= 1.\,7\,\Omega/\text{km}$.
Earth fault current at 3.3 kV is neutral point earthing resistance limited to only 30 A. This resistance swamps all other sequence components if the motor is earthed by only the cable armour.

Touch voltage $= 30 \times 1.7 \times 75/1000 = 3.83\,\text{V}.$

This is well below the continuous allowable IEC 60364 dry condition 50 V limit.

12.4.7.3 *400 V distribution cable*

Current carrying capacity
Consider the supply to a 400 V small power and lighting distribution board with a load of 38 kW (including an allowance for future extensions. Note that low voltage switchboards should be specified with a level of equipped and unequipped spare ways to cater for such future extensions).

$$
\text{Full load current} = \frac{35 \times 10^3}{1.73 \times 400} = 55\,\text{A}
$$

Derating factors
Derating factors associated with cables laid in air, touching, on cable tray apply. The current rating is based upon a specified ambient temperature (30°C) shielded from direct sunlight. For XLPE cable the maximum continuous conductor operating temperature is taken as 90°C and the maximum conductor short circuit temperature as 250°C. For six to eight cables laid touching on horizontal tray the group derating factor based upon IEC 60287 and the table given below $= 0.72$.

$$
\text{Necessary cable rating} = \frac{55}{0.72} = 76.5\,\text{A}.
$$

Table 12.9   Touching cables on tray horizontally or vertically installed

Number of cables	2	3	4/5	6/8	≥9
Horizontal derating factor	0.85	0.78	0.75	0.72	0.70
Vertical derating factor	0.80	0.73	0.70	0.68	0.66

Table 12.10   Current carrying capacity and voltage drop – Multicore cable having thermosetting insulation, non armoured (Copper conductors) (BS7671 Table 4E2A and 4E2B)

Installation (in air, clipped direct to a cable tray)	Conductor area, mm²	Twin cable single phase, AC or DC		3 or 4 core cable three phase AC	
		Current rating (A)	Approximate volt drop per amp per metre (mV)	Current rating (A)	Approximate volt drop per amp per metre (mV)
	1.5	24	31	22	27
	2.5	33	19	30	16
	4	45	12	40	10
	6	58	7.9	52	6.8
	10	80	4.7	71	4.0
	16	107	2.9	96	2.5

From manufacturers' data sheets $16\,mm^2$ cable (current rating = 95 A) may be selected and allows a margin for power factor.

Short circuit rating

For short circuits on an LV system the protective device must clear the fault limiting the maximum conductor temperature (250°C for XLPE insulation). For cables of $10\,mm^2$ and greater cross-sectional area the maximum fault clearance time, t, is based upon:

$$t = \frac{K^2 A^2}{I^2_{SC}}$$

where t = fault clearance time (s)
   A = cable conductor cross-sectional area (mm^2)
   I_{SC} = short circuit current (A)
   K = 143 for XLPE insulation

The breaking capacity of the protective device must be at least equal to the highest current produced by a short circuit at the installation location. For a 3.58 kA fault level the maximum fault clearance time, $t = 143^2 \times 16^2 / 3580^2 = 0.41$ second.

Protection by a 63 A MCB to IEC 157-1 (now superseded by IEC 60947-2) the fault clearance time would be 0.01 second.

Voltage drop (V_d)
For a 16 mm^2 cable V_d =2.6 mV/A/m (Table 12.10).
Therefore for a 55 A full load current and 20 m cable run

$$V_d = 2.6 \times 55 \times 20$$
$$= 2.86\,V$$
$$= 0.72\%$$

Earth loop impedance
For 16 mm^2 cable the loop impedance = 6.42 milli Ω/m (from manufacturer's data).
Cable armour cross-sectional area = 44 mm^2.
For 20 m cable length loop impedance, Z_S = 0.1284 Ω.
Single phase short circuit current, I_{SC} = V/Z_S =230/0.1284 = 1791 A.

$$I_{SC} = \frac{KA}{\sqrt{t}} \quad \text{where } K = 54 \text{ for steel wire armour and XLPE insulation.}$$

From the formula, maximum operating time of protective device,

$$t_{max} = \frac{(54^2 \times 44^2)}{1791^2}$$
$$= 1.75 \text{ seconds}$$

For a 63 A MCB to IEC 157-1 (IEC 60947-2) the tripping time for this fault level will be 0.014 second.

Figure 12.1 illustrates typical symmetrical short circuit current/time duration ratings for PILC-, PVC- and XLPE-insulated cables.

12.5 CALCULATION OF LOSSES IN CABLES

12.5.1 Dielectric losses

Cables of the same conductor diameter, insulation material and similar construction from different manufacturers will have similar small dielectric losses which may be compared when buying cable during the tender adjudication stage. The larger the conductor diameter, the greater the losses for a given insulating dielectric material. Dielectric losses in XLPE-insulated cables will be appreciably lower than in oil-filled paper-insulated types which have a higher capacitance per unit length. For example, consider circuits containing 3 single core or 3 core 132 kV cables:

Oil-filled cables, 240 mm^2 to 800 mm^2	Dielectric losses typically 4.5 to 9.5 W/m
XLPE cables, 240 mm^2 to 800 mm^2	Dielectric losses typically 0.7 to 1.2 W/m

12.5.2 Screen or sheath losses

Screen or sheath/metallic layer losses will be proportional to the current carried by the cables and will be approximately the same for standard cables of the same types and size. If the cables are to be installed on systems with high earth fault levels the sheath or metallic layer cross-sections will have to be increased. In particular care should be taken regarding possible future network expansion and interconnections which might involve increasing fault levels over the life time of the cable installation. Losses may be reduced in the case of circuits employing single core cables by single point bonding on short cable routes (<500 m) and cross bonding on longer routes (see Section 12.4.5).

Cable losses may be calculated and compared at the tender adjudication stage from the maximum permissible loss angle value in accordance with IEC Standards and the maximum current carrying capacity of the cable. Some Tenderers base their calculations on the loss angle value obtained during cable type tests and the specified cable current rating required. This will give results appreciably lower than the permissible maximum value.

Where the costs due to cable losses are to be evaluated this should be specified at the tender or enquiry stages so that the manufacturers can state the actual maximum losses and not the maximum permissible losses. For example, 132 kV circuits containing 3 single core or 3 core 240 mm^2 to 800 mm^2 standard XLPE or oil-filled cables will have sheath (or screen) losses of the order of 1.0 to 10 W/m.

12.6 FIRE PROPERTIES OF CABLES

12.6.1 Introduction

This is an area of increasing public and legislative concern, and therefore of increasing interest to engineers. There have been major advances in the fire performance of cables in recent years, and Table 12.11 lists some of the relevant standards.

12.6.2 Toxic and corrosive gases

It is recognized that conventional flame retardant cables having sheathing based upon PVC type materials evolve considerable quantities of halon acid gases such as hydrogen chloride upon burning. Such materials are not therefore suitable for use in confined spaces where the public are likely to travel, and moreover the fire in the ENEL power station at La Spezia in 1967 showed that in certain circumstances PVC cables will burn completely and contribute to the spread of a fire. Materials have now been developed for cable oversheaths and bedding which are normally free of halogen based compounds. They consist of a mixture of inorganic filler such as aluminium hydroxide and polymers such as ethylene vinyl acetate, acrylates and ethylene propylene rubbers. Cables manufactured with

Table 12.11   Standards relating to fire properties of cables

Standard	Description
IEC 60331	Fire resisting characteristics of electric cables.
IEC 60332	Tests on electric cables under fire conditions. Test methods and flame propagation of power and control/communication cables. Note the identical EN60332 and equivalent national standard BSEN60332 supersede EN50265 and BS 4066.
IEC60754	Test of gases evolved during combustion of electric cables.
IEC61034	Measurement of smoke density of cables burning under defined conditions. Identical EN61034 and national equivalent BSEN 61034 supersede EN50268 and BS7622.
BS6387	Performance requirements for cable required to maintain circuit integrity under fire conditions.
BS6724	Electric cables. Thermosetting insulated, armoured cables for voltages of 600/1000 V and 1900/3300 V, having low emission of smoke and corrosive gases when affected by fire.
BS7211	Electric cables. Thermosetting insulated, non-armoured cables for voltages up to and including 450/750 V, for electric power, lighting and internal wiring, and having low emission of smoke and corrosive gases when affected by fire.
BS 7835	Specification for cables with cross-linked polyethylene or ethylene propylene rubber insulation for rated voltages from 3800/6600 V up to 19 000/33 000 V having low emission of smoke and corrosive gases when affected by fire.
EN50267	Common test methods for cables under fire conditions. Tests on gases evolved during combustion of materials from cables. Apparatus. BSEN50267 is identical and supersedes BS6425. Similarly French standard NF C 20-454 is superseded.

such materials are known as 'Low Smoke and Fume' (LSF) and have acid gas evolution less than 0.5% in comparison to 25–30% for PVC compounds.

IEC 60754-1 specifies a method of determining the amount of halogen acid gas, other than hydrofluoric acid, evolved during combustion of halogen based compounds. The method essentially measures the existence of halogen acid greater than 0.5%, the accuracy limit for the test. Therefore cables tested having less than the 0.5% limit are generally termed 'zero halogen' or 'low smoke zero halogen' (LS0H).

12.6.3 Smoke emission

Normal cable sheathing compounds also give off dense smoke when burned and this is of particular concern in underground transport system installations. The generation of large amounts of smoke obscures vision and reduces the ease with which the fire brigade is able to bring members of the public to safety in the event of a fire. LSF cables therefore play an important part in reducing this danger to a minimum. London Underground Limited (LUL) have developed a test of practical significance which has been designed to measure the density of smoke emission from cables and it has now been adopted by British and IEC Standards. This defines the standard absorbance produced across the opposite faces of a test cubicle and is popularly known as the 3 m cube test. Paris Metro (RATP) adopts the French Standard UTE C20-452 on smoke emission which determines under experimental conditions the specific optical density of smoke produced by burning material. This slightly different approach is generally known as the NBS smoke chamber test.

12.6.4 Oxygen index and temperature index

'Oxygen index' is the minimum concentration of oxygen in an oxygen/ nitrogen mixture in which the material will burn. As air contains approximately 21% oxygen it is stated that a material with an oxygen index greater than about 26% will be self extinguishing. In general, a particular oxygen index value offers no guarantee of resistance to the spread of flames. In practice materials having identical oxygen indices may have widely different burning properties especially if base polymers or additives are of different types.

The 'temperature index' of a material is the minimum temperature at which the material supports combustion in air containing 21% oxygen when tested under controlled conditions. The test is useful for the comparison of similar materials but no correlation with flammability under other fire conditions is implied.

Oxygen and temperature indices are to some extent inter-related. The engineer specifying such cable requirements should *not* pick out the most favourable parameters from different manufacturers' literature and expect them to comply. For example, a high oxygen index using a particular combination of materials may result in a slightly less favourable temperature index rating. In some cases where manufacturers have been requested to provide cables with a temperature index of 280°C or above this requirement was only met at the expense of other important parameters such as tensile properties and water permeability. Acceptable values of oxygen index and temperature index recommended by leading manufacturers and specified for LSF compounds would be:

- Oxygen index equal to or greater than 30.
- Temperature index equal to or greater than 260°C.

12.6.5 Flame retardance/flammability

12.6.5.1 *Single core cables*

Flame retardant cables meet the requirements of IEC 60332 Part 1. (For European use EN 60332 is identical to the IEC standard). These tests define the cable performance under fire conditions. The tests are carried out on a single length of cable supported vertically in a draught free enclosure with a burner applied to the lower end of the cable. After a specified time the heat source is removed and the cable should not continue to burn after a stated length of time. The extent of charring at the top of the cable is also defined.

12.6.5.2 *Cables in bunches or groups*

Single cables which pass the test mentioned in Section 12.6.4.1 above may not necessarily pass the test when grouped together in vertical racks, where propagation of the fire takes place. Propagation of fire depends upon a number of factors, but is in particular a function of the total volume of combustible material in the cable run.

The tests involved in this category attempt to simulate group cabling installation conditions and are generally covered in IEC 60332 Part 3. The IEC standards define three categories for grouped cables, A, B and C which are related to the volume of combustible (organic) material per metre.

LSF power cables manufactured by leading companies should be covered by the IEC Standards mentioned above. An important feature of the construction of all of the cables which relates to flame retardance is the cable armour. For example, XLPE insulation as a material on its own is not flame retardant. Provision of the cable armour separates the insulated cores from air for combustion even after the sheath has been destroyed.

12.6.6 Fire resistance

Fire resistance is the term used to define cables which can maintain circuit integrity for a specified period of time during a fire.

Such cables have to conform to a severe test in which the middle portion of a 1200 mm long sample of cable is supported by two metal rings 300 mm apart and exposed to a flame of a tube type gas burner. Simultaneously the rated voltage of the cable is applied throughout the test period. Not less than 12 hours after the flame has been extinguished the cable is re-energized and no failure must occur. There are may variations of time and temperature and also impact tests to simulate falling debris and application of a water deluge after the flame has been extinguished.

Two typical types of fire resistant cables are described below:

1. Mineral insulated (MICC or Pyrotenex) cables complemented with an outer LSF covering and rated 500/750 voltage. IEC 60702 (EN60702 in Europe) covers mineral insulated cables but testing in UK is to BS 6387 (NF C32-070 CR1 in France). The outer LSF covering would be required to meet BS 6724 (NF C32-200) as far as behaviour in a fire is concerned. Cable type BS 6387 – CWZ has a 3 hour resistance to fire up to 950°C.
2. Lapped mica/glass tape to be covered by an extruded cross-linked insulation, armoured, LSF sheathed. Rated voltage 600/1000 V and to meet BS 6387 types CWZ or lower temperature performances type A/B/SWX.

Trends in the development of building safety regulations in various countries (e.g. the Construction Products Directive (CPD) in Europe) may mean that use of cables with a defined level of fire resistance in all buildings may be mandated in the future.

12.6.7 Mechanical properties

Achieving good mechanical qualities in a cable material is finely balanced by the requirement of maintaining good low smoke/toxic gas emission and reduce flame propagation.

All cable materials must possess reasonable tensile strength and elongation properties with good resistance to abrasion, where the oversheath should not suffer cracks or splits during installation. Leading manufacturers have now formulated compounds for LSF cables which have similar properties to existing standard sheathing materials. Cables must also have acceptable tear resistant properties and where used for cable sheathing provide adequate protection in a wet environment.

Testing requirements for mechanical properties are defined in IEC 60229 (or EN 60811 for European countries; BS 6469-99-2 covers UK tests not covered in 60811).

12.7 CONTROL AND COMMUNICATION CABLES

12.7.1 Low voltage and multicore control cables

A wide range of cables exists for a multitude of specific applications. Open terminal substation control cables are usually multicore 600/1000 V PVC-insulated copper conductor types laid in concrete troughs from the substation control building to the switchgear. Within a high security substation building LSF cables may be specified both within and between equipment. Such cable cores are described by the individual conductor cross-sectional area (mm^2) together with the number of individual strands and associated strand diameter (mm) making up the conductor core. Control cables are generally armoured when laid direct in ground. The type of multicore cable screen and armour will determine the flexibility of the cable and associated bending radius. In general steel wire armoured cables have a smaller bending radius than steel tape armour types.

Some standard specification data and standard sizes for such control cables are detailed in Table 12.12.

Care must be taken to ensure adequate cable conductor cross-sectional area when selecting sizes for association with substation current relays located some distance from their associated CT. At the same time the traditional practice of standardizing upon 2.5 mm^2 cables for substation control and relay cubicles is now outdated and the terminations onto modern low current consumption electronics are often incapable of accommodating such a large conductor size. Insulation levels for certain applications such as pilot wire cables (pilots) associated with overhead line or feeder cable differential protection schemes must also be clearly specified. Such cables may require an enhanced insulation to counter induced voltages from the parallel power circuit.

Table 12.12   Multicore control cable technical particulars

Description	Parameter (Units)
VOLTAGE	V
CORES	
Number of cores	
CONDUCTOR	
Cross-sectional area	mm^2
Material	
INSULATION	
Nominal thickness	mm
Minimum thickness	mm
Diameter of insulation mm	
ARMOUR BEDDING	
Type	
Nominal thickness	mm
ARMOUR	
Number of wires or tapes	
Diameter of wire or thickness of tape	mm
Galvanized steel to ... Standard	
Armour resistance	ohm/km
OUTER COVERING	
Material	
Minimum average thickness	mm
COMPLETED CABLE	
Overall diameter	mm
Weight per metre	kg/m
Maximum drum length	m
CAPACITANCE	
Of each conductor to earth per km	pF/km
Of each core at 20°C	pF/km
MINIMUM BENDING RADIUS	
Around which cable may be laid	m (or as a function of overall cable diameter)
At terminations	m
MAXIMUM DC RESISTANCE	
Of each core at 20°C per km	ohm/km

Table 12.13   Standard multicore cable sizes

Conductor cross-sectional area & configuration	1.5 mm^2 (1/1.38 mm or 7/0.50 mm) 2.5 mm^2 (1/1.78 mm or 7/0.67 mm) 4 mm^2 (7/0.85 mm)
Number of cores	2, 3, 4, 5, 7, 12, 19, 27, 37 and 48
Core identification	Colour for up to 4 cores Numbers above 4 cores

12.7.2 Telephone cables

Telephone cables likely to be encountered by the transmission and distribution engineer are PE- or PVC-insulated and may be specified as unfilled (standard underground situations) or filled (submarine, high humidity environment or cables laid in waterlogged ground) with a gel to prohibit the ingress of water.

Table 12.14   Standard telephone cable sizes

Description	Characteristic (units)	Test
Standard sizes	0.4, 0.5, 0.6, 0.9 and 1.13 (mm)	
Wire insulation thickness	0.3 mm PE or 0.6 mm PVC	2 kV DC
	0.5 mm PE	5 kV DC
	0.8 mm PE	10 kV DC
Inner sheath thickness	up to 1.2 mm PE	5 kV DC
	about 1.8 mm PE	15 kV DC
Core identification	Standard colour code	

Major trunk route cables are often installed with a dry nitrogen gas system. The lay of the cable and a standard colour coding scheme assists in the identification of the correct 'pair' or circuit in a multicore telephone cable. For such communication cables the characteristic impedance is also important since maximum power transfer is achieved when impedance matching is achieved. Maximum loop impedance for a telephone circuit is typically 1000 Ohms. Attenuation in telephone cables is normally measured in dB assuming a 600 Ohm impedance. Of particular practical interest to power installation and civil services engineers is the pulling strength capability of such relatively small cables in order for the maximum distance between cable draw pits or pulling chambers to be determined in an early stage of the design. Since such 'hard wire' small signal cables are susceptible to electromagnetic interference from adjacent power cables adequate screening must be provided. Where feasible control and communication (C&C) copper telephone type cables should be laid at the following minimum distances adjacent parallel power cables:

HV single core cables	>500 mm
HV multicore power cables	>300 mm (add a physical barrier between power and C&C cables if spacing <150 mm)
External MV and LV power cables	>50 mm (add a physical barrier between power and C&C cables if spacing <25 mm)
Internal MV and LV power cables	>50 mm (use separate trunking if this spacing not possible)

Some standard sizes and specification data for telephone cables are detailed in Tables 12.14 and 12.15.

12.7.3 Fibre optic cables

12.7.3.1 *Introduction*

In order to improve the rate of transmission of information and the amount of information that can be transmitted over a given channel path the widest possible bandwidth must be employed. In order to achieve the necessary bandwidths

Table 12.15   Telephone cable technical particulars

Description	Parameter (units)
PAIRS	
Number of pairs	
CONDUCTOR	
Diameter	mm
INSULATION	
Type	
Thickness – nominal	mm
– minimum	mm
INNER SHEATH	
Thickness – nominal	mm
– minimum	mm
ARMOUR	
No. of wires or tapes	
Diameter of wire or thickness of tape	mm
OUTER COVERING	
Material	
Minimum average thickness	mm
Anti termite/worm protection additives	
COMPLETE CABLE	
Overall diameter	mm
Weight per metre	kg/m
Maximum drum length	m
Tensile strength	N
Maximum distance between pulling chambers	m (for ducted systems)
MAXIMUM DC LOOP RESISTANCE	
per km of conductor at 20°C	ohm/km
MINIMUM INSULATION RESISTANCE	
per km of cable at 20°C	ohm/km
CAPACITIVE UNBALANCE	
between any two pairs at audio frequency	pF
MAXIMUM MUTUAL CAPACITANCE	
per km of cable	nF
MAXIMUM CROSS TALK	
Under balanced cable conditions	
– Audio pairs at 1 kHz (say)	dB
– Carrier pairs at 60 kHz (say)	dB
NOMINAL IMPEDANCE	
Audio pairs at 1 kHz (say)	ohm
Carrier pairs at 60 kHz (say)	ohm

higher and higher frequencies have been employed. Power line carrier systems at frequencies of tens of kHz have been used by superimposing a modulated radio carrier on the overhead transmission line phase conductors. Alternatively, high quality telephone circuits may be installed or rented from the local telephone company. Microwave radio links using radio frequency coaxial cable feeders (see IEC 60096 – Radio-frequency cables) between transceivers and aerials are also used to carry more information. The latest development of this trend is for data transmission and digitized speech or other digitized analogue signals to be

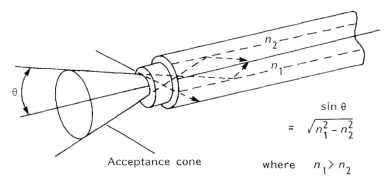

$$sin\ \theta = \sqrt{n_1^2 - n_2^2}$$

where $n_1 > n_2$

Acceptance cone

Figure 12.13   Light entering the fibre within the boundary of the acceptance cone will be propagated through the fibre

transmitted over fibre optic cable where the bandwidth is more than adequate at infra red frequencies.

Some of the advantages of fibre optic cable are summarized below:

- Fast reliable communications over long distances ($>$1Gbit/s over 100 km).
- Low transmission loss or signal attenuation.
- High privacy or security since signals carried are immune from remote detection.
- Wide bandwidth availability hence large data handling capacity.
- Signals unaffected by electromagnetic interference (EMI), radio frequency interference (RFI) and lightning noise. The cable may therefore be installed, without special screening, adjacent to power cables and overhead lines.
- No dangerous voltages are employed or induced in such cables. Therefore they may be installed in hazardous environments which require intrinsic safety.
- Complete electrical isolation between terminations is achieved. This avoids voltage gradients and ground loop problems encountered with hard wire cable solutions.
- Small overall cable diameter, light weight and flexible nature makes for easy installation. However, it is very important to note carefully that fibre optic cables must *not* be handled roughly (excess pulling tension, over clamping to cable tray, etc.) nor without conforming to the correct installation specification.

12.7.3.2 *Fibre optic cable principles*

An optical fibre cable consists of a very pure thin optical strand of silica glass material surrounded by an optical cladding of lower refractive index. The infra red light radiation in the frequency range 10^{14} Hz passes down the fibre by a series of total internal reflections (Fig. 12.13). Single or mono mode fibres have very dense and exceptionally small (5 μm) internal core diameters. They offer the greatest information carrying capacity of all fibre types and support

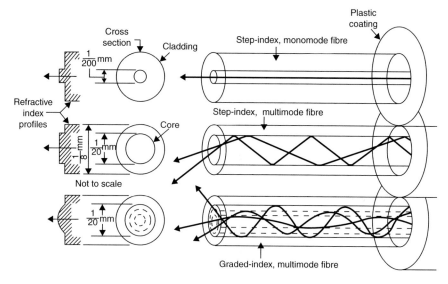

Figure 12.14   Mono- and multimode fibres

the longest transmission distances. Multimode fibres have wider cores (50 μm) with either an abrupt or graded refractive index outer profile. The graded outer cladding allows longer transmission distances at higher bandwidths (Fig.12.14).

A fibre optic cable communication system always consists of a transmitter light source (laser diode, light emitting diode (LED) or pin diode in order of cost) pulsed by electronic circuitry at the required data rates, the fibre optic cable and a detector at the receiving end which decodes the light pulses back into electronic signals. A transmitter and receiver are built into a single electronic circuit ('chip') for two way duplex communication. 820 nm to 850 nm wavelengths are used for low data rate communication but other wavelengths (1300 nm, 1550 nm) may be used for long distance systems. This is lower in frequency than white light radiation and cannot be seen by the naked eye.

12.7.3.3 *Optical budget*

The transmission system is given an 'optical budget' as a level of attenuation which should not be exceeded if correct reception by the detector equipment at the remote end of the link is to be ensured. Four factors are involved:

1. Light source power.
2. Fibre loss (dependant upon fibre size and acceptance angle or matching of light source to the cable).
3. Receiver sensitivity.
4. Jointing or splicing, coupling and connector losses.

Consider a 200 µW light emitting diode transmitter power source and a receiver sensitivity of 2 µW.

Optical power budget (dB) = $10 \log_{10}$ (available infra red light input)/(required infra red light output) = $10 \log_{10} 200/2 = 20$ dB.

Fibre optic cable end connector attenuation loss = 3 dB.

Fibre optic cable splice loss (3 joints at 2 dB each) = 6 dB.

Fibre attenuation (3 dB/km).

Therefore maximum cable length allowable is approximately 3 km before a regenerator is necessary to boost the signal. The receiver bandwidth capability may be traded off against its sensitivity for some applications. Modern joints have reduced losses by a factor of ten in the last 10 years such that good manufacturers should be able to offer splice losses of only 0.2 dB.

12.7.3.4 *Terminology*

The following terms refer to the make up and type of cable and fittings:

Armour: Extra protection for a fibre optic cable to improve the resistance to cutting and crushing. The most common form is galvanized steel wire as for power cables.

Bifurcator: An adaptor with which a loose tube containing two optical fibres can be split into two single fibre cables.

Buffer: Material surrounding the fibre to protect it from physical damage.

Cladding: The outermost region of an optical fibre, less dense (lower refractive index) than the central core. Acts as an optical barrier to prevent transmitted light leaking away from the core.

Core: The central region of an optical fibre, through which signal carrying infra red light is transmitted. Manufactured from high density silica glass.

Loose tube: A type of cable in which one or more optical fibre is/are laid loosely within a tube.

Moisture barrier: A layer of protection built into the cable to keep moisture out.

Multimode fibre: An optical fibre which allows the signal carrying infra red light to travel along more than one path.

Primary coating: A thin plastic coating applied to the outer cladding of an optical fibre. Essential in protecting the fibre from contamination and abrasion.

Sheath: The outer finish of a cable. Usually an extruded layer of either PVC or PE.

Single mode fibre: An optical fibre so constructed that light travelling along the core can follow only one path. (Also called 'monomode'.)

Step index or step index profile: A measurement shown in diagrammatic form illustrating how the quality of glass used in this type of optical fibre graduates, in clearly defined steps, from the highest to the lowest. The shift from one level of density or refractive index to another causes to light to be totally internally reflected back into the core as it travels along the fibre.

Strain member: Part of an optical fibre cable which removes any strain on the fibres. Commonly used materials include steel and synthetic yarns.

Tight buffered: A cable in which the optical fibres are tightly bound.

The following terms refer to transmission characteristics:

Analogue link: Fibre optic cables cannot be easily used to transmit analogue data directly in analogue form because light source variations, bending losses in cables, connector expansion with temperature, etc. introduce distortion. The analogue signal is normally converted to a digital form in an analogue to digital (A/D) converter; accurately determined by the number of bits used, multiplexing the digital bits into one stream in a multiplexer (MUX) and using a pulsed transmission approach.

Attenuation: A term which refers to a decrease in transmission power in an optical fibre. Usually used as a measurement in decibels (dB), for example low attenuation means low transmission loss.

Bit/s: Bits per second. Basic unit of measure for serial data transmission capacity (kbit/s, Mbit/s, Gbit/s, etc.).

Bit error rate: The frequency at which the infra red light pulse is erroneously interpreted. Usually expressed as a number referenced to a power of 10 (e.g. 1 in 10^5).

Dark fibre: Unused or spare fibre perhaps in a multi fibre cable.

Data rate: The capability to accurately transmit data in a specified rate range.

Drop and insert: Simplest extension to a point to point optical fibre link. Extends the link along its length from one 'drop' point (node) to the next. Incoming light energy is split between the receiving port at the insert into the link and also to the ongoing output port.

Frequency shift keying: A form of modulation whereby the frequency of an optical carrier system is varied to represent digital states '0' and '1'. Useful in schemes where 'handshaking' is employed to recognize transmitter and receiver.

Handshaking: A predefined exchange of signals or control characters between two devices or nodes that sets up the conditions for data transfer or transmission.

Minimum output power: The amount of light, typically measured in microwatts, provided into a specific fibre size from the data link's light source.

Modem: A contraction of the term 'MOdulator–DEModulator'. A modem converts the serial digital data from a transmitting terminal into a form suitable for transmission over an analogue telephone channel. A second unit reconverts the signal to serial digital data for acceptance by the receiving terminal.

Multiplexer (MUX): Employed in pairs (one at each end of a communication channel) to allow a number of communications devices to share a single communications channel. Each device performs both multiplexing of the multiple user inputs and demultiplexing of the channel back into separate user data streams.

Photodetector: Device at the receiving end of an optical link which converts infra red light to electrical power.

Pulse width distortion: The time based disparity between input and output pulse width.

Receiver sensitivity: The amount of infra red light typically measured in microwatts or nanowatts required to activate the data link's light detector.

Regenerators: Devices placed at regular intervals along a transmission line to detect weak signals and re-transmit them. Seldom required in a modern fibre optics system. Often wrongly referred to as 'repeaters'.

12.7.3.5 *Cable constructions and technical particulars*

Recent standards covering the fibres themselves and the total fibre optic cable make up and accessories are:

IEC 60793 Optical Fibres

Part 1 – Measurement Methods and Test procedures; generic specification establishing uniform requirements for geometrical, optical, transmission, mechanical and environmental properties of optical fibres.

Part 2 – Product Specifications, Class A (multimode) and Class B mono mode fibres.

IEC 60794 Optical Fibre Cables

Parts 1 & 2 covering Generic and Product Specifications.

IEC 60874 Connectors for Optical Fibres and Cables

IEC 60875 Fibre Optic Branching Devices

Figure.12.15 shows a variety of basic fibre optic cable constructions. Table 12.16 indicates typical technical parameters to be considered when specifying a cable for a particular application. Fibre optic cables may be buried underground using the transmission wayleave between substations. Power cables are typically supplied in 500 m to 1000 m drum lengths. This is short for fibre links and introduces the need for a large number of fibre optic cable joints at the corresponding power cable joint if the fibre is introduced as part of the overall power cable makeup. In addition trenching operations are expensive and a cheaper technique is to specify the fibre optic cable as part of the overhead line earth cable in new installations. A further even better alternative is to wrap the fibre optic cable around the overhead line earth cable. This allows greatly increased lengths of fibre cable to be installed free of joints. Installation equipment has been developed to cater for such installations without outages.

12.8 CABLE MANAGEMENT SYSTEMS

12.8.1 Standard cable laying arrangements

It is necessary to be able to accurately locate buried cables, and indeed all buried services (fresh water piping, gas mains, foul water piping, etc.), in order to avoid damage when excavations are taking place. Reference should be made to drawings that have been regularly updated to reflect the current

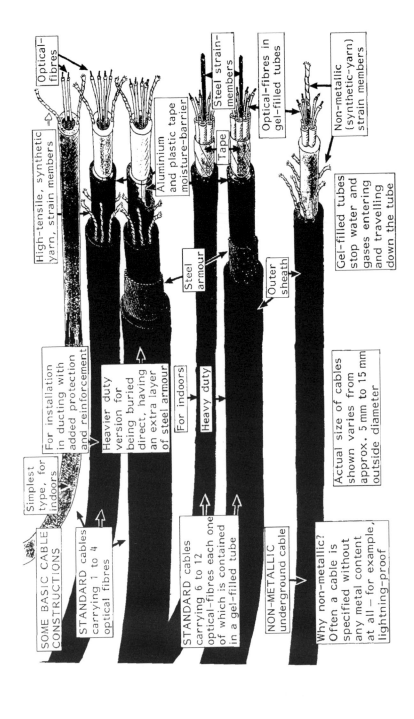

Figure 12.15   Fibre optic cable constructions

Table 12.16   Fibre optic cable technical parameters

Description	Parameter (units)
CABLE TYPE/REFERENCE	
FIBRE	
Reference	
Material	
Number	
DIMENSIONS	
Core	
Cladding	µm
Primary Coating	µm
	µm
MAXIMUM ATTENUATION AT	
– 850 nm	
– 1300 nm (or as required)	dB/km
	dB/km
MINIMUM BANDWIDTH AT	
– 850 nm	
– 1300 nm (or as required)	MHz km
	MHz km
NOMINAL NUMERICAL APERTURE	
PROTECTIVE LAYER (PRIMARY COATING)	
Nominal thickness	
Minimum thickness	mm
	mm
ARMOUR BEDDING (where applicable)	
Type	
Nominal thickness	
	mm
ARMOUR	
No. wires or tapes	
Diameter of wire or thickness of tapes	No.
	mm
OUTER COVERING (SHEATH)	
Material	
Average thickness	
	mm
COMPLETED CABLE	
Overall diameter	
Weight per metre	mm
Maximum drum length	kg
	m
MAXIMUM TRANSIENT WITHSTAND TENSION	N
MAXIMUM MECHANICAL WITHSTAND TENSION	N
MINIMUM BENDING RADIUS	
For cable laying	
At cable-termination point	mm
	mm

status of buried services. In addition to studying these 'as built' drawing records digging should only take place after the necessary 'permit to work' authorization for excavations in a specific area or route has been obtained. It is useful for contractors to adopt a standard cable laying arrangement in road verges. Figure 12.3 shows such a typical standard arrangement.

Oil-filled cable tanks are best buried in purpose built pits if being installed adjacent to roadways in order to avoid collision damage by vehicles. Figures 12.17a and 12.17b show typical above and below ground oil pressure vessel installation arrangements, trifurcating joint and cable sealing ends.

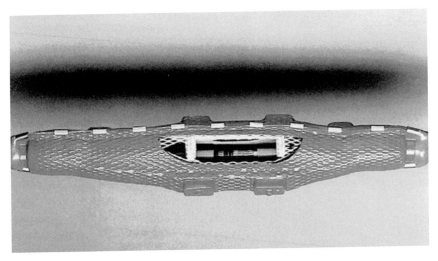

Figure 12.16   12.7/22 kV, 3 core, 120 mm², XPLE, LSF, straight cable joint (note other types also exist for this cable)

It is also normal practice to identify the cable route by cable markers attached to adjacent permanent walls or buildings. Such markers detail the distance horizontally from the marker to the line of the cable route and the depth of the cables at that location. Unfortunately such markers are often removed in the intervening period between one installation and the next so that reliance has to be placed upon existing record drawings and permit to dig systems. Further security to the existing services installation is achieved by placing cable tiles and/or marker tape above the cables. A cross section of direct buried 132 kV oil-filled cables with both cable tile protection and marker tape is shown in Fig.12.5. When further cables are to be placed in the same wayleave as used by an existing installation a mechanical digger can be used to carefully dig down to the depth of the cable tiles and then hand digging used to expose the existing cable area. In this way damage to existing cable outer sheaths is minimized.

12.8.2 Computer aided cable installation systems

Integrated computer aided design and drawing (CADD) packages are available for the engineer to assist in:

- The optimum routing of cables (shortest route, least congested route, segregated route, etc.).
- Selection of cable sizes for the particular environmental and electrical conditions.

Such programs should have the following advantages and facilities:

- Improved ability to respond to change requests.
- Greater speed in locations of services and repairing faults

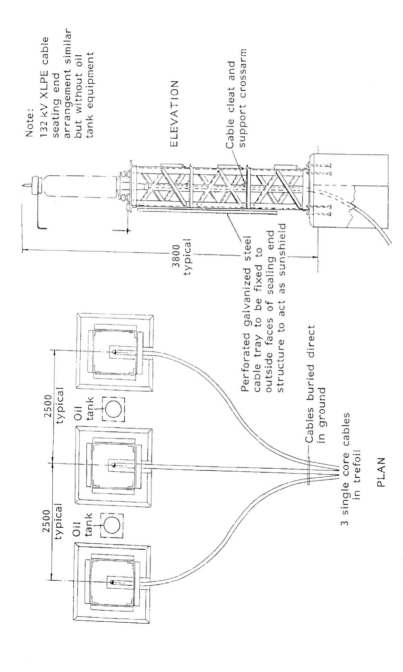

Note:
132 kV XLPE cable
sealing end
arrangement similar
but without oil
tank equipment

ELEVATION

Cable cleat and
support crossarm

Perforated galvanized steel
cable tray to be fixed to
outside faces of sealing end
structure to act as sunshield

3800
typical

Cables buried direct
in ground

2500
typical

2500
typical

Oil
tank

Oil
tank

3 single core cables
in trefoil

PLAN

Figure 12.17   (a) Typical sealing and arrangement

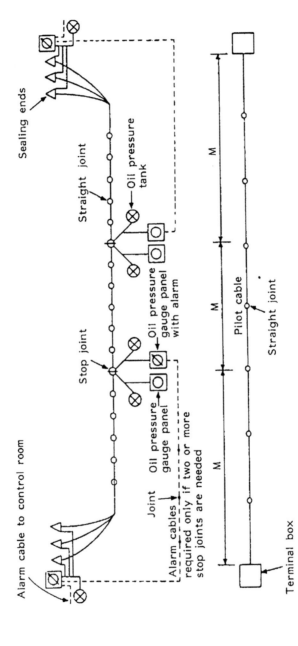

Figure 12.17   (b) Typical oil feed and alarm system arrangement

- Automatic or computer assisted rapid cable route design and selection in accordance with carrier selection, accommodation, segregation and separation rules.
- Accurate cable sizing for the given route and data base record.
- Consistent accuracy in design and drawing quality. Production of a drawing register to hold drawing numbers, sheet numbers, titles, revisions, etc.
- Automatic calculation of material quantities, including glands, termination equipment, cable tiles, cable ties, sand surround, etc. Production of procurement schedules resulting from the design.
- Production by direct printing of cable installation cards which are able to withstand the rigours of on-site handling.
- Overall progress monitoring and control with up to date information on the number of cables scheduled, routed, design approvals, drummed, shipped, installed, etc.

One such program is the CAPICS (Computer Aided Processing of Industrial Cabling Systems) suite primarily intended to assist in major power installations developed by Balfour Kilpatrick Ltd., Special Products Division, Renfrew, Scotland. A typical cable installation card from such a program is shown in Fig. 12.18 and a design schedule in Fig. 12.19. Other proprietary systems are available.

CAD systems also improve the quality of buried cable installation information. The drawing files should be arranged in 'layers' such that each layer is used for different buried services (water, telephone, lighting, power, foul water, etc.). The composite drawing then consists of merging the different files for a full services co-ordination drawing. The program should be integrated with a data base to keep records of duct routes, duct occupancy, drawpit locations, cable information, etc. This may then be used to tie in with a 'permit to work' scheme and records updated accordingly as the work proceeds. Examples of the output from such a system based upon AUTOCAD is given in Figs 12.20 and 12.21.

12.8.3 Interface definition

As an important general principle it is essential to define as accurately as possible the technical and physical interfaces between different subcontracts within an overall transmission and distribution project. This should be done at the earliest stage in the project. Since cable installation works are often carried out by specialist contractors particular attention must be paid to this by the design engineer. Detail down to the supply of termination materials at the interface point must be given so that materials and construction plant orders may be correctly placed in a timely manner. Lack of definition will only lead to inefficiency and costly disputes at a later stage.

In addition to a description of the subcontract package terminal points an interface drawing such as that shown in Fig. 12.22 is often even more useful.

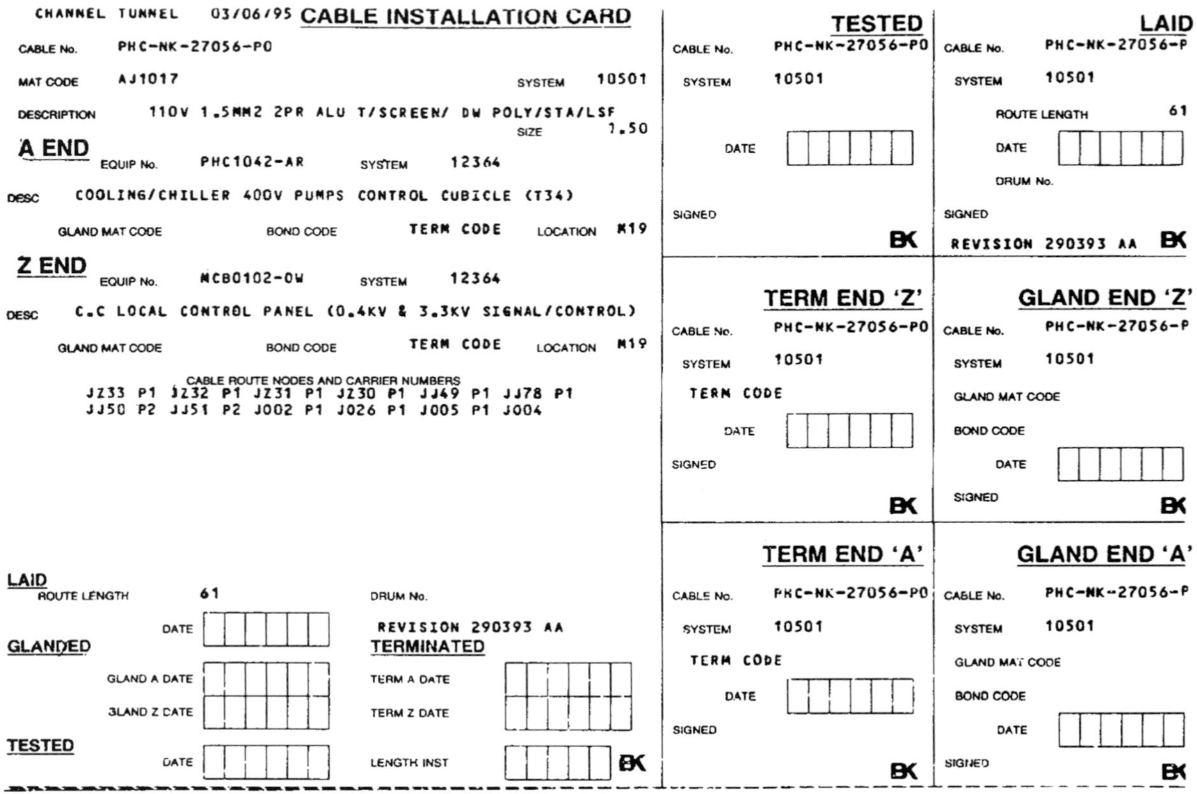

CHANNEL TUNNEL 03/06/95 CABLE INSTALLATION CARD

CABLE No. PHC-NK-27056-PO

MAT CODE AJ1017 SYSTEM 10501

DESCRIPTION 110V 1.5MM2 2PR ALU T/SCREEN/ DW POLY/STA/LSF

SIZE 1.50

A END EQUIP No. PHC1042-AR SYSTEM 12364

DESC COOLING/CHILLER 400V PUMPS CONTROL CUBICLE (T34)

GLAND MAT CODE BOND CODE TERM CODE LOCATION M19

Z END EQUIP No. MCB0102-OW SYSTEM 12364

DESC C.C LOCAL CONTROL PANEL (0.4KV & 3.3KV SIGNAL/CONTROL)

GLAND MAT CODE BOND CODE TERM CODE LOCATION M19

CABLE ROUTE NODES AND CARRIER NUMBERS

JZ33 P1 JZ32 P1 JZ31 P1 JZ30 P1 JJ49 P1 JJ78 P1
JJ50 P2 JJ51 P2 J002 P1 J026 P1 J005 P1 J004

LAID ROUTE LENGTH 61 DRUM No.

DATE

REVISION 290393 AA

GLANDED

GLAND A DATE

GLAND Z DATE

TERMINATED

TERM A DATE

TERM Z DATE

TESTED

DATE

LENGTH INST

BK

TESTED

CABLE No. PHC-NK-27056-PO

SYSTEM 10501

DATE

SIGNED

BK

LAID

CABLE No. PHC-NK-27056-P

SYSTEM 10501

ROUTE LENGTH 61

DATE

DRUM No.

SIGNED

REVISION 290393 AA BK

TERM END 'Z'

CABLE No. PHC-NK-27056-PO

SYSTEM 10501

TERM CODE

DATE

SIGNED

BK

GLAND END 'Z'

CABLE No. PHC-NK-27056-P

SYSTEM 10501

GLAND MAT CODE

BOND CODE

DATE

SIGNED

BK

TERM END 'A'

CABLE No. PHC-NK-27056-PO

SYSTEM 10501

TERM CODE

DATE

SIGNED

BK

GLAND END 'A'

CABLE No. PHC-NK-27056-P

SYSTEM 10501

GLAND MAT CODE

BOND CODE

DATE

SIGNED

BK

Figure 12.18 Cable installation card

```
BK-CAPICS        PROJECT NO. HC1001        PROJECT NAME  CHANNEL TUNNEL                              PRINT DATE 07/09/93  PAGE NO  1
PROGRAM COPCDS                             CABLE DESIGN SCHEDULE - SELECT CBL.SYS                    WEEK NO.  3693
SYSTEM NO. 70164      SOUTH - GENERAL - CROSSOVER ADIT UNDERSEA
                                                                                                                       EQP.DRW.NO.
CABLE        RC SYSTEM E SYST. GLAND AREA TERM EQUIPMENT    EQUIPMENT TITLE

MVT-N6-05885-   CC 70164   A 77026   M27 N      MVT2709-HO UNDERSEA X/O DOOR CONTROL PANEL XO 2708 (WAS MVT8401-OW)
CCTS G/P1/HO               Z 50165   M27 N      MVT8451-CR UK UNDERSEA CROSSOVER DOOR JUNCTION BOX CRS1
                                    REMARKS
              S R D I   U HCPH  PMD  AX6125 1000V 4CR CU/XLPE/STA/LSF
CSA  LGTH DRUM NO  P R  A   A3 ROUTE R164 P1 RI63 P1 RI15 P1 PI17 P1 RN12 P1 RN10 P1 RA11 P1 RA10 P1 RA02 P1
35.00  128                           RA08 P1 RA09

MVT-N6-05985-   CC 70164   A 77026   M27 N      MVT2709-HO UNDERSEA X/O DOOR CONTROL PANEL XO 2708 (WAS MVT8401-OW)
CCTS G/P1/HO               Z 50165   M27 N      MVT8401-CR UK UNDERSEA CROSSOVER DOOR JUNCTION BOX CR1
                                    REMARKS
              S R D I   U HCPH  FMD  AX6125 1000V 4CR CU/XLPE/STA/LSF
CSA  LGTH DRUM NO  P R  A   A3 ROUTE R164 P1 RI63 P1 RI15 P1 PI17 P1 RN12 P1 RN11 P1 RN10 P1 RA11 P1 RA10 P1 RA03 P1
35.00  106                           RA04 P1 RA12

MVT-NK-05885-   CC 70164   A 77026   M27 N      MVT2709-HO UNDERSEA X/O DOOR CONTROL PANEL XO 2708 (WAS MVT8401-OW)
                           Z 50165   M27 N      MVT8451-CR UK UNDERSEA CROSSOVER DOOR JUNCTION BOX CRS1
                                    REMARKS
              S R D I   U HCPH  PMD  AJ1044 48V 1.5MM2 25PR ALU T/SCREEN DW POLY/STA/LSF
CSA  LGTH DRUM NO  P R  A   A3 ROUTE R164 C1 RI63 C1 RI15 C1 RI17 C1 RN12 C1 RN11 C1 RN10 C1 RA11 C1 RA10 C1 RA02 C1
1.50  128                            RA08 C1 RA09

MVT-NK-05886-   CC 70164   A 77026   M27 N      MVT2709-HO UNDERSEA X/O DOOR CONTROL PANEL XO 2708 (WAS MVT8401-OW)
                           Z 50165   M27 N      MVT8451-CR UK UNDERSEA CROSSOVER DOOR JUNCTION BOX CRS1
                                    REMARKS
              S R D I   U HCPH  PMD  AJ1044 48V 1.5MM2 25PR ALU T/SCREEN DW POLY/STA/LSF
CSA  LGTH DRUM NO  P R  A   A3 ROUTE R164 C1 RI63 C1 RI15 C1 RI17 C1 RN12 C1 RN11 C1 RN10 C1 RA11 C1 RA10 C1 RA02 C1
1.50  128                            RA08 C1 RA09

MVT-NK-05887-   B9 70164   A 77026   M27 N      MVT2709-HO UNDERSEA X/O DOOR CONTROL PANEL XO 2708 (WAS MVT8401-OW)
                           Z 50165   M27 N      MVT8451-CR UK UNDERSEA CROSSOVER DOOR JUNCTION BOX CRS1
                                    REMARKS
              S R D I   U HCPH  PMD  AJ1018 110V 1.5MM2 7PR ALU T/SCREEN/ DW POLY/STA/LSF
CSA  LGTH DRUM NO  P R  A   A3 ROUTE R164 C1 RI63 C1 RI15 C1 PI17 C1 PN17 C1 PN11 C1 RN10 C1 PA11 C1 PA10 C1 RA02 C1
1.50  128                            RA08 C1 RA09

MVT-NK-05888-   CC 70164   A 77026   M27 N      MVT2709-HO UNDERSEA X/O DOOR CONTROL PANEL XO 2708 (WAS MVT8401-OW)
                           Z 50165   M27 N      MVT8451-CR UK UNDERSEA CROSSOVER DOOR JUNCTION BOX CRS1
                                    REMARKS
              S R D I   U HCPH  PMD  AJ1016 110V 1.5MM2 4PR ALU T/SCREEN DW POLY/STA/LSF
CSA  LGTH DRUM NO  P R  A   A3 ROUTE R164 C1 RI63 C1 RI15 C1 PI17 C1 PN12 C1 PN11 C1 RN10 C1 RA11 C1 PA10 C1 RA02 C1
1.50  128                            RA08 C1 RA09

MVT-NK-05889-   B8 70164   A 77026   M27 N      MVT2709-HO UNDERSEA X/O DOOR CONTROL PANEL XO 2708 (WAS MVT8401-OW)
                           Z 50165   M27 N      MVT8451-CR UK UNDERSEA CROSSOVER DOOR JUNCTION BOX CRS1
                                    REMARKS
              S R D I   U HCPH  PMD  AJ1018 110V 1.5MM2 7PR ALU T/SCREEN/ DW POLY/STA/LSF
CSA  LGTH DRUM NO  P R  A   A3 ROUTE R164 C1 RI63 C1 RI15 C1 PI17 C1 PN12 C1 PN11 C1 RN10 C1 PA11 C1 PA10 C1 RA02 C1
1.50  128                            RA08 C1 RA09
```

Figure 12.19 Design schedule

TRANSLINK JOINT VENTURE

UK TERMINAL

FILE

Reference No. 4272

APPLICATION FOR PERMIT TO DIG

☑ New Application
☐ Revision

APPLICANTS   SECTION 1

To N. MORGAN (Permit Controller) I request a PERMIT TO DIG as it is proposed to

disturb the ground on 26/10/93 (Date) For the purpose of WATERMAIN TEE REMOVAL

in the area of SUBSTATION 'R'   Applicant's Name A. NOTHER

Name of Company/Section/Subcontractor ACME WATER COMPANY

Contact point and telephone No. (0303) 276431

*Approved (Section Engineer/Supervisor) A Nother.

(Copy of working drawing to be attached)

*From Approved List

[right margin, rotated] Application Valid from 26.10.93. N. M. for 7 days only

INVESTIGATION (SECTION 2)

The above area has been reproduced on the attached drawing (A3/A4 sheet) Ref. No. 4272 All services indicated are to be located by hand digging or by other approved methods as stated below, PRIOR TO A PERMIT TO DIG BEING ISSUED.

NOTES: (i) THE AREA IS TO BE SCANNED THOROUGHLY — | Investigation/ Excavation requires a W.A.D. |

(ii) BEWARE WATERMAIN BELOW. ALSO HV DUCTS BT DUCTS TO THE NORTH AND S.U.D. TO THE SOUTH.

APPLICANTS   SECTION 3 | W.A.D. No. |

COMPLETED INVESTIGATION (SECTION 3)

The area defined has been fully investigated by *myself, and all services listed above in Section 2 have been located & identified. Any deviations or additions are on attached sheet.

I would like to nominate as the PERMIT HOLDER. I agree to abide by the UK Terminal Procedure for Permit to Dig plus all the conditions listed on the 'Permit to Dig' form.

Signed   Date

*Applicant

PERMIT TO DIG (SECTION 4)

PERMIT TO DIG No. has been allocated to *you, commencing on for a period of

days, and is subject to the conditions below.

CONDITIONS OF USE: Possession of this Permit does not exempt the Permit Holder from his responsibility to proceed in a safe and responsible manner.

Signed (Permit Controller)   Date

Pink Copy – File	White Copy – Safety	Blue Copy – Applicant	Green Copy – Permit Holder Machine Driver

*Applicant

HOLD POINT   DO NOT BACKFILL PRIOR TO SURVEY

SURVEY DETAILS (SECTION 5)

I confirm that the services installed in this Permit consisting of

have been surveyed by myself on (Date)

Surveyed by (Land Surveyor)   Date

PERMIT CLOSURE (SECTION 6)

Applicant   I confirm that the work is now complete and is correctly shown on the attached survey

Drawing No.

Signed   Date

Permit Controller   I acknowledge receipt of the attached survey which will be added to my records and the Permit is now closed.

Signed   Date

Form 130/28026D

Figure 12.20   Permit application

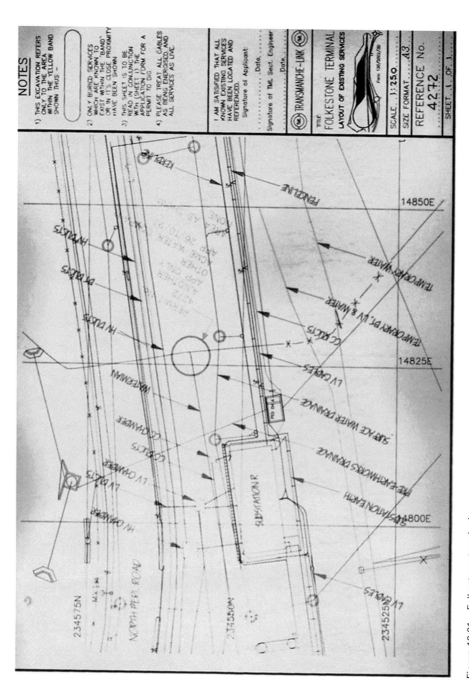

Figure 12.21 Folkestone terminal

434 Cables

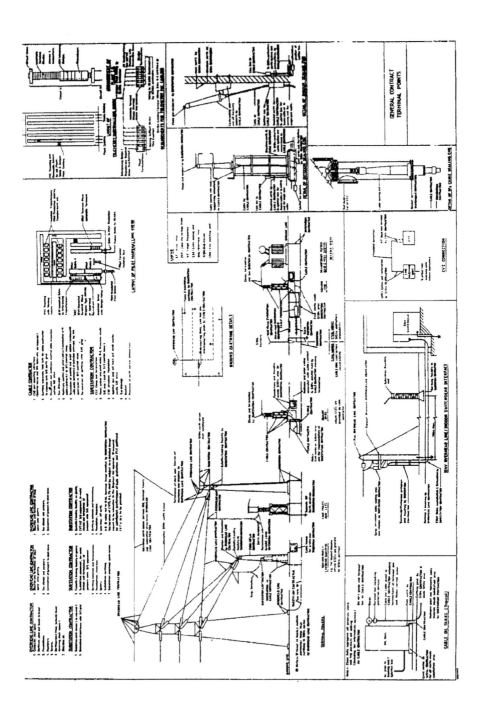

Figure 12.22 Interface drawing, general contract terminal interface points

REFERENCES

1. BICC Publication No. 597D, Paper Insulated Power Cables, Solid Type Lead Sheathed, for voltages up to and including 33 kV to BS 6480 1969.
2. BICC Publication No. 598D, PVC Insulated Power Cables, for voltages up to and including 3.3 kV to BS 6346 1969.
3. BICC Publication No. 808, XLPE Insulated Power Cables, for voltages up to and including 3.3 kV to BS 5467 1977.
4. E. W. G. Bungay & D. McAllister, BICC Electric Cables Handbook, BSP Professional Books, 3rd Edition, 1997.S

13 Switchgear

13.1 INTRODUCTION

Switchgear is a general term covering switching devices and their combination with associated control, measuring, protective and regulating equipment. The term covers assemblies of such devices and equipment with associated interconnections, accessories, enclosures and supporting structures intended for use in connection with transmission and distribution networks. The different types of air, oil, vacuum and SF_6 switchgear together with the theory of arc interruption are already well-covered in standard reference books.[1–4] This chapter therefore concentrates on the description of various switching phenomena under different practical circuit conditions and then relates these basic principles to the different switchgear designs currently available on the market. In particular this chapter is intended to assist the reader in specifying switchgear for particular applications.

13.2 TERMINOLOGY AND STANDARDS

The descriptions of different types of switchgear, intended for different duties, are listed in Table 13.1. It is important not to be too lax with the terminology. For example, the use of the term isolator to describe a switch will not on its own sufficiently describe the required capability of the device.

A **circuit breaker** is intended to switch both load and short circuit currents. Unlike a fused device it enables supplies to be quickly restored after operation on short circuit and is the most expensive form of switchgear. It is not primarily intended for frequent operation although vacuum and SF_6 breakers are more suited to load switching duties than older switchgear types.

A **contactor** is operated other than by hand and is intended for switching loads under normal and overload conditions. It is designed for frequent operation but

Table 13.1   Explanation of commonly used switchgear terminology

Terminology	Description
Circuit breaker	A mechanical switching device, capable of making, carrying and breaking currents under normal circuit conditions and also making, carrying for a specified time and breaking currents under specified abnormal circuit conditions such as those of short circuit.
Contactor	A mechanical switching device having only one position of rest, operated otherwise than by hand, capable of making, carrying and breaking currents under normal circuit conditions including operating overload conditions.
Current limiting circuit breaker and current limiting fuselink	A circuit breaker with a break time short enough to prevent the short circuit current reaching its otherwise attainable peak value. Similarly a fuselink, during and by its operation in a specified current range, limits the current to a substantially lower value than the peak value of the prospective current.
Disconnector	A mechanical switching device which provides, in the open position, an isolating distance in accordance with the specified requirements. A disconnector is intended to open or close a circuit under negligible current conditions or when there is no significant voltage change across the terminals of each of its poles. It is capable of carrying rated current under normal conditions and short circuit through currents for a specified time. Also sometimes known as a no-load isolator. It is important to clarify the term isolator or disconnector. It can apply as follows: (a) off-circuit isolator – capable of switching 'dead' (non-energized) circuits only; (b) no-load isolator – capable of switching under 'no-load' (negligible current flow) conditions only. Ensure when specifying such a device that it is capable of switching any applicable no-load charging current.
Earthing switch	A mechanical switching device for earthing parts of a circuit, capable of withstanding for a specified period current under abnormal conditions such as those of a short circuit, but not required to carry current under normal circuit conditions. An earthing switch may have a short circuit making capacity either to act as a 'fault thrower' at the end, say, of a long distribution feeder or to cater for inadvertent operation of a live circuit to earth.
Fuse switch	A switch in which a fuselink or a fuse carrier with a fuse-link forms the moving contact. Such a device may be capable of closing onto a fault ('fault-make' and the fuse will operate).
Moulded case circuit breaker (MCCB)	A circuit breaker having a supporting housing moulded insulating material forming an integral part of the circuit breaker. Also note the miniature circuit breaker (MCB) which is of the current limiting type (see Chapter 11).
Switch	A mechanical switching device capable of making, carrying and breaking current under normal circuit conditions. This may include specified operating overload and short-term, short circuit current conditions. If so specified breaking full load rated current. It may then be called a 'fault-make, load break' switch or isolator.

has a limited short circuit current carrying and switching capability. It is therefore often backed up by fuses or a circuit breaker.

A **disconnector** provides in the open condition a specific isolating distance. It has only a very limited current switching capability and is not intended for frequent use or for breaking full load current.

A **switch** is used for switching loads but is not suitable for frequent operation. Switches may be manual or motor operated, and have a short circuit current-making capability but no breaking capability and must therefore be used in combination with a short circuit interrupting device (usually fuses). Where the fuse and switch are in combination in series the unit is called a **switch fuse**; where the fuse forms part of the moving contact of the switch it is termed **a fuse switch**.

Some useful IEC standards covering switchgear are detailed in Table 13.2.

It is important to distinguish between the terms metal-clad and metal-enclosed as applied to switchgear and control gear. **Metal-enclosed** refers to complete switchboards, except for the external connections, with an external metal enclosure intended to be earthed. Internal partitions may or may not be incorporated and where installed need not be metallic. **Metal-clad** refers to metal-enclosed switchgear and control gear in which the components (each main switching device, outgoing way, busbar system, etc.) are arranged in compartments separated by earthed metal partitions. The partitions and busbar/feeder shutters or covers should be carefully specified to various **IP** levels of protection.

Gas insulated switchgear (GIS) has all live parts contained in SF_6 gas-tight enclosures. Three phase busbar systems may use steel or aluminium enclosures. The busbars are physically arranged in trefoil formation largely to cancel out the resultant stray magnetic fields and any associated enclosure eddy current losses. The enclosure may also be sectionalized with insulating parts to further reduce such losses. Single phase busbar arrangements normally use lighter aluminium alloy or stainless steel corrosion proof enclosures. The illustration on p. 449 shows a modern 400 kV GIS indoor substation installation in Saudi Arabia (courtesy Reyrolle Switchgear and successors).

13.3 SWITCHING

13.3.1 Basic principles

13.3.1.1 *General*

Figure 13.1 shows a typical switching arrangement with a source impedance $R_S + j\omega L_S$ and downstream impedance from the circuit breaker to the fault $R_L + j\omega L_L$. The shunt impedances (capacitance and insulation resistance of machines, switchgear, cables and overhead lines) may normally be ignored when calculating short circuit currents since they are several orders of magnitude greater than the series impedances involved. In addition, it should be noted that series resistance values are only some 1% to 3% of the inductive reactance for

Table13.2 Useful IEC switchgear standards

IEC standard	Title	Notes
60038	IEC standard voltages	Applies to AC transmission and distribution with standard frequencies of 50 and 60 Hz and nominal voltages above 100 V.
60050-441	International Electrotechnical Vocabulary. Switchgear, controlgear and fuses	
60059	IEC standard current ratings	
60060	High-voltage test techniques	Various parts, covering definitions, test requirements and measuring systems
60296	Fluids for electrotechnical applications – Unused mineral insulating oils for transformers and switchgear	
60364-5-53	Electrical installations of buildings – Part 5-53: Selection and erection of electrical equipment – Isolation, switching and control	Low voltages less than 1 kV AC
60376	Specification of technical grade sulphur hexafluoride (SF_6) for use in electrical equipment	Properties, methods and tests for new and unused SF_6.
60427	Synthetic testing of high-voltage alternating current circuit breakers	Applies to circuit breakers covered by IEC 62271-100. General rules for circuit breaker testing.
60466	AC insulation-enclosed switchgear and controlgear for rated voltages above 1 kV and up to and including 38 kV	
60470	High-voltage alternating current contactors and contactor-based motor-starters	Includes conditions for compliance with operation, behaviour and dielectric properties and associated tests. Applicable to vacuum contactors.
60480	Guide to the checking of sulphur hexafluoride (SF_6) taken from electrical equipment	Checks for the condition of SF_6.
60529	Degrees of protection provided by enclosures	IP categorisation
60694 (supersedes IEC 1208)	Common clauses for high-voltage switchgear and control gear standards	Essential cross reference when reading IEC 62271
60859	Cable connection for gas-insulated metal-enclosed switchgear for rated voltages of 72.5 kV and above	Complements and amends other switchgear and cable termination Standards.

(Continued)

Table13.2 Continued

IEC standard	Title	Notes
60947 (supersedes IEC 157)	Low-voltage switchgear and control gear	Applies to equipment intended to be connected to circuits operating at rated voltages less than 1 kV. Standard is broken down into several parts.
61958	High-voltage prefabricated switchgear and controlgear assemblies – Voltage presence indicating systems	
62271-100 (supersede IEC56)	High-voltage switchgear and controlgear – High-voltage alternating current circuit breakers	Applicable to indoor and outdoor circuit breaker installations up to and including 60 Hz on systems having voltages above 1000 V. Covers operating devices and auxiliary equipment.
62271-102 (supersedes IEC129, IEC 1128 and IEC 1129)	High-voltage switchgear and current controlgear – Alternating disconnectors and earthing switches	Indoor and outdoor installations. Also covers operating devices and auxiliary equipment.
62271-105 (supersedes IEC 420)	High-voltage switchgear and controlgear Alternating current switch-fuse combinations	Apply to combinations for use on three phase distribution systems at rated voltages between 1 and 52 kV. Functional assemblies of switches including switch disconnectors and current limiting fuses thus able to interrupt: – load breaks up to rated breaking current – overcurrents up to rated short circuit breaking current of the combination by which automatic interruption is initiated.
62271-107 (supersedes IEC 420)	High-voltage switchgear and controlgear Alternating current fused circuit-switchers for rated voltages above 1 kV up to and including 52 kV	
62271-200 (supersedes IEC 298)	High-voltage switchgear and controlgear AC metal-enclosed switchgear and controlgear for rated voltages above 1 kV and up to and including 52 kV	Service conditions, applicable terms and rated characteristics. Rules for design and construction, type and routine tests. General information on selection of devices, tenders, transport, erection and maintenance. Supplements IEC 60694, 'Common concepts for high voltage switchgear and control gear standards'.
62271-203 (supersedes IEC 517)	High-voltage switchgear and controlgear Gas-insulated metal-enclosed switchgear for rated voltages above 52 kV	Covers requirements for SF_6 switchgear.

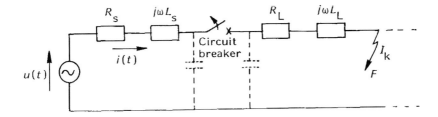

$i(t)$ = circuit breaker current

I_k = rms value of symmetrical short circuit current

$u(t)$ = source voltage

R_s = source resistance

L_s = source inductance

R_L = load resistance between circuit breaker and fault

L_L = load inductance between circuit breaker and fault

F = fault

Notes: Traditionally the sub-transient reactances of the source contributing machines have been used to determine the fault current and circuit breaker-rated short-circuit current selection.

Thus by ignoring the decrement of the AC component which occurs after the sub-transient stages correctly leads to a conservative approach to circuit breaker ratings.

Figure 13.1   Simplified network representation for a circuit breaker operation under short circuit conditions

generators and transformers and some 5% to 15%, depending upon construction, for high-voltage overhead lines. Short circuit currents in a high-voltage network are therefore practically totally inductive with a power factor of some $\cos \phi = 0.07$. At lower voltages resistance values become more important and may be investigated depending upon the accuracy of the analysis required (see Chapter 1, Section 1.4.2.5).

Initially the circuit breaker shown in Fig. 13.1 is closed and carries the fault current $i_s(t)$. The relay protection senses the fault and initiates a circuit breaker trip. Figure 13.2 shows the behaviour of the short circuit current $i_s(t)$, through the circuit breaker and the voltage across the circuit breaker $u_{cb}(t)$. At a point in time, t_1, the circuit breaker contacts begin to part and arcing occurs across the contacts. The arc is extinguished by the particular circuit breaker arc quenching mechanisms used and involves stretching the arc and rapid cooling. In modern SF_6 or vacuum circuit breakers the current is interrupted at the next

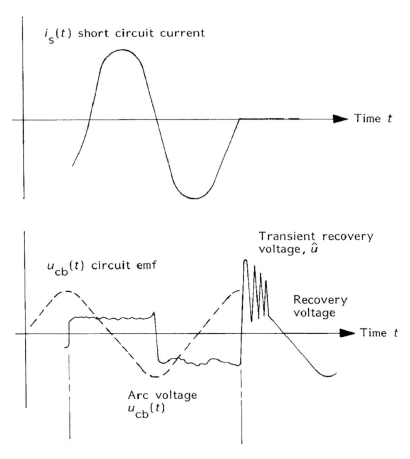

Figure 13.2   Short circuit, $i_s(t)$ through the circuit breaker and characteristic waveform for recovery voltage, $u_{cb}(t)$ across circuit breaker upon fault clearance

or next but one current zero (2 cycles breakers) at time t_2. Older, oil or air circuit breakers take slightly longer (typically 4 cycles) before the arc is extinguished. The arc duration in modern breakers is therefore relatively short ($\sim 10\,\mathrm{ms}$) and this coupled with the low arc voltage leads to low energy dissipation during the circuit breaker operating process. However, for the UK transmission system (between $132\,\mathrm{kV}$ and $400\,\mathrm{kV}$) breakers are tested for longer arcing times (arcing windows) – characteristically $\leq 20\,\mathrm{ms}$ – to cover for the conditions brought about by earth faults.

The characteristic waveform of the recovery voltage is shown in Fig. 13.2. A high frequency voltage oscillation, known as the 'transient recovery voltage' (TRV), fluctuates about the power frequency recovery voltage waveform. Its behaviour is determined by the circuit parameters and the associated rapid redistribution of energy between the network component electric and magnetic fields. If the power factor of the faulted circuit is high (i.e. resistance is a significant proportion of the total fault impedance) then the circuit or power source voltage at

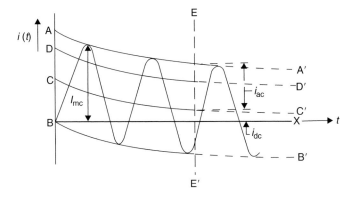

AA' BB'	= envelope of current wave
BX	= normal zero line
CC'	= displacement of current-wave zero-line at any instant
DD'	= rms value of the AC component of current at any instant, measured from CC'
EE'	= instant of contact separation (initiation of the arc)
I_{mc}	= first maximum peak short circuit current
i_{ac}	= peak value of the AC component of current at instant EE'
$\dfrac{i_{ac}}{\sqrt{2}}$	= rms value of the AC component of current at instant EE'
i_{dc}	= DC component of current at instant EE'
$\dfrac{i_{dc} \times 100}{i_{ac}}$	= percentage value of the DC component
ϕ	= phase angle (short circuit current lags voltage across the circuit breaker by approximately 90°)
δ	= closing angle related to u

Note: Inpractice some decrement of the AC component can occur

Figure 13.3   Determination of making and breaking current, and of percentage DC component

current zero will be low. At low power factors (predominantly inductive or capacitive circuits) the circuit voltage at current zero will be high and result in a tendency for the arc to re-strike. This is the basic reason why inductive and capacitive circuits are more difficult to interrupt than resistive circuits. The circuit breaker must, therefore, be designed to withstand the transient recovery voltage. Whether or not the arc extinguishes after the first current zero depends upon establishing adequate dielectric strength across the circuit breaker contacts faster than the rate of rise of TRV and the peak TRV involved. Repeated dielectric or thermal breakdown of the circuit breaker insulating medium between the contacts is also reduced by efficient and rapid thermal quenching.

The short circuit current is characterized by a degree of asymmetry resulting from an AC component (contained in envelope AA'/BB') and a decaying DC component (line CC') as shown in Fig. 13.3. The exact response depends upon the instant the switching of the AC waveform takes place and the relative R, L and C circuit parameters involved. The response may be calculated by solving the differential equations for the network involved and the solution for the current response $i(t)$ takes the form:

$$i(t) = \frac{u\{\cos[\omega t + (\delta - \phi)] - \cos(\delta - \phi)e^{-t/\tau}\}}{\sqrt{(R^2 + X^2)}}$$

where $i(t)$ = circuit breaker short circuit current
        R = resistance
        X = reactance
      $u(t)$ = source voltage ($\hat{u}$ = peak source voltage)
        δ = closing angle related to voltage across circuit breaker = 0
        ϕ = phase angle
        τ = time constant = $L/R = X/\omega R$

13.3.1.2 DC component

The exponential decay time constant of the DC component, τ, is often taken as 45 ms and is a typically representative value in IEC 62271-100 based on a power factor of 0.07 (but special time constants 120 ms <52 kV, 60 ms 72.5 – 400 kV and 75 ms 550 kV and above are also provided for, and high dc components are now being experienced in UK especially around London and on 132 kV system). Ranges of time constant, τ, for cables, high-voltage transmission lines and generators are shown in Fig. 13.4.

13.3.1.3 AC component

The AC component itself may, or may not, be subject to decay. This depends upon whether or not the sub-transient or transient reactances of source generators form a significant part of the total impedance of the overall fault circuit.

For short circuits on a distribution system where transformers whose kVA rating is low in relation to the capacity of the system are interposed between the short circuit and sources of generation, the AC component decay is negligible.

Where the short circuit is close to sources of generation the sub-transient and transient reactances of the machines will form a significant part of the total fault circuit impedance. The AC component delay will therefore be more appreciable. Consider a generator circuit breaker located on the high voltage side of a large (say, 1000 MVA) power station generator transformer. Theoretically the circuit breaker could experience an asymmetrical short circuit current which decays more rapidly than the DC component. During the initial decay period, when current asymmetry is a maximum, it is possible for no current zeros to occur and short circuit interruption to be delayed as a result. In practice, the arc resistance at the fault and across the circuit breaker contacts will reduce the DC time

Percentage DC component
(Note: 100% = BC in Fig. 13.3)

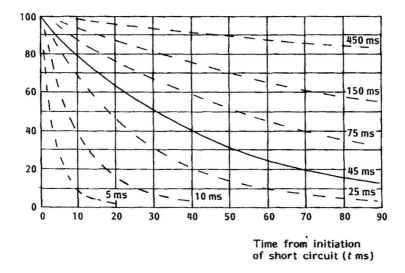

Cable time constant   = 5 ms
Transformers time constant   = 25 – 150 ms
Overhead lines time constant = 10 – 60 ms
Generators time constant   = 50 – 450 ms
Standard IEC time constant   = 45 ms

Figure 13.4   Percentage DC component in relation to time t (ms) and different circuit component time constants τ (ms)

constant and result in a faster DC component decay period. Generator circuit breakers located on the low voltage side of the generator transformer may be presented with an even more severe case of slow DC decay in comparison with the AC component. The attenuating effect of the series connected low L/R transformer impedance ratio is not present and the fault levels will be at least an order of magnitude higher without the limiting effect of the transformer reactance. It is therefore very important to specify circuit breakers for such duties carefully. IEEE Standard C37.013—1993, 'IEEE Standard for AC High Voltage Generator Circuit Breakers Rated on a Symmetrical Current' gives very useful guidance. Generally generator circuit breakers for the larger generators are isolated phase design. Faults will thereby mostly be phase – earth not phase – phase and this can assist interruption as there will be two 'healthy' phases with no dc component.

13.3.1.4 *Circuit breaker short circuit current ratings*

The 'rated short circuit breaking current' is the highest short circuit current which the circuit breaker is capable of breaking and is specified in terms of the AC and DC components. The AC component is termed the 'rated short circuit current' and is expressed in kA rms. The DC component (unless the breaker is non-standard and subject to special agreements between manufacturer and purchaser) is characterized in accordance with the IEC 62271-100 negligible AC component decrement and short circuit power factor of 0.07. Traditionally the sub-transient reactances of the source generators contributing to the fault have been used to determine the fault current. By ignoring the decrement of the AC component which occurs after the sub-transient stage, this correctly leads to a conservative selection of the circuit breaker-rated short circuit current. Tables 13.3 and 13.4 give typical circuit breaker short circuit ratings ranging from 10 to 50 kA and normal current ratings ranging from 400 to 4000 A. The short circuit currents impose large electromechanical forces on the switchgear busbars and contacts. The circuit breaker mechanism has to be designed to be able to close onto the peak value of short circuit current with full asymmetry and carry the fault current for 1 or 2 s without the contacts overheating, parting or damage occurring. Mechanisms are usually of the stored energy or externally fed power-operated type using pneumatic, hydraulic, spring or solenoid systems. Current practice tends to solenoid plus permanent magnets.[5] Because it is necessary to overcome the large forces present when closing onto a fault, manual-operated circuit breakers, where the closing force is dependent upon positive operation by the operator, have generally been discontinued except at the lowest voltage and short circuit levels.

If the operating time of the breaker from the instant of trip initiation to the instant of contact separation is known (the minimum opening time), it is possible to determine the 'actual rated short circuit breaking current'. From fault inception IEC 62271-100 allows 10 ms relay operating time followed by breaker opening time say 22–35 ms and then the first major loop after contact separation. Generally breaker opening times are 22–35 ms, arcing times 10–23 ms and therefore total break times of 32–58 ms. So consider a circuit breaker with 35 ms minimum opening time which has a rated rms short circuit current of 25 kA and auxiliary tripping power supply. The minimum time interval from the instant of the fault to the instant when the arcing contacts have separated in all three poles will be the relay operating time, 10 ms, plus the minimum opening time to total 45 ms.

Therefore in Fig. 13.3:

$$EE' = 45 \, \text{ms.}$$

$$\text{at } i_{ac} = \text{peak value of the AC components at instant } EE' = 25 \cdot \sqrt{2}$$
$$= 35.35 \, \text{kA}$$

$$i_{ac}/\sqrt{2} = \text{rms value of the AC component (which corresponds to the rated short circuit current of 25 kA)}$$

$$i_{dc} = \text{DC component of current at instant } EE'$$

From Fig. 13.4, using a delay time constant of $\tau = 45 \, \text{ms}$, after 45 ms: i_{dc}

$$i_{dc}/i_{ac} \cdot 100\% = 35\%.$$

Therefore $i_{dc} = i_{ac} \cdot 0.35 = (25 \cdot \sqrt{2}) \cdot 0.35 = 12.37 \, \text{kA}.$

Table 13.3 Co-ordination table of rated values for circuit breakers, 3.6 to 72.5 kV

Rated Voltage U(kV)	Rated short circuit breaking current Isc(kA)	Rated normal current In(A)	Rated normal current In(A)	Rated normal current In(A)	Rated normal current In(A)	Rated normal current In(A)	Rated normal current In(A)	Rated normal current In(A)	Rated normal current In(A)
3.6	10	400							
	16		630						
	25				1250	1600		2500	
	40				1250	1600		2500	4000
7.2	8	400							
	12.5	400	630		1250				
	16		630		1250	1600			
	25		630		1250	1600		2500	
	40				1250	1600		2500	4000
12	8	400							
	12.5	400	630		1250				
	16		630		1250	1600			
	25		630		1250	1600		2500	
	40				1250	1600		2500	4000
	50				1250	1600		2500	4000
17.5	8	400	630		1250				
	12.5		630		1250				
	16		630		1250				
	25				1250				
	40				1250	1600		2500	
24	8	400	630		1250				
	12.5		630		1250				
	16		630		1250				
	25				1250	1600		2500	
	40					1600		2500	4000
36	8		630						
	12.5		630		1250				
	16		630		1250	1600			
	25				1250	1600		2500	
	40					1600		2500	4000
52	8			800					
	12.5				1250				
	20				1250	1600	2000		
72.5	12.5			800	1250				
	16			800	1250				
	20				1250	1600	2000		
	31.5				1250	1600	2000		

Note: This table was derived from IEC56, which has been superseded by IEC62271 plus IEC 60694. These newer standards do not define selected current ratings for each voltage level, but simply state that the value of rated normal current and the rms value of the AC component of rated short circuit breaking current must both be selected from the R10 series specified in ISO 3. This series is 1–1.25–1.6–2.0–2.5–3.15–4–5–6.3–8 and their products by 10^n, where n is an integer. The figures in this table comply with that requirement, and remain a useful guide to typical manufacturing switchgear ranges.

Table 13.4   Co-ordination table of rated values for circuit breakers, 123 to 765 kV

Rated voltage U(kV)	Rated short circuit breaking current ISC(kA)	Rated normal current In(A)	Rated normal current In(A)	Rated normal current In(A)	Rated normal current In(A)	Rated normal current In(A)	Rated normal current In(A)
123	12.5	800	1250				
	20		1250	1600	2000		
	25		1250	1600	2000		
	40			1600	2000		
145	12.5	800	1250				
	20		1250	1600	2000		
	25		1250	1600	2000		
	31.5		1250	1600	2000	3150	
	40			1600	2000	3150	
	50				2000	3150	
170	12.5	800	1250				
	20		1250	1600	2000		
	31.5		1250	1600	2000	3150	
	40			1600	2000	3150	
	50			1600	2000	3150	
245	20		1250	1600	2000		
	31.5		1250	1600	2000		
	40			1600	2000	3150	
	50				2000	3150	
300	16		1250	1600			
	20		1250	1600	2000		
	31.5		1250	1600	2000	3150	
	50			1600	2000	3150	
362	20				2000		
	31.5				2000		
	40			1600	2000	3150	
420	20			1600	2000		
	31.5			1600	2000		
	40			1600	2000	3150	
	50				2000	3150	4000
550	40				2000	3150	
800	40				2000	3150	

Note: This table was derived from IEC56, which has been superseded by IEC62271 plus IEC 60694. These newer standards do not define selected current ratings for each voltage level, but simply state that the value of rated normal current and the rms value of the AC component of rated short circuit breaking current must both be selected from the R10 series specified in ISO 3. This series is 1–1.25–1.6–2.0–2.5–3.15–4–5–6.3–8 and their products by 10^n, where n is an integer. The figures in this table comply with that requirement, and remain a useful guide to typical manufacturing switchgear ranges.

Now the actual rms value of the asymmetric current characterized by i_{dc} and i_{ac}

$$\sqrt{\{(i_{ac}^2/2) + i_{dc}^2\}} = \sqrt{\{(25 \times \sqrt{2})^2/2 + (12.37)^2\}} = \sqrt{(25^2 + 12.37^2)} \text{ kA}$$

$$= 27.89 \text{ kA rms}$$

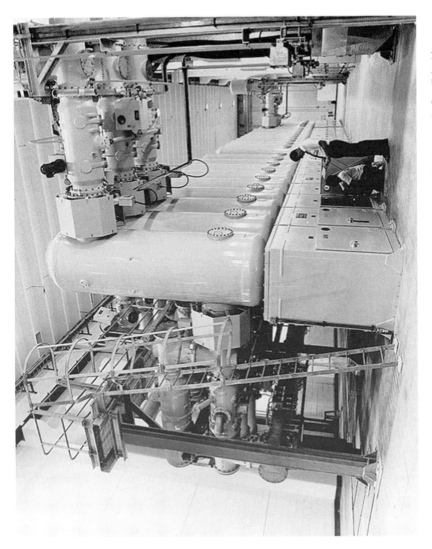

Figure 13.5   Modern 440 kV gas insulated switchgear (GIS) installation undergoing final commissioning tests in Saudi Arabia

Thus whilst the circuit breaker would be offered by the manufacturer as having a rated short circuit current of 25 kA rms, it would actually have a rated short circuit breaking current of 27.89 kA rms and 47.72 kA peak ($i_{dc} + i_{ac}$).

The rated short circuit current is often referred to by the manufacturers as 'symmetrical breaking capacity' and the rated short circuit breaking current as the 'asymmetrical breaking capacity'.

In addition to the breaking short circuit current capability, the circuit breaker is also called upon to 'make' short circuit current, that is to close onto a short circuit. In these circumstances the circuit breaker must be able to latch successfully whilst subject to the magnetic forces associated with the peak value of the first half cycle of fault current. The first short circuit current peak, I_{max}, is given by $I_{max} = I_k\sqrt{2}(1 + e^{-t/\tau})$ where I_k is the rms value of the symmetrical short circuit current. For full asymmetry some 10 ms after the short circuit begins and using the 45 ms time constant, $I_{max} = I_k\sqrt{2}(1 + e^{-t/\tau})$ $\sqrt{2}(1 + e^{-10/45}) = 2.55I_k$.

For readers interested in mathematics it is worth analysing the circuit shown in Fig. 13.1 by hand using Laplace transforms or by computer using a numerical analysis program with short time steps.

Where concentrations of induction motors exist (such as in refineries or in a petrochemical complex) their contribution on a system to the fault levels may be introduced into computer aided network analysis in order to determine more accurately the required circuit breaker fault rating characteristics. Normally induction motors only form part of the network load and it should be noted that such motor time constants are relatively short. Consequently there is a rapid decay in their fault current contribution and it is not normally necessary to increase circuit breaker interrupting capacity. The motor contribution is more important when determining the required making capacity of the circuit breaker and there may be a case for increasing the ratio of 2.5 between I_{max} and $i_{ac}/(\sqrt{2})$. The manufacturer's advice may be sought in such cases since their standard breaker may well be able to comply with such making capacity requirements. Alternatively, a circuit breaker may be selected with slightly higher than necessary interrupting capacity in order to obtain the required making capacity.

13.3.2 Special switching cases

13.3.2.1 Current chopping

Extinction of the arc can normally only occur at current zero. There was a tendency in some circuit breakers (notably MV vacuum and air blast types) in the 1970s to have such good dielectric strength that at low current levels they were capable of extinguishing the arc before current zero occurred. In reactive circuits this can lead to very serious high-voltage spikes which can be several per unit system peak power frequency voltage as the energy transfer takes place (see Fig. 13.6). These temporary overvoltages can cause the breakdown of insulation. This might especially be the case if such circuit breakers are used in conjunction with older equipment such as transformers with relatively

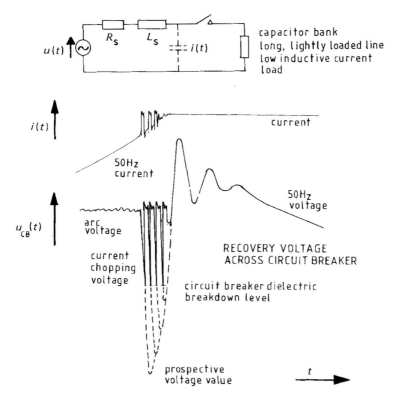

Figure 13.6   Overvoltages caused by current chopping before natural current zero

poor insulation whilst switching magnetizing currents or when switching line charging (capacitive) currents. Surge suppression devices may be fitted on the load side of the circuit breaker to mitigate the problem.

With higher voltage SF_6 switchgear the circuit breaker contact arc voltage is normally constant at higher currents and the arc energy is removed by rapid convection cooling effects. At lower inductive currents the arc tends to be extinguished by arc extension and by turbulence within the circuit breaker contact chamber. The arc current can become unstable as energy is exchanged between source and load reactances such that a high frequency voltage oscillation occurs. The oscillation may be such as to allow the rapidly oscillating current waveform to pass through a current zero before its natural power frequency zero occurs. Again this may lead to a current chopping phenomena.

The interruption of capacitive currents (such as when disconnecting open circuit cables, capacitor banks or long, lightly loaded overhead lines) may also lead to current chopping. Considerable efforts have been made by switchgear manufacturers to provide designs suitable for re-strike-free operation.

SF_6 transmission breakers are designed to be re-strike free and IEC 62271-100 now includes a test for line charging current and capacitor bank interruption that does not allow any re-strikes.

13.3.2.2 *Pole factors*

A three pole circuit breaker will not trip all three phases simultaneously. The first pole to clear will experience the highest transient recovery voltage and the associated power frequency recovery voltage for this first phase will begin to appear after the second pole has interrupted the current flow as shown in Fig. 19.2, Chapter 19. During this interval the transient recovery voltage (TRV) waveform contains high-voltage spikes. The severity of the voltage transient is a function of the network earthing and sequence impedances. For most systems (depending upon the earthing practice) the ratio of X_0/X_1 varies approximately between 0.6 and 3 with average values in the range 1.5 to 2. The pole factor is the ratio of the power frequency recovery voltage to the corresponding phase voltage after the current interruption. In solidly earthed systems the highest first pole-to-clear factors occur with three phase faults. First pole-to-clear factors for different fault conditions are listed in Table 13.4a.

13.3.2.3 *Overvoltages on high voltage overhead line energization*

The exchange of energy between source and load impedances during switching is of particular significance on systems at voltages above 245 kV (IEC Range II). In addition, it should be noted that on long unloaded transmission lines the Ferranti effect may cause the power frequency receiving end voltage to be higher than the sending end voltage such that switching overvoltages occur throughout the system. Such switching overvoltages, which may be as high as 3 per unit,

Table 13.4a   First pole-to-clear factors for different fault conditions

Fault condition	Pole factor
Unearthed three phase fault close up to circuit breaker	Pole factor = 1.5 For systems with earthed or impedance earthed neutrals (as commonly found in systems below 245 kV) the first pole-to-clear factor for all three phase faults is 1.5.
Unearthed three phase fault with source positive and zero sequence impedances X_1 and X_0 together with significant transformer or overhead line impedance (Y_1, Y_0) downstream beyond the fault location:	Pole factor $= 1.5 \dfrac{2 \frac{X_0 + Y_0}{X_1}}{1 + 2 \frac{X_0 + Y_0}{X_1}}$
	For X_0/X_1 in the range 0.6 to 3 this give pole factors between 0.8 and 1.3.
Earthed three phase fault	Pole factor $= 1.5 \dfrac{2 \frac{X_0}{X_1}}{1 + 2 \frac{X_0}{X_1}}$

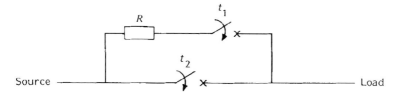

Figure 13.7 Line energization by means of a circuit breaker with a closing resistor

consist of a power frequency and transient high frequency component. The latest IEC insulation co-ordination recommendations are based on switching impulse levels greater than 2 per unit. They may be minimized and reduced to this level by:

1. Inductive reactor compensation connected between phase and earth at the sending and receiving ends to compensate for capacitive charging currents and reduce power frequency surges.
2. Use of inductive voltage transformers (rather than CVTs) connected at the ends of line or cable circuits. These have the effect of helping to discharge the line or cable capacitive currents.
3. Cancellation of the trapped charge effect by using control circuitry such that the circuit breaker only closes during that half cycle of the supply voltage which has the same polarity as the trapped line charge.
4. Energizing the line by initially switching with a series resistance at time t_1 and then short circuiting the resistor out in one (or more) stages some 6 to 10 ms later at time t_2 (Fig. 13.7).

13.3.3 Switches and disconnectors

Switches are capable of load breaking, that is interrupting currents up to their continuously rated current value. Disconnectors have negligible current interrupting capability and are only used in the off-load condition.

LV switches ($<$1000 V) are normally air insulated, and MV types air, oil or SF_6 insulated. Switches have spring tripping to ensure fast action. Both switches and disconnectors must be designed to be thermally stable and suitable (i.e. a specific temperature rise must not be exceeded) for their continuous current rating and for short time (1 or 3 s) through fault current rating conditions. Figure 13.8 shows a 4000 A, 52 kV disconnector undergoing temperature rise tests in the factory before release to site. The numerous small wires are attached to thermocouples mounted on the disconnector. The long copper busbar connections between the disconnector assembly and the current injection cables assist in the thermal isolation between the source current and the disconnector assembly under test.

Figures 13.9a–13.9c show various types of open terminal disconnector for transmission and distribution applications.

Figure 13.8   4000 A, 54 kV disconnector undergoing factory temperature rise tests

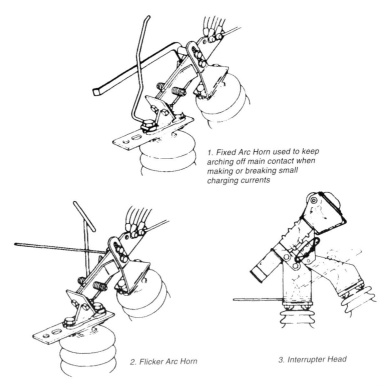

1. Fixed Arc Horn used to keep arching off main contact when making or breaking small charging currents

2. Flicker Arc Horn

3. Interrupter Head

Figure 13.9(a)   Distribution 'rocking arc' disconnector-pole-mounted applications up to 36 kV for sectionalising and isolating apparatus in rural electrification schemes. Fixed contact and moving blades are designed for smooth operation. Contact material is hard drawn high conductivity copper, silver plated over the contact area. The units are designed to break line charging or load currents within specified limits. Courtesy Hawker Siddeley Switchgear

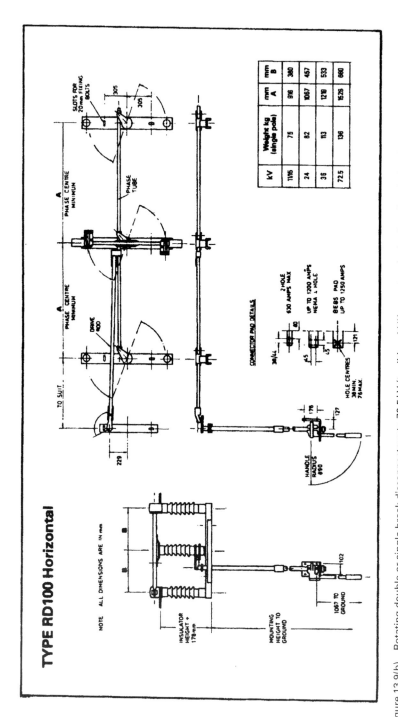

Figure 13.9(b)   Rotating double or single break disconnectors to 72.5 kV (traditional UK open terminal substation designs). The units may incorporate earth switches, manual or power operation, use cap and pin or solid core insulators, have facilities for padlocks, mechanical key type or electrical interlocks and removable interrupter heads. Courtesy Hawker Siddley Switchgear

Figure 13.9(c)   Double break disconnectors

Switches must also be capable of fault making. This is essentially a mechanical design problem. When closing onto a short circuit the device is subjected to the maximum peak value of short circuit current occurring in the first half cycle after fault initiation. Electromagnetic forces are then at their maximum and the switch must reliably close under these conditions. IEC standards assume the peak value of making current to be 2.5 times the rms value of rated short circuit current for MV switches and between 1.7 and 2.2 times for LV switches.

Tables are enclosed giving details of the major technical particulars and guarantees that need to be specified for the following substation components:

Table 13.5 Disconnectors and earthing switches
Table 13.6 Busbars (see also Chapter 18)
Table 13.7 Post type insulators (see also Chapter 6)

13.3.4 Contactors

A contactor is designed for frequent load switching but not for short circuit interruption. It has a relatively light operating mechanism intended for many

Figure 13.9(d)   Pantograph disconnector

Table 13.5   Open terminal disconnector and earthing switch technical particulars and guarantees checklist

Item	Characteristics
Manufacturer	
Type reference	
Description	Centre post swivel, pantograph, etc.
Number of breaks per pole	
Type of contacts	
Contact surface material	
Type of operating mechanism	
Normal rated current	A
Rated short time current (3 or 1 s)	kA rms
Charging current breaking capacity	A
Magnetizing current breaking capacity	A
Rated power frequency withstand voltage	kV
Lightning impulse withstand voltage	
(a) to earth	kV
(b) across isolating distance	kV
Rated switching impulse withstand voltage	
(a) to earth	kV
(b) across isolating distance	kV
Total weight of three phase isolator complete	kg
Air gap between poles of one phase	mm
Period of time equipment has been in commercial operation	Years
Names of supply utility references	

Table 13.6   Busbar technical particulars and guarantees checklist

Item	Characteristics
Manufacturer	
Material	
Overall diameter or dimensions	mm
Nominal cross-sectional area	mm²
Maximum rated current	A
Maximum working tension of main connections	kN/mm²
Resistance of conductors per 100 m at . . . . . . °C	ohms
Tensile breaking stress of conductor material	kN/mm²
Maximum permissible span length	m
Maximum sag under own weight of maximum span	mm
Fixed clamps – catalogue reference and drawing no.	
Flexible clamps – catalogue reference and drawing no.	
Bimetallic joints – catalogue reference and drawing no.	

Table 13.7   Post-type insulator technical particulars and guarantees checklist

Item	Characteristics
Manufacturer	
Insulator type and material	
IEC Standard Reference	
Maximum working vertical load	
(a) compression	kN
(b) tension	kN
Maximum horizontal working load	kN
Minimum failing load (torsion)	kN
Minimum failing load (bending)	kN
Minimum failing load (tension)	kN
Shed profile (reference drawing no.)	
Greatest diameter	mm
Number of units in one insulator	
Length of overall complete post	mm
Weight of complete post	kg
Electrostatic capacity	pF
1 min power frequency withstand voltage (dry)	kV
1 min power frequency withstand voltage (wet)	kV
Lightning impulse withstand voltage (dry, 1.2/50 μs wave)	kV
Switching impulse withstand voltage (wet)	kV
Minimum guaranteed creepage distance	mm
Protected creepage distance	mm
Period of time equipment has been in commercial operation	years
Names of supply utility references	

Notes: A similar format may be used for hollow or post-type insulators for small oil volume (SOV), air blast or SF_6 switchgear, and hollow insulators for CTs and CVTs.

thousands of reliable operations between maintenance inspections. In comparison a circuit breaker is designed to make and interrupt short circuit currents and has a powerful fast acting operating mechanism. Contactors, when used in conjunction with series fuses, give excellent short circuit protection and switching performance. Low-voltage contactors are normally air break types, and

Table 13.8 Contactor ratings and parameters

Rated voltage	Includes rated operating voltage and insulation voltage level.
Rated current	The rated thermal current is the maximum current the contactor can carry for 8 h without exceeding temperature rise limits.
Rated frequency (of supply)	Normally 50 or 60 Hz
Rated duty	This defines the number of duty cycles ranging from uninterrupted duty (contactor closed for an indefinite period) to an intermittent duty of 1200 operating cycles per hour.
Making and breaking capacities	The rated making/breaking capacity defines the value of current under steady state conditions which the contactor can make/break without welding or undue erosion of the contacts and is defined in accordance with the contactor utilization category.
Utilization category	The utilization category depends upon switching requirements. Four categories AC-1 to AC-4 are available. An onerous duty (AC-4) would be switching off motors during starting (plugging) conditions. A typical duty (AC-3) would be for the starting and switching off squirrel cage induction motors during running conditions. The AC-3 category allows for making and breaking capability of 8x the rated operational current at 0.35 power factor (pf) and, under conditions of specified electrical endurance (see below), a making capacity of 6 x breaking capacity 1 x rated operational current at 0.35 pf.
Mechanical and electrical endurance	These define the conditions under which a contactor shall make a number of specified operating cycles without repairs or replacements. For LV contactors such minimum requirements might be specified as: • Mechanical endurance (i.e. off load) 1×10^6 operating cycles • Electrical endurance (i.e. on load) AC-3 or AC-4 operating cycles For MV contactors such minimum requirements might be specified as: • Mechanical endurance (i.e. off load) 25 000 operating Cycles • Electrical endurance (i.e. on load) 5000 operating cycles

MV contactors air, oil or SF_6 insulated. Contactor ratings are based on a number of parameters as described in Table 13.8.

Vacuum contactors have been commercially available since the mid-1960s and SF_6 since the late 1970s. The advantages over air break contactors are smaller physical size, less floor space and floor loading, and their long-term reliability and reduced maintenance. Vacuum, SF_6 and air break contactors with series fuses (corrugated elements) are available for duties in terms of motor size:

3.6 kV—up to ~2000 kW
7.2 kV—up to ~4000 kW
12 kV—up to ~2000 kW

For larger motor sizes and at higher voltages it is necessary to use a circuit breaker for direct on-line (DOL) starting.

When designing contactor control circuits care must be taken to ensure that repeated closing/tripping (pumping) of the contactor is avoided. All control contacts in the 'hold in' circuits should have a positive action with a definite make/break differential in such devices as pressure switches, level switches, etc. A detail not to be overlooked is the circuitry surrounding an economy resistor used to reduce the current consumption of the contactor coil once closed. This resistor can be bridged by a normally closed contactor auxiliary contact. If the economy resistor should then fail (become open circuit) the contactor closing command could then set up a 'hunting' action. The contactor closes and is then immediately opened as the resistor bridging contact opens, this action being repeated rapidly. A latching feature with DC tripping should be considered.

13.4 ARC QUENCHING MEDIA

13.4.1 Introduction

Modern open terminal high-voltage switchgear is primarily based on the SF_6 gas circuit breaker interrupting medium. Vacuum and small oil volume (SOV) switchgear designs are also available and economic up to about 145 kV. Air blast circuit breakers are no longer produced by any major manufacturers.

The options available for indoor equipment at the medium-voltage level consist of metal-enclosed equipment employing phase-segregated small oil volume, vacuum and SF_6 circuit breakers. Medium-voltage bulk oil circuit breakers based on designs perfected in the 1950s are still produced and have a very satisfactory history of reliable use. However, the additional regular maintenance costs, weight and need to incorporate oil catchment and fire detection/suppression systems carefully into the building and civil services designs has reduced their popularity.

Three phase or phase-segregated gas insulated switchgear (GIS) is now used over the complete medium and high-voltage spectrum. Layouts employing such GIS equipment may occupy only one-ninth the volume of an open terminal equipped outdoor substation site.

Detailed switchgear design studies and choice of interrupting medium should give consideration to circuit breaker duties including capacitive and inductive switching, short circuit transient recovery voltage (TRV), rate of rise of recovery voltage (RRRV) and any special needs associated with system parameters. The circuit breaker specifications should also consider the methods of isolation and earthing in order to ensure these are compatible with the particular electricity supply utility operations and maintenance practice. A comparison of circuit breakers with different arc quenching mediums is given in Table 13.9.

Table 13.9 Circuit breaker comparison

	Small oil volume (SOV)	SF$_6$ (GIS, open terminal)	Vacuum	Air blast	Bulk Oil (Indoor – metal-clad or enclosed)	GIS (Indoor – metal-clad or enclosed)	Vacuum (Indoor – metal-clad or enclosed)
MAINTENANCE	**Advantages** Can be economic at low interrupting capacities up to approximately 145 kV.	**Advantages** Low maintenance. Considerable choice from many manufacturers.	**Advantages** Low maintenance. Vacuum 'bottles' easy to replace.	**Advantages** Relatively simple to maintain circuit breaker itself.[1] Still found in installations built up to late 1970s at the highest voltage levels.	**Advantages** Available up to 36 kV for use in extending existing switchboards or where equipment is a supply utility standard. Obsolescent for outdoor open terminal installations. Low cost SOV equipment remains available.	**Advantages** Low maintenance. Compact, small site area, available up to highest voltage levels. At distribution voltage levels below 36 kV many designs considered 'maintenance free' for life time of equipment.	**Advantages** Low maintenance. Vacuum 'bottles' easy to replace. Popular up to 36 kV and especially at 12 kV. Lightweight and compact designs available. Vertical or horizontal housing.
	Disadvantages Maintenance after fault clearance.	**Disadvantages** Special care in handling SF$_6$.	**Disadvantages** Limited availability. May be found for open terminal designs up to 72.5 kV were used in conjunction with SF$_6$ insulation systems. Spare vacuum 'bottle' holding required.	**Disadvantages** Main and standby compressor systems required.	**Disadvantages** Regular oil maintenance required. Civil and building services requirements to be considered in overall installation costs.	**Disadvantages** Special care in handling SF$_6$.	**Disadvantages** Eventual 'bottle' replacement.

(Continued)

Table 13.9 Continued

	Small oil volume (SOV)	SF₆ (GIS, open terminal)	Vacuum	Air blast	Bulk Oil (Indoor – metal-clad or enclosed)	GIS (Indoor – metal-clad or enclosed)	Vacuum (Indoor – metal-clad or enclosed)
OPERATIONS	**Advantages** Overhead connections possible for most circuit configurations. Low cost in good environmental conditions. Substation extensions independent of particular switchgear manufacturer.	**Advantages** Overhead connections possible for most circuit configurations. Economic in good environmental conditions. Substation extensions independent of particular switchgear manufacturer.	**Advantages** Overhead connections possible for most circuit configurations. Low energy, lightweight operating mechanisms. Substation extensions independent of particular switchgear manufacturer.	–	**Advantages** –	**Advantages** Good performance independent of atmospheric pollution when properly housed. Available in phase segregated and non-phase segregated form. Full protection against live parts.	**Advantages** Good performance independent of atmospheric pollution when properly housed. Full protection against live parts.
	Disadvantages Large outdoor switchyard site areas. Extensive civil trench and foundation works.	**Disadvantages** Large outdoor switchyard site areas. Extensive civil trench and foundation works.	**Disadvantages** Large outdoor switchyard site areas. Extensive civil trench and foundation works.	–	**Disadvantages** Bulk oil now largely obsolete. SOV equipment requires regular oil checks.	**Disadvantages** Special cable connections or bus duct/through wall bushings maybe required at higher voltages. Simple modular housing recommended. Insulation co-ordination more difficult.	**Disadvantages** Not available at highest voltages. Simple modular housing recommended. Check switching transient performance.

Note (1): Whilst as a general principle it is true that maintenance is simple, where you have multi-head breakers skilled personnel are required to match each interruptor head operating time to keep the spread time within a few ms (generally less than 5 ms) and also the spread time between phases to <10 ms. The skilled personnel who used to carry out this work are fast disappearing from the industry.

13.4.2 Sulphur hexafluoride (SF$_6$)

SF$_6$ gas is stable and inert up to about 500°C, it is incombustible, non-toxic, odourless and colourless. Figure 13.10 gives a comparison of the dielectric breakdown strength of SF$_6$ gas with both air and transformer oil as a function of gas pressure. SF$_6$ gas possesses excellent insulating properties when pressurized in the range 2 to 6 bar and has a dielectric strength some 2.5 to 3 times that of air at the same pressure. The gas is about 5 times heavier than air with a molecular weight of 146 and specific gravity of 6.14 g/l. At normal densities the gas is unlikely to liquefy except at very low operating temperatures less than −40°C and equipment may be fitted with heaters if this is likely to be a problem, or , as an alternative to heaters, mixed gases are also used below −40°C (SF$_6$ + N$_2$ or SF$_6$ + CF$_4$). Industrial SF$_6$ gas used in circuit breakers and bus systems is specified with a purity of 99.9% by weight and has impurities of SF$_6$ (0.05%), air (0.05% O$_2$ plus N$_2$), 15 ppm moisture and 1 ppm HF. Absorbed moisture leaving the switchgear housing and insulator walls leads to the moisture content of the SF$_6$ gas stabilizing at between 20 and 100 ppm by weight.

Gases at normal temperatures are good insulators but the molecules tend to dissociate at the elevated arc plasma temperatures (~20 000°K) found during the circuit breaking process and become good conductors. SF$_6$ gas also dissociates during the arcing process and is transformed into an electrically conductive plasma which maintains the current until the next or next but one natural power frequency current zero. SF$_6$ gas has proven to be an excellent arc quenching medium. This arises not only from its stability and dielectric strength but also its high specific heat, good thermal conductivity and ability to trap free electrons.

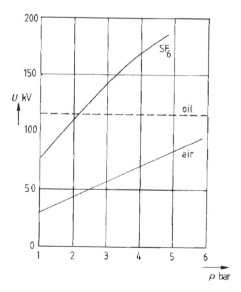

Figure 13.10   SF$_6$ air, and oil dielectric breakdown strength as a function of pressure

It cools very rapidly (few μs) and the sulphur and fluorine ions quickly recombine to form stable insulating SF_6. Such properties all assist in the removal of energy from the arc during the circuit breaking process.

At voltage levels below 36 kV the equipment is often of a 'sealed for life' variety. Higher voltage equipment may be opened for inspection and maintenance after several thousand switching or tens of short circuit operations. SF_6 leakage rates from high-voltage GIS should be less than 0.5% per annum. Secondary dissociation products formed during the arcing process may remain in a gaseous state (mainly SOF_2 but also SO_2F_2 and HF) in very small concentrations. Non-conductive fluorides and sulphides (e.g. WF_6 and CuF formed from the reaction with circuit breaker contact materials) may also be recondensed in very small quantities on the walls of the equipment and form a dust deposit. Standard health and safety precautions (gloves, dust mask, goggles, well-ventilated room, etc.) must therefore be executed when carrying out such maintenance procedures.

At voltages up to about 15 kV and for lower breaking currents both circuit breakers and contactors can use the rotating arc principle. Instead of moving cold gas (air, SF_6 or oil gas bubble) into the arc, the arc is made to rotate under the action of a magnetic field produced by the load or short circuit current. This stretches and moves the arc in the gas to create cooling and eventual arc extinction.

Puffer type SF_6 circuit breakers employ a piston attached to the moving contact to force cool gas into the arc in order to cool and extinguish it. The advantages of these types of SF_6 circuit breaker may be summarized as:

1. Complete isolation of the interrupter from atmosphere and contaminants.
2. Absence of oil minimizes fire risk.
3. Generally, up to 36 kV the interrupter is considered sealed for life and maintenance free.
4. Overall maintenance requirements are low and involve attention to the mechanism.
5. The equipment does not require a heavy operating mechanism, dead and live weight is low. The equipment therefore tends to be compact. This offers civil works savings for indoor metal-enclosed or GIS designs.

Figures 13.11 and 13.12(a)–(c) show SF_6 circuit breaker interruption methods using arc rotation and puffer principles. The latter are gradually being replaced by auto puffer designs – see Fig. 13.12(d).

13.4.3 Vacuum

Vacuum interrupter tubes or 'bottles' with ceramic and metal casings are evacuated to pressures of some 10^{-6} to 10^{-9} bar to achieve high dielectric strength. The contact separation required at such low pressures is only some 0 to 20 mm

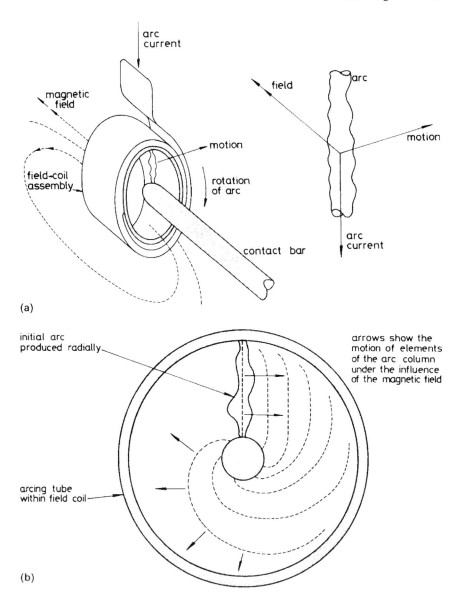

Figure 13.11   (a) Arc motion in a magnetic field. The arc has been extended towards the coil axis by transverse movement of the contact bar. It is shown after transfer to the arcing tube and consequent production of the magnetic field. It is thus in the best plane to commence rotation. (b) Development of a spiral arc. The arc is motivated to move sideways but this tendency is limited by the shape of the electrodes and it develops quickly into a spiral with each element tending to move sideways as shown by the arrows. (c) An actual high speed photograph taken through a porthole. The direction of rotation is in line with that shown in (a) and (b) but as seen from the opposite side of the field coil. A typical speed of rotation is 3000 revolutions per second depending upon the value of arcing current

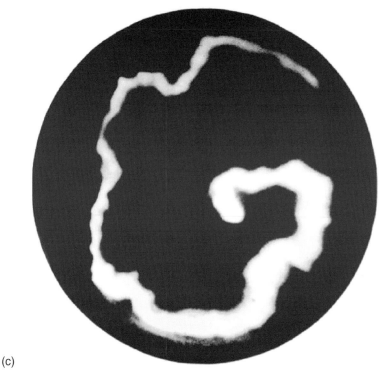

(c)

Figure 13.11   *Continued*

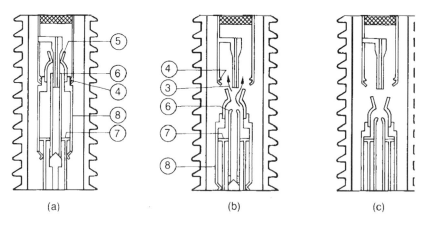

(a)   (b)   (c)

Figure 13.12   SF$_6$ puffer principle of operation (a) Interruption components in CLOSED position; (b) quenching process as arc is gas-blasted; (c) OPEN position

and low energy mechanisms may be used to operate the contacts through expandable bellows. Figure 13.13 shows a cut away view of such a device. The engineering technology required to make a reliable vacuum interrupter revolves around the contact design. Interruption of a short circuit current

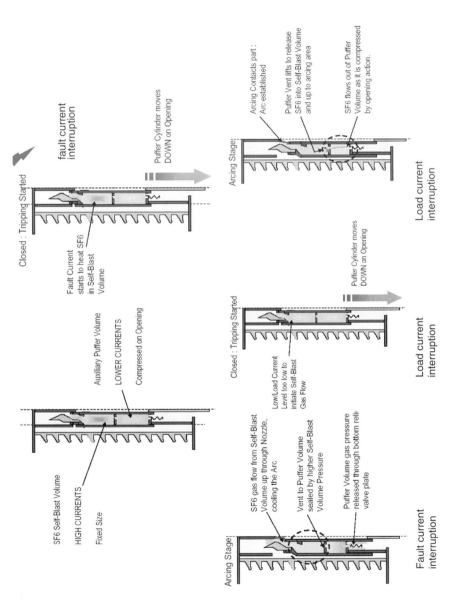

Figure 13.12(d)   Auto-puffer circuit breaker interruption method (courtesy of ABB)

involves the initial formation of a conductive path between the contacts which very rapidly becomes a high grade insulator normally after the first current zero. The conductive path consists of contact metal vapour. The arc is extinguished when the current falls to zero. Conducting metal vapour condenses on metallic screens or sputter shields inside the vacuum tube walls within a few μs and the dielectric strength is restored to form an open circuit. The shields prevent metal vapour deposits from reducing the overall dielectric strength of the insulated vacuum interrupter casing. Arcing times are of the order of half a cycle (10 ms at 50 Hz). To avoid overheating the contact system must be designed to allow the arcing to move to different points over the contact surface area by utilizing its own magnetic field and by using special contact materials. In this way current chopping and the associated transient overvoltages are avoided except at the lowest levels of current interruption (a few amps). The life time of such devices is very long (typically 20 000 switching and a hundred short circuit operations) before replacement is required. The upper voltage range for vacuum interrupters is extended in some designs by surrounding the vacuum bottles and busbars with SF_6.

The advantages of the vacuum circuit breaker or contactor are:

1. Complete isolation of the interrupter from atmosphere and contaminants.
2. Absence of oil minimizes fire risk.
3. Maintenance requirements are low and involve attention to the operating mechanism.
4. Very compact metal-enclosed designs are available. Where necessary the designs may incorporate a series fuse to improve short circuit capability and still render the units safe should the contact surfaces weld under loss of vacuum conditions.

13.4.4 Oil

Mineral oil has good dielectric strength and thermal conductive properties. Its insulation level is, however, dependent upon the level of impurities. Therefore regular checks on oil quality are necessary in order to ensure satisfactory circuit breaker or oil-immersed switch performance. Carbon deposits form in the oil (especially after heavy short circuit interrupting duties) as a result of decomposition under the arcing process. Oil oxygen instability, characterized by the formation of acids and sludge, must be minimized if cooling properties are to be maintained. Insulation strength is particularly dependent upon oil moisture content. The oil should be carefully dried and filtered before use. Oil has a coefficient of expansion of about 0.0008 per °C and care must be taken to ensure correct equipment oil levels. The physical properties of some switchgear and transformer insulating oils available from the Shell Company are listed in Table 13.10.

Oil insulating properties may be assessed by measurement of electric strength, volume resistivity or loss angle. Bulk oil circuit breakers have given

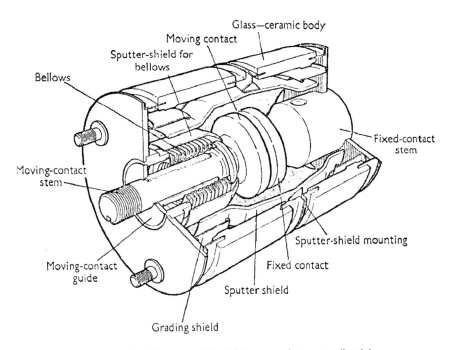

Figure 13.13 Constructional features of the 11 kV vacuum interrupter 'bottle'

Table 13.10 Details of some insulating oils

Oil designation	Mass density at 15°C (kg/1)	Flash point (open cup) (°C)	Pour point (°C)	Kinematic viscosity at 40°C (cSt)	Kinematic viscosity at 100°C (cSt)	Viscosity index
Shell Diala Oil A	0.886	148	<−50	9.4	2.3	50 (approx.)
B	0.872	163	−42	13.2	2.9	45
C	0.872	163	−42	13.2	2.9	45
D	0.864	149	−45	9.5	2.4	50
Shell Diala Oil AX	0.886	148	<−50	9.4	2.3	50 (approx.)
BX	0.872	163	−42	13.2	2.9	45
DX	0.864	149	−45	9.5	2.4	50
(oxidation inhibited versions of A, B and D grades)						

years of reliable service and it should be noted that oil is not a poor or less suitable extinction medium. However, oil circuit breakers for new installations may now be considered obsolete as a result of the maintenance burden necessary to keep the oil in good condition. In addition fire suppression/detection features must be included in the overall building services and civil engineering substation design when using oil circuit breakers. Small oil volume circuit breakers are still available from European manufacturers at distribution voltage levels for relatively low short circuit duties and where short circuit breaking times are not

Figure 13.14   66 kV open terminal substation in the Middle East showing oil-filled cable sealing ends and pressure tanks, rotating post insulators and small oil volume (SOV) circuit breakers

critical to system stability. They offer phase-segregated design thus eliminating the risk of inter phase faults within the interrupting chamber. The small oil volume also greatly reduces the fire risk.

The oil-assisted arc interruption process is difficult to model and the practical design of circuit breaker heads is complex because of the need to cope with the three liquid oil/gas/arc plasma phases. Oil viscosity varies greatly with temperature and gas bubble pressure evolution during short circuit interruption may vary between one and several hundred bar. In comparison, a similar-rated vacuum or SF_6 circuit breaker does not require such a complex mechanical assembly. A typical small oil volume circuit breaker installation and head are shown in Figs 13.14 and 13.15.

13.4.5 Air

Air circuit breakers are normally only used at low voltage levels but are available with high current ratings up to 6000 A and short circuit ratings up to 100 kA at 500 V. The physical size of such units, which contain large arc chutes, quickly makes them uneconomic as voltages increase above 3.6 kV. Their simplicity stems from the fact that they use ambient air as the arc quenching

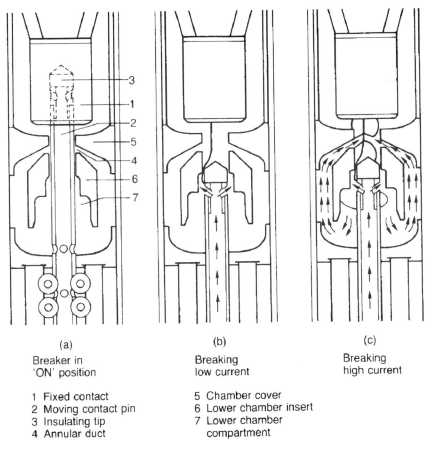

(a)	(b)	(c)
Breaker in 'ON' position	Breaking low current	Breaking high current

1 Fixed contact
2 Moving contact pin
3 Insulating tip
4 Annular duct

5 Chamber cover
6 Lower chamber insert
7 Lower chamber
  compartment

Figure 13.15   Arc extinction in a small oil content circuit breaker (T breaker)

medium. As the circuit breaker contacts open the arc is formed and encouraged by strong thermal convection effects and electromagnetic forces to stretch across splitter plates (Fig. 13.16). The elongation assists cooling and deionization of the air/contact metallic vapour mixture. The long arc resistance also improves the arc power factor and therefore aids arc extinction at current zero as current and circuit breaker voltage are more in phase. Transient recovery voltage oscillations are also damped thus reducing overvoltages. Arc products must be carefully vented away from the main contact area and out of the switchgear enclosure. As explained in Chapter 11, many MCB and MCCB low-voltage current limiting devices are only designed to have a limited ability to repeatedly interrupt short circuit currents. Care must therefore be taken when specifying such devices. Figure 13.17 shows 400 V air circuit breaker with fully repeatable high short circuit capability as typically found in a primary substation auxiliary supply switchboard.

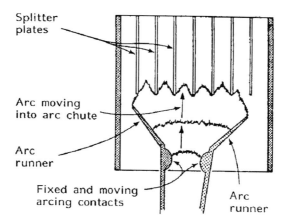

Figure 13.16   LVAC air circuit breaker arc extension across splitter plates

The air blast circuit breaker uses a blast of compressed air across the contacts to assist the interrupting process. Rapid fault clearance times (~2 cycles) largely independent of the short circuit current involved are possible because of the permanent availability of a given blast of compressed air through a nozzle formed in the main fixed contact. The arc is stretched by the air blast and heat removed by forced convection. It is important that the compressor supplies sufficient air to ensure that the arc extinction is still eventually achieved even if not at the first or second current zero. Current chopping when interrupting low currents is usually overcome by paralleling the arc with a resistance connected across the main contacts. Final interruption is then achieved by a fast acting switch in series with the main contacts. Such breakers have high rated current and short circuit current capabilities. They are reliable and reasonably maintenance free. However, they tend to be very noisy in operation (not good for use in substation sites adjacent to built-up areas), require a reliable compressed air plant and have high dynamic loads. This will increase the maintenance burden in comparison with other types of arc interrupting circuit breakers. However, note that the actual operating mechanisms for many circuit breakers use compressed air.

A typical arrangement is shown in Fig. 13.18. Such breakers have been superseded by SF_6 designs which are now available up to 800 kV.

13.5   OPERATING MECHANISMS

13.5.1   Closing and opening

Off-load or off-circuit disconnectors and associated earthing switches may be hand-operated or motor driven through mechanical linkages with the required mechanical advantage to ensure satisfactory operation.

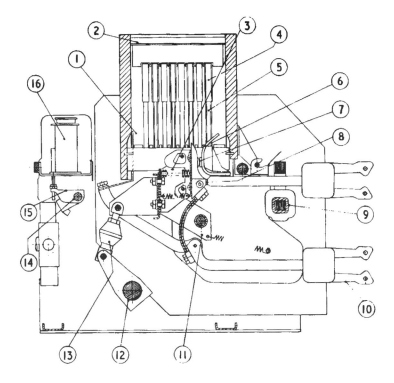

1. Arc runner
2. Heat exchanger for rapid cooling of arc
3. Moving contact carrier
4. Insulating high temperature refractory layer covering top half of de-ionisation plates
5. De-ionisation plates
6. Arcing horns
7. Sintered arcing contacts
8. Main contacts (silver)
9. Current-transformers controlling tripping devices
10. Isolating contacts
11. Location pin of moving contact assembly
12. Main operating shaft to closing mechanism
13. Insulating connecting rod (asbestos glass fibre)
14. Main trip rod
15. Trip pawl for tripping devices
16. Magnetic thermal tripping devices

Figure 13.17   Withdrawable 400 V air circuit breaker (courtesy of Brush Electrical Engineering Co Ltd)

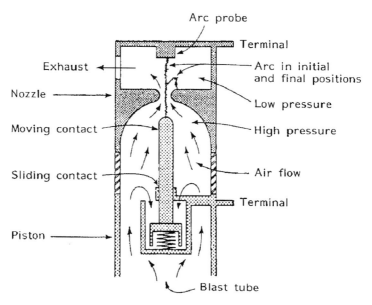

Figure 13.18   Air blast circuit breaker arcing process

For circuit breakers it is essential to use a more positive operating mechanism that does not rely upon operator strength or technique. Manual or motor-driven operating mechanisms which compress a spring for energy storage are used at medium voltages for bulk oil, SOV and SF_6 switchgear. Vacuum switchgear generally needs lower energy mechanisms for the smaller travel distances involved. The sizing of the DC supply necessary to control an MV switchboard is described in Chapter 4.

At higher voltage and short circuit levels pneumatic- or hydraulic-driven mechanisms are used to provide the necessary power to overcome the circuit breaker operating restraining forces. Individual supply systems involve independent hydraulic or compressed air installations in each circuit breaker. Spring operating mechanisms are also available to the highest voltages, normally single pole design mechanisms, one per phase.

Group systems are often specified to feed a group of air blast circuit breakers from the same (usually duplicated) compressed air source. The schematic for a decentralized compressed air group supply system is shown in Fig. 13.19. The air receivers should be equipped with renewable type air filters, safety valves, and blow down valves. Air driers may be used but are not always necessary, depending on the pressure reduction ratio and the ambient temperature variation. Compressed air should be fed in to the HP storage tanks at the bottom of the tank and extracted from the top. The air in the tanks will always be 100% humid and the system should be designed to allow water droplets in the air to settle to the bottom of the tank before use. The pressure reduction will dry the air sufficiently for use in the breaker. A lock-out feature should be

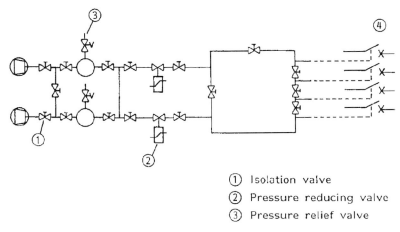

① Isolation valve
② Pressure reducing valve
③ Pressure relief valve

Figure 13.19   Schematic showing decentralized, duplicated compressed air installation feeding a set of four air blast circuit breakers

incorporated into the design such that if air pressure in the receiver falls below that suitable for reliable circuit breaker operation closing or opening is prevented and an alarm raised. The design capacity of such a plant is based on the number of circuit breakers in the group, the number of realistic switching cycles being considered (usually based on three complete close and trip operations unless there is an auto-reclose requirement), the air quantity used by the circuit breaker mechanism and arc extinction process together with an estimate for pipeline and circuit breaker system leakage losses. It is important to consider the necessity of good compressed air system maintenance. The author has visited several compressed air installations in developing countries where compressors are continually running until failure occurs because of excessive leakages due to a lack of spare parts.

The closing spring type of mechanism is specified for automatic spring recompression after each circuit breaker closing operation. It is therefore immediately available for the next time it is required. Should the motor drive supply fail, a handle is supplied with the switchgear so that the spring mechanism may be charged by hand. The closing spring may be released either locally by a hand-operated latch or remotely via an electromagnetic latch powered from a reliable AC or DC supply used during the closing action.

The opening spring is normally arranged to be charged during the closing operation. In this way the breaker is always ready for circuit breaking duties even under auxiliary supply failure. Figure 13.20 shows the closing and opening spring mechanisms on a YSF type Yorkshire Switchgear SF_6, 24 kV, 400 A circuit breaker. The meter shown in the photograph is monitoring the SF_6 gas pressure. The opening spring is automatically unlatched by an indirect shunt trip coil which is in turn energized via the circuit breaker protection scheme. Built-in direct release schemes are also found on LVAC air circuit breakers.

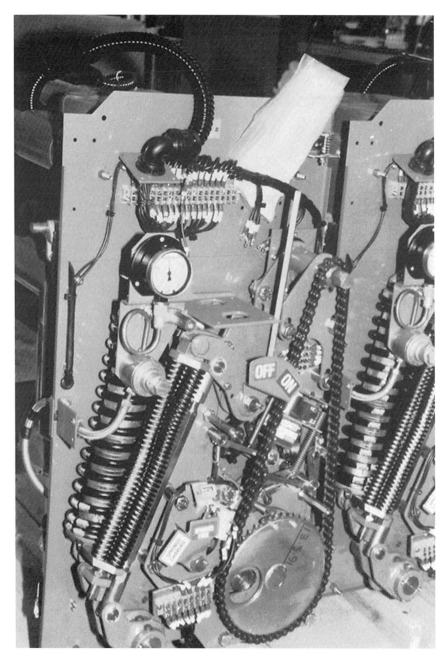

Figure 13.20   24 kV, 400 A SF$_6$ circuit breaker operating mechanism (courtesy of Yorkshire Switchgear)

13.5.2 Interlocking

Since disconnectors must not be operated on load, interlocks between circuit breakers and the associated disconnectors must be incorporated into the overall substation design. For example, disconnectors should be so interlocked that they cannot be operated unless the associated circuit breaker is open. In a duplicate busbar system the interlocking must prevent the simultaneous closing of two busbar disconnectors unless the busbars are already electrically connected through the bus coupler disconnectors and associated bus coupler circuit breaker (if installed). Similarly, circuit breakers must be interlocked so that except under maintenance conditions it is not possible to close the breaker unless the selected busbar and circuit disconnectors are already closed. Such interlocks ensure safe operation of equipment under all service conditions by detecting and logically processing the switching status of all the switchgear involved. Interlocking may be achieved by the use of mechanical key exchange box or linkage systems which enforce the correct sequence of switching operations. Alternatively, electromagnetic bolt systems may be employed which inhibit disconnector movement. Electrical interlocks should function so as to interrupt the operating supply.

For withdrawable switchgear such interlocks would involve:

1. The circuit breaker cannot be inserted into the switchgear cubicle housing unless the 'isolated' position has been selected.
2. The circuit breaker cannot be closed unless it is in the fully 'engaged' or 'isolated' position.
3. The circuit breaker assembly cannot be coupled to or decoupled from the cubicle busbar and feeder circuits unless it is in the 'open' position.
4. The circuit breaker cannot be inserted into any position other than that selected on the selector mechanism (busbar earth, feeder earth, normal service).
5. The busbar shutters.

13.5.3 Integral earthing

A useful maintenance feature for indoor MV withdrawable switchgear is to use the circuit breaker switch itself to earth either the circuit or busbars. The switch may be moved into different positions within the cubicle to achieve this as shown in Fig. 13.21(a). An alternative to this, used by most European manufacturers in both withdrawable and fixed breaker design switchgear, is to use fixed fault making earthing switches mounted in the cubicles, as illustrated in Figure 13.21(b). This ABB switchgear incorporates the REF542plus control unit, which not only provides inbuilt solid-state system protection (overcurrent, earth fault and distance protection, if required) but also a programmable logic to prevent switch maloperation. Mechanical interlocking is available as an option, and some utilities still prefer this, in view of the difficulty of ensuring compliance with IEC 61508.

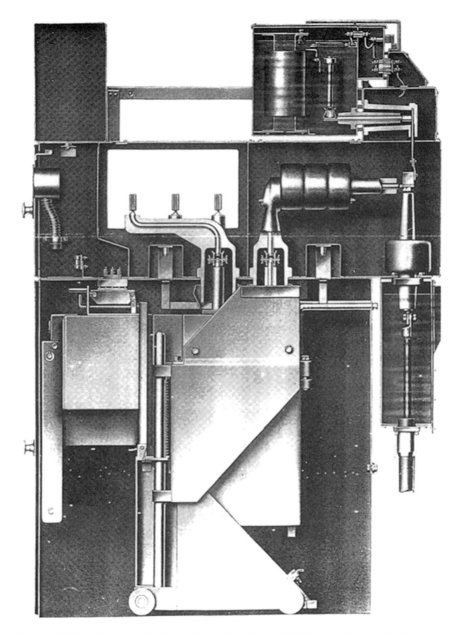

Figure 13.21(a)   Cross section through an 11 kV, 400 A vertical isolator, horizontal withdrawal circuit breaker. The circuit breaker carriage is shown for connection between the circuit and busbars. The carriage may be moved to different positions for the integral earthing through the circuit breaker of either the circuit or busbars

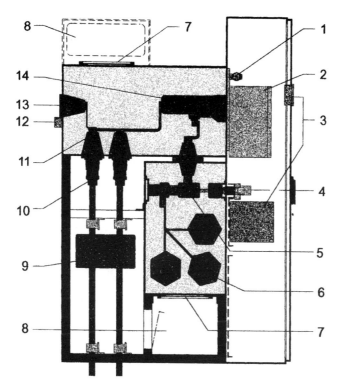

1   Density sensor

2   Circuit-breaker operating mechanism

3   Multifunction Protection and Switchgear Control unit
    REF542 plus

4   3-position switch operating mechanism

5   3-position switch

6   Busbar

7   Pressure relief disk

8   Pressure relief duct (optional)

9   Ring type current transformer

10 Cable plug

11 Cable socket

12 Measuring sockets for capacitive voltage indicator system

13 Test socket

14 Circuit-breaker

☐   SF$_6$ gas

Figure 13.21(b)   Cross section through a 24 kV 1250 A non-withdrawable SF$_6$ circuit
breaker, with SF$_6$ insulated busbars and integral earthing switch (courtesy of ABB)

13.6 EQUIPMENT SPECIFICATIONS

13.6.1 12 kV metal-clad indoor switchboard example

13.6.1.1 *Scope of the work*

This is a practical example of MV switchboard definition. Drawing 16218-10-EE0001 (Fig. 13.25) is the key 132/11 kV system single line diagram. The work involves connecting a large pharmaceutical production complex, with its own combined heat and power (CHP) generation, into the electrical supply utility 132 kV grid system. The total work for this project involves double circuit overhead line tee connections and a small two transformer 132/11 kV substation. The substation includes the design, supply installation, testing and commissioning of a new 11 kV, 9 panel switchboard which is covered in this example.

13.6.1.2 *System parameters*

In order to specify the equipment correctly it is first essential to determine and accurately define the environmental and electrical system requirements. These are detailed Tables 13.11 and 13.12 with appropriate explanations. Note that it is very important to consider standardization issues when defining parameters for new equipment. For example, the 30 V DC tripping supply meets with that already used elsewhere on the large pharmaceutical site by the purchaser. In a similar way there are great advantages in using similar switchgear, relays, etc. in order to minimize spares holdings and ensure maintenance crews are not confronted with a wide variety of equipment.

13.6.1.3 *Fault level and current ratings*

The maximum and minimum fault level at the interconnection point should be specified in writing by the electrical supply utility. Even without this information the high impedance (30%) 132/11 kV transformer impedance will greatly reduce the fault contribution from the electrical supply utility. Assuming a negligible source impedance the fault level, F_{MVA} on the 132/11 kV transformer secondary side on a 100 MVA base will be:

$$F_{MVA} = \frac{100 \cdot S}{Z\%}$$

where S = MVA rating of equipment
F_{MVA} = Three phase fault level in MVA on the secondary side of one transformer
$Z\%$ = % impedance expressed as a percentage at a stated MVA rating $S\%$,
F_{MVA} = 100 · 30 (MVA)/30 (%)
= 100 MVA

Table 13.11 Climatic conditions

Parameter	Value	Notes
Seismic factor	0.05 g	No special precautions necessary. (see Chapter 17)
Atmospheric pollution	Heavy	Affects outdoor equipment creepage distances.
Soil thermal resistivity	30°C cm/W	Affects cable ratings.
Soil temperature (at cable laying depth)	9°C	Affects cable ratings.
Soil resistivity	30 Ohm-m at average 1 m depth 50 Ohm-m at 2 to 10 m depth falling gradually to 200 Ohm-m at 30 m depth	Affects the design of the substation earthing system (see Chapter 8).
Maximum daily average temperature	21.6°C (July)	Use 40°C max. ambient temperature for switchgear rating. This is a normal standard to which switchgear is designed without cost penalties to the Purchaser.
Maximum annual absolute temperature	33.9°C	
Minimum daily average temperature	0°C	
Minimum annual absolute temperature	−15.6°C	
Annual average relative humidity	85%	Specify anti condensation heaters in the switchboard.
Annual number of lightning strikes to ground.	0.6 per km²/year	132/11 kV transformers will be protected by 132 kV surge arresters and cable sealing end spark gaps. 11 kV switchboard will be connected by cables from the transformers. The standard 12 kV switchgear impulse rating of 75 kV to be checked with supply utility as adequate (see Chapter 19, Section 19.5 and Chapter 9).
Altitude	3 m above ordinance datum	No high altitude insulation co-ordination problems to consider.

For two transformers in parallel the effective fault limiting impedance halves and the maximum possible fault level contribution from the electrical supply utility becomes 200 MVA = $200/\sqrt{3} \cdot 11\,kV$ = 10.5 kA.

In addition to this the local generation contribution must be added. Further, since this is a large industrial complex with many induction motors the short-term motor contribution must also be considered. A full study using computer analysis typically as described in Chapter 1 is worthwhile and gives an 11 kV switchboard fault level of 360 MVA or approximately 19 kA under maximum generation and electrical supply utility fault contribution conditions. In accordance with Tables 13.3 and 13.4 and allowing for some possible future fault level increases a standard switchboard-rated short circuit breaking current of 25 kA is selected.

Table 13.12   Electrical system parameters

Parameter	Value	Notes(1)
Nominal service voltage	11 kV ± 5%	Use 12 kV rated voltage equipment, standard voltage levels to IEC 38, current ratings to IEC 59 and switchgear to IEC 56.
Impulse voltage withstand	75 kV	As per IEC 56
Number of phases	3	
Frequency	50 Hz ± 2%	
System earthing	Resistance earthed	Interconnections with nearby electrical supply utility generation sites via the 132 kV OHL earth and cable feeder armour connections to on-site generation earths. This leads to a diversification through the alternative paths of the maximum fault current.
System fault level	25 kA	As per IEC56. Calculations show that this rated short circuit breaking current may be used to meet system requirements without cost penalty to the purchaser.
DC systems	30 V DC +10%, −15%	For DC motor-driven auxiliaries, tripping, indicating lamps and controls. Switchgear design to operate down to 85% nominal voltage and trip relays, coils, etc. to operate reliably down to 80% of nominal voltage. Note this is a standard value already used elsewhere on the purchaser's pharmaceutical site.

Note 1: As this is taken from a real example, the IEC references as used are quoted. For current equivalents see Table 13.2.

The current rating of the incoming circuit breakers and busbars is matched to the transformer full load current with an allowance for possible future short-term overload. For the particular type of switchgear that the purchaser wishes to use, a 2000 A busbar is available as a manufacturer's standard and is therefore specified. Ring feeder and interconnector circuit breaker current ratings are derived from requirements and load growth projections covered by system studies. The next greater standard rating is then specified.

13.6.1.4 *General requirements*

Table 13.13a may now be completed as shown. The IP 34 enclosure rating is a manufacturer's normal protection and perfectly adequate for an indoor installation. The local controls are specified after discussion with the purchaser's maintenance and engineering staff.

Table 13.13a   Typical MV switchgear accessories enquiry data sheet – general requirements

General requirements	Description
(1) Rated voltage and system	12 kV, three phase, three wire, 50 Hz
(2) Switchgear type	Nominated or preferred manufacturer's name and catalogue reference
	Fully extendable
	9 panel metal-clad
	Vertical isolation, horizontal withdrawal
	SF_6, vacuum, etc. circuit breakers
	Integral earthing
(3) Rated lightning impulse withstand (1.2/50 μs)	75 kV peak
(4) Rated 1 min power frequency withstand	28 kV rms
(5) Short circuit rms breaking current	25 kA
(6) Short time current duration	3 s
(7) Supply voltage of opening and closing devices and auxiliary circuits	30 V DC (often 110 V DC for primary substation switchboards)
	240 V, one phase AC
(8) Busbar rating	2000 A
(9) Degree of enclosure protection	IP 34
(10) Operating mechanism	Motor wound spring charged
	Shunt trip coil
(11) Panel mounted control switches	CB trip/close
	CB local/remote control
	Heater selector on/off
(12) Panel mounted indication lamps and colour	CB isolated (amber)
	CB closed (red)
	CB open (green)
	Trip circuit healthy (white)
(13) Panel mounted fuses/MCBs/links	Yes
(14) Applicable standards	IEC 62271

13.6.1.5 *Particular requirements*

After the system studies mentioned in Section 13.6.1.3, the Protection, Instrumentation and Metering Drawing 16218-10-EE-0003 (Fig. 3.27) is prepared using the standard symbols shown in Drawing 16218-10-EE-0002 (Fig. 3.26). The switchboard uses a relatively complex protection scheme which arises from the local generation and interconnection to the grid. Another point of interest in this example is the use of *Is*-Limiters (see Chapter 11). The pharmaceutical complex has existing, older, 11 kV, 250 MVA fault-rated switchgear. Since the interconnection to the Grid increases the fault level to approximately 360 MVA this older switchgear must be protected or replaced. The use of the *Is*-Limiters (items IS1 and IS2 on drawing 16218-10-EE-0001) avoids expensive switchgear replacement costs. The switchgear particular technical requirements may then be tabulated and clearly defined as shown in Table 13.13b. Items such as cable termination detail spare parts and special tools should not be left to chance and may be specified and ordered with the switchgear.

During an open competitive enquiry for such MV switchgear a technical specification should accompany the tables and drawings of general and particular

Table 13.13b Typical MV switchgear accessories enquiry data sheet – particular requirements

Parameter	1	2	3	4	5	6	7	8	9
Circuit description	Ring main feeder 1	Ring main feeder 2	Inter-connector 1	Incomer 1	Bus section	Incomer 2	Inter-connector 2	Ring main feeder 3	Ring main feeder 4
Current rating	630 A	630 A	1250 A	2000 A	2000 A	2000 A	1250 A	630 A	630 A
Ammeter scale reading	0–500 A	0–500 A	–	–	–	–	–	0–500 A	0–500 A
Voltmeter scale reading	–	–	–	–	–	–	–	–	–
Ammeter selection	Yes	Yes	–	–	–	–	–	Yes	Yes
Voltmeter selection	–	–	–	–	–	–	–	–	–
Current transformer data									
Core A Class	1/5P	1/5P	X	X	X	X	X	1/5P	1/5P
Ratio	630/5	630/5	1200/5	2000/5	2000/5	2000/5	1200/5	630/5	630/5
Core B Class	X	X	X	1/5P	X	1/5P	X	X	X
Ratio	2000/5	2000/5	2000/5	2000/5	2000/5	2000/5	2000/5	2000/5	2000/5
Core C Class	X	X	X	X	X	X	X	X	X
Ratio	2000/5	2000/5	2000/5	2000/5	2000/5	2000/5	2000/5	2000/5	2000/5
Core D Class				X		X			
Ratio				2000/5		2000/5			
Integral earthing	Feeder	Feeder	Feeder	Cable	Busbar	Cable	Feeder	Feeder	Feeder
Voltage transformer data									
1st secondary Class			6P (5 limb open delta)	1.0		1.0	6P (5 limb open delta)		
Ratio			11/0.11 kV	11/0.11 kV		11/0.11 kV	11/0.11 kV		
2nd secondary Class			1.0/3P				1.0/3P		
Ratio			11/0.11 kV				11/0.11 kV		
No. and type of trip coils	1-shunt 30 V DC	1-shunt 30 V DC	1-shunt 30 V DC	1-shunt 30 V DC	1-shunt 30 V DC	1-shunt 30 V DC	1-shunt 30 V DC	1-shunt 30 V DC	1-shunt 30 V DC
No. and type of closing coils	1-spring release 30 V DC	1-spring release 30 V DC	1-spring release 30 V DC	1-spring release 30 V DC	1-spring release 30 V DC	1-spring release 30 V DC	1-spring release 30 V DC	1-spring release 30 V DC	1-spring release 30 V DC
Spring charge motor	240 V AC	240 V AC	240 V AC	240 V AC	240 V AC	240 V AC	240 V AC	240 V AC	240 V AC
Protection relays	(a) 2 × IDMTL o/c + instantaneous high set & 1 × IDMTL	(a) 2 × IDMTL o/c + instantaneous high set & 1 × IDMTL	(a) Pilot wire protection Type……	(a) 2 × IDMTL o/c & 1 × IDMTL e/f Type……	(a) High impedance busbar protection with main	(a) 2 × IDMTL o/c & 1 × IDMTL e/f Type…… (b) 1 × IDMTL	(a) Pilot wire protection Type……	(a) 2 × IDMTL o/c + instantaneous high set & 1 × IDMTL	(a) 2 × IDMTL o/c + instantaneous high set & 1 × IDMTL

	(1)	(2)	(3)	(4)	(5)	(6)	(7)	(8)
Protection	e/f Type ...... (c) High impedance busbar protection with main and check zones, BB'E'	e/f Type ...... (c) High impedance busbar protection with main and check zones, BB'E' (d) Intertrip remote CB direct from CB auxiliary contacts	(c) High impedance busbar protection with main and check zones, BB'E'	1 and check zones, BB'E' & 'F' (c) High impedance busbar protection with main and check zones, BB'E'	directional e/f & 2 x IDMTL directional o/c Type ...... (c) High impedance busbar protection with main and check zones, BB'E'	(b) 1 × IDMTL directional e/f & 2 x IDMTL directional o/c Type ...... (c) High impedance busbar protection with main and check zones, BB'F' (d) Intertrip remote CB direct from CB auxiliary contacts	e/f Type ...... (c) High impedance busbar protection with main and check zones, BB'F'	e/f Type ...... (c) High impedance busbar protection with main and check zones, BB'F'
Other equipment	Remote monitoring pulse transmitter energy metering Range ...... Type	Remote monitoring pulse transmitter energy metering Range ...... Type	Alarms: (a) Protection operated (b) Substation general alarm (c) ...... spares			Remote monitoring pulse transmitter energy metering Range ...... Type	Remote monitoring pulse transmitter energy metering Range ...... Type	
Selector switch			– synchronising on/off – manual/auto – logic selector switch – synchronising override on/off switch		– synchronising on/off – manual/auto – logic selector switch – synchronising override on/off switch	– synchronising on/off – manual/auto – logic selector switch – synchronising override on/off switch		
Cable Type	1 x 3c, Cu XLPE/SWA/PVC	1 x 3c, Cu XLPE/SWA/PVC	6 x 1c,Cu XLPE/AWA/PVC	2 x 3c, Cu XLPE/SWA/PVC	6 x 1c, Cu XLPE/AWA/PVC	2 x 3c, Cu XLPE/SWA/PVC	1 x 3c, Cu XLPE/SWA/PVC	1 x 3c, Cu XLPE/SWA/PVC
Section	up to 185 mm²	up to 185 mm²	up to 630 mm²	up to 300 mm²	up to 630 mm²	up to 300 mm²	up to 185 mm²	up to 185 mm²
Termination	Dry type	Dry type	Dry type	Dry type	Dry type	Dry type	Dry type	Dry type
Glands	Yes	Yes	Yes	Yes	–	Yes	Yes	Yes
Anti condensation heater	Yes	Yes	Yes	Yes	–	Yes	Yes	Yes

Loose equipment: (a) 1 earthing switch operating handle (if applicable); (b) 1 set of special tools: (c) 2 raise/lower handles (if vertical isolation, horizontal withdrawal type switchgear); (d) 2 closing mechanical maintenance operating devices; (e) 1 test jumper set (if applicable); (f) 1 wooden tool box complete with full set of maintenance tools; (g) 1 wall mounted key cabinet including one set of brass padlocks; and two sets of keys Installation: To be floor mounted on unistrut channel.

requirements. Such a specification would give background details to the manu-facturer (site location, access and temporary storage facilities) and also request details on particular switchgear mounting arrangements, whether metal-enclosed/metal-clad equipment is required, any particular results arising from the system studies, the factory test requirements, etc. A useful proforma and checklist covering indoor switchgear is given in Table 13.14 Parts A and B.

13.6.2 Open terminal 145 kV switchgear examples

13.6.2.1 *System analysis*

It is essential to involve an element of system analysis performed by senior and experienced systems engineers before switchgear at the higher voltage levels can be correctly specified. This will need particularly to take into account the network earthing, any likely switching overvoltages and also to give more gen-eral attention to insulation co-ordination. This is particularly important when specifying ZnO surge arresters and insulation levels associated with GIS. The number of outage conditions to be considered makes computer-assisted load flow and fault level system analysis (also if applicable harmonic checks) almost essential. However, the simple sums explained in Chapters 1 and 25 and in other chapters throughout this book are still important and may be used as a guide as to the order of magnitude expected from the computer generated results.

The earthing arrangements used throughout the network and values of posi-tive, negative and zero sequence impedances will determine the ratio of fault current to three phase fault current. In highly interconnected solidly earthed sys-tems the ratio of Z_0/Z_1 may be less than unity. The three phase symmetrical earth fault level may not therefore be the worst case fault condition. The ratio Z_0/Z_1 is therefore a measure of the 'effectiveness' of the system earthing and may also vary over the system. For the calculation of modern circuit breaker breaking currents it is worthwhile taking the sub-transient values of generator reactance into account. For older oil circuit breakers and when determining the minimum currents available to operate protection relays it is more usual to consider tran-sient reactances since the sub-transient reactance effects usually disappear before oil circuit breakers have operated. When determining the circuit breaker making currents or the maximum through-fault current for protection purposes the sub-transient reactance values should be used. The mechanical forces asso-ciated with short circuits increase with the square of the current. In practice, most computer programs now allow both transient and sub-transient values to be entered. One should not get carried away with the idea that all the available computing power will in any way improve results since the raw input data may only be best estimates in the first place.

Figure 13.22 shows a hand calculation of peak symmetrical fault current, peak asymmetrical fault current and maximum short circuit current based on contributions from the electrical supply utility grid, local generation and local industrial plant motors to determine the switchboard fault levels.

Table 13.14   Indoor metal-enclosed/metal-clad switchgear technical particulars and guarantees

Item	Characteristic
Part A	
General	
Manufacturer	
Type reference	
Metal-enclosed/metal-clad	
Vacuum/SF$_6$/SOV/bulk oil	
Rated voltage	kV
Rated 1 min power frequency withstand voltage	kV
Rated lightning withstand voltage	kV
Rated frequency	
Rated normal current	
(a) busbars	A
(b) feeders	A
Busbar spout shutters	Yes/no
Shutter material	Steel......
Independent shutter locking facility	Yes/no
Degree of protection (IEC 60529)	
(a) enclosure	IP......
(b) partitions	IP......
(c) shutters	IP......
Circuit breaker	
(a) fixed or withdrawable?	
(b) horizontal or vertical isolation	
(c) horizontal or vertical withdrawal for maintenance	
Busbar	
(a) material	Cu/Al......
(b) cross section	mm^2
(c) insulation material	PVC/LSF/resin/etc.
(d) fire certification (IEC 60466, etc.)	
Dimensions and weights	
Minimum clearances in air	
(a) phase to phase	mm
(b) phase to earth	mm
(c) across circuit breaker poles	mm
Minimum creepage distances	
(a) phase to phase	mm
(b) phase to earth	mm
(c) across circuit breaker poles	mm
Overall dimensions of each circuit breaker panel type	
(a) feeder (...A)	Height/width/depth (mm × mm × mm)
(b) incomer (...A)	Height/width/depth (mm × mm × mm)
(c) bus-section (...A)	Height/width/depth (mm × mm × mm)
Space necessary for circuit breaker withdrawal for maintenance	mm
Weight of whole panel complete with all fittings as in service	kg
Maximum shock load imposed upon floor when operating under worst case conditions	N (tension or compression)
Method of earthing	
(a) busbars	
(b) incomers or feeders	
Is earthing device capable of:	
• closing onto a fault?	Yes/no
• withstanding short circuit current for 1 or 3 s?	Yes/no
Is mechanical interlocking provided for earthing and isolation facilities?	Yes/no

(Continued)

Table 13.14   Continued

Item	Characteristic
Maximum number of CT windings that may be accommodated	
Has switchgear been subjected to internal arc tests to IEC62271-200	Yes/no
Period of time switchgear has been in satisfactory commercial operation	Years
Names of supply utility references	
Are floor plates, rails or carrier trolleys included for supporting switchgear during maintenance?	Yes/no
Recommended maintenance tools	
Special tools	
Part B	
Type tests	
Testing authority	
Test certificate report reference and date	
Short time withstand current	
(a) 1 s	kA rms
(b) 3 s	kA rms
Rates short circuit breaking current	
(a) symmetrical	kA rms
(b) asymmetrical	kA rms
Rated short circuit making current	kA peak
First phase to clear factor	
Rated transient recovery voltage (TRV) at 100% rated short circuit breaking current	kV peak
Rate of rise of transient recovery voltage (RRRV)	kV/μs
Rated line charging breaking current	A
Rated cable charging breaking current	A
Rated out-of-phase breaking current	A
Rated duty cycle	
Is the circuit breaker re-strike free?	Yes/no and comments
Performance data	
Opening time	ms
Making time	ms
Method of closing	Solenoid latch – spring-charged coil, etc.
Solenoid closing coil (or other method) current	A
Solenoid closing coil (or other method) voltage	V (DC assumed)
Method of tripping	Solenoid latch – spring-charged coil, etc.
Trip coil current	A
Trip coil voltage	V (DC assumed)
Type of arcing contact or arc control	
Type of main contact	
Material of contact surfaces	
Does the magnetic effect of load current increase main contact pressure?	
Number of breaks per phase	mm
Length of each break	mm
Length of operating mechanism stroke	A
Current chopping level	
Number of operations allowable before maintenance is recommended as necessary:	
• at rated asymmetrical breaking current	
• at rated symmetrical breaking current	
• at rest normal current	

(*Continued*)

Table 13.14   Continued

Item	Characteristic
• at no load	
• overall maintenance regime	
Quantity of interrupting medium in each three phase circuit breaker	
• oil	l
• SF$_6$ density (normally taken as a pressure reading at ...... °C)	Bar
Vacuum or SF6 monitoring facilities	Yes/no
Is a facility provided to measure contact wear without dismantling	(useful for vacuum interrupters)
Notes	

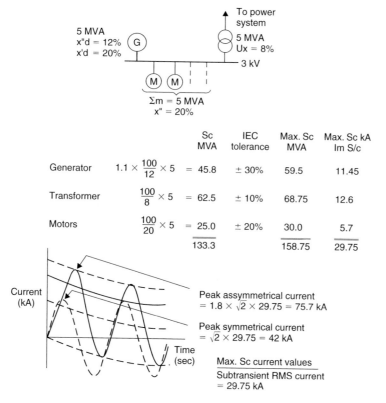

Short circuit current levels

		Sc MVA	IEC tolerance	Max. Sc MVA	Max. Sc kA Im S/c
Generator	$1.1 \times \frac{100}{12} \times 5 = 45.8$		± 30%	59.5	11.45
Transformer	$\frac{100}{8} \times 5 = 62.5$		± 10%	68.75	12.6
Motors	$\frac{100}{20} \times 5 = 25.0$		± 20%	30.0	5.7
		133.3		158.75	29.75

Current (kA)

Time (sec)

Peak assymmetrical current
= 1.8 × √2 × 29.75 = 75.7 kA

Peak symmetrical current
= √2 × 29.75 = 42 kA

Max. Sc current values
Subtransient RMS current
= 29.75 kA

Figure 13.22   Hand calculation of peak symmetrical fault current, peak asymmetrical fault current and maximum short circuit current levels

13.6.2.2 *Technical particulars and guarantees*

Tables 13.15 and 13.16 are proforma checklists based on IEC 62271-100 and IEC 60044 for use when specifying open terminal circuit breakers and pedestal

Table 13.15   Open terminal circuit breaker technical particulars and guarantees checklist

Item	Characteristic
General	
Manufacturer	
Type of reference	
Interrupting medium (small oil volume, SF6, vacuum, etc.)	
Number of phases	
Frequency	Hz
Rated voltage	kV
Lightning impulse withstand voltage	kV peak
Switching impulse withstand voltage	
(a) to earth	kV peak
(b) across open breaker	kV peak
Power frequency withstand voltage	
(a) breaker closed	kV
(b) breaker open	kV
Rated normal current	A
Type tests	
Testing authority	
Test certificate report reference	
Short time withstand current	
(a) 1 s	kA rms
(b) 3 s	kA rms
Rated short circuit breaking current	
(a) symmetrical	kA rms
(b) DC component	%
Short circuit making current	kA peak
Rated operating duty cycle	
First phase to clear factor	
Rated transient recovery voltage at 100%	
Rated short circuit breaking current	kV peak
Rated small inductive breaking current	A
Rated line charging breaking current	A
Rated cable charging breaking current	A
Rated out-of-phase breaking current	A
Rated characteristic for short line/close-up faults	A
Maximum guaranteed switching overvoltage	kV
Constructional features	
Is a series break incorporated?	Yes/no
Is a device used to limit transient recovery voltage	Yes/no
Method of closing (spring coil, hydraulic, pneumatic, etc.)	
Solenoid closing coil current and voltage	A V(DC)
Method of tripping (spring coil, hydraulic, pneumatic, etc.)	
Trip coil current and voltage	A V
Is the circuit breaker trip free?	Yes/no
Minimum clearances in air	
(a) between phases	mm
(b) phases to earth	mm
(c) across interrupters	mm
(d) live parts to ground level	mm
Number of breaks per phase	
Dimensions and weights	

(Continued)

Table 13.15   Continued

Item	Characteristic
Weight of circuit breaker complete	kg
Maximum shock load imposed uponfloor/foundation	kgf (tension/compression)
Quantity of oil/gas in complete breaker	l
Routine pressure test on circuit breaker tank	Bar
Nominal interrupting head gas pressure at °C	Bar
Period of time equipment has been in commercial operation	Years
Names of supply utility references	

Table 13.16   MV circuit breaker failure mode statistics

Item	% of total failures
Operating mechanism	50
Control circuit	
• Switches, relays	10
• Other	15
Switching system	
• Contacts, insulating medium	<1
• Insulation (cable terminations, etc.)	15
• Live parts	9

mounted CTs. They cover the items described in this chapter, Chapters 5 and 19. Note the special cases for rated out-of-phase breaking current, rated cable charging breaking current, rated capacitor breaking current and rated small inductive breaking current have all been included in the circuit breaker checklist. Since the manufacturer will match one of his standard items of equipment to the specification it is worth noting the statistics associated with different circuit breaker failure modes. When checking tenders it is important to give these areas emphasis and to seek advice from other users of the same switchgear.

13.6.3 Distribution system switchgear example

An MV fused contactor may be used to protect and control the switching of an MV/LV (6.6./0.38 kV) distribution transformer. This is a feasible and less expensive option than using a circuit breaker at this voltage level. The HRC fuse limits the prospective short circuit current and therefore the severity of the fault. Note that, as explained in Fig. 4.6, the standby earth fault (SBEF) CT is located in the earth connection of the star point of the LV (low voltage, 0.38 kV) transformer winding. This is connected to an earth fault relay arranged to trip the contactor. The design takes advantage of the maximum LV fault current, when referred to the HV (high voltage, 6.6 kV) side of the transformer, being well within the interrupting capability of the contactor.

This is a result of the fault limiting effect of the transformer impedance. A star point CT and earth fault relay are used because faults in the transformer

LV winding and LV switchboard most often occur initially as earth faults. This is especially so in the case of LV switchboards with all insulated busbars. The SBEF relay is sensitive to the detection of earth faults well down into the LV transformer winding.

The 6.6 kV fuses provide effective protection against phase-to-phase and phase-to-earth faults occurring on the 6.6 kV HV side of the transformer. The fuses also give 'coarse' protection against faults on the LV side of the transformer. Striker pins should be specified on the fuses in an arrangement whereby they initiate a contactor trip of all three phases.

Careful co-ordination of the protection characteristics of the HV fuses, LV SBEF and LV switchboard outgoing circuit protection is necessary in order to ensure correct protection discrimination. In this example an extremely inverse IDMTL relay characteristic has been used to grade best with the HV transformer and LV switchboard fuse characteristics. Allowances must be made for the permissible tolerances in the operating characteristics of the fuses and relay and for the contactor tripping time.

The contactor should be specified as the latched type in order to ensure that it remains closed under conditions of system disturbances. The overall installation will therefore involve the specification of a suitable auxiliary DC supply (see Chapter 4).

The 0.38 kV switchboard has fuse switch units. MCBs could be specified after careful assessment of the required switching capability under short circuit conditions, their compatibility in the overall protection setting co-ordination, and whether they are required repeatedly to break short circuits without replacement (see IEC 60947-2). The larger outgoing circuits, other than motor circuits, must be provided with a means to detect earth faults and to switch off selectively.

13.6.4 Distribution ring main unit

Three phase low voltage (~400 V) supplies for domestic consumers are normally derived from distribution transformers (~100 to 1000 kVA rating) which are fed via medium voltage (12 or 24 kV) ring main circuits from primary substations. The supply to the high voltage side of the distribution transformer must be arranged:

- to give adequate protection;
- to cater for normal operational switching;
- to provide switching facilities to isolate faulty parts of the circuit and to allow normal maintenance, replacements, extensions and testing.

In order to meet these requirements it is usual to fit switches at each transformer tee-off point on the MV ring. Savings can be made by utilizing fault make/load break ring main switches rather than circuit breakers in these positions, although this option in not applicable in high security networks where each cable section has unit protection. The transformer may be protected by a switch fuse, circuit breaker or contactor depending upon the protection philosophy

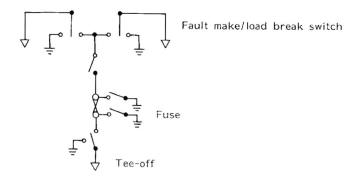

Fault make/load break switch

Fuse

Tee-off

Figure 13.23   Ring main unit schematic

adopted by the electricity supply utility. The arrangement is shown in Figs 13.23 and 13.24.

The fundamental requirements of the ring main unit are as follows:

1. ring main switches;
   - continuously carry ring full load current;
   - make and break ring full load current;
   - carry the full system fault current for a 1 or 3 s design criteria;
   - make onto a full system fault current.
2. Tee-off switches
   - continuously carry tee-off circuit full load current;
   - make and break full load current of the tee-off circuit (including magnetizing inrush currents, motor starting overloads, etc.);
   - make and break the full system fault current.
3. Environment
   - The ring main units must be designed to match the environmental and any special pollution requirements. Often in the UK the units are outdoor types. On the continent such units are more often enclosed in packaged substation housings.
4. Impulse levels
   - Normally an impulse level of 75 kV is sufficient. When connected to overhead line distribution circuits an impulse level of 95 kV may be specified depending upon impulse co-ordination design.
5. Insulation and earthing
   - air, oil, SF_6 or vacuum insulation may be adopted;
   - in practice it is necessary to ensure that if fuses are used for transformer tee-off protection they are easily accessible for replacement;
   - fuse access must only be possible after each side of the fuse has been isolated from the busbars and transformer;
   - in addition, an auxiliary earth must also be applied to both sides of the fuse in order to discharge any static which may have built up on the fuse connections before any maintenance takes place;

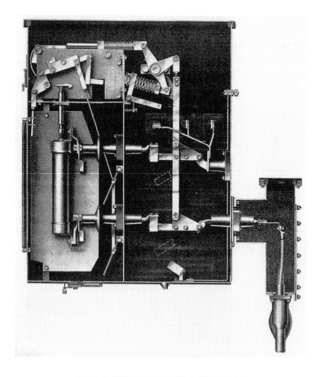

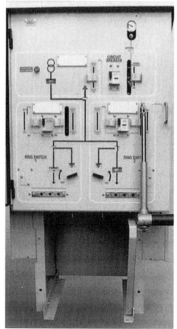

Figure 13.24   Type HFAU oil-filled switch fuse (*above*) and SF$_6$ ring main unit (*left*)

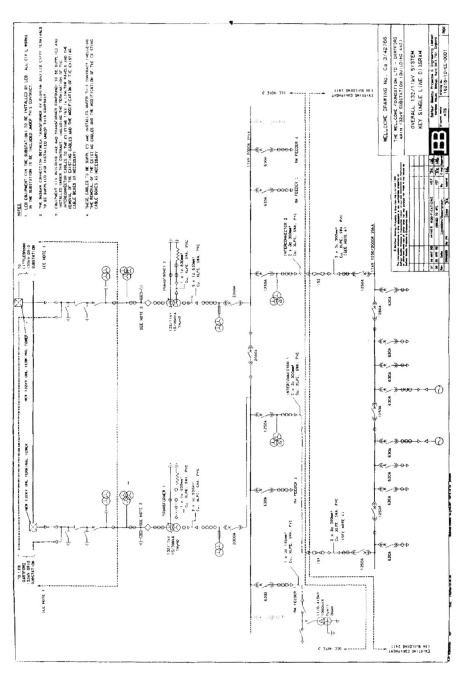

Figure 13.25   12 kV metal-clad indoor switchgear example – single line diagram

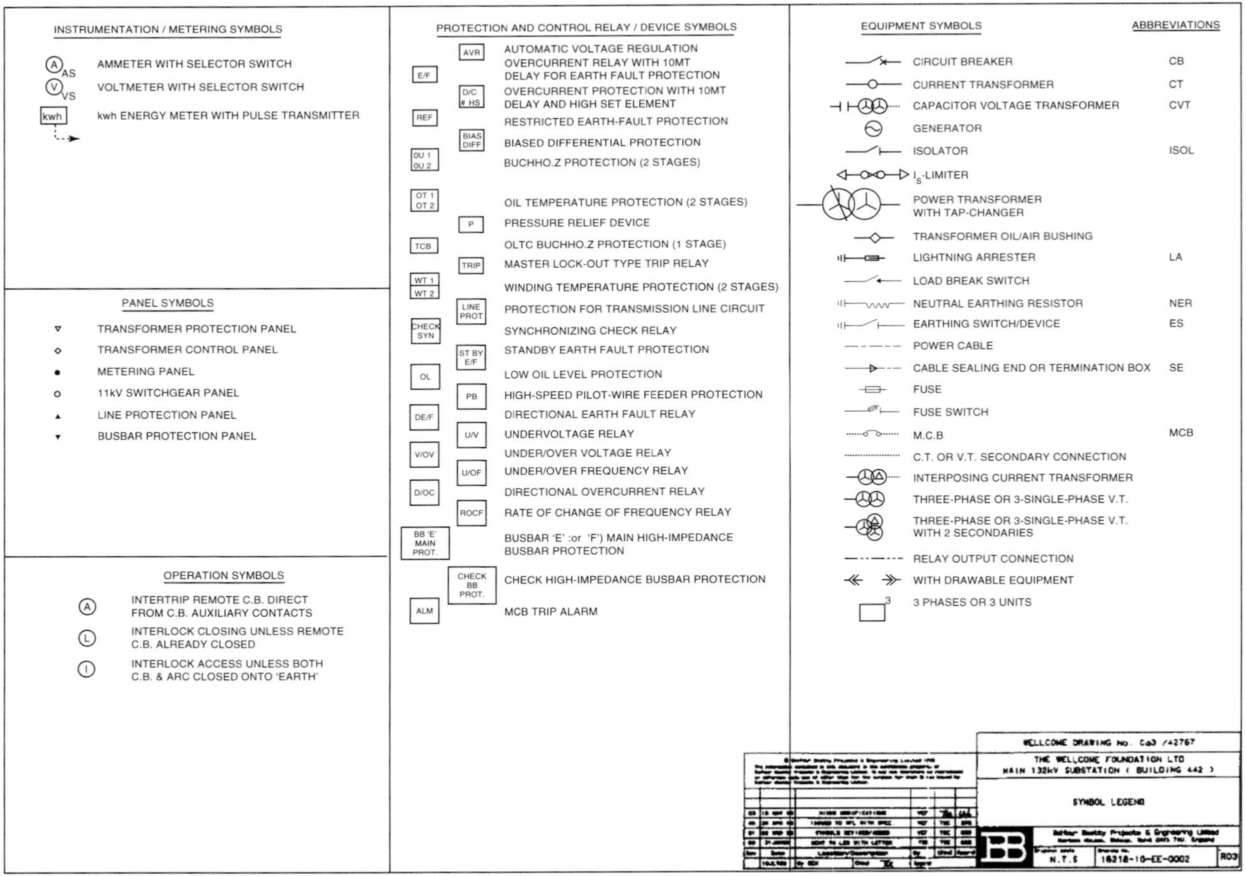

Figure 13.26   12 kV metal-clad interior switchgear example – symbols

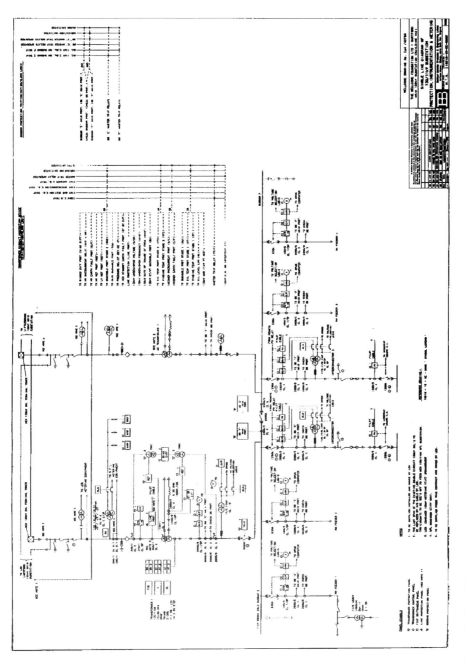

Figure 13.27   12 kV metal-clad indoor switchgear example – protection, metering and instrumentation

● fuse changing must be arranged to be quick and simple and capable of being performed in all weathers even if the ring main unit is of the outdoor type.

6. Test facilities and interlocks
   ● provision should be made for access to the ring main cable terminations for test purposes;
   ● suitable interlocks and labelling must be incorporated to prevent maloperation;

7. Extensibility
   ● When required and specified the designs should allow for future extensions to the busbars.

It should be noted, however, that such extensibility is not a normal feature and unless specified at the outset a separate RMU switchboard will be required for any extra switches needed in the future.

8. Maintenance
   ● For maintenance the whole unit has to be shut down in one go, including incoming cables. In contrast a circuit breaker switchboard can have its individual breakers maintained on a circuit by circuit basis.

Figure 13.24 shows cutaway views of a typical outdoor oil ring main and switch fuse unit of which many thousands have been installed throughout the world with a very high reliability record. More modern SF_6 insulated units follow the same layout principles. Heat shrink terminations would normally be used with XLPE cables.

REFERENCES

1. R. T. Lythall, *The J&P Switchgear Book*, Newnes-Butterworth.
2. C. H. Flurscheim (ed), *Power Circuit Breaker Theory*, The Institution of Electrical Engineers, Power and Energy Series No.1.
3. Ryan and Jones, *SF₆ Switchgear*, The Institution of Electrical Engineers, Power and Energy Series No.10.
4. A. Greenwood, *Vacuum Switchgear*, The Institution of Electrical Engineers, Power and Energy Series No.18.
5. E. Dullni, H. Fink and C. Reuber, *A Vacuum Circuit Breaker with Permanent Magnetic Actuator and Electronic Control*, CIRED 199.

14 Power Transformers

14.1 INTRODUCTION

Excellent text books are already available dealing with the design theory and operation of power transformers.[1] This chapter therefore concentrates on highlighting certain important aspects of:

1. Voltage selection – Calculation of transformer voltage ratio, specification of insulation levels, examples of voltage regulation, rating, tap ranges and impedance calculations.
2. Thermal aspects – Specification of temperature rise and ambient conditions. Some comments are made on constructional features of different types of transformer in common use together with the purpose and selection of accessories. A review of the relevant IEC Standards and summary of the parameters to be specified by the user when detailing a transformer for a particular application are given.

14.2 STANDARDS AND PRINCIPLES

14.2.1 Basic transformer action

The phasor diagram for a single phase transformer with a 1:1 turns ratio supplying an inductive load of power factor $\cos \theta_2$ is shown in Fig. 14.1. The transformer no-load current, I_0 consists of the physically inseparable magnetizing current and core loss components. The primary magnetizing current, I_m is lagging the primary induced emf, E_1 by 90°. The primary core loss component, I_c consists of hysteresis and eddy current components. The hysteresis loss is proportional to the frequency of operation and the peak flux density while the eddy current loss is a function of the frequency, rms flux density and the thinness of

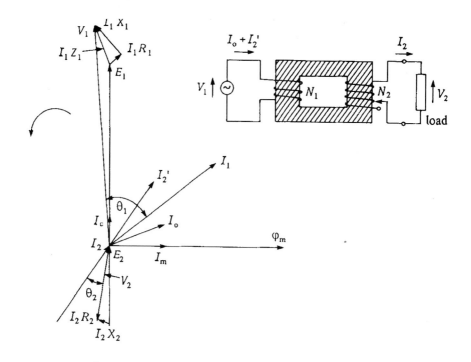

I_o = no-load primary current
I_m = primary magnetizing current
I_c = primary core loss current (iron loss component)
$I_1 R_1$ = primary resistance voltage drop
$I_1 X_1$ = primary reactance voltage drop
$I_1 Z_1$ = primary impedance voltage drop
I_1 = total primary current $(I_o + I_2')$
I_2 = secondary load current
I_2' = load component of total primary current (reflected secondary current)
φ_m = maximum peak value of magnetic flux (linkage and leakage flux)
V_1 = primary terminal voltage
E_1 = primary induced emf
V_2 = secondary terminal voltage
E_2 = secondary induced emf
$\cos \theta_1$ = primary total load power factor
$\cos \theta_2$ = secondary load power factor

- Voltage drops shown divided between primary and secondary sides
- I_2 is not in phase with I_1 because of primary no-load current, I_o
- $\theta_1 > \theta_2$ because of inductive nature of the transformer

Figure 14.1   Phasor diagram for single phase 1:1 turns ratio transformer supplying an inductive load of lagging power factor $\cos \theta_2$

the core laminations. Normally the magnetizing current is much larger than the core loss component and in power transformers the no-load current, I_0 is almost equal to I_m. Typically the no-load current, I_0, represents some 1.5% of full load current for small distribution transformers and may be less than 0.75% for large high voltage transformers. The no-load current is small because the primary links with its own magnetic field and electromagnetic theory explains that this will induce a back-emf to oppose the voltage applied externally to the coil. The open circuit transformer therefore acts as a highly inductive choke with a power factor of some 0.15 lagging.

Virtually the whole magnetic field created by the primary is attracted into the steel core and is encircled by the secondary winding. If the magnetic field is considered common to both primary and secondary transformer windings the actual field strength (in theory at least) becomes of no importance and only the four variables of voltage and coil winding turns remain giving the fundamental transformer expression:

$$V_1/V_2 \sim N_1/N_2 \tag{14.1}$$

Under load conditions the voltage induced in the secondary winding coil drives a current into the load. In addition, the secondary current also produces its own magnetic field which acts to oppose (and thus reduce) the original field in the steel core laminations. This in turn reduces the field in the primary and allows more current to flow until a turns balance is reached. The total primary and secondary load current, I_1 and I_2 produce equal and opposite magnetic fields in the core so the overall effect is to leave the magnetic field unchanged from what it was before the load was applied to the secondary coil.

This leads, at large load currents when the primary current, I_1, is much greater than the no-load current, I_0, to the second fundamental expression:

$$N_1 \cdot I_1 = N_2 \cdot I_2 \tag{14.2}$$

It should be noted that the magnetic flux levels in the core do not rise in proportion to the load current. The magnetic field due to the secondary is always balanced by that due to the primary current. The net magnetizing flux is due only to the magnetizing current and magnetic flux levels do not therefore reach very high levels under abnormal short circuit conditions.

Combining the two equations gives:

$$V_1 \cdot I_1 = V_2 \cdot I_2 \tag{14.3}$$

14.2.2 Transformer equivalent circuit

The transformer equivalent circuit shown in Fig. 14.2 is a fundamental basis for transformer calculations involving voltage drop or regulation under various load conditions (short circuit currents, tap settings, power factor, load currents, etc.).

The magnetizing circuit is taken as a shunt-connected impedance (inductance to represent the setting up of the magnetic field and resistance to represent heat losses in the core). As an approximation this equivalent circuit assumes the no-load current, I_0, to be sinusoidal and the core flux constant at all loads. In practice, the non-linear core material flux density/magnetizing force (B/H) curve, means that even for a sinusoidal-applied voltage a slightly distorted magnetizing

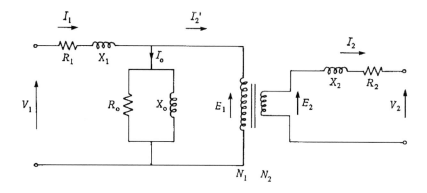

V_1 = primary terminal voltage
E_1 = primary induced emf (theoretical)
V_2 = secondary terminal voltage
E_2 = secondary induced emf (theoretical)

$$E_1/N_1 = E_2/N_2 \qquad\qquad V_1/N_1 \sim V_2/N_2$$

I_o = vector sum of primary magnetizing and core loss currents
I_1 = total primary current $(I_o + I_2')$
I_2 = secondary load current
I_2' = load component of total primary current (reflected secondary current)

X_o & R_o = magnetizing and core loss reactive and resistive components
X_1 & R_1 = primary winding reactive leakage and coil resistance
X_2 & R_2 = secondary winding reactance and resistance

N_1 = primary coil number of turns
N_2 = secondary coil number of turns

Figure 14.2   Transformer equivalent circuit

current results. The magnetizing current is rich in harmonics which must be kept
in check by keeping the flux density within specified limits. During transformer
energization a 'transient' current inrush rich in second harmonic will result. The
magnitude of this inrush depends upon the instant of switching and the residual core flux. Transients may be more than two times full load current with
significant decay over periods between 5 and 50 cycles depending upon transformer rating. This effect can be detected by transformer protection relays in a
manner whereby the presence of the second harmonic component is used as a
restraint feature. In this way the relay can be used to differentiate between a true
fault and inrush current and avoid anomalous tripping.

The two resistances, R_1 and R_2, represent the ohmic resistances of the
primary and secondary windings. The two inductances, X_1 and X_2, which are
not independent, represent the leakage reactance in a realistic transformer. In
practice, not all the magnetic field of the primary is linked with the secondary
coil. Leakage results in a slightly lower secondary voltage than the simple turns
ratio theory predicts and the greater the load current the greater the deviation

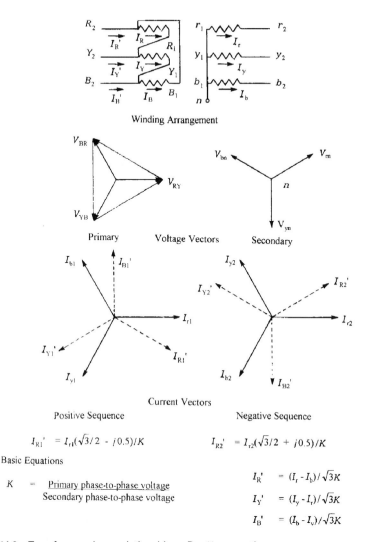

$$I_{R1}' = I_{r1}(\sqrt{3}/2 - j0.5)/K \qquad I_{R2}' = I_{r2}(\sqrt{3}/2 + j0.5)/K$$

Basic Equations

$$K = \frac{\text{Primary phase-to-phase voltage}}{\text{Secondary phase-to-phase voltage}}$$

$$I_R' = (I_r - I_b)/\sqrt{3}K$$

$$I_Y' = (I_y - I_r)/\sqrt{3}K$$

$$I_B' = (I_b - I_y)/\sqrt{3}K$$

Figure 14.3   Transformer phase relationships – Dyn11 connections

from the ideal. In addition, further losses occur from magnetostriction whereby the physical dimensions of the core laminations change by a few parts in a million in a complex pattern with the flux cycle. This in turn causes hum at audible even harmonics of the supply frequency.

14.2.3 Voltage and current distribution

Vector representation of voltage and currents in transformer windings allows the practising engineer to visualize the relationships involved. These relationships are shown for Dy11 and Yd11 vector group transformer connections in Figs 14.3 and 14.4, respectively. The length of the vector is made equal to the maximum or

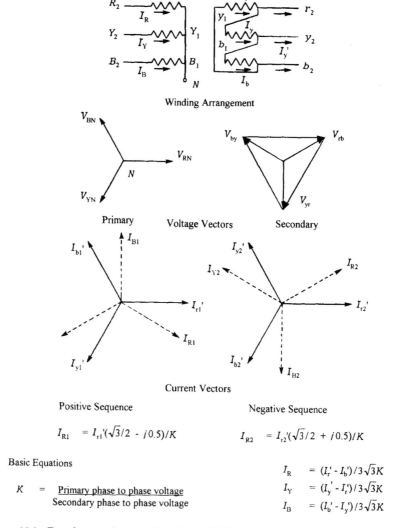

Figure 14.4   Transformer phase relationships –YNd11 connections

rms value of voltage or current. The convention is for the arrow heads to point away from the source of generation towards the load.

14.2.4 Transformer impedance representation

The systems engineer, as opposed to the transformer designer, is chiefly concerned with the representation and characteristics of a given power transformer in the transmission and distribution network. In fault and load flow studies transformers are represented in the network diagrams by their equivalent

Rating (MVA)	Minimum percentage impedance (%)
Up to 0.630	4.0
0.631 to 1.25	5.0
1.251 to 3.15	6.25
3.151 to 6.30	7.15
6.301 to 12.50	8.35
12.501 to 25.00	10.0
25.001 to 200.00	12.5

Notes: (a) Clause 4.3 of IEC 60076-1 gives preferred transformer ratings as being in the R10 series of ISO 3(1973). That is, integral multiples of 100, 125, 160, 200, 315, 400, 500, 630, 800, 1000 …, but only up to 10 MVA
(b) Impedance ranges for higher voltage transformer units are typically 150 kV: 12% to 15% and 275 kV, 15% to 20%
(c) Base is transformer rating at nominal tap. Tappings may cause variations of approximately ± 10%
$\%Z_{\text{auto-transformer}} = \%Z$ for equivalent two winding MVA × (HV – LV)/HV
where $\%Z$ for equivalent two winding MVA = Auto-transformer MVA × (HV – LV)/HV

Figure 14.5   Typical two winding transformer ratings and minimum positive or negative percentage impedances (Z_1 and Z_2) together with derivation for auto-transformer impedances

impedances. The positive and negative sequence impedances of two winding power transformers are equal and equivalent to the ordinary leakage impedance used in three phase calculations. Transformer impedance is usually expressed as a percentage reactance (a transformer is highly inductive) on the base of the transformer rating. For transformers with dual or triple ratings (see paragraph 14.4.6 for cooling codes, for example ONAN/ONAF or ONAN/ONAF/OFAF) the correct rating base and tap position must be clearly detailed when specifying the transformer impedance required. Typical values are given in Fig. 14.5.

There is a move in city centre primary substation design towards direct conversion between the highest and lowest distribution voltages (say, 132 kV to 11 kV) rather than via an intermediate voltage level (e.g. 132 kV to 66 kV to 11 kV). In such cases the transformer impedance must be carefully specified to limit the secondary fault level and still maintain good voltage regulation as described later in this chapter. Three winding transformers also have equivalent positive and negative sequence impedances and may be represented in an impedance network by three, rather than a single, impedances as shown in Fig. 14.6. Typical auto-transformer/two winding transformer impedances may be estimated for rough fault calculation purposes as shown in Fig. 14.5 if actual rating plate data is unavailable.

Transformer zero sequence impedances will vary over a wide range depending upon the winding vector grouping and neutral point earthing of both transformer and/or source generators within the system. As explained in Section 14.2.1, a turns balance is normally produced within the transformer windings. However, under fault conditions the zero sequence impedance is a result of the extent to which the configuration allows the zero sequence current in one winding to be balanced by equivalent ampere turns in another winding. The zero sequence impedance of star/star transformers is also dependent upon the

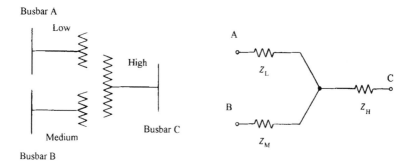

Primary windings represented by Low and Medium voltage designations
Secondary winding represented by High voltage designation.

The required positive or negative sequence impedance diagram is the three terminal
arrangement where:

$$Z_L = (Z_{LM} + Z_{LH} - Z_{MH})/2$$

$$Z_M = (Z_{LM} + Z_{MH} - Z_{LH})/2$$

$$Z_H = (Z_{LH} + Z_{MH} - Z_{LM})/2$$

Z_{LM} = impedance of L & M with H open circuited
Z_{LH} = impedance of L & H with M open circuited
Z_{MH} = impedance of M & H with L open circuited

Impedances are usually quoted in terms of Z_{LM}, etc. and must be converted to a
common voltage base (if quoted in ohms) or a common MVA base (if quoted in % or
pu).

Figure 14.6   Impedance representation of three winding transformers

core configuration. Several examples of the distribution of zero sequence currents (represented by arrows with no vector significance but indicating the
magnitude by the number of arrows involved with a given phase) in typical
transformer windings for different typical vector groupings are shown in Fig.
14.7 together with approximations of zero sequence impedance magnitudes.

14.2.5 Tap changers

14.2.5.1 Introduction

The addition of extra turns to the secondary winding as shown in Fig. 14.1
allows a change in output voltage from, say, V_2 to $V_2 + \Delta V$ as the primary to
secondary turns ratio is decreased. In transmission systems the control of the

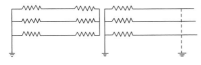

(a) Starstar (YNyn) transformer vector grouping with primary and secondary star points together with source generator solidly earthed.

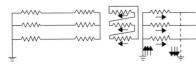

(b) Starstar (Yyn) transformer vector grouping with secondary star point and source generator solidly earthed.

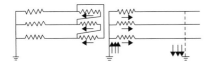

(c) Starstar with delta tertiary transformer vector grouping with secondary star point and source generator solidly earthed.

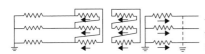

(d) Deltastar (Dyn) transformer vector grouping with secondary star point and source generator solidly earthed.

(c) Deltastar with delta tertiary transformer vector grouping with secondary star point and source generator solidly earthed.

$$Z_0 = Z_1 = Z_2$$

Primary and secondary ampere-turns balance. The transforme primary star point and the source generator are solidly earthed such that zero sequence currents arising from a fault on the secondary side of the transformer may flow in the primary circuit. The overall zero sequence impedance is that of the transformer and generator transferred to the same MVA or voltage base.

$$Z_0 \gg Z_1$$

Without primary star point earthing there is no path for zero sequence currents on the primary side of the transformer. Therefore zero sequence fault currents on the secondary side are relatively small. The transformer connection approximates to an open circuit for Zero sequence components. The actual value for Z_0 depends upon the transformer magnetic circuit arising from three or five limb construction.

$$Z_0 \sim Z_1$$

The delta tertiary winding allows a 'trapped' circuiting flow of zero sequence currents arising from a secondary fault. There is no zero sequence return circuit path back to the generator. Therefore the transformer primary winding and generator do not carry zero sequence currents. Z_0 is low but depends upon leakage flux of secondary and tertiary windings. Z_1 depends upon leakage flux of primary and secondary windings. Z_0 is therefore of same order of magnitude as Z_1.

$$Z_0 = Z_1 = Z_2$$

The zero sequence secondary fault current is induced in the delta primary windings. The primary and secondary zero sequence ampere-turns are balanced and Z_0 equals the leakage impedance. No zero sequence currents flow in the generator as there is no earth return circuit on the primary side of the transformer. They circulate and are 'trapped' in the delta primary.

$$Z_0 < Z_1$$

The zero sequence secondary fault current is induced in the delta primary and tertiary windings. Z_0 is low and normally less than the leakage impedance.

Figure 14.7   Zero sequence impedance approximations for different transformer vector groups and different system earthing configurations

voltage may be achieved by varying the transformer ratios or the effective number of turns in service by using taps. There is a practical limit to the number of separate winding tap positions that can be accommodated arising from the physical size of the tap changer required and tapping winding insulation between adjacent steps. Transformer voltage control is therefore characteristically by means of small step changes in voltage. Tap changers may be motor driven or manually operated via a switch. Alternatively, the change in turns ratio may involve physically and manually changing tapping connections. Such arrangements may be found on the smaller distribution dry type transformers. Tap changer switches may be mounted separately on the side of the tank with their own separate oil insulation. This is intended to allow for easier maintenance. Alternatively, the tap changer may be mounted in the main transformer tank in order to reduce costs and result in a compact transformer design.

14.2.5.2 Tap changer types and arrangements

Tap changers may be:

1. Off-circuit – The tap change may only be carried out when the transformer is not energized. Off-circuit tap changers are usually relatively simple switches mounted close to the winding tappings. The switches are under oil and are designed to change position only when the transformer is de-energized. There is consequently no breaking of current flow. The tap changer is operated by a handle, or wheel, from the outside of the tank in most transformers.
2. Off-load – The tap changer may be operated when the circuit is energized but not when the circuit is drawing load current.
3. On-load – The tap changer may be operated under load conditions. An on-load tap changer has a much more difficult duty than the off-circuit type. An international survey on failures in large power transformers (CIGRE Working Group Study Committee 12, Electra, Jan. 1983, No. 88, pp. 21–48) showed that tap changers were the source of some 40% of transformer faults. As the name implies, the on-load tap changer may change tapping position with transformer load current flowing.

On-load tap changer selection is best completed in conjunction with the manufacturer unless some standardization policy by the electrical supply utility dictates otherwise.

On-load tap changer manufacturer's requirements are:

1. General
   - Reliability
   - Minimal maintenance
   - Lowest cost
   - Electrical supply utility preferences.
2. Technical
   - Dielectric strength
   - Overload and fault current capability
   - Breaking capacity
   - Electrical and mechanical life expectancy
   - Service and processing pressure withstand capability.

There are three basic tapping arrangements (see Fig. 14.8) and each have their own advantages and disadvantages depending upon the application. In addition,

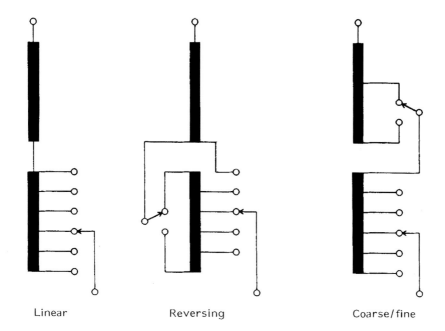

Figure 14.8   Basic arrangements of tapped windings

the connection point for the taps depends upon whether the transformer is a double wound unit or an auto-transformer.

The *linear arrangement* is generally applied for smaller tapping ranges and results in a relatively simple tap changer. It is restricted to smaller ranges because of difficulties which can arise from bringing out a large number of tapping leads from the winding and also owing to impulse voltages being developed across a large number of tapping turns.

For larger tapping ranges the *reversing arrangement* can be used. The changeover selector allows the taps to be added or subtracted from the main winding, effectively halving the number of connections and giving a larger tapping range from a smaller tap winding. A disadvantage of the reversing arrangement is that on the position with the minimum number of effective turns the total tapping winding is in circuit resulting in higher copper losses in the transformer.

A *coarse/fine arrangement* incorporates some of the advantages of the reversing arrangement but exhibits lower copper losses on the minimum tap position. The main disadvantage of the coarse/fine scheme is the cost of providing separate coarse and fine tapping windings.

14.2.5.3 *Electrical connections to the main winding*

For double wound transformers (see Fig. 14.9) the tappings may be applied to either star- or delta-connected windings. The most common connection is at

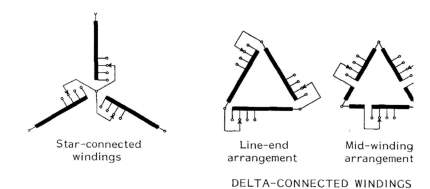

| | DELTA-CONNECTED WINDINGS | |
| Star-connected windings | Line-end arrangement | Mid-winding arrangement |

Figure 14.9   OLTC arrangements in double wound transformers

the neutral end of star-connected HV windings. This results in the most economical tap winding and allows a low voltage class, three phase tap changer to be used. When tappings are applied to delta-connected windings the lack of a neutral leaves the choice of connecting the tappings at either the line end or in the middle of the main winding. The line-end connection requires the tap changer to be fully insulated from the system voltage. The mid-winding connection can be used to reduce the dielectric stresses but other parameters such as transformer impedance, winding insulation level and economic considerations will affect the final choice.

When tappings are applied to auto-transformers the choice of connection is even more involved. From a dielectric viewpoint the ideal position for the tappings is at the neutral end. As with double wound transformers this has the advantage of a smaller, lower cost tap changer. In core form transformers the physical disposition of the windings greatly affects the transformer impedance. The neutral tap changer position therefore results in a possibly advantageous, low impedance variation over the tapping range. However, since the operation of auto-transformers differs from double wound types this advantage may be outweighed by other considerations as shown in Fig. 14.10. The main disadvantage of the neutral end tapping connection arises from the fact that reducing the turns of the LV circuit also reduces turns in the HV circuit and therefore a larger tapping winding is required for a given voltage change. For auto-transformers with high ratios this is not a great disadvantage and the neutral end connection can usually prove the most economic solution for auto-transformer ratios above 2.5/1. The disadvantage becomes more pronounced on transformers with low ratios and large tapping ranges. This is one of the reasons why the majority of the 400/132 kV auto-transformers on the UK National Grid system have neutral end tap changers whereas the line-end tap changer is much more common for 275/132 kV units.

Another effect of varying the turns in the LV and HV auto-transformer circuits simultaneously is that the volts per turn, and therefore the flux density of the transformer core, varies over the tapping range. Consequently, the voltage

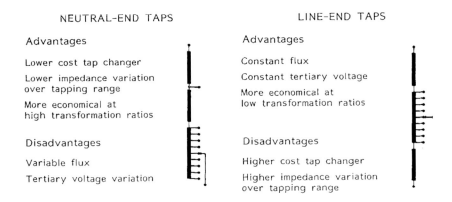

NEUTRAL-END TAPS	LINE-END TAPS

NEUTRAL-END TAPS

Advantages

Lower cost tap changer

Lower impedance variation
over tapping range

More economical at
high transformation ratios

Disadvantages

Variable flux

Tertiary voltage variation

LINE-END TAPS

Advantages

Constant flux

Constant tertiary voltage

More economical at
low transformation ratios

Disadvantages

Higher cost tap changer

Higher impedance variation
over tapping range

Figure 14.10   Auto-transformer tapping arrangements

induced in any auxiliary winding such as a tertiary will vary with tap position. The line-end connection has the advantage of constant flux density and therefore constant tertiary voltage over the tapping range. Also, the change in turns ratio is achieved in a more cost effective manner particularly for lower transformer ratios. The main disadvantages are the higher cost of the line-end tap changer and the higher impedance variation over the tapping range resulting from the preferred disposition of the windings.

14.2.5.4 *Dielectric stresses*

Figure 14.11 illustrates the critical stresses of a three phase neutral-end coarse-fine tap changer. The most onerous stresses appear during the transformer dielectric tests and their magnitude depends upon the transformer design parameters and the tapping position during the tests. The magnitude and precise location of maximum stresses is determined at the design stage through numerical simulation.[5]

14.2.5.5 *Tap changer duty*

Figure 14.12 illustrates the switching sequence of the tap changer selector and diverter switch. The purpose is to transfer connection from the selected tap to a preselected adjacent tapping without interrupting the power supply to the load.

During the short time that the transfer switch is in transit between contacts M1 and M2 the load is carried by a transition impedance. With the exception of some units in North America (which use reactors) the transition impedance is nowadays normally a resistor. The transition resistors (RT) are designed according to the step voltage and rated through current, and the very fast transfer time in the order of tens of milliseconds means that the transition resistors need only be short time rated. The main contacts M1 and M2 are called upon

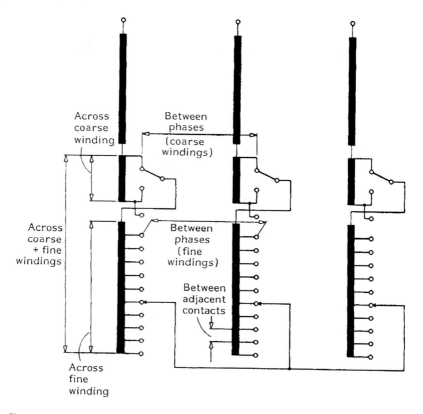

Figure 14.11   On-load tap changer critical stresses

to carry the full load current continuously. The diverter switch contacts T1 and T2 must be capable of sustaining arc erosion and mechanical duty resulting from making and breaking full load current. The arcing of these contacts produces gases which saturate the adjacent oil and a barrier must be provided to separate this oil from the main transformer oil.

14.2.5.6 *In-tank tap changers*

Figure 14.13 illustrates the physical arrangement of the in-tank tap changer. The leads from the tapping winding are connected to the selector contacts within the main transformer oil. The diverter switches are enclosed in an oil-filled insulating cylinder which is piped to its own conservator. The oil contact with the diverter switch is therefore isolated ensuring that degradation products resulting from the switching process do not contaminate the transformer oil.

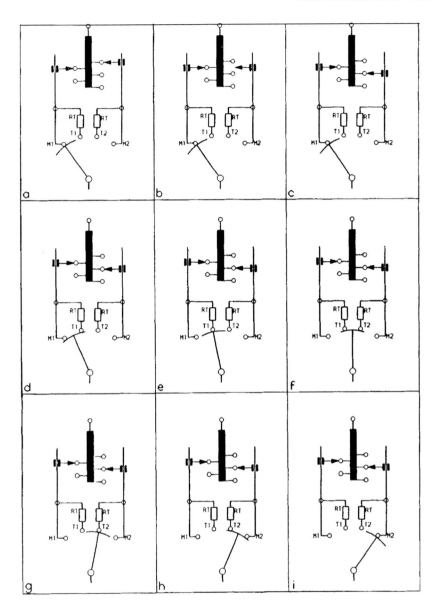

Figure 14.12   Operating sequence of tap selector and divertor switch

Maintenance is confined to the diverter compartment and the selector contacts are considered maintenance free. Access to the selector contacts and separation of the selector oil from the transformer oil to increase the selectivity of dissolved gas-in-oil analysis may be specified. In these cases it is necessary either to separate the in-tank tap changer from the transformer by a barrier board as

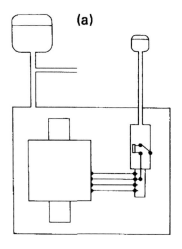
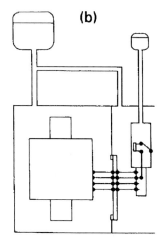

(a)

(b)

In-Tank OLTC
Separate Diverter Oil
Common Selector Oil

Barrier Board OLTC
Separate Diverter Oil
Separate Selector Oil

Figure 14.13   In-tank tap changers

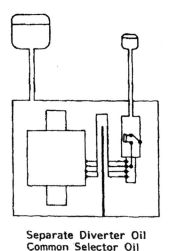
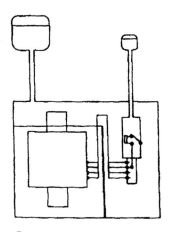

Separate Diverter Oil
Common Selector Oil

Retains oil over windings
when switch is drained

Figure 14.14   Weir-type tap changer

shown in Fig. 14.13b or alternatively to supply a separate bolt-on tap changer. If access to the selector contacts without dropping the transformer oil level below the level of the windings is specified then a weir may be fitted inside the transformer as shown in Fig. 14.14. The tap changer oil may then be drained independently of the main transformer oil.

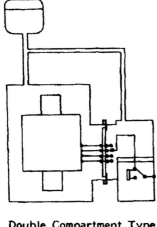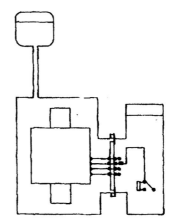

**Double Compartment Type
Separate Selector Oil
Separate Diverter Oil**

**Single Compartment Type
Common Selector & Diverter Oil**

Figure 14.15   Bolt-on tap changers

14.2.5.7 *Bolt-on tap changers*

The two main types of bolt-on tap changers are shown in Fig. 14.15.

The *double-compartment* type separates the selector contacts from the diverter switch forming two main compartments. This system allows the tap changer manufacturer to separate the mechanical drives to the selector and diverter mechanisms. The diverter may be operated by a spring-loaded device at switching speed and the selector can be driven directly from the output shaft of the motor drive mechanism at slower speeds. This is the traditionally preferred type for larger transformers built in the UK.

The *single-compartment* bolt-on tap changer utilizes selector switches which combine the function of selection and transfer in one mechanical device. The fact that arcing products are in contact with insulation subjected to high voltages limits the application of single-compartment tap changers to the lower ratings.

14.2.6 Useful standards

The principle reference for power transformers is IEC 60076. A number of other standards are also relevant, but work is in hand to supersede many of them by sections of 60076, thus making one easier reference. Reference to the IEC index, accessible via the internet, will enable the current status to be checked as work progresses. See also Table 14.1.

Table 14.1   Useful standards

Standard	Description	Notes
IEC 60076	Power transformers	
	Part 1 – General	Covers scope, service conditions, definitions
	Part 2 – Temperature rise	
	Part 3 – Insulation levels, dielectric tests and external clearances in air	
	Part 4 – Guide to the lightning impulse and switching impulse testing of power transformers and reactors	Supplements part 3, and supersedes IEC 60722
	Part 5 – Ability to withstand short circuit	
	Part 6	Work in progress to supersede IEC 60289
	Part 7	Work in progress to supersede IEC 60354
	Part 8 – Application guide for power transformers	Assists purchaser in power transformer selection. Supersedes IEC 60606
	Part 10 – Measurement of transformer and reactor sound levels	Supersedes IEC 60551
	Part 11 – Dry type power transformers	For ratings up to and including 36 kV. Supersedes IEC 60726
	Part 12	Work in progress to supersede IEC 60905
IEC 60085	Electrical insulation – thermal evaluation	Considers the thermal evaluation of insulation materials and of insulation systems, their interrelationship and influence of service conditions. Relation to stability in service
IEC 60137	Bushings for alternating voltages above 100 V	
IEC 60156	Insulating liquids – determination of the breakdown voltage at power frequency – test method.	Describes conventional tests to reveal the extent of physical pollution by water and other suspended matter, and the advisability of carrying out drying and filtration treatment before introduction into apparatus
IEC 60214	Tap changers	Applies to on-load resistor and reactor types
	Part 1 – Performance requirements and test methods	The majority of tap changer faults are of mechanical origin, so the mechanical endurance test is important, as is the service test duty (a measure of contact wear). Dielectric strength, overload and fault current capability tests are also covered. A routine pressure and vacuum test for oil-filled compartments is included but withstand levels are not specified and left to the tap changer manufacturer

(Continued)

Table 14.1   Continued

Standard	Description	Notes
	Part 2 – Application guide	Supersedes IEC 60542, and to be read in conjunction with IEC 60076. Provides useful field service advice
IEC 60270	High voltage test techniques – partial discharge methods	
IEC 60289	Reactors	Includes coverage of shunt, current limiting and neutral earthing reactors. Will eventually be superseded by future 60076-6
IEC 60296	Specification for unused mineral insulating oils for transformers and switchgear	
IEC 60354	Loading guide for oil-immersed power transformers	Provides recommendations for specifications and loading of power transformers complying with IEC 60076 from the point of view of operating temperatures and thermal ageing. Will eventually be superseded by future 60076-7
IEC 60599	Mineral oil-impregnated electrical equipment in service – guide to the interpretation of dissolved and free gases analysis	
IEC 60616	Terminal and tapping markings for power transformers	
IEC 60905	Loading guide to dry type power transformers	Permits calculation of loadings for naturally cooled types complying with IEC 60076-11 in terms of rated current. Will eventually be superseded by future 60076-12

14.3 VOLTAGE, IMPEDANCE AND POWER RATING

14.3.1 General

The correct specification of transformer voltages, impedance(s) and kVA rating(s) are described in this section.

14.3.2 Voltage drop

As shown in Fig. 14.2 there is an internal voltage drop in a transformer under secondary load conditions. The volt drop is due to the leakage reactance and the winding resistance. Rather than express the impedance in ohms per phase the normal convention with transformers is to express the impedance as a percentage value referred to the kVA (or MVA) rating of the transformer.

The change in transformer terminal voltage from no load to full load is the *regulation* of the transformer. This change corresponds with the volt drop appearing at full load. Several formulae are available to calculate volt drop, the more accuracy required the more complex the formula. The following is adequate for most purposes:

$$\Delta U = [(R \cdot p)^2 + (X \cdot q)^2]^{1/2} \div 100\%$$

where X = leakage reactance (%)
    R = winding resistance (%)
    p = power factor, cos ϕ (in %)
    q = sin ϕ (in %)
    ΔU = % volt drop at full load

For example, a transformer with a leakage reactance of 10%, a resistance of 0.5% and supplying a load at 0.85 (85%) power factor will have the following full load volt drop:

$$\Delta U = [(0.5 \cdot 85)^2 + (10 \cdot 53)^2]^{1/2} \div 100\%$$
$$\sim 5.3\%$$

Notice that the formula includes the winding resistance which is small compared with the leakage reactance but may be included to retain accuracy.

14.3.3 Impedance

The short circuit impedance or internal impedance is a main parameter for a transformer. Extreme values are limited by design factors; the lowest value by the minimum physical distance between windings, the highest by the effects of the associated high leakage flux.

For any given rating and voltage the size and weight of a transformer are functions of its percentage reactance. A small percentage reactance means a large main flux requiring larger iron cross-section. As reactance is increased the iron cross-section decreases, iron loss decreases but copper loss increases. The ratio of copper loss to iron loss is appreciably increased and the total loss increased slightly. High reactance has the disadvantage of a large voltage drop (requiring a large tapping range to compensate and maintain secondary voltage) and a large amount of reactive power consumed within the transformer itself. For larger transformer ratings a high reactance may, however, be considered desirable because it limits the short circuit current and therefore maintains the rating of associated system switchgear. Some compromise must be arrived at between these conflicting requirements and minimum values are shown in Fig. 14.5.

For three phase systems the zero sequence impedance of the transformer is also of importance since it determines the magnitude of fault currents flowing between the neutral of a star-connected winding and earth during phase-to-earth faults. The transformer zero sequence impedance is dependent upon the core configuration (3 or 5 limb for core type transformers) and whether or not a delta-connected auxiliary winding is fitted (refer to Fig. 14.7 and Section 14.5.2.3).

14.3.4 Voltage ratio and tappings – general

A transformer intended to connect, for example, a 132 kV system to a 20 kV system may, at first sight, simply require a voltage ratio of 132/20 kV. In practice, this may not be the most appropriate ratio to specify to the manufacturer since the following aspects need to be taken into account:

1. The 132 kV system voltage is not constant and may vary as much as ±10% from the nominal value.
2. Volt drop on load will depress the voltage at the 20 kV terminals.

To accommodate these effects virtually every practical transformer will need tappings to allow selection of different voltage ratios to suit different circumstances. In some situations, where the transformer regulation and the primary voltage variations are small, a change from one tapping to another would be very infrequent, if ever, in the transformer life. In such cases 'off-circuit' or 'off-load' tappings are adequate.

In the majority of transmission system applications, system voltage control is achieved by changing transformer taps and an 'on-load' tap changer facility is needed for frequent changes in tapping without removing the transformer from service.

14.3.5 Voltage ratio with off-circuit tappings

In domestic and industrial distribution systems, transformers stepping down from 11 kV to 3.3 kV or 0.415 kV will normally be satisfactory without on-load tap changers. Such transformers will usually have impedances of around 4% to 6% giving a full load volt drop at 0.85 pf of 3% or 4%. In many cases the primary voltage will be fairly well controlled to, say, ±3% of the nominal value. Combining the primary voltage variation effect with the transformer regulation effect gives an overall reasonably satisfactory 9% to 10% voltage variation on the secondary terminals. The voltage ratio is usually chosen to give approximately nominal secondary voltage at full load. Thus a ratio of 11 kV to 433 V is commonly chosen to feed a 415 V system. Distribution transformer off-circuit tappings giving −5.0%, −2.5%, 0%, +2.5% and +5.0% variation in ratio are conventionally specified and will be adequate for the majority of situations. The middle tap of a transformer is referred to as the 'principal tap'.

The role of the off-circuit tap changer is then to match the transformer to the circumstances of the installation. For example, an 11 kV/433 V transformer close to a main 33/11 kV feeder substation may see an 11 kV voltage level biased towards the high side – for example, 11.3 kV ± 2%. On the other hand, a remote 11 kV/433 V transformer may see an 11 kV voltage biased towards the low side – for example, 10.8 kV ± 3%. In the former case the +2.5% tapping would be used (giving a ratio 11.28 kV/433 V), and in the latter the −2.5% tapping would be used (giving a ratio 10.7 kV/433 V). For standardization, both transformers use the same specification; only the tapping is changed for the particular service condition.

14.3.6 Voltage ratio and on-load tappings

The procedure for specifying voltage ratio and tapping range for on-load tap changes is quite involved and often causes problems. Section 14.3.9 gives an example of some of the factors to be considered.

IEC 60076-1 defines three categories of voltage variation for transformers with tappings. They are defined as follows:

- Constant flux voltage variation (CFVV) – The tapping voltage in any untapped winding is constant from tapping to tapping. The tapping voltages in the tapped winding are proportional to the tapping factor.
- Variable flux voltage variation (VFVV) – The tapping voltage in the tapped winding is constant from tapping to tapping. The tapping voltages in any untapped winding are inversely proportional to the tapping factor.
- Combined voltage variation (CbVV) – The transformer is specified using both principles, but applied to different parts of the range. This approach is particularly used in instances of a large tapping range.

Particular care must be taken in attempting to specify the impedance variation with tapping, as this can restrict design. Refer to IEC 60076-1 for extensive guidance.

14.3.7 Basic insulation levels (BIL)

The amount of insulation applied to the winding conductors is usually influenced by the impulse voltage rating of the winding rather than by the power frequency voltage rating. Impulse voltages due to lightning or switching activity appearing at the terminals of the transformer stress the winding insulation and this effect may be reduced by the application of surge arresters. The factors involved in correct specification of the transformer basic insulation level are explained in Chapter 9. See also IEC 60076-3 which gives particular guidance on insulation levels for transformers.

14.3.8 Vector groups and neutral earthing

Three phase windings of transformers will normally be connected in a delta configuration, a star (wye) configuration, or, less commonly, in an interconnected star (zig-zag) configuration as shown in Fig. 14.16. The vector grouping and phase relationship nomenclature used is as follows:

- Capital letters for primary winding vector group designation.
- Small letters for secondary winding group designation.
- D or d represents a primary or secondary delta winding.
- Y or y represents a primary or secondary star winding.
- Z or z represents a primary or secondary interconnected star winding.
- N or n indicates primary or secondary winding with an earth connection to the star point.
- Numbers represent the phase relationship between the primary and secondary windings. The secondary to primary voltage displacement angles are given in accordance with the position of the 'hands' on a clock relative to the mid-day or twelve o'clock position. Thus 1 (representing one o'clock) is $-30°$, 3 is $-90°$, 11 is $+30°$ and so on.

Therefore a Dy1 vector grouping indicates that the secondary red phase star voltage vector, V_{rn}, is at the one o'clock position and therefore lags the primary red phase delta voltage vector, V_m, at the twelve o'clock position by $30°$, i.e. the one o'clock position is $30°$ lagging the primary twelve o'clock position for conventional anti-clockwise vector rotation.

Similarly a Dyn11 vector grouping indicates that the secondary red phase voltage leads the primary voltage by $30°$, i.e. the eleven o'clock position leads the twelve o'clock position by $30°$. The secondary star point is earthed.

Yy0 would indicate $0°$ phase displacement between the primary and secondary red phases on a star/star transformer.

Dz6 would indicate a delta primary interconnected star secondary and $180°$ secondary-to-primary voltage vector phase displacement.

The system designer will usually have to decide which vector grouping arrangement is required for each voltage level in the network. There are many factors influencing the choice and good summaries of the factors of most interest to the manufacturer can be found in Ref. (1). From the user's point of view, the following aspects will be important:

1. Vector displacement between the systems connected to each winding of the transformer and ability to achieve parallel operation.
2. Provision of a neutral earth point or points, where the neutral is referred to earth either directly or through an impedance. Transformers are used to give the neutral point in the majority of systems. Clearly in Fig. 14.16 only the star or interconnected star (Z) winding configurations give a neutral location. If for various reasons, only delta windings are used at a particular voltage level on a particular system, a neutral point can still be provided by a purpose-made transformer called a 'neutral earthing transformer' or 'earthing compensator transformer' as shown in Fig. 14.16 and also as described in Chapter 4.

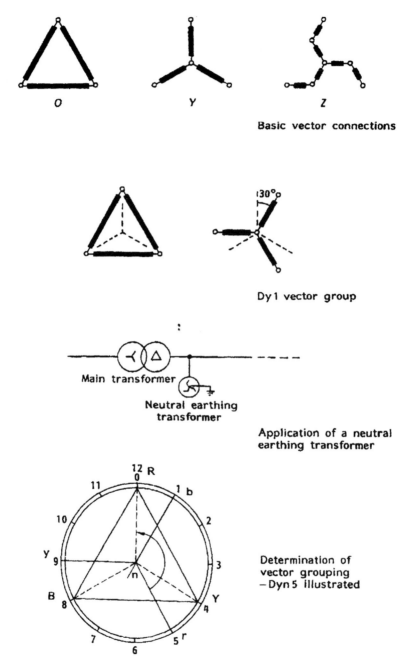

Basic vector connections

Dy1 vector group

Application of a neutral
earthing transformer

Determination of
vector grouping
—Dyn5 illustrated

Figure 14.16   Winding arrangements

3. Practicality of transformer design and cost associated with insulation requirements. There may be some manufacturing difficulties with choosing certain winding configurations at certain voltage levels. For example, the interconnected star configuration is bulky and expensive above about 33 kV. Of considerable significance in transmission systems is the cost and location of the tap changer switchgear as explained in Section 14.2.5.
4. The Z winding reduces voltage unbalance in systems where the load is not equally distributed between phases, and permits neutral current loading with inherently low zero-sequence impedance. It is therefore often used for earthing transformers.

14.3.9 Calculation example to determine impedance and tap range

14.3.9.1 *Assumptions and data*

It is required to calculate the impedance and tap changer range for a star/star auxiliaries transformer.

- The voltage variation on the primary side is 132 kV ± 10%.
- The maximum allowable voltage variation on the secondary side is 21 kV − 0%, +5%.
- The maximum allowable 3.3 kV voltage is 3.54 kV (+7.5%).
- The maximum transformer load is anticipated to be initially 31.9 MVA at 0.9 pf and increased to an ultimate future figure of 38.3 MVA at 0.9 pf.
- The maximum allowable secondary 21 kV side fault current is 12.5 kA.
- Maximum primary side 132 kV source fault level = 2015 MVA.

14.3.9.2 *Rating calculation*

The HV principal tapping voltage is 132 kV.

The LV no-load voltage at principal tap is chosen as 22.05 kV. This voltage should be adequately high to cater for the on-load voltage drop and also adequately low to avoid over-voltage problems under specific load rejection conditions.

The initial LV maximum current $= 31.9/\sqrt{3} \cdot 22.05 = 0.835$ kA

According to IEC 60076, rated power equals the product of no-load voltage and rated LV current.

Thus the initial required transformer-rated power $= \sqrt{3} \cdot 22.05 \cdot 0.835 = 32.54$ MVA

The ultimate LV maximum current $= 38.3/\sqrt{3} \cdot 22.05 = 1.003$ kA

The ultimate required transformer-rated power $= \sqrt{3} \cdot 22.05 \cdot 1.003 = 39.09$ MVA

The auxiliaries power transformer rating chosen was 35/40 MVA ONAN/ ONAF.

14.3.9.3 *Network impedance representation*

Figure 14.17 shows the system configuration comprising of two 132/21 kV auxiliary power transformers (the details for which we are investigating), 21 kV cable network, 21/3.3 kV and 3.3/0.4 kV transformers and 3.3 kV and 0.4 kV loads. In this calculation the base MVA is chosen as 10 MVA. Base voltages are 132 kV and 21 kV. It is necessary to determine the auxiliary transformer impedance required to limit the fault level to the specified 12.5 kA.

The system configuration is reduced to an impedance network for making and breaking duties as shown in Figs. 14.18a and 14.18b. For readers wishing to work through such an impedance estimation the following network parameters may be used:

21 kV, 3c, XLPE, 18 700/22 000 V, 120 mm^2 copper cable

resistance @ 90°C = 0.196 ohm per km
reactance = 0.108 ohm per km
capacitance = 0.25 μF per km or 78.5 micro mho per km @ 50 Hz

The per unit values may be obtained from this data as follows:

Base impedance = (21 kV)^2/10 MVA = 44.1 ohm
pu resistance = 0.196/44.1 = 0.0044 pu per km
pu reactance = 0.108/44.1 = 0.0024 pu per km
pu susceptance = (78.5 · 10^{-6}) · 44.1 = 0.00346 pu per km

The effect of cable capacitance is negligible and may therefore be ignored in a simple hand calculation.

Motor contribution to fault level

When a fault occurs near an induction motor the motor will contribute to the fault current. The motor may be represented as a voltage source behind a reactance. This reactance can be obtained using contribution factors.

For breaking duty calculation, two contribution factors apply:

1. $X_m = 1.5 X_d''$ for a motor with a rating above 250 hp
2. $X_m = 3.0 X_d''$ for a motor with a rating below 250 hp

where X_m is the effective motor reactance during the fault period and X_d'' is the sub-transient reactance – see, for example, IEEE-recommended practice for electrical power distribution for industrial plants. It may be assumed here that $X_d'' = 0.9 X_d$ where X_d is the transient reactance. The motor starting current is typically specified as six times full load current for 400 V motors and four times full load current for 3.3 kV motors. The transient motor reactance $X_d = 1/(\text{starting current})$. Using this information the effective reactance, for

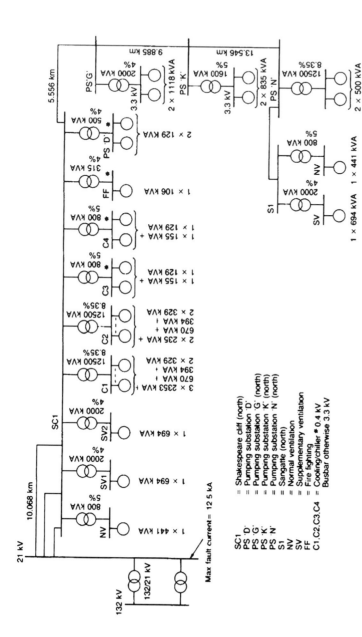

Figure 14.17   System configuration

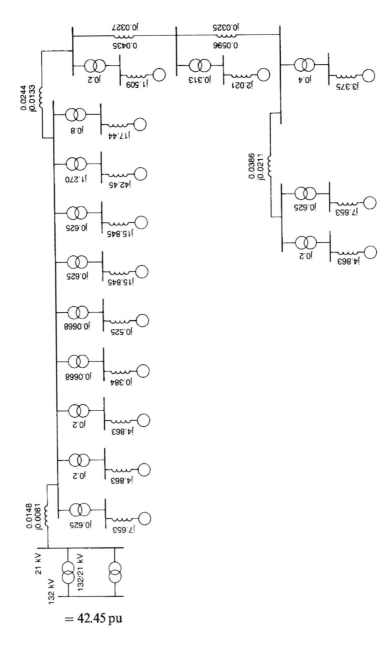

Figure 14.18a   System impedances on 10 MVA base for breaking duty

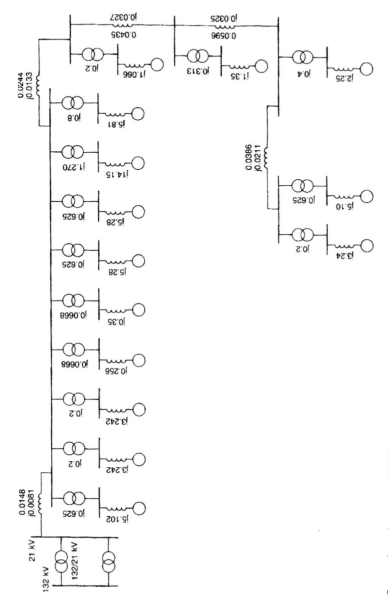

Figure 14.18b   System impedances on 10 MVA base for making duty

the breaking duty calculation, of all the motor loads shown in Fig. 14.17 may be calculated.

For example, the 3.3 kV, 441 kVA motor reactance

$$X_m = 1.5X_d'' = 1.5 \cdot 0.9 \cdot X_d$$
$$= 1.5 \cdot 0.9 \cdot 10\,\text{MVA}/(0.441\,\text{MVA} \cdot 4)$$
$$= 7.653\,\text{pu}$$

The 400 V, 106 kVA motor reactance with a starting current $6 \times$ flc (full load current) becomes:

$$3X_d'' = 3.0 \cdot 0.9 \cdot 10\,\text{MVA}/(0.106\,\text{MVA} \cdot 6)$$

For making duty calculation the sub-transient reactance should be used giving contribution reactances of 5.102 pu and 14.15 pu for the 3.3 kV and 400 V motors detailed above, respectively.

14.3.9.4 *Transformer impedance*

Figure 14.19 shows the reduced network equivalent impedance for breaking and making duties. Chapter 1 describes simplifications which may be used for hand network reduction calculations. Neglecting the system resistance the reactance of the auxiliaries transformers is calculated as follows:

Base current $= 10\,\text{MVA}/(\sqrt{3} \cdot 21\,\text{kV}) = 0.275\,\text{kA}$

The maximum allowable fault current is 12.5 kA, therefore:

$$I\,(\text{pu}) = 12.5/0.275 = 45.45\,\text{pu}$$

The maximum fault current occurs when the voltage (HV side) is 1.1 pu (+10%) and the voltage at the 3.3 kV side is 1.075 pu (+7.5%). The break fault current from the motors (see Fig. 14.19a):

$$= \frac{1.075}{0.1709} \times \text{base current at 21 kV}$$
$$= \frac{1.075}{0.1709} \times 0.275$$
$$= 1.729\,\text{kA}$$

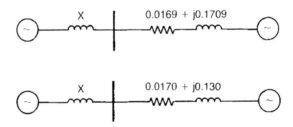

Figure 14.19   Equivalent circuit for breaking duty (top) and making duty (bottom)

The make fault current from the motors (see Fig. 14.19b):

$$= \frac{1.075}{0.130} \times 0.27$$
$$= 2.274 \text{ kA}$$

The maximum allowable fault current with the two auxiliaries transformers operating under the minimum impedance condition in parallel $= 12.5 - 2.274 = 10.226 \text{ kA}$. (Note: The transformers in this example were specified at the outset with an internal design to cater for later unforeseen load growth by the addition of oil pumps and forced oil cooling to give a possible future 45 MVA OFAF rating.)

Since the transformers have an LV-rated voltage of 22.5 kV, the transformer LV-rated current

$$45 \text{ MVA}/(\sqrt{3} \cdot 22.5 \text{ kV}) = 1.155 \text{ kA}$$

The maximum source fault level on the 132 kV primary side of the transformers is 2015 MVA. If X_T is the transformer impedance on a 45 MVA base, then:

$$\frac{1.1 \times 1.155}{\dfrac{45}{2015} + \dfrac{X_T}{2}} = 10.226 \text{ kA}$$

or

$$X_T = 0.204 \text{ pu}$$
$$= 20.4\%$$

say 21% at maximum negative tap. Therefore the nominal impedance is chosen as 23%.

14.3.9.5 Tap range calculation

In this calculation the equivalent circuit shown in Fig. 14.20a is used. The pu voltage on the LV side of the auxiliary 132/21 kV transformer is given by:

$$\text{LV (pu)} = \frac{\text{HV (pu)}}{a} \times 1.071 - DV$$

where DV = the voltage drop across the transformer due to the load current
          a = HV turns/HV turns at principal tap
          $1.071 = 22.5/21$

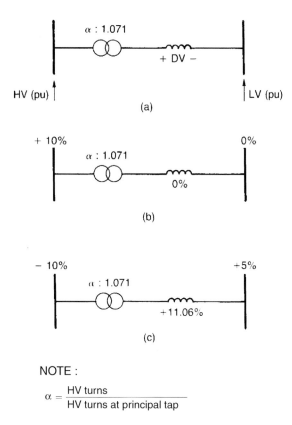

Figure 14.20   Transformer regulation: (a) equivalent circuit; (b) no load equivalent circuit; (c) full load equivalent circuit

(a) *Highest tap*
The highest tap on the 132 kV windings occurs when:
- the 132 kV side voltage is at a maximum ($+10\%$)
- under no-load conditions ($DV = 0$). The no-load equivalent circuit is shown in Fig. 14.20b
- the 21 kV side voltage is at its nominal ($\sim 0\%$)

$$\text{LV (pu)} = \frac{\text{HV (pu)}}{a} \times 1.071 - DV$$
$$1 = \frac{1.1}{a} \times 1.071 - 0$$
$$a = 1.178$$

The change in the winding turns to compensate therefore equals $(a - 1) = 1.178 - 1 = 0.178$ or 17.8%. With 1.25% per tap step, 18.75% is the nearest highest tap position required.

(b) *Lowest tap*

The full load equivalent circuit is shown in Fig. 14.20c. The lowest tap on the 132 kV windings occurs when:

- the 132 kV side voltage is at a minimum (-10%)
- under full ultimate load conditions (39.09 MVA loading)
- the 21 kV side voltage is at its maximum ($+5\%$)

$$DV = K(V_R \cos \phi + V_X \sin \phi) + [K^2(V_X \cos \phi - V_R \sin \phi)^2]/200$$
$$+ [K^4(V_X \cos \phi - V_R \sin \phi)^4]/8 \times 10^6$$

(*J & P Transformer Book*, 11th Edition, p. 112)

where V_R = percentage resistance voltage at full load = 1% (assumed)
V_X = percentage leakage reactance voltage at full load = 23%
$\cos \phi$ = power factor of load = 0.9, $\sin \phi$ = 0.436
K = actual load/rated load = 39.09/45 = 0.869

The voltage regulation

$$DV = 0.869(1 \cdot 0.9 + 23 \cdot 0.436) + [0.869^2(23 \cdot 0.9 - 1 \cdot 0.436)^2]/200$$
$$+ [0.869^4(23 \cdot 0.9 - 1 \cdot 0.436)^4]/8 \times 10^6$$
$$= 9.496 + 1.550 + 0.012$$
$$DV = 11.06\%$$

The lowest tap is therefore:

$$\text{LV (pu)} = \frac{\text{HV (pu)}}{a} \times 1.071 - DV$$
$$1.05 = \frac{0.9}{a} \times 1.071 - 0.1106$$
$$a = 0.831$$

The change in the winding turns to compensate therefore equals $(a - 1) = 0.831 - 1 = -0.169\%$ or -16.9%. With 1.25% per tap step, -17.5% is the next lowest tap position required. To include one spare tap, -18.75% is a preferred lowest tap position.

14.3.9.6 *Conclusions*

The nominal impedance of the auxiliaries transformers in this example should therefore be 23% in order to limit the fault level to 12.5 kA under the worst case conditions (two transformers in parallel, maximum source fault level and maximum motor contribution). For satisfactory operation of the 21 kV system a tap range of -18.75% to $+18.75\%$ should be specified when placing the order for the transformers.

14.4 THERMAL DESIGN

14.4.1 General

Heat is mainly produced in a transformer due to the passage of load current through the resistance of the winding conductors (load loss), and due to heat production in the magnetic core (no-load loss). Additional but less significant sources of heat include eddy current heating in conductors and support steel structures, and dielectric heating of insulating materials.

Transformer thermal design is aimed at removing the generated heat effectively and economically so as to avoid deterioration of any of the components of the transformer due to excessive temperature. In oil-immersed transformers the core and windings are placed in a tank filled with mineral oil. The oil acts as the primary cooling medium since it is in close contact with the heat-producing components. Dry type transformers, where the windings are resin cast, may be specified for particular low fire risk applications with ratings up to about 10 MVA.

14.4.2 Temperature rise

Heat is produced directly in the windings from the I^2R losses in the conductors. Insulation usually consists of paper tape wound around the conductor. This gives the required insulation of the conductor from its neighbouring turns. The paper tape is saturated with oil since the whole winding is immersed in the bulk of oil inside the transformer tank. This gives insulation from other windings, and from the earthed parts of the transformer structure.

Heat generated in the conductor must firstly be conducted through the paper tape insulation and then into the bulk of oil. From there the heat is conducted and convected away from the winding eventually to be dissipated into the air surrounding the transformer. In order to avoid damage to the insulation, the maximum service temperature must be limited. The basis for 'normal life expectancy' of oil-immersed transformers with oil-impregnated Class 105 (previously defined as class A) paper insulation is that 'the temperature of the insulation *on average* shall not exceed 98°C'.

In practice, not all parts of a winding operate at the same temperature since some parts are cooled more effectively than others. The part of the winding which reaches the hottest temperature is known as the 'hot spot'. The hot spot location in the winding is not precisely known, although infrared imaging techniques may be used if a fault is suspected. Modern transformers incorporate fibre optic devices to couple the temperature transducers to the recording apparatus in order to obtain sufficient insulation. *Direct* hot spot thermal fibre optic probes are located at the calculated hot spot position. *Distributed* thermal sensor fibre optic probes more accurately map the temperature image of the transformer but in practice they are more difficult to install. Conventional temperature probes are not suitable for direct attachment to conductors which may be at a high voltage above earth. Therefore the 'average temperature' of a complete winding is normally determined by measuring its change in resistance above a reference temperature. Research and development tests have established that the hot spot temperature is about 13°C above the average winding temperature in typical naturally cooled transformers. Measurement of average winding temperature therefore allows the hot spot to be deduced, at least in an empirical way.

When a transformer is unloaded the conductor temperature is virtually the same as the ambient temperature of the air surrounding the transformer. When load current is passed, the conductor temperature rises above ambient and eventually stabilizes at an elevated value (assuming the load current is constant). The total temperature of the hot spot is then given as:

hot spot temperature = ambient temperature
+ average winding temperature rise
+ hot spot differential

The basis of the IEC specification for thermal design, with the transformer at full load, is to assume an *annual average* temperature of 20°C. On average, over a year therefore, the limit of 98°C is achieved if:

98°C ⩾ 20°C + average winding temperature rise + 13°C

Therefore the average winding temperature rise should be ⩽65°C and this forms the basis of the IEC specification for 65°K *average winding temperature rise* (70°K for cooling type OD – see paragraph 14.4.6).

There are also IEC requirements for the temperature rise of the insulating oil when the transformer is at full load. The specified rise of 60°C ensures that the oil does not degrade in service and is compatible with allowing the average winding temperature to rise by 65°C.

14.4.3 Loss of life expectancy with temperature

Insulating materials are classified by a statement of the maximum temperature at which they can be operated and still be expected to give a satisfactory life span. Operation at a moderately elevated level above the maximum recommended temperature does not result in immediate insulation failure. However, it will result in shortened life span. The law due to Arrhenius gives the estimated life span as:

Loss of life expectancy = $A + B/T$
A and B are empirical constants for a given material
T is the absolute temperature in °K

For the particular characteristics of Class A transformer insulation, the Arrhenius law results in a halving of life expectancy for every 6°C above the temperature for normal life. (Conversely, life expectancy is increased for a temperature reduction, but this effect can only be applied in a limited way since life spans beyond about 40 years would be influenced by factors other than temperature alone.)

The Arrhenius effect allows for periods of operation with the insulation above its specified 'normal life' temperature provided these periods are balanced by periods of lower temperature where the life is above normal. This effect may be utilized in normal transformer specifications since, in operation, the hot spot temperature will fluctuate both with variations in ambient temperature and variations in loading level.

In the summer the ambient temperature may rise to 40°C so that the hot spot at full load would rise to $40 + 65 + 13 = 118°C$. In the winter, however, at say 0°C, the hot spot at full load would only reach $0 + 65 + 13 = 78°C$. So long as the annual average temperature was not above 20°C, the overall life expectancy would remain normal due to the additional life gained in the winter counterbalancing the increased loss of life in the summer.

The Arrhenius effect can also be applied to allow overloading of a transformer. Consider, for example, a 24 hour period in which the transformer is loaded to 75% of its rating for all but 2 hours when it is loaded to 120% of its rating. The period at 120% load has no overall detrimental effect on the life of the transformer since the increased loss of life in the 2 hours of overload is balanced by the slower-than-normal ageing at 75% load.

The overloading with no loss in life described above can be extended one step further to cover emergency conditions when a definite loss in life is tolerated in order to meet an abnormal, but critical, system operational requirement. Thus a transformer could be operated at, for example, 200% normal load for, say, 2 hours with an additional loss in life of, say, 5 days in that 2 hours (assuming other parts of the transformer such as bushings or tap changer contacts, to which the effect does not apply, can also accept a 200% load for this time).

Refer to IEC 60354 for a more comprehensive guide to oil-filled transformer overload values and durations, and IEC 60905 for dry type transformers.

14.4.4 Ambient temperature

Since ambient temperature has an important influence on transformer performance and internal temperature such environmental details must be included in the transformer enquiry specifications. The IEC reference ambient temperature is given in four components as follows:

maximum: 40°C
maximum averaged over a 24 hour period: 30°C
annual average: 20°C
minimum: $-25°C$

In some parts of the world the first two values may not be exceeded but the annual average is often above 20°C. In Middle East desert areas the first three temperatures may all be exceeded by 10°C.

If any of the IEC reference ambient temperatures are exceeded by the site conditions the permitted internal temperature rises are adjusted to restore the basic thermal equation for normal life expectancy. For example, if the annual average temperature was 25°C instead of 20°C, the permitted average winding rise is reduced to 60°C to restore the 98°C total hot spot temperature. Note that the correct annual average temperature to use when specifying transformers is a 'weighted' value given as follows:

$$Ta^1 = 20 \log_{10} 1/N \left[\sum_1^N 10^{Ta/20} \right]$$

where Ta^1 = weighted annual ambient temperature
         Ta = monthly average temperature
         N = month number

The weighted value is designed to take proper account of the Arrhenius law.

14.4.5 Solar heating

The sun provides an additional source of heat into the transformer which must be taken into account in tropical climates. The additional temperature rise of the oil in the transformer will be small, typically 2°C or 3°C, for most transformers. It is only for small transformers, such as pole-mounted units, where the exposed surface area is large compared with the volume, that the effects become significant. In these circumstances it may be necessary to subtract 5°C or even 10°C from the permitted winding temperature rise at full load in order to maintain normal life expectancy. Even for large transformers where the effect of solar heating on internal temperature rise is negligible, the manufacturer should be advised of exposure to tropical solar radiation since the operation of other components such as temperature gauges, electronic control modules, and gaskets, may be adversely affected.

14.4.6 Transformer cooling classifications

In the simplest cooling method, the heat conducted to the oil from the windings and core is transmitted to the surrounding air at the tank surface. In practice, only the smallest distribution transformers, for example 10 kVA pole mounted, have enough tank surface area to dissipate the internal heat effectively (see Fig. 14.21a). As the transformer size increases the surface area for heat dissipation is deliberately increased by attaching radiators to the tank. A 1000 kVA

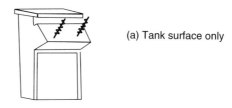

(a) Tank surface only

(b) Radiators on tank

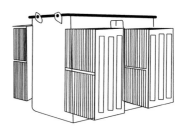

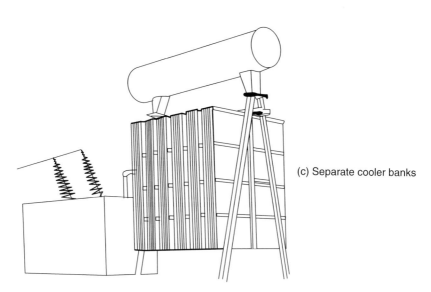

(c) Separate cooler banks

Figure 14.21   Cooling arrangements

Figure 14.22   225/21 kV, 35/40 MVA, ONAN/ONAF transformer at Coquelles substation, France

hermetically sealed transformer with radiators is shown in Fig. 14.21b. A 200 kVA pole-mounted transformer with radiator tubes is shown in Fig. 17.13, Chapter 17. As the transformer rating increases still further the number of radiators required becomes too large all to be attached to the tank and separate cooler banks are used as indicated in Fig. 14.21c.

In the cooling method described above, no moving parts are used. As the oil is warmed inside the tank, the warmer oil rises to the top of the tank and into the tops of the radiators. As the oil cools, it falls to the bottom of the radiator and then back into the bottom of the tank. This sequence then repeats itself, giving a 'natural' circulation of cooling oil.

Increased cooling efficiency is obtained by fitting fans to the radiators to blow cooling air across the radiator surfaces. Figure 14.22 shows cooling fans on a 225/21 kV, 35/40 MVA, ONAN/ONAF transformer.

A further increase in efficiency is achieved by pumping the oil around the cooling circuit, thereby boosting the natural circulation. The oil is forced into closer contact with the winding conductors to improve the heat extraction rate. In practice, baffles and cooling ducts direct the oil into the heat-producing areas.

The IEC cooling classification codes allow the desired type of cooling to be simply specified. The codes indicate the primary cooling medium, i.e. the medium extracting the heat from the windings and core, and the secondary cooling medium, i.e. the medium which removes the heat from the primary cooling medium. The type of cooling method (how it is circulated) can also be specified. The following codes are used:

Kind of cooling medium	Code
Mineral oil	O
Water	W
Air	A
Non-flammable oil	L

Kind of Circulation	
Natural	N
Forced	F
Forced directed liquid	D

The coding method is to specify, in order, the primary cooling medium, how it is circulated; the secondary cooling medium, how it is circulated.

For example, an oil-immersed transformer with natural oil circulation to radiators dissipating heat naturally to surrounding air is coded as ONAN. Adding fans to the radiators changes this to ONAF, and so on.

Notice that a dry type transformer, with heat dissipation directly (but naturally) to the surrounding air uses only a two letter code, namely AN.

14.4.7 Selection of cooling classification

Choosing the most appropriate method of cooling for a particular application is a common problem in transformer specification. No clear rules can be given, but the following guidance for mineral oil-immersed transformers may help. The basic questions to consider are as follows:

1. Is capital cost a prime consideration?
2. Are maintenance procedures satisfactory?
3. Will the transformer be used on its own or in parallel with other units?
4. Is physical size critical?

ONAN

This type of cooling has no mechanical moving parts and therefore requires little, if any, maintenance. Many developing countries prefer this type because of reliability, but there is an increasing cost penalty as sizes increase.

ONAF

A transformer supplied with fans fitted to the radiators will have a rating, with fans in operation, of probably between 15% and 33% greater than with the fans not in operation. The transformer therefore has an effective dual rating under ONAN and ONAF conditions. The transformer might be specified as 20/25 MVA ONAN/ONAF. The increased output under ONAF conditions is reliably and cheaply obtained.

Applying an ONAN/ONAF transformer in a situation where the ONAF rating is required most of the time is undesirable since reliance is placed on fan operation. Where a 'firm' supply is derived from two transformers operating in parallel on a load-sharing basis the normal load is well inside the ONAN rating and the fans would only run in the rare event of one transformer being out of service. Such an application would exploit the cost saving of the ONAF design without placing too much emphasis on the reliable operation of the fans. Note that fans create noise and additional noise mitigating precautions may be needed in environmentally sensitive areas.

OFAF

Forcing the oil circulation and blowing air over the radiators will normally achieve a smaller, cheaper transformer than either ONAF or ONAN. Generally speaking, the larger the rating required the greater the benefits. However, the maintenance burden is increased owing to the oil pumps, motors and radiator fans required. Application in attended sites, with good maintenance procedures, is generally satisfactory. Generator transformers and power station interbus transformers will often use OFAF cooling.

ODAF/ODWF

These are specialized cooling categories where the oil is 'directed' by pumps into the closest proximity possible to the winding conductors. The external cooling medium can be air or water. Because of the design, operation of the oil pumps, cooling fans, or water pumps is crucial to the rating obtainable and such transformers may have rather poor naturally cooled (ONAN) ratings. Such directed and forced cooling results in a compact and economical design suitable for use in well-maintained environments.

14.4.8 Change of cooling classification in the field

Transformers may be specified with future load requirements in mind such that the design may allow for the future addition of oil pump or air fan equipment. As loads increase in distribution systems and transformers become overloaded, a relatively cheap increase in rating can be obtained by converting ONAN transformers to ONAF by fitting radiator fans. The manufacturer should always be consulted with regard to fan types and number, and the actual rating increase for a particular transformer.

If considered in the initial design specification it may be practical to fit oil circulation pumps to obtain a higher OFAF rating at some future date as the load demand increases. The rating increase is dependent on the internal design of the cooling circuit. Fitting oil pumps to a transformer not having such cooling ducts only supplements the natural oil circulation past the bulk of the winding assembly, and has very little improvement in cooling efficiency.

14.4.9 Capitalization of losses

Although transformers are very efficient machines increasing attention is paid to minimizing the cost of losses in electrical systems over the lifetime of the plant. A transformer manufacturer can build a lower loss transformer if required but this usually results in the use of more materials, or more expensive materials, with the end result of a higher initial purchase cost. Even so, the total cost of buying *and* operating the transformer over a life of, say, 25 years can be less for an initially more expensive, but low loss, transformer. Refer to Chapter 22, Section 22.2, for an introduction to financial and economic assessments.

When a number of manufacturers have been asked to bid for a particular transformer contract, a choice can be made on the basis of *total* cost, that is the capital cost, plus the cost of supplying the losses over an anticipated life span. To assign a cost to the losses can be an elaborate procedure. Note that the basis for costing losses must be advised to the manufacturer at the time of inviting quotations in order that the manufacturer can optimize capital cost and the cost of losses to give a competitively priced transformer. In most cases the consultant or electrical supply utility will simply specify separate capitalizing factors for the load and no-load losses and typical figures for UK transmission transformers are: no-load loss capitalization rate £4000/kW; load loss capitalization rate £650/kW.

The transformer manufacturer will then simply arrive at the capitalized price as:

$$\text{capitalized cost} = \text{selling cost} + 4000 \times \text{no-load losses (kW)}$$
$$+ 650 \times \text{load loss (kW)}$$

It is sometimes useful for the supplier to provide alternative designs (e.g. a high loss and low loss design) to illustrate the variation in prime and capitalized costs as a function of transformer losses. The methods of capitalizing losses have been the subject of numerous studies. Very precise calculations are not considered to be justified since the accuracy of the results may only be of the same order as the assumptions made regarding the evolution of the parameters.

In some special cases the user may specify some capitalizing formulae to be applied and an example is detailed below. Tenders may be requested by the purchaser that guarantee the losses quoted by the manufacturers. No-load losses should be quoted at a given reference temperature and voltage. Load losses must be carefully quoted at a given rating and tap position. Auxiliary losses (fans, etc.) must be detailed at each level of cooling. For reactors only the total guaranteed losses are normally required. If £*NET* is the net difference in evaluated cost of losses (given here in UK£) between test and guaranteed losses then:

$$£NET = (NL_t - NL_g)EVAL_{NL} + (LL_t - LL_g)EVAL_{LL}$$
$$+ (AL_t - AL_g)EVAL_{AL}$$

if £*NET* > 0, then:

$$£PEN = 1.15 \times £NET$$

where

NL_t = tested no-load losses (kW)
NL_g = guaranteed no-load losses at time of tender enquiry (kW)
$EVAL_{NL}$ = no-load loss evaluation factor (£/kW)
LL_t = tested load losses (kW)
LL_g = guaranteed load losses at time of tender enquiry (kW)
$EVAL_{LL}$ = load loss evaluation factor (£/kW)
AL_t = tested auxiliary losses (kW)
AL_g = guaranteed auxiliary losses at time of tender enquiry (kW)
$EVAL_{AL}$ = auxiliary loss evaluation factor (£/kW)
$£PEN$ = price adjustment (in £) to be deducted from the base price if £*NET* > 0.

The intent of this approach is to balance the potential estimated savings against the known capital transformer cost, and also to compensate for the uncertainty of predicted parameters such as the cost of money, inflation of energy costs, future electric plant construction costs, predicted load and load growth rates and uncertainty as to the exact lifetime of the transformer.

14.5 CONSTRUCTIONAL ASPECTS

14.5.1 Cores

Cores are constructed from an iron and silicon alloy which is manufactured in a way to enhance its magnetic properties. Basic cold-rolled flat alloy sheets known as 'cold-rolled grain-oriented silicon steel' have been used since the 1960s. Since this time improved quality control producing better grain orientation and thinner sheets ('Hi-B steels') have reduced no-load losses by some 15% compared to conventional cold-rolled grain-oriented silicon steel types. The magnetic core is made up from several thin sheets, 0.3 mm to 0.23 mm thick, of the core metal. Each sheet has a thin coating of insulation so that there is no conduction path from sheet to sheet. This technique is used to minimize eddy currents in the core metal. (If the core were a solid block of metal these eddy currents would produce excessive heating.) Surface laser-etched 0.23 mm steels are now also being used by leading transformer manufacturers. This results in a further 15% loss reduction and such treatment may be justified as a result of the electrical supply utility's loss capitalization formulae. It is important for the transmission and distribution systems engineer to specify the flux density in conjunction with the manufacturer before ordering transformers. If the flux density is too high the transformer may go into saturation at the most onerous tap setting. Typical values for modern cold-rolled grain-oriented silicon steel transformers should not exceed 1.7 Tesla (Wb/m²) without manufacturers' advice.

Rapidly cooled, thin (typically 0.025 mm thick) amorphous ribbon steel cores with lower magnetic saturation limits of about 1.4 Tesla can reduce the losses in the core compared to cold-rolled grain-oriented silicon steel by up to 75%. Distribution transformers using this material are widely used in North America.

A continuous magnetic circuit is obtained by avoiding air gaps or non-magnetic components at joints. Core lamination clamping bolts are no longer used in modern designs. The laminations are held together by the hoop stress of the windings, by fibreglass or banding or by pinching the yoke between external clamps.

14.5.2 Windings

14.5.2.1 *Conductors and insulation*

Transmission and distribution oil-immersed power transformer windings are usually made of copper to reduce load losses. Winding insulation in oil-immersed transformers is a cellulose paper material. High voltage transformers are vacuum impregnated with high quality, extremely clean, hot mineral-insulating oil with a water content less than 2 ppm (parts per million) and air content less than 0.2%. As ratings increase the winding conductors are connected in parallel (to reduce eddy current loss) and transposed (to avoid leakage flux circulating currents).

Aluminium has a higher specific resistance than copper and therefore requires a larger cross-section for a given current rating. It is not therefore generally used in power transformers. However, aluminium has certain advantages over copper when used as foil windings in dry type cast resin distribution transformers. Aluminium has a coefficient of expansion of approximately 24×10^{-6} °K^{-1}, compared to 17×10^{-6} °K^{-1} for copper, and this is more similar to the resins used. The short circuit thermal withstand time tends to be greater for aluminium compared to copper in an equivalent design and aluminium foil eddy current losses are lower. Modern designs using epoxy and fibreglass resins, vacuum moulded to the conductors, having high thermal conductivity (0.5 W/m °K) and extremely high electrical strength (200 kV/mm) generally allowing higher conductor temperatures than Class A. The impulse withstand tends to be lower for dry type transformers but this depends upon strip or foil winding construction and the resins used. Some manufacturers offer 95 kV BIL designs whereas IEC 60076-11 requires 75 kV BIL for 12 kV systems.

14.5.2.2 *Two winding (double wound)*

This is the basic transformer type with two windings connecting a higher voltage system to a lower voltage system. This type is the normal arrangement for step-down transformers in distribution and subtransmission systems, and for generator transformers.

14.5.2.3 *Three winding*

14.5.2.3.1 *General*
There are situations where, for design reasons, or because a third voltage level is involved, that a third winding is added. The impedance representation of three winding transformers is detailed in Section 14.2.4.

14.5.2.3.2 *Delta tertiary*
A star/star transformer is often supplied with a third (delta-connected) winding for one or more of the following reasons:

- To reduce the transformer impedance to zero sequence currents and therefore permit the flow of earth fault currents of sufficient magnitude to operate the protection. (See Section 14.2.4 and Fig. 14.7.)
- To suppress the third harmonics due to the no-load current in the earth connection when the neutral is earthed. These harmonics have been known to induce disturbances in neighbouring low voltage telecommunication cables.
- To stabilize the phase-to-phase voltages under unbalanced load conditions (e.g. a single phase load between one phase and neutral). Without a tertiary winding the current flowing in the uncompensated phases is purely magnetizing and, by saturation, causes deformation of the phase voltages and displacement of the neutral point. The addition of a delta tertiary winding balances the ampere turns in all three phases eliminating such phenomena.
- To enable overpotential testing of large high voltage transformers to be carried out by excitation at a relatively low voltage. This requirement depends upon the transformer manufacturers' test bay capabilities. However, for such test purposes the tertiary may only need to be of a very low rating and connected only for the factory tests.
- To provide an intermediate voltage level for supply to an auxiliary load where a tertiary winding offers a more economical solution than a separate transformer.

Because of these apparent advantages a general view that such a tertiary winding is essential has flourished. However, this is not the case and the tertiary involves an increase in transformer cost of approximately 6% to 8% with a corresponding increase in losses of some 5%. The cross-section of the tertiary winding is usually determined by fault withstand considerations.

With normal three limbed core type star/star transformers satisfactory earth fault current is normally available. Typical values of zero sequence impedance are as follows:

Single phase transformers	5000% to 10 000%
Shell type or five limb core type transformers	1000% to 5000%
Three limb core type transformers	50% to 100%

Therefore the reduction in zero sequence impedance by the addition of a tertiary winding is not as significant in three limb transformers as in the other units.

The development of low loss cold-rolled grain-orientated steel together with improved core construction methods has reduced the magnetizing currents in modern transformers. The delta tertiary should only be specified when very small ripple voltages (>2% to 3% of the fundamental) or overvoltages of the order of a few percent due to out-of-balance loads cannot be tolerated. However, for a bank of single phase transformers, five limbed core types and shell types with an unearthed primary neutral, a delta tertiary should be specified for low zero sequence impedance and 'trapping' triple harmonics considerations.

14.5.2.4 *Auto-transformers*

The basic transformer principle can be achieved using a single winding (per phase). If a tap is made part way down the winding, this can be the low voltage terminal just as though this were a separate winding.

By eliminating the second winding, an auto-transformer is potentially cheaper than a two winding counterpart. In practice, such cost savings only apply for voltage transformation ratios of up to about 3:1 if adequate power transfer is to be achieved. Thus, for example, a transformer with a voltage ratio of 275/132 kV will be a straightforward auto-transformer choice; a ratio of 275/66 kV would, however, probably favour a double wound arrangement.

Both high voltage and low voltage systems have the same neutral (auto-transformers are usually star connected) and this would often be undesirable except in transmission systems where solid earthing of neutrals is common at all voltage levels.

14.5.3 Tanks and enclosures

14.5.3.1 *Oil preservation*

The transformer oil acts as an insulation and heat transfer medium. To keep good insulating properties the oil must be dry and free from contaminants. The transformer tank has to be strong enough to take the mass of oil (plus the core and windings) and to allow lifting and haulage into position.

The simplest way of keeping the oil in good condition is to seal the oil inside the tank and not permit any contact with the atmosphere. However, the oil volume changes as the transformer heats up and it is necessary to allow for this expansion in the tank design. The following methods may be used depending on the rating of the transformer, its location and the particular policy of the manufacturer:

1. Sealed rigid tank – Oil expansion is catered for by not completely filling the tank with oil. The space above the oil is filled with a dry inert gas, such as dry nitrogen, which has no chemical reaction with the oil. Pressure changes within the tank are relatively large and require a stronger tank than normal. This type of tank construction is common in the USA up to quite large ratings of, say, 50 MVA, but rather uncommon in Europe above about 10 MVA at 33 kV.
2. Sealed expandable tank – There is a limit to the size of transformer that can successfully use this technique. Ratings up to 2 MVA at 11 kV are normally satisfactory. The tank is completely filled with oil but the surfaces are flexible (usually corrugated) in order to allow for changes in the effective volume of oil with temperature. Distribution transformers can be specified as hermetically sealed units with a semi-flexible tank of corrugated appearance to take up the forces of oil expansion on heating.
3. Positive pressure nitrogen – This method applies to very large transformers and has the advantage of a sealed rigid tank system but with pressure changes minimized by a venting and topping-up method using an external nitrogen supply. The nitrogen is kept at a small pressure above atmospheric by gas supply cylinders attached to the tank and operating through pressure sensing valves. Careful maintenance of the gas supply and supply valves is necessary.
4. Conservator (with breather) – This method may be used for virtually any size of transformer. The main tank is completely filled with oil and changes in volume are allowed by an expansion tank (conservator) mounted above the main tank. The conservator has a vent to the atmosphere. To avoid excessive moisture intake, an air drying device is used in the vent. Such drying is either by silica gel crystals, or more effectively, by a refrigerant drier unit. Careful oil maintenance is necessary especially at transmission voltages. The refrigerant drier is widely used at 275 kV and 400 kV in UK (see Section 14.6.6 for more details).
5. Conservator (with diaphragm seal) – This method is used by some manufacturers to give the advantages of an expansion tank system but without contact with the atmosphere. The expansion tank contains a flexible synthetic rubber diaphragm which allows for oil expansion, but seals the oil from the atmosphere (see Fig. 14.23). In theory, oil maintenance is less than with the breather systems but a disadvantage is that any moisture and contaminants trapped in the oil during manufacture may be sealed in for life.

The quality of tank welding, gasketing and painting must be carefully specified and inspected prior to release from the manufacturer's works. Such

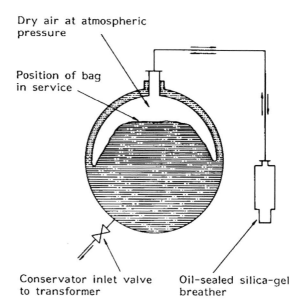

Dry air at atmospheric pressure

Position of bag in service

Conservator inlet valve to transformer

Oil-sealed silica-gel breather

Figure 14.23   Arrangements using expansion bag in the conservator

precautions will avoid oil leakage due to poor welds or gasket assembly and ensure suitable paint thickness, finish and application procedures for harsh environments.

14.5.3.2 *Dry type transformer enclosures*

Dry type transformers have some physical protection around them to keep personnel away from live parts, and to protect the core end windings from dust, water ingress, condensation, etc. A sheet steel enclosure or, more simply, an open steel mesh surround may be specified for indoor applications depending upon the IP classification required.

14.5.4 Cooling plant

Oil cooling is normally achieved by heat exchange to the surrounding air. Sometimes a water jacket acts as the secondary cooling medium. Fans may be mounted directly onto the radiators and it is customary to use a number of separate fans rather than one or two large fans.

Oil pumps for OFAF cooling are mounted in the return pipe at the bottom of the radiators. The motors driving the pumps often use the transformer oil as their cooling medium.

With ODAF cooling, the oil-to-air coolers tend to be compact and use relatively large fan blowers. With this arrangement the cooling effectiveness is very dependent on proper operation of the fans and oil pumps since the small amount of cooling surface area gives relatively poor cooling by natural convection alone.

Water cooling (ODWF) has similar characteristics to the ODAF cooling described above and is sometimes found in power station situations where ample and well-maintained supplies of cooling water are available. Cooling effectiveness is dependent upon the flow of cooling water and therefore on proper operation of the water pumps. Natural cooling with the out-of-service water pumps is very limited. Operational experience has not always been good, with corrosion and leakage problems, and the complexity of water pumps, pipes, valves and flow monitoring equipment. The ODAF arrangement is probably favourable as a replacement for the ODWF designs. Double wall cooler pipes give added protection against water leakage. The inner tube carries the water and any leakage into the outer tube is detected and causes an alarm. This more secure arrangement is at the expense of slightly reduced heat transfer for a given pipe size.

Normal practice with cooling plant is to duplicate systems so that a failure of one need not directly affect operation of the transformer. Two separate radiators or radiator banks and duplicate oil pumps may be specified. In the larger ODAF cooling designs there may be four independent unit coolers giving a degree of redundancy. The transformer may be rated for full output with three out of the four coolers in service.

Dry type transformers will normally be naturally air-cooled (classification AN) or incorporate fans (classification AF).

14.5.5 Low fire risk types

14.5.5.1 *General*

Mineral oil-immersed transformers present a potential fire hazard. The spreading of a fire resulting from a transformer fault is limited by including a bund and oil-catch pit arrangement in the civil installation works. In sensitive locations such as inside buildings, power station basements, offshore oil rigs, underground railways, etc. dry type construction or non-flammable oil-cooled transformers may be specified. The possible instances where a transformer may be involved in a fire may be considered to fall into three categories:

- An internal fault leads to ignition and subsequent burning of the materials within the transformer. However, note that transformers are normally protected by overcurrent devices which should clear arcing faults in 0.5 seconds or less and even in poorly protected cases in less than approximately 4 seconds. Under such conditions a cast resin transformer should not allow any small flames produced to be sustained for, say, longer than 45 seconds.
- The transformer is housed in an enclosed space involving traditional building materials (wood, etc.) which could ignite and engulf the transformer in

flames with temperatures of 800°C to 900°C. The contribution of the transformer to the fire should be severely limited and should not emit toxic smoke or fumes, and visibility should not be greatly impaired due to the transformer smoke contribution.

- The transformer is housed in an enclosure in which a fire involving hydrocarbon fuels or plastic materials (oil, polythene, etc.) occurs engulfing the transformer in flames with temperatures in excess of 1000°C.

14.5.5.2 *Dry type transformers*

Dry types are available up to 10 MVA and 36 kV with cast resin or conventional dry type Class 180 (previous categorization H) insulation temperature limits. The fully cast resin-encapsulated transformer units have the following advantages:

- Unaffected by humidity, dust, etc.
- Relatively simple assemblies using few insulating materials and less prone to electrostatic stress.
- High thermal time constant and superior short circuit withstand giving good overload performance often better than conventional air-cooled types.
- Avoids non-biodegradable problems associated with polychlorinated biphenyls (PCBs) or Askarels which are now banned from use.

A cast resin 1 MVA 21/0.4 kV transformer with a sheet metal IP 21 enclosure is shown in Fig. 14.24. The interconnections for the delta configuration are clearly visible together with 'off-circuit' tap connections and links.

14.5.5.3 *Non-flammable liquids*

There are a number of non-flammable liquids available, with little to choose between them. Some types are not as good as mineral oil in heat conduction or in lubrication properties, so there may be some minor design differences between a mineral oil-immersed unit and a non-flammable liquid-immersed unit. Some older transformers may be filled with Askarel or similar non-flammable fluid. This fluid is not permitted now for new installations in most countries due to its high level of toxicity.

Both dry type and non-flammable liquid-immersed types will cost more than an equivalent mineral oil-immersed unit. Possible reduced civil works must be taken into account when assessing overall dry type transformer installation costs.

14.5.5.4 *Fire protection*

Transmission system transformers are generally only protected to the extent that oil spillage from a burst tank is contained. The oil drains through adjacent

Figure 14.24   1 MVA, 21/0.4 kV, cast resin transformer

stonework and is held in a bund with a wall surround approximately 300 mm high. The draining and cooling of the oil through the stone chippings into the bund is intended to extinguish flame and the wall prevents pollution to natural drainage. Additional protection for outdoor installations may be offered by temperature sensors located above the transformer which initiate a water spray or foam system to extinguish the fire. Indoor installations may use CO_2 or a modern replacement for halon gas.

14.5.6 Neutral earthing transformers

The single zig-zag-connected winding is all that is needed to provide the neutral earthing facility as shown in Fig. 14.16c. A secondary earthing transformer winding is often specified to provide a substation low voltage auxiliary power source. An explanation of such earthing transformer selection is given in Chapter 4, Section 4.4.6.

14.5.7 Reactors

14.5.7.1 *General*

Reactors have single windings and are intended to provide inductive reactance. Shunt reactors are connected to the system to provide an inductive load for the purposes of compensating the capacitive loads of cables and lightly loaded overhead lines.

Series reactors are connected in series with a circuit in a system to reduce fault currents, or in some instances to balance the impedance between two parallel paths.

An air-cored coil will have a relatively low inductive reactance and above about 145 kV the impulse withstand requirements limit their use. Such reactors are often suitable for series reactors or reactive compensation schemes. Air-cored construction offers the cheapest solution up to certain MVAr and voltage ratings. Figure 14.25 shows a 60 MVAr air-cored reactor associated with the Channel Tunnel single phase 25 kV traction load-to-three phase 132 kV supply reactive compensation balancer scheme. Higher values of reactance can be achieved by introducing a magnetic core but with a deliberate air gap in series with the steel components. Shunt reactors will often use this method.

Steel tank reactors are normally oil-filled, with similar cooling requirements to oil-filled transformers. They are more expensive than an equivalent air-cored units, but have much lower external magnetic fields (see Section 14.5.7.2), which may be important in urban or industrial areas.

14.5.7.2 Assessment of acceptable levels of magnetic fields

The World Health Organization published Environmental Health Criteria 69 in 1987 and considers that the magnetic field strength not considered to produce any biological effects is about 0.4 milli Tesla for 50 Hz or 60 Hz. The maximum rate of change of magnetic field for some heart pacemakers to remain synchronous is 40 milli Tesla/second. This has been calculated to correlate to a 50 Hz field strength of 0.12 milli Tesla and 0.1 milli Tesla at 60 Hz. Warnings are also given concerning metallic implants but without any specific restriction levels. These figures are of the same magnitude as other guidelines concerning exposure to static and time varying electromagnetic fields and radiation. A conservatively safe value of exposure and reference level currently under consideration is 40 milli Tesla divided by the frequency. At 50 Hz this corresponds to 0.8 milli Tesla.

It is obviously necessary to take these values into account especially when dealing with large air-cored reactors. An estimation of the field strength should be made at the design stage and practical precautions taken to screen or fence off areas on site. Particular attention must be paid to the level of magnetic field strength likely to result in adjacent public areas and staff working areas. Adequately screened or oil-filled reactors may have to be specified if safe levels cannot otherwise be achieved.

The civil works must take into account the effects of induced currents in reinforcement and switchyard fencing. Reinforcement must be segregated into short sections and loops avoided by use of spacers made of insulating materials. The switchyard fence close to high field sources must also be divided into short sections and precautions taken with earthing arrangements in order to avoid circulating currents.

Figure 14.25   60 MVAr air-cored reactor

14.5.7.3 *Underground transformers*

In dense urban areas where substation sites are difficult to obtain, distribution transformers may be directly buried or installed in underground chambers. Small units up to 1 MVA may directly buried, and are typically ON/AF types in Europe or solid insulation AN/AF in the USA, in each case with air ducts leading to a radiator on the surface. Particular care has to be taken to minimise the effect of soil drying; use of thermal backfill is one means to achieve this. Maintenance facilities are minimal, and problems have arisen in some cases with tank corrosion of directly buried units.

In exceptional cases special designs have been used for transformers up to 100 MVA in underground substations, but these are in fully accessible rooms, with carefully built arrangements for allowing future replacement.

14.6 ACCESSORIES

14.6.1 General

The basic transformer assembly of windings, core, tank and terminations is supplemented by a number of accessories for monitoring, protection and safety purposes.

Some accessories are optional and will not necessarily be justified on every transformer; others are important to the safety or operation of the transformer and will therefore be of a mandatory nature.

The following sections briefly describe some accessories available.

14.6.2 Buchholz relay

A Buchholz relay is connected in the oil feed pipe connecting the conservator to the main tank. The relay is designed to:

- Detect free gas being slowly produced in the main tank, possibly as a result of partial discharging. Under such conditions the relay may be set to give an alarm condition after a certain amount of gas has evolved. Examples of incipient faults include broken down core bolt insulation on older transformers, shorted laminations, bad contacts and overheating in part of the windings.
- Detect a sudden surge movement of oil due to an internal transformer fault. Under such conditions the relay is normally set to trip the high and low voltage transformer circuit breakers. Examples of such oil surge faults include earth faults, winding short circuits, puncture of bushings and short circuits between phases.
- Provide a chamber for collection and later analysis of evolved gas. Chemical analysis of the gas and transformer oil can give maintenance staff an indication as to the cause of the fault.

Buchholz relays are considered mandatory for conservator type transformers since they are protective devices. They should be installed in accordance with the manufacturers' instructions since a certain length of straight oil piping is required either side of the relay to ensure correct operation.

From time to time maloperation of Buchholz relays is reported. This is often due to vibration effects and a relay designed for seismic conditions may overcome the problem.

14.6.3 Sudden pressure relay and gas analyser relay

On non-conservator type transformers, the useful protective and gas analysis feature of the Buchholz relay cannot be provided. In its place, a sudden pressure relay detects internal pressure rises due to faults, and gas devices can be used to detect an accumulation of gases.

Sudden pressure relays are normal accessories for sealed transformers. Gas analyser devices tend only to be used on large important transformers.

14.6.4 Pressure relief devices

A pressure relief device should be regarded as an essential accessory for any oil-immersed transformer. Very large transformers may require two devices to adequately protect the tank. Violent pressures built up in the transformer tank during an internal fault could split the tank and result in the hazardous expulsion of hot oil. In order to avoid tank rupture resulting from the high pressures involved in an internal transformer fault a quick acting pressure relief device is specified and used to give a controlled release of internal pressure.

Older transformers may have been fitted with a rupturing diaphragm type device where the excess pressures breaks a fragile diaphragm and allows oil to be discharged. Not only does this not reseal but the overall operating time may be too slow to protect the tank against splitting.

14.6.5 Temperature monitoring

A correctly specified and loaded transformer should not develop excessive temperatures in operation. Oil and winding temperature is monitored in all but small (say, less than 200 kVA) distribution transformers.

Apart from the facility to monitor temperature (useful during controlled overloading), an important feature of the winding temperature indicator is to initiate automatic switch-on and switch-off of cooling fans and oil circulation pumps. In this way a dual rated transformer with a cooling classification of, for example, ONAN/ONAF will automatically switch from ONAN to ONAF (and back) according to the transformer loading conditions. The winding temperature monitor on an oil-immersed transformer simulates the temperature by using an oil temperature sensor and injecting additional heat into the sensor from a current transformer connected to one of the transformer terminals. In this way the winding temperature monitor registers a temperature above that of the oil by an amount that is dependent on the load current of the transformer. This arrangement is usually calibrated on site and is used to indicate the hot spot winding temperature.

The oil temperature monitor is usually a capillary type thermometer with the sensor placed in the vicinity of the hottest oil in the tank (i.e. at the top of the tank just prior to entering the radiators).

Both oil and winding temperature monitors are fitted with contacts which can be set to operate at a desired temperature. Such contacts are used for alarm (and possibly trip) purposes and also to operate auxiliaries as noted above.

Alarm and trip temperature settings are usually advised by the manufacturer. Note that it will usually be necessary to modify the settings if the transformer is used for controlled overloading since winding and oil temperature are allowed to reach higher temperatures during overloading than during normal operation.

Dry type transformers incorporate thermistor probes through the resin, usually by the low voltage winding hot spot area. Negative temperature coefficient thermistors are available with a resistance range from some $2000\,\Omega$ at 200°C to $1\,M\Omega$ at ambient temperature with an accuracy of some $\pm3\%$. Settings for alarm and trip conditions may be made using an electronic control device.

14.6.6 Breathers

As noted in Section 14.5.3, breathers are placed in the vent pipes of conservators to dry the air entering the conservator as the volume of oil contracts on transformer cooling.

Traditional breathers use the moisture absorbing properties of silica gel crystals. These crystals need replacement when they become saturated with moisture. Replacement is indicated by a change in colour of the crystals from blue to pink.

An alternative technique is to continuously extract moisture dissolved in the transformer oil by freezing the moisture out of the air by passing it over refrigerating elements and then evaporating it off to the atmosphere. This approach is used in the 'Drycol' device shown in Fig. 14.26. The oil is kept particularly dry and researches have shown an improved life span of the transformer. The 'Drycol' device is a standard fitting on 275 kV and 400 kV transformers in UK.

14.6.7 Miscellaneous

14.6.7.1 Core earth link

The magnetic core is earthed to the transformer tank at one point only in order to prevent a metallic path for circulating currents. It is useful to specify that

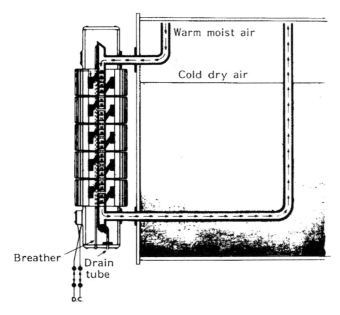

Figure 14.26   Principle of the Drycol breather – showing direction of air flow during a drying period (courtesy of GEC Transformers Ltd)

the earth is made through a removable link so that on-site tests can be made to check for other core earths that may have been produced by rough handling of the transformer or a fault in service.

Some manufacturers may have difficulty with this facility and it is worth ensuring during manufacture that a truly accessible link is being provided.

14.6.7.2 *Oil level gauge*

The simplest arrangement is a sight glass, but on many transformers a dial type indicator will be used. The correct oil level should be shown on the indicator for a range of ambient temperatures appropriate to the site. Low oil level contacts can be used to provide an alarm.

Transformers with expandable tanks, or with diaphragm seals, do not have a free oil level. In the case of the diaphragm seal it may be possible in some designs to attach the float of a conventional oil level indicator to the diaphragm seal in order to provide an alarm for seal breakage.

14.6.7.3 *Tap changer accessories*

An on-load tap changer mechanism should be equipped with an oil surge detection relay to indicate a fault within the tap change compartment if this is separate from the main tank. Gas evolution during tap changing is not abnormal so if a conventional Buchholz relay is applied the gas collection facility is not used. With the double-compartment tap changer arrangement the diverter and selector are located in different compartments and full Buchholz protection is possible.

14.6.7.4 *Oil sampling valve*

Routine maintenance of transformers includes the testing of oil for moisture, contaminants and possibly also for dissolved gas content. An oil sampling valve is therefore a necessary accessory on most transformers. Oil sampling is not common on sealed distribution transformers which may be regarded as 'sealed for life'.

14.6.8 Transformer or...ng details

14.6.8.1 *Specifications*

The primary objective of the purchaser's specification for a transformer should be to state precisely the duty the transformer is to perform and the conditions under which it will operate. Having specified these, the details of the design and construction should, as far as possible, be left up to the manufacturer. To specify intimate details in the design, such as current density or flux density, merely ties the hands of the designer and could well result in an inefficient or unnecessarily expensive design.

The purchaser's main concern is to ensure that as far as possible every new transformer purchased is capable of performing under the specified operating conditions for a service life in the order of 40 years. Key activities in ensuring this are the contract specification for the transformer, quality assurance during design and manufacture, effective testing before leaving the manufacturer's works and on site and appropriate maintenance and diagnostic testing in service (see Chapters 19 and 22). Most electrical supply utilities now rely on functional or technical specifications which cover the requirements of their own particular company. These are reinforced with an approval procedure which includes a design review to ensure product compatibility with their particular needs.

The best use should therefore be made of recognized specifications typically as described in Section 14.2.6.

14.6.8.2 *Rated power and rated voltage*

A major item which is sometimes subject to confusion is the *rated power* of the transformer. Taking into account the anticipated loading conditions, IEC

60076-10 gives examples for determining the required rated power of a transformer for a given set of loading conditions.

Transformers are assigned a rated power for each winding which refers to a continuous loading. This is a reference value for guarantees and tests concerning the load losses and temperature rises. A two winding transformer has only one value of rated power (identical for both windings). For multi-winding transformers the rated power of each winding should be stated. In the case of a three winding transformer, the rated power and voltage of each winding must be known in order to determine steady state operation.

The rated power for the three phase case given by IEC definitions is:

$$\text{Rated power} = \sqrt{3} \times \text{rated winding voltage} \times \text{rated winding current}$$

The rated secondary voltage is given as the no-load voltage on the secondary side of the transformer with rated voltage applied to the primary winding. This differs from the full load secondary voltage by the amount of voltage drop through the short circuit impedance of the transformer. According to this definition the rated apparent power is the power input, taken by the primary from the supply, and not the power output delivered to the load (which is smaller due to the secondary voltage drop on full load). This differs from the ANSI definition which defines rated power as the output supplied at rated secondary voltage.

14.6.8.3 *System parameters*

It is important that the transformer designer receives sufficient information on the system to determine the conditions under which the transformer will operate. The following information must be supplied together with any special conditions relating to the installed location, increased clearances and creepages due to atmospheric pollution, etc.:

• Range and variation of system voltage and frequency
• Required insulation levels of line and neutral terminals
• System fault level
• Altitude if in excess of 1000 m.

14.6.8.4 *Technical particulars and guarantees*

Table 14.2 Parts 1 to 3 are intended to assist the reader in covering all the points raised in this chapter when specifying oil-immersed power transformers. Another similar set of guidance – particularly covering electrical aspects – is given in annexes A and B of IEC60076-1. Details of test requirements are covered in more detail in Chapter 19. Environmental conditions together with general power system technical details (primary and secondary nominal

Table 14.2   Part 1 – Power transformer technical particulars and guarantees

Item	Characteristics
General:	
Manufacturer	
Design standard	IEC 60076, etc.
Indoor/outdoor	
Type:	Generator, station, distribution, etc.
Isolation: Separate winding	
Auto	
Booster	
Tertiary winding	
Construction: Core	
Shell	
3 or 5 limb	
Cooling: Oil immersed	ONAN/ONAF/OFAF, etc.
Hermetically sealed	
Dry type cast resin	AN/AF, etc.
Mounting; skid/wheel etc.	
Number of phases	1 or 3, etc.
Rating: ONAN	MVA
ONAF	MVA
OFAF	MVA
Other	MVA
Ratio: (No load, principle tap)	kV
Impedance voltage:	Specify HV/LV, HV/Tertiary, LV/Tertiary
Full load, principle tap,.........rating	% (note the ONAN, ONAF etc. rating must be clearly indicated)
Tolerances	$\pm$ ....%
Full load, lowest voltage tap (min. No. ....turns), rating	%
Tolerances	$\pm$ ....%
Full load, highest voltage tap (max. No. ....turns), rating	%
Tolerances	$\pm$ ....%
Short circuit impedance at:	
Lowest tap (min. No. turns) & tolerances	Ω/phase $\pm$ .... Ω
Highest tap (max. No. turns) & tolerances	Ω/phase $\pm$ .... Ω
Vector group:	
Tap changer:	
Manufacturer	
Type reference	
Type: Off-circuit	
Off-load	
On-load	
Automatic/manual	
Range	+% −%
Number of steps	
Location: HV or LV side	
Required for parallel operation?	Yes/no
Terminal Arrangements:	(Details of bushings, bushing CTs,
HV	cable box arrangements, segregation,
HV neutral	dry type terminations, numbers of
Tertiary (if required)	cables to be accommodated, etc.)
HV earthing	
LV earthing	
Tertiary earthing (if required)	

(Continued)

Table 14.2   Continued

Item	Characteristics
Noise:	(Maximum A-weighted sound pressure levels measured to IEC 60076-10)
Transformer only	dBA
Transformer plus cooling at each cooling stage	dBA
Flux density:	
Lowest tap	Tesla
Principle tap	Tesla
Highest tap	Tesla
Impulse withstand:	(1.2/50 μs waveform)
HV	kVp
LV	kVp
Tertiary	kVp
Neutral terminals	kVp
Power frequency withstand:	(1 minute)
HV	kV rms
LV	kV rms
Tertiary	kV rms
Neutral terminals	kV rms

Note: Details of system parameters and environmental conditions (voltage and frequency variations, altitude if in excess of 1000 m, temperature, etc.) also to be specified.

voltages and tolerances, frequency, earthing arrangements, BIL, etc.) and a single line diagram showing the basic protection arrangements must also be included with the tender specifications.

Table 14.2 Part 1 covers the main electrical aspects. Remember to quote the no-load voltage ratio and to allow for regulation. The required percentage impedance must also be specified at a known transformer rating and tap. This is especially important for dual rating transformers (ONAN/ONAF, etc.) in order to avoid confusion. Such items as flux density are best determined in conjunction with the manufacturer. Any loss capitalization formula must be included at the enquiry stage in order to allow the transformer manufacturer to optimize the design and for the purchaser to obtain competitive prices on an equal basis.

Table 14.2 Part 2 covers accessories and physical details which are necessary in order to make arrangements for transport to site and also to allow civil design to proceed.

Table 14.2 Part 3 covers possible overload requirements and transformer tests. Type and special tests are expensive and may not be necessary if similar units made by the same manufacturer are already in satisfactory service with test certification. In any event only one unit of each type being manufactured under a contract will normally require such special testing.

Table 14.3 details noise levels that leading manufacturers are able to conform to without serious cost penalties to the purchaser.

Table 14.2   Part 2 – Power transformer accessories and physical details, particulars and guarantees

Item	Characteristics
Fittings and equipment: (Yes/no)	
On-load tap changer switch	
Lifting lugs	
Jacking lugs	
Holding down bolts	
Wheel flanges for rail mounting	
Skid underbase	
Conservator	
Drain valve	
Filter valve	
Oil cooling system valve	
Oil sampling device	
Thermometer pocket	
Oil level indicator	
Silica gel breather	
Pressure relief device with alarm and trip contacts	
Earth terminals	
Oil level indicator with alarm contact	
Oil temperature indicator with alarm and trip contacts	
Oil temperature cooling system initiation contacts	
Gas/oil operated relay alarm and trip (gas and surge) contacts	
Surge arresters on HV side as part of overall substation design	
Surge arrester counters on each phase	
Cooling system fault relay	
Rating plate	
Bushing CTs to suit overall protection design	
Neutral CTs to suit overall protection design	
Auxiliaries supply voltage:	
1 phase	V Hz
3 phase	V Hz
Transformer weights and dimensions:	
Thickness of transformer tank:	
Top	mm
Sides	mm
Bottom	mm
Radiator and/or cooling tubes	mm
No. coolers or cooling banks	No.
Rating of each cooler or cooler bank	
Overall length	mm
Overall width	mm
Overall height	mm
Overall largest dimensions for transportation	L mm $\times$ W mm $\times$ H mm
Oil weights and capacities:	
Total quantity of oil required	litres
Filling medium of tank for shipment	(note some transformers are shipped under dry nitrogen)
Filling medium of coolers for shipment	
Oil quantity required to cover windings for shipment	litres
Total capacity of conservator	litres

(Continued)

Table 14.2   Continued

Item	Characteristics
Quantity of oil in the conservator between the highest and lowest visible levels	litres
Weight of core and windings	kg
Weight of each cooler complete with oil	kg
Total weight of transformer	kg
Weight of heaviest piece of transformer for shipment/transport	kg
Details of transport to site limitations	(Site location, lifting facilities at port and site, road and rail limitations, etc.)
Impact recorders	Yes/no or number required
Additional information:	
Brochure and technical detail references	

Table 14.2   Part 3 – Power transformer test details, particulars and guarantees

Item	Characteristics
Loading:	
Overloads	(to IEC 60354 or IEC 60905)
Temperature rise at rated output	
Windings	°C (65°C typical for oil-filled power transformer)
Top oil	°C (60°C typical for oil-filled power transformer)
Maximum ambient temperature	°C
Type of load	(to assist in determining I_{eq} rms – see Chapter 26)
Transformer test data:	
No-load loss at principal tap	kW
No-load loss at tapping with maximum loss	kW (only applicable with variable flux – e.g. auto-transformer with neutral end taps)
Fixed losses at full load and principal tap	kW
Load loss	kW
Maximum winding hot spot temperature at full load	°C (also specify applicable ambient temperature)
Maximum observable oil temperature at full load	°C (also specify applicable ambient temperature)
Maximum current density in winding at full load	
HV	A/mm^2
LV	A/mm^2
Magnetising current at principal tap	% of full load current (HV winding for transmission or distribution transformer)
Efficiency at principal tap, full load, unity power factor	%
Efficiency at principal tap, full load, 0.8 pf	%
Inherent regulation at principal tap an full load	
unity pf	%
0.8 pf	%

(Continued)

Table 14.2   Continued

Item	Characteristics
Winding resistance at principal tap (per phase)	
HV	Ω @ .... °C
LV	Ω @ .... °C
Auxiliary losses	kW
Factory test requirements:	(Yes/no)
Ratio test on all taps, polarity and voltage vector relationship	
Measurement of winding resistance at each tap	
Measurement of insulation resistance	
Induced overvoltage withstand test	
Separate source voltage withstand test	
Measurement of no-load loss and current	
Measurement of impedance voltage (principal tapping), short circuit impedance and load loss at rated current	
Measurement of power absorbed by cooling fans	
Functional test of auxiliaries	
Pressure test on radiators	
Zero phase sequence impedance measurement	
Relief valve pressure test	
Test of core assembly resistance to earth	
Tests and certificates for all bushings	
Dye penetrometer test for all welds	
Polarity tests on CTs	
Insulation resistance and voltage withstand for all motors, auxiliaries, controls and alarms	(Test voltage to be agreed with manufacturer)
Oil leakage test on tanks, conservators, piping, tap changer and disconnect chambers	
Verification of flux density	
Vacuum test on main tank	
Core magnetization	
Additional test requirements:	(These may not be necessary
Lightning impulse test	depending upon whether the
Heat run test	manufacturer has had similar
Bushing current transformer tests (accuracy, ratio, etc.)	transformer designs type tested in the past. Such tests are
Partial discharge test	relatively expensive and the tender
Bushing partial discharge test	enquiry should list such tests
Audible sound level test (to prove compliance with specified requirements, NEMA Specification TR1 Table 1–2, or similar)	separately and allocate provisional sums against each test should they be required. Only one transformer of each type being ordered need be subjected to such tests)

Table 14.3 Audible sound levels for oil-immersed power transformers

Column 1 – Class* OA, OW and FOW Ratings
Column 2 – Class* FA and FOA First-stage Auxiliary cooling**
Column 3 – Straight FOA* Ratings, FA* FOA* Second-stage Auxiliary Cooling‡‡

Equivalent two-winding rating/

Average Sound Level††, Decibels	350 kV 1	2	3	450, 550, 650 kV BIL 1	2	3	750 and 825 kV BIL 1	2	3	900 and 1050 kV BIL 1	2	3	1175 kV BIL 1	2	3	1300 kV BIL and Above 1	2	3
57	700																	
58	1000																	
59																		
60	1500			700														
61	2000			1000														
62	2500			1500														
63	3000			2000														
64	4000			2500														
65	5000			3000														
66	6000			4000			3000											
67	7500	6250#		5000	3750#		4000	3125#										
68	10000	7500		6000	5000		5000	3750										
69	12500	9375		7500	6250		6000	5000										
70	15000	12500		10000	7500		7500	6250										
71	20000	16667		12500	9375		10000	7500										
72	25000	20000	20800	15000	12500		12500	9375		12500								
73	30000	26667	26667	20000	16667	20800	15000	12500		15000								
74	40000	33333	33333	25000	20000	26667	20000	16667		20000			12500					
75	50000	40000	41667	30000	26667	33333	25000	20000	20800	25000	16667		15000					
76	60000	53333	53333	40000	33333	41667	30000	26667	26667	30000	20000	20800	20000	16667			16667	
77	80000	66667	66667	50000	40000	53333	40000	33333	33333	40000	26667	26667	25000	20000	20800		20000	
78	100000	80000	83333	60000	53333	66667	50000	40000	41667	50000	33333	33333	30000	26667	26667	20000	26667	20800
79		106667	100000	80000	66667	83333	60000	53333	53333	60000	40000	41667	40000	33333	33333	25000	33333	26667
80		133333	133333	100000	80000	100000	80000	66667	66667	80000	53333	53333	50000	40000	41667	30000	40000	33333
81			166667		106667	133333	100000	80000	83333	100000	66667	66667	60000	53333	53333	40000	53333	41667
82			200000		133333	166667		106667	100000		80000	83333	80000	66667	66667	50000	66667	53333
83			250000			200000		133333	133333		106667	100000	100000	80000	83333	60000	80000	66667
84			300000			250000			166667		133333	133333		106667	100000	80000	106667	83333
85			400000			300000			200000			166667		133333	133333	100000	133333	100000
86						400000			250000			200000			166667	133333		133333
87									300000			250000				166667		166667
88									400000			300000				200000		200000
89												400000				250000		250000
90																300000		300000
91																400000		400000

*Classes of cooling (see 2.6.1 of American National Standard C57. 12.00–1980)
**First- and second-stage auxiliary cooling (see TR 1 0.02)
†The equivalent two-windings 55°C or 65°C rating is defined as one-half the sum of the kVA rating of all windings
#Sixty-seven decibels for all kVA rating equal to this or smaller
††For intermediate kVA ratings, use the next larger kVA rating
‡‡For column 2 and 3 ratings, the sound levels are with the auxiliary cooling equipment in operation

REFERENCES

1. S. Austen Stigant and A. C. Franklin, *J & P Transformer Book*, 10th Edition, Newnes-Butterworths.
2. D. J. Allan, 'Power transformers – the second century', *IEE Power Engineering Journal*, Vol. 5, No. 1, January 1991.
3. H. Grant, Application of on-load tap changers to power transformers, IEE Discussion Meeting, Edinburgh, 1993.
4. R. O. Kapp and A. R. Pearson, the performance of star–Star transformers, *IEE Journal*, 1955.
5. J. Sturgess, Virtual valuations – using finite-element models to test power transformers can prevent embarrassing failures, *IEE Power Engineer*, April/May 2005.

15 Substation and Overhead Line Foundations

15.1 INTRODUCTION

The design of overhead line tower or substation gantry structure foundations must be such as to safely sustain and transmit to the ground the combined dead load, imposed load and wind load in such a manner as not to cause any settlement or movement which would impair the stability of the structure or cause damage. The settlement is a result of the transfer of load from the structure to the soil layers. Essentially settlement must be minimized to an acceptable level for the design life of the structure and adequate factors of safety applied to ensure this. Foundation design requires information on the properties of the soil and in particular its compressibility, moisture content, plasticity characteristics, friction between soil particles and for fine soils its undrained shear strength. This chapter describes typical soil investigations and foundation design. Such design is the responsibility of the civil engineer. The details described in this chapter are intended to give the transmission and distribution electrical engineer an appreciation of the factors involved.

15.2 SOIL INVESTIGATIONS

Ground investigations are carried out by geotechnical experts using boreholes, trial pits and penetrometer tests.

Investigations take the form of *in situ* and laboratory tests. *In situ* tests include standard penetration tests to provide data on the relative density to sand for the more coarse-grained soils.

Laboratory investigations on soil samples taken from boreholes or trial pits will measure grain size, density, shear strength, compressibility, chemical composition and moisture content such that the soil can be categorized.

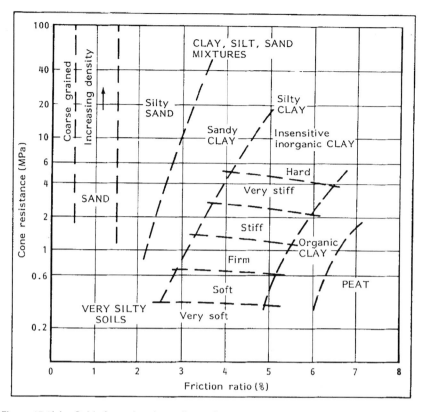

Figure 15.1(a)   Guide for estimating soil type from penetrometer testing

Figures 15.1a and 15.1b give a useful guide for estimating soil types based on cone penetrometer end resistance and friction ratio. Examples of Middle East and UK substation soil penetrometer site investigations are given in Figures 15.2a to 15.2c. Soil chemical test results, grain size distribution, consolidation and plasticity are given in Figures 15.3a to 15.3d, respectively.

Guidance on geotechnical design is provided in Eurocode 7 (EN1997).

15.3 FOUNDATION TYPES

The results from the soil investigation allow the civil engineer to decide which type of foundation will most economically support the structure. The actual practical solution will take into account the access requirements for a piling rig, availability and transportation of materials.

Large concrete raft type platforms as shown diagrammatically in Fig. 15.4a are used where the upper layers of soil have relatively low bearing capacities.

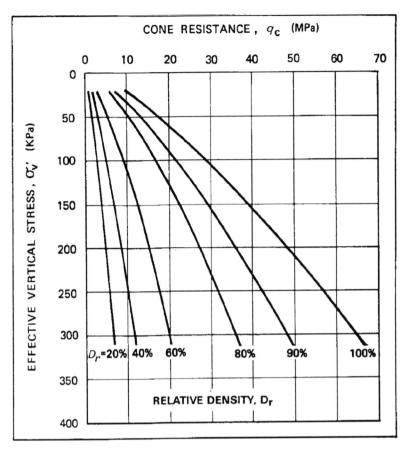

Cone End Resistance (q_c) (MPa)	Compaction of Fine Sand	S.P.T. (N)	Relative Density D_r (%)	Angle of Internal Friction Φ (degrees)
<2	very loose	<4	<20	<30
2-4	loose	4-10	20-40	30-35
4-12	medium dense	10-30	40-60	35-40
12-20	dense	30-50	60-80	40-45
>20	very dense	>50	80-100	>45

Figure 15.1(b)   Guide for estimating soil type from penetrometer testing

This type of foundation will evenly distribute the load over a wide area thus avoiding potential bearing capacity failure and ensuring that any settlement will be acceptable and even. Figure 15.4b shows a close-up view of 'pad and chimney' foundation steelwork during concrete pouring operations.

Pile foundations, Figures 15.5a and 15.5b, are necessary for very poor soils and are used where the weight of the structure is likely to cause bearing

RECORD OF BOREHOLE No: EB1

START DATE: 7TH JANUARY 1992

DRILLING METHOD: CABLE PERCUSSION

EQUIPMENT: PILCON WAYFARER

CASING DETAILS: 200mm to 17.80m / 150mm to 27.50m

BOREHOLE DIA: 200mm to 19.50m / 150mm to 30.00m

DRILL FLUID:

SHEET 1 OF 3
ORIENTATION: VERTICAL
CO-ORDINATES E 549174 N 182717
GROUND LEVEL 1.65 m.O.D.

Date & Time	Casing Depth (m)	Water Level - m (Flush Return %)	SAMPLE/CORE RECOVERY Depth (m)	Type Total %	No. RQD (SCR)	Core Size	Fracture	DESCRIPTION OF STRATA	Depth -m (Thickness)	Level m.O.D	Strata Symbol
7/1								MADE GROUND: Reinforced concrete.	(0.30) 0.30	1.35	
	NONE	DRY	0.50 - 1.00 C(2)	B	1			MADE GROUND: Very loose, dark grey, sandy (fine to coarse) subangular fine to coarse gravel (including ash, cinders and brick) and firm brown silty slightly sandy (fine to coarse) clay with some to much subangular fine and medium gravel (including flint, brick and ceramics).	(0.50) 0.80	0.85	
	1.50	DRY	1.30 1.50 - 2.00 (0.50m Rec)	D P	2 3			Firm, grey and brown, slightly organic very silty CLAY. (ALLUVIUM)	1.30	0.35	
			2.00 - 2.50	B	4			Soft to firm, grey slightly organic silty to very silty CLAY. (ALLUVIUM)	(1.40)		
	2.50	DRY	2.50	U [10]	5				2.70	-1.05	
			3.00 3.00 - 3.50	D B	6 7			Firm, brown, clayey, locally very clayey, fibrous and amorphous PEAT, with occasional roots and wood fragments.			
	3.50	DRY	3.50	U [14]	8			(PEAT DEPOSITS)			
			4.00	D	9			Below 4.00m: Becoming slightly clayey, amorphous peat.	(3.10)		
	4.00	DRY	4.50	U [10]	10						
			5.00	D	11						
			5.50	D	12			Below 5.50m: Becoming fibrous and amorphous peat.			
			5.80	D	13				5.80	-4.15	
	6.00	DRY	6.00	U [14]	14			Soft to firm, blue, green grey, locally slightly organic very silty CLAY, with occasional pockets of brown amorphous peat, and occasional gravel (noted by the driller).	(0.90)		
			6.50 6.50 - 7.00	D B	15 16			(ALLUVIUM)			
7/1 8/1	6.00 6.00	DRY 2.80	2.80	W	17				6.70	-5.05	
	7.00	3.10	6.70 7.00 - 7.50 C(7)	D B	18 19			Loose, grey brown, slightly sandy (medium and coarse) subangular to subrounded fine to coarse GRAVEL. (TERRACE GRAVEL)			
			7.75	D	20				(6.50)		
	8.50	2.70	8.50 - 9.00 C(14)	B	21			Below 8.50m: Medium dense.			
			9.25	D	22						
8/1								(Continued on Sheet 2).			

REMARKS
1. Concrete was broken out by hand from ground level to 0.30m (2 hours).

Logged by: CRR   Date: B4/1
Checked by:
Approved by:

SCALE 1 : 50 (When reduced to A4)

FOUNDATION & EXPLORATION SERVICES

BARKING REACH POWER STATION
GROUND INVESTIGATION
PHASE II

CONTRACT No 2085
FIGURE 2.0

Figure 15.2(a)   Soil investigation borehole profile (courtesy of Foundation and Exploration Services)

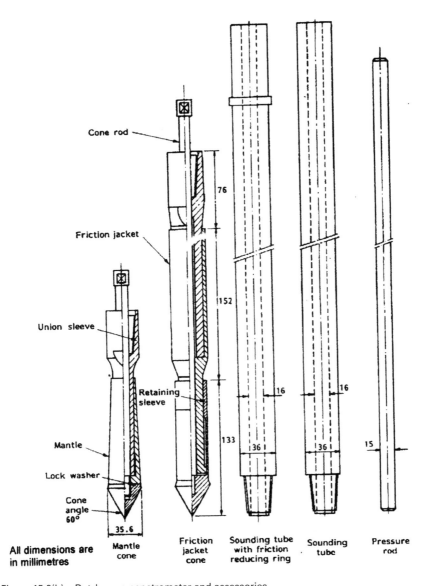

Cone rod

Friction jacket

Union sleeve

Retaining sleeve

Mantle

Lock washer

Cone angle 60°

76

152

16

16

133

36

36

15

35.6

All dimensions are in millimetres

Mantle cone

Friction jacket cone

Sounding tube with friction reducing ring

Sounding tube

Pressure rod

Figure 15.2(b)   Dutch cone penetrometer and accessories

capacity failure or excessive settlement of the upper layers. The pile foundation transmits the load to the lower and more stable areas of ground. Several piles may be required for each tower foundation depending upon the load capacity. Bored, cast-*in-situ* piles typically utilize temporary steel casings which are bored into the ground and removed during concreting. Alternatively, the piles can be steel or precast concrete sections which are driven into the ground. Less common are galvanized steel screw anchor piles which are screwed into the ground until the required resistance is achieved. They are then left in the ground

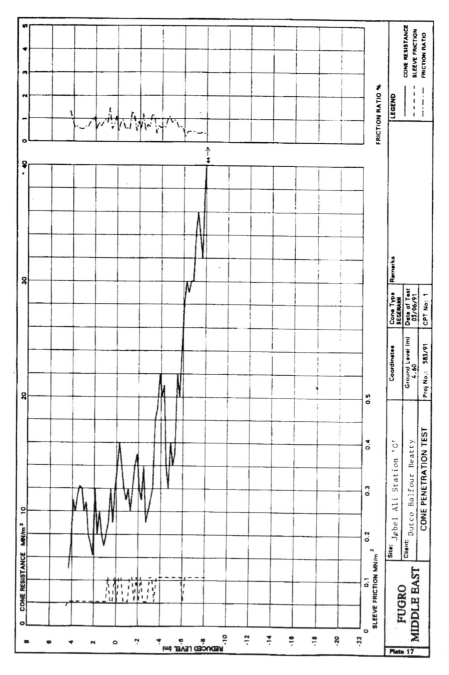

Figure 15.2(c)   Dutch cone penetration test

SUMMARY OF CHEMICAL ANALYSES

Borehole/Sample No.	Depth m	Date/Time of Sampling	Date/Time of Testing	Sulphate Content (as SO₃)			Chloride Content (as Cl)		pH Value	Organic Content % (Dry Mass)	Carbonate Content % (By Mass)	Remarks
				Soil % (Dry Mass)	2:1 Water:Soil Extract g/l	Groundwater g/l	Soil % (Dry Mass)	Groundwater g/l				
EB1/1	0.5			0.01	–	–	–	–	9.6	–	–	
EB3/7	2.6			–	–	–	–	–	–	70.2*	–	* Determined by loss on ignition at 450°C.
EB3/9	3.1			0.59	0.45	–	–	–	5.7	–	–	
EB3/13	5.0			–	–	–	–	–	7.2	1.1	–	
EB3/15	6.0			0.03	–	0.06	–	–	7.2	–	–	
EB3/17	6.5			0.01	–	–	–	–	7.6	–	–	
EB5/4	1.6			0.01	–	–	<0.01	–	8.8	69.5*	–	
EB5/7	3.0			–	–	–	–	–	–	–	–	
EB5/14	6.0			–	–	0.08	–	0.375	7.4	–	–	
EB5/20	8.45			0.02	–	–	0.05	–	8.4	–	–	
EB7/7	1.4			0.75	1.78	–	–	–	8.1	–	–	
EB7/21	6.2			0.02	–	–	–	–	8.3	–	–	
EB8/6	3.0			–	–	–	–	–	–	66.5*	–	
EB8/12	6.6			–	–	–	–	–	–	1.4	–	
EB10/8	4.5			–	–	–	–	–	–	34.2	–	
EB10/12	6.0			–	–	0.05	–	–	7.1	–	–	
EB11/7	3.5			0.42	1.41	–	–	–	7.3	–	–	
EB11/8	4.0			–	–	–	–	–	–	63.7*	–	
EB11/12	6.5			–	–	–	–	–	–	.1	–	

* EB5/15 & 45 See Sheet 2

FOUNDATION & EXPLORATION SERVICES

Compiled by	Date	Checked by	Date	Approved by	Date	For Contractor	Date	For Engineer	Date
MAI	3/2	SR			4/2				

BARKING REACH POWER STATION GROUND INVESTIGATION PHASE II

Sheet 1 of 7   CONTRACT No 2085   FIGURE 288

Figure 15.3(a)   Soil chemical test results (courtesy of Foundation and Exploration Services)

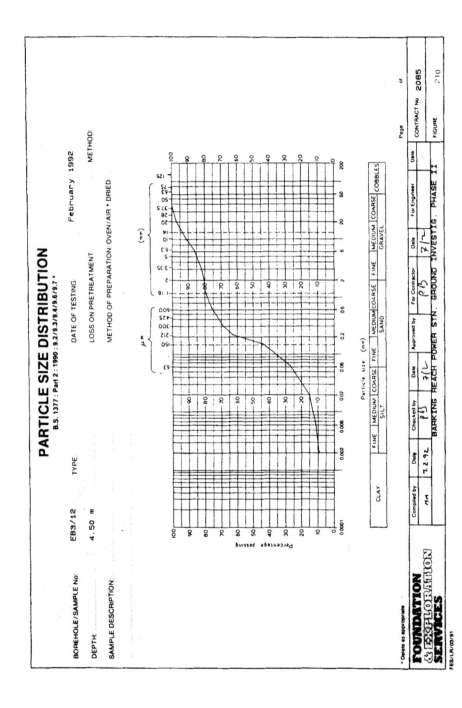

Figure 15.3(b)   Soil grain size distribution

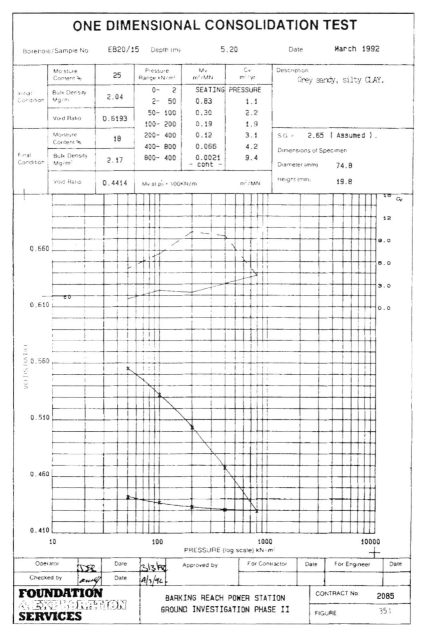

Figure 15.3(c)   Soil consolidation test results

and the tower attached directly to the screw anchor stubs. They are more expensive than traditional piling methods because of the material costs but they provide an extremely fast method of construction. The pile installation machine consists of a hydraulic unit attached to a JCB, Poclain or similar type of digger.

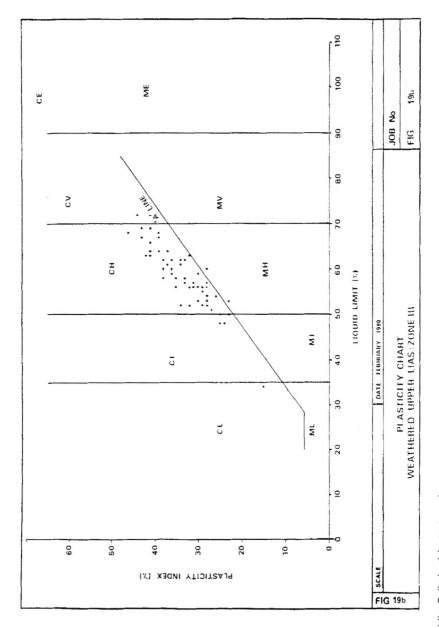

Figure 15.3(d)   Soil plasticity test results

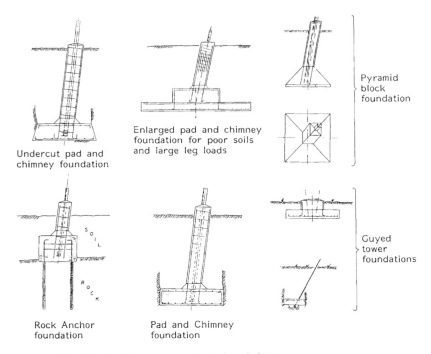

Undercut pad and
chimney foundation

Enlarged pad and chimney
foundation for poor soils
and large leg loads

Pyramid
block
foundation

Rock Anchor
foundation

Pad and Chimney
foundation

Guyed
tower
foundations

Figure 15.4(a)   Overhead line power and pole foundations

Figure 15.5b shows an auger for forming bored cast-*in-situ* concrete piles at a
400 kV substation site.

Denser soils with high bearing capacity only require small rectangular
foundations or strip footings to transfer adequately the structure weight to the
soil. Undercutting arrangements ensure stability in uplift conditions (see Fig.
15.6a and 15.6b).

Hard ground or rock conditions require anchor foundations where reinforcement is grouted into predrilled holes in the rock.

Simple distribution pole foundations may not require concrete. The pole is
lowered into a predrilled hole with compacted soil backfill.

15.4 FOUNDATION DESIGN

The wind loading on the towers will result in uplift or compression forces being
transmitted through the tower legs to the foundations. Terminal or heavy angle
towers will have tower leg foundations remaining in uplift or compression
although a check is necessary to ensure broken wire conditions do not reverse
the effect. Straight line or light angle towers can have the loading reversed
depending upon the wind direction and therefore the foundations for each tower
leg must be capable of restraint in both modes (see Fig. 15.7). EN 50341

Figure 15.4(b)   Pad and chimney foundation steelwork – concrete pour in progress

requires 'special attention … to be paid to the interaction of loads resulting from active soil pressures and the permanent weight of foundation and soil'. It also warns of the buoyancy effect of ground water on soil and foundations.

EN50341-3-9 also has specific requirements on precautions relating to uplift.

15.5 SITE WORKS

15.5.1 Setting out

Accurate setting out is essential in substation and overhead line work in order to match the prefabricated assemblies and structures to the foundation holding-down arrangements.

Figure 15.5(a)   Auger for forming bored cast-*in-situ* concrete piles (courtesy of Stent Foundations, Balfour Beatty)

15.5.2 Excavation

The specifications issued with the tender documentation associated with the works must include details for shoring up trenches and excavations in order to safeguard workers from collapse. In addition, it should be noted that cable trench depth, cable surround and backfill material have a direct effect upon cable ratings. In a similar way foundation depths may affect the final height of overhead line supporting structures and associated clearances. Therefore if cable trench or overhead line structure excavation depths differ from those specified then the civil engineers must be made aware that the electrical characteristics may be compromised.

15.5.3 Piling

Sample load checks on a random selection of piles must be included in the civil works specifications in order to prove that the predicted load bearing capability has been obtained in practice. This is particularly important for bored piles. For driven piles or screw anchor foundations the torque converter or resistance limit is set on the pile driving machine and this must be regularly checked for correct calibration.

Figure 15.5(b)   Auger for forming cast-*in-situ* concrete piles – 400 kV Sellindge substation, Kent

The load tests measure settlement under pile loading and recovery on removal of the test load. For a satisfactory proof load test on a trial pile, loads at 2 × working load should not result in excessive settlement. Routine checks during construction works on actual foundation piles at 1½ × working load are typical.

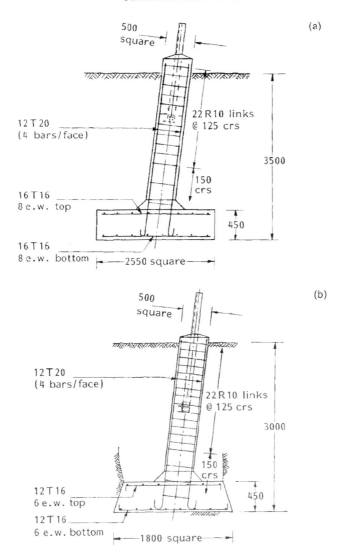

Figure 15.6(a) Standard pad foundation; (b) undercut pad foundation

15.5.4 Earthworks

An economic design will attempt to match cut and fill earthwork quantities. This might be possible by using the 'cut' to prepare the substation level switchyard. As long as cable or line entries to the substation are still possible the resulting 'fill' may then be used to form embankments around the switchyard thereby reducing the environmental impact.

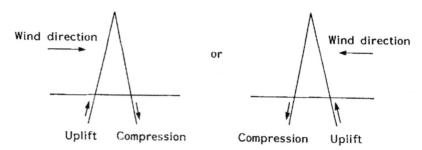

Figure 15.7   Uplift and compression on tower foundations

Soils must be deposited in shallow layers. Voids are reduced and the ground consolidated by compacting each layer in turn. The final soil density of the 'fill' should be within 90% of the optimum density obtained by a standard Proctor compaction test.

15.5.5 Concrete

15.5.5.1 *Concrete strength*

Concrete specification is the responsibility of the civil engineer. The electrical engineer needs to appreciate the fundamentals and the terminology. Concrete design is based on the required characteristic strength measured in N/mm^2. Concrete gains strength over time as the concrete 'cures'. Concrete cube samples are taken from each batch and measured for strength in a materials laboratory often after 7 and 28 days. Results from the tests taken after 7 days will give a good guide to the 28-day characteristic strength. The curing period will of course impose constraints upon the timing for the installation of switchgear and steelwork structures on the foundations. Concrete may be quoted in terms of its aggregate size and 28-day characteristic cube strength as, say, 30/20, meaning $30\,N/mm^2$ strength and 20 mm nominal maximum aggregate size. Figure 15.8 shows concrete site investigations taking place during construction of the Yanbu-Medina 380 kV transmission line in Saudi Arabia. The 'slump cone' gives details of the nature and consistency of the concrete. The cube moulds are used to form the concrete cubes for determination of concrete strength.

The overhead lines standard EN50341 refers to Eurocode 2 (EN 1992) for concrete foundation design.

15.5.5.2 *Concrete durability*

Concrete durability depends upon the degree of exposure, the concrete grade (or strength) and the cement content. A high density, alkali-resistant concrete

Figure 15.8   Concrete cube and slump cone samples being prepared –Yanbu-Medina
380 kV transmission line, Saudi Arabia (courtesy of Balfour Beatty Projects and Engineering)

will better resist the effects of moisture penetration. Since concrete is a porous
material reinforcement bars within the concrete will be subject to possible
corrosion if safeguards are not taken. Adequate concrete 'cover' to the reinforcement should be included in the specifications in order to reduce moisture or
salts penetrating to the rebar. Should the rebar corrode it will expand and the
considerable forces involved cause the concrete to crack.

15.5.5.3 *Concrete slump*

The amount of water present in a mix has a large influence on the strength and durability of concrete. Too much water will not only decrease strength and durability but could also be the cause of shrinkage cracks as the concrete dries out. The optimum water content of the mix should be determined by trial mixes thereby allowing the slump to be defined. This desired slump can then be checked on site by means of a slump test.

15.5.5.4 *Concrete curing*

Curing is the stage of concrete construction where chemical reactions under controlled conditions ensure that the concrete correctly gains its design strength. Concrete will gain strength rapidly at first and an initial 'set' takes place within a few hours and a good strength after 3 days. About 60% of final strength is gained after 7 days and full strength is assumed to have been achieved after 28 days although the process goes on for many months. The concrete should be kept damp during the curing process by using modern chemical sprays, or the more traditional wet sacking. Concrete surfaces must be protected from dry windy conditions or intense sunlight by polythene sheeting or sacking screens in order to avoid rapid surface evaporation.

15.5.5.5 *Cracks*

Uncontrolled cracking of a concrete member can seriously affect its structural integrity. Cracks in reinforced concrete members should not generally exceed 0.3 mm in width in order to avoid possible deterioration of the reinforcing steel. Cracking must, therefore, be controlled. This is initially done by the designer in his prudent spacing of reinforcement and also on site by adequate curing as described in Section 15.5.5.4. On large-volume concrete pours insulated shutters and adequate cooling may also be necessary in order to prevent thermal cracking.

15.5.5.6 *Site supervision*

The electrical engineer may be called upon to assist with site supervision and should look out for the following points regarding concrete foundation works:

1. Materials, including rebar, should be checked for cleanliness and grading.
2. Sand and aggregate storage conditions must be such as to keep materials clean. Materials must be clearly identified in order to avoid accidental mixing of wrong components. A batching plant (even if not fully automated)

should have a concrete hardstanding and bunkers for the easy delivery of materials by lorry and segregation of different grades of aggregates. Cement must be stored in a damp-proof building and used in the order in which it is delivered – first in, first out.

3. Volume or weight measurement of the different concrete components may be used to obtain the required mix. Weight measurement is to be preferred since the volume of a fixed amount of fine aggregate can vary considerably depending upon its moisture content. Full bags rather than part bags of cement should be used in any one mix. The mixing of structural concrete must always be done by machine.

4. Trials on the specified mix should be done in advance before full construction work begins on site in order to prove that the local materials match up to the expectations of the specifications. Such tests must form part of the overall concrete cube strength programme that should continue throughout the construction period. It is essential that additional water is not added during the work since excess water will produce a porous concrete with low strength.

5. Skilled carpenters should be used to prepare firm shuttering for concrete placement. Concrete compaction must be achieved with the aid of mechanical vibration equipment. This also allows even mixing between the different pours of concrete into the foundation. Such a process should be as continuous as possible in order to avoid weak joints between the pours.

6. The curing process must be controlled by ensuring that the concrete is kept moist.

15.5.6 Steelwork fixings

Overhead line towers connect to stubs which form an integral part of the foundation. Stubs or fixing bolts are locked into the foundation reinforcement as part of the design.

Large items of substation plant such as power transformers may merely sit on the foundation and no special fixing arrangements are required. In environmentally sensitive areas vibration pads may be required between the transformer and the base, and then it is *essential* that no rigid fixing exists between the plant and the concrete mass. Stability is achieved by the weight of the transformer itself. For smaller transformers with wheel fixings arrangements are made to lock the wheels in position. Larger transformers tend to be skid mounted and steel runner plates may form part of the foundation design such that the transformer may be slid into position without damage to the concrete surface (but again, special precautions may be needed to avoid damaging vibration pads).

Substation steelwork structures or switchgear may be connected to the foundations either by setting fixing bolts into the concrete foundation or by leaving pockets in the foundation for future grouting-in of fixing bolts. Accurate setting out is essential when the fixing bolts form an integral part of the foundation. A wooden template should be used to hold the bolts in the correct position during concrete pours. Pockets left in the foundation allow more flexibility. Bolts may be adjusted in position in the pockets and then grouted-in in their final position after matching with the switchgear or steelwork. It is essential that the correct bolts are used and that all connections are correctly tightened.

16 Overhead Line Routing

16.1 INTRODUCTION

Before any design or planning work on an overhead line is contemplated the national and regional authorities must first be consulted in order to ensure no statutory regulations are being contravened with regard to:

- safety factors on supporting structures and conductors
- maximum conductor working temperatures
- clearances between accessible ground and conductors
- minimum separation between the overhead line and railways, telegraph lines and pipelines
- planning permissions.

The logical sequence for the design and planning of the routing of an overhead line is shown in Fig. 16.1. It is assumed in this example that the client, or his consultant, carries out the preliminary routing and includes this information in a tender specification such that competitive tenders may be received from a variety of design and construct contracting organizations. The contractor will then carry out the detailed line routing and profile work. By careful preliminary routing the effect on the environment may be minimized. In addition, the client is in a position to narrow the choice of the tower and span design to the most economic. At the same time the client is able to take into account a strategy for minimum maintenance costs.

16.2 ROUTING OBJECTIVES

The preliminary routing work determines the physical constraints involved and allows the establishment of the least-cost solution for the overhead line.

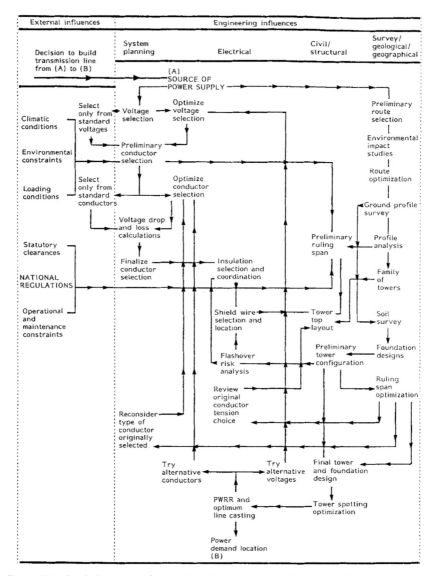

Figure 16.1   Logical sequence for overhead line design, planning and routing

Estimated quantities for the towers, foundations and conductors may then be included in tender documentation for the supply and/or erection of the overhead line.

The detailed routing survey and profile allows the towers to be located in the most economic manner. It will take into account proximity restrictions and maintenance of specified design parameters such as electrical clearances, wind spans, angles of deviation, etc.

16.3 PRELIMINARY ROUTING

16.3.1 Survey equipment requirements

1. Good maps for the expected line route and adjacent areas to a suitable scale; typically 1:10 000 for cross-country work.
2. Good survey quality compass and compass bearing monocular. These greatly assist orientation and road or river crossing locations and projection of sections.
3. Theodolite and level may be worthwhile but they are not essential for the preliminary survey.
4. 100 m tapes.
5. Hammers, identification paint and pegs.
6. Ranging rods to enable checks and recording of location relative to centres of proposed angle towers.

16.3.2 Aerial survey

Simple aerial survey photographs, or if available, satellite photographs, greatly aid the routing designer and reduce the amount of time taken for the ground survey. The proposed route is indicated on the photographs in conjunction with maps of the area if these are available. However, corridor mapping can now be provided as a complete service using LiDAR (light detection and ranging) in conjunction with GPS (geographical positioning system) satellite systems. Specialist contractors can provide CAD (computer aided design) drawings of the full route plan and profile, on the basis of a helicopter overflight of the planned route (see Section 16.4.3).

16.3.3 Ground survey

The exact route may differ considerably from that proposed by studying maps and aerial photographs. This is because of the difficulty in obtaining wayleaves, overcoming local planning requirements and ensuring that any specific local landowner requirements that may be accommodated are considered.

16.3.4 Ground soil conditions

It may be possible to route the overhead line such that the chosen ground conditions favour low foundation costs. However, in practice huge savings are unlikely since considerable deviations are likely to be necessary and this will in turn increase the overall materials cost of the overhead line route.

Figure 16.2   400 kV single circuit twin conductor overhead line crossing the Zagros
Mountains in Iran (courtesy of Balfour Beatty)

16.3.5 Wayleave, access and terrain

Figure 16.2 shows an example of a 400 kV single circuit overhead line crossing extremely mountainous terrain in the Zagros mountains of Iran between
Reza Shah Kabir Dam and Arak. Overhead lines often cover areas without

Figure 16.3   Helicopter assisted conductor stringing – Hong Kong (courtesy of Balfour Beatty)

good communications access. In such cases the construction of overhead lines is greatly assisted by the use of helicopters. Figure 16.3 shows helicopter assisted conductor stringing for the China Light and Power Company in Hong Kong.

16.3.6 Optimization

16.3.6.1 *Practical routing considerations*

The sending and receiving ends of the transmission line from existing or future substations or tee-off points are first established and are usually well defined. The straight line between these two points must then be investigated to see if this really represents the cheapest solution. In practice, wayleave availability, access, ground conditions, avoidance of populated or high atmospheric pollution coastal or industrial areas, difficulties for tower erection and maintenance almost always require deviations from the straight line option. Further economic considerations involving parameters which are difficult to equate in purely financial terms such as impacts of the line on the environment must be considered. If a 400 kV tower costs approximately £100 000 to design and detail then the use of a limited number of standard designs may well prove cheaper than having a large number of special tower types necessary to achieve the more direct line route.

Lines should not be routed parallel to pipelines or other similar services for long distances because of possible induced current effects. Where this cannot be avoided minimum distances of, say, 10 m should be maintained between the vertical projection of the outer phase conductor of a 145 kV overhead line and the pipeline. Similarly, precautions should be taken with regard to proximity to gas relief valves or hydrants. Gaz de France sets threshold levels of maximum AC-induced currents in pipelines at $100 \, A/m^2$. Corrosion effects from AC should be negligible because of the current reversal but research shows a small polarizing effect which could lead to corrosion in the very long term. Oil companies also require minimum clearances between overhead line counterpoise (if installed) and buried steel pipes of, say, 3 m for the first 5 kA of earth fault current plus 0.5 m for each additional kA.

16.3.6.2 *Methodology*

Once the terminal points for the line have been established they are linked on the maps avoiding the areas mentioned in Section 16.3.6.1. Angle or section towers may be provided near the terminal points in order to allow some flexibility for substation entry and slack spans or changes to the future substation orientation and layout.

The proposed route is then investigated by walking or driving along the whole of the route. Permissions to cross private property must first be obtained from the landowner. The purpose of this thorough investigation is to ensure that the route is feasible and what benefits could accrue from possible changes.

The feasible preliminary route is then plotted on the maps to at least 1:10 000 scale. The approximate quantities of different tower types (suspension, 30° angle towers, 60° angle towers, terminal towers, etc.), conductor and earthwire established. Suspension towers will often account for more than 80% of the total number of towers required on the overhead line route and quantities must be optimized and accurately assessed. Ground conditions are recorded during the field trip in order to estimate the different tower foundations required (piled – and, if necessary, is access for a piling rig possible? – screw anchor, 'normal', rock, etc.). At the same time an estimate of the difficulties likely to be encountered in obtaining the required tower footing resistance and the need for a counterpoise, tower earth rods, etc., is made.

The cost of the line is proportional to the tower steel and foundation loads. The tower weight, W, may be approximated from Ryles formula:

$$W = C \cdot h \cdot \sqrt{M}$$

where W = tower weight of steel (kg)
   C = constant
   h = tower overall height (m)
   M = tower overturning moment under maximum loading conditions at ground level (kg/m)

Tower heights and their overturning moment are established for a variety of basic spans. This then allows the engineer to concentrate on the total number of intermediate towers required. Such an iterative procedure is, of course, entirely suited to computer analysis. However, the computer will place towers in inconvenient or impossible locations without the knowledge resulting from the field survey described. The costs of suspension insulator strings, fittings and foundations are then added to the estimated number of towers in order to derive the basic span and the first approximation to the cheapest overhead line routing solution.

A constant ratio is applied to each basic span in order to obtain the wind span. This constant (typically $1.1\times$) is necessary to allow for some flexibility over uneven ground. A factor of $2\times$ over the basic span may be used as a guide to the weight span (which does not greatly affect tower design) and this will allow for tower spotting and wind spans to be optimized.

The average span is the basic span multiplied by an efficiency factor which takes into account the nature of the ground and varying span lengths envisaged from flat to hilly terrain. The estimated quantities for materials may be derived from the average span. Finally, the technical specifications for the overhead line are drawn up for use in tender documentation.

16.4 DETAILED LINE SURVEY AND PROFILE

16.4.1 Accuracy requirements

The objective is to draw up a plan and section so that further refinement of the tower distribution may be made. The party carrying out this work will depend upon the type of contract being let by the electricity supply authority in charge of the works. An explanation of the suitability of different types of contract for different areas of transmission and distribution work is explained in Chapter 22.

The required accuracy should be to $\pm 0.5\,\mathrm{m}$ in the horizontal plane and to $\pm 0.1\,\mathrm{m}$ in the vertical plane. Greater accuracy is possible from survey data but in practice cannot always be easily transferred to the profile (but see comment in Section 16.3.2; LiDAR survey claims 'centimetre precision' and can be provided in a CAD form compatible with profiling). The location of angle and terminal towers is best specified in a contract document rather than allow a complete free hand to the overhead line contractor. This is because access may be an important parameter for the electricity supply authority if maintenance costs for the line are to be kept down.

The vertical profile ground line is surveyed from one angle or terminal tower to the next. When national maps of good quality are available the vertical survey data may be cross-referenced to bench marks of a known level. Horizontal survey dimensions to tower centrelines are checked against the 1:10 000 map and differences investigated until resolved on site. In hilly terrain side slopes in excess of $\pm 0.3\,\mathrm{m}$ must be recorded together with all major features (angles of deviation, other power overhead lines or cable routes, underground services, roads, rail, river and pipeline crossings, buildings adjacent to the wayleave, etc.).

16.4.2 Profile requirements

16.4.2.1 *Vertical and horizontal scales*

In order to keep the drawings to a manageable size the detailed survey drawings are scaled to typically 1:200 vertical and 1:2000 horizontal or as necessary in hilly terrain. On sloping sites it will be necessary to ensure that foundation depths are not compromised and individual tower legs may have to be adjusted to correct to the tower centre profile level. The profiles, whether computer generated or not, should be on graph type paper with a grid background. This greatly eases the reading of span lengths or clearances even when photocopy prints have slight distortions.

16.4.2.2 *Templates*

Traditionally the design line profile was prepared by hand using a sag template made from perspex ($\sim$3 mm thick) with all the engraving on the back using the same scales as the ground profile. A template would show:

- the maximum sag condition curve (usually at maximum temperature but could be under in extreme loading conditions)
- the minimum sag condition (usually at minimum temperature without ice loading)
- basic span and cases up to, say, $\pm$20% above and below the basic span.

Normally the parabolic approximation will suffice for distribution work unless special long spans or hilly terrain with slopes $>$15° are envisaged. For transmission lines a full catenary calculation may be more appropriate (see Section 16.4.3). Using the parabolic approximation the tension for any equivalent span is then given by:

$$EA\alpha\left(t_2 - t_1\right) + \left[W_1^2 \cdot g^2 \cdot L^2 \cdot \frac{EA}{24T_1^2}\right] - T_1 = \left[W_2^2 \cdot g^2 \cdot L^2 \cdot \frac{EA}{24T_2^2}\right] - T_2$$

where E = modulus of elasticity MN/m^2
A = conductor cross-sectional area mm^2
α = coefficient of linear expansion per °C
t_1 = initial temperature °C
t_2 = final temperature °C

W = weight of conductor and may include wind and/or ice loadings
W_1 = initial conductor unit effective weight kg/m
W_2 = final conductor unit effective weight kg/m
g = gravitational constant (1 kgf = 9.81 N) 9.81 m/sec^2
L = span length m
T = conductor tension N
T_1 = initial conductor tension N
T_2 = final conductor tension N
S = sag m
B = ice weight constant (refer BSEN 50341 or national standard)
x = conductor diameter mm
y = radial thickness of ice mm
p = wind pressure N/m^2

The sag for different span lengths is then derived from:

$$S = \frac{W \cdot g \cdot L^2}{8T} \text{ m}$$

Ice weight per unit length = $By(y + x)$ kg/m

Wind load = $\dfrac{p(2y + x)}{1000}$ N/m (see Chapter 18, Section 5)

Effective conductor weight = $\sqrt{[(\text{weight of conductor + ice})^2 + \dfrac{(\text{wind load})^2}{g^2}]}$ kg/m

More than one technically acceptable solution for tower locations is always available and therefore the final test of acceptability is based on cost. It is essential that tower fittings, extensions and foundations are taken into account. In addition, clearances must not be infringed.

16.4.3 Computer aided techniques

It is now normal practice to use computer aided techniques to prepare the overhead line profile and with modern computer tools the sag/tension relationship may be calculated using full catenary equations. Chainage, level, vertical and horizontal angles may all be transferred directly from a modern theodolite via a portable computer to an office power line survey and computer aided drafting and design (CADD) facility. This eliminates the need for completion by the surveyor of a field record book and any transcription errors that might occur. In addition, some packages allow details of type of ground, ownership, etc., also to be recorded electronically during the survey. As a further feature such surveys may be linked into co-ordinates derived from geostationary 'geographical information system' satellites (see Section 16.3.2). It has been estimated that power utilities using such systems can achieve survey and data transfer time savings of between 40% and 50% and savings of 50% and nearly 80% for line design involving poles and towers, respectively (*Electricity International*, Vol. 3, March/April 1991).

Once field data has been transferred to the CADD tool the ground line profile may be automatically produced with all the annotations that the surveyor has included in the field. Overhead line structures may be 'spotted' at any point along the profile manually by the engineer or automatically by the computer and strung with any conductor type. The software library containing conductor, pole and tower information will then be used to calculate sag and tension for the given conductor, uplift forces on any structures and ensure ground clearances are not infringed at a user specified temperature.

Typical computer profiles may be generated using OSL® Software on an IBM® compatible personal computer workstation. This particular package also allows 3-D representation to detail such items as terminal tower downleads and jumper clearances. (OSL is a trademark of Optimal Software Ltd and produces a suite of programs including Powerline, Powercard, PowerSite, Towerline, TowerCad, Towerlog, TowerSite, PoleLine, PoleCad, PoleLog and PoleSite. Optimal Software is an Integraph Third Party Software Partner. IBM is a trademark of International Business Machines. Intel is a trademark of Intel Corporation.) Another suitable program is PLS-CADD.

17 Structures, Towers and Poles

17.1 INTRODUCTION

This chapter describes the basic input data required and terminology used for the design of substation steel structures, overhead line towers and poles.

The industry is currently revising its approach to the general concepts of tower design. The loadings and related strengths required for overhead line design have normally been determined by Statutory Instruments, the client's or the consulting engineer's specifications. The international standards applicable are IEC 60826, 'Design criteria for overhead lines', EN 50341 'Overhead lines exceeding 45 kV' and EN 50423 'Overhead electrical lines exceeding AC 1 kV up to and including AC 45 kV'. IEC 60826 gives clear and straightforward guidance using many of the graphs and tables directly from BS8100 and in its National Normative Aspects (NNAs) EN 50341-3-9 points to loading derived from the same standard for use in the UK. The same criteria are not necessarily applicable in all territories, where, for example, averaging time for wind speeds may be an issue. See also report IEC/TS 61774 (Overhead lines – meteorological data for assessing climatic loads).

The newer ENs offer the alternatives of a probabilistic approach to design or an empirical approach based upon national/local experience. Such issues are the responsibility of the specialist structural engineer. Therefore this chapter gives very basic examples to allow the electrical engineer to understand the fundamental principles and terminology involved rather than the specific methodologies.

It should also be noted that open-terminal substation equipment support structures are nowadays being fabricated more and more from aluminium alloy angle rather than from galvanized steel. The structures may be welded up and drilled to tight tolerances in the factory. The prefabricated structures are light weight and may be transported directly to site. Although there is an initial higher capital materials cost this is largely offset by not having to provide special corrosion protection finishes. In addition the aluminium alloy material has a low resistivity. Therefore earth connections from the substation earth mat to the base of the support structures are normally sufficient. Additional copper tapes to the 'earthy ends' of the insulator supports are not specifically required.

There is a wide range of applicable standards – Table 17.1 lists a selection.

Table 17.1   A selection of overhead line design standards

Reference	Description
IEC 60383	Insulators for overhead lines with a nominal voltage above 1000 V
IEC 60471	Dimensions of clevis and tongue couplings on string insulator units
IEC 60720	Characteristics of line post insulators
IEC 60797	Residual strength of string insulator units of glass or ceramic material for overhead lines after mechanical damage to the dielectric
IEC 60826	Design criteria for overhead transmission lines
IEC/TR 60828	Loading and strength of overhead transmission lines
IEC/TR 61774	Meteorological data for assessing climatic loads for overhead lines
EN 12465	Wood poles for overhead lines – durability requirements
EN 12479	Wood poles for overhead lines – signs, methods of measurement and densities
EN12509	Timber poles for overhead lines – determination of modulus of elasticity, binding strength, density and moisture content
EN12510	Wood poles for overhead lines – strength grading criteria
EN12511	Wood poles for overhead lines – determination of characteristic values
EN14229 (see Note 3)	Wood poles for overhead lines – requirements
EN 12843	Precast concrete masts and poles
EN 50341	Overhead electrical lines exceeding AC 45 kV. Part 3 covers all the different National Normative Aspects (NNAs)
EN50423	Overhead lines AC 1 to 45 kV – based on 50341 but provides specific simplifications or/changes
Eurocode 1 – EN1991	Basis of design and actions on structures; 1991-1-3 covers snow loads; 1991-2-4 covers wind loads
Eurocode 2 – EN1992	Design of concrete structures – 1992-3 covers concrete foundations
Eurocode 3 – EN1993	Design of steel structures
Eurocode 7 – EN1997	Geotechnical design
Eurocode 8 – EN1998	Design provision for earthquake resistance of structures
BS1990	Wood poles for overhead power and telecommunication lines
BS3288 – 2	Insulator and conductor fittings for overhead power lines – specification for a range of fittings. Other parts of this standard are superseded, or becoming so.
BS7354	Code of practice for the design of high-voltage open-terminal stations
BS8100	Lattice towers and masts. Part 1 is a Code of practice for loading; Part 2 is a guide to Part 1; Part 3 is a CoP for strength assessment of tower and mast members; Part 4 covers the loading of guyed masts

Notes: 1. See Table 18.1 for standards relating to conductors.
2. A much more extensive schedule of related standards is provided in EN 50341-1
3. At the time of writing, EN14229 has only been issued as a draft for public comment.

17.2 ENVIRONMENTAL CONDITIONS

17.2.1 Typical parameters

In order to match both the mechanical and electrical characteristics of the overhead line or substation arrangement the environmental conditions and climatic details must first be collected and analysed. The following parameters are required (for installations in the EU refer to the Eurocodes 1, 2, 3, 5, 7 and 8 (EN1991–1998), although where there is conflict EN 50341 prevails):

Maximum ambient shade temperature	°C
Minimum ambient shade temperature	°C
Maximum daily average temperature	°C
Maximum annual average temperature	°C
Maximum wind velocity (3-second gust)	km/hr
Minimum wind velocity (for line rating purposes)	km/hr
Solar radiation	mW/sq m
Rainfall	m/annum
Maximum relative humidity	%
Average relative humidity	%
Altitude (for insulation level)	m
Ice (for loading conditions)	
Snow (for loading conditions)	
Atmospheric pollution	light, medium, heavy, very heavy
Soil type	clay, alluvial rock, etc.
Soil temperature at depth of cable laying	°C
Soil thermal resistivity	°C m/W
Soil resistivity	ohm-m
Isokeraunic level	thunderstorm days or lightning flashes to ground per km^2
Seismic factor	

17.2.2 Effect on tower or support design

17.2.2.1 *Wind load*

It is normal practice to consider wind loads on structures due to a 3-second gust that occurs over a 50-year period. This basic wind speed figure is to be obtained from meteorologists. On overseas work it may be difficult to obtain data as records may not have been accurately kept over such a period. The wind load is related to the wind speed in accordance with the code of practice applicable to the country where the work is being carried out. In the EU the relevant information is set out in the NNAs of EN50341 and 50423. It describes procedures for calculating wind loads on both structures and conductors, with

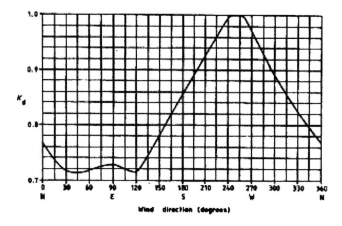

Figure 17.1 G_b Wind direction factor (from EN50341)

considerable variation in detail between individual countries. The following example uses the UK NNAs.

The site reference wind speed, V_r is the mean hourly wind speed at a level 10 m above the effective height of obstructions 'appropriate to the site terrain' and is given by:

$$V_r = \gamma_v \, K_d \, K_R \, V_B \tag{17.1}$$

where V_B(m/s) = Basic wind speed – obtainable from wind maps in
              BS8100 Part 1
        K_d = Wind direction factor – see Fig. 17.1, or site records
        K_R = Terrain roughness factor – see Table 17.2 or site records
        γ_v = Partial safety factor on wind speed = 1 for reliability
              level I (50 years)

The variation of wind speed with height for sites in level terrain is given by

$$V_z = V_r(\{z - h_e\}/10)^\alpha \quad \text{for } z \geqslant 10 + h_e \quad \text{or } (V_r/2)(1 + z)/ \tag{17.2}$$
$$(10 + h_e) \quad \text{for } z < h_e$$

where V_z = Mean wind speed at height z in metres above ground level
        α = Power law index of speed with height – see Table 17.2
        h_e = Effective height of surface obstructions – see Table 17.2

The dynamic pressure q_z at height z shall be taken as:

$$q_z = (\rho_a/2)Vz^2 \tag{17.3}$$

where ρ_a = The density of air, taken as 1.22 kg/m^3 for UK

Table 17.2 G_B terrain characteristics (from BSEN 50341-3-9)

Category	Terrain description	Terrain roughness factor K_R	Power law index of variation of wind speed with height α	Effective height h_e (m)
I ($Z_0 = 0.003$ m)	Snow covered flat or rolling ground without obstructions; large flat areas of tarmac; flat coastal areas with off-sea wind	1.20	0.125	0
II ($Z_0 = 0.01$ m)	Flat grassland, parkland or bare soil, without hedges and with very few isolated obstructions	1.10	0.140	0
III ($Z_0 = 0.03$ m)	Basic open terrain, typical UK farmland, nearly flat or gently undulating countryside, fields with crops, fences or low hedges, isolated trees	1.00	0.165	0
IV ($Z_0 = 0.10$ m)	Farmland with frequent high hedges, occasional small farm structures, houses or trees	0.86	0.190	2
V ($Z_0 = 0.30$ m)	Dense woodland, domestic housing typically covering 10% to 20% of the plan area	0.72	0.230	10

Notes: 1. Z_0 is the terrain aerodynamic roughness parameter.
2. The lower (smoother) of any two possible categories should be adopted in case of doubt or where environs may change.
3. The terrain description should apply to environs extending several km upwind from the site.
4. Higher (rougher) categories that occur within only a few km upwind may not be sufficiently extensive to develop an 'equilibrium wind profile' and should generally be ignored.
5. In urban areas ($Z_0 = 0.8$ m) where towers rise above the general level of surrounding buildings, category V should be used. Specialist advice should be sought where adjacent high buildings could affect design.

It is considered good practice to apply a gust factor at this stage, to derive the 'gust wind pressure' q_z', where:

$$q_z' = q_z(1 + K_{com} G_b)$$

K_{com} = a combination factor to take account of the improbability of maximum gust loading on both conductors and towers occurring simultaneously. It may be conservatively taken as 1.0

G_b = the basic gust factor for the support, depending on the height of the support. See Fig. 17.2.

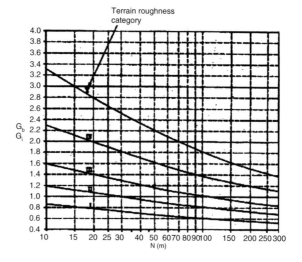

The basic gust response for towers, G_b, Insulators and Fittings, G_i, is given by the maximum of G_x or G_y where

$$G_x = K_1\, K_2\, (3{,}976/K_R - 2{,}485)$$

$$K_1 = (1+\alpha/2)\,(10/H)^\alpha$$

$$K_2 = \{2/S_1 + 2/S_1{}^2\,(e^{-S1} - 1)\}^{0.5}$$

$$S_1 = (H/100{,}8)\,(10/H)^\alpha$$

$$G_y = K_3\, K_4\, K_5\, (3{,}976/K_R - 2{,}485)$$

$$K_3 = (1+\alpha/2)\,(10/K_6)^\alpha$$

$$K_4 = \{2/S_2 + 2/S_2{}^2\,(e^{-S2} - 1)\}^{0.5}$$

$$K_5 = (K_6/H)^\alpha\,\{1-(1-K_6/H)^2\}\,/\,\{1-(1-K_6/H)^{\alpha+2}\}$$

$$S_2 = (K_6/100{,}8)\,(10/H)^\alpha$$

$$K_6 = H/10 \text{ but not less than } 10\text{m}$$

Figure 17.2   Basic gust response factor for towers (G_b) and insulators and fittings (G_i) (from BSEN 50341)

17.2.2.2 Wind loading example

The wind load on a structure is calculated by considering the areas of the structure exposed to the wind. The maximum wind load in the direction of the wind for each panel of the body, P_{tw}, may be taken as:

$$P_{tw} = q_z{}'\, A_s C_n K_\theta \quad \text{or}$$

$$P_{tw} = q_z\, A_s C_n (1 + K_{com}\, G_b) K_\theta \tag{17.4}$$

where

A_s = the total area projected normal to the face of the members in the tower face, for the panel considered, or, for poles, the face area of the pole over the panel considered.

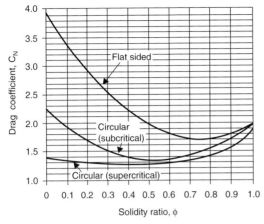

The drag coefficient for towers composed of flat-sided members, C_{Nf}, subcritical circular section members, C_{NC}, and supercritical circular-section members, $C_{NC'}$:

$$C_{Nf} = 1{,}76\,C_1\,[1 - C_2\,\phi + \phi^2]$$

$$C_{NC} = C_1\,(1 - C_2\,\phi) + (C_1 + 0{,}875)\phi^2$$

$$C_{NC'} = 1{,}9 - \sqrt{\{(1 - \phi)\,(2{,}9 - 1{,}14C_1 + \phi^2)\}}$$

where C_1 = 2,25 for square towers
         = 1,9 for triangular towers
       C_2 = 1,5 for square towers
         = 1,4 for triangular towers
       ϕ = solidity ratio

Figure 17.3 Overall normal drag coefficients, C_n, for towers (EN 50341-3-9)

C_n = the overall drag (pressure) coefficient, which for lattice towers is dependent upon the 'solidity ratio' which takes account of the open nature of the structure (see Fig. 17.3). The calculation of this ratio is complex (see EN50341) – the figure typically lies between 0.1 and 0.6.
K_θ = the wind incidence factor (unity for a full facing wind)

Thus if the basic wind speed for a UK location is determined as 24 m/s, the wind direction factor 0.9, the terrain roughness factor 0.86 (farmland with high hedges), the site reference wind speed is 18.58 m/s from equation (17.1). Then from equations (17.2) and (17.3) the dynamic pressure on a tower panel 40 m above the ground would be:

$$(1.22/2)(18.58[(40 - 2)/10]^{0.19})^2 = 333.8\,\text{N/m}^2$$

From Fig. 17.3, assuming a solidity factor of 0.2 the overall drag coefficient would be 2.9 for a flat sided structure and the maximum wind load on a panel of projected area 40 m² is given in equation (17.4) as:

$$333.8 \times 40 \times 2.9(1 + 1 \times 1.7) \times 1 = 104.5\,\text{kN}$$

17.2.2.3 *Wind load on a substation gantry*

The design of substation gantry busbar support structures is treated similarly to towers. It must take into account adequate height in order to allow for the maximum conductor sag condition. This will normally occur at maximum ambient temperatures and full busbar load current. A check could also be made for extreme circumstances at no load with maximum ice build-up. The mechanical loading on the gantry will include:

- wind loading (as described in Sections 17.2.2.1 and 17.2.2.2);
- maximum conductor tension.

The combined effect would be calculated as in Section 17.2.4 following. The conductor tension must allow for the weight of the insulator strings. Also it should be noted that the maximum conductor tension will occur under minimum temperature, minimum sag conditions.

17.2.2.4 *Wind load on a conductor example*

Consider an aerial conductor (for example, a moderately smooth earth wire forming a substation lightning screen) with a length of 150 m and diameter 25 mm. Assume the same wind conditions, height, etc. as in Sections 17.2.2.1

Table 17.3   Safe design tension taking account of terrain turbulence (*from Electra 186 October 1999*)

Terrain category	Terrain characteristics	$(H/w)_{adm}$ *(m)*
I	Open, flat, no tree, no obstruction, with snow cover, or near/across large bodies of water; flat desert	1000
II	Open, flat, no obstruction, no snow; e.g. farmland without any obstruction, summer time	1125
III	Open, flat or undulating with very few obstacles, e.g. open grass or farmland with very few trees, hedgerows and other barriers; prairie, tundra	1225
IV	Built-up with some trees and buildings, e.g. residential suburbs; small towns; woodlands and shrubs. Small fields with bushes, trees and hedges	1425

Note: 1. This table uses H/w, the ratio of horizontal tension in the span to conductor weight per unit length, as the tension parameter. It is important to note that this horizontal tension refers to the initial tension, before wind and ice loading and before creep, at the average temperature of the coldest month on the site of the line.
2. Valid for homogeneous aluminium and aluminium alloy conductors Ax (AAC and AAAC), bi-metallic aluminium conductors Ax/Ay (ACAR) and steel reinforced aluminium conductors A1/Syz (ACSR).

and 17.2.2.2. The wind loading on the conductor is determined from the formula:

Wind loading, $p_{cw} = lq_zD_cC_c(1 + G_c)\sin^2\psi$

where l = length of conductor (m)
        D_c = diameter of conductor (m)
        q_z = wind pressure = 333.8 N/m² (from Section 17.2.2.2)
        C_c = overall drag coefficient for the conductor – see Table 17.3
  Note 1. This needs the effective Reynold's number to be calculated.
        2. That ice loading can affect this.
        G_c = conductor gust response factor, the product of the length factor
           and the height factor (see Figs 17.4 and 17.4(a))
        Ψ = the angle of wind incidence to the conductor (assumed 90°)

Thus wind load = $150 \times 333.8 \times 0.025 \times 1.2(1 + [0.75 \times 1.4]) \times 1^2$
               = 3.12 kN

17.2.3 Conductor loads

17.2.3.1 *Conductor tensions*

The starting point for all conductor sag/tension calculations is the clear definition of the bases and conditions upon which the minimum factor of safety at which the

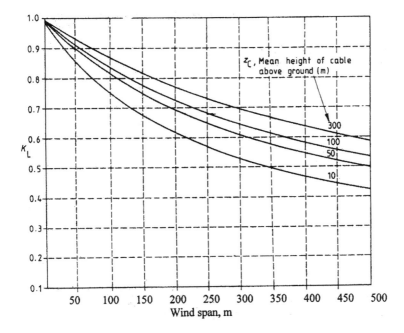

Figure 17.4   Length factor K_L for conductor wind loads (EN 50341-3-9)

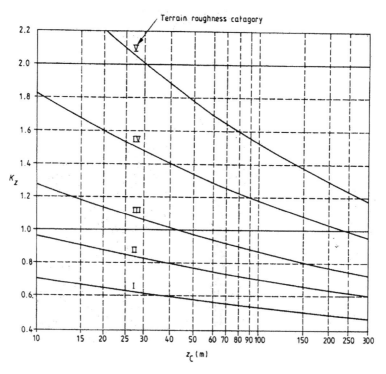

Figure 17.4(a)   Conductor height factor (EN 504341-3-9)

conductor is allowed to operate are set. The following are typical requirements with values given for UK conditions in respect of transmission lines:

- Maximum working tension (MWT)
  Conductor tension shall not exceed 50% of its breaking load (factor of safety of 2) at, say,

Temperature	$-6°C$
Cross wind pressure	$383\,N/m^2$
Radial ice thickness	$12.7\,mm$

- Everyday stress (EDS)
  Conductor tension shall not exceed 20% of its breaking load at, say,

Temperature	$16°C$
Cross wind pressure	–
Radial ice thickness	–

The figure of 20% has been used above to serve this example, but derivation of safe design tension must take account of a number of factors. Research into the effect of aeolian vibrations[7] over the life of a conductor has shown that the effect of the local terrain on turbulence is relevant, resulting in recommendations for safe design temperature depending on terrain – see Table 17.3.

Usually either MWT or EDS will be the critical basis for calculations and the other condition will then automatically be met. Often there is a particular span length above which the one basis is critical and below which the other one is. The tension, T, in the conductor for a given sag, S, is given by the formula:

$$T = \frac{W \cdot g \cdot L^2}{8S} \, N$$

where W = weight of conductor per unit length (kg/m)
L = span of the conductor (m)
g = gravitational constant (1 kgf = 9.81 N)
S = sag (m)

This is based on the parabolic curve shape for the conductor which, for less than 300 m spans or high span-to-sag ratios (sag is less than 10% of span and generally level topography), is very close to the more mathematically correct catenary formula. (See paragraph Section 18.5.2.5). The sag/tension formula is given in Chapter 16, Section 16.4.2.2.

17.2.3.2 Conductor tension example

A conductor is to have a maximum working tension of 65.95 kN at a temperature of −6°C with 12.7 mm of radial ice and a wind load of 383 N/m². Calculate the sag at 20°C in a span of 400 m. The sag/tension formula is given in Chapter 16, Section 16.4.2.2:

$$EA\alpha(t_2 - t_1) + (W_1^2 g^2 L^2 \, EA/24T_1^2) - T_1 = (W_2^2 g^2 L^2 EA/24T_2^2) - T_2$$

The conductor data is:
x = conductor diameter   = 28.62 mm
E = modulus of elasticity   = 69 × 10³ MN/m²
A = total conductor cross-sectional area   = 484.5 mm²
α = coefficient of linear expansion   = 19.3 ×10⁻⁶/°C
W_2 = final weight of conductor alone   = 1.621 kg/m

The calculation is given in Table 17.4.

17.2.3.3 Short circuit loadings

Under short circuit conditions lateral mechanical attraction or repulsion forces will occur between the different phase conductors (see Reference 5). The effect of conductor movement during short circuits is erratic and difficult to calculate. Such movement is taken into account in the overall design by allowing adequate clearances between the phase conductors. The conductor short circuit forces are usually ignored in the structural design of overhead line towers or substation gantries because of the very short durations of the faults.

Table 17.4 Sag and tension calculation

Parameter	Formula	Calculation	Results and units
Ice load	$B\sqrt{(y + x)}$	$2.87 \cdot 10^{-3} \cdot 12.7(28.62 + 12.7)$	1.505 kg/m
Total vertical load	Ice load + conductor weight, W_2	$1.505 + 1.621$	3.126 kg/m
Wind load	$p(2y + x)/10^3$	$383(2 \cdot 12.7 + 28.62) \times 10^{-3}$	20.69 N/m
Effective conductor weight, W_1	$\sqrt{[(\text{weight of conductor} + \text{ice})^2 + (\text{wind load}/g^2)]}$	$\sqrt{[(3.126)^2 + (20.69/9.81)^2]}$	3.771 kg/m
Tension @ 20°C, T_2	$EA\alpha(t_2 - t_1) + (W_1^2 g^2 L^2 EA/24T_1^2) - T_1 = (W_2^2 g^2 L^2 EA/24T_2^2) - T_2$	$[69 \cdot 10^3 \times 484.5 \cdot 19.3 \cdot 10^{-8}(20 - \{-6)\}] + [(3.771^2 \cdot 9.81^2 \cdot 400^2 \cdot 69 \cdot 10^3 \cdot 484.5)/24 \cdot 65\,951^2] - 65\,951 = [(1.621^2 \cdot 9.81^2 \cdot 400^2 \cdot 69 \cdot 10^3 \cdot 484.5)/24 \cdot T_2^2] - T_2$ $16\,775 + 70\,123 - 65\,951 = [56\,358 \cdot 10^9/T_2^2] - T_2$ $T_2^2[T_2 + 20\,947] = 56\,358 \cdot 10^9$ This gives an equation of the form: $x^3 + ax^2 = k$ which may be solved by an iterative process. With $T_2 = 32\,500$ N then error difference is very small and sufficiently accurate	32.5 kN
Sag @ 20°C, S	$\dfrac{WgL^2}{8T}$	$(1.621 \cdot 9.81 \cdot 400^2)/8 \cdot 32\,500$	9.8 m

17.2.3.4 *Ice loading*

The build-up of ice on conductors will increase effective conductor weight, diameter and wind loading. Local experience must be used in the application of ice loads to structural design. As an example, EN 50341-3-9 calls for a uniform ice load on all spans of 5 kN/m^3 to be considered for UK designs, or 9 kN/m^3 in case of wind and ice.

17.2.3.5 *Seismic loads*

The application of seismic loads in structural design is a specialist subject. Eurocode 8 covers it in detail, but the following gives a simplistic overview.

The acceleration due to a seismic event is categorized as a fraction of the gravitational constant, g. This may be given for both horizontal and vertical effects over a frequency spectrum. Table 17.5 details such acceleration factors for what is loosely described as a $0.2 \, g$ seismic event. Such an event refers to a surface wave travelling outwards from the epicentre exercising both horizontal and vertical forces on equipment. For simplicity, loadings on substation structures could be taken as an equivalent horizontal load.

An example of an analysis on the stability of a small distribution transformer under $0.2 \, g$ seismic conditions is given below. Consider the transformer with the physical characteristics given in Table 17.6.

Table 17.5 Frequency spectrum – acceleration factors for 0.2 g seismic event

Frequency	Horizontal acceleration	Vertical acceleration
0–0.64 Hz	0.15 g	
0.64–2.35 Hz	0.15 linear to 0.4 g	
2.35–10 Hz	0.4 g	2/3 of horizontal acceleration
10–30 Hz	0.4 linear to 0.2 g	
30 Hz and above	0.2 g	

Table 17.6 Cast resin dry type transformer physical details and calculated stability results under 0.2 g seismic conditions

Transformer (cast resin type)	Weight (kg)	Height/width ratio	Calculated uplift per wheel (% nominal weight)	Calculated uplift per wheel (kgf)
160 kVA	895	1.35	8.7	78
315 kVA, 21/0.4 kV	1613	1.15	4.5	77
500 kVA, 21/0.4 kV	2063	1.34	8.6	177
1000 kVA, 21/3.3 kV	3780	1.34	8.6	323
1600 kVA, 21/0.4 kV	4600	1.22	6.2	283
1600 kVA, 21/3.3 kV	5470	1.22	6.2	337
2000 kVA, 21/3.3 kV	6600	1.22	6.2	409

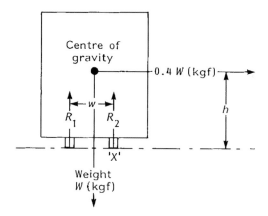

Figure 17.4(b)   Overturning force on transformer

Transformer weight $= W$ (kgf)
Height from base to transformer centre of gravity $= h$ (m)
Width of transformer wheel mounting points $= w$ (m)
Vertical wheel mounting transformer reaction forces $= R_1$ and R_2 (kgf)

Assume wheel mounted transformer sliding is prevented, then taking moments about point 'X' for the $0.4\,g$ horizontal seismic factor and assuming no vertical effects (Fig. 17.4(b)):

1. $0.4Wh + R_1w = \dfrac{Ww}{2}$

$$R_1 = W\left(\frac{1}{2} - 0.4\,\frac{h}{w}\right)$$

The transformer overturns if $R_1 \leq 0$, i.e.

$$\frac{h}{w} \geq \frac{0.5}{0.4} \geq 1.2$$

Again taking moments about point 'X' for the maximum upward acceleration of $(2/3 \times 0.4\,g)$, upward reaction $= (1 + 0.27)W$ kgf:

2. $0.4Wh + R_1w = 1.27\left(\dfrac{Ww}{2}\right)$

$$R_1 = W\left(0.635 - 0.4\,\frac{h}{w}\right)$$

The transformer overturns if $R_1 \leq 0$, i.e.

$$\frac{h}{w} \geq \frac{0.635}{0.4} \geq 1.59$$

At maximum downward acceleration of $-(2/3 \times 0.4\,g)$, the downward reaction $= (1 - 0.27)W$ kgf:

3. $0.4Wh + R_1 w = 0.73 \left(\dfrac{Ww}{2} \right)$

$$R_1 = W \left(0.365 - 0.4\,\dfrac{h}{w} \right)$$

The transformer overturns if $R_1 \leqslant 0$, i.e.

$$\frac{h}{w} \geqslant \frac{0.365}{0.4} \geqslant 0.912$$

The worst case for transformer stability is therefore at the maximum downward acceleration (ground cyclic movement falling away beneath the transformer) and the uplift on the rear wheels $= (0.365 - 0.4\,h/w) \times 100\%$ of nominal transformer weight. If not held down to resist overturning the transformers will slide because the coefficient of friction, μ, between the transformer steel wheels and the plinth is unlikely to be better than the 0.4 required in practice (for steel on steel $\mu = 0.25$). In this example all but one of the distribution transformers have an h/w ratio $\geqslant 1.2$ and will therefore overturn without the effect of vertical acceleration effects provided:

1. they are prevented from sliding, and
2. restraining effects of connecting cables or busbar trunking are ignored.

The transformers may be restrained by fixing arrangements to resist the following forces at *each* wheel:

1. $0.2\,W$ longitudinally
2. $0.2\,W$ uplift $\qquad$ where W = nominal weight of the unit
3. $0.5\,W$ deadweight

Switchgear and control or relay cubicles will need to be looked at on an individual basis in the same way.

17.2.3.6 Combined loads

The simultaneous application of individual worst case loads is unlikely to occur in practice and the simple arithmetic addition of all such load cases would lead to an uneconomic and over-engineered solution. The individual loads are therefore factored to arrive at a sensible compromise. For example, wind load plus ice load is often taken as full ice loading plus wind load at, say,

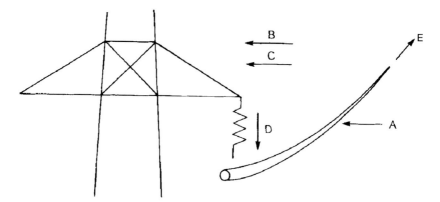

Figure 17.4(c)   Forces on an overhead line tower

50% basic wind speed. Similarly, wind load plus seismic load is normally taken as full earthquake load plus 50% wind load.

The forces involved on an overhead line tower are shown in Fig. 17.4(c).

A = horizontal conductor wind load, wind span × wind pressure (see Section 17.2.2.4) (note that for simplicity the wind effect on insulators, fittings, etc. has not been considered in Section 17.2.2, but in practice a designer will consider this, and the standards give suitable guidance)

B = horizontal structure wind load (see Section 17.2.2.2)

C = component of wind loading due to direction of wind and effective structure area normal to the wind (see equation (17.4).

D = vertical conductor weight span × conductor weight, including weight of insulator strings and fixings

E = longitudinal loads due to conductor tension. This will occur under uneven loading, such as for broken wire conditions or at a terminal tower with a slack span on one side of the tower entering to a substation gantry

These forces will, in general, result in a turning moment causing compression on one side of a tower and tension ('uplift') on the opposite side.

Having calculated the forces on structure and conductors, standards such as EN50341 apply a 'partial factor', which depends on the selected reliability level and takes account of possible modelling inaccuracies and uncertainties in the disturbing 'actions'. Such factors will depend on different national experience. EN50341-3-9 recommends for the UK a factor of 1.0 in calculating wind loads, with or without ice, provided the design period is no more than 50 years, but specifies a factor 1.5 when calculating static maintenance and construction loads, and 2.0 on conductor tension when conductors are being pulled by powered winches. Partial factors are also applied to the various material properties such as resistance of steel cross sections and of bucking of sections, compressive concrete strength, etc.

17.3 STRUCTURE DESIGN

17.3.1 Lattice steel tower design considerations

17.3.1.1 *General*

It should be noted that structural design is not an exact science, but BS5950 covering the subject is comprehensive and will, in general, lead to an economic design. The standard is based on material yield strengths with factors to take account of dynamic and static loads. Eurocode 3 (design of steel structures) is applicable, supplemented by EN50341, with different national criteria provided in EN50341 Part 3.

The l/r (l = length and r = radius of gyration) slenderness ratios for different steel member sections are obtained from tables in BS5950, Volume 1, Section – Properties Member Capacities. Obviously the longer and thinner the steel section the less load it will be able to take before failure along its axis. An equal steel angle will have a radius of gyration equal in both planes whereas an unequal angle will have different radius of gyration values in the x and y planes and for the design of a steel column the minimum value of 'r' should normally be taken. This gives a higher l/r ratio and correspondingly lower design compressive strength to work with.

132 kV substation gantries for an incoming overhead line are shown in Figs 17.5(a) and 17.5(b). Figures 17.6 and 17.7 show a typical tower general arrangement and associated steel member schedule. Tables 17.7 and 17.8 give typical allowable stress capabilities for steel struts with yield strengths, $Y_s = 245 \, \text{N/mm}^2$ and $Y_s = 402 \, \text{N/mm}^2$ respectively. Note these figures are for

Figure 17.5(a)   Oman – 132 kV line entry gantry

Figure 17.5(b)   132 kV gantry bolted connected – Channel Tunnel, Folkestone

example only; designers should refer to BSEN 10025 (hot rolled steel), and the Eurocodes or other equivalent national structural standards. Figure 17.8 gives dimensions and properties of light equal angle steel sections from BS5950, Structural Use of Steelwork in Buildings. Figure 17.9 gives bolt capacities for standard 4.6 and 8.8 grade metric bolts.

Steel lattice transmission tower design is based on compression formulae such as those in Tables 17.7 and 17.8 for leg members with different length/radius of gyration (l/r) values. The self-supporting tower design uses steel angle columns supported by stress-carrying bracing and redundant members. Higher strength steels show their greatest advantage in the lower l/r range where it can be seen from the tables that the allowable stress is not so sensitive to the slenderness of

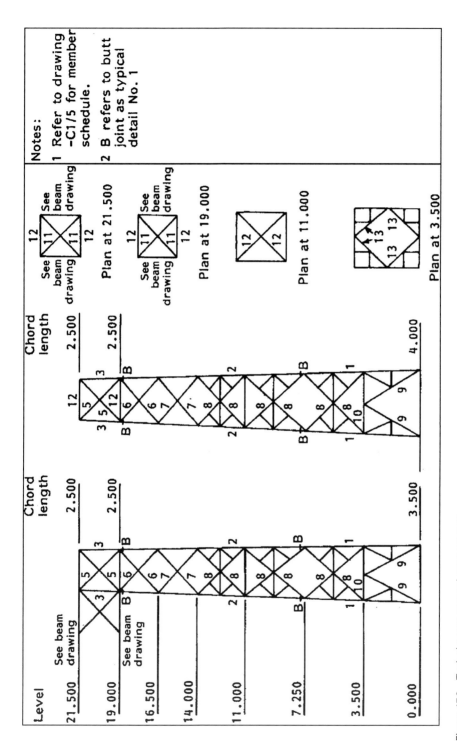

Figure 17.6   Typical tower general arrangement

Member	Section (Equal angle)	Steel Type (JIS)	No. of Bolts (per end)	Dia. mm	Gusset Plate	Joint Type	Typical Detail	Remarks
MAIN MEMBERS								
1	130×9	C4	6 Top / 8 Bottom	24	8	Butt	1/12/17	Foundation stubs to be 2000 long with 4 No. cleats
2	120×8	C4	4 Top / 6 Bottom	24	8	Butt	1/10/17	Thicker plate at beam connection beam details.
3	120×8	C2	4	20	8	Butt	1/10	ditto
BRACING								
5	90×6	C2	3	20	6		10	Thicker plate at beam connection – see details
6	80×6	C2	3	16	6		4	ditto
7	80×6	C2	2	20	6		4	
8	65×6	C2	3	16	6		4/15/16	
9	90×7	C2	3	20	7		12/16	
10	70×6	C2	2	16	6		4/18	
11	65×6	C2	1	16	6		4	
12	65×5	C2	1	16	6		10/14	
13	65×5	C2	1	16	6		18	

NOTES
1. This drawing to be read in conjunction with drg C1/4 and the specification.
2. Steel type refers to steel class in accordance with JIS standard
3. Members not detailed to be as follows:

Length	Section	Bolt
1800	50 × 5	C21M16
2200	60 × 5	C21M16
2500	65 × 5	C21M16

Figure 17.7   Typical tower member schedule

Table 17.7 Typical strut formulae for maximum allowable stress, U_c N/mm², on gross section for axial compression. $Y_S = 245$ N/mm²

l/r	0	1	2	3	4	5	6	7	8	9
0	245	245	245	245	245	245	245	245	245	244
10	244	244	244	244	244	243	243	243	243	242
20	242	242	241	241	241	240	240	239	239	238
30	238	237	237	236	236	235	234	234	233	232
40	231	231	230	229	228	227	226	225	224	223
50	222	221	220	219	218	216	215	214	212	211
60	209	208	206	205	203	202	200	198	196	195
70	193	191	189	187	185	183	181	179	177	175
80	173	171	169	167	165	163	161	159	157	155
90	153	151	149	147	145	143	141	139	137	135
100	133	131	129	128	126	124	122	121	119	117
110	116	114	112	111	109	108	106	105	103	102
120	101	99	98	97	95	94	93	91	90	89
130	88	87	86	84	83	82	81	80	79	78
140	77	76	75	74	73	72	72	71	70	69
150	68	67	67	66	65	64	63	63	62	61
160	61	60	59	59	58	57	57	56	55	55
170	54	54	53	52	52	51	51	50	50	49
180	49	48	48	47	47	46	46	45	45	44
190	44	43	43	43	42	42	41	41	41	40
200	40	39	39	39	38	38	38	37	37	37
210	36	36	36	35	35	35	34	34	34	33
220	33	33	33	32	32	32	31	31	31	31
230	30	30	30	30	29	29	29	29	28	28
240	28	28	28	27	27	27	27	26	26	26
250	26	0	0	0	0	0	0	0	0	0

Note: This table is provided for illustration only. The figures are derived from BS 449, which has been superseded by BS5950.

the member involved. The equivalent area of the member and the radius of gyration are looked up in tables for standard steel sections. See, for example, Fig. 17.8 taken from BS449 for a 60 × 60 × 6 mm angle. The minimum radius of gyration is 11.7 mm for the weakest axis.

An accurate analysis procedure is necessary to take into account the tension in the conductors, ice and wind effects. In particular, account must be taken of broken wire conditions (unequal loading on either side of the tower) and also the effect of the insulator strings. Only a very brief introduction to the principles involved is given here.

Such tower design is carried out by specialist structural engineers and is outside the scope of work purely for electrical engineers. Computer techniques are nearly always employed for all but the simplest structures. Table 17.9 lists some of the factors that have to be taken into account in an accurate calculation, which must allow for bending forces, compression and tension forces, flexural buckling and lateral buckling.

Table 17.8   Typical strut formulae for maximum allowable stress, U_C N/mm², on gross section for axial compression. Y_S = 402 N/mm²

l/r	0	1	2	3	4	5	6	7	8	9
0	402	402	402	402	402	402	402	401	401	401
10	401	401	400	400	400	399	399	398	398	397
20	397	396	396	385	384	382	381	380	380	378
30	389	388	387	386	385	384	382	381	380	378
40	377	375	373	371	369	367	365	363	360	358
50	355	353	350	347	344	341	337	334	330	327
60	323	319	315	311	306	302	298	294	289	285
70	280	276	271	267	262	258	253	249	244	240
80	236	231	227	223	219	215	211	207	204	200
90	196	193	189	186	183	179	176	173	170	167
100	164	161	159	156	153	151	148	146	143	141
110	139	136	134	132	130	128	126	124	122	120
120	118	116	114	113	111	109	108	106	105	103
130	102	100	99	97	96	95	93	92	91	89
140	88	87	86	85	84	83	81	80	79	78
150	77	76	75	74	73	73	72	71	70	69
160	68	67	67	66	65	64	64	63	62	61
170	61	60	59	59	58	57	57	56	55	55
180	54	54	53	53	52	51	51	50	50	49
190	49	48	48	47	47	46	46	45	45	45
200	44	44	43	43	42	42	42	41	41	40
210	40	40	39	39	39	38	38	38	37	37
220	37	36	36	36	35	35	35	35	34	34
230	34	33	33	33	32	32	32	32	31	31
240	31	31	39	30	30	30	30	29	29	29
250	28	0	0	0	0	0	0	0	0	0

Note: This table is provided for illustration only. The figures are derived from BS 449, which has been superseded by BS5950.

17.3.1.2 *Adequacy of steel angle in compression*

Consider a tower main leg of length 4 m to be designed to carry a maximum compressive force of 400 kN. A 120 × 120 × 8 mm high yield stress equal angle steel, Y_s = 402 N/mm², equivalent area of 1876 mm² and minimum radius of gyration, r_{min} = 23.8 mm is to be used and checked for adequacy in this application.

Compressive stress on member = $400 \cdot 10^3/1876$ = 213 N/mm².

From Table 17.8 the associated l/r ratio must not be greater than 85 for this condition. Therefore the maximum unsupported length of the leg must not exceed 85 × 23.8 = 2023 mm and therefore a brace support must be provided at, say, mid-length of the leg using this type of steel in this application in compression.

17.3.1.3 *Bolted connections*

There are basically three types of bolt connector in common use:

- ISO metric black hexagonal bolts, screws and nuts to EN 24016 (strength grade 4.6).
- ISO metric precision hexagonal bolts, screws and nuts to EN 24016 (strength grade 8.8).
- High strength friction grip bolts and associated nuts and washers to EN 143399 (grades 8.8 and 10.9 and 10.9 with wasted shank) – these are more usually considered for buildings and bridges rather than towers.

The term 'black bolt' is not sufficiently precise to be used without clarification as to the exact bolt grade being described. Both EN24016 grade 4.6 bolts and sometimes 8.8 precision bolts are supplied in the 'black' condition. The term 'precision' refers to the bolt shank dimensional tolerance. Normally an allowance of 2 mm over the bolt size is made for the bolt hole. A 22 mm diameter hole would therefore be drilled for an M20 bolt.

The nomenclature used for bolt types gives the yield strength (Y_s) and ultimate tensile strength (UTS). Consider a bolt with grade '$x \cdot y$'. The first number, x, in the bolt grade is a tenth of the ultimate tensile strength expressed in kgf/mm^2 (note this is not in N/mm^2 although 1 kgf $\sim$ 10 N). The second number, y, is the ratio of the yield strength to UTS $\times$ 10.

For example, a 4.6 bolt has

$$Y_s = 4 \cdot 6 = 24\,\text{kgf/mm}^2 \approx 240\,\text{N/mm}^2$$
$$\text{UTS} = 4 \cdot 10 = 40\,\text{kgf/mm}^2 \approx 400\,\text{N/mm}^2.$$

Similarly a high tensile 8.8 bolt has

$$Y_s = 8 \cdot 8 = 64\,\text{kgf/mm}^2 \approx 640\,\text{N/mm}^2$$
$$\text{UTS} = 8 \cdot 10 = 80\,\text{kgf/mm}^2 \approx 800\,\text{N/mm}^2.$$

There are three main aspects of bolted joints to be considered:

- Bearing – the stress on the inner surface of the bolt hole imparted by the bolt. The thicker the plates being bolted together the larger the bearing area and the lower the bearing stress.
- Reduction in steelwork material and cross-sectional area due to the presence of the bolt holes.
- The bending and prying effects of tension in the bolts.

Checking the effect of bolt holes for connecting
steel members together
If structural members are bolted together then an allowance has to be made for the reduction in steel bulk and therefore stiffness due to the holes required for the bolted connection. Steel plates may be connected together by bolted

EQUAL ANGLES

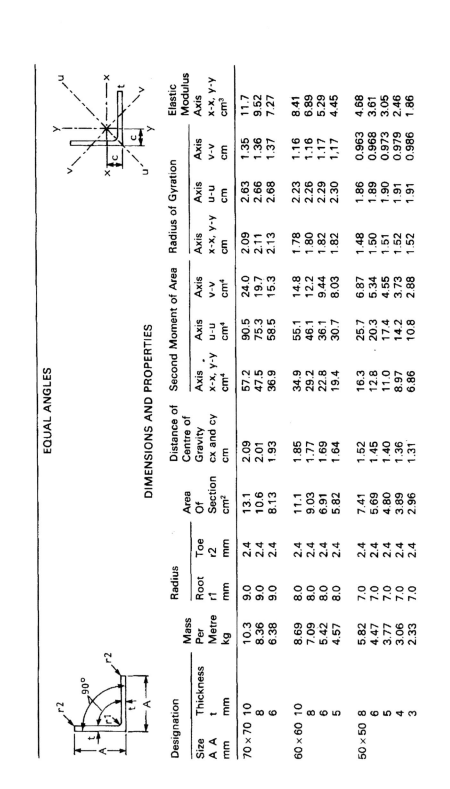

DIMENSIONS AND PROPERTIES

Designation Size A A mm	Thickness t mm	Mass Per Metre kg	Radius Root r1 mm	Radius Toe r2 mm	Area Of Section cm²	Distance of Centre of Gravity cx and cy cm	Second Moment of Area Axis x-x, y-y cm⁴	Second Moment of Area Axis u-u cm⁴	Second Moment of Area Axis v-v cm⁴	Radius of Gyration Axis x-x, y-y cm	Radius of Gyration Axis u-u cm	Radius of Gyration Axis v-v cm	Elastic Modulus Axis x-x, y-y cm³
70 × 70	10	10.3	9.0	2.4	13.1	2.09	57.2	90.5	24.0	2.09	2.63	1.35	11.7
	8	8.36	9.0	2.4	10.6	2.01	47.5	75.3	19.7	2.11	2.66	1.36	9.52
	6	6.38	9.0	2.4	8.13	1.93	36.9	58.5	15.3	2.13	2.68	1.37	7.27
60 × 60	10	8.69	8.0	2.4	11.1	1.85	34.9	55.1	14.8	1.78	2.23	1.16	8.41
	8	7.09	8.0	2.4	9.03	1.77	29.2	46.1	12.2	1.80	2.26	1.16	6.89
	6	5.42	8.0	2.4	6.91	1.69	22.8	36.1	9.44	1.82	2.29	1.17	5.29
	5	4.57	8.0	2.4	5.82	1.64	19.4	30.7	8.03	1.82	2.30	1.17	4.45
50 × 50	8	5.82	7.0	2.4	7.41	1.52	16.3	25.7	6.87	1.48	1.86	0.963	4.68
	6	4.47	7.0	2.4	5.69	1.45	12.8	20.3	5.34	1.50	1.89	0.968	3.61
	5	3.77	7.0	2.4	4.80	1.40	11.0	17.4	4.55	1.51	1.90	0.973	3.05
	4	3.06	7.0	2.4	3.89	1.36	8.97	14.2	3.73	1.52	1.91	0.979	2.46
	3	2.33	7.0	2.4	2.96	1.31	6.86	10.8	2.88	1.52	1.91	0.986	1.86

Size	Thickness											
40 × 40	6	3.52	6.0	2.4	4.48	1.20	6.31	9.98	2.65	1.19	1.49	0.770	2.26
	5	2.97	6.0	2.4	3.79	1.16	5.43	8.59	2.26	1.20	1.51	0.773	1.91
	4	2.42	6.0	2.4	3.08	1.12	4.47	7.09	1.86	1.21	1.52	0.777	1.55
	3	1.84	6.0	2.4	2.35	1.07	3.45	5.45	1.44	1.21	1.52	0.783	1.18
30 × 30	5	2.18	5.0	2.4	2.78	0.918	2.16	3.41	0.917	0.883	1.11	0.575	1.04
	4	1.78	5.0	2.4	2.27	0.878	1.80	2.85	0.754	0.892	1.12	0.577	0.85
	3	1.36	5.0	2.4	1.74	0.835	1.40	2.22	0.585	0.899	1.13	0.581	0.649
25 × 25	5	1.77	3.5	2.4	2.26	0.799	1.21	1.90	0.524	0.731	0.915	0.481	0.711
	4	1.45	3.5	2.4	1.85	0.762	1.02	1.61	0.430	0.741	0.931	0.482	0.586
	3	1.11	3.5	2.4	1.42	0.723	0.803	1.27	0.334	0.751	0.945	0.484	0.452

Figure 17.8 Dimensions and properties of steel equal angle (BS 5950)

4.6 BOLTS IN MATERIAL GRADE 43 AND 50

Diam of Bolt mm	Tensile Stress Area mm²	Tensile Cap kN	Shear value Single Shear kN	Shear value Double Shear kN	Bearing value of bolt at 435N/mm and end distance equal to 2 × bolt diameter Thickness in mm of Plate Passed Through 5	6	7	8	9	10	12.5	15	20	25	30
12	84.3	16.4	13.5	27.0	26	31	0	0	0	0	0	0	0	0	0
16	157	30.6	25.1	50.2	34	41	48	55	0	0	0	0	0	0	0
20	245	47.8	39.2	78.4	43	52	60	69	78	87	0	0	0	0	0
22	303	59.1	48.5	97.0	47	57	67	76	86	95	120	0	0	0	0
24	353	68.8	56.5	113	52	62	73	83	94	104	131	0	0	0	0
27	459	89.5	73.4	147	58	70	82	94	106	117	147	176	0	0	0
30	561	109	89.8	180	65	78	91	104	117	131	163	196	0	0	0

Values printed in bold type are less than the single shear value of the bolt. Values printed in ordinary type are greater than the single shear value and less than the double shear value. Values printed in italic type are greater than the double shear value. Bearing values are governed by the strength of the bolt

8.8 BOLTS IN MATERIAL GRADE 43

Diam of Bolt mm	Tensile Stress Area mm²	Tensile Cap kN	Shear value Single Shear kN	Shear value Double Shear kN	Bearing value of bolt at 435N/mm and end distance equal to 2 × bolt diameter Thickness in mm of Plate Passed Through 5	6	7	8	9	10	12.5	15	20	25	30
12	84.3	37.9	31.6	63.2	27	33	38	44	49	55	69	0	0	0	0
16	157	70.7	58.9	118	36	44	51	58	66	73	93	110	147	0	0
20	245	110	91.9	184	46	55	64	73	82	92	115	138	184	0	0
22	303	136	114	227	50	60	70	81	91	101	127	152	202	253	0
24	353	159	132	265	55	66	77	88	99	110	138	166	221	276	0
27	459	207	172	344	62	74	86	99	112	124	155	186	248	310	373
30	561	252	210	421	69	82	96	110	124	138	173	207	276	345	414

Values printed in ordinary type are less than the single shear value of the bolt. Values printed in bold type are greater than the single shear value and less than the double shear value. Values printed in italic type are greater than the double shear value. Bearing values are governed by the strength of the plate

Figure 17.9 Bolt capacities

Table 17.9   Data used in the design of Lattice steel towers (per EN 50341-1)

Symbol	Description
A	Cross-sectional area; gross cross-sectional area of bolt
A_{eff}	Effective cross-sectional area
A_{net}	Net cross section at holes
A_S	Tensile stress area of bolt
b	Nominal width
b_{eff}	Effective width of the leg
c	Distance between batten plates
d	Bolt diameter
d_o	Hole diameter
E	Modulus of elasticity
e_1	End distance from centre of hole to adjacent end in angle
e_2	Edge distance from centre of hole to adjacent edge in angle
F	Concentrated horizontal load
f_u	Ultimate tensile strength
f_{ub}	Ultimate tensile strength for bolt
f_y	Yield strength
f_{yd}	Design yield strength
i	Radius of gyration about the relevant axis
L	System length
$M_{c\,Rd}$	Design moment of resistance in bending
M_{sd}	Bending moment at cross section
m	Number of angles
N	Axial force
N_d	Compression force; force in the compression member
$N_{R,d}$	Design value of the buckling resistance
N_{sd}	Design value of the tensile or compressive force at cross section
P	Spacing of two holes in the direction of load transfer
p	Spacing of two holes, measured perpendicular to the member axis
S_d	Tension force; force in the supporting member (tension or compression)
s	Staggered pitch, spacing of centres of 2 consecutive holes
t	Thickness
W_{eff}	Effective cross section modulus
γ_{M1}	Partial safety factor for resistance of member in bending or tension or to buckling
γ_{M2}	Partial safety factor for resistance of net section at bolt holes
γ_{Mb}	Partial safety factor for resistance of bolted connections
λ	Slenderness ratio for the relevant buckling load
λ_{eff}	Effective slenderness
λ^-	Non-dimensional slenderness for the relevant buckling load
λ_p	Ratio of width to thickness (b/t)
p	Reduction factor
χ	Reduction factor

connections with forces acting in shear across the bolt diameter. Friction grip bolts are normally only used in rigid frame structures where high shear loads and moment loads are involved. In a pinned three dimensional truss structure, such as a steel lattice tower, the design will involve only very slight bending moments. High strength friction grip bolts would not therefore normally be used in a lattice tower structure to clamp the plates together.

Consider 22 mm diameter bolt holes in each right angle $120 \times 120 \times 8$ mm steel gantry leg face for M20 bolts to withstand a 350 kN maximum tensile force. From Fig. 17.9 an M20 grade 8.8 bolt has an allowable shear value of

91.9 kN. Therefore at least four No. M20 grade 8.8 bolts or at least nine No. M20 grade 4.6 bolts are required. Larger bolts will reduce bearing stresses. It is important to notice that if a design has been formulated around grade 8.8 bolts then they should not be replaced at a later date by a bolt of a lesser grade. Some engineers design on the basis of grade 4.6 bolts and specify grade 8.8 bolts in the material schedule if no major cost disadvantage ensues.

There are standard edge distances and back marks for bolt drilling in standard section steel members. For example, a bolt centre should not be less than $1.4 \times$ hole diameter from the edge of the member in the direction of the load, and a minimum distance of not less than $1.25 \times$ hole diameter in the direction normal to the load (BS5950). The application of such precautions takes into account the reduction in steel bulk due to the bolt holes. More conservative guidelines are also given in EN50341, which takes account of whether or not the shear plane passes through the threaded portion of the bolt.

17.3.1.4 *Bracing*

The calculation to confirm the adequacy of a steel brace is similar to that given for the tower leg in Section 17.3.1.1.

A 3.5 m long mild steel, $Y_s = 245 \, \text{N/mm}^2$, $60 \times 60 \times 6 \, \text{mm}$ angle brace, equivalent area 691 mm^2 and $r_{min} = 11.7 \, \text{mm}$ (Fig. 17.8) is to be designed to carry a maximum compressive force of 80 kN.

Experience would show that this is rather a slender steel section for the proposed load. Compressive stress $= 80 \times 10^3/691 = 116 \, \text{N/mm}^2$. From Table 17.7 the l/r ratio must not be greater than 110. Therefore the maximum unsupported length of the brace must not exceed $110 \times 11.7 = 940 \, \text{mm}$ and therefore the brace must have additional supports at $3500/940 \approx$ four points along its length when using this type of steel in this application.

17.3.1.5 *Analysis*

The structural analysis may be carried out by:

- computer
- graphical methods

It is normal to use computer methods to carry out the analysis and often the complete design with simple hand calculations as shown in Sections 17.3.1.2 to 17.3.1.4 only to check certain results. The tower or gantry structure is designed to have members either in compression or tension. The computer checks each element to ensure that it is capable of withstanding the applied loads. The checks are carried out in accordance with standard codes of practice applicable to the country involved or as specified by the design or consulting engineer. A most useful reference is the *Steel Designers Manual*.

Figure 17.10   Double circuit 400 kV tower undergoing type tests

17.3.2 Tower testing

New tower designs may be type tested at special open air laboratories. Figure 17.10 shows a double circuit 400 kV tower undergoing tests at Chels Combe Test Station in the UK.

17.4 POLE AND TOWER TYPES

17.4.1 Pole structures

Pole structures are especially used for economic household distribution at voltage levels of 380/415 V and 20/11 kV where planning permission allows such arrangements in place of buried cables. Such pole structures are also used at the lower transmission voltage levels, typically at up to 145 kV but also with multipole and guyed (stayed) arrangements at voltages up to 330 kV. Low-voltage designs are based on matching the calculated equivalent pole head load to the particular type and diameter of wood, steel or concrete to be employed. At higher voltages specific designs are used in order to select optimum size and relative cost. Some examples of different pole arrangements are given in Fig. 17.11.

Wood poles must be relatively straight and defects such as splits and shakes are unacceptable. There are various national standards, and Table 17.1 lists some European standards. Commonly used soft woods such as fir, pine and larch require impregnation with creosote, anti-termite repellents if to be used in tropical countries and similar chemicals to prevent decay. Some hardwoods may not need chemical treatment but these are becoming very expensive and their use is considered by some to damage the environment. The poles are usually

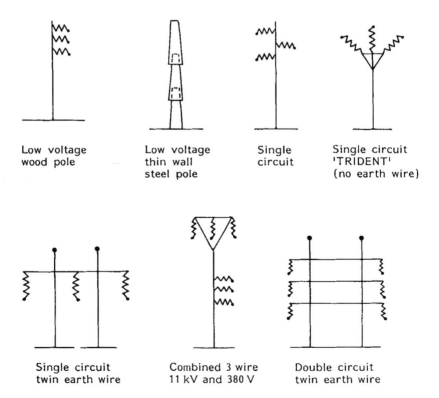

Low voltage
wood pole

Low voltage
thin wall
steel pole

Single
circuit

Single circuit
'TRIDENT'
(no earth wire)

Single circuit
twin earth wire

Combined 3 wire
11 kV and 380 V

Double circuit
twin earth wire

Figure 17.11   Typical pole arrangements

direct buried with a depth of burial normally equal to one sixth of pole length. If this is insufficient to resist the design turning moment, supplementary blocking or 'Permasoil' is used to provide further resistance to overturning.

Tubular poles are available in single column circular sections, octagonal shapes made from folded steel and stepped/swaged sections. Very thin steel wall sections, slightly conical in shape, are also available. The poles are shipped with the smaller sections inside the larger ones for compactness. They are then erected on site by sliding one section over the other to form the pole as shown in Fig. 17.11.

Concrete poles are available in prestressed spun or unstressed cast concrete. Light fibreglass poles are also available for light head loads. Figure 17.12 shows a typical 115 kV twin pole single circuit line in Saudi Arabia with the oil wells in the distance, and Fig. 17.13 is a pole-mounted 11/0.415 kV distribution transformer.

17.4.2 Tower structures

Steel lattice towers are generally used at the higher voltage levels where longer spans, high wind loads, ice loads and heavy conductors make the use of wood or light steel poles impractical. Figure 17.14 gives some examples of typical tower outlines for single and double circuit configurations with single and double earth wires. In order to standardize, towers are categorized typically to fulfil the following duties:

Suspension towers	straight line and deviation angles up to about 2°
10° angle or section tower	angles of deviation up to 2° or at section positions also for heavy weight spans or with unequal effective negative weight spans
30° angle	deviation angles up to 30°
60° angle	deviation angles up to 60°
90° angle	deviation angles up to 90°
Terminal tower	terminal tower loading taking full line tension on one side of tower and none or slack span on other – typically at substation entry

The terminology adopted to describe such towers varies and must be clearly described in order to avoid confusion. For example, a double circuit 30° angle tower for twin conductor use could be described in short form as D30T, but the 'T' can be confused between 'twin' and 'triple'. Many users would restrict the description to D30. Similarly, a double circuit terminal tower for twin conductor use could be described as DTT, or more commonly just DT. Extensions are described with a further addition; for example D30E6 describes a dual circuit tower with a 30° maximum angle and a six metre body extension.

In addition to the conductor and insulator set loadings, tower design must take into account shielding angles (lightning protection). Further clearances must be maintained as the insulator sets swing towards the earthed tower

Figure 17.12   115 kV twin pole, single circuit line – Saudi Arabia

Figure 17.13   Pole-mounted 11/0.415 kV transformer and LV distribution

structure under certain wind conditions. Figure 18.18 indicates the physical size of a 400 kV cap and pin glass insulator suspension set undergoing maintenance on the Lydd/Bolney line in Southern England. Figure 17.15 shows 'ducter' (low resistance) measurements being taken on an overhead line clamp during refurbishment work on the Elland/Ferrybridge overhead line in Yorkshire. The tension insulator set is in the foreground.

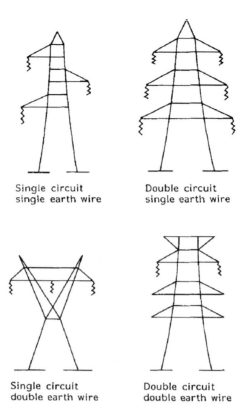

Single circuit
single earth wire

Double circuit
single earth wire

Single circuit
double earth wire

Double circuit
double earth wire

Figure 17.14   Typical tower outlines

Figure 17.15   400 kV cap and pin insulator set. Low resistance 'ducter' testing on connections in progress

Figure 17.16   500 kV double circuit tower erection in progress, China

REFERENCES

1. Calculations of wind and ice loadings – an example of the application of IEC reports, 826-2 and 826-4, *Electra* 132, October 1990, Working Group 06 (Line reliability and security) of Study Committee 22.
2. Calculation of overhead line tower loadings, wind loading only, Paper 22–84 (WG06)14 being CIGRE paper 22–87 (SC)01 now publication IEC 60826-2.
3. Overhead distribution lines – some reflections on design, *IEE Proceedings*, Vol. 133, Pt. C, No. 7, November 1986.
4. Overhead lines – loading strength: the probabilistic approach viewed internationally, *IEE Power Engineering Journal,* September 1991.
5. Mechanical effects of high short circuit currents in substations, *IEEE Transactions on Power Apparatus and Systems,* Vol. PAS-94, No. 5, September/October 1975.
6. *Steel Designers Manual,* Crosby, Lockwood, Staples, Metric Edition.
7. Safe design tension with respect to aeolian vibrations – part 1, *Electra* 186, October 1999 – TF 22.11.04.

18 Overhead Line Conductor and Technical Specifications

18.1 INTRODUCTION

Overhead lines are, in essence, air-insulated cables suspended from insulated supports with a power transfer capacity approximately proportional to the square of the line voltage. Overhead lines are cheaper in initial capital cost and are generally more economic than cable feeders. For the transmission of equivalent power at 11 kV a cable feeder would cost some 5 times the cost of a transmission line, at 132 kV 8 times and at 400 kV 23 times. Such comparisons must, however, be treated in more depth since they must take into account rights of way, amenity, clearance problems, planning permissions associated with the unsightly nature of erecting bare conductors in rural and urban areas, and ongoing maintenance requirements.

18.2 ENVIRONMENTAL CONDITIONS

In order to match both the mechanical and electrical characteristics of the overhead line conductor to the environmental conditions climatic details must first be collected and analysed. The parameters required are as described in Chapter 17, Section 17.2.1.

- Temperature
  The maximum, minimum and average ambient temperature influences conductor current rating and sag. For temperate conditions typically 20°C with 55°C temperature rise. For tropical conditions 35°C or 40°C with 40°C or 35°C temperature rise. Maximum conductor operating temperature should not exceed 75°C for bare conductors to prevent annealing of aluminium. Conductor

temperatures up to 210°C are possible with 'GAP' conductor (see Section 18.3.2).

- Wind velocity
  Required for structure and conductor design. Electrical conductor ratings may be based on cross wind speeds of 0.5 m/s or longitudinal wind speeds of 1 m/s.

- Solar radiation
  Required for conductor ratings but also for fittings such as composite insulators which may be affected by exposure to high thermal and ultraviolet (UV) radiation. Typical values of 850 W/m^2 and 1200 W/m^2 may be assumed for temperate and tropical conditions respectively.

- Rainfall
  Important in relation to flooding (necessity for extension legs on towers), corona discharge and associated electromagnetic interference, natural washing and insulator performance.

- Humidity
  Effect on insulator design.

- Altitude
  Effect on insulation and conductor voltage gradient.

- Ice and snow
  Required for design of conductor sags and tensions. Build-up can also affect insulation as well as conductor aerodynamic stability.

- Atmospheric pollution
  Effect on insulation and choice of conductor material (IEC 60815-1 and -2 – Guide for the selection of insulators in respect of polluted conditions – drafts for public comment).

- Soil characteristics
  Electrically affecting grounding requirements (soil resistivity) and structurally the foundation design (weights, cohesion and angle of repose).

- Lightning
  Effect on insulation levels and also earth wire screening arrangements necessary to provide satisfactory outage performance.

- Seismic factor
  Effect on tower and foundation design.

- General loadings
  Refer also to IEC 60826 (Design criteria for overhead lines), EN50341 (Overhead lines exceeding AC45 kV – supersedes 60826 for European use) and BS8100 (Loading and strength of overhead transmission lines).

18.3 CONDUCTOR SELECTION

18.3.1 General

The selection of the most appropriate conductor size at a particular voltage level must take into account both technical and economic criteria as listed below:

1. The maximum power transfer capability must be in accordance with system requirements.

2. The conductor cross-sectional area should be such as to minimize the initial capital cost and the capitalized cost of the losses.
3. The conductor should conform to standard sizes already used elsewhere on the network in order to minimize spares holdings and introduce a level of standardization.
4. The conductor thermal capacity must be adequate.
5. The conductor diameter or bundle size must meet recognized international standards for radio interference and corona discharge.
6. The conductor must be suitable for the environmental conditions and conform to constructional methods understood in the country involved (such as IEC, BS, etc.).

18.3.2 Types of conductor

The international standards covering most conductor types are IEC 61089 (which supersedes IEC 207, 208, 209 and 210) and EN 50182 and 50183 (see Table 18.1). For 36 kV transmission and above both aluminium conductor steel reinforced (ACSR) and all aluminium alloy conductor (AAAC) may be considered. Aluminium conductor alloy reinforced (ACAR) and all aluminium alloy conductors steel reinforced (AACSR) are less common than AAAC and all such conductors may be more expensive than ACSR. Historically ACSR has been widely used because of its mechanical strength, the widespread manufacturing capacity and cost effectiveness. For all but local distribution, copper-based overhead lines are more costly because of the copper conductor material costs. Copper (BS 7884 applies) has a very high corrosion resistance and is able to withstand desert conditions under sand blasting. All aluminium conductors (AAC) are also employed at local distribution voltage levels.

From a materials point of view the choice between ACSR and AAAC is not so obvious and at larger conductor sizes the AAAC option becomes more

Table 18.1   Relevant national and international standards

Number	Title	Comment
IEC 61089	Round wire concentric lay overhead electrical stranded conductors	Supersedes IEC 207 (AAC), 208 (AAAC), 209 (ACSR) and 210 (AACSR)
EN 50182	Conductor for overhead lines: round wire concentric lay stranded conductor	Supersedes IEC 61089 for European use. BSEN 50182 identical
EN 50183	Conductor for overhead lines: aluminium–magnesium–silicon alloy wires	
BS 183	Specification for general purpose galvanized steel wire strand	For earth wire
BS 7884	Specification for copper and copper–cadmium conductors for overhead systems	

attractive. AAAC can achieve significant strength/weight ratios and for some constructions gives smaller sag and/or lower tower heights. With regard to long-term creep or relaxation, ACSR with its steel core is considerably less likely to be affected. Jointing does not impose insurmountable difficulties for either ACSR or AAAC types of conductor as long as normal conductor cleaning and general preparation are observed. AAAC is slightly easier to joint than ACSR. The characteristics of different conductor materials are given in Table 18.2.

Figure 18.1 illustrates typical strandings of ACSR. The conductor, with an outer layer of segmented strands, has a smooth surface and a slightly reduced diameter for the same electrical area.

Historically there has been no standard nomenclature for overhead line conductors, although in some parts of the world code names have been used based on animal (ACSR – UK), bird (ACSR – North America), insect (AAAC – UK) or flower (AAAC – North America) names to represent certain conductor types (see Table 18.3a).

Aluminium-based conductors have been referred to by their nominal aluminium area. Thus, ACSR with 54 Al strands surrounding seven steel strands, all strands of diameter $d = 3.18$ mm, was designated 54/7/3.18; alu area $=$ 428.9 mm^2, steel area $= 55.6$ mm^2 and described as having a nominal aluminium area of 400 mm^2. In France, the conductor total area of 485 mm^2 is quoted and in Germany the aluminium and steel areas, 429/56, are quoted. In Canada and USA, the area is quoted in circular mils (1000 circular mils $=$ 0.507 mm^2).

Within Europe standard EN50182 has co-ordinated these codes while permitting each country to retain the actual different conductor types via the National Normative Aspects (NNAs). Table 18.3b explains the EN 50182 designation system.

The development of 'Gap type' heat-resistant conductors offers the possibility of higher conductor temperatures. The design involves an extra high strength galvanized steel core, and heat-resistant aluminium alloy outer layers, separated by a gap filled with heat-resistant grease. To maintain the gap, the wires of the inner layer of the aluminium alloy are trapezoid shaped. Depending on the alloys used, temperatures of up to 210°C are possible, with a current carrying capacity of up to twice that of hard-drawn aluminium. This offers particular value where projects involve upgrading existing circuits.

18.3.3 Aerial bundled conductor and BLX

Power failures on open wire distribution systems (up to 24 kV) under storm conditions in the 1960s led to various distribution companies investigating what steps could be taken to increase service reliability. At low voltage levels aerial bundled conductor (ABC) is now becoming rapidly more popular because of improved reliability and the low installation and maintenance costs compared to conventional open wire pole distribution. For short distribution

Table 18.2   Characteristics of different conductor materials

Property	Unit	Annealed copper	Hard-drawn copper	Cadmium copper	Hard-drawn aluminium	Aluminium alloy (BS3242)	Galvanized steel
Relative conductivity	(%)	100	97 (average)	79.2 (minute)	61 (minute)	53.5	—
Volumetric resistivity @ 20°C	(Ω mm^2/m)	0.01724 (std.)[a]	0.01771 (average)[b]	0.02177 (maximum)	0.02826 (maximum)	0.0322 (std.)	—
Mass resistivity @ 20°C	(Ω kg/km)	0.15328	0.15741	0.19472	0.07640	0.08694	—
Resistance @ 20°C	(Ωmm^2/km)	17.241	17.71	21.77	28.26	32.2	—
Density	(kg/m^3)	8890	8890	8945	2703	2703	7780
Mass	(kg/mm^2/km)	8.89	8.89	8.945	2.703	2.703	7.78
Resistance temperature coefficient @ 20°C	(per °C)	0.00393	0.00381	0.00310	0.00403	0.0036	—
Coefficient of linear expansion	(per °C)	17×10^{-6}	17×10^{-6}	17×10^{-6}	23×10^{-6}	23×10^{-6}	11.5×10^{-6}
Ultimate tensile stress (approximately) BS values	(MN/m^2)	255	420	635	165	300	1350
Modulus of elasticity	(MN/m^2)	100×10^3	125×10^3	125×10^3	70×10^3	70×10^3	200×10^3

Notes: [a]For calculation this figure may be taken as 0.17241379. [b]Value assumed UTS of 420 MN/m^2.

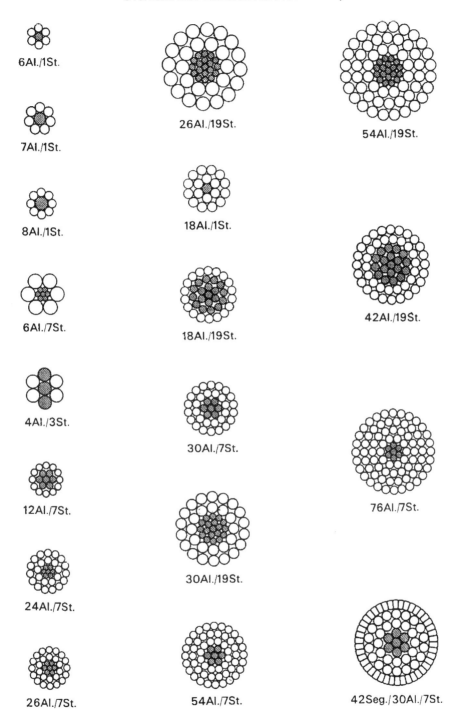

6Al./1St.

7Al./1St.

8Al./1St.

6Al./7St.

4Al./3St.

12Al./7St.

24Al./7St.

26Al./7St.

26Al./19St.

18Al./1St.

18Al./19St.

30Al./7St.

30Al./19St.

54Al./7St.

54Al./19St.

42Al./19St.

76Al./7St.

42Seg./30Al./7St.

Figure 18.1   Conductor arrangements for different CSR combinations

Table 18.3(a)   Typical properties of some ACSR conductors

Historic Code name	Stranding	Alu area (mm²)	Steel area (mm²)	Diameter (mm)	Mass (kg/km)	Breaking load (kN)	Resistance (ohm/km)
Horse	12/7/2.79	73.4	42.8	13.95	538	61.2	0.3936
Lynx	30/7/2.79	183.4	42.8	19.53	842	79.8	0.1441
Zebra	54/7/3.18	428.9	55.6	28.62	1621	131.9	0.0674
Dove	26/3.72 +						
	7/2.89	282	45.9	23.55	1137	99.88	0.1024

Table 18.3b   Conductor designation system to EN50182:2001

1. A designation system is used to identify stranded conductors made of aluminium with or without steel wires.
2. Homogeneous aluminium conductors are designated ALx, where x identifies the type of aluminium. Homogeneous aluminium-clad steel conductors are designated yzSA where y represents the type of steel (Grade A or B, applicable to class 20SA only), and z represents the class of aluminium cladding (20, 21, 30 or 40).
3. Composite aluminium/zinc coated steel conductors are designated ALxlSTyz, where ALx identifies the external aluminium wires (envelope), and STyz identifies the steel core. In the designation of zinc coated steel wires, y represents the type of steel (Grades 1 to 6) and z represents the class of zinc coating (A to E).
4. Composite aluminium/aluminium-clad steel conductors are designated ALxlyzSA, where ALx identifies the external aluminium wires (envelope), and yzSA identifies the steel core as in 2.
5. Conductors are identified as follows:
   (a) a code number giving the nominal area, rounded to an integer, of the aluminium or steel as appropriate;
   (b) a designation identifying the type of wires constituting the conductor. For composite conductors the first description applies to the envelope and the second to the core.

EXAMPLES:

16-AL1:	Conductor of AL1 aluminium with an area of 15.9 mm², rounded to 16 mm².
587-AL2:	Conductor of AL2 aluminium with an area of 5869 mm², rounded to 587 mm².
401-AL1128-ST1A:	Conductor made of AL1 aluminium wires around a core of ST1A zinc coated steel wires with a Class A zinc coating. The integer area of AL1 wires is 401 mm² and that of the ST1A wires 28 mm².
401-AL1128-A20SA:	Conductor made of AL1 aluminium wires around a core of grade A, class 20 aluminium-clad steel wires. The integer area of AL1 wires is 401 mm² and that of the A20SA wires 28 mm².
65-A20SA:	Conductor made of grade A, class 20 aluminium-clad steel wires with an area of 65 mm².

lines, where voltage drop is not a limiting factor determined by the line react-ance, the ABC installation is some 160% of the cost of the equivalent open wire construction at 24 kV. For longer lines and higher currents where the line reactance is important the cost differential diminishes. The initial capital cost of the cable alone is, however, up to twice the cost of the equivalent open wire conductor. Environmentally it could be argued that the ABC end product is marginally more pleasing. However, for the 10–14 kV distribution levels the use of ABC is more problematic due to the requirement of employing underground cable joining techniques at high level (especially difficult in maintenance situations).

There are two distinct ABC systems in use. One system uses a self-supporting bundle of insulated conductors where all conductors are laid up helically and where tension is taken on all conductors which are of hard-drawn aluminium. An alternative system is where all conductors are insulated and the hard-drawn aluminium phase conductors are laid up around an aluminium alloy neutral which has greater tensile strength and acts as a catenary wire to support the whole bundle. The insulation material may be polyvinylchloride (PVC), linear polyethylene (PE) or cross-linked polyethylene (XLPE).

A comparison of the advantages and disadvantages between ABC and conventional open wire distribution construction is given in Table 18.4.

With ABC construction core identification and the need to distinguish between neutral and phase conductors is essential and in practice such overhead line emergency work is often carried out in poor light. One, two or three prominent ribs are introduced along the length of the core insulation to identify the phases, and multiple low profile ribs along the neutral so that conductors may be identified by touch irrespective of the position of the neutral in the bundle.

For medium voltage distribution lines, (10 kV to 25 kV) BLX conductors are now preferred in many countries.[14] These are stranded, alloyed aluminium conductors, with cross-section from 50 mm^2 to 150 mm^2, covered with a cross-linked and UV resistant polymer. The conductor is grease filled during manufacture to make the covered cable watertight and corrosion proof. The use of this conductor greatly reduces transient faults, while allowing reduced clearances and thus minimizing wayleave problems. It does not present the HV jointing problems of bundled conductor, and the insulation life is much longer than on previously used PVC or Nylon covered conductors.

18.3.4 Conductor breaking strengths

It may come as a surprise to many readers that the declared breaking strength (sometimes referred to as the ultimate tensile strength) of a conductor has no unique value. The value depends upon the method of calculation employed as stipulated in the National or IEC Standards to which the conductor material is supplied. Differences quoted in breaking strengths for a given conductor configuration are *not* due to the material itself but to this calculation methodology. Table 18.5 shows the results of calculations for two different ACSR conductors using different calculation methodologies.

The IEC and BS values listed in Table 18.5 are fairly close for these two conductors. Such anomalies have presented a dilemma to manufacturers of conductors and fittings as to how to decide whether test results were applicable to the conductors or to the fittings. The calculation of conductor behaviour under changing loading conditions (wind, ice) and temperature is related to breaking strengths, and the design of fittings (tension clamps, repair sleeves,

Table 18.4   Comparison between ABC, BLX and open wire distribution

Property	Pros (ABC vs open wire)	Construction (ABC vs open wire)	BLX features
1. Ultimate tensile strength	Higher for neutral catenary wire support. Simple fittings. Less stock/stores holdings.	Possible support/fittings failure prior to bundle failure	As equivalent open wire aluminium alloy conductors
2. Current ratings	–	Lower – however note design at distribution voltage level is usually based on voltage drop rather than on current carrying capacity	–
3. Voltage regulation	Lower AC reactance (typically −25%)	Higher DC resistance (typically +15%)	Reactance will be affected by decision whether or not to reduce spacing.
4. Earth loop impedance	–	Line lengths will be less (typically −15%) because of higher DC resistance. This is an important point in PME systems	–
5. Short circuit ratings	–	Thermal limits on *both* conductor and insulation mean more attention to protection is required	Short circuit temperatures of 200°C allowable.
6. Costs			
(a) Fittings	Same	Same	1.3 × bare conductor
(b) Conductor	–	1.6 to 2×	1.8 × bare conductor
(c) Poles and stays	−10%	–	0.9 × bare conductor
(d) Labour *pole top* refurbishment	−36%	problems with jointing at HV	total new works cost for HV is approx equivalent to bare conductor
new works	−25%	problems with jointing at HV	
under eaves refurbishment	−17%	–	
new works	−22%	–	
(e) Maintenance	lower costs at LV	problems with jointing at HV	0.9 × bare conductor

etc.) must also be related to these values. Hence it is necessary in any overhead line specification to state clearly which standard calculated breaking strengths are to be based on in order to avoid disputes at a later date. If in doubt it is suggested that the IEC values should be used.

Table 18.5   Calculated conductor breaking strengths according to some different standards

Calculation standard	ACSR conductor 50/8 breaking strength (kN)	ACSR conductor 380/50 Breaking strength (kN)
BS215, Pt. 2. 1970	16.81	120.96
ASTM B2 32-74(Class A)	17.45	121.62
ASTM B232-74 (Class B)	16.91	119.74
ASTM B232-74 (Class C)	16.67	114.97
NF C34 120 1968 (R)		144.21
NF C34 120 1968 (N)		118.75
DIN 48 204 (declared)	17.09	123.14
DIN 48 204 (theoretical area)	16.83	120.80
DIN 48 204 (calculated area)		120.72
CSA C49 1–75	17.19	123.40
IEC 209 (now IEC 61089)	16.87	120.71
EN 50182	16.81	121.30

Notes: Conductor ACSR	48-AL1/8-ST1A	382-AL1/49-ST1A
Old reference	50/8	380/50
Stranding	6/1/3.2	54/7/3.0
Steel area	8.042 mm^2	49.480 mm^2
Aluminium area	48.255 mm^2	381.703 mm^2
Total area	56.297 mm^2	431.183 mm^2

18.3.5 Bi-metal connectors

Where an aluminium conductor is terminated on a copper terminal of, say, an isolator a special copper/aluminium joint is necessary to prevent the formation of a corrosion cell. A termination of this type usually comprises of an aluminium sleeve compressed onto a copper stalk with an insulating disc separating the two surfaces which are exposed to the atmosphere (see Fig. 18.2). The two dissimilar materials are generally welded together by friction welding as this process ensures a better corrosion resistance at the interface. An additional protection is afforded by the use of an anticorrosion varnish. When using such fittings it is always recommended that the aluminium component is above the copper one. Even slight traces of copper on aluminium have a disastrous effect on the aluminium material.

18.3.6 Corrosion

Since overhead lines are erected in different climatic conditions throughout the world a knowledge of their performance has been built up over the years. Aluminium conductors have good corrosion behaviour essentially resulting from the formation of an undisturbed protective surface oxide layer which prevents further corrosion attack. ACSR is known to suffer from bi-metallic corrosion which is noticeable as an increase in conductor diameter due to corrosion products in the steel core known as 'bulge corrosion'.

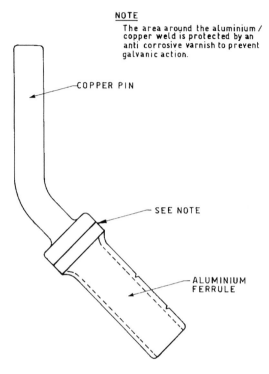

NOTE
The area around the aluminium /
copper weld is protected by an
anti corrosive varnish to prevent
galvanic action.

COPPER PIN

SEE NOTE

ALUMINIUM
FERRULE

Figure 18.2   Bi-metal connector

Early problems associated with deterioration of the steel cores used in ACSR conductors have been resolved over the years by the use of high temperature greases. These greases prevent the onset of any galvanic corrosion between the galvanized steel core and the outer aluminium wires. They have a high drop point which allows continuous operation of the conductor at 75°C and full service life protection. AAAC will obviously offer superior corrosion resistance than ungreased ACSR. Conductors that are not fully greased are not recommended for corrosive areas. The resistant properties of ACSR also depend upon the number of layers of aluminium surrounding the steel core. The conclusions of research carried out in the late 1960s showed that:

- Pure aluminium had the best corrosion resistance under the majority of environmental conditions.
- Smooth body conductors were the most corrosion resistant, especially if the inner layers were greased.
- Small diameter wires were most susceptible to corrosion damage and to failure. Thus for a given conductor area it is preferable to have fewer larger diameter strands.
- The overall corrosion performance of aluminium alloy conductors depends upon the type of alloy used.

For very aggressive environments the following order of preference is suggested:

- Aluminium conductor fully greased.
- Aluminium conductor with alumoweld core fully greased.
- ACSR fully greased.
- Aluminium alloy conductor fully greased.
- Aluminium conductor with alumoweld core ungreased.
- ACSR with greased core.

18.4 CALCULATED ELECTRICAL RATINGS

18.4.1 Heat balance equation

A conservative approach to the conductor thermal current rating in wind, ignoring any voltage regulation considerations, is given by the following simplified heat balance equation as valid for stranded conductors:

Heat generated (I^2R conductor losses) = heat lost by convection (watts/km)
+ heat lost by radiation (watts/ km)
− heat gained by solar radiation (watts/km)

$$= H_C + H_R - H_S$$
$$I^2 R_{20}\{1 + \alpha(t + \theta)\} = 387\,(V \cdot d)^{0.448} \cdot \theta$$
$$+\, \Pi \cdot E_C \cdot s \cdot d\{(t + \theta + 273)^4 - (t + 273)^4\} - \alpha_S \cdot S \cdot d \text{ (watts/km)}$$

where I = current rating, amps
R_{20} = resistance of conductor at 20°C
α = temperature coefficient of resistance per °C (for ACSR at 20°C, $\alpha = 0.00403$)
t = ambient temperature, °C
θ = temperature rise, °C (t_1 = initial temperature and t_2 = final temperature)
α_S = solar absorption coefficient – depends upon outward condition of the conductor and varies between 0.6 for new bright and shiny conductor to 0.9 for black conditions or old conductor. Average value of 0.8, say, may be taken for initial design purposes.
S = intensity of solar radiation, watts/m^2
d = conductor diameter, mm
V = wind velocity normal to conductor, m/s
E_C = emissivity of conductor – differs with conductor surface brightness. Typical values are 0.3 for new bright and 0.9 for black aluminium, ACSR or AAAC conductor. Average value = 0.6, say.

$$s = \text{Stefan–Boltzmann's constant} = 5.7 \times 10^{-8} \text{ watts/m}^2$$
$$\Pi = \text{pi, constant } (22/7) = 3.141592654\ldots$$

For design purposes 0.5 or 0.6 m/s wind speeds are often taken. Higher wind speeds would lead to higher ratings. Also it is possible to allocate different ratings for different seasonal conditions; this is specifically provided for in UK utility procedures.[16] In practice, the heat balance is a highly complex process but the above equation is adequate for many purposes. Further research is still going on in this field using deterministic models with values based on experience without attempting to correlate wind speed with air temperature or solar radiation. Also a more sophisticated probabilistic approach, taking account of available statistical data, is now available.[15] This is valid since practical measurements have shown that in almost all cases the conductor temperature is lower than that predicted by other methods, and the approach – defined as the General Approach in the European standard EN 50341, (which is applicable for new build lines only) – has been adopted in the UK.

18.4.2 Power carrying capacity

Approximate economic power transfer capacity trends for different line voltages based on power transfer being proportional to the square of the line voltage are given in Figs 18.3a and b for transmission voltages up to 500 kV. In practice, the capacity will be limited over long distances by the conductor natural impedance (voltage regulation) as well as by conductor thermal capacity. Depending upon the required electrical load transfer, the number of overhead line conductors of a particular type used per phase may vary.

Conductor configurations are shown in Fig. 18.4.

Therefore under the following specific tropical conditions (40°C ambient temperature, 0.894 m/s wind speed, 100 mW/cm² solar radiation and 35°C temperature rise) the calculated ratings for typical ACSR twin conductors at 230 kV would be:

$2 \times 200\,\text{mm}^2$ (nominal) – 1052 A (419 MVA)
$2 \times 300\,\text{mm}^2$ (nominal) – 1296 A (516 MVA)
$2 \times 400\,\text{mm}^2$ (nominal) – 1558 A (620 MVA)
$2 \times 500\,\text{mm}^2$ (nominal) – 1742 A (694 MVA)
$2 \times 600\,\text{mm}^2$ (nominal) – 1890 A (753 MVA)

A typical set of power transfer curves for the $2 \times 400\,\text{mm}^2$ conductor case are given in Fig. 18.5. The optimum rating for the particular line length is given by the intersection of the regulation curves for, say, 0.9 power factor (pf) with either the thermal limit or the voltage regulation lines, whichever does not infringe the voltage or current limit specified for the line. It should be noted that adequate technical performance is usually judged upon the load flow

VOLTAGE / POWER	11kV	33kV	66kV	132kV	220kV	275kV	330kV	400kV	500kV
5 MVA a)	100-DOG	25-GOPHER							
b)	300-GOAT	25-GOPHER							
10 MVA a)	300-GOAT	50-RABBIT	25-GOPHER						
b)	INADEQUATE	50-RABBIT	25-GOPHER						
25 MVA a)		200-PANTHER	75-RACCOON						
b)		200-PANTHER	75-RACCOON						
50 MVA			200-PANTHER	200-PANTHER					
100 MVA				2x150-WOLF					
200 MVA					250-BEAR				
200 MVA									
300 MVA					2x175-LYNX	400-ZEBRA			
400 MVA					2x250-BEAR	2x175-LYNX			
500 MVA					2x400-ZEBRA	2x250-BEAR (2X BATANG)			
600 MVA						2x350-ANTELOPE or BISON			
700 MVA							2x350-ANTELOPE or BISON		
800 MVA							3x300-GOAT	2x400-ZEBRA	
1000 MVA								3x250-BEAR or DOVE	
1200 MVA								3x400-ZEBRA	
1800 MVA								4x400-ZEBRA or 4xCROW	3x450-ELK 4x(2B2)DOVE
2000 MVA									4x300-GOAT

Notes:

1. Numbers refer to nominal Alu area e.g. 100mm²

2. For voltages up to and including 66kV, conductor size is governed by thermal rating and/or voltage drop - surface gradients are normally acceptable.

3. (a) is thermal rating
   (b) is rating for 10% voltage drop with power factor = 0.9 over a distance of 10km. Other ratings would apply for other assumptions.

4. For voltages 132kV and above conductor size is also governed by surface gradient and electrical stability of systems. Ratings of lines are affected by equipment in substations.

5. Typically minimum conductor sizes would be
   132kV   1 x 14.2mm
   275kV   2 x 19.3mm
   400kV   4 x 18mm
   There are variations as function of construction and altitude of line.

6. Table prepared for tropical conditions. For temperate conditions ratings would be 20 to 30% higher.

7. Complexity of presentation exemplified by CEGB which have prepared 75 pages of tables for 6 conductor types.

PTD/GO/SS/MARCH 1987

Figure 18.3(a)   Approximate conductor sizes (ACSR) for power transfer capabilities

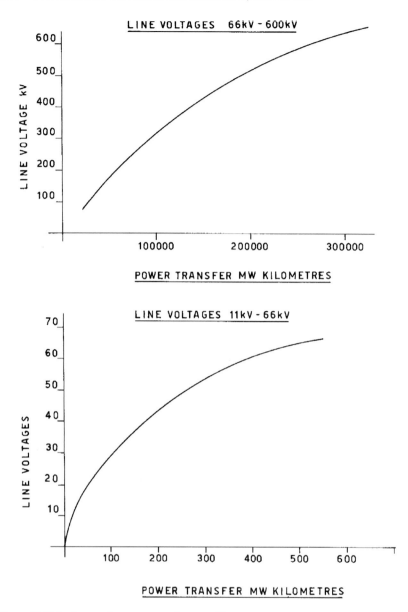

Figure 18.3(b)   Economic power transfer capacities

under single circuit outage conditions. In comparison, economic loadings do not normally consider outages and are based on normal operating conditions.

Calculated ratings for typical ACSR conductors at lower voltage levels of 11, 33 and 66 kV overhead lines using different conductors over different distances are given in Table 18.6.

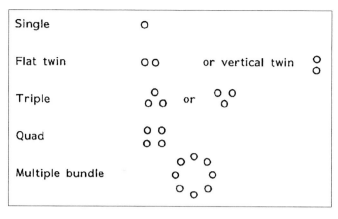

Figure 18.4 Typical conductor configurations

18.4.3 Corona discharge

High voltage gradients surrounding conductors (above about 18 kV/cm) will lead to a breakdown of the air in the vicinity of the conductor surface known as corona discharge. The effect is more pronounced at high altitudes. Generally, the breakdown strength of air is approximately 31 kV peak/cm or 22 kV rms/cm. This is a useful guide for the selection of a conductor diameter or conductor bundle arrangement equivalent diameter. Corona discharge and radio interference noise generated cause problems with the reception of radio communication equipment and adversely affect the performance of power line carrier signals.

At higher voltage levels, and certainly at voltages of 400 kV and above, interferences due to the corona effect can be the dominant factor in determining the physical size of the conductor rather than the conductor thermal rating characteristic. Increasing the conductor diameter may be necessary in order to reduce the surface stress to acceptable levels. Obviously there is a limit with regard to the practical size, strength and handling capability for conductors. The bundling of conductors as described in Section 18.4.2 assists in the effective increase in overall conductor diameter and hence leads to lower stress levels.

The surface voltage gradient may be determined from Gauss's theorem showing that an increase in radius or equivalent radius leads to a reduction in surface voltage gradient:

$$V_g = \frac{Q}{(2 \cdot \pi \cdot \varepsilon_0 \cdot r)}$$

where V_g = voltage surface gradient (volts/cm)
  Q = surface charge per unit length (coulomb/m)
  r = equivalent radius of smooth conductor (cm)
  ε_0 = permittivity of free space = $1/\{36 \cdot \pi \cdot 10^9\}$(F/m)

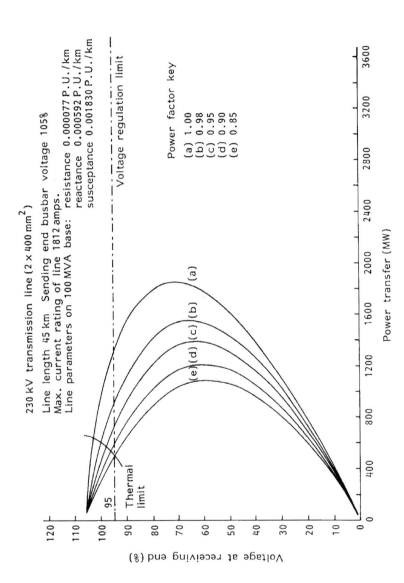

230 kV transmission line (2 × 400 mm²)

Line length 45 km Sending end busbar voltage 105%
Max. current rating of line 1812 amps.
Line parameters on 100 MVA base: resistance 0.000077 P.U./km
                                   reactance 0.000592 P.U./km
                                   susceptance 0.001830 P.U./km

Voltage regulation limit

Power factor key

(a) 1.00
(b) 0.98
(c) 0.95
(d) 0.90
(e) 0.85

Thermal limit

Voltage at receiving end (%)

Power transfer (MW)

Figure 18.5   Power transfer curves

Table 18.6   Typical load carrying capacity of distribution lines

Line voltage (kg)	Conductor equivalent configuration spacing (mm)	ACSR conductor code	AAC conductor code	MW capacity based upon 5% regulation			
				8 (km)	16 (km)	24 (km)	32 (km)
11	1400	Sparrow	Iris	0.95	0.49	0.33	0.25
		Raven	Poppy	1.4	0.7	0.47	0.35
		Linnet	Tulip	3.00	1.5	1.00	0.75
				16 (km)	32 (km)	48 (km)	64 (km)
33	1500	Quail	Aster	5.00	2.50	1.70	1.25
		Penguin	Oxlip	6.70	3.35	2.20	1.70
		Linnet	Tulip	8.35	4.18	2.80	2.10
		Hen	Cosmos	11.50	5.75	3.80	2.90
				32 (km)	64 (km)	96 (km)	128 (km)
66	3000	Quail	Aster	12.50	6.25	4.18	3.14
		Linnet	Tulip	16.00	8.00	5.32	3.99
		Hen	Cosmos	18.40	9.18	6.12	4.59

In practical terms this may also be expressed as follows:

$$V_g = \frac{U_p}{[(d/2)\log_e(2D/d)]} \text{ kV/cm}$$

where V_g = voltage surface gradient (kV/cm)
U_p = phase voltage (kV)
d = diameter of single conductor (cm)
D = distance between phases for single phase line or equivalent spacing for three phase lines (cm)

For the three phase line configuration, $D = \sqrt[3]{D_{ry} \cdot D_{yb} \cdot D_{br}}$ where D_{ry}, D_{yb} and D_{br} are the spacings between the different phases r, y & b.

Consider a 132 kV single circuit Zebra ACSR line with conductor diameter 28.62 mm and spacings as shown in Fig. 18.6.

$$D_{ry} = \sqrt[2]{6^2 + 1.8^2} = 6.26 \text{ m}$$

$$D_{yb} = \sqrt[2]{7^2 + 1.8^2} = 7.23 \text{ m}$$

$$D_{br} = \sqrt[2]{1^2 + 3.6^2} = 3.74 \text{ m}$$

$$D = \sqrt[3]{D_{ry} \cdot D_{yb} \cdot D_{br}} = 5.53 \text{ m} = 553 \text{ cm}.$$

$$V_g = U_p/[(d/2)\log_e(2D/d)] = 132/\sqrt{3}/[(2.86/2)\log_e(2 \cdot 553/2.86)]$$

$$= 76.2/[1.43 \cdot \log_e(386.71)]$$

$$= 76.2/1.43 \cdot 5.96$$

$$= \underline{8.94 \text{ kV/cm}} \text{ which is within the } 18 \text{ kV/cm criteria.}$$

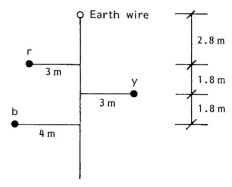

Figure 18.6   Corona discharge calculation example – 132 kV zebra conductor spacing

Table 18.7   Some reception classifications

Signal-to-noise ratio (dB)	Subjective impression of reception quality
32	Entirely satisfactory
27	Very good, background unobtrusive
22	Fairly good, background evident
16	Speech easily understood, background very evident
6	Difficulty in understanding speech
0	Noise swamps speech

Radio frequency interference (RFI) noise is measured in decibels above 1 microvolt per metre (dB > 1 V/m) from comparative equations of the form:

$$\text{RFI} - \text{RFI}_0 = 3.8 \, (E_{\text{mean}} - E_{0 \, \text{mean}}) + 40 \log_{10} d/d_0 + 10 \log_{10} n/n_0$$
$$+ \, 30 \log_{10} D_0/D + 20 \log_{10} (1 + f_0{}^2)/(1 + f^2)$$

where RFI = calculated radio noise (dB > 1 μV/m)
E_{mean} = calculated mean voltage gradient (kV/cm)
d = conductor diameter (cm)
n = number of subconductors in bundle
D = distance between phase and measuring antenna (m)
f = frequency (Hz)

The suffix '0' refers to the same quantities obtained from measurements. Acceptable noise levels depend upon the quality of service required and is described in terms of an acceptable signal-to-noise or signal plus noise-to-noise ratio. Some reception classifications are given in Table 18.7.

Thus if a signal has a field strength of say 60 dB > 1 μV/m and a fairly satisfactory reception is required then the noise from the adjacent overhead line should not exceed 60 − 22 = 38 dB > 1 μV/m. Audible noise is generally not considered to be a controlling factor at voltage levels below 500 kV. However, more research is being carried out in this area.

18.4.4 Overhead line calculation example

If we wish to transfer 40 MVA over a distance of 70 km then we can calculate the conductor and tower size using the simple hand calculations given below. Such calculations are a useful check to the more normally used computer generated solutions and allow an understanding of the basic principles involved.

Assume Lynx ACSR overhead line under the following tropical conditions operating at 132 kV:

Maximum operating temperature	75°C
Maximum ambient air temperature	40°C (temperature rise = 35°C)
Lynx conductor maximum resistance	0.1441 ohm/km
Lynx conductor diameter	19.53 mm
Emissivity	0.6
Solar absorption coefficient	0.8
Solar radiation intensity	1000 W/m^2
Wind velocity	1 mph = 0.447 m/s

Effective wind velocity = actual wind velocity $\cdot$ p/760 $\cdot$ 293/(273 + t)
$$= 0.447 \cdot 760/760 \cdot 293/313 = 0.418 \, \text{m/s}$$
$$= 41.8 \, \text{cm/s (assuming normal atmospheric pressure)}$$

Load current at 132kV for 40 MVA power transfer
$$= 40 \cdot 10^6 / \sqrt{3} \cdot 132 \cdot 10^3$$
$$= \underline{175 \, \text{A}}$$

The conductor thermal rating capability is first determined, ignoring any voltage drop considerations, by comparing the 175 A load current requirement and the rating of the conductor derived from the heat balance equation detailed in Section 18.4.1 (or use the more complex approach referred to there).

$$I^2R = 13.8(t_2 - t_1) \cdot 10^{-4} \cdot (V \cdot d)^{0.448}$$
$$+ \Pi \cdot E_C \cdot s \cdot d \cdot (T_2^4 - T_1^4)$$
$$- \alpha_S \cdot S \cdot d \text{ (watts/cm)}$$

$$I^2 0.1441 \cdot 10^{-5} = 13.8(75 - 40) \cdot 10^{-4} \cdot (41.8 \cdot 1.953)^{0.448}$$
$$+ \Pi \cdot 0.6 \cdot 5.7 \cdot 10^{-12} \cdot 1.953 \cdot ([273 + 75]^4$$
$$- [273 + 40]^4) - 0.8 \cdot 1000 \cdot 10^{-4} \cdot 1.953$$
$$= 0.347 + 20.99 \cdot 10^{-12} \cdot (348^4 - 313^4) - 0.156$$
$$= 0.347 + 0.106 - 0.156$$
$$I^2 = 0.297/0.1441 \cdot 10^{-5}$$
$$= 206.107 \cdot 10^3$$
$$I = \sqrt{206.107 \cdot 10^3}$$
$$= \underline{454 \, \text{A}}$$

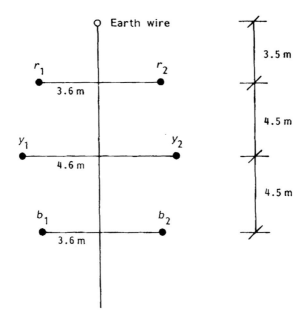

Figure 18.7   Calculation example – 132 kV Lynx conductor spacing

The conductor type is therefore more than adequate on thermal considerations for the load required.

A check is then made for any corona discharge limitations. Assume a conductor configuration as shown in Fig. 18.7 and calculate for only one circuit, r_1, y_1, b_1:

$$D_{ry} = \sqrt[2]{4.5^2 + 1^2} = 4.61\,\text{m}$$

$$D_{yb} = \sqrt[2]{4.5^2 + 1^2} = 4.61\,\text{m}$$

$$D_{br} = 9\,\text{m}$$

$$D = \sqrt[3]{D_{ry} \cdot D_{yb} \cdot D_{br}} = \sqrt[3]{4.61 \cdot 4.61 \cdot 9} = 5.76\,\text{m}.$$

$$\begin{aligned}
V_g &= U_p/[(d/2)\log_e(2D/d)] = 132/\sqrt{3}/[(1.953/2)\log_e(2 \cdot 576/1.953)] \\
&= 76.2/[0.977 \cdot \log_e(589.86)] \\
&= 76.2/0.977 \cdot 6.38 \\
&= \underline{12.22\,\text{kV/cm}}
\end{aligned}$$

which is within the 18 kV/cm criteria and Lynx conductor is therefore acceptable from both a corona and current carrying capacity point of view.

If capacitive reactance is ignored the voltage drop, V_d, for a line length, l, is calculated from the usual formula:

$$V_d = I(R\cos\phi + X\sin\phi) \cdot l$$

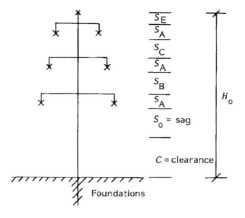

Overall tower height $H_0 = C + S_0 + 3S_A + S_B + S_C + S_E$

Figure 18.7(a)   Basic span – overall tower height and clearances

If the load at the end of the line is given in kVA then for a three phase system the load current, $I = kVA/\sqrt{3} \cdot U$ where U equals the line voltage in kV.

The main practical problem is now to obtain accurate values for the line reactance. Some typical line reactance values are given in the Table 18.8.

The value of $(R \cos\phi + X \sin\phi)$ is approximately constant for overhead line configurations with conductor sizes above 150 to 200 mm². Therefore very large conductors are necessary to improve any voltage drop problems if such conductor sizes prove to be inadequate. In such circumstances consideration should be given to increasing the transmission voltage level.

It is useful to introduce the concept of allowable kVA km for a given voltage drop for a variety of overhead line configurations and different conductors. For a 10% voltage drop then:

$$0.1\,U = \frac{\sqrt{3} \cdot kVA}{\sqrt{3} \cdot U} \cdot l \cdot (R \cos\phi + X \sin\phi) \cdot 10^{-3}$$
$$kVA \cdot l = 100 \cdot U^2/(R \cos\phi + X \sin\phi) \text{ with length, } l \text{ in km.}$$

Tables may be prepared based on this equation for different conductors at different power factors giving the allowable kVA km for a given % voltage drop.

18.5 DESIGN SPANS, CLEARANCES AND LOADINGS

18.5.1 Design spans

The general parabolic sag/tension equation is explained in Chapter 17 Section 17.2.3. In order to design suitable tower dimensions for an overhead line it is

Table 18.8   ACSR conductors inductive reactance, Ω/km (equivalent spacings given) (*IEE Proc*, Vol. 133, Pt. C, No. 7, November 1986)

Equivalent Al area (mm²)	25	30	50	100	125	150	175	200	250	300
Stranding	6/1/2.36	6/1/3.35	6/1/4.1	6/4.72 + 7/1.57	30/7/2.36	30/7/2.59	30/7/2.79	30/7/3.0	30/7/3.35	30/7/3.71
Current (temperate) A	157	242	311	371	429	482	528	579	664	755
Current (tropical) A	130	198	253	299	343	384	419	457	521	587
R Ω/km (20°C)	1.093	0.5426	0.3622	0.2733	0.2203	0.1828	0.1576	0.1362	0.1093	0.08911
R Ω/km (75°C)	1.317	0.6539	0.4365	0.3294	0.2655	0.2203	0.1899	0.1641	0.1317	0.1074
0.3 m spacing (415 V)	0.298	0.276	0.263	0.253	0.239	0.233	0.229	0.224	0.217	0.211
1.4 m spacing (11 kV)	0.395	0.373	0.360	0.350	0.336	0.330	0.326	0.321	0.314	0.308
1.5 m spacing (33 kV)	0.399	0.377	0.364	0.355	0.340	0.335	0.330	0.325	0.318	0.312
3.0 m spacing (66 kV)	0.442	0.420	0.408	0.398	0.384	0.378	0.373	0.369	0.362	0.356
3.6 m spacing (110 kV)	0.454	0.432	0.419	0.410	0.395	0.390	0.385	0.380	0.373	0.367
4.9 m spacing (132 kV)	0.473	0.451	0.439	0.429	0.415	0.409	0.404	0.400	0.393	0.386

necessary to calculate the conductor sags and tensions. The maximum conductor tension (which will occur at minimum temperature) is evaluated in order to ensure a sufficient mechanical strength margin for the particular conductor. The sag is calculated in order to fix the tower height. The ruling condition for the conductor has to be determined based on either the maximum working tension (MWT), the everyday stress (EDS) or, potentially, the maximum erection tension (MET). The conductor has to be designed such that the maximum anticipated loads do not exceed 50% of the breaking load at $-6°C$ (MWT condition) and 20% at, say, an everyday temperature of 16°C (EDS condition).

18.5.1.1 *Basic span*

The optimum spacing of towers and their height becomes a financial exercise.

With short spans and low towers the total number of towers and associated fittings will be large to cover a certain route length but less steel per tower will be necessary. If long spans are used then the conductor sag between tower points becomes greater and fewer stronger, higher towers and fittings, but with correspondingly more steel, are necessary to ensure correct clearances. The extent of labour associated with a variable number of towers for a given route length will also be important.

The overall height of the tower (see Fig. 18.7(a))

$$H = C + S_O + 3S_A + S_B + S_C + S_E$$

where C = statutory clearance to ground
S_A = length of insulator suspension set
S_B, S_C, S_E = vertical distances between crossarms and conductor above or to earth wire
S_O = sag of conductor (proportional to square of span)

Given the mechanical loading conditions and phase and earth wire conductor types an evaluation of the basic span may be made as follows. Assume an arbitrary length in a flat area over, say, 100 km. Inevitably there will be some angle/section towers whose positions will be fixed beforehand. From experience let this number be N_0. If L is the basic span and l the span length then the number of suspension towers will be the next integer from $[(100 \cdot L/l) + l - N_0]$.

1. Conductors and earth wire – costs for supply and installation.
2. Insulators – selection depending upon mechanical loading and pollution levels such that S_A may be defined.
3. S_B, S_C, S_E – a function of the still air clearance co-ordinated with the insulation level.

4. Tower weight (W) – lengthy designs may be omitted at this stage by using the formulae such as that by P. J. Ryle:

Approximate weight of tower $W = K_1 \cdot H \cdot \sqrt{M_t}$

Approximate base width $K_2 \cdot \sqrt{M_t}$

where　H = overall tower height
　　　　M_t = ultimate overturning moment (OTM) at the base of the tower.

This must be the largest OTM corresponding to the highest loading conditions affecting one leg of the tower and taken as the sum of the transverse and longitudinal moments due to conductor tension, tower and conductor wind loadings. For convenience the OTM due to wind on the structure as a proportion of all other loads may be accepted as:

~25% (20 to 30%) for intermediate towers
~10% (7 to 15%) for small angle (10° to 30°) towers
~8% (5 to 10%) for 60° towers

K = constant:

Tower type	K_1	K_2
Conventional mild steel	0.008	0.30

With a knowledge of suspension and tension tower weights, supply and installation costs may be assessed.

5. Foundations – depends upon soil properties and a site visit is necessary to assess the situation. However, an assessment of uplift and compression loads may be made since an approximate base width is calculable.

The summation of the costs involved will then give an indication of the approximate total cost. By varying the span length l (with its influence on S_O and associated quantities), cost vs span may be evaluated and plotted. Such curves as shown in Fig.18.8 are in practice normally very flat at the bottom. Experience shows that a span selected slightly greater than the minimum derived from such an initial analysis gives an overall optimum choice. From a recent international survey the supply and installation costs of overhead lines may be broken down as in Table 18.9.

Table 18.9　Relative supply and installation costs for overhead lines

Description	up to 150 kV	150–300 kV	>300 kV
Conductors	31.6	31.5	34.1
Earth wires	4.1	3.5	3.9
Insulators	8.8	9.3	6.9
Towers	36.0	36.0	36.4
Foundations	19.5	19.7	18.7

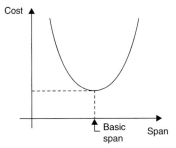

(a) Cost/span plot to determine most economic basic span.
The basic span is the horizontal distance between centres
of adjacent supports on level ground.

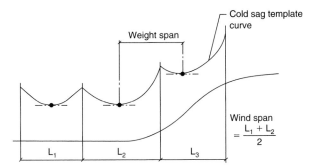

The wind span is half the sum of adjacent horizontal span lengths
supported on any one tower.
The weight span is the equivalent length of the weight of
conductor supported at any one tower at minimum temperature
in still air.

(b) Wind and Weight Span

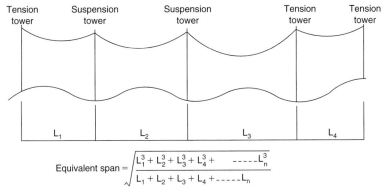

$$\text{Equivalent span} = \sqrt{\frac{L_1^3 + L_2^3 + L_3^3 + L_4^3 + \quad - - - - - L_n^3}{L_1 + L_2 + L_3 + L_4 + - - - - - L_n}}$$

The equivalent span is used for determination of sag in spans for which the tension in any
section length is that which would apply to a single span equal to the equivalent span.

(c) Equivalent Span

Figure 18.8   Overhead line spans

This breakdown is only approximate and gives average values for many lines and practices encountered by Balfour Beatty throughout the world and varies according to line voltage, conductor configuration and the design of the supporting structure. In addition, an allowance has to be made for the routing survey, land clearance, erection and similar incidentals. Basic spans might be approximately 365 m at 230 kV and 330 m at 132 kV. The minimum allowable ground clearance between phase conductors and earth is derived from specified conductor clearance regulations for the country involved, in still air at maximum conductor temperature. Survey figures for the proportion of tower costs compared to the overall line costs ranged from 8% to 53% with ACSR, but from 25% to 45% with AAAC.

18.5.1.2 *Wind span*

The wind span is half the sum of the adjacent span lengths as shown in Fig. 18.8b. At 230 kV this might be 400 m under normal conditions and 300 m under broken wire conditions. Correspondingly, at 132 kV typical values are 365 m and 274 m respectively.

18.5.1.3 *Weight span*

The weight span is the distance between the lowest points on adjacent sag curves on either side of the tower as also shown in Fig. 18.8b. It represents the equivalent length or weight of conductor supported at any one tower at any time. For design purposes, it is the value under worst loading conditions (minimum temperature in still air) which gives the greatest value. A tower at the top of a hill may be heavily loaded and it is usual to assume a weight span which can reach up to twice the value of the basic span. In fairly level terrain a value of 1.6 to 1.8 may be adopted.

The ratio of weight span to wind span is also important since insulators on lightly loaded towers may be deflected excessively thus encroaching electrical clearances. A ratio of weight span to wind span of approximately 1.5–2 is often considered acceptable. This ratio is easily computed with the use of the 'cold' template. When plotting tower positions, the engineer must be aware of the maximum weight span and of such ratios. Typical weight span values at 230 kV and 132 kV are given below:

230 kV		132 kV	
Suspension towers			
−750 m	Normal conditions	680 m	Normal conditions
565 m	Broken wire conditions	510 m	Broken wire conditions
Tension towers			
750 m	Normal conditions	680 m	Normal conditions
750 m	Broken wire conditions	680 m	Broken wire conditions

18.5.1.4 *Equivalent span*

The equivalent span is defined as a fictitious single span in which tension variations due to load or temperature changes are nearly the same as in the actual spans in a section. The mathematical treatment to obtain the equivalent span is based on parabolic theory and there is no similar concept using full catenary equations. For sagging the overhead line conductors the tension appropriate to the equivalent span and the erection temperature as shown in Fig. 18.8c is used. Erection tensions are calculated from final tensions making an allowance for creep. This is equated to a temperature shift which is applied to final tensions.

18.5.1.5 *Creep*

Creep is a phenomenon which affects most materials subjected to stress. It manifests itself by an inelastic stretch (or permanent elongation) of the material in the direction of the stress. Certain materials such as aluminium are more susceptible than others. For example steel suffers only a limited amount of creep. The increase in conductor length resulting from inelastic stretch produces increased sags which must be taken into account in the overhead line design and installation process so as not to infringe clearances.

Some mathematical models have now been evolved to help the engineer assess the effects of creep and examples are given here (see also IEC technical report 61597 – calculation methods for stranded bare conductors):

$$\varepsilon = K \sigma^\beta e^{\phi\theta} t^{\mu/\sigma^\delta} \text{ mm/km} \qquad\qquad \text{Formula 1}$$

and,

$$\varepsilon = K \sigma^\beta e^\phi t^\mu \text{ mm/km} \qquad\qquad \text{Formula 2}$$

where ε = permanent inelastic elongation (creep)
              K = constant
              σ = average stress in conductor
β, ϕ, μ, δ = creep indices obtained by test
              e = natural logarithm base = 2.7182818…
              t = time in hours
              θ = temperature in °C

Since the total inelastic strain can be considered as the result of geometric settlement of the strands and of the metallurgical creep thereafter, the derivation of the constants and of the indices is of prime importance. In the UK it has been decided that tests should be carried out in such a way that the geometric settlement would be taken into account in the constants and indices and that the formulae above would give the total creep. Typical values for the constants involved in the above equations are given in Tables 18.10a to d.

Table 18.10a   Creep coefficients for ACSR conductors (Formula 1)

Conductor stranding	Al/steel	Al/steel area ratio	Process	K	φ	β	μ	δ
54	7	7.71	HR	1.1	0.0175	2.155	0.342	0.212 7
			EP	1.6	0.0171	1.418	0.377	0.1876
48	7	11.4	HR	3.0	0.0100	1.887	0.165	0.011 6
30	7	4.28	EP	2.2	0.0107	1.375	0.183	0.036 5
26	7	6.16	HR	1.9	0.0235	1.830	0.229	0.080 21
24	7	7.74	HR	1.6	0.0235	1.882	0.186	0.007 71
18	1	18.0	EP	1.2	0.0230	1.502	0.332	0.133 1
12	7	1.71	HR	0.66	0.0115	1.884	0.273	0.147 4

Note: Industrial processing of aluminium rod: HR = hot rolled; EP = extruded or Properzi.

Table 18.10b   Creep coefficients for AAAC conductors (Formula 2)

K	φ	β	μ
0.15	1.4	1.3	0.16
Not available	Not available	Not available	Not available

Table 18.10c   Creep coefficients for AAC conductors (Formula 2)

	K						
	Number of make up wires						
Process	7	19	37	61	φ	β	μ
Hot rolled	0.27	0.28	0.26	0.25	1.4	1.3	0.16
Extruded or Properzi	0.18	0.18	0.16	0.15	1.4	1.3	0.16

Table 18.10d   Creep coefficients for ACAR conductors (Formula 2)

Process	K	φ	β	μ
Extruded or Properzi	0.04 + {0.24 m/(m + 1)}	1.4	1.3	0.16

Notes: m = aluminium area/aluminium alloy area.

When applying the technique of creep evaluation the designer must forecast reasonable conductor history. Typical conditions might be as shown in Table 18.11 where t_{III} and t_{IV} represent the periods for which compensation should be made. Figure 18.9 illustrates an acceptable procedure for creep assessment. As an illustration of the steps to be followed consider the following example.

1. The EDS is to be 20% of the UTS of the conductor at 20°C.
2. The maximum stress occurs when the conductor is subjected to a wind of 50 kg/m² at 0°C, no ice.
3. The maximum operating temperature is 70°C.

Table 18.11   Typical conditions

Stage	Stress	Temperature	Time
I	Running out	Average ambient	Time for running out
II	Pretension (if provided)	Average ambient	As decided by design office
III	Stress at given temperature	Mean yearly temperature + 5°C	t_{III}
IV	Stress at given temperature	Maximum operating conductor temperature	t_{IV}
V	Maximum stress	Temperature corresponding to maximum stress condition	t_{IV}

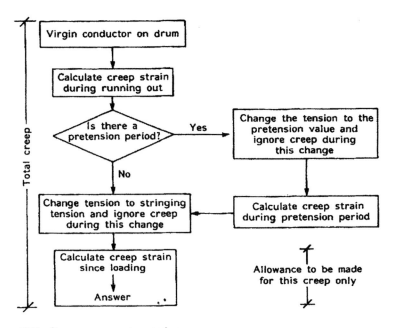

Figure 18.9   Creep assessment procedure

4. Accept a span length of 400 m. (In practice, three values should be taken: a maximum and a minimum span both deduced from the profile, and a basic span. The span which gives the highest value of creep strain is selected as a basis for creep compensation.)
5. Creep strain to be calculated for a period of 30 years.
6. Conductor is manufactured from aluminium rod obtained by the Properzi method.

Some decisions based on experience are then necessary regarding the duration of the maximum and minimum stresses, and values may then be inserted in a tabular format as shown in Table 18.12a.

If we consider the general change-of-state sag/tension equation the influence of creep strain and temperature are both linear (see Section 18.5.2.5).

Table 18.12a

Stage	Stress	Temperature	Time
I	20% UTS	20°C	1 hour
II	Nil (no pretension)	Not applicable (no pretension)	Not applicable (no pretension)
III	Calculate by program	25°C	257 · 544 hours[a]
IV	Calculate by program	70°C	2628 hours[a]
V	Calculate by program	0°C	2628 hours[a]

Notes: [a] The period for which compensation should be made.

Table 18.12b   Typical creep values of stranded conductors (IEC report 61597)

Type of conductor	Estimated creep after 10 years Mm/m	Equivalent temperature difference °C
A1	800	35
A2, A3	500	22
A1/A2, A1/A3	700	30
A1/Sxy	500	25

$$W^2 \cdot L^2/24 \cdot (1/T_2^2 - 1/T_1^2) = 1/EA(T_2 - T_1) + \alpha(\theta_2 - \theta_1)$$
$$+ (\varepsilon_2 - \varepsilon_1) \cdot 10^{-6}$$

It is therefore possible to express creep strain, ε, by an equivalent temperature change, i.e.

$$\varepsilon = \alpha\Delta\theta_e \times 10^6$$

where α is the coefficient of thermal expansion per °C. This simplifies the conductor sag calculation, because the elongation due to creep can be simulated by a temperature difference using the appropriate coefficient of linear expansion, as illustrated in Table 18.12b.

This concept is equally applicable if lifetime creep is predicted using other techniques. For example:

- Using creep values from conductor creep tests made at actual mechanical and temperature conditions over a long time(normally more than 2 months) and extrapolating the creep curve up to 10, 30 or 50 years. Normally the final sag calculation is made using the 10 year figure, because the additional creep between 10 to 50 years is relatively small and a reasonable part of that may have elapsed from the time of stringing up to the time of clamping the conductor.
- Using creep values from accelerated conductor creep tests made at higher mechanical tension. The creep value at a certain time will then correspond to what is known to apply under real conditions after (say) 30 years.

- Using creep compensations made with the help of sag and tension charts. As an illustration, with Zebra conductor it has been assessed that creep strain at the end of 10 years (t = 87 600 hours) is $\varepsilon = 616$ mm/km giving

$\Delta\theta_e = 32°C$ approximately, then:

- maximum design temperature of conductor = 50°C, say
- equivalent temperature corresponding to creep, $\Delta\theta_e = 32°C$
- temperature for evaluating sag at time, t, and corresponding to the maximum design temperature of the conductor when no pretension or overtension regime are applicable, $\theta + \Delta\theta_e = 82°C$

Clearly, this will result in a penalty in the height of all towers. An alternative would be to reduce the sag at sagging time resulting in a temporary overtension in the conductor, resulting in an overdesign penalty on the angle towers. By applying several combinations of temperature correction or pretension the designer is able to aim for the least onerous solution.

A detailed study of the references given at the end of this chapter concerning this subject is recommended. The following guidelines may be useful:

1. At the start of the computer run, a number of time intervals should be defined in which the temperature is assumed to remain constant. The temperature can vary between one time interval and another.
2. If the tension remains reasonably constant throughout an interval, as could be the case during running out and pretensioning of a conductor, the creep at the end of that interval is obtained directly from the relevant equation:

$$\varepsilon = K\sigma^\beta e^{\phi\theta} t^{\mu/\sigma^\delta} \text{ mm/km} \quad \text{or} \quad \varepsilon = K\sigma^\beta e^\phi t^\mu \text{ mm/km}$$

18.5.1.6 Catenary equations for sloping spans

The following equations have been found useful and are summarized for convenience. The basic catenary equations may be found in most university mathematics textbooks. The symbols used are explained in Fig. 18.9a.

Put the origin at the lowest point of the catenary curve.

$$L = x_1 + x_2$$
$$\tan\phi = h/L$$
$$s = \text{suspended length}$$
$$\theta = \text{temperature differential between initial and final conditions}$$

The catenary equations are as follows and may be simplified with $c = T_0/W$.

$$y = T_0/W \cdot \cosh\{Wx/T_0\} = c \cdot \cosh x/c$$

$$s = T_0/W \cdot \sinh\{Wx/T_0\} = c \cdot \sinh x/c$$

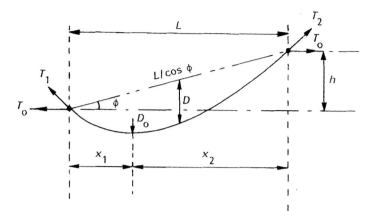

Figure 18.9(a)   Catenary equations for sloping spans

$$\therefore h = c(\cosh \{[L - x_1]/c\} - \cosh x_1/c) \tag{1}$$

$$= 2c \sinh L/2c \cdot \sinh ([L/2] - x_1)/c) \tag{1'}$$

$$S = c(\sinh x_1/c + \sinh \{[L - x_1]/c\} \tag{2}$$

(a) Distance of lowest point of curve from supports:
Left-hand support from equation (1'):

$$\sinh ([L/2] - x_1)/c = h/2c \sinh L/2c$$
$$\therefore x_1 = L/2 - c \sinh^{-1} \{h/(2c \sinh L/2c)\} \tag{3}$$
$$x_1 \sim L/2 - T_0/W \cdot h/L = L/2 - T_0/W \cdot \tan \phi$$

Right-hand support

$$x_2 = L - x_1$$
$$\therefore x_2 = L/2 + c \sinh^{-1} \{h/(2c \sinh L/2c)\} \tag{4}$$

(b) Sag:
The sag is measured at mid-span.

$$\therefore D = (c \cosh x_1/c + h/2) - c \cosh ([L/2] - x_1)/c$$
from (1), $D + h = c\{\cosh\{[L - x_1]/c\} - \cosh ([L/2] - x_1)/c\} + h/2$
$$\therefore D = c\{\cosh [L/2c] \cdot \cosh ([L/2] - x_1)/c \sim \sinh [L/2c] \cdot \sinh$$
$$([L/2] - x_1)/c - \cosh ([L/2] - x_1)/c\} - h/2$$

from (1')

$$D = c(\cosh[L/2c] - 1) \cdot \sqrt{\{1 + h^2/(4c^2 \sinh^2 L/2c)\}} \tag{5}$$

$$D \sim (L2/8c) \cdot 1/\cos \phi$$

(c) Change of state:
Suspended length
from (2)

$$s = 2c \cdot \sinh L/2c \cdot \cosh\{[L/2 - x_1]/c\}$$

$$\therefore \quad \text{from (1')}$$

$$s = 2c \cdot \sinh L/2c \cdot \sqrt{\{1 + h^2/(4c^2 \sinh^2 L/2c)\}}$$

$$= \sqrt{\{(4c^2 \sinh^2 L/2c) + h^2\}}$$

$$\sim \{L^2 + 2L^2/6 \cdot (L/2c)^2 + h^2\}^{1/2}$$

$$\therefore s \sim (L/\cos\phi) \cdot (1 + L^2 \cdot \cos^2 \phi/24c^2)$$

Temperature change
Let

s_f = length at final slope
s_i = length at initial slope
$s_f = s_i\{1 + [(T_f - T_i)/AE] + [(t_f - t_i)\alpha]\}$

where

$$(t_f - t_i) = \theta°$$

$$\therefore \sqrt{\{(4 c_f^2 \sinh^2 L/2c_f) + h^2\}} = \sqrt{\{(4c_i^2 \sinh^2 L/2c_i) + h^2\}} \cdot$$
$$\{1 + [(T_{of} - T_{oi})/AE] + [\theta \cdot \alpha]\}$$

by squaring and transformation:

$$c_f^2 \sinh^2 L/2c_f - c_i^2 \sinh^2 L/2c_i = 1/2 \{4c_i^2 \sinh^2 (L/2c_i) + h^2\} \cdot$$
$$\{(T_{of} - T_{oi})/AE] + [\theta \cdot \alpha]\}$$

$$\therefore \{L^2 \cdot \cos^2\phi/24\} \cdot \{1/c_f^2 - 1/c_i^2\} \sim \{[(T_{of} - T_{oi})/AE] + [\theta \cdot \alpha]\}$$

(d) Tension at supports:
Tangential tension

$$T_1 = T_0 \cosh x_1/c = T_0 \cosh\{[L/2 - (L/2 - x_1)]/c\}$$
$$= T_0\{\cosh L/2c \cdot \cosh(L/2 - x_1)/c - \sinh L/2c \cdot \sinh(L/2 - x_1)/c\}$$
$$\therefore \quad T_1 = T_0\{\cosh L/2c \cdot \sqrt{[1 + h^2/(4c^2 \sinh^2 L/2c)]} - h/2c\} \tag{6}$$

similarly

$$T_2 = T_0 \cosh (L - x_1/c)$$
$$= T_0\{\cosh L/2c \cdot \sqrt{[1 + h^2/(4c^2 \sinh^2 L/2c)]} + h/2c\}$$

Vertical component of tension

$$T_{1\text{vert}} = T_1 \sin \phi$$

but $T/W = s \operatorname{cosec} \phi$

$$\therefore \quad T_{1\text{vert}} = Ws_1 = T_0 \sinh x_1/c$$

$$T_{1\text{vert}} = T_0\{\sinh L/2c \cdot \cosh(L/2 - x_1)/c - \cosh L/2c \cdot \sinh(L/2 - x_1)/c\}$$

$$= T_0\{\sinh L/2c \cdot \sqrt{[1 + h^2/(4c^2 \sinh^2 L/2c)]} - h/2c \coth L/2c\}$$

$$T_{1\text{vert}} \sim W\{L/2 - c \tan \phi\}$$

similarly

$$T_{2\text{vert}} = T_0\{\sinh L/2c \cdot \sqrt{[1 + h^2/(4c^2 \sinh^2 L/2c)]} + h/2c \coth L/2c\}$$

$$T_{2\text{vert}} \sim W\{L/2 + c \tan \phi\}$$

18.5.2 Conductor and earth wire spacing and clearances

18.5.2.1 *Earth wires*

Where there is a risk of a direct lightning strike to the phase conductors, transmission lines are provided with overhead earth (or ground) wires to shield them and also to provide a low impedance earth return. The degree of shielding of the overhead line phase conductors from lightning strikes is determined by the shielding angle afforded by the earth wire(s) running over the overhead line. A single earth wire is considered to afford a 30° shielding angle as shown in Fig. 18.10a. Where lines are erected in areas of high lightning activity, or with supporting structures with wide horizontal spacing configurations, two earth wires may be provided to permit a lower shielding angle and superior protection. A 0° shielding angle arrangement is shown in Fig. 18.10b. The vertical spacing between the earth and phase conductors must be such as to ensure sufficient clearance to prevent mid-span flashovers under transient conditions. The sagging should be arranged so as to ensure that the vertical mid-span clearance between the phase and earth conductors is about 20% greater than at the supports. Galvanized stranded steel presents a low cost earth wire material. Where severe pollution exists or where protection schemes demand a low impedance earth path, ACSR or other materials may be used.

In the UK the original 132 kV overhead lines were designed with a 45° angle of protection and gave satisfactory cover. When this angle was applied to the higher 400 kV overhead lines it was found advantageous to reduce the angle of protection to 30° in order to reduce the number of strikes.

The calculation of lightning behaviour of overhead lines is complex but the electrogeometric model is a convenient way of visualizing the process. Assume a cloud at height H above the ground with a stepped leader originating from point 'O' as shown in Fig. 18.11a. If the distance OE to the earth wire

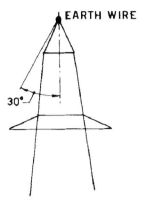

**(a) TYPICAL 132 kV DOUBLE CIRCUIT TOWER
WITH 30° SHIELD ANGLE**

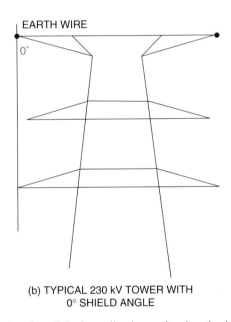

**(b) TYPICAL 230 kV TOWER WITH
0° SHIELD ANGLE**

Figure 18.10   Protections from lightning strikes by overhead earth wires

is less than the distance OC, the strike is more likely to hit the earth wire than the phase conductors. If a line EB is drawn at a tangent to the circle centre, O, then by construction:

$$\sin \alpha = DB/EB = OA/OE = (H - h)/H$$

If it is assumed that $H = 2h$ then $\sin \alpha = (2h - h)/2h = 0.5$ and $\alpha = 30°$. This angle of protection is often adopted on the basis that the cloud height is likely to be twice the height of the earth wire.

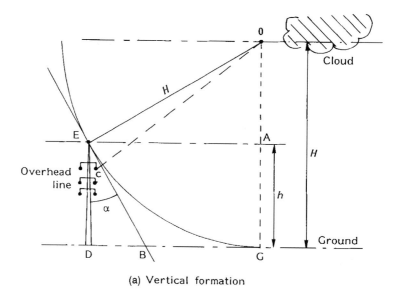

(a) Vertical formation

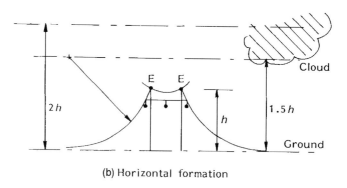

(b) Horizontal formation

Figure 18.11   Overhead line earth wire lightning screen protection

When considering conductors arranged in horizontal formation as shown in Fig. 18.11b it is customary to assume a cloud level at 1.5 to 2 h. If, for example, $H = 1.5\,h$ then $\sin\alpha = (1.5\,h \times h)/1.5\,h = 0.5/1.5 = 0.333$ and $\alpha \sim 20°$.

18.5.2.2 Earthing counterpoise

A lightning strike on the earth wire will be dissipated into the ground after passing through the transmission tower structure and foundations. Wave propagation along electrical lines obeys classical wave propagation theory. Wave reflections will occur at points of discontinuity such as points of changing

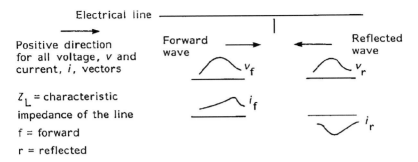

Figure 18.12   Wave propagation along electrical lines

impedance. The basic equations for the arrangement shown in Fig. 18.12 are given below.

The voltage and current along the line are the vector sum of the forward and reflected waves at any time, t.

$$\bar{v} = \bar{v}_f + \bar{v}_r \tag{a}$$

$$\bar{i} = \bar{i}_f + \bar{i}_r \tag{b}$$

$$\bar{v}_f = \bar{Z}_L \bar{i}_f, \quad \bar{v}_r = -\bar{Z}_L \bar{i}_r \tag{c}$$

therefore

$$\bar{Z}_L \bar{i} = \bar{v}_f - \bar{v}_r, \quad \bar{Z}_L \bar{i} + \bar{v} = 2\bar{v}_f \tag{d}$$

For a line terminated on an impedance, $\bar{Z}_T$, the relationships between voltage, $\bar{v}$ and current, $\bar{i}$, at the receiving end is

$$\bar{v} = \bar{Z}_T \bar{i}$$

Combining with equations (a) to (d) allows resolution in terms of the incident wave

$$\bar{v} = 2\bar{Z}_T \cdot \bar{v}_f / (\bar{Z}_L + \bar{Z}_T)...(e) \qquad (v) - \text{instantaneous voltage}$$
$$\bar{v}_r = (\bar{Z}_T - \bar{Z}_L) \cdot \bar{v}_f / (\bar{Z}_L + \bar{Z}_T)...(f) \qquad (r) - \text{reflected voltage}$$
$$\qquad\qquad\qquad\qquad\qquad\qquad\qquad\qquad (f) - \text{forward voltage}$$

For a line with open circuit at the receiving end ($\bar{Z}_T = $ infinity) then $\bar{v}_f = \bar{v}_r$ and $\bar{v} = 2v_f$ illustrating possible voltage doubling effects. Consider the example shown in Fig. 18.13.

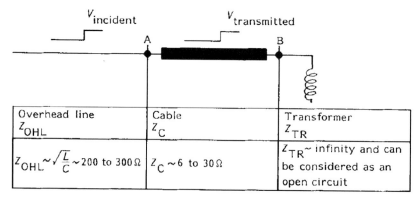

Figure 18.13   Cable and transformer characteristic impedances

At the interface, A, between the overhead line and the cable there is an impedance mismatch. The incident wave, $v_{incident}$ will be transmitted through the cable ($\bar{v}_{transmitted}$) and reflected ($\bar{v}_{reflected}$) in accordance with equation (e):

$$\bar{v}_{transmitted} = 2\,\bar{Z}_C \cdot \bar{v}_{incident}/(\bar{Z}_{OHL} + \bar{Z}_C)$$
$$\bar{v}_{reflected} = (\bar{Z}_C - \bar{Z}_{OHL}) \cdot \bar{v}_{incident}/(\bar{Z}_C + \bar{Z}_{OHL})$$

The transmitted voltage, $\bar{v}_{transmitted}$ is fully reflected at the interface, B, between the cable and the effectively open circuit transformer impedance. This process continues between points A and B in the circuit with multiple reflections and wave distortion. The BIL (Basic Insulation Level) of all the equipments (cables, terminations, transformer bushings, etc.) has to be specified to match the maximum anticipated voltages. In the above example consider a 132 kV overhead line, cable and transformer combination with:-

$$\bar{v}_{incident} = 830\,kV$$
$$\bar{Z}_C = 10\,\Omega$$
$$\bar{Z}_{OHL} = 220\,\Omega$$

Surge voltage entering the cable, $v_{transmitted} = 2Z_C \cdot \bar{v}_{incident}/(\bar{Z}^{OHL} + \bar{Z}_C)$
$$= 2 \cdot 10 \cdot 830/(220 + 10)$$
$$\sim 72\,kV$$

If the equipment BIL is specified as 650 kV (a standard IEC value for 145 kV rated equipment) then the maximum voltage magnification allowed is $650/72 = 9$ times. This value will assist in the determination of suitable protection equipment (see also Chapter 9).

Structures having a high footing impedance (or surge impedance) will cause the development of extremely high potentials during the lightning strike conditions. This may in turn be greater than the phase-to-neutral insulation of the line and cause a 'back flashover' to the phase conductor. In order to minimize this effect the tower footing impedance is specified to a low value (typically less than 10 ohms). This is achieved by connecting the tower footing to bare counterpoise conductors laid in the ground.

Such conductors may radially project from the base of the tower. Alternatively, a continuous counterpoise is sometimes direct buried and connected to each tower along the line length. Earth rods may also be used at the tower base to try and reduce the footing impedance. National and international regulations require touch potentials to be kept within defined limits. Since all transmission lines have slight leakage from phase conductors to earth it is essential to ensure proper tower earthing.

Supporting steelwork (crossarms) associated with wooden pole lines may not be directly connected to earth thus saving the cost of earth conductors and electrodes. The pole itself acts as an insulator. If the conductors are struck by lightning then auto-reclose switchgear and protection should be provided to clear the fault and restore the supply.

18.5.2.3 *Distribution voltage level clearances*

For open wire construction at distribution voltage levels (380 V–24 kV) the earth or neutral wire is normally placed at the bottom (nearest the ground) of the conductor set so as to minimize the danger caused by poles, ladders, etc. touching the wires from underneath. Clearances for conductors are usually the subject of statutory regulations (e.g. in the UK the Statutory Instruments 2002, No. 2665, The Electricity Safety, Quality and Continuity Regulations). Because of the insulated nature of ABC, ground clearances may be reduced in comparison with open wire construction under certain circumstances. Typical minimum clearances used in UK and Europe are indicated below:

18.5.2.4 *Transmission voltage level clearances*

There are no universally agreed clearances as they depend upon insulation level, pollution, span, type of overhead line construction, etc. Table 18.13a details some examples of typical clearances for overhead transmission lines, derived from the guidance in EN50341. Table 18.13b indicates the legal minimum clearances in UK (see also IEC60071, Parts 1, 2 and 3 for more details). Chapter 17, Section 17.4.2 comments on physical clearances around insulator sets.

18.5.2.5 *Overhead line calculation example – sag/tension*

With the information given in Section 18.5 we can now continue the calculation from Section 18.4.3. Having determined the acceptability of the chosen conductor for the electrical load it is now possible to determine the correct tower dimensions in order to obtain adequate clearances. We need to calculate

Table 18.13a   Examples of typical overhead line clearances (based on maximum conductor temperature or the load – EN 50341)

Clearance consideration	Clearance (m) from line with highest system voltage of:		
	52 kV	145 kV	245 kV
To ground in unobstructed countryside	5.6	6.2	6.7
To rockface or steep slope	3.1	3.2	3.7
To trees which cannot be climbed	0.6	1.2	1.7
To trees which *can* be climbed	2.1	2.7	3.2
To buildings with fire-resistant roofs and roofs with slope >15° to horizontal	3.1	3.2	3.7
To buildings with fire-resistant roofs and roofs with slope <15° to horizontal	5.1	5.2	5.7
Horizontal clearance to buildings	2.6	3.2	3.7
To fire sensitive installations	10.6	11.2	11.7
To antennae, lamp posts, etc. which cannot be stood upon	2.6	3.2	3.7
Line crossings of minor roads, railways and waterways	2.6–6.6 depending on nature of road, railway (e.g. electrified or not) and waterway (e.g. with structures or not)	3.2–7.2 depending on nature of road, railway (e.g. electrified or not) and waterway (e.g. with structures or not)	3.7–7.7 depending on nature of road, railway (e.g. electrified or not) and waterway (e.g. with structures or not)

Note: Equivalent clearance guidance for lines <52 kV is given in EN50423.

Table 18.13b   Minimum height above ground for overhead lines in UK (Electricity regulation 1988, amended 1990)

Nominal system voltage Vn (kV)	Clearance (m)
33 < Vn ⩽ 66	6.0
66 < Vn ⩽ 132	6.7
132 < Vn ⩽ 275	7.0
275 < Vn ⩽ 400	7.3

Note: These clearances are minimum, but are calculated at maximum likely line temperatures.

the maximum conductor tension at minimum temperature and ensure this is within the capability of the ACSR Lynx conductor. In addition, we need to calculate the maximum sag, at maximum temperature, to ensure correct tower height. Parabolic equations are used to simplify the arithmetic in this hand calculation example – the error compared to catenary calculations is less than 0.5% for spans less than 300 m and sag less than 5%. This means it is usually quite satisfactory for distribution circuits, but for good design of many transmission lines the more accurate calculations are justified. Assuming then the following conditions:

Minimum tropical temperature = 0°C (no ice)
Everyday tropical temperature = 20°C
Basic span   = 330 m
Wind pressure   = 680 N/m²
Lynx breaking load   = 79.8 kN
Lynx mass   = 0.842 kg/m
Lynx overall diameter   = 19.53 mm

(a) Check for ruling tension condition
MWT factor of safety (with wind and ice loads)   = 2.5 ×, say
EDS factor of safety (no wind, no ice)   = 5 ×, say
At 0°C, with wind: tension = breaking load/2.5   = 79.8/2.5
                                                        = 31.92 kN

At 20°C, no wind: tension = breaking load/5 = 79.8/5 = 15.96 kN
Wind load = $p\{y + x\}$

where p = wind pressure = 680 N/m²
         y = radial thickness of ice = 0 mm
         x = conductor diameter = 19.53 mm

The effective weight of the conductor, W_1, under maximum loading conditions is derived from the resultant of wind load and the weight of the conductor itself:

$$W_1 = \sqrt{[(\text{weight of conductor} + \text{ice})^2 + (\text{wind load}/g)^2]}$$
$$= \sqrt{[(0.842 \cdot 9.81)^2 + (680 \cdot 0.01953)^2]}$$
$$= \underline{1.594} \text{ kg/m } (\sim15.64 \text{ N/m} = 0.01564 \text{ kN/m})$$

At 0°C, T_1 = 31.92 kN so calculate T_2 at 20°C from the general change-of-state equation for tensions in conductors (using the parabolic calculations):

$$EA\alpha(t_2 - t_1) + (W_1{}^2 \cdot g^2 \cdot L^2 \cdot EA/24\,T_1{}^2) - T_1 = (W_2{}^2 \cdot g^2 \cdot L^2$$
$$\cdot EA/24\,T_2{}^2) - T_2$$

where E = modulus of elasticity   = 84 × 10³ MN/m²
         A = cross-sectional area of   = 226.2 mm²
             conductor
         α = coefficient of linear expansion = 19.3 × 10⁻⁶ per °C
         $t_2 - t_1$ = temperature differential   = 20°C (use negative sign
                                                      for temperature fall)
         W_1 = effective weight of   = 15.64 N/m
             conductor at conditions
             which produce tension T_1

W_2 = final conductor unit effective weight alone at T_2 conditions ... = 8.26 N/m

g = gravitational constant ... = 9.81 m/s²

L = basic span length ... = 330 m (in practice probably a bit on the long side for this conductor)

T_1 = initial known conductor tension based upon ruling situation of most onerous condition with a factor of safety on minimum temperature or every day temperature ... = 31920 N

T_2 = final required conductor tension (N)

$$84 \times 10^3 \times 10^6 \cdot 226.2 \times 10^{-6} \cdot 19.3 \times 10^{-6} \cdot (20 - 0) + \{15.64^2 \cdot 330^2 \cdot 84 \times 10^3 \times 10^6 \cdot 226.2 \times 10^{-6}/(24 \cdot 31920^2)\} - 31920$$
$$= 8.26^2 \cdot 330^2 \cdot 84 \times 10^3 \times 10^6 \cdot 226.2 \times 10^{-6}/(24 \cdot T_2^2) - T_2 - 3887.3$$
$$= (5.8823 \times 10^{12}/T_2^2) - T_2$$

This may be solved by trial and error or by using a simple calculator subroutine with an initial approximation for T_2 as $T_2 = \sqrt[3]{5.8823 \times 10^2} \sim$ 18052 N. Further iterations give T_2 = 19440 N = 19.44 kN. (Alternatively, using the Newton–Raphson iteration in a spreadsheet package will give accurate results). Since this value is greater than the permissible value at 20°C of 15.96 kN, the latter value must be used as the sagging basis. *Therefore the limiting condition becomes 15.96 kN.*

Under such circumstances a check is advisable as to the MWT resulting from this sagging basis since the MWT will affect the design of the tension structures. We use the same general change-of-state equation as before but put:

W_1 = 8.26 N/m

t_2 = 0°C

t_1 = 20°C

T_1 = 15960 N

W_2 = 15.64 N/m

Thus

$$84 \times 10^3 \times 10^6 \cdot 226.2 \times 10^{-6} \cdot 19.3 \times 10^{-6} \cdot -20$$
$$+ \{8.26^2 \cdot 330^2 \cdot 84 \times 10^3 \times 10^6 \cdot 226.2 \times 10^{-6}/(24 \cdot 15960^2)\} - 15960$$
$$= 15.64^2 \cdot 330^2 \cdot 84 \times 10^3 \times 10^6 \cdot 226.2 \times 10^{-6}/(24 \cdot T_2^2) - T_2 - 201.18$$
$$= (2.1089 \times 10^{13}/T_2^2) - T_2$$

Taking an initial approximation for T_2 as $T_2 = \sqrt[3]{2.11 \times 10^{13}} \sim 27630\,\text{N}$. Further iterations give T_2 at $0°C = 27695\,\text{N}$. Hence a value of $27.70\,\text{kN}$ may be used as a value for checking the strength of towers.

(b) Next check the sag at maximum temperature

Having established that the everyday temperature, no wind, is the ruling condition the maximum sag (at maximum temperature $= 75°C$) is calculated in order to determine the height of the tower arm for the lowest phase conductor. In this way adequate ground clearance is ensured. The parabolic approximation for the conductor shape is again used, in this case with $t_2 = 75°C$ and $t_1 = 20°C$, for this simple hand calculation (used for illustration here, but see earlier comment on use of spreadsheets). In areas liable to flood, extension legs may have to be applied to the towers to ensure clearances under maximum water height conditions. In the change-of-state equation $W_1 = W_2 = 8.26\,\text{N/m}$ since only the bare weight of conductor needs to be taken into account.

$$EA\alpha(t_2 - t_1) + (W_1^2 \cdot g^2 \cdot L^2 \cdot EA/24\,T_1^2) - T_1 = (W_2^2 \cdot g^2 \cdot L^2 \cdot EA/24\,T_2^2) - T_2$$

$84 \times 10^3 \times 10^6 \cdot 226.2 \times 10^{-6} \cdot 19.3 \times 10^{-6} \cdot (75 - 20)$
$+ \{8.26^2 \cdot 330^2 \cdot 84 \times 10^3 \times 10^6 \cdot 226.2 \times 10^{-6}/(24 \cdot 15960^2)\} - 15960$
$= 8.26^2 \cdot 330^2 \cdot 84 \times 10^3 \times 10^6 \cdot 226.2 \times 10^{-6}/(24 \cdot T_2^2) - T_2$
$- 27302.48 = (5.8823 \times 10^{12}/T_2^2) - T_2$

$T_2 \sim 12202\,\text{N} \sim 12.2\,\text{kN}$, say.

The sag is determined from the equation,

$\text{sag} = (W \cdot g \cdot L^2)/8T$
$\quad\quad = (0.84 \cdot 9.81 \cdot 330^2)/8 \cdot 12202$
$\quad\quad = 9.21\,\text{m}$

Note for comparison if catenary equations had been used the

$\text{sag} = c\,(\cosh L/2c - 1)$

where $c = T_0/W = 12202/8.26 = 1477.24$
$\quad\quad \text{sag} = 1477.24\,(\cosh 330/2 \times 1477.24 - 1)$
$\quad\quad\quad\quad = 9.22\,\text{m}$, i.e. very little difference with the parabolic approximation.

(c) Determine the lowest conductor fixation point

For a $132\,\text{kV}$ line typical minimum phase conductor-to-ground clearance might be $7\,\text{m}$. Lowest conductor connection point on the tower (assuming approximately flat terrain) becomes:

$\text{sag} + \text{suspension insulator set string and fitting length}$
$\quad\quad + \text{allowable clearances} = 9.22 + 2\,(\text{say}) + 7 = 18.22\,\text{m}$

When estimating the level above ground of the attachment point of the insulator, no allowance has been made for creep. In such a case, the erection sag would have to be smaller than calculated so as to allow for a margin of sag increase due to creep. Designs made on this basis should be conscious of the fact that should the worst loading conditions occur soon after erection, the maximum calculated tension would be exceeded certainly for at least the first meteorological cycle.

18.5.3 Broken wire conditions

It is essential that the structures supporting the overhead line are capable of withstanding unequal loads. Suspension and tension structures must be designed for the vertical and transverse loadings plus the unbalanced longitudinal forces due to the simultaneous breakage of up to two complete phase conductors or one earth wire, whichever is the more onerous. The towers themselves are usually designed such that no failure or permanent distortion occurs when loaded with forces equivalent to $2\times$ the maximum simultaneous vertical, transverse or longitudinal working loadings for suspension towers and $2.5\times$ for tension towers. Under broken wire conditions the towers must be capable of withstanding typically $1.25\times$ the maximum simultaneous resulting working loadings.

18.5.4 Conductor tests/inspections

Conductors from a reputable manufacturer's standard product range should already have full type test certification for electrical and mechanical properties (e.g. to IEC 61089 or EN 50182 for ACSR and BS183 for galvanized earth wire). Tests at the manufacturer's works, in addition to any routine requirements in standards, may typically include those shown in Table 18.14.

Table 18.14 Overhead line conductor test requirements

	Hard-drawn aluminium wire	Complete ACSR or AAAC	Steel wires	Earth wires (galvanized steel)
Appearance and finish	Yes	Yes	Yes	Yes
Diameter	Yes	Yes	Yes	Yes
Resistivity	Yes			
Tensile test	Yes		Yes	
Wrapping test	Yes		Yes	
Lay ratio		Yes		Yes
Weight per metre		Yes		Yes
Grease weight per metre		Yes		
Breaking strength and resistance		Yes	Yes	Yes
Stress determination at 1% elongation			Yes	
Torsion test			Yes	
Thickness of galvanizing			Yes	Yes

18.6 OVERHEAD LINE FITTINGS

18.6.1 Fittings related to aerodynamic phenomena

18.6.1.1 *Dampers and aeolian vibrations*

Aeolian vibrations are characterized by their high frequency (5 to 30 or even up to 60 Hz) and low amplitude (one to two conductor diameters). They occur most frequently in winds of laminar flow in the range 0.5–10 m/s. Techniques have been evolved for the estimation of the amplitudes by the 'energy balance principle' in which the wind input energy is equated to the energy dissipated by the conductor and fittings. Also an estimation of conductor lifetime based on their 'endurance capability' is possible to allow design, selection and installation of vibration dampers. The most common type of damper is the Stockbridge type which comprises of two hollow masses attached together by means of a flexible connection. The whole assembly is clamped to the conductor close to the suspension point. The fixing distance between the suspension point and the damper is a function of the diameter and tension of the protected conductor. Typical arrangements are shown in Fig. 18.15. Other types of dampers such as buckles and festoons have also proved successful.

18.6.1.2 *Aerodynamic dampers and galloping*

Galloping is a low frequency (0.1 to 1 Hz) high amplitude($\pm$several metres) motional phenomena which can affect both single and bundled conductors. Galloping occurs mostly in the vertical plane with oscillations of 1, 2 or 3 half wavelengths per span. It may involve two different mechanisms:

1. In the absence of ice it can arise with large diameter conductors (>40 mm). The phenomena may be controlled by using smooth body conductors.
2. Ice-initiated oscillations which occur at near freezing temperatures associated with freezing rain, wet snow or hoar frost. The deposits modify the shape of the conductors into unstable aerodynamic profiles at moderate wind speeds. The ice deposits necessary to initiate such oscillations may be very small and difficult to detect by eye.

Some of the solutions which have given success in limiting galloping include:

- removing spacers from twin conductor configurations
- addition of pendulum detuners, perforated cylinders, aerodynamic dampers, interphase spacers and air flow spoilers.

Another form of galloping (of short duration) which can cause flashovers arises from ice shedding. The sudden fall of ice from conductors releases stored potential energy and the conductors violently rebound. A provisional solution to this problem has been to modify the lengths of the crossarms so that no two or three

Span length	Number of dampers
Up to 500 m Over 500 m	1 damper/span end 2 dampers/span end

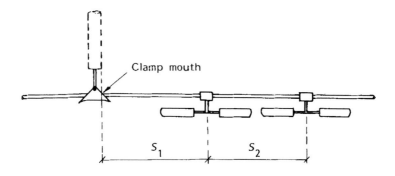

Calculation for positioning of dampers

$S_1 = 0.282d \sqrt{(T/M)}$ metres   where d = conductor diameter mm
$S_2 = 0.242d \sqrt{(T/M)}$ metres T = tension kN
                                            M = mass kg/km

Applicable to conductor and earth wire tension and suspension positions.

If armour rods are used at suspension clamps then dampers must be placed at least 100 mm from the ends of the armour rods.

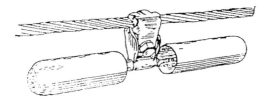

Figure 18.15   Stockbridge damper quantity and positioning principles

phases are in the same vertical plane. In the UK the middle crossarm on double circuit towers has been offset by 1.5 m on the new families of towers.

18.6.1.3 *Subspan oscillations and spacers or spacer-dampers*

Subconductor oscillations are restricted to bundles when pairs of subconductors lie in almost the same horizontal plane. The oscillations usually take the form of lateral, anti-phase motions (0.5–3 Hz) although other modes such as twisting or snaking of the bundle occur. Horizontal motion causes stress reversals at or near to the suspension points or spacer clamps. This type of oscillation is initiated by low turbulent wind flow normal to the conductor bundle with velocities in the range 5–24 m/s. Early experience indicated that there was a reduced likelihood of subspan oscillation when the ratio of bundled conductor spacing to conductor diameter exceeded 15. In addition, the installation of conductor spacers with rubber inserts in the clamps or of spacer-dampers with rubber or elastomeric damping materials in the clamping arm hinges at suitable locations helped to reduce the problem.

18.6.1.4 *Armour rods and armour grip suspensions*

Reinforcement of conductors to prevent fatigue occurring at suspension points is achieved by means of galvanized steel rods. The rods are preformed into a helix approximately 1–1.5 m long and wrapped round the conductor at the most vulnerable suspension points. The effective increase in conductor diameter caused by the addition of the rods also tends to reduce the amplitude of vibrations and increase conductor fatigue life.

A development of armour rod protection has been the design of armour grip suspension (AGS) units which use rubber inserts. AGS units are specified for attachment of conductors involving fibre optic cables.

18.6.1.5 *Joints and repair sleeves*

Compression type joints are normally specified to have an ultimate tensile strength (UTS) of at least 95% of that of the conductor for spans under tension. Many new specifications now require joints with a strength equal to 100% of the declared conductor breaking load. All joints must have the same current carrying capacity as the conductor in order to avoid hot spots.

Repair sleeves are special fittings which can be installed over a damaged portion of a conductor. The maximum number of broken strands for such a repair must be closely defined. The sleeves may be of the preformed or compression type and designed to restore the mechanical and electrical properties of the conductor.

18.6.2 Suspension clamps

These are shaped to form a fully articulated support for the conductors. They are curved at the ends in the vertical plane to allow the conductor to take up the maximum angle of inclination as it leaves the clamp as caused by sag. Clamps must be made of materials to suit the duty and avoid galvanic corrosion cells forming between dissimilar metals.

18.6.3 Sag adjusters

These consist of pivoted clamping plates with adjustment holes to allow the sag of the conductor to be regulated after the initial erection in steps of, say, 10 mm over a 300 mm range.

18.6.4 Miscellaneous fittings

These include:

Tower or pole anti-climbing guards
Climbing steps
Danger plates
Tower or pole number plates
Phase plates
Line circuit identifications
Aircraft warning spheres

18.7 OVERHEAD LINE IMPEDANCE

18.7.1 Inductive reactance

18.7.1.1 *Basic formula, geometric mean radius and geometric mean distance*

The inductance of an overhead line is derived from its self-inductance and the mutual inductance between the different phase conductors. For a simple go and return two wire single phase AC system the inductance:

$L = 0.2(0.25 + \log_e d/r)$ mH/km (see Figure 18.15a)

The expression may be reduced to a single term by taking a hypothetical value for the conductor radius which still gives the same value for the inductance. Thus:

$L = 0.2(\log_e d/\text{GMR})$ mH/km

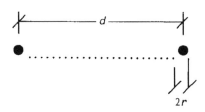

Figure 18.15(a)

where d = separation between conductor axes (mm)
            r = radius of conductor (mm)
        GMR = geometric mean radius of conductors (mm)

For solid conductors the GMR $\sim 0.78r$ and for stranded conductors it varies according to the number and size of the strands. Table 18.15 gives values of GMR for ACSR and all copper or all aluminium stranded conductors of conductor radius, r.

18.7.1.2 Three phase formula

For a three phase system the mutual effects of one conductor on another modify the formula and the expression under symmetrical operating conditions becomes:

$L = 0.2(\log_e \text{GMD/GMR} + K)$ mH/km line to neutral inductance
K = correction factor for steel core in ACSR ($K = 0$ for single material conductors), conductor diameter = $2r$, phases: r = red, y = yellow and b = blue

GMD = geometric mean distance between the r, y and b phases

$$= \sqrt[3]{(d_{ry} \cdot d_{yb} \cdot d_{br})}$$

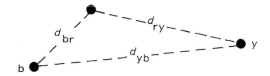

Figure 18.15(b)

Table 18.15   GMR values as function of conductor radius, r

All aluminium or all copper conductor		ACSR	
Number of strands	GMR	Number of Al strands	GMR
7	$0.726r$	6	$0.768r$
19	$0.758r$	12	$0.859r$
37	$0.768r$	26	$0.809r$
61	$0.772r$	30	$0.826r$
91	$0.774r$	54	$0.810r$
127	$0.776r$		
169	$0.779r$		
Solid	$0.779r$		

Table 18.16   Typical overhead line zero/positive sequence reactance ratios

Earth wire	Overhead line	X_0/X_1
–	Single circuit	3.5
–	Double circuit	5.5
Galvanized steel	Single circuit	3.5
Galvanized steel	Double circuit	5.0
Non-magnetic	Single circuit	2.0
Non-magnetic	Double circuit	3.0

18.7.1.3 *Positive and negative sequence reactance*

The positive and negative sequence inductive reactances (X_1 , X_2) of a three phase overhead line are equal and for a frequency, f Hz, become:

$$X_1 = 2 \cdot \pi \cdot f \cdot 10^{-3}[0.2(\log_e \text{GMD/GMR} + K)] \text{ ohms/km line to neutral}$$
$$\text{inductive reactance}$$

18.7.1.4 *Zero sequence reactance*

Zero sequence reactance (X_0) is complicated to calculate. The value depends upon the position and materials used for the earth wires and the log of the square root of the ground resistivity. Typical values for the ratio of zero-to-positive sequence reactance for double and single circuit overhead lines are given in Table 18.16.

18.7.2 Capacitive reactance

For short overhead lines at higher voltages the shunt capacitance is usually ignored if data is not immediately available. For computer modelling involving overhead line or cable impedances if a shunt reactance value is known it might as well be used since the computer is doing all the normally time consuming mathematical manipulations anyway. It is important to notice that a long, lightly loaded overhead line may have a receiving end voltage higher than the sending end voltage due to capacitance effects. The expression for line-to-neutral capacitance, C, is in the form:

$$C = 1/(18 \log_e \text{GMD}/r) \text{ μF/km}$$

The earth plane modifies this expression and still more complex formulae are applicable to bundled conductor configurations.

18.7.3 Resistance

Resistance values for different conductor materials are given in Table 18.2. The effective resistance of high voltage overhead lines is usually negligible in comparison to the inductive reactance and is therefore often neglected in simple hand calculations for load flow or fault current analysis. Only at low and medium voltages does the resistance become significant for short circuit calculations.

18.8 SUBSTATION BUSBAR SELECTION – CASE STUDY

18.8.1 Introduction

This case study describes the steps to be taken for the selection of a 132 kV primary substation catenary busbar. It highlights the use of the topics discussed in this chapter. A photograph showing the actual busbar arrangement at the Channel Tunnel, 132/21/25 kV 240 MVA firm capacity main HV substation near Folkestone, South East England, which resulted from this study, is shown in Fig. 18.16.

18.8.2 Conductor diameter/current carrying capacity

240 MVA at 132 kV corresponds to a busbar rating of approximately 1050 A. The climatic conditions (say, 1 mph wind speed and temperate climate temperature rise) are specified and the conductor diameter which will achieve this rating determined from the heat balance equations. For simplicity such figures are normally found from tables in manufacturers' literature such as the *BICC Overhead Conductor Design Book* and *Prescot Aluminium Company Catalogue*. Correction factors have to be applied to the current rating figures given in such tables depending upon whether all aluminium, aluminium alloy or ACSR is being considered. In this case suitable conductor diameters for both current rating and voltage gradient considerations are:

All aluminium conductor	33.53 mm
Aluminium alloy conductor	35.31 mm
ACSR	34–35 mm depending upon aluminium-to-steel area correction factors.

18.8.3 Conductor selection on weight basis

Various standard conductor types which meet the current rating criteria are next investigated. Usually the client will wish to minimize spares stock holdings and will request that the designer tries to standardize. A selection is detailed in Table 18.17.

In addition, ACSR busbar conductor of suitable diameter and current rating may be investigated (see Table 18.18).

The availability and pricing of conductor at the time of purchase is next considered. In this instance UK conductors had an advantage and therefore North American and European conductors were not considered further. Amongst the

Figure 18.16   'Tarantula' 132 kV busbar

Table 18.17   Physical parameters of some AAC and AAAC cables

All aluminium conductors (AAC)		All aluminium alloy conductors (AAAC)	
TARANTULA	British to BS215 Part 1: 1970 [a]	**ARAUCARIA (EN code 821-AL3)**	British to BS3242: 1970[a]
Stranding	37/5.23	Stranding	61/4.14
Overall diameter	36.61 mm	Overall diameter	37.26 mm
Weight	2.192 kg/m	Weight	2.266 kg/m
Breaking load (UTS)	120 kN	Breaking load (UTS)	230 kN
COLUMBINE	USA to ASTM B231: 1985	**800 (EN code 802-AL3)**	German to DIN[a] 48201-6: 1981
Stranding	61/3.78	Stranding	91/3.35
Overall diameter	34.02 mm	Overall diameter	36.85 mm
Weight	1.887 kg/m	Weight	2.219 kg/m
Breaking load (UTS)	104.1 kN	Breaking load (UTS)	224 kN
800 (EN code 802-AL1)	German to DIN 48291-5[a] 1981	**ASTER 851 (EN code 851-AL4)**	French to NFC [a] 34-125: 1976
Stranding	91/3.35	Stranding	91/3.45
Overall diameter	36.85 mm	Overall diameter	37.95 mm
Weight	2.219 kg/m	Weight	2.474 kg/m
Breaking load (UTS)	118.39 kN	Breaking load (UTS)	273.9 kN

Note: [a]All these individual European standards are supersede by EN 50182, where the same conductors appear with EN codes under the National Normative Aspects (NNAs). TARANTULA is not specifically listed in EN50182, (the largest size being CICADA (628-AL1)), but the standard makes provision for such larger sizes to be coded as necessary.

Table 18.18   Physical and electrical parameters of some ACSR cables

Aluminium conductor steel reinforced (ACSR)

DIPPER	USA size to ASTM B232: 1986
Stranding	45/4.4 + 7/2.93
ACSR current rating correction factor	0.967
Equivalent current	1086 A
Required conductor diameter	34.54 mm
Actual conductor diameter	35.19 mm (and therefore acceptable)
Weight (greased)	2.352 kg/m
Breaking load (UTS)	160.63 kN
680/85 (EN code 679-AL1/86-ST1A)	German to DIN 48204:1984[a]
Stranding	54/4.0 + 19/2.4
ACSR current rating correction factor	0.942
Equivalent current	1115 A
Required conductor diameter	35.56 mm
Actual conductor diameter	36.0 mm (and therefore acceptable)
Weight (greased)	2.648 kg/m
Breaking load (UTS)	210 kN
CROCUS 865 (EN code 717-AL1/148-ST6C)	French to NFC 34-120:1976 [a]
Stranding	66/3.72 + 19/3.15
ACSR current rating correction factor	0.9104
Equivalent current	1153 A
Required conductor diameter	36.07 mm
Actual conductor diameter	38.07 (and therefore acceptable)
Weight (greased)	3.317 kg/m
Breaking load (UTS)	319 kN

Note: [a]All these individual European standards are supersede by EN 50182, where the same conductors appear with EN codes under the National Normative Aspects (NNAs)
Note: Current rating correction factor for ACSR = $\sqrt{(m/[m+1])}$ where m is the aluminium/steel area.

UK conductors TARANTULA (AAC) and ARAUCARIA (AAAC) are both suitable from a current carrying point of view. TARANTULA is lighter. ACSR conductors of suitable current carrying capacity introduce a weight penalty to the designer and therefore a reasonable choice for this application would be the TARANTULA.

A final check can be made on the current carrying capacity of TARANTULA under the assumed conditions:

From Section 18.4.1 we have $I^2R + H_S = H_C + H_R$

For TARANTULA $R_{20} = 0.03627\ \Omega/km$ $\alpha = 0.00403\ \Omega/°C$

$$R_{20} = R_0(1 + \alpha \cdot 20)   \qquad \therefore R_0 = R_{20}/(1 + \alpha \cdot 20)$$

$$R_{75} = R_0(1 + \alpha \cdot 75)   \qquad \therefore R_{75} = R_{20} \cdot (1 + \alpha \cdot 75)/(1 + \alpha \cdot 20)$$

$$= 0.04371\ \Omega/km$$

Heat gained by solar absorption $H_S = \alpha_S \cdot S \cdot d$

where α_S = solar absorption coefficient, say 0.9

S = intensity of solar radiation, say 1050 Watts/m^2

d = conductor diameter = 36.61 mm

$$H_S = 0.9 \cdot 1050 \cdot 36.61 \times 10^3 \times 10^{-3}$$

$$= 34596.45\ W/km$$

Heat lost by convection $H_C = 387(V \cdot d)^{0.448}\theta$

where V = wind speed (1 mph) = 0.44 m/s
       d = conductor diameter = 36.61 mm
       θ = temperature rise = 40°C
      $H_C = 387(0.44 \cdot 36.61)^{0.448}\ 40$
         = 53768.81 W/km

Heat lost by radiation $H_R = \pi \cdot E_C \cdot s \cdot d\{(t + \theta + 273)^4 - (t + 273)^4\}$
where E_C = emissivity of conductor, say 0.9
      s = Stefan–Boltzmann's constant = 5.7×10^{-8} W/m^2
      t = ambient temperature = 35°C
      $H_R = \pi \cdot 0.9 \cdot 5.7 \times 10^{-8} \cdot 36.61\{(35 + 40 + 273)^4$
         $- (35 + 273)^4\}$
         = 33436.45 W/km

Hence I^2R = 53768.81 + 33436.45 − 34596.45 = 52608.81 W/km
and $I = \sqrt{(52608.81/0.04371)}$ = 1097 A which meets the design
      requirement

From this check the reader can obtain an impression of the relative importance of the assumptions. For example, if a wind speed of 0.6 m/s is adopted (a value frequently used on the European continent) the current carrying capacity of TARANTULA becomes 1178 A. If the intensity of solar radiation, S, is taken as 1200 W/m^2 then I = 1044 A. The economic effect of a more sophisticated probabilistic approach, as set out in Ref. (15), and referred to earlier in 18.4.1, becomes more apparent.

18.8.4 Conductor short circuit current capability

The short circuit capability is a function of the duration and magnitude of the short circuit current. The main consideration is that the conductor must not lose a significant amount of tensile strength due to annealing. The fault duration (typically specified as not to exceed 1 or 3 seconds and of course much shorter if satisfactory busbar protection operates correctly to clear the fault) is insignificant in comparison with the cooling time of the conductor. Available information suggests maximum temperatures of 210°C for copper and 200°C for aluminium. A formula for short circuit current rating assuming adiabatic conditions is given below. When dealing with ACSR conductors in accordance with IEC 60865-1, only the aluminium component is considered for R, α, W and S.

$$I_{SC} = \sqrt{(W \cdot S \cdot J \cdot \log_e[1 + \alpha\{t_2 - t_1\}])/(R_{20}[1 + \alpha\{t_1 - 20\}] \cdot \alpha)}$$

where t_1 = temperature before fault occurs (maximum operating
         temperature) = 75°C

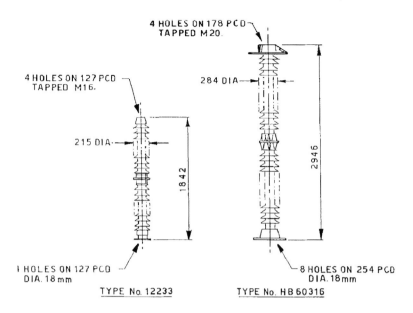

4 HOLES ON 178 PCD TAPPED M20.

284 DIA

2946

4 HOLES ON 127 PCD TAPPED M16.

215 DIA

1842

I HOLES ON 127 PCD
DIA. 18 mm

TYPE No. 12233

8 HOLES ON 254 PCD
DIA.18mm

TYPE No. HB 60316

Technical details			132 kV	275 kV
Electrical values, kV	Power frequency flashover voltage	Dry	520	870
		Wet	380	620
	Impulse flashover voltage	Positive	850	1450
		Negative	1050	1650
	Power frequency withstand voltage (one minute)	Dry	470	790
		Wet	330	570
	Impulse withstand voltage		750	1300
	Power frequency puncture voltage		PUNCTURE-PROOF	

Type No.		12233	HB60316
Dimensions mm			
Diameter		215	284
Total creepage distance - minimum		3353	7500
Protected creepage distance -minimum		1150	3420
Mechanical failing loads-minimum			
Cantilever - upright	kN	4.5	9.0
- underhung	kN	3.5	—
Tension	kN	90	90
Torsion	Nm	4500	12500
CEGB Application		132 kV Upright and Underhung	275 kV Upright
Approximate Weight of unit kg		110	240

Figure 18.17(a)   Typical porcelain post insulators – at the lower voltages post insulators are largely polymeric

t_2 = maximum permissible conductor temperature = 200°C
W = weight/m of conductor in kg = 2.192 kg/m for
    TARANTULA conductor
S = specific heat of aluminium = 214 calories/kg/°C

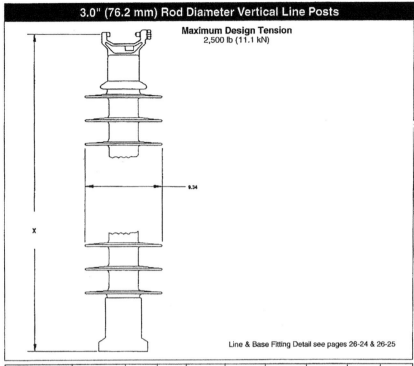

| 3.0" (76.2 mm) Rod Diameter Vertical Line Posts |

Maximum Design Tension
2,500 lb (11.1 kN)

9.34

Line & Base Fitting Detail see pages 26-24 & 26-25

Selection Guide Typical Line Voltage, kV				Catalog # with 5" Bolt Circle & Vert. Clamptop	"X" Length Inches (mm)	No. of Sheds	Dry Arc Distance inches (mm)	Leakage Distance inches (mm)	(1) 60 Flashover ANSI		(1) Critical Flashover ANSI		(2)RCL pounds (kN)	Net Weight pounds (kg)
69	115	138	161						Dry-kV	Wet-kV	Pos-kV	Neg-kV		
				523005-1205	38.0 (965)	10	29 (737)	77 (1956)	295	250	445	540	2500 (11.1)	54 (24.5)
				523006-1205	43.2 (1097)	12	34 (864)	93 (2362)	345	295	530	620	2500 (11.1)	59 (26.8)
				523007-1205	48.5 (1219)	14	39 (991)	108 (2743)	395	335	615	705	2500 (11.1)	64 (29.1)
				523008-1205	53.8 (1367)	16	45 (1143)	124 (3150)	445	380	695	785	2500 (11.1)	69 (31.4)
				523009-1205	59.2 (1504)	18	50 (1270)	140 (3556)	495	420	780	870	2500 (11.1)	74 (33.6)
				523010-1205	64.5 (1638)	20	55 (1397)	156 (3962)	545	465	865	950	2335 (10.4)	79 (35.9)

Notes: (1) Tests in accordance with ANSI C29.1-1982.
(2) RCL is the maximum cantilever continuous load at which the post should be applied.

Figure 18.17(b)   Polymeric post insulation (Ohio Brass™)

$$J = 4.18 \text{ Joules per calorie}$$
$$\alpha = \text{temperature coefficient of resistance at } 20°C$$
$$= 0.00403 \text{ ohms/}°C$$
$$R_{20} = \text{resistance/m of TARANTULA conductor at } 20°C$$
$$= 0.03627 \times 10^{-3} \text{ ohms/m}$$
$$T = \text{duration of fault} = 1 \text{ second (in this case)}$$

With these parameters $I_{SC} = 66.93$ kA which is well in excess of the 25 kA, 1 second design criteria. TARANTULA conductor therefore meets the short circuit requirements.

It should be noted however that the assumption used above, which ignores the steel conductor component, is conservative. A more accurate approach, which takes account of the heat capacity of the steel core but not its conductivity, is now available.[17] The same Electra report also makes the point that if the highest short-time conductor temperature is reached, a negligible decrease in strength can occur, and the strength and lifetime of a conductor depend on its whole operational history. Consequently, users may validly specify other maximum temperatures than those set out in IEC 60865-1.

18.8.5 Conductor support arrangements

18.8.5.1 *Loading conditions*

The substation land area and any planning constraints are major factors in determining busbar supporting gantry heights and span lengths in addition to electrical clearance criteria. In this example the gantry height is limited to 8.5 m in order to avoid observation of the substation from the nearest housing estate, about 1 km away. In addition, 132 kV overhead line entry to the substation was replaced with approximately 14 km of underground cable feeder for aesthetic reasons.

The busbar conductor maximum sag is limited to 1/60th of the ruling span length at temperate climate (16°C EDS) as a design criteria in order to minimize movement and strain on equipment connected to the busbar via 'down droppers'.

The maximum working tension is determined under certain design criteria conditions listed below. These are based on Grid Company Standards and the particular environmental data for the substation area.

1. Dead weight of busbar catenary conductor + down droppers + insulator sets: under simultaneous transverse wind load at −5.6°C
2. Dead weight of busbar catenary conductor + down droppers + insulator sets: under simultaneous transverse wind load and short circuit attraction forces at −5.6°C

Ice loads are taken for a 12 mm radial ice thickness buildup (specific gravity of ice = 0.913).

The wind force on the busbar conductor, $F = C_f \cdot q \cdot A_e$ newtons, is derived from the basic wind speed as explained in Section 17.2, Chapter 17 using British Standard, CP3, Chapter V, Part 2, 1972.

The loads to be considered then become:

a = vertical load = dead weight (inc. insulator sets and down droppers) + ice

b = transverse load = wind load + short circuit load

c = resultant load = $\sqrt{(a^2 + b^2)}$

18.8.5.3 *Short circuit force*

The force F_{SC} on the busbar conductor is based on the maximum force resulting from a short circuit (Ref. CEGB design memo 099/68).

The force $F_{SC} = (k \cdot 0.2 \cdot I_{SC}^2)/d$ N/m

where k = stress factor (1.6 for 15 m to 30 m spans)
 I_{SC} = maximum three phase short circuit current, kA
 d = conductor spacing, m

For a 21.65 kA short circuit design criteria and 2.95 m conductor spacing, $F_{SC} = 50.84$ N/m = 5.18 kg/m.

18.8.5.4 *Equivalent span*

Tension variations due to altered loading and/or temperature variations for any given span may be determined from the equivalent span formula and the sag/tension formula derived from simple parabolic equations.

$$\text{Ruling span} = \sqrt{(L_1^2 + L_2^2 + L_3^2 + \cdots + L_N^2 + /(L_1 + L_2 + L_3 + \cdots + L_N)}$$

Where $L_1, L_2, L_3, \ldots L_N$ are the actual spans within a section under the same horizontal tension as that for the equivalent span (see Fig. 18.8c). The sag in any actual span then becomes:

$$\text{Sag} = (\text{actual span/equivalent span})^2 \times \text{sag in equivalent span}$$

18.5.5.5 *Closest approach*

The closest approach distance between the main busbar conductors is based on the maximum temperature condition under wind and short circuit loads. It is derived from the swing angle,

$$\theta_{SA} = \tan^{-1}(T/W)$$

where T = transverse load (arithmetic sum of wind and short circuit forces) kg/m
 W = conductor weight (including down droppers and insulator sets) kg/m

Checks are made to ensure adequate phase clearance. Horizontal distance between busbar phase conductors,

$$D_H = \text{sag} \times \sin\theta_{SA}$$

Figure 18.18   400 kV Lydd-Bolney (UK) overhead line–quad conductor glass insulators showing arcing horns and stockbridge dampers

For sag $= 0.78$ m in this particular case (calculated as described in Section 18.8.5.4)

$T =$ short circuit force + wind force $= 5.18 + 4.623 = 9.8$ kg/m

$W = 7.2073$ kg/m

$\theta_{SA} = 53.7°$
$D_H = 0.78 \times \sin 53.7° = 0.63$ m

Closest approach distance $=$ phase spacing – conductor short circuit in-
swing
$= 2.95 - 2 \times 0.63$ (as a conservative simplification assume two phases swing towards each other)
$= 1.69$ m

This exceeds the preferred phase-to-phase working clearance of 1.5 m at 132 kV and is therefore satisfactory.

Figure 18.19   Vibration damper replacement, 400 kV Thames crossing reinsulation

Figure 18.20   Conductor stringing, 500 kV double circuit, quad conductor, Daya Bay, China

Figure 18.21   Conductor running out in progress. Tensioner and drums at 60 degrees angle tower. United Arab Emirates – 132 kV East Coast Transmission

18.8.5.6 *Forces acting on post insulators*

If post insulators are used as intermediate busbar supports between gantries or as tie points on slack span down droppers, a check has to be made to ensure their capability to withstand horizontal forces. Technical details for typical 132 and 275 kV post insulators are given in Fig. 18.17.

Considering the 132 kV insulator in Fig. 18.17 the minimum mechanical failing load (cantilever upright) is 4500 N. If these insulators are to be used as busbar mid-span supports over a total span length of 47 m between substation gantries and with a short circuit lateral force of 50.84 M/m (see Section 18.8.5.3) the short circuit loading will be approximately:

$$(47/2) \cdot 50.84 = 1195\,\text{N}$$

This particular post insulator is therefore capable of withstanding horizontal forces due to substation busbar short circuits. Such an intermediate post insulator busbar support allows, in this case, for a reduction in the height of the substation gantries. In this way, the visual impact of the substation on the environment may be reduced.

REFERENCES

1. J. Bradbury, P. Dey, G. Orawski and K. H. Pickup, 'Long term creep assessment for overhead line conductors', *Proc. IEE,* Vol. 122, No. 10, October 1975.
2. CIGRE WG 22-05, 'Permanent elongation of conductors, predictor equations and evaluation methods', *Electra*, No. 75, March 1981.
3. P. J. Ryle, 'Steel tower economics', *Journal IEE* (1945), Vol. 93, Part II, pp. 263–284.
4. CIGRE WG 22-09, 'International survey of component costs of overhead transmission lines', *Electra*, No. 137, August 1991.
5. C. O. Boyse and N. G. Simpson, 'The problem of conductor sagging on overhead transmission lines', *Journal IEE*, Vol. 91, Part II, No. 21, June 1944.
6. J. Bradbury, G. F. Kuska and D. J. Tarr, 'Sag and tension calculations for mountainous terrain', *IEE Proc.*, Vol. 129, Pt. C, No. 5, Sept. 1982.
7. Ch. Avril, 'Construction des lignes aeriennes a` haute tension', *Editions Eyrolles*, 61 boulevard St Germain, Paris V, 1974.
8. CIGRE Group 33, WG01, 'Guide to procedures for estimating the lightning performance of transmission lines', CIGRE, Paris.
9. The Secretary of State for Energy in relation to England and Wales, Statutory Instruments 1988, No.1057, ELECTRICITY – The electricity Supply Regulations, 1988.
10. CIGRE WG 22-01, 'Aeolian vibrations on overhead lines', CIGRE paper 22-11, 1970.
11. CIGRE WG 22-04, 'Endurance capability of conductors', Final Report, July 1988. CIGRE, Paris.
12. EPRI, 'Transmission line reference book: Wind induced conductor motions based on EPRI research project 792', Electric Power Research Institute (1979). 3412 Hillview Avenue, Palo Alto, California, USA.
13. J. G. Allnutt, S. J. Price and M. J. Tunstall, 'The control of subspan oscillation of multi-conductor bundled transmission lines', CIGRE 1980, paper 22-01.
14. R. Hart, 'HV Overhead lines – the Scandinavian experience', *IEE Power Engineering Journal*, June 1994.
15. CIGRE WG 22-12, 'The thermal behaviour of overhead conductors – sections 1 & 2', *Electra*, No. 144, October 1992.
16. Electricity Association, 'Engineering recommendation P27: Current rating guide for high voltage overhead lines operating in the UK distribution system'.
17. CIGRE WG 22-12, 'The thermal behaviour of overhead conductors – section 4', *Electra*, No. 185, August 1999.

19 Testing and Commissioning

19.1 INTRODUCTION

This chapter describes the testing and commissioning necessary before power transmission and distribution equipment is ready for service. Consideration is given under the following headings:

- *Quality assurance*: From the start of manufacture to the end of its useful life equipment is subjected to quality controls (QC) by a series of planned and controlled inspections and checks to achieve quality assurance (QA). Particularly useful frameworks for such quality assurance schemes and quality control checks which ensure that the equipment is 'fit for purpose' are defined in ISO 9000. (Note ISO 9000 is published as a European norm EN 9000 and most sections of BS5750 have been superseded by the national equivalent BSEN 9000).
- *Works inspection and testing*: Works tests are intended to check the design characteristics of the equipment against specified standards. This may involve tests to destruction in order to gauge the extreme capabilities of equipment (e.g. new overhead line tower designs).
- *Site inspection and testing*: Site tests are less severe and cover both the individual items of plant and the complete system.
- *Testing and commissioning methods*: Whatever the method, it is essential that commissioning staff have a good understanding of the equipment involved and the modes of operation of the system. Safety of personnel and plant during commissioning is also vital; this is best addressed by good planning and procedural and safety documentation. The planning must take account of the risk of slippage in the earlier parts of the project programme, since a squeeze on commissioning time must be avoided at all costs. There should also be a proper budget for commissioning – 1% to 1.5% of the electrical system cost is a typical allowance which will be more than off-set by increased reliability and reduced operating costs resulting from adequate commissioning.

19.2 QUALITY ASSURANCE

19.2.1 Introduction

Quality assurance is a system of documented procedures employed by an organization to ensure that the end product (or 'deliverable') (be it design, equipment in various stages of manufacture, an individual finished item of equipment or a complete system) is fit for purpose. Such documentation is called a quality plan. The plan sets out in detail methods of quality control, periodic self-assessments against a known set of criteria, maintenance of a register or log of results together with a method of feedback to correct any deficiencies. It should therefore be possible using such a scheme to achieve traceability throughout the life cycle of a design or an item of equipment. Traceability will cover:

- the original Contract requirements or project brief;
- control of all documentation;
- control of the design/manufacture/installation interfaces;
- system changes;
- checks to ensure that the end design, manufacture or installation meets the contract specification.

A quality assurance system will therefore require the provision of the following items:

- Policy on quality throughout the organization.
- Organization for executing the policy.
- Defined procedures for executing the policy.
- Defined organizational responsibilities.
- Defined system of verification checks and responsibilities.
- Assigned authority.
- System of training programmes for all staff throughout the organization.
- Controls for design and design change including:
  - safety factors
  - reliability
  - maintenance
  - reviews
  - traceability.

For example, consider the case of an overhead line conductor that shows signs of outer strand breakage some months after installation. The cause of the damage must be ascertained before doubts can be cleared and a final solution to the problem found. In such a case the problem must lie with the original material and/or the handling and installation methods and/or the environmental conditions. An analysis of the quality control records throughout

Table 19.1   Useful quality and reliability standards

Standard	Description
IEC 60300-1	Dependability management systems. Describes the concepts and principles of dependability management systems. Identifies the generic processes in dependability for planning, resource allocation, control, and tailoring necessary to meet dependability objectives. Deals with the dependability performance issues in the product life-cycle phases concerning planning, design, measurements, analysis and improvement. Dependability includes availability performance and its influencing factors: reliability performance, maintainability performance and maintenance support performance. Aims at facilitating co-operation by all parties concerned (supplier, organization and customer) and fostering understanding of the dependability needs and value to achieve the overall dependability objectives.
IEC 60300-2	Provides guidelines for dependability management of product design, development, evaluation and process enhancements. Life cycle models are used to describe product development or project phases. Applicable for detailed planning and implementation of a dependability programme to meet specific product needs.
IEC 60300-3	Dependability management – application guide. Sixteen very practical parts (some still in draft).
IEC 60812	Analysis techniques for system reliability – Procedure for failure mode and effects analysis (FMEA).
IEC 61025	Fault tree analysis (FTA)
IEC 61160	Design Review – Procedure for this essential element in achieving quality in design.
IEC 61882	Hazard and operability studies (HAZOP studies) – Application guide.
ISO 9000	Quality management systems. Fundamentals and vocabulary: Describes the fundamentals of quality management systems and defines the terms used in the ISO 9000 series of standards.
ISO 9000-1	Quality management and quality assurance standards. Guidelines for selection and use Clarifies the principal quality concepts and provides guidance on the selection and use of the ISO 9000 family of standards on quality systems that can be used for internal quality management purposes and for external quality assurance purposes.
ISO 9000-2	Quality management and quality assurance standards. Generic guidelines for the application of ISO 9001, ISO 9002 and ISO 9003.
	Describes the fundamentals of quality management systems and defines the terms used in the ISO 9000 series of standards.
ISO 9001	Quality management systems – Requirements QMS requirements where an organization needs to demonstrate ability to provide product that meets customer/regulatory requirements and aims to enhance customer satisfaction.
ISO/IEC 9003 computer	Software engineering. Guidelines for the application of ISO 9001:2000 to software.
ISO 9004	Quality management systems – Guidelines for performance improvements. Provides guidelines beyond the requirements of ISO 9001. It is not a guide to the implementation of ISO 9001. Not intended for certification, regulatory or contractual use.

Note: The European standards in the series EN9000 are identical to the corresponding ISO standards, as are the British standards of the BSEN 9000 series. The latter supersedes BS5750.

the life cycle of the project (the QA data pack) will assist in determining the cause.

19.2.2 Inspection release notice

After design, manufacture or installation the product or deliverable should be inspected for conformance and passed as being substantially complete before being released for the next phase of the work. All outstanding items not meeting the specified requirements (often recorded on a 'snags' or 'punch list') should be included in the 'inspection release notice' (again, different Organizations have specific names for this, e.g. IRN or QICC (= quality, inspection and change control) document). In some cases these outstanding items will have different levels of criticality. They may, therefore, be categorized as requiring immediate attention before a release notice may be granted or simply recorded as in need of rectification before some future quality hurdle may be passed. This is often done by a simple 'a', 'b' or 'c' notation.

19.2.3 Partial acceptance testing

The distribution and transmission engineer will be involved with system engineering where a variety of individual items are installed together in order to form an overall power system. Such systems may be broken down into commissioning 'lots' in order to simplify the testing and commissioning process. For example, substation commissioning lots could be broken down into LVAC supplies, DC supplies, switchgear bays, local/remote control, SCADA, etc. for 'partial acceptance testing' (PAT) and certification.

19.2.4 System acceptance testing

Before a substation may be handed over to the client after design, supply, installation and partial acceptance testing, the complete system must be shown to function correctly as an integral whole. This involves satisfactory 'system acceptance testing' (SAT), commissioning and certification of all the various interfaces between the partial acceptance test lots. In cases of substation extension work this will also include verification of correct interfacing between new and existing equipment.

19.2.5 Documentation and record systems

During design, manufacture and installation inspection systems must give up-to-date, precise and detailed records of all activities. Design documents are issued with various categorizations depending upon the level of approval. There are a number of such categorization systems – the following is one example:

Revision	*Status*	*Comments*
A	PRE or VLO	Initial 'preliminary' design status document issue following normal internal drawing office checking and approval procedures.
B	VLA	VLO status document has been circulated within the organization internally and modified if necessary to incorporate any additional comments. This checking stage is very useful on multidiscipline projects where many interfaces between various subprojects exist. The document is again 'validated' and approved at this status before any further issue for checking by the client or directly to site for installation. An overall approval is usually added to the document at this stage from a senior member of the originator or main contractor organization. This gives assurance as to the overall multidiscipline or multi-department checking and validity of the document.
E (say)	DES or APR	'Descriptive' or 'approved' status document which has gone through the VLO and VLA status checks described above. The document or drawing gives input or background information relating to the project but which is not directly required for manufacture or installation purposes.
1	EXN	'Execution' status document or drawing which has gone through the VLO and VLA status checks described above and is released to site or to the shop floor for execution.
4 (say)	ASB	'As built' status document or drawing that has been issued on the basis that it incorporates any site or shop floor modifications that have taken place during commissioning or manufacture to make the original EXN status document out of date.

For example, during the commissioning phase small site modifications to equipment wiring or software are often necessary. These must be rigorously controlled and recorded at the time the changes are made. This is even more essential if software changes relate to safety related equipment and IEC 61508 applies. The changes to schematic diagrams may be recorded in the form of 'red marked-up' drawings where the changes are shown clearly in red crayon on the original EXN status paper print. One copy of such up-to-date drawings must be retained on site and another sent back to the responsible party for formal updating and issued to the client upon completion of the works or during the maintenance period. Advanced on-site CAD communication facilities may allow direct input of such updates from the field into the master CAD drawing, but extreme care must be taken to ensure proper authorization, revision number control and issue sequence.

When possible results obtained during commissioning tests should be inserted on test sheets rather than just go/no go ticks. These test sheets then provide a useful record for future service testing. They should form a complete set of testing and commissioning records that are included in the QA data pack for the project.

Operation and maintenance manuals are an essential part of the documentation for the whole project. The production of such documentation is often given insufficient resource by contracting organizations even though it forms a definite contract requirement. The supply of the operation and maintenance manuals (O & MM) should therefore be tied to the release of moneys to the contractor as an incentive for their timely production. A first-class guide for the content and presentation of the operation and maintenance manuals is contained in BS4884, Parts 1 and 2. Briefly the O & MM should contain:

- Factory and site test certificates for each item of plant and reference to relevant design calculations and the QA data pack.
- Complete drawing list appropriate to the individual sections of the works.
- Maintenance instructions for all plant including preventative and corrective maintenance procedures. For electronic or individual circuit cards details should be provided for fault tracing to the card involved or to the components upon the card.
- Maintenance and inspection schedules for all items of plant giving details of type of work required on a weekly, monthly, annual, etc. basis.
- Proformas of the required maintenance record sheets for all items of plant.
- Standard handbooks produced by manufacturers making it absolutely clear as to which variant, option or alternative assemblies are applicable to the actual plant supplied under the transmission and distribution project.

Appendix A gives a useful checklist for items to be included in a commissioning partial or system acceptance test record.

Appendix B gives a useful flow diagram for the routing of drawings for approval under a substation design and construct contract.

19.3 WORKS INSPECTIONS AND TESTING

19.3.1 Objectives

Works tests are one of a series of tests for part or all of an installation:

- Type tests are used to prove the basic design – they are normally done at special test stations and may, in the extreme, involve tests to destruction.

- Works or factory tests are production tests used to confirm that manufacture is in line with design.
- Site tests are less severe and are intended to check the whole installation, (i.e. plant and associated cabling and connections) thus confirming delivery and erection were correct.
- Maintenance tests are periodic tests carried out during the lifetime of the installation to identify any developing problems and to confirm performance characteristics. They may be a repeat of the 'Site Tests'.

19.3.2 Specifications and responsibilities

Before placing an order for equipment it is essential that the required level of definition has been achieved. Such definition is the responsibility of the client or his engineer. Factory tests are carried out by the manufacturer and may be witnessed by an inspector from the client's and/or the engineer's organization. It is absolutely no use placing a loosely worded order with a contractor and expecting him later to include a wide range of tests which he has not catered for in his pricing. The level of detail to be specified may therefore range from virtually nothing (leaving all type and routine test requirements up to a reputable contractor) or clearly specifying detailed requirements with the order. Test requirements are often an area of dispute with a contractor and should therefore be finalized before an order is placed. One solution for minimizing type test cost disputes is to allow in the contract a priced bill of quantities for the work and avoid spending money if some equipment is shown already to have satisfactory type test certification. Indeed, major items of plant will normally have obtained type test certificates before they are marketed at all.

Often the satisfactory completion of type and/or routine testing, together with all associated documentation in a QA data pack, is a requirement before the equipment may be released from the factory for delivery to the client.

19.3.3 Type tests

Type tests are intended to:

- verify the design concepts of the plant;
- demonstrate that the plant performance complies with the relevant specification and manufacturer's guarantees;
- demonstrate the mechanical and electrical characteristics of the plant (e.g. weatherproofing, ability to operate safely in a particular environment or within a particular transmission network).

Such tests are expensive to carry out and might for instance include the making and breaking capacity of switchgear or the temperature rise of transformer windings. They are normally carried out by a specialist body independent of the manufacturer. Failure to satisfy the type test requirements may cause the plant under test to be damaged either during such tests or later when in service in the field.

Type tests are usually demonstrated on the first item of plant supplied on an order (for example, on a particular transformer design not previously made by the particular manufacturer) or on the first of a range of standard equipment to be supplied on an order (for example, on standard switchgear). Following a successful type test programme, the plant is usually provided with a certificate of type test issued by a recognized authority. It is not usual to repeat such type tests unless a fundamental change is made to the design or in the event of a particular client requiring the tests to be repeated in his presence.

19.3.4 Routine production tests

Following satisfactory type testing, each subsequent item of plant is subjected to a series of agreed production tests. These are intended to demonstrate that this item has similar design characteristics to the prototype and that it has been properly manufactured.

Examples of type and routine tests are given in Section 19.5.

19.4 SITE INSPECTION AND TESTING

19.4.1 Pre-commissioning and testing

The object of site inspection and commissioning tests is to ensure that the plant has been correctly installed and that it will function in a safe and proper manner. Tests at site are intended to verify the operation and to check that the plant has not been damaged in transit, during erection or even during the site tests themselves. Site tests are therefore less demanding than those carried out in the works. Before site work commences (and preferably before the contract for the work is agreed and signed) the client and/or his engineer and the contractor should study and agree the local regulations (or where necessary national alternatives) applicable to the safety of the workforce and the installation of plant. These regulations must then be strictly implemented. (Note that an international contractor would be unwise to apply safety standards significantly lower than those applicable in his home country, even if local standards permit that).

Transmission and distribution installations should be subjected during erection on site and upon completion of installation to a series of formal, witnessed and recorded inspections as part of the quality control procedures. Functional tests on the plant should be carried out as soon as possible in order to enable any difficulties to be solved without affecting the overall project completion date. The following recommendations will apply to all installations:

- An agreed inspection work programme (which may require outages to be arranged) to be detailed in advance on a working programme and finalized at weekly 'look ahead' client/engineer/contractor review meetings. In the case of energizing a new substation several subcontractors may be involved (e.g.

overhead line, power line carrier/communications and substation contractors) as well as different parts of the client organization (projects, operations, maintenance, protection, etc.). The co-ordination necessary should not be underestimated.

- Inspection to be carried out and witnessed by suitably trained persons. The testing and commissioning period is a good time to organize training programmes for the client's future operations and maintenance personnel.
- All inspections to be formally recorded and issued to the engineer or client for formal approval.
- Inspection shall, as far as is practicable, be independent of maintenance, construction and operation activities.
- Whenever possible a single point of integrative responsibility for all the inspection, testing and commissioning activities should be appointed in each of the client, engineer and contractor organizations. In this way outage times should be minimized and co-ordination of all activities ensured. Contractors will request extra monies from the client for substantiated idle time resulting from badly co-ordinated working practices.

19.4.2 Maintenance inspection

During plant shutdowns for overhaul and maintenance and before re-energization, the transmission and distribution plant will be subjected to certain inspection and testing procedures. This also applies to plant, such as a cable, that has been de-energized for a long period of time. Planned maintenance schedules should be drawn up by the electricity supply utility before the installation contractor has completed his work. Typical Transformer, On-Load Tap Changer and Substation Inspection Proforma Reports are given in Tables 19.2–19.4.

19.4.3 On-line inspection and testing

On-line (or 'in service') inspection and testing is normally limited to visual, external, physical examinations in order to ensure that the plant is in a safe condition. Considerable success has been achieved using infra-red detectors for inspecting overhead lines and open terminal substation busbars for hot spots caused by faulty terminations. In addition, 'live line' washing techniques are available for cleaning overhead line or open terminal substation insulators. Purified water with a high resistance value is used in a fine spray jet from a well-earthed nozzle. Functional testing of protection and trip schemes may require special switching arrangements initially to reconfigure the power system network. Such planned shut downs of the plant to be tested and network reconfiguration ensure continuity of supply to consumers while the testing takes place. The electricity supply utility will formulate such planned outage schemes at different times of the year (depending upon the load demand or the requirements of consumers) for different maintenance scenarios.

Table 19.2   Transformer Inspection Test Report

Test	Remarks and notes	Test	Remarks and notes (when required)
1. External examination		3) Internal Examination	
Oil level		Sludge	
Max temp/max load		Core	
Off load tap changer		Core clamps	
Insulators		Windings	
Cable boxes		Connections	
Oil leaks		Off load tap changer contacts	
Breathers		Tank and radiators	
Conservator		Gaskets	
Buchholz		Painting	
Painting		Internal bushing CTs (if fitted)	
Cooling equipment			
Pressure relief devices		4) Insulation Tests	
Earthing connections		HV to earth	
Silica gell		LV to earth	
External CTs (i.e. SBEF CT)		HV to LV	

Test	Transformer	Tapchanger	Conservator	Test	Remarks and notes (when required)
2) Oil sampling and testing				5) Instrumentation	
Drain valve condition				Oil temperature	
Oil appearance				Winding temperature	
Oil acidity				Others	
Oil 'crackle' test					
Oil electric strength					

Remarks:-

SUBSTATION:-	TRANSFORMER TEST REPORT	BB PROJECTS AND ENGINEERING
TRANSFORMER:-	Client:-	Drg. No.
By:- Checked:- Approved:- Date:-	Job No.	Sheet: .... of ....

Table 19.3 On-load Tap Changer Inspection Test Report

Test	Remarks and notes	Test	Remarks and notes
Oil appearance		Limit switches	
Oil acidity		Auxiliary contacts	
Oil 'crackle' test		Small wiring	
Oil electric strength		Contact box	
Oil leaks		Voltage relay	
Breathers		Time delay relay	
Selector Contacts		Line drop compensators	
Diverter contacts		Other relays	
Diverter resistors or reactors		Tap position indicator (local)	
Mechanism - general		Other instruments	
Worm drives		Interlocks	
Shafts and couplings		Heaters	
Clutches		Painting	
Motors		Tap operations meter reading	

Remarks:-

SUBSTATION:- TRANSFORMER:-	TRANSFORMER TEST REPORT	BB PROJECTS AND ENGINEERING
By:- Checked:- Approved:- Date:-	Client:- Job No.	Drg. No. Sheet ..... of .....

Table 19.4 Substation General Visual Inspection Report

ITEM	OK	Action taken	Action required	Remarks
1) General structures				
Gates and doors				
Padlocks				
Ground				
Drainage				
Birds and vermin				
Access				
Building				
Roof				
Guttering and spouts				
Windows				
Paintwork				
Fixed locks				
Name plates				
Danger notices				
HVAC				
Trenches and covers				
Doors and panic bolts				
2) Safety items				
Safety rules				
Treatment for shock				
Log book				
Key cabinet				
Telephone				
Fire detection/supression				
Substation drawings				
Portable equipment				
3) Transformers				
General condition				
Oil level				
Max Temperature reading				
Oil pumps				
Fans				
Oil leakage				
Breathers				
Tap changer operations				

ITEM	OK	Action taken	Action required	Remarks
4) HV Switchgear				
General condition				
Noise				
Gas pressure/oil level				
Spring mechanism				
Pneumatic mechanism				
Hydraulic mechanism				
instruments				
Relays				
Indication lamps				
Padlocks				
Heaters				
DC supply				
AC supply				
Marshalling kiosk				
Labelling				
Tools and accessories				
5) Outdoor switchyard				
General condition				
Noise				
Busbars & labelling				
Connections				
Surge arresters/counters				
Bushings/insulators				
General housekeeping				
Arcing horns				
6) Earthing				
Connections				
Earth bars and leads				
Neutral earthing				
7) Cables				
General condition				
Cable boxes/terminations				
Oil or gas pressures				
Cleats and supports				

19.5 TESTING AND COMMISSIONING METHODS

19.5.1 Switchgear

19.5.1.1 *Introductions*

This section should be read in conjunction with Chapter 13, where the principles and effects of different switching actions under different network conditions are explained. The following standards cover switchgear testing:

IEC 62271-100	High voltage switchgear and control gear – HV alternating current circuit breakers
IEC 62271-203	High voltage switchgear and control gear – Gas-insulated metal-enclosed switchgear for rated voltages of 72.5 kV and above
IEC 60060	High voltage test techniques
IEC 60427	Synthetic testing of high voltage alternating current circuit breakers
IEC 60480	Guide to the checking of sulphur hexafluoride (SF_6) taken from electrical equipment
IEC 60694	Common specifications for high voltage switchgear and control gear standards. This includes a specification for the maintenance of circuit breakers, superseding IEC 1208.

19.5.1.2 *Type tests*

- Dielectric measurement.
- Temperature rise tests.
- Making and breaking tests.
- Mechanical endurance.

19.5.1.2.1 *Dielectric measurements*

Dielectric measurements confirm that switchgear of a given voltage rating is able to withstand the voltage stresses specified as likely to occur in service from switching operations or lightning surges.

One-minute power frequency voltage withstand tests are applied to the switchgear. A voltage in excess of the switchgear-rated voltage is used in order to confirm adequate clearances and insulation strength between phases, across open contacts and between phase and earth. The tests also cover the withstand capability of such items as switchgear closing and opening resistors, GIS designs and grading capacitors. An adequate margin between the withstand voltage levels and temporary overvoltages that may arise during switching operations has to be allowed for when determining the test voltage to be used.

In addition, the performance of asymmetric electrode shapes under different high test voltage polarity has to be investigated.

The dielectric strength of air is a function of pressure and temperature (although other factors such as humidity, dust, radiation and in particular, the electrode shape can significantly affect the breakdown strength). Therefore correction factors are applied to type test voltages used in the laboratory in order to simulate a variety of environmental conditions. For example, larger switchgear clearances have to be allowed for in designs for switchgear operation at high altitude where the air density is reduced. Similarly, the type test voltage is increased to compensate for the higher withstand voltage of air at sea level. SF_6 dielectric strength also varies with density and type tests should be conducted under the least favourable conditions.

For outdoor equipment the tests may be performed under both dry and wet conditions. The wet conditions attempt to take into account the effects of water cascading down the switchgear porcelain insulators as a method of confirming adequate insulator profiles and creepage distances. The performance of insulators under polluted or low temperature ice conditions also requires assessment and is the subject of special agreements between the switchgear/insulator manufacturers and the client or his engineer.

Impulse withstand tests using both positive- and negative-going waveforms attempt to prove the ability of the switchgear to withstand surges arising from both lightning and switching conditions. Switching surges are normally only relevant to equipment designed for rated voltages in excess of 300 kV, insulation Range II (refer to Chapter 9). In this insulation category the overall switchgear dimensions are largely determined by the necessary switching and lightning surge clearances. A standard test involves the application 15 times of test voltages (normally using a standard 1.2/50 μs impulse voltage wave shape to simulate lightning strike conditions and a 250/2500 μs wave shape for switching surges) between each phase and earth in turn with the circuit breaker closed and remaining phases earthed. In addition, test voltages are also applied across each set of open circuit breaker contacts. The overall test is considered satisfactory if not more than two flashovers occur during any one series of 15 tests provided that these discharges only occur in self-restoring insulation (i.e. air, oil or SF_6 gas). A breakdown of solid insulation normally results in an inability to recover and constitutes a test failure. Additional tests to verify this may be carried out if required.

At rated voltages in excess of 300 kV a bias test is applied to the switchgear. A power frequency is applied to one terminal of the circuit breaker and a lightning impulse applied to the opposite terminal in order to prove the design under conditions of a lightning strike whilst the switchgear is in the open condition and the waveform peaks are at opposite polarity.

Partial discharge tests are not normally specified as part of a complete circuit breaker assembly type test. However, partial discharge tests are particularly applicable to switchgear designs involving components with solid insulation. The insulation should be discharge free at rated voltage since continued internal breakdown will lead to chemical and thermal damage and eventual insulation failure. However, very low intensity discharges are unavoidable in practice and minimum acceptable levels, based on practical experience, are imposed in order to confirm long insulation and switchgear life. An attempt may be made using oscillograph traces to determine the voltage levels for discharge inception and extinction. An analysis of the traces allows assessment of whether the breakdown is a result of voids in the insulation, corona discharge, etc. Alternatively, the maximum discharge level in pico-coulombs may be established during the power frequency overvoltage testing.

External discharges will lead to electromagnetic interference. Corona breakdown may be seen visually in a darkened laboratory and the inception and extinction voltages recorded. Radio noise levels resulting from external discharges are measured in dB above a reference level in μV. Table 19.5 details circuit breaker-rated lightning impulse and switching overvoltage levels to IEC 62271 and IEC 60694.

19.5.1.2.2 *Temperature rise tests*

The correct power frequency-rated current is passed through the closed switchgear contacts. Plots of temperature against time are recorded in order to determine the constant ultimate temperature rise at various temperature sensor locations on the switchgear. Standard acceptable temperature rise limits are specified for the various switchgear components. These are based on experience of safe working temperatures (ambient plus maximum temperature rise) which give reliable long-term switchgear performance. If temperature limits are exceeded insulating/interrupter oil may begin to lose its properties (a widely used rule of thumb states that every 10°C increase in operating temperature cuts insulation life in half). Paper insulation in particular suffers a reduction in life at high temperatures. Further, mechanical parts may expand with loss of contact pressure, oxidation of contact surfaces may occur together with increases in contact resistance and the resulting further temperature rise will eventually lead to arcing and insulation breakdown.

When standard manufacturer's switchgear is mounted in custom-built panels special attention must be paid to allow for adequate heat transfer such that temperature rise limits are not exceeded. Low voltage switchgear manufacturers often give details of busbar sizes and enclosure dimensions necessary in order to achieve this adequate heat transfer and dissipation. It is essential that the actual enclosure to be used in practice is also used during the type testing if it is to give meaningful results for the particular application. Acceptable temperature rise is based on an average ambient temperature of 35°C and maximum ambient temperature of 40°C. Suitable derating factors must be applied to the current rating of equipment specified for use in hot climates with ambient temperatures exceeding these values.

19.5.1.2.3 *Making and breaking tests*

IEC 62271 in conjunction with IEC 60694 defines the various circuit breaker making and breaking type test duties performed at rated voltage, rated current

Table 19.5   Circuit breaker rated lighting impulse and switching overvoltage levels to IEC 62271 and IEC 60694.

Rated voltage (kV rms)	Rated lighting Impulse withstand voltage (kV peak)	(A) Maximum permissible switching overvoltage to earth (kV peak)	(A) Maximum permissible switching overvoltage to earth (per unit, pu)	(B) Maximum (permissible switching overvoltage to earth (kV peak)	(B) Maximum permissible switching overvoltage to earth (per unit, pu)
12	60	29.5	3	24.5	2.5
	75	39.2	4	24.5	2.5
17.5	75	43	3	35.7	2.5
	95	57	4	35.7	2.5
24	95	59	3	49	2.5
	125	74	3.8	49	2.5
36	145	88	3	73	2.5
	170	112	3.8	73	2.5
52	250	149	3.5	106	2.5
72.5	325	207	3.5	148	2.5
100	380	246	3	204	2.5
	450	286	3.5	204	2.5
123	450	302	3	251	2.5
	550	352	3.5	251	2.5
145	550	356	3	297	2.5
	650	415	3.5	297	2.5
170	650	417	3	348	2.5
	750	487	3.5	348	2.5
245	850	540	2.7	400	2
	950	600	3	400	2
	1050	600	3	400	2
300	950	637	2.6	490	2
	1050	735	3	490	2
362	1050	710	2.4	592	2
	1175	800	2.7	592	2
420	1300	790	2.3	688	2
	1425	895	2.6	688	2
525	1425	900	2.1	858	2
	1550	985	2.3	858	2
765	1800	1125	1.8	1125	1.8
	2100	1250	2	1250	2

Note: pu maximum permissible switching overvoltage = peak value (kV peak)/{rated value (kV rms) x √2/3}

and short circuit current. The test plant is of a special design to cater for the short circuit powers involved. Full three phase testing is impracticable for ratings above about 8 GVA at 275 kV where synthetic tests to IEC 60427 are used to simulate field conditions. Such tests are therefore normally carried out at accredited and independent short circuit test laboratories after initial design and test work has been completed by the manufacturer. At the end of the test sequences the circuit breaker mechanical and insulating parts must be in essentially the same condition as before the test duty. In some cases (such as sealed-for-life GIS and vacuum switchgear) dismantling of the breaker is necessary in order to assess the effect of the tests visually.

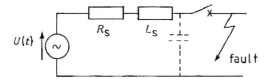

Equivalent circuit

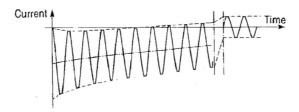

Short circuit current decays from initial
symmetrical value to steady state short
circuit level

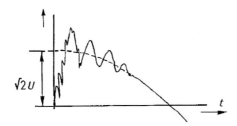

Recovery voltage across circuit breaker.
(difference between supply and line side
recovery voltages)

Figure 19.1   Close-up short line fault

(a) Terminal faults

Close-up terminal fault tests consist of subjecting the circuit breaker in stages
to 10%, 30%, 60% and 100% rated symmetrical three phase short circuit currents and a 100% asymmetrical short circuit through its contacts to simulate a
close-up fault such as might inadvertently occur in practice due to earths being
left on busbars after maintenance. A 3-minute interval is allowed between each
break or opening (O) operation and such an accurately timed duty cycle with
three breaks is often expressed as 'O–3–O–3–O'. Normally the time interval is
not less than 2 minutes and not greater than 10 minutes (Fig. 19.1).

Close and open (CO) tests at full short circuit current confirm the making capability of the circuit breaker. A typical duty cycle would be designated 'O–3–CO–3–CO'. This is intended to demonstrate that the closing mechanism, contacts and interrupter design and closing speed are sufficient to overcome short circuit forces. The most onerous condition for such forces is under maximum asymmetry in one phase. The additional stresses placed upon the circuit breaker as a result of auto-reclose protection schemes and its ability to withstand them are simulated by varying the time interval between the first open and close/open operation. For example, a typical test duty cycle would become 'O–15 sec–CO–3–CO'. The test equipment identifies any tendency for the dielectric to break down after the first or second current zero as a result of the rate of rise and peak value of the transient recovery voltage (TRV). Since all three poles of the circuit breaker do not interrupt the fault simultaneously the first phase-to-clear power frequency recovery voltage, as explained in Chapter 13, experiences a voltage rise up to 1.5 times the normal-rated voltage to earth (Fig. 19.2).

(b) Line faults
At the higher voltage levels (above 72.5 kV) switching surges, caused by an exchange of energy between the impedances on the source and faulted line side of the circuit breaker, can lead to oscillatory high transient recovery voltages. This presents a severe test case since multiple restriking of the arc could occur leading to circuit breaker failure. The short line fault test circuit simulates the impedances and uses a duty cycle 'O–3–O–3–O' in a series of breaking tests at 90% and 75% of the rated short circuit breaking current and appropriate prospective TRV.

(c) Capacitive switching
A sequence of capacitive current switching test duties consists of breaking 100% and 20% to 40% of the rated capacitive charging current of the overhead line or cable (typically at rated voltages above 72.5 kV for overhead line charging currents and above 24 kV for cable circuits). (Note these tests may be particularly important for circuit breakers controlling harmonic filters, power factor correction installations, or even very extensive cable networks).

(d) Out-of-phase tests
These determine the ability of the circuit breaker to interrupt circuit connections between, for example, out-of-phase generators under asynchronous conditions. The resultant recovery voltage (up to 2.5 times the normal-rated power frequency voltage) is much higher than that for the close-up terminal fault case with a fault current of only some 25% of the rated short circuit breaking current. Standard test duties consist of two opening operations (O and O) at 20% to 40% of the rated out-of-phase breaking current followed by one close–open (O and CO) test at the full rated out-of-phase breaking current. In accordance

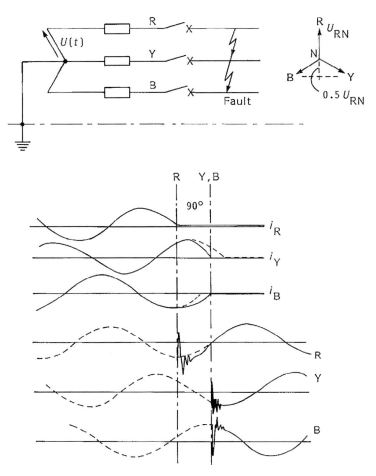

Figure 19.2   Power frequency recovery voltages for a three-phase fault in a solidly earthed system. R-phase, first pole to clear. Recovery voltage across first pole to clear $= 1.5 \times 2 \times 2$ phase voltage (transient voltage doubling and 1.5 first pole to clear factor)

with the recommendations of the now superseded IEC 56 (no equivalent recommendation is given in the later IEC 62271-100) a 10 ms DC component decay results in a current peak 1.8 times the symmetrical short circuit peak.

(e) Short time currents
Short time withstand current tests demonstrate the ability of circuit breakers, busbars, disconnectors, CTs, etc., to withstand through-fault current without damage. The initial peak test current may also be arranged to attain that of a specified fully asymmetrical condition. The standardized time durations of one or three seconds have generally been derived from what is considered to be a realistic time for backup protection to operate in practical situations.

19.5.1.2.4 *Mechanical endurance*

The circuit breaker or switching device is put through a series of up to 2000 open–close operations and at the end of the test undue wear and the following are checked:

Closing/opening and speed	Time spread between pole operations	Operating device times re-charging times
Control circuit power consumption	Trip circuit power consumption	Auxiliary contact condition
Duration of opening and closing command impulses	Enclosure and interrupter head tightness	Gas densities (or pressures)
Resistance of main circuit	Main contact condition	Rigidity of structure

Additional tests over a wide temperature range may also be carried out.

19.5.1.3 *Routine tests*

- Power frequency withstand.
- Voltage withstand on control and auxiliary circuits.
- Measurement of resistance of main circuit.
- Mechanical operating tests.
- Visual checks.

19.5.1.4 *Site tests*

It is important not to overstress equipment by inappropriate repetition of type or routine tests under inadequate site conditions. However, it is necessary to confirm that no damage has occurred during transport and erection of the equipment. Depending upon the relative amount of site assembly it may be necessary to ask the manufacturer to witness site erection and/or to assist in the switchgear commissioning process.

It is particularly important that the dielectric strength of the total assembly of GIS components that have been delivered to site separately and then bolted together are assessed. This is intended to ensure that wrong fastenings, damage during transit or handling, storage, presence of foreign bodies, etc., are avoided. AC voltage tests are especially sensitive in detecting contamination such as conducting particles. DC voltage tests are not recommended and existing cable test specifications are not applicable to metal-enclosed GIS. One-minute AC withstand test voltages for GIS site tests are detailed in Table 19.6.

The voltages are applied between each phase conductor, one at a time, and the enclosure, the other phase conductors being connected to the earthed enclosure. The insulation between phase conductors should not then be subjected to separate site dielectric tests. Impulse tests consisting of three impulses of each polarity may also be carried out at the levels detailed in Table 19.6.

Table 19.6   Site dielectric test voltages for GIS

Rated voltage (kV rms)	Lightning impulse voltage (kV)	Switching impulse voltage (kV)	AC voltage (kV rms)
72.5	260	208	112
100	360	288	148
123	440	352	184
145	520	416	220
170	600	480	260
245	760	608	316
300	840	680	380
362	940	760	450
420	1040	840	520
525	1140	940	620
765	1440	1140	750

Switchgear site tests generally include:

General checks	Assembly to manufacturer's drawings
	Tightness of terminal connections, piping, junctions and bolted joints
	Painting and corrosion protection
	Cleanliness
Electrical circuit checks	
Insulation checks	Dielectric strength of insulating oils (see IEC 60296) and level, SF6 quality, humidity content (see IEC 60376) filling pressure or density except for sealed apparatus
Mechanical tests	Operating circuits (hydraulic, pneumatic, spring charged)
	Consumption during operation
	Verification of correct rated operating sequence (recharging, etc.)
Time quantities	Closing and opening times
	Operation and control of auxiliary circuits
	Recharging time of operating mechanism after specified sequence
	Checks on specific operations
Electrical tests	Dielectric tests
	Resistance of main circuit

A typical MV circuit breaker maintenance inspection report proforma is shown in Table 19.7.

19.5.2 Transformers

19.5.2.1 *Introduction*

This section should be read in conjunction with Chapter 14. IEC60076 divides transformer tests into type, routine and special tests.

Table 19.7   Typical Proforma MV Circuit Breaker Maintenance Inspection Report

Item	As found	As left	Remarks / work done	Item	As found	As left	Remarks / work done
Main contacts				Screws			
Arc contacts				Operating mechanism			
Switch tanks				Plug contacts			
Arc fingers				Auxiliary switches			
Oil				Auxiliary wiring			
Insulators				Shutters			
Fixings				Voltage transformer			

SUBSTATION:- CIRCUIT BREAKER:- By:- Checked:- Approved:- Date:-	SUBSTATION CIRCUIT BREAKER MAINTENANCE INSPECTION REPORT Client:- Job No.	**BB** PROJECTS AND ENGINEERING Drg. No. Sheet...... of ....

Table 19.8   Standard dielectric measurement impulse voltage levels for oil-immersed transformers

System highest voltage (kV rms)	Impulse voltage level (1.2/50 μs waveform, kV peak)	Insulation to earth
3.6	45	
7.2	60	
12.0	75	
17.5	95	
24	125	uniform
36	170	
52	250	
72.5	325	
100	450	
100	380	
123	450	
145	550	
170	650	graded
245	900	
300	1050	
362	1175	
420	1425	

19.5.2.2 Type tests

- Dielectric measurement.
- Temperature rise test.
- Noise level measurement.
- Tap changer tests.

The dielectric tests confirm that the transformer is capable of meeting the specified breakdown insulation levels. This is achieved by the application in the factory of impulse waveforms in accordance with Clauses 12 and 14 of IEC 60076-3 from specially designed impulse generator equipment. The normal 1.2/50 μs impulse voltage wave shape has a steep fronted 1.2 μs rise time (with 30% tolerance) between 10% and 90% of the peak voltage and a nominal wave tail of 50 μs (with 20% tolerance) between peak voltage and 50% of the peak voltage. In North America wave shapes of 2.5/40 μs may be specified. Standard impulse levels for oil-immersed transformers are given in Table 19.8.

Temperature rise tests confirm that the transformer insulation, oil, core and windings do not exceed the specified limits. The temperature rise tests should be conducted when the transformer operates at its full MVA rating on the tapping corresponding to maximum losses. Since the tests involve running the transformer in the factory at its maximum rating special test arrangements have been devised to minimize the power consumption involved. If two similar transformers are available a 'back-to-back' arrangement may be used. The two low voltage windings are connected in parallel and energized while the high voltage windings are connected in opposition. The transformer taps are set to provide the full load circulating current.

Noise level measurements are becoming more frequently requested. They may be performed to IEC 60076-10 which defines the methods to be used for transformers, reactors and associated cooling equipment. It is important to specify noise levels that may be economically achieved in practice. The 1974 American NEMA Specification TR1 gave useful guidance for attainable levels from reputable manufacturers for different sized transformers. Also see BEMA 227 and in the UK Electricity Supply Industry publication 989907. Where noise is critical it may be necessary to construct a special transformer enclosure.

On-load tap changer type tests are defined in IEC 60214. Where tap changer assemblies are destined to work in parallel with each other the satisfactory parallel operation should be demonstrated in the factory before despatch.

19.5.2.3 Routine tests

- Winding resistance measurement.
- Voltage ratio check.
- Phase relationship check.
- Impedance voltage.
- Load loss.
- No-load loss and current.
- High voltage tests.
- On-load tap changer functional tests.

Some clients also require transformer tank pressure tests in order to check for gasket or weld leakage. Insulators, oil and transformer auxiliaries (winding and oil temperature measurement devices, pressure relief devices, Buchholz relays, etc.) have their own separate routine tests. At site:

- Check for physical damages upon arrival of the transformer on site. A shock recorder is sometimes specified to record possible damaging knocks during transit.
- For larger transformers which arrive oil-filled – filter, circulate and dry the insulating oil using a mobile oil treatment plant. If delivered dry nitrogen filled, they must be filled with oil or synthetic fluid on site. Oil drum seals must be checked before use.
- Confirm the suitability of the insulating oil with checks for physical pollution by water and other suspended matter in accordance with IEC 60156 and IEC 60296 using a portable oil test set. The oil must be allowed to settle in a clean, specifically dimensioned test chamber before testing commences. The more sophisticated oil test sets automatically ramp up the voltage between two spherical electrodes immersed in the oil in the test chamber until breakdown occurs. After each breakdown all traces of carbon must be removed by thoroughly cleaning the test cell and refilling with a new oil sample.
- The interpretation of transformer oil gas analysis is described in IEC 60599. This is a test used during regular transformer maintenance or after a Buchholz gas or surge fault as a guide to the state of the transformer insulation.

- A simple method of oil moisture content may be obtained from an oil 'crackle test'. A test tube of oil is carefully heated over a bunsen burner flame. If moisture is present (down to very low levels of approximately 60 ppm) a 'crackling' sound will be audible. Samples may also be tested in a site laboratory using the 'Karl Fischer' test which can detect moisture levels in the oil down to 3 ppm. New oil should approximately have only 1% of dissolved air by volume and a moisture content of less than 5 ppm. A more sophisticated approach takes account of the fact that the dielectric breakdown value of insulating oils is proportional to the relative saturation (RS) of water in oil rather than the concentration in ppm. (For example, at 10°C only 36 ppm of water can be dissolved in the oil (i.e. 100% RS), whereas when the temperature increases to 90°C the amount of water that can be dissolved rises to 600 ppm). The oil RS should not exceed 50%.
- Check the voltage ratio over a variety of tap positions.
- Complete a phase relationship check to confirm the markings on the transformer terminals. This may be achieved by connecting together a primary and secondary phase, applying a low voltage to the primary and recording the resulting primary, secondary and primary-to-secondary phase voltages (see Chapter 26, Section 26.4).

Proforma transformer and transformer tap changer site inspection test sheets are included in Tables 19.2 and 19.3.

19.5.2.4 *Special tests*

Special tests are made at the request of the client or engineer. It should be appreciated that such tests may put considerable strain on the transformer and should not be called for unless

- type certification from the manufacturer for similar transformers made to the same design is unavailable;
- the special tests actually simulate conditions that could really occur in practice.

The simulation of an incoming surge which has been chopped by the breakdown of co-ordinating arc gaps or surge arresters is achieved by the 'chopped wave' test in a particular sequence to IEC 60076, with chopping times of the order of 2–6 μs and the peak voltage applied at least equal to the specified full wave test. As an alternative, partial discharge measurements may be made on higher-rated voltage transformers. IEC 60076-4, 'Guide to the lightning and switching impulse testing of power transformers and reactors', gives explanatory comments on the procedures.

Information is given on wave shapes, test circuits including test connections, earthing practices, failure detection methods, measuring techniques and interpretation of results.

Transformer zero sequence impedance affects the earth fault level in the power system and measurement in the factory is therefore a frequent special test requirement.

Transformer short circuit tests allow the manufacturer to prove to the client or engineer that the unit is capable of withstanding the stresses involved. Such tests are, however, costly and time consuming and therefore rarely completed in practice. It is possible to check the manufacturer's calculations and to request verification of the computer models used by comparing the results with tests carried out by reputable manufacturers on trial units in the past.

No-load harmonic component tests may be important in applications where the transformers are associated with telecommunication circuits.

Special tests may be requested to determine the power requirements for the auxiliaries such as motorized fans and oil pumps. Such requirements represent losses and may be included in loss capitalization equations. For dual (ONAN/ONAF) ratings the fans are considered to operate only rarely and would not normally therefore be included as losses in such economic assessments.

19.5.3 Cables

This section should be read in conjunction with Chapter 12. Dimensional and mechanical tests are performed at the factory. This section deals with site testing.

19.5.3.1 *High voltage DC testing*

High voltage testing is carried out in order to determine the electrical strength of cable insulation. Site tests are performed by applying a predetermined high voltage to the insulation. Particular care must be taken to ensure that the insulation is not subjected to excessive high voltages as it may become overstressed with a consequent future breakdown or reduction in its working life. DC site test voltages, regardless of insulation type, are used to ensure that the cable, cable joints and terminations are correctly made and installed as indicated in Table 19.9. Throughout the tests the conductors not under test are connected to earth. If withstood for a given period of time it is considered that the insulation will perform safely at the rated voltage.

Three core armoured cables and single core cables protected by aluminium armour wires, tapes or sheaths may experience corrosion if the outer PVC sheath deteriorates or if it is mechanically damaged during installation such that water or moisture can penetrate. Sheath integrity is verified by a 5 kV 3-minute test applied between armouring and earth. This test may be enhanced (and sheath damage is easier to locate) by specifying that the cable outer PVC sheath is covered with a graphite coating.

Table 19.9   DC voltages for cable pressure and insulation resistance testing

For HV pressure testing (duration 3 minutes)

Cable voltage designation (kV – AC)	Test voltage (kV – DC)	Cable voltage designation (kV – AC)	Test voltage (kV – DC)
0.75	insulation resistance test only	–	–
0.6/1.0			
1.9/3.3	7	8.7/15	30
3.0/3.0	10	10/10	30
3.8/6.6	13	13.7/22	44
6.0/6.0	18	15/15	45
6.35/11	22	19/33	65

For insulation resistance test

System voltage	Test voltage
Below 1000 V AC	500 V DC
1000 V–4.6 kV	2500 V DC
Above 4.6 kV	5000 V DC

Notes: (a) Test voltages are applied continuously.
(b) Test voltages should be slowly raised to the required value over a period of about 1 minute and the test period starts once the full voltage is reached. In this way the capacitive and absorption currents will have decreased and circuit conditions stabilized such that true leakage current may be measured.
(c) IEC Standards specify maintenance tests after installation should be 70% of permissible factory test DC voltages.

19.5.3.2 AC tests

When an AC voltage is applied to a cable a leading charging current, I_c, flows. Also, because the cable insulation is not a perfect dielectric a small leakage current, I_r, flows in phase with the applied voltage. This leakage current causes losses in the dielectric which generate heat. The total no-load current flowing is the vector sum of leakage and quadrature currents. The equivalent circuit of a composite insulation may be represented by a parallel combination of resistance and capacitance as shown in Fig. 19.3 where M1 and M2 are different materials and the potential drop in each is given by:

$$V_{M1} = I_r\,R1$$
$$V_{M2} = I_c/(j\omega C1) \quad \text{and} \quad V_{M1} \neq V_{M2}$$

From Fig. 19.3

$$I_r = I_c \tan \delta$$

The dielectric losses of a cable are given by:

$$P = VI_r$$
$$= VI_c \tan \delta$$
$$= V^2\omega C \tan \delta$$

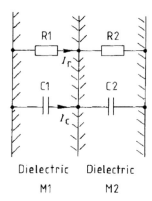

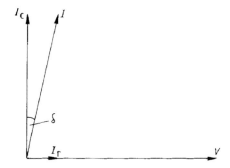

Figure 19.3   Dielectric losses

and shows that the dielectric losses are influenced by the frequency, capacitance, the square of the voltage and the 'loss factor', $\tan \delta$. Therefore during AC testing a considerable amount of power is absorbed in the insulation, causing heating and accelerating the ageing process. Therefore AC testing is used as a 'go/no go' test to determine whether or not the insulation can withstand the applied voltage. On small test items the capacitance is sufficiently low so that the leakage current is significant and any avalanche current effects at insulation breakdown are apparent. For long cables high values of capacitance are involved and the capacitive current may be several hundred times greater than the leakage current. Therefore large AC test sets are involved.

19.5.3.3 *Comparison of DC and AC testing*

The voltage distribution in the test sample is completely different under DC and AC steady state test conditions. The DC test stresses the insulation material in a non-homogeneous structure in proportion to the insulation resistance of each element of the path. During such steady state DC tests there are no

Table 19.10   DC and AC insulation test comparison

DC test	AC test
No dielectric loss	Dielectric loss resulting in possible thermal flashover
Lower temperature resulting in higher insulation resistance	
Reduced thermal flashover	
Flashover depending on the profile and polarity of the insulation	
No capacitive current and therefore voltage distribution is different from operating conditions	Large test set required to supply capacitive currents.
	Note: The winding to frame capacitance of a 30 kV generator may draw a capacitive current of 3 A which would require a 30 kVA test set.
Tests governed by surface resistivity.	Tests governed by permittivity of insulation material.
Flashover may occur due to surface contamination, dust, humidity, moisture, etc.	

capacitive currents and dielectric losses. This results in less heating of the insulation and reduces the risk of thermal flashover. The DC test set size is a function of the voltage because of the small leakage currents involved.

The AC test will stress the various materials in proportion to their dielectric constant. It will subject the material to a vibrating mechanical force more nearly simulating AC equipment operating conditions. Such AC tests are better suited to being performed in the manufacturers' works.

Irrespective of whether DC or AC testing is carried out the insulation will retain a high voltage charge upon completion of the test. For safety reasons this must be discharged through an impedance to earth. Because of the capacitance and dielectric absorption effects it takes considerably longer to discharge static voltage after a DC high voltage test and the earth connection should be maintained for at least 5 minutes. Table 19.10 gives an overall comparison of DC and AC tests.

A typical proforma HV pressure test sheet for cables and substation plant is given in Table 19.11.

19.5.3.4 *Fibre optic cables*

The transmitters and receivers used in fibre optic communication are usually digital devices that convert analogue or digital electrical signals to digital light signals and vice versa. These devices are routinely tested after manufacture. Following installation their performance will be demonstrated during the end-to-end system commissioning tests. Periodic routine maintenance testing should also be carried out in accordance with the manufacturer's recommendations.

Detailed guidance on basic test and measurement procedures for fibre optic interconnecting devices and passive components is provided in IEC 61300-1.

Table 19.11   Typical Proforma HV pressure test sheet for cables and substation plant

Description of cable / plant tested	Working voltage (kV)	Conductors tested	Conductors earthed	D.C. Voltage (kV)	A.C. Voltage (kV)	Duration (Minutes)	Leakage current (mA)	Insulation Resistance (MΩ)	Remarks

SUBSTATION:-	PRESSURE TESTING OF HV CABLES AND PLANT	BB PROJECTS AND ENGINEERING	
By:-	Client:-		Sheet .... of ....
Checked:-	Job No.		
Approved:- Date:-		Drg. No.	

Part 2 of standard IEC 61300 contains 48 basic test and measurement procedures, while Part 3 has 40 parts covering examinations and measurements.

The fibre optic transmission path is a crucial element in the performance of the communication system, and it must be thoroughly tested following initial installation, during maintenance or if re-routing or other modification work is carried out. Problems relating to the transmission path are usually caused by unacceptable reduction or total loss of the light signal reaching the receiver. The cause may be reflective or non-reflective events. Test equipment is available to quickly determine and diagnose any specific problems.

Typically, a fibre optic cable testing unit can be connected to the fibre at the sending end using suitable adapters and connection cables. The test unit is set up to match the refractive index of the fibre, the wavelength of the transmission signal and the mode of signalling used. The unit will send repetitive pulses down the line and record the strength and timing of all the returning signals and display the result as a gradually reducing magnitude/distance plot.

Where the light signal is reflected by connectors or mechanical splices, a spike results. Where there is a loss greater than the normal progressive attenuation such as the loss due to a fusion splice, there will be a step reduction in the trace. The test unit can usually be set up to provide a tabular print-out of abnormalities in the transmission path to facilitate reconciliation with known features of the path. Should an unknown change be revealed or if a known change is abnormally large it can be investigated. Benchmarking fibre optic routes during commissioning is useful for subsequent comparison during maintenance testing.

Optical power meters can be used to identify whether a fibre is in service and the direction of the signal flow. Typically, the meter applies only a slight bend to the fibre and the negligible additional insertion loss will not interfere with the normal signal traffic within the fibre. Use of such a device to positively identify a fibre before cutting and splicing during modification work can eliminate the risk of cutting the wrong fibre.

Optical fibres often carry light at frequencies invisible to the human eyes but nevertheless powerful enough to damage them. Tracing light leakage at poor connectors, fibre breaks or escaping from tight bends or breaks in the fibre's jacket can therefore present serious safety risks. Cards are available which convert any infra-red light shining upon them into visible light to help in detection of leakage, but a safer alternative is to replace the normal invisible light source with a low power visible red laser to enable any leakage to be readily and safely identified. Where connectors must be microscopically examined to identify dirt, pits or scratches, it is also preferable to interpose a miniature video camera or videoscope to eliminate any risk to eyesight.

When testing fibre optic paths, communication between testers working at locations quite remote from one another can be a problem. Talk sets, which use spare fibres for voice communication, can be obtained that can also be used by personnel testing control or protection schemes operating between the same remote locations.

19.5.4 Protection

19.5.4.1 *Introduction*

Refer to Chapter 10 for an explanation of the terminology used in this section. It is essential to test protection equipment on site in order to ensure the following:

- That the relays are correctly installed.
- That the relay equipment has not been damaged during transit and that any packaging restraint has been correctly removed.
- That the relay equipment is correctly connected and wired up in accordance with the approved drawings.
- That the relays and associated trip coils operate within the required margins and are set to the required settings.

In general this requires the following types of test to be carried out:

1. Physical check on all wiring and connections to ensure cabling conforms to the approved schematic diagrams and that all connections are secure and correctly tightened and labelled. In particular, the integrity of current transformer circuits should be checked as an open circuit CT will not only be likely to fail but also creates lethal high voltages. Check on all fuses, MCBs, links, test switches and earthing terminals, etc. and ensure that the relay and instrument cases are correctly earthed.

   Check that the DC supply voltage polarity is correct in order to avoid damage to voltage transient suppression diodes connected across relay coils and that coils are not continuously energized from the positive DC supply.

   Carefully inspect the relay cases or racks for dirt and condensation that could affect operation.

2. Insulation resistance measurements on all AC and DC circuits taking into account the manufacturers' recommendations and the need to ensure that a high voltage 'Megger' is not used for readings where transistor equipment is involved. Records to be maintained for future reference.

3. Secondary current and voltage injection tests to ensure that the relays are in good working order and that the correct relays with the correct characteristics have been shipped. IDMTL overcurrent relays are now available with a variety of characteristics but most older electromechanical relays have particular characteristics not alterable. The correct sequence of operation of AC and DC auxiliary relays, tripping, impedance relay starters and measuring elements is also determined. Check that where provided trip circuit supervision is functioning correctly for the correct circuits involved. Check that the relay 'flag' indicators continue to operate correctly at reduced (80%) voltage and are correctly reset.

4. CT magnetization curve checks are important when a number of CTs are installed on site and errors in identification are possible. Also for balanced

protection schemes correctly matched CTs must be employed. It is important to take measurements to determine the 'knee point' of the CT.

5. Primary injection CT ratio, polarity checks and phase continuity tests from the CTs back to the relay equipment.
6. Confirmation of the directional sense of directional relay elements under load and simulated load conditions.
7. Check that the relays are all adjusted to the correct settings. An initially low back-up earth fault relay setting is preferred before energization to limit the extent of damage should a fault occur on new equipment.

19.5.4.2 *Primary injection tests*

19.5.4.2.1 *Objectives*
The object of carrying out primary injection tests is to ensure that the entire circuit is correct:

- CT ratios.
- Polarity of one CT group with another.
- Continuity of entire circuit (CT, secondaries, relay coils, circuits, etc.).
- Phasing.

Primary injection requires a portable, heavy current single phase injection transformer operating from the local mains supply with several low voltage heavy current windings and taps to supply a convenient proportion of the CT-rated primary current. Windings may be connected in series or parallel according to the current required. The transformer is usually rated at about 10 kVA allowing currents up to 1000 A. For transformer-balanced protection schemes it is possible to circulate sufficient current for protection test purposes between a pair of parallel transformers by selecting different tap settings.

Primary injection tests may be extended where necessary to cover control, metering and instrumentation checks.

19.5.4.2.2 *CT polarity*
The polarity of CTs is standardized. If the relay scheme includes distance, directional, differential or restricted earth fault relays, the polarity of the main CTs must be checked. Primary and secondary polarity markings are placed on the units before they leave the factory but it may be necessary to carry out a 'flick' test on site:

1. Connect a low voltage battery through a current limiting resistor and then to an analogue moving coil centre zero DC voltmeter. When the switch is closed note the direction of 'kick' of the voltmeter pointer (Fig. 19.4a).
2. Next connect the CT as shown in Fig. 19.4b. Again observe the direction of 'kick' of the voltmeter. If it is the same direction as for the initial test as

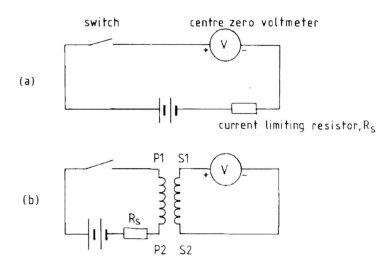

Figure 19.4   CT polarity 'flick' test

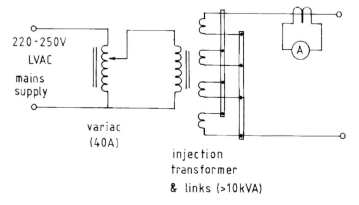

Figure 19.5   Primary injection test set

described in 1. above then the polarity is correct. The direction of the instantaneous injected current from primary terminal P1 to P2 generates a secondary current through the external circuit from S1 to S2. The series resistor should be high enough to keep the resultant current as low as practicable. The core should be demagnetized after the test by injecting an AC power frequency current through the secondary and gradually reducing the value to zero.

Primary injection testing using a test circuit as shown in Fig. 19.5 will also ascertain CT polarity. A temporary short circuit is placed across the phases of the primary circuit and a rated circulating current generated from the primary injection test set. The ammeter connected in the residual CT secondary circuit will give the 'spill' current of a few milliamps with correct polarity

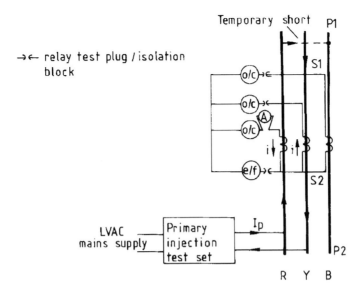

Figure 19.6   CT ratio check

CT connections and a reading proportional to twice the primary current if they are incorrectly connected. The ammeter should therefore initially be set on a high current range to avoid instrument damage. It is also advisable temporarily to short out any low setting earth fault relay element connected in the residual circuit in order to avoid overheating until the polarity check has been made.

When measuring spill currents in balanced protection schemes a test should be made both with and without large secondary burdens.

19.5.4.2.3 *CT ratio*
The ratio of a set of three phase CTs is measured using the circuits as shown in Fig. 19.6. Current is passed through the primary conductors and measured on the test set ammeter, A1. The secondary current is measured on ammeter A2, and the ratio of the value on A1 to that on A2 should approximate to the ratio detailed on the CT nameplate and drawings. In this example, a special test tool is inserted into the relay case to bridge and carry the ammeter A2 connections. It is also often convenient to measure the secondary output from the CTs at links located, for example, in the circuit breaker marshalling kiosk in the switchyard or in the switchgear itself.

19.5.4.2.4 *Magnetizing curves*
It is recommended that an auto-transformer of at least 8 A rating is used when testing 5 A secondary-rated CTs. As the magnetizing current will not be sinusoidal a moving iron or dynamometer ammeter should be used. It is often found that 1 A or less secondary-rated CTs have a knee-point voltage higher than the

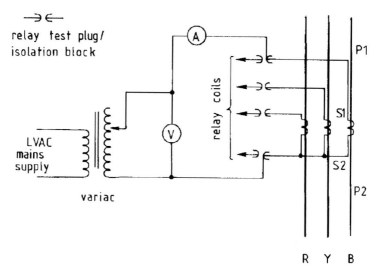

Figure 19.7   Test arrangement for CT magnetization curves

local mains supply. In such cases the required voltage may be obtained using an interposing step-up transformer. A typical test arrangement is shown in Fig. 19.7. Generally, the applied voltage should be raised slowly until the magnetizing current is seen to rise rapidly for a small voltage increase. This indicates the approximate knee point or saturation flux level of the CT. The magnetizing current should then be recorded for a few levels of secondary voltage as the voltage is reduced to zero.

19.5.4.3 *Secondary injection tests*

Secondary injection tests allow relay characteristics to be confirmed at the relatively low secondary CT and relay power levels. The purpose of such testing on site is not to repeat the factory tests but to confirm that the correct relay type has been installed and that it has not suffered any damage during transit. Test blocks and sockets should be specified for relay circuits as shown in Fig. 19.8 to facilitate the testing process. However, as a rule it is best to include as much of the secondary circuit from the CT to the relay as possible in the testing process. Relays may have non-linear current coil impedances and this can cause test waveforms to be distorted if the injection supply voltage is fed directly to the coil. Further, the presence of harmonics in the test waveform may affect the relay operating sensitivity and give unreliable results. High quality injection test sets are therefore designed to avoid such distortion. The injection transformer should incorporate secondary tappings corresponding to the relay current settings. This reduces harmonics in the test circuit which are due to saturation of the relay magnetic circuit. The sets offer control of the current supply by an adjustable series reactance which keeps the power dissipation

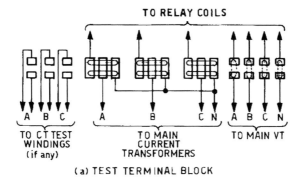

(a) TEST TERMINAL BLOCK

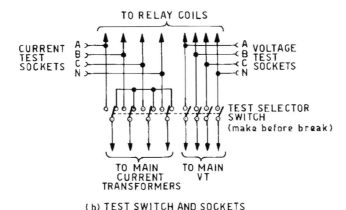

(b) TEST SWITCH AND SOCKETS

Figure 19.8   Test block and sockets

small and the equipment light and compact. Test sets use precision voltage and current meters and electronic timers as an integral part of the portable test set equipment.

A typical secondary injection overcurrent test set arrangement is shown in Fig. 19.9.

When using the test set the injection current should first be approximately set with the relay coil short circuited out in order to prevent excessive heating. The short is then removed for final current adjustment.

19.5.4.4 *Relay tests*

Reference should be made to the manufacturers' specific literature for site relay testing and commissioning. Some guidelines for specific types of relays are included here together with typical test sheets for reference purposes. Overall commissioning tests may well involve some duplication of results from both secondary and primary injection. The secondary injection tests should still be carried out as they act as a useful reference for future routine maintenance checks.

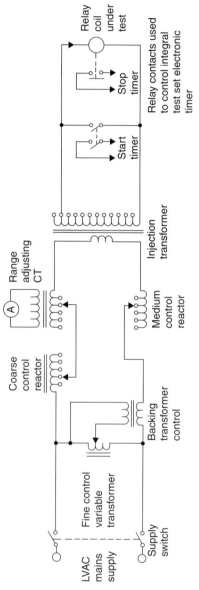

Figure 19.9 Typical circuit diagram of secondary injection overcurrent test set

19.5.4.4.1 *Instantaneous overcurrent relays*

- Measure the minimum current that gives relay operation over a range of current settings.
- Measure the minimum current at which resetting takes place.
- Check the correct functioning of the relay plug-bridge CT short circuiting device.

19.5.4.4.2 *IDMTL relays*

- Test as for instantaneous relays but also check the minimum operating and resetting currents at maximum, minimum and chosen settings.
- Measure the operating time at suitable values of current to check the time/current curve at two or three points with unity time multiplier setting.

19.5.4.4.3 *Differential relays*

- Test current is fed to the relay from one set of CTs at a time so that each 'end' is checked in turn.
- Careful attention must be paid to the characteristics of the matching CTs and the lead burdens between the CTs and the protection relay. Stabilizing coils and phase matching CTs must be checked and set to the required values.
- After proving each 'end' separately overall checks should be carried out under simulated primary 'in-zone' and 'out-of-zone' fault conditions in order to prove protection stability under through fault conditions. Often it is not possible to carry out an exhaustive range of tests and it may be necessary to compromise with, for example, three phase, earth fault and phase-to-phase 'out-of-zone' fault simulations to prove stability and possibly three phase and earth fault 'in-zone' tests to verify correct protection operation.
- Switching operations on the power circuit feeding each side of a transformer should be performed in order to prove the stability of the protection under magnetizing inrush conditions.
- Full load current must be passed through the main CT primary windings in order to check the stability of the full transformer differential protection scheme. If it is not possible to circulate sufficient current by altering the main transformer tap settings on two parallel transformers then a test generator will be required. The primary full load current may be made to circulate through the transformer by putting a three phase short circuit on one side of the transformer external to the protection. The machine should then be coupled to the other side of the transformer, the generator run slowly up to speed and the excitation slowly raised until full load current flows through the transformer primary windings. It should be appreciated that only 12% or less of the transformer winding-rated voltage will have to be generated to make full load primary currents circulate. If circulating current around two parallel transformers with different tap settings is not possible or if a generator is not available then protection stability checks will have to be checked when the

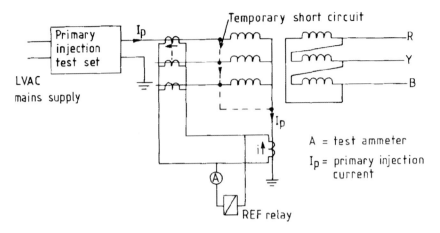

Figure 19.10   Transformer restrictive earth fault (REF) relay stability check

transformer is first put into service on load. It is advisable to measure the spill current through the relay operating coils during the load test with the tap changer set on its maximum or minimum tap setting (choose the greatest ratio away from the neutral ratio). This spill current, expressed as a percentage of the load current used in the test, indicates the minimum amount of bias the relay needs to maintain stability for through faults.

- Restricted earth fault protection scheme testing and commissioning should follow the same principles of 'in-zone' operation and 'out-of-zone' stability as for transformer-balanced differential protection schemes as described above. For example, if the line CTs are not installed in the bushings of the transformer apply a temporary short circuit connection on the transformer side of the CTs and circulate current through each phase and neutral CT in turn as shown in Fig. 19.10.

- Busbar protection schemes must be tested to confirm high speed operation, ensure absolute stability for external faults, and to give complete discrimination between the different protection zones. Secondary injection should therefore be performed on all relays with voltage and current measurements being taken for high impedance schemes. The correct CT overlap and any associated CT secondary circuit supervision relay scheme operation verified. Primary injection stability and operation tests must be considered essential for all possible fault conditions as part of the busbar protection commissioning procedures.

19.5.4.4.4 *Pilot wire protection*
- Identify pilot wire cores and check continuity, loop resistance and correct polarity of connections at each end of the pilot wire circuit. Note that systems

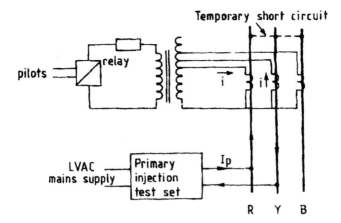

Figure 19.11   Pilot wire protection relay has fault sensitivity check.
Note: for earth fault sensitivity check circulate current through one CT only at a time

often require pilots to be crossed such that the cores are connected to different relay terminals at each end of the pilot circuit.
- Check pilot wire supervision system if fitted for correct identification of open and/or short circuit pilot wire conditions.
- If using rented pilots inform the telephone company of the times and types of tests you intend to perform in advance.
- If additional barrier insulation transformers are used check correct connections.
- Check overcurrent and earth fault sensitivity and correct summation transformer connections by primary injection through each CT primary in turn (see Fig. 19.11).
- Stability is best checked with load current flowing in the feeder since primary injection test sets will not normally have sufficient power to circulate the required current along the feeder lengths.

19.5.4.4.5 *Directional relays*
- The most satisfactory directional relay check is to use appreciable load current of known direction. Relay contact closing operation should be checked when the load current is in the operating direction and remain open when the current is in the reverse direction.
- A variac and phase shifting transformer may be used in order to vary the voltage magnitude and phase angle to the relay directional element. A variac is also required to vary current magnitude to the relay. The relay is then tested to ensure that it possesses a 180° directional characteristic at various levels of current and voltage and that the characteristic angle has been correctly set.
- Directional overcurrent relays are normally fed with current from the appropriate overcurrent phase and a polarizing voltage from the other two quadrature phases in order to obtain a 90° system angle connection. This results in

the required basic relay characteristic angle between relay current and voltage when operating at unity power factor. Care must be taken over the chosen characteristic angle since there is a danger of maloperation if used incorrectly. For example, the 60° or 30° connections should not generally be used with transformer feeders whereas theoretically the 90° connection will be satisfactory. The directional unit operates essentially instantaneously in modern solid state relays and in less than 10 ms for electromechanical types which is negligible compared to the overall relay operating time. Also check that the relay is rendered inoperative when the directional contact is open.

- Directional earth fault protection relays may be fed with residual earth fault current and polarized with either residual voltage or current. Therefore under normal load conditions these types of relay will not operate. The directional characteristics of the relay are therefore confirmed on site again by using a phase shifting test arrangement between the inputs to the relay.

19.5.4.4.6 *Distance relays*
- Generally the relay manufacturer's test set should be used.
- Distance protection relay settings are derived from the basic positive and negative sequence line data, line lengths, system data concerning the presence of parallel feeders and transformers within the reach of the relay and system fault level and load flow data. It is convenient to set up a small program on a portable computer/calculator to convert from the test set values and relay front panel settings to line R and X values such that the relay operating characteristic may be drawn out on graph paper during the commissioning tests.
- Test each phase in turn at the adopted setting and appropriate line angle in order to confirm the correct operation and accuracy of the relay.
- Check that the relay operates in the correct direction sense.
- When possible if the correct order of load and power factor can be obtained on the system complete a series of load tests to confirm the direction of operation.

19.5.4.4.7 *Protection channel tests*
- Tests should be carried out in conjunction with the telecommunication engineers.
- Check power line carrier transmitter frequency and power output into a correctly matched dummy load and then into the power line via the associated matching unit.
- Carry out power line or fibre optic channel attenuation tests both with the associated power feeder de-energized and earthed, isolated and then fully energized over the channel frequency spectrum and at the particular carrier frequencies to be used in the protection scheme. The tests should attempt to simulate phase-to-earth and phase-to-phase faults in order to ensure the signals are recovered at the receiving end under these conditions. The attenuation

Client	Order Reference	
Contractor	Date	
Substation	By	Checked

Circuit name and reference No.
CT type and reference Nos.
Relay type, manufacturer and reference Nos.

1) Insulation resistance (see separate test report ref....................................)

2) CT Magnetising Curves (see separate test report ref..............................)

3) Secondary Injection Relay Test Results

Plug or DIP switch setting	Overcurrent Minimum operating current		Plug or DIP switch setting	Earthfault Minimum operating current
	R	B		
50%				10%
75%				20%
100%				30%
125%				40%
150%				50%
175%				60%
200%				70%
No plug				No plug

Plug setting	Current	TSM	O/C Operating time	Plug setting	Current	TSM	E/F Operating time

4) Primary Injection

Injection Phases	Injection current	R	Y	B	E/F
R - Y					
Y - B					
Y - Earth					

5) Indication flags/lamps, trips and alarms
Flags and indicator operation..
Trips proved...
Correct relay operation annunciator alarms...

6) Protection arrangement (sketch)

7) Comments

Figure 19.12   IDMTL overcurrent and earth fault protection test report

should be from transmitter output at one substation end to receiver input at the other remote substation.

- Confirm that the channel attenuation falls within the specified attenuation budget.

Client		Order Reference	
Contractor		Date	
Substation		By	Checked

Circuit name and reference No.
CT types and reference Nos.
Relay type, manufacturer and reference Nos.

1) **Insulation resistance** (see separate test report ref................................)

2) **CT Magnetising Curves** (see separate test report ref................................)

3) **Secondary Injection Relay Test Results**

Plug or DIP settings   Stability without stabilising shunts   Stability with stabilising shunts

4) **Primary Injection**
Injection Phases   Injection current   Fault setting
   R - Y
   Y - B
   Y - N
   N - Y

5) **Indication flags/lamps, trips and alarms**

Flags and indicator operation..
Trips proved...
Inter trips proved...
Correct relay operation annunciator alarms..

6) **Protection arrangement (sketch)**

7) **Comments**

Figure 19.13   Transformer restricted earth fault (REF) protection test report

- Confirm power line carrier line traps, line couplers and filters are correctly matched and tuned if attenuation falls outside recommended values. Note that power line carrier traps have very low 'Q' values in comparison to normal radio frequency coils. Check the signal strengths being received on the remote side of line traps to assess their attenuation at the power line carrier frequency.

Client	Order Reference	
Contractor	Date	
Substation	By	Checked

Circuit name and reference No.
CT type and reference Nos.
Relay type, manufacturer and reference Nos.

1) Insulation resistance and continuity (with all interconnecting wiring and interposing CT connected)
    LV1 CT secondary
    HV CT secondary
    Relay pilots
    Relays

2) CT Magnetising Curves (see separate test report ref..............................)

3) Secondary Injection Bias Characteristic Tests
       % Bias   Bias Tap   Operating current
                                       R   Y   B

4) Primary Injection - Ratio / balance checks
4.a) CT Ratio and Phase compensation (see protection arrangement sketch below)
    HV CT ratio   HV CT phase connection
    LV1 CT ratio   LV1 CT phase connection

4.b) Interposing CT Ratio and Phase compensation (see protection arrangement sketch below)
    Interposing HV CT ratio   Interposing HV CT phase connection
    Interposing LV1 CT ratio   Interposing LV1 CT phase connection

  Note:- An LV2 connection applies to transformers with an active tertiary winding with external connections
       supplying an auxiliary load.

4.c) HV to LV1 balance (with three phase through fault test connection and transformer on nominal tap)
    HV primary amps
    HV secondary amps
    LV1 primary amps
    LV1 secondary amps

    Relay pilot "spill" amps   R phase........   B phase........   Y phase ......
    Relay bias amps

Note:- For a three way differential protection scheme test both LV circuits (LV1 and LV2) under the through fault
    test condition with transformer on nominal tap.

5) Primary Injection - Fault settings (primary injection of appropriate CTs associated with the scheme)
    LV1 E/F primary amps setting
    LV1 phase fault primary amps setting

    HV E/F primary amps setting
    HV phase fault primary amps setting

6) Indication flags/lamps, trips and alarms
Flags and indicator operation...
Trips proved..
Correct relay operation annunciator alarms.......................................

7) Protection arrangement sketch and comments

Figure 19.14   Transformer bias differential protection test report

- Check receiver sensitivity and selectivity. Add an adjustable attenuator in the signal path and confirm that a security margin exists. A new line may have clean insulators, etc., and be less noisy than after a few years in service. A margin of at least 10 dB should therefore be available.
- Check the system under loss of AC supply and especially during changeover from mains AC to standby DC supply.

19.5.4.5 *Test sheets and records*

Figures 19.12–19.14 are typical proforma relay setting schedules and site commissioning test sheets for overcurrent and earth fault IDMTL relays and transformer differential protection.

APPENDIX A:

COMMISSIONING TEST PROCEDURE REQUIREMENTS

1 Programme of activities
   1.1   Physical Limits, Description and Role
   1.2   Needs and Constraints
         (Environmental Requirements, Services, Spares, Consumables and Test Equipment)
   1.3   Performances to be Achieved
         (Functional Tests and Acceptance Criteria)
   1.4   Task List
         (Including Test Report Format)
   1.5   Sequence of Events for Tests
   1.6   Personnel Involved in Tests and Organogram
   1.7   Provisional Test Planning

2 Test procedures
   2.1   Needs and Constraints
         (Temporary devices such as earthing sticks and checklist for application and removal, drawing checklist, etc.)
   2.2   Safety
         (Permit to Work, Applicable Safety Rules, Specific Safety Precautions)
   2.3   Sequence of Events for Tests
         (Specific Test Details)
   2.4   Measurement Instruments Used
         (List of Test Gear, Serial Numbers and Confirmation of Calibration)

3 Test Reports
    3.1   Purpose and Objectives
    3.2   Analysis and Presentation of Results (Proforma Test Sheets to be used where possible)
    3.3   Conclusions and Follow-up

APPENDIX B:

DRAWINGS, DIAGRAMS AND MANUALS

1 Construction Drawing Routing Scheme (Fig.19.15)

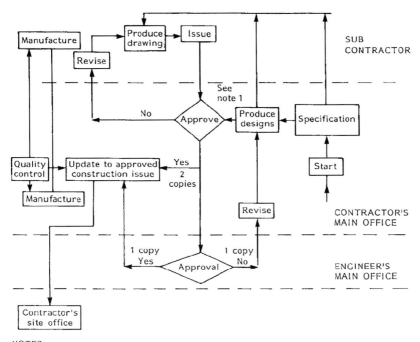

NOTES

1   The contractor is responsible for incorporating sub–contractor's drawings into the detailed designs and also for ensuring designs fully comply with the specification requirements.

Figure 19.15   132/11 kV substation and ancillary distribution works drawings, diagrams and manuals construction drawing routing scheme

2 'As-Built'/Record Drawing Routing Scheme (Fig. 19.16)

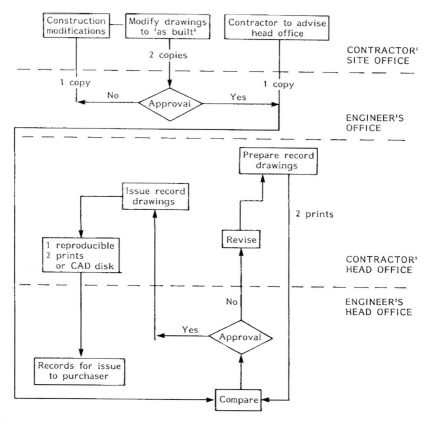

Figure 19.16   132/11 kV substation and ancillary distribution works as built/record drawing

20 Electromagnetic Compatibility

20.1 INTRODUCTION

This chapter describes how to apply the latest Electromagnetic Compatibility (EMC) Standards to Transmission and Distribution Equipment.

The subject of EMC comes under EMC Directive 89/336 EEC for all European Union countries. This is being replaced by a new version, 2004/108/EC which is similar in many respects. In the USA the Federal Communication Commission (FCC) standards apply, elsewhere European or IEC standards are commonly used. Both sets of standards are similar and many of the European standards are identical to the corresponding IEC standards.

In brief, the EU EMC directive requires that all electrical and electronic equipments constructed for use in any EU country must ensure:

- The electromagnetic disturbance generated by the apparatus does not exceed a level allowing radio and telecommunication's equipment and other apparatus to operate as intended (i.e. to limit the electromagnetic noise from the equipment).
- The apparatus has an adequate level of intrinsic immunity to electromagnetic disturbance in the intended environment enabling it to operate satisfactorily (i.e. to ensure adequate screening and noise immunity in the operational environment).

The effect of conducted electromagnetic disturbances on power systems is discussed separately in Chapters 24 and 25 (Power quality).

Many good theoretical and descriptive books related to EMC already exist (see References 1 to 9). This chapter describes the application of the theory to real practical transmission and distribution examples and typical measurements that may be made on site.

Table 20.1   UK national standards

Document number	Document title
BS 5049[a]	Radio interference characteristics of overhead power lines and high voltage equipment
BSEN 61000	Electromagnetic compatibility – British standard is identical to EN 61000 and based on IEC 61000[c]
BSEN 50121	Railway applications – Electromagnetic compatibility
NRPB-GS11[b]	Guidance as to restrictions on exposures to time varying electromagnetic fields and the 1992 recommendations

Notes: (a) BS5049 part 1 to 3 are equivalent to CISPR parts 1 to 3.
(b) NRPB has subsequently issued new advice in document 2 of NRPB Vol. 15 (Advice on limiting exposure to electromagnetic fields). This recommends adoption CISPR exposure guidelines issued in 1998[10] which in turn recommend a reduction factor of 5 in the basic exposure restrictions for members of the public compared with workers.
(c) BSEN 61000 is very broad. See Table 20.3 for further details.

Table 20.2   Generic Standards for apparatus

Document number	Document title
EN 61000-6-1	Immunity for residential, commercial and light-industrial environments
EN 61000-6-2	Immunity for industrial environments
IEC 61000-6-3	Emission standard for residential, commercial and light-industrial environments
EN 61000-6-4	Emission standard for industrial environments
EN 61000-6-5	Immunity for power station and substation environments

Note: The EN standards are identical to the corresponding and similarly numbered IEC standards.

20.2 STANDARDS

Tables 20.1–20.3 list current EMC national and international standards which are relevant to transmission and distribution projects. It is the requirement of the EU Commission that all standards will be common throughout the Community. To this extent various committees and working parties have been convened and tasked to meet this requirement. Where such common standards are not in existence, national standards will be used.

Engineers concerned with substations in a railway environment should note that the applicable series of standards (EN 50121 EMC in railway applications) has not been harmonized. The implications of this are discussed in Section 20.3.

Although the effects of non-ionizing electromagnetic radiation (NIEMR) on workers does not as yet come under the auspice of an EU Directive (the EMF Directive 2004/104/EC has been published and will come into force in the future), the list of standards includes reference to the National Protection Laboratory guidance document for Exposure of Humans to Electromagnetic

Table 20.3   International standards (for equipment)

Document number	Document title
EN 55022[a]	Limits and Methods of Measurement of Radio Interference Characteristics of Information Technology
EN 55011[a]	Radio Frequency Limits and Methods of Measurement of Electromagnetic Disturbance Characteristics of Industrial, Scientific and Medical Radio Frequency Equipment
EN 55024[a]	Information Technology Equipment. Immunity Characteristics. Limits and Methods of Measurement
IEC 61000-1[b]	EMC – General
IEC 61000-2[b]	EMC – Environment
IEC 61000-3[b]	EMC – Limits
IEC 61000-4[b]	EMC – Testing and Measurement Techniques
IEC 61000-5[b]	EMC – Installation and Mitigation Guidelines
IEC 61000-6[b]	EMC – Generic Standards
CISPR 18	Radio Interference Characteristics of Overhead Power Lines and High Voltage Equipment

Note: (a) EN 55011 is based on CISPR 11, EN 55022 is based on CISPR 22 and EN 55024 on CISPR24.
(b) All the IEC 61000 series of standards are published in a number of subsidiary parts (e.g. 61000-3-7 EMC-limits-Assessment of emission limits for fluctuating loads in MV and HV power systems – basic EMC publication).

Fields. As the measurement of the fields to which these documents refer is similar in some respects to those related to EMC testing, this chapter will also address the relevance of these requirements.

Guidance on the choice and applicability of generic EMC standards is available in CENELEC Report R110-002.[11]

20.3 COMPLIANCE

The EMC directive has legal implications on both its implementation and the meeting of the standards set. Substations are considered as fixed installations as defined in the Guidelines to the Directive [8]. Thus, where the directive applies, the transmission and distribution engineer has a liability to ensure that compliance with the directive is met and maintained throughout the operating life of the apparatus. In many other territories compliance would be seen as good practice.

Compliance with product specific or generic standards, will normally be achieved through testing supplied equipment to harmonized EMC standards, with the provision of an EMC 'Declaration of Conformity' (DoC). Note however that for railway substations compliance is necessarily through a Technical Construction File (TCF) as EN 50121 is not a harmonized standard.

For a complete installation, proof of compliance with the required criteria, as summarized in Section 20.1, may be achieved by preparing firstly an 'EMC Management Plan', which will show how the question of meeting EMC requirements will be organized, worked out and documented throughout the design, construction and operation of the project. Secondly, an 'EMC Control Plan' will include DoC's, TCFs, risk analyses measurement results, etc. In general, installations can be proven to comply with the essential protection requirements of the EMC Directive where it can be demonstrated that compliant products are installed to good engineering practice.

20.4 TESTING

20.4.1 Introduction

Testing of apparatus can fall into two distinct categories:

- Individual apparatus tests.
- In-site apparatus/system testing.

The concepts of individual apparatus testing, for both emissions and immunity, are well defined in the various test standards. Table 20.2 lists the Generic standards which are used to conduct these tests. As the requirements are accepted internationally, the methods and procedures are not discussed in this chapter. Suffice it to say that testing is required to be performed in test houses that have been audited and accredited as suitable to conduct the tests identified in the relevant standards.

The Generic Standards are to be used where no product standard exists. In many cases this situation will arise within typical power and distribution designs. Depending on the installation location of the system being designed, the responsible engineer will have to decide which Generic Standard he or she elects to follow.

Within these standards there are references to the general international standards which are listed in Table 20.3.

Once the different equipments have been approved against their individual test standards, it is not acceptable for the design engineers to conclude that all EMC responsibilities have been met. It is the interconnection of the various equipments together which will still require a confirmatory assessment and testing. As all systems and their associated interconnections are different, there may be existing conditions which are different to those considered in the standards.

However, with the design engineer's knowledge of the standards required for his various system equipments, judgemental on-site testing can be performed to ensure the requirements of the EU Directive, as paraphrased in Section 20.1, are achieved.

20.4.2 Magnetic field radiated emission measurements

In the power distribution environment, magnetic field emission measurements are normally conducted over the frequency range 20 Hz to 50 kHz where

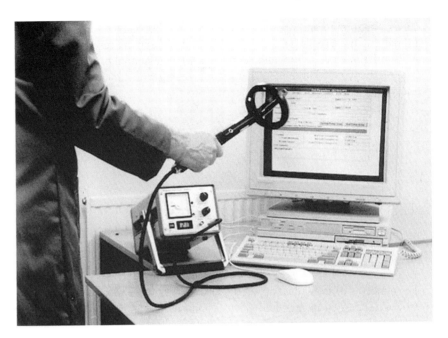

Figure 20.1   Magnetic field portable measurement equipment

installation design and health effects are considered. As a transmission and distribution network could be spread over large distances it is necessary to evaluate the 'worse case' condition. This is normally performed using hand-held, battery-driven equipment. Typical equipment is shown in Fig. 20.1. It should be noted that EMC product approval magnetic field emission requirements are limited to lighting equipment and industrial equipment containing wanted radio frequency sources and special test facilities or test ranges are required.

The test engineer would move around the area under investigation with the antenna probe held in a fixed orientation. The probe consists of three orthogonal coils, so the displayed information is the sum of the fields being radiated at that particular point irrespective of their frequencies. The reading displayed on the unit is calibrated and shown in dBpicoTesla and is therefore directly related to the standard. The design engineer can deduce two criteria from these results:

- If the displayed information is 40 dB, i.e. 100 times, below the design level, he or she may decide that further testing is not required.
- If the displayed level is closer to the design level than the 40 dB, or if more definitive measurements are required, he or she can place the specialist measurement equipment in the defined 'worse case' area.

Figure 20.2 shows a typical layout of test equipment suitable for the detailed measurements.

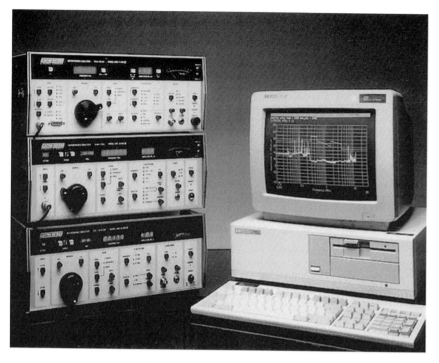

Figure 20.2   EMC emissions test equipment layout. Highly accurate receivers and frequency/amplitude plot recorded on computer

Measurements are made whilst the loop antenna is fixed in each of its polarizations. The receiver is then stepped through the defined frequency range. Data, on the levels of radiated energy for each frequency step, are gathered and stored within the computer. The data can be retrieved and presented in graphical format for analysis against the design limits at a later date. A typical plot of field measurements with no substation equipment energized is shown in Fig. 20.3a. Figure 20.3b shows the radiated fields with the substation equipment energized.

It must be noted that, although for convenience the peak levels of radiated energy are normally checked against the specification or design limit, measurements related to the NRPB document have to be calculated and weighting factors taken into consideration for the different harmonic frequencies involved.

20.4.3 Electric field radiated emission measurements

Power distribution and transmission technologies involve frequencies normally from 50 Hz and up to the 25th harmonic but some associated electronic equipment may contain high frequency sources that could cause interference to radio systems. The EU Directive is, in particular, concerned with the interference to communication equipment and requires measurements in the frequency range

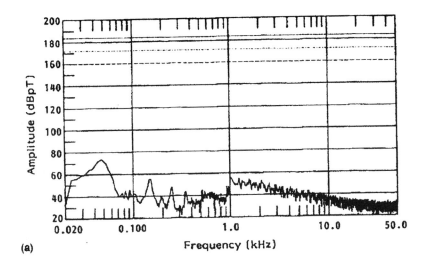

(a)

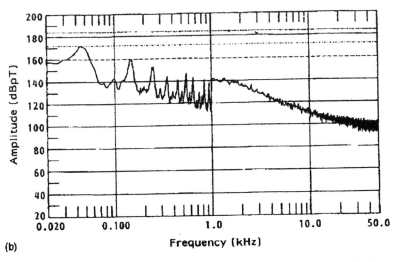

(b)

Figure 20.3   (a) Ambient magnetic field measurement. (b) equipment magnetic field measurement

30–1000 MHz. Also, with the development of more sophisticated electronic control equipment in the power industry, the reduction of radiated higher frequency electric fields becomes more critical; e.g. clock frequencies and fast leading switching pulse waveform edges used by microprocessor equipments must not be degraded. In substations or power stations, it is more important to ensure that electronic systems have adequate immunity to the local disturbance fields and there is a lower level of requirement to control emissions from supplied equipment and switching devices.

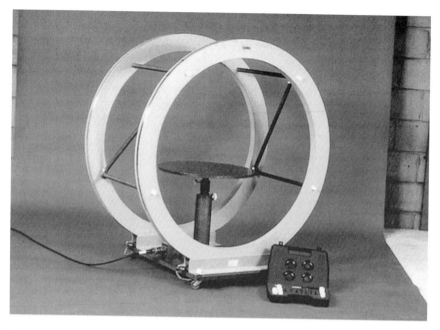

Figure 20.4   Helmholtz coils (for creating an environment free of the earth's magnetic field)

As with the measurement of magnetic fields, initial ambient electric field data may be gathered using portable equipment. Care must be taken in the use of this equipment in order to avoid the human body affecting the results. Use of this type of equipment for anything other than 'worse case' location identification (e.g. for product compliance with relevant standards), can only be performed with the antenna probes mounted on wooden tripods on a calibrated test site prior to results being taken.

Having identified the 'worse case' locations, specialist equipment testing can now be performed. Whilst equipment suitable for these tests is similar to that used for the magnetic field testing, the main area of change is in the type of antennae. The antennae used relate to the frequency bands in which the testing is to be performed and use of an antenna outside its operation frequency band must not be carried out.

Data gathered during these tests can be evaluated later and checked against the design limits. Typical plots obtained during the testing would be similar to those shown in Fig. 20.3 with amplitude measured in dBμV/m against frequency.

20.4.4 Conducted emission measurements

The levels of harmonic current emissions and flicker are set out in IEC 61000-3-2 and 61000-3-3. These have been issued as harmonized European standards

EN61000-3-2 and EN611000-3-3 and are being incorporated as part of the EMC directive. The consequent levels of harmonic and unbalance voltages which can be reflected back into the supply network are controlled by the different distribution networks. Typically, IEC 60034-1 (rotating electrical machines) allows for no more than 1% unbalance on polyphase voltage systems and that the instantaneous peak harmonic voltage is less than 5% of the fundamental peak voltage. See also Chapters 24 and 25.

The EU Directive defines the conducted transients which an apparatus is allowed to generate into the local environment. Table 20.3 lists the appropriate international standards. These ensure the general requirements, as paraphrased in Section 20.1, are achieved.

The concepts of these tests are well defined in the various test standards. As these requirements are accepted internationally, the methods and procedures are not discussed in this chapter. Suffice it to say that testing should be performed by persons or test houses who will have been audited and approved as suitable to conduct the tests identified in the relevant standards.

20.4.5 Immunity testing

Once the apparatus has been installed within the confines of the design engineer's system, it becomes more difficult to perform radiated and conducted immunity measurements. It may be argued that, providing the equipment operates within its own confines and under all definable conditions, further testing is not required. Problems of taking this argument to its final conclusion occur when systems are shown to be susceptible in operation. It may now be necessary to conduct localized immunity testing in order to satisfy legal and contractual requirements.

Testing at this level can only be achieved by using the principles set out in the various product EMC standards. With conducted immunity testing, provided the relevant cables can be accessed, test methods and procedures identified in these documents can be followed.

Radiated immunity testing presents a larger problem. It is forbidden, by the Wireless Telegraphy Act, to transmit signals into the environment unless a licence is obtained. Further, if a licence is granted, then it will only cover a defined and stipulated frequency, which is of no use to the design engineer. Other test methods which will approximate to the radiated immunity testing must be used. One such method, which was developed for the aircraft industry so as to overcome this particular problem, is known as Bulk Current Injection. In this method relatively low levels of the interference signals can be injected onto the system cable form with the use of a current transformer. Similarly, the interference signal may be coupled into the cable form using a galvanic connection. These test methods are covered in the associated apparatus test standard IEC 61000-4-6.

20.5 SCREENING

20.5.1 Introduction

In the design of any apparatus and the subsequent system, screening will be used to overcome potential but possible underevaluated EMC concerns. Care in the use of the screening of both cables and equipment must be taken. Poor engineering of screening may cause more problems than they resolve. Textbooks have been written which investigate the theory of screening and provide calculated examples of achieving design limits (see References 3 and 7).

General points which should be considered in deciding on the level and type of screening are as follows.

20.5.2 The use of screen wire

All wires including the screens are at some point terminated. The termination of the screen will be critical to its operating efficiency. Control and communication cable screens should only be connected at one end, thereby reducing the probability of circulating currents. Where this conflicts with safety 'step and touch' potential requirements the bonding of screens may take place within the run lengths of the cables and/or secure 'gapping' of the screen may be incorporated (Fig. 20.5). If the cable is terminated at a multiway connector then the connector should be enclosed in a metal housing; this housing acts as a continuation of the cable screening. Where screened connectors cannot be used and wire 'tails' are used to terminate the cable screens the tails *must* be kept short, less than 3 cm, and taken to known clean earth points (Fig. 20.5).

20.5.3 The use of screen boxes and Faraday enclosures

Any aperture in the box will, unless properly screened, degrade the effectiveness of the box. Areas of particular note include lids which must be fitted with EMC gaskets and close spacing of fixing screws (Fig. 20.6). The entry ports of any cables are of importance. Where screen cables are being brought directly into the unit, continuity of the screening at the entry gland must be maintained by a 360 peripheral glanded connection.

Where unscreened cables are being used each wire must be taken into the box via EMC filters (Fig. 20.6a).

Figure 20.6b shows the schematic diagrams of three typical in-line, or feedthrough, filters used in EMC. The general rules that can be taken in the design of transmission and distribution projects are that:

- In-line feed through capacitors will short circuit high frequency signals to earth.

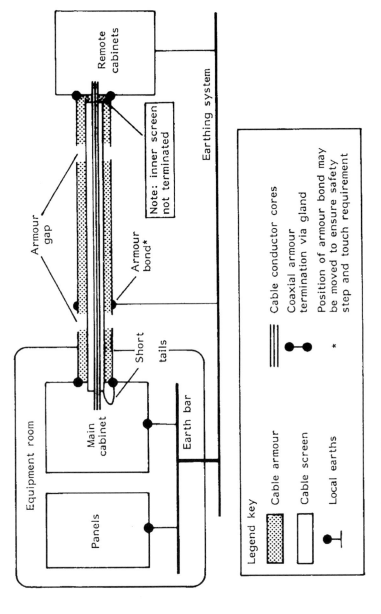

Figure 20.5   Screened and armoured screen earthing arrangements

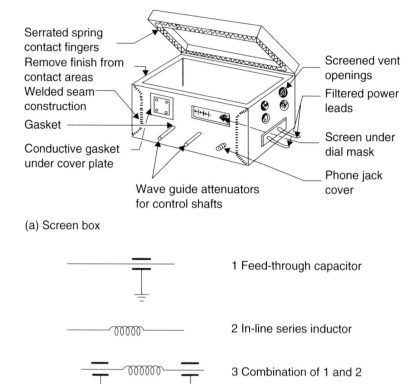

Serrated spring contact fingers
Remove finish from contact areas
Welded seam construction
Gasket
Conductive gasket under cover plate

Screened vent openings
Filtered power leads
Screen under dial mask
Phone jack cover

Wave guide attenuators for control shafts

(a) Screen box

1 Feed-through capacitor

2 In-line series inductor

3 Combination of 1 and 2

(b) In-line and feed-through filters

Figure 20.6   (a) Screening and filtering of boxes. (b) typical EMC suppression filters

- In-line inductors will act as high impedances to the fast rise edges of signal and therefore prevent their transmission into the culprit circuitry. The high frequency performance of the inductor may be limited by the self-capacitance of the windings.

20.5.4 The use of screen floors in rooms

If a substation design requires the use of screen metal false floors within a control room containing electronic equipment, it is essential that the design engineer is conversant with the reasons for this type of floor. If it is required simply to provide a safety earth for both the users and the equipment, then the construction of the floor need only take into account its need to maintain low frequency, low impedance paths. This will include the installation of earth connections to the floor with particular care being taken to ensure continuity is

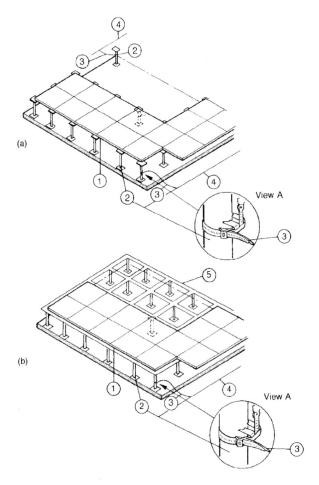

Figure 20.7   (a) Typical stringerless screen floor construction. (b) typical stringer screen floor construction. Key: 1, screen floor tiles; 2, mounting pillars; 3, earth connection wire; 4, main earth wire; 5, metal support stringers

maintained when screen tiles are removed due to maintenance or installation of new equipments. The same care needs to be taken where tiles are cut to allow for the passage of cables and air ducts. For this type of floor a 'stringerless' construction may be used where the tiles are mounted on individual corner pillars (Fig. 20.7a). View A shows a typical method of connecting the pillars to an earth termination wire. Maintenance may be kept to a minimum with standard cleaning of the tiles and pillar heads only when a tile is removed.

Where the floor is to be used to screen the equipment in the room from any conducted or re-radiated noise associated with cables under the floor, then different design criteria are required from those described above. It is recommended that such a requirement will force the design engineer into using a 'stringer' floor construction (Fig. 20.7b). Should the above 'stringerless' construction be used then each pillar must be taken to the common ground point. The area above the screen floor should now be considered as the interior of a screen box or Faraday enclosure. Hence cables entering this area must be glanded and filtered in a manner as described above. Where air ducts are set in the floor, the apertures must be constructed as an EMC wave guide. Figure 20.8 shows a 'honeycomb' construction of a typical ducting aperture. These have known filtering characteristics and are ordered and installed such as to maintain EMC security over the frequency ranges being considered in the room's design parameters.

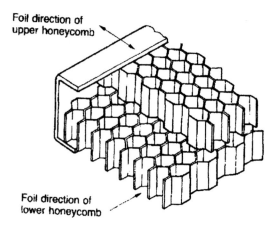

Figure 20.8   Typical honeycomb ducting construction

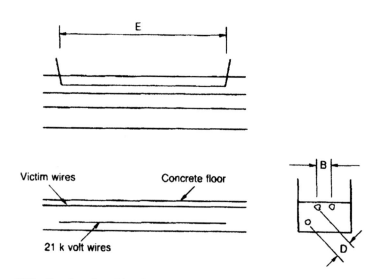

Figure 20.9   Drawing of corridor showing cable layout

Absorption losses can be calculated from

$$A_{dB} = 3.34E^{-3} \times t \times \sqrt{(f \times G \times \mu)} \text{ dB}$$

Reflective losses can be calculated from

$$R_{dB} = 20 \log_{10} ((0.462/r)\sqrt{\frac{\mu}{f \times G}} + 0.136 \times r\sqrt{\frac{f \times G}{\mu}} + 0.354) \text{ dB}$$

where μ = permeability of shielding material relative to copper
t = thickness of shielding material (mils)
r = distance from source to shielding material (mm)
f = frequency in Hz
G = shielding material conductivity relative to copper

Many shielding manufacturers, and in particular manufacturers of shielded wires, will have standard graphs for the above calculations and these should be used whenever possible.

20.7 CASE STUDIES

20.7.1 Screening power cables

20.7.1.1 *Concern*

The general cable configuration is shown in Fig. 20.11.
Due to system design criteria the following conditions have been forced on the installation of the equipment:

- Two sensitive signal cables transmitting information to nearby receiver coils are placed within 1.0 metres of a three phase 21 kV cable for a distance of

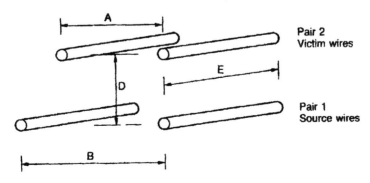

Figure 20.11   Cable layout for calculation

10 metres. As the signal cables are operating on the principle of radiating energy signals to the receiver coils, they cannot be screened or in any other way protected from the noise. The operational frequency band of the victim system is 7 kHz to 12 kHz.

- The 21 kV power cable is feeding multiple systems including noise generators such as pumps, motors and transformers.

20.7.1.2 Requirements

Noise which may be present on the 21 kV cable must not be induced on the signal carrying wires or radiated such that it may interfere with the receiver coils.

20.7.1.3 Solution

Due to the low operational frequency of the signal cables and to ensure the above requirement is achieved, it is necessary to place the 21 kV cable in a solid metal pipe. A pipe having an internal diameter suitable for the 75 mm 21 kV cable and having a wall thickness of 7. 6 mm was found to be available.

20.7.1.4 Known factors and assumptions

The following initial data needed to be determined and, as this is an evaluation to ensure proposed solutions are acceptable, engineering assumptions are made.

1. What thickness of steel tube will be used?
   7.6 mm – Due to existing availability
2. What is the sensitivity of the receiving system?
   125 μA/m at 7–12 kHz
   This sensitivity is obtained from the victim's data sheets.
      It is decided that any induced field at 10 kHz should not exceed 1/100th, 40 dB, of this value.
3. What is the minimum current in the signal cables?
   150 mA
   This figure is obtained from the victim's data sheets.
4. How far are the signal cables from the 21 kV cable?
   0.945 m
   Obtained from detailed system drawings.
5. How far apart are the signal cables?
   1.7 m
   Obtained from detailed system drawings.
6. What is the harmonic current at 10 kHz associated with the 21 kV cable?
   As no details are available use 0.1% of full load

7. What is the impedance of the signal cables?
   50 Ω, 5 Ω and 0.5 Ω
   The victim's data sheets showed the minimum impedance to be 50 Ω. Calculations at 5.0 Ω and 0.5 Ω have been used to ensure a safety margin.
8. Due to the close proximity of the interference source with the victim it is assumed that the two sets of cables are parallel. This allows for standard equations to be used in the initial calculations.

20.7.1.5 *Proof of suitability for conducted noise in victim wires*

The mutual inductance between sets of long parallel pairs of straight wires can be calculated from the equation provided in Section 20.6.3:

$$M = \frac{(2 \times A \times B \times E)/10^7}{D^2} \text{ Henrys}$$

From the above equation it is possible to calculate the voltage and current induced in Pair 2, i.e. the victim wires:

$$V = \frac{M \times I_1}{T} \times 10^9 \text{ volts}$$

$$I_2 = \frac{V}{Z} \text{ amps}$$

where I_1 = maximum current flowing in Pair 1
  T = rise time of the maximum current flowing in Pair 1 in nanoseconds
  Z = impedance of the Pair 2 victim circuits in ohms

Using the above equations and data the worse case current can be estimated at 10 kHz:

At $Z = 50$ ohms $I_2 = 68$ μamps
At $Z = 5$ ohms $I_2 = 680$ μamps
At $Z = 0.5$ ohms $I_2 = 6.8$ mamps

Thus the worse case induced current at 10 kHz is 6.8 mamps. The acceptable induced current at 10 kHz is 1.5 mamps.
Thus the attenuation required by the proposed 7.6 mm wall steel pipe at 10 kHz is:

$$20 \log_{10} \frac{1.5}{6.8} = -13 \text{ dB}$$

The negative sign is applied to indicate the figure is to be applied as a reduction to the incident field.

The attenuation must be provided by the sum of the rejection and absorption losses where:

μ = permeability of steel = 1000
t = the thickness of the steel in mils = 300
r = the source to shield distance in mm = 35.56
f = the frequency in Hz = 10 000
G = the conductivity relative to copper = 0.1

Reflection losses can be calculated as:

$$R_{dB} = 20 \log_{10} ((0.462/r)\sqrt{\frac{\mu}{f \times G}} + 0.136 \times r\sqrt{\frac{f \times G}{\mu}} + 0.354) \text{ dB} =$$

$$R_{dB} \approx 0 \text{ dB}$$

Absorption loss can be calculated as:

$$A_{dB} = 3.34E^{-3} \times t \times \sqrt{(f \times G \times \mu)}$$

$$A_{dB} = 1002 \text{ dB}$$

Therefore the total attenuation provided by the steel pipe at 10 kHz will be 1002 dB.

Thus using a steel pipe with a wall thickness of 7.6 mm provides sufficient attenuation to meet the objectives defined above.

20.7.1.6 Proof of suitability for radiated noise in victim receiver coils

The magnetic field set up by the 21 kV cable at 10 kHz is:

$$H = \frac{I}{28 \times \pi \times r} \text{ amps/metre}$$

(Formula taken from references)
where r = the radial distance in metres between the 21 kV cable and the receiving coils (this was found to be 1.09 metres when measured on design drawings)

I = the maximum current flowing in the 21 kV cable at 10 kHz

Thus $H = 43.7$ mamps/metre.
The acceptable level of magnetic field strength is:

1.25 μamps/metre

The level of shielding required is:

$$S_{dB} = 20 \log_{10} \frac{1.25 E - 6}{43 E - 3}$$

$$= -91 \text{ dB}$$

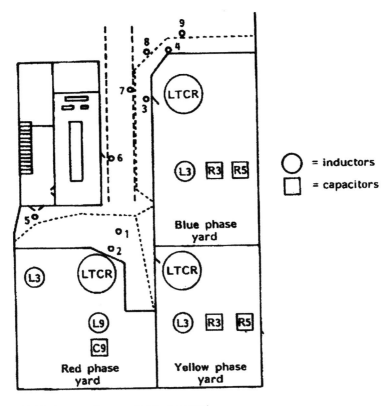

Figure 20.12   Layout of power distribution yard

Thus using a steel pipe with a wall thickness of 7.6 mm provides sufficient attenuation (1002 dB) to achieve the objective of protecting the receiving coils from the 21 kV noise.

20.7.2 Measurement of field strengths

The case study investigated in this section deals with the initial measurement of magnetic fields within the confines of a substation switchyard. Identical methods could be used for the measurement of both magnetic and electric field strengths in any area, and is particularly useful in the measurement of field strengths produced by long distribution cables.

Figure 20.12 shows the basic layout of the switchyard containing static VAr compensation air core reactors and capacitor banks. It is required to determine the levels of magnetic field around the various perimeter fences to ensure those levels prescribed by the NRPB guidelines are not exceeded.

Using the portable test equipment shown in Fig. 20.1 the engineer determines the points where the maximum total field is monitored. Due to the wideband operation of the equipments it must be remembered that the measurements show the fields at the fundamental and all harmonic frequencies. Points 1 to 4 in the figure are typically where the maximum field strengths will be observed, this being due to the large inductors within the yard and their location with respect to the fencing.

In the particular case study the levels monitored were:

Position	Level mT
1	0.54
2	0.95
3	0.38
4	0.40
5	0.035
6	0.032
7	0.088
8	0.139
9	0.122

The measurement observed at position 2 was above the 0.8 mT reference level proposed by the NRPB (Note: current NRPB levels may be lower). With this apparently high level it is considered advisable to take detailed measurements at point 2 using the specialist measurement equipment. Figure 20.3b shows the graph obtained from these measurements. A number of issues can be determined from this result. Firstly, it is predominantly the fundamental, 3rd to 17th harmonics which are being radiated. Secondly, none of the levels reach the 0.8 mT level, i.e. 178 dBpT.

The NRPB guidance document recommends that:

Where exposure occurs at more than one frequency, the exposure can be considered to be less than the effective reference level if

$$\Sigma R_f \leqslant 1$$

where R_f is the ratio of the measured value to the reference level in the appropriate unit at the frequency f.

Using the information taken from Fig. 20.3b, and using the above equation, it can be shown that the summation is approximately 2.3 mT, i.e. an unsatisfactory condition. Even if the fundamental frequency component is removed from the equation the total field density is above 1.8 mT. Restrictions can be placed on the access of personnel within the area such that they are kept in a narrow corridor around the building. Further measurements were taken at position 5 with the results shown in Fig. 20.13. If the total field density is now recalculated we find the answer is approximately 0.11 mT, i.e. a satisfactory condition.

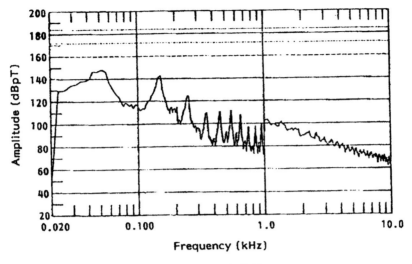

Figure 20.13   Measurement of equipment magnetic field

REFERENCES

1. Donald R. J. White, *Electrical Noise and EMI Specifications*, Vol. 1, 1971.
2. Donald R. J. White, *Electromagnetic Interference Test Methods and Procedures*, Vol. 2, 1980.
3. Donald R. J. White, *Electromagnetic Interference Control Methods and Techniques*, Vol. 3, 1973.
4. Donald R. J. White, *Electromagnetic Interference Test Instrumentation Systems*, Vol. 4, 1980.
5. Donald R. J. White, *Electromagnetic Interference Predictions and Analysis Techniques*, Vol. 5, 1972.
6. Donald R. J. White, *Electromagnetic Interference Specifications, Standards and Regulations*, Vol. 6, 1975.
7. Donald R. J. White and Michel Mardiguian, *EMI Control Methodology and Procedures*, 1985.
8. C. R. Paul, *Introduction to Electromagnetic Compatibility*, 1992.
9. H. W. Ott, *Noise Reduction Techniques in Electronic Systems*, 1998.
10. European Commission, *European Commission Guidelines on the application of council directive 89/336/EEC 3 May 1989 on the approximation of the laws of the member states relating to electromagnetic compatibility and Amendments 92/31/EEC, 93/68/EEC, 93/97/EEC*, 1997.
11. CENELEC, *Guide to Generic standards*, Report R110–002: 1993
12. ICNIRP, *Guidelines for limiting exposure to time-varying electric, magnetic, and electromagnetic fields (up to 300°GHz)*, Health Phys, Vol. 74, 1998, pp. 494–522.

21 Supervisory Control and Data Acquisition

21.1 INTRODUCTION

Complex and dispersed power systems necessitate large manpower resources for control, maintenance and management functions. Such resources may be reduced or employed more efficiently with the help of computer systems. This chapter describes the basic interfaces, software and hardware necessary for transmission and distribution power system supervisory control and data acquisition (SCADA). Programmable logic controllers (PLCs) and substation bay controllers may be used for local automatic control functions. These controllers are described together with a practical interlocking application example. This chapter also introduces traditional power line carrier communication and signalling methods. Communication via fibre optic links is mentioned separately in Chapter 12. This chapter goes on to describe a centralized power transmission network control system and covers the very important subject of software management. Such management is essential if software development is to be achieved within quality, time and budget constraints.

21.2 PROGRAMMABLE LOGIC CONTROLLERS (PLCs)

21.2.1 Functions

Programmable logic controllers (PLCs) were initially developed for discrete control applications in machine and materials handling production engineering environments. The on-going development of PLCs for the control and monitoring of industrial systems has increased their capabilities from simple hard wired logic elements (NAND, NOR gates) to advanced functions using software-controlled microprocessors for piping and instrumentation diagram (P&ID)

algorithms, floating point arithmetic, network communication and multiple processor configurations for parallel processing. Modern PLCs are capable of handling power system local control automation requirements. IEC 61131 is rapidly becoming the internationally recognized standard for configuring PLCs. PLCs evolved to become bay controllers, which are basically items of relay grade hardware that are directly connected to instrument transformers (CT/VT) in a substation and have built in binary input and output modules together with a large logic function library that can then be programmed like a PLC. The significant difference between a PLC and a bay controller is the fact that bay controllers are built with substation requirements in mind such as EMC, high making and breaking capacities, added relaying functionality, etc.

21.2.2 PLC selection

21.2.2.1 *Control and monitoring specifications*

The development of a control system may be divided into various stages as shown in the project development life cycle diagram (Fig. 21.1). A management decision, based on timing and resource availability, is made as to the best stage to obtain competitive tenders for remaining design, supply and installation work from specialist contractors. The first step is to carefully detail the system to be controlled together with possible future expansion requirements. This initial description must carefully detail the hardware and software interfaces and addresses such questions as the physical location of devices, supervisory control connections, motor or actuator loads and physical enclosure protection.

The second step is to define the operational control requirements in a concise and accurate descriptive form. At this second stage, it is essential that full consultation is made with operatives and maintenance crews as well as the engineers in order to ensure the correctness of the descriptions and definitions of the user's wishes. These descriptive control requirements are then converted in a particular format as a sequence of logical events.

A specification (sometimes termed Functional Design Specification or FDS) is next prepared for both the hardware and software. The hardware specification should cover the following points:

- conformity requirements with any existing systems
- communication gateway (RS232/485, etc. or fibre optics glass/plastic) and associated 'protocols' (the protocol is the transmitting/receiving data exchange rules which govern the message format, timing and error checking)
- the input and output devices to be connected either directly or via interposing accessories
- power supply requirements
- codes and applicable standards
- installation environment (enclosure protection, temperature, humidity, etc.)

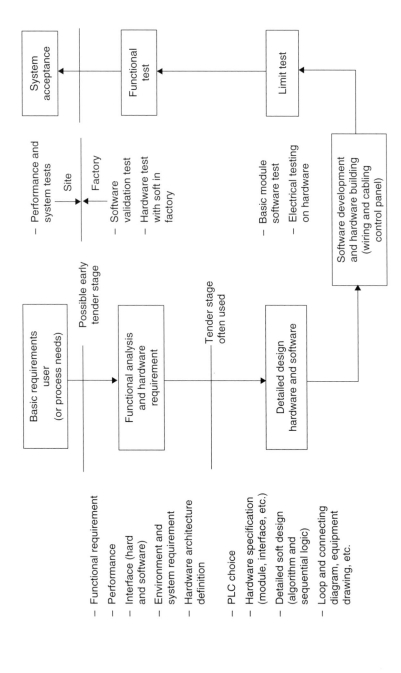

Figure 21.1   PLC system development lifecycle

- factory and site testing requirements
- documentation and quality assurance (QA) requirements.

The software specification provides a complete and definitive statement of what the control system has to do but not, at this stage, how to do it. It provides the basis for the system design and implementation, and includes both the descriptive and logical sequences and functions taking into account:

- functions to be implemented (Boolean or sequential, P&ID functions, maths, etc.)
- data exchanges (type of information per actuator or motor, analogue values, commands to be exchanged, etc.)
- complete input/output listing
- system software (redundancy or self-diagnostic)
- support structure (programming tool giving access to PLC for on or off-line testing and diagnostics)
- factory and site testing requirements
- documentation and QA requirements.

A search is next made for the PLC system that is best fitted to these carefully defined needs at the most competitive price. The size of the PLC system is determined from what tasks it is required to perform by defining the input/output requirements, the memory size and spare capacity. Other requirements include the piping and instrumentation (P&ID) loop control, floating point maths to perform the calculations and special functions such as time delays. This is the stage for selection from the various options for the most appropriate technical solution.

21.2.2.2 *Technical solutions*

The options will comprise of a list of hardware and software components which will implement the functions specified in the system specifications. Once the choice is made detailed design of the PLC system follows. The various algorithm and logical sequences, time delays, fault treatments and data exchange tables used have to be validated through the detailed software design document. Such a document may use:

- logic blocks or ladder logic in line with ISA Standards;
- organograms if development is in the specific pseudo software language used by a particular PLC supplier;
- IEC 61131–3, programming languages for programmable controllers.

The 'response time' of the PLC is the time it takes to translate a change on an input to effect an output. This is not the same as the 'scan time' which is only one of the response time elements involved. The response time takes into account:

- the input/output update times
- the times to process counters, timers and mathematical instructions
- communication times if the PLC is part of a network control system.

A typical example would be a PLC scan of 1000 instructions in 10 ms and a response time of 35 ms per 1000 instructions.

21.2.2.3 *Communication links*

Local automatic primary substation control will invariably involve more than one PLC. The integrated control system will require data to be passed from the switchgear to the associated PLC, from one PLC to another and also to the overall supervisory management system. Correct communication links are therefore the key to the fully automated system and will be the source of problems at the commissioning stage if not properly defined. In the past manufacturers have introduced their own communications protocols and formats such that a variety of software/hardware communication standards exist. It was due to this the standard organizations in Europe and North America started working on open standards. This work has culminated in a set of standards called IEC 61850. It is therefore essential for the user to define the communication standards to be used at the outset (or refer to the open standard) when preparing the specification and enquiry documents. There are three principal communications network arrangements as shown in Fig. 21.2.

The ring and bus/optic star arrangements are the most widely used. Twisted pair copper, coaxial and fibre optic cables link the network together with the choice depending upon transmission protocol used, quantity and rate of

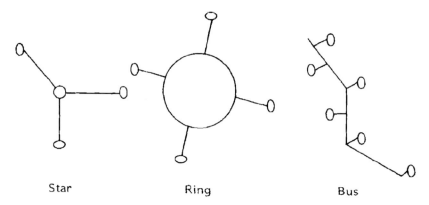

Star   Ring   Bus

Figure 21.2   Communication network arrangements

information being exchanged, length of circuits and cost for the particular application. Other special considerations for the system specification include:

- system resilience requirements
- alarm reporting
- operator/graphics station
- electromagnetic compatibility
- documentation
- interface requirements.

Figure 21.3 shows a typical switchgear control, metering and alarm interface. In this example, a separate marshalling cabinet is proposed with jumpers between the switchgear equipment connections and the SCADA control termination blocks. This arrangement has the advantage of greatly simplifying testing and maintenance by allowing easy access to all the interface points in one cubicle. The disadvantage of the dedicated separate marshalling cubicle is the added expense, the space requirements and the introduction of additional connections into the system.

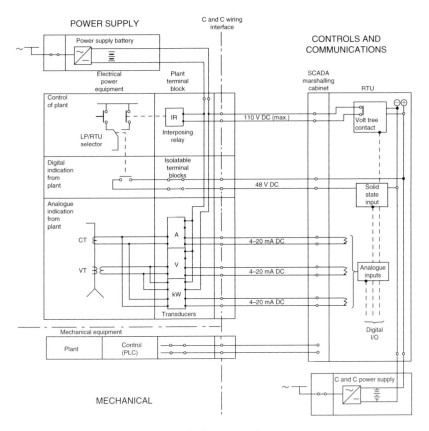

Figure 21.3   Interface for remote control of power equipment

21.2.3 Application example

21.2.3.1 *User requirements*

Figure 21.4 is a single line diagram of a switchboard with two incoming circuit breakers, A and B, and a bus section circuit breaker, C. In this example the simplified user requirements are:

- Automatic control for closing and opening circuit breakers A, B and C.
- Remote monitoring of circuit breaker A, B and C positions.
- Display of incoming circuit current and voltage.
- Control modes to be either local manual or local automatic by PLC.

A brief introduction to the basic requirements for a remote control system is also provided in this example.

21.2.3.2 *Input and output requirements*

The analogue and digital input and output requirements are defined.

(a) *Analogue inputs*: The analogue inputs are described by standard loop current transducers as shown in Table 21.1.

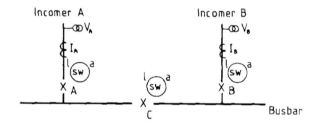

A, B, C   Circuit breakers

(sw)   local / automatic mode switch

Figure 21.4   Single line diagram for PLC application example

Table 21.1   Voltage and current data for example

Description	Sense	Range/engineering units
Voltage incoming circuit A	4–20 mA	0–500 V
Current incoming circuit A	4–20 mA	0–100 A
Voltage incoming circuit B	4–20 mA	0–500 V
Current incoming circuit B	4–20 mA	0–100 A

(b) *Digital inputs*: The digital inputs are defined by logical '0' and '1' condition states. Some simplifications have been introduced into this example (e.g. no maintenance or earth positions for the circuit breakers have been introduced) since the purpose is to explain the basic design steps to be followed rather than describe a complex case (see Table 21.2).

(c) *Digital outputs*: While defining digital outputs, it is very important to consider their duty range. If they are expected to act on the process, i.e. operate a circuit breaker, the switching load and breaking loads may have to considered or suitable interposing relays added externally (see Table 21.3).

Table 21.2   Digital inputs for example

Description	Requirement	Logic
Mode switch circuit A	Automatic	1
Circuit breaker A position	Open	1
Circuit breaker A position	Closed	1
Circuit breaker A condition	Faulty	1
Mode switch circuit B	Automatic	1
Circuit breaker B position	Open	1
Circuit breaker B position	Closed	1
Circuit breaker B condition	Faulty	1
Mode switch circuit C	Automatic	1
Circuit breaker C position	Open	1
Circuit breaker C position	Closed	1
Circuit breaker C position	Faulty	1

Table 21.3   Digital outputs for example

Description	Requirement	Logic
Circuit breaker A	Open command	1
Circuit breaker A	Close command	1
Incomer A voltage display	Binary coded decimal (BCD)	4 digits
Incomer A current display	Binary coded decimal (BCD)	4 digits
Circuit breaker B	Open command	1
Circuit breaker B	Close command	1
Incomer B voltage display	Binary coded decimal (BCD)	4 digits
Incomer B current display	Binary coded decimal (BCD)	4 digits
Circuit breaker C	Open command	1
Circuit breaker C	Close command	1

21.2.3.3 *System specification*

Normal and abnormal operating conditions are defined together with any system operating constraints in conjunction with maintenance, operations and engineering staff:

(a) Normal condition:
- Circuit breaker changeover control is under automatic mode.
- The left-hand side of the switchgear busbar is fed from incomer A with circuit breaker A closed and the bus section circuit breaker C open.
- The right-hand side of the switchgear busbar is fed from incomer B with circuit breaker B closed and the bus section circuit breaker C open.

(b) Abnormal condition:
- Power input failure or circuit breaker A faulty on incomer A. The whole busbar (both right- and left-hand sections) shall be fed from incomer B with the bus section circuit breaker C closed.
- Power input failure or circuit breaker B faulty on incomer B. The whole busbar (both right- and left-hand sections) shall be fed from incomer A with the bus section circuit breaker C closed.

(c) Operating constraints:
- Bus section circuit breaker C must not be closed when circuit breakers A and B are both in the closed position. Such a constraint may typically be due to fault level restrictions on the switchgear with the two incoming supplies paralleled.
- When circuit breaker A or B or C is under local mode, the automatic control is disabled. This is a safety constraint.
- When power supply is restored after failure on incomer circuits A or B the system should remain in its current configuration awaiting operator intervention. This ensures positive action and status acknowledgement by operations personnel.

(d) Communications requirements between PLC and communications controller:
- Master/slave configuration where the PLC/bay controller local to the switchgear is the slave and the communications controller, which has intelligence and interfaces between the communications network and the PLC, is the master for remote control purposes.
- Define the communications network protocol (e.g. 'Modbus, IEC 61850').
- Define the serial link format (e.g. RS 485, Optic bus glass-plastic).

(e) Remote control operation:
Usually, the central control centre (CCC) is comprised of duplicated mini-computers as the central processor associated with various man/machine interfaces. These interfaces include such items as a hand-dressed (Fig. 21.5)

Figure 21.5   Hand dressed conventional mimic panel. The panel consists of individual engraved tiles to display the substation single line diagram, switch position and key metering

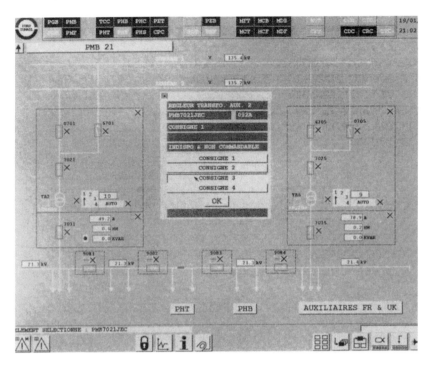

Figure 21.6   VDU display of 132/21 kV substation single line diagram

or automatically updated (Fig. 21.6) mimic displays, operator consoles/ keyboards, visual display units (VDUs), data loggers, event recorders, telephone, public address and radio speech communications. The CCC acquires information from the communications controller or remote telemetry units (or 'remote terminal units') (RTUs) associated with interrogation scan. Each RTU has a unique address code and is accessed in turn for a given period of time when information requests or control signals may be sent and information received. In order to avoid large amounts of data overloading the system during a fault (e.g. a busbar fault would create a multitude of circuit breaker status changes, network load flow alterations, metering and alarm indications) information is prioritized. Further 'front-end processors' are used for data acquisition in order to free the main computer for data processing.

21.2.3.4 *Detail design to tender enquiry stage*

From the foregoing a logic block diagram is next prepared as shown in Fig. 21.7. This should then be fully detailed into an overall descriptive and

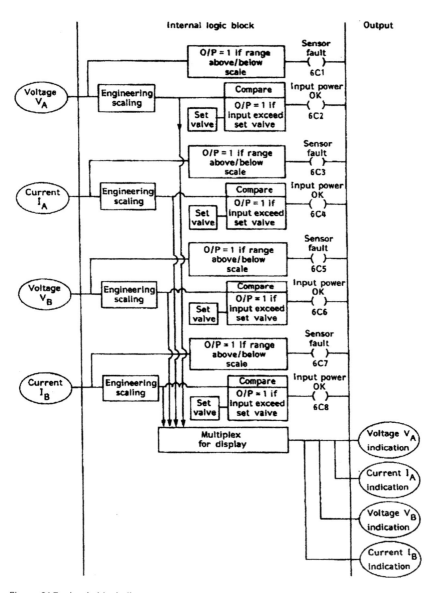

Figure 21.7   Logic block diagram

technical specification such that enquiries may be launched with different manufacturers to supply equipment for the particular application. Figure 21.8 shows the PLC control cubicle for the Channel Tunnel 21 kV network automatic interlocking. This is designed to ensure that the UK and French unsynchronized Grid supplies are not paralleled inadvertently by incorrect switching sequences. The PLCs are in the top left-hand corner of the cubicle.

Figure 21.8   Typical PLC control cubicle for switchgear interlock control

21.3 POWER LINE CARRIER COMMUNICATION LINKS

21.3.1 Introduction

The use of transmission lines as communications channels has obvious advantages to the electrical supply utility since it saves investing in additional

Table 21.4   IEC standards for power line carrier

Standard	Description
IEC 60353	Line traps for AC power systems
IEC 60481	Coupling devices for power line carrier systems
IEC 60495	Recommended values for characteristic input and output quantities of single sideband power line carrier terminals
IEC 60663	Planning of (single sideband) power line carrier systems

dedicated communications radio, hard wire or fibre optic cable links. System Control and Data Acquisition (SCADA) requires a communications network to transmit the information back to a central control centre, (CCC). There is a fundamental relationship and trade-off between the amount of information that may be transmitted over a given communications circuit, the speed of transmission and the bandwidth of the communications channel involved. The larger the bandwidth the faster a greater amount of information may be transmitted. Hence the bandwidth is the limiting factor for the signalling speed upon which the telecontrol system response times are based. Power line carrier circuits operate at only a few hundred kHz carrier frequency with signalling speeds restricted as a consequence to approximately 600 Baud when a single (4 kHz) channel is shared with speech, and up to 9600 Baud (analogue) or 28 kbs (digital) on an equivalent data only channel. With modern digital equipment it is possible to attain signalling speeds >64 Kbit/s using a good quality line (e.g. better than 30 dB signal-to-noise ratio).

The subject is well covered by IEC Standards as listed in Table 21.4. The data transmission is performed by modulating the carrier frequency using audio frequency shift keying with modem (modulator/demodulator) interface units. Higher carrier frequencies (and hence larger bandwidths and signalling speeds) are not possible because of the stray capacitance (and hence high losses and attenuation) involved in overhead line power circuits. Telecontrol systems designed around power line carrier communication links are therefore specified with rather slow five second response times. The response time here is defined as the time between a change of state occurring at an outlying substation and it being announced at the CCC. This is one of the major reasons why fibre optic cable communications links are taking over from power line carrier-based systems. The other key reason is the immunity of fibre optic links from electromagnetic interference.

21.3.2 Power line carrier communication principles

21.3.2.1 Modulation

Power line carrier systems amplitude modulate the carrier frequency. Full amplitude modulation (AM) has a frequency spectrum of sidebands symmetrical about the carrier frequency as shown in Fig. 21.9. These sidebands contain all the information being transmitted and the carrier frequency is only the bearer of the

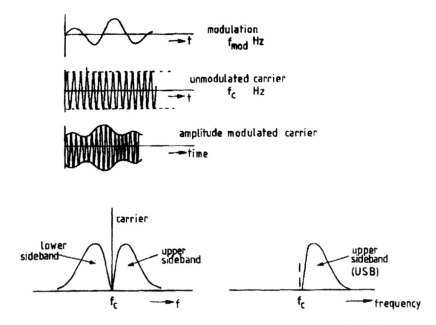

Figure 21.9   Amplitude modulation of a radio frequency carrier and associated double sideband full carrier (AM) or single sideband reduced carrier (SSB) spectrum

messages. It is therefore possible to achieve savings in transmitter power without degrading signal performance by reducing the power of the carrier frequency and by deleting one of the sidebands. This is known as single sideband (SSB) transmission. The carrier is not fully suppressed because it is used to synchronize the remote end receiver with the corresponding transmitter. The transmitter and receiver circuits therefore tend to be slightly more complex than those used for normal broadcast AM transmission because of the filtering and accurate synchronizing involved. The lower ranges ($\sim$30 kHz to 200 kHz) of carrier frequencies are used on long transmission lines and the higher ranges ($\sim$200 kHz to 500 kHz) on shorter lines. This helps to offset the attenuation effects of long lines with high frequency transmission. Computer simulations may be used to optimize the best frequency band to employ on a given overhead line taking into account interference from any adjacent circuits.

Each power line carrier path can carry one audio frequency (AF) channel. This requires a minimum bandwidth of some 4 kHz. The lower $\sim$2 kHz end of this base band is often reserved for speech which requires a bandwidth from approximately 300 kHz to 2000 Hz. It is possible to use the speech band for teleprotection, thereby freeing up bandwidth for data transmission and the channel is often used as a telephone system for the electrical supply utility. Dialing pulses may be transmitted by shifting a pilot signal frequency and detecting the shift pattern at the receiving end. An override facility is normally also provided for emergency/maintenance purposes whereby the telephones are connected directly via a front panel jack socket into the speech circuits.

The remainder of the channel bandwidth, 2000 Hz to ~3480 Hz is available for telecontrol, teleprotection and telegraph transmission using frequency shift keying (FSK). This form of modulation has many advantages over on-off keying of the carrier frequency and provided that the wanted signal (mark or space) is slightly stronger than any interfering signals the information will be correctly received. The main difficulty is that the use of automatic gain control is very limited and the time constant must be short. This is because the mark and space (logical '0' and '1') frequencies are only separated by a few tens of Hertz and may fade independently of one another. A strong mark may be followed by a weak space especially under power fault conditions. A pilot signal, added in the spectrum outside the audio base band (e.g. at 3600 Hz), is therefore used for supervision of the power line carrier channel and regulation of the receiver automatic gain control (agc). The Baud is the shortest single signal unit in a signalling code and may be expressed as the reciprocal of the time of the shortest signal element. For example if the shortest signal element were 20 ms in length then the data transmission speed would be $1/0.02 = 50$ Bauds. The bandwidth of the telecontrol channel is determined by the frequency shift speed. A 200 Baud telecontrol channel shifting ± 90 Hz occupies 360 Hz of bandwidth and the frequencies used are selected from CCITT standard recommended channels as described in IEC 60481 and 60663.

Power line carrier schemes are used in conjunction with overhead line distance protection direct intertripping/blocking or permissive intertripping/blocking as described in Chapter 10. It is, of course, essential that such signals are correctly transmitted and received over the very transmission line that the protection scheme is attempting to protect from the consequences of a prolonged fault. During the fault noise will be generated that could degrade the teleprotection signal. Therefore the power line carrier teleprotection signal is boosted to maximum power and all other signals may be disconnected (speech and telecontrol) in order to improve the reliability under fault conditions.

21.3.2.2 *Circuit configurations*

It is not usual to find power line carrier installations on distribution lines at voltages less than 36 kV. This is because such lines tend to have many tee-off points which would attenuate the signals and necessitate the installation of many power frequency rated filters or 'line traps'. Also short power lines may employ pilot wire protection and the telecontrol system requirements may be able to use spare pilot cable cores.

The high frequency carrier signal is coupled to the overhead transmission line via high voltage coupling capacitors of value around 5000 pf. These act as a low impedance (few hundred ohms) at carrier frequencies but as an open circuit at power frequency ($\sim 0.6\,M\Omega$ @ 50 Hz) thus isolating the radio equipment from the power equipment. In addition coupling filters and transformers are necessary to match the power line carrier transmitter output impedance to the overhead line and thereby ensure maximum power transfer.

The carrier frequencies must not be effectively short circuited to ground through earthing switches at substations or through the neutrals of power transformers. Each power line carrier overhead line transmission circuit must therefore be effectively isolated at radio frequency from the substation busbars, transformers and switchgear.

This is achieved by 'line traps' which are parallel inductance and capacitance (L, C) tuned circuits. These line traps are inserted in series with, and at the end of, the transmission line to act as high impedance at the carrier frequency and prevent such frequencies entering the substation busbars. The line trap coil has a low impedance at 50 Hz in order to minimize power frequency losses. Surge diverters are connected across the tuned circuit to prevent damage against surges. The traps are specified to carry rated current and to withstand short circuit conditions.

Figure 21.10 shows phase-to-earth and phase-to-phase coupling arrangements between the power line carrier radio frequency equipment and the power frequency overhead line. Phase-to-earth coupling requires only half the equipment necessary for the phase-to-phase method. If a power system fault occurs on the phase being also used as a teleprotection or telecontrol channel the power line carrier signal will be considerably degraded and an assessment has to be made as to the security of the system under these conditions. For double circuit transmission lines it is possible to arrange the power line carrier protection intertripping for one circuit to be transmitted over the adjacent circuit. In this way the teleprotection channel does not signal over the actual line it is protecting. A diagrammatic representation of this commonly used arrangement is shown in Fig. 21.11a. Figure 21.11b shows a Middle East 145 kV substation overhead line bay with the incoming gantry and Power Line Carrier line traps mounted on CVTs associated with the distance protection scheme. The CVTs are used to couple the power line carrier signal to the overhead line.

21.4 SUPERVISORY CONTROL AND DATA ACQUISITION (SCADA)

21.4.1 Introduction

The term Supervisory Control and Data Acquisition (SCADA) refers to the network of computer processors that provide control and monitoring of a remote mechanical or electrical operation (e.g. management of a a power distribution grid or control of mechanical processes in a manufacturing plant). Typically in the past a SCADA-based system would encompass the computers and network links which manage the remote operation via a set of field located programmable logic controllers (PLCs) and remote telemetry units (RTUs). The PLCs or RTUs would be connected to field transmitters and actuators and would convert analogue field data into digital form for transmission over the network.

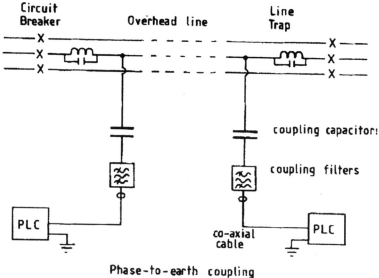

Phase-to-earth coupling

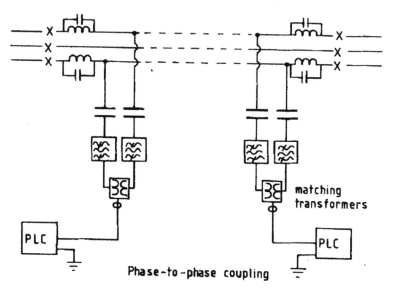

Phase-to-phase coupling

Figure 21.10   Power line carrier coupling arrangements

In many instances today, with respect to high voltage substations or even medium voltage substations, the substation control devices are not necessarily RTUs or PLC's, but IEDs (intelligent electronic devices) that serve the purpose of protection, local and remote control. These IEDs provide a means to

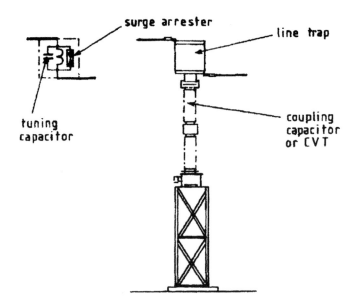

surge arrester

line trap

coupling
capacitor
or CVT

tuning
capacitor

Figure 21.11(a)   CVT and line trap installation arrangement

acquire and transmit the analogue and binary input data to the control system via communication links.

SCADA systems are in essence a real-time operating database that represent both the current and past values or status of the field input/output points (tags) used to monitor and control the operation.

Relationships can be set up within the database to enable functional (or computed) elements to be represented which provide operators with a logical representation of the remote operation. This representation enables the whole operation to be monitored and controlled through a central point of command whereby concise information is available in a clear schematic and textual form typically on graphics workstations.

The supervisory functions of SCADA systems present plant operators with a representation of the current and historical states by means of hierarchical graphic schematics, event logs and summaries. These screens also identify all abnormal conditions and equipment failures which require operator acknowledgement and remedial actions. The control functions enable specified items of plant to be controlled by issuing direct commands, by instigating predetermined control sequences, or by automatically making a programmed response to a particular event or status change.

SCADA systems do not usually handle the collation of statistical data for management information purposes. However, a SCADA system usually exists in an integrated computer hierarchy of control and as such interfaces usually exist to other computer-based systems.

Figure 21.11(b)   Qaboos 132 kV substation, Oman-OHL incomers with surge arresters, CVTs/line traps and cable sealing ends

21.4.2 Typical characteristics

It is convenient to describe SCADA systems by considering their typical characteristics in relation to input/output, modes of control and interfaces with operating personnel.

21.4.2.1 Plant input/output

Typically a SCADA system interfaces with plant over a wide geographical area via PLCs and other RTU/bay control equipment local to the plant. The number and types of the input/output points, the nature of the local equipment connected. There are two basic modes of capture of input data which may be used by the central processing facility of the SCADA. These are:

(i) scheduled capture, whereby the local units are polled on a regular basis and all input data is transferred; or

(ii) change of state capture, whereby only input data which has changed is transferred.

The input and output data are held centrally in a real time database. By holding historical as well as current data it is possible to provide facilities for analysis and reporting of trends. This facility is often particularly important for systems where most of the data is analogue rather than discrete.

In the database input and output data is usually grouped into functional units or elements. For example several input/output points might be grouped to provide the complete representation of an electrical circuit breaker. Frequently such groups of plant input points are transformed to calculate computed points which are also stored in the database. For example a single computed point might represent the status of a number of associated circuit breakers.

A typical use of such computed points is in the management of alarms. In many cases alarms are categorized at least into major and minor alarms. Each alarm is itself likely to be a computed point usually computed from an input value and a trip level. Some major alarms may also be computed from combinations of minor alarms. Complex strategies for predicting alarm conditions may also be used.

21.4.2.2 *Control modes*

Control of plant associated with a SCADA system may be either local or remote. Local control exercised automatically, for example by a local PLC, or may be by local mechanical or electrical controls (automatically or manually operated).

Remote control via the SCADA may be instigated by an operator or may be automatic. Automatic controls can be initiated by time (scheduled control) or events (change of state control). In both cases control frequently involves initiating a pre-programmed sequence of actions which are then automatically carried out.

One advantage of distributed control with the control systems located locally in substations and reporting to a higher level SCADA system is the lower dependence on troublesome communication links. In modern computer-controlled stations, the SCADA system sends out a simple remote control command. Any sequential control sequence is executed locally.

21.4.2.3 *Operator interface*

The operator interface for a modern SCADA system should be designed to provide the maximum support to the operator in his role of monitoring and controlling the plant. In order to achieve this considerable use is made of sophisticated real time graphics to display current and retrospective input output values and trends.

A well-designed operator interface can provide considerable support in alarm management. Where there is a potential for large number of alarms it is

particularly important that they are grouped, classified and displayed in a coherent fashion which enables the operator to concentrate on the more important alarms. Often the facility to filter out minor or consequential alarms or to acknowledge them in groups for later response can be valuable on its own.

Nowadays graphical displays are usually Windows based. There is an increasing trend to using a mouse or a touch sensitive screen to supplement a keyboard for most operator input. Graphical displays should incorporate a hierarchy of displays from high level overall plant schematics to tables of associated input/output points at the lowest levels. Banner display of important information about key events is often used. In order to supply all this functionality it is common to use multiple screens at single operator positions.

An important security facility which is required in many SCADA systems is the definition of different classes of user with access to different functions or facilities. The distinction may simply be between supervisor and operator or between supervisor terminal and operator terminal or may involve several levels of access. Most systems provide the facility to set up or configure the SCADA database. This is usually necessary for the installation and commissioning of the system but should not be available to the ordinary operator.

21.4.3 Design issues

The throughput of the system is one of three main issues in any design. This will be dictated by:

(i)   the magnitude and number of the field input/outputs;
(ii) the data capture time required, which is usually dictated by the time constraints of the process being monitored and/or the time taken to respond to an event;
(iii) whether any sophisticated schemes are used for data compression;
(iv) whether a deterministic communications protocol is required guaranteeing a response in a specified time;
(v)   what level of integrity is expected of the data communications;
(vi) what physical media for data transmission is acceptable in the particular application.

Certain key plant input/output and associated operation may be deemed as being high integrity. For such input/output redundancy needs to be considered in one or more areas of a SCADA system to minimize the effects of failure. Typical redundancy may include:

- dual links for plant input/output to two or more PLC or RTUs;
- dual communication links handling dialogues with the main supervisory processors;
- redundancy in the main supervisory processors provided by either employing a standby fault tolerant processor or by having two or more processors

providing a shadowing function. In this case a 'standby' processor would shadow all operations of a 'normal' processor and watchdog mechanisms would enable a switchover to occur if any communication failure or data integrity errors are detected.

As the loading in terms of the number of plant input/output points increase the processing power required increases. The central architecture of a SCADA system may require several processors each dedicated to specific operations. A typical partitioning would include:

- *Front End Processors* (*FEP*). FEP are dedicated to handling data acquisition from field RTU and PLC equipment;
- *Graphics Workstations*. Where a number of operator positions are required a distributed client-server-based architecture spreading the load between a main supervisory processor and two or more graphics workstations should be provided.
- *Main Supervisory Processor* (*MSP*). One or more MSPs provide centralize control and representation of the field input/output (plant status) by means of one more databases. The central processor will perform functions such as data logging, handling of control sequences, maintenance of logical (functional) equipment states.

21.4.4 Example (Channel Tunnel)

The Channel Tunnel Engineering Management System (EMS) employs a SCADA system configured to manage remote equipment via 26 000 direct input/output points and a further 7000 computed points. The equipment under the EMS control is the Fixed Equipment located in the two terminals in Folkestone in the UK and Coquelles in France and in the three tunnels (Running Tunnel-North, Running Tunnel-South and the Service Tunnel). The Fixed Equipment manages the following:

- Electrical Distribution:
  - Connections to National Grids (225 kV and 132 kV)
  - Supply to 25 kV Overhead Catenary System
  - Tunnel distribution of 21 kV and 3.3 kV supplies
  - Terminal and Tunnel Lighting.
- Mechanical Systems:
  - Normal and Supplementary ventilation systems
  - Tunnel Cooling
  - Pumping
  - Fire Fighting equipment.

Figure 21.12 shows the RTU equipment located in the 178 equipment rooms located between the service and running tunnels. These handle the data

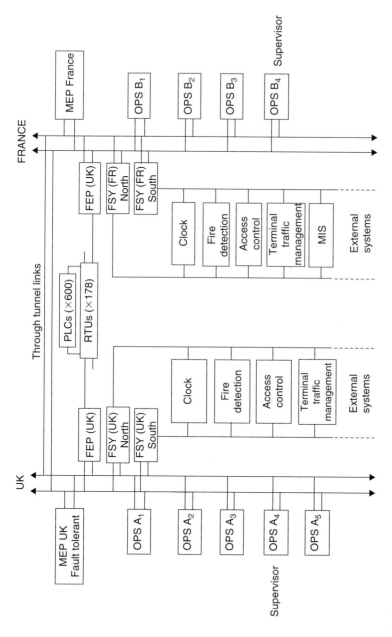

Figure 21.12 Arrangement of RTUs in Channel Tunnel SCADA system

acquisition and control of the 26 000 input/output points via 600 PLCs. When an input/output point changes state the new status is sent to both the French and UK control centres using a drop insert connection to the RTUs. The input/output states are handled simultaneously by main EMS processors (MEPs) in both UK and French control centres. The MEPs are DEC VAX processors running identical SCADA application software. The machines operate in a normal/standby mode. The normal machine is the master and handles all operator dialogues from both the UK and French operator positions. The standby processor whilst maintaining data compatibility with the normal processor also monitors the health of the normal processor, site networks and through tunnel point-to-point links. If any failures are detected then a switchover will occur and the standby machine will move to a normal status.

Three dedicated FEPs are provided in each terminal. Two of these FEPs handle communications with RTUs. The other four FEPs (two in each terminal) provide dual redundant links to a number of external systems such as fire detection and access control.

Data integrity is provided as follows:

- Certain plant input/output has links to two different RTU processors.
- All RTUs communicate with both the French and UK control centres. In addition input/output states received in the French control centre are routed to the UK control centre by the through tunnel links. Similarly the UK control centre transmits input/output states to the French control centre. In a full availability operation each MEP receives two identical messages which are filtered accordingly.
- Redundant on site networks.

Dedicated operator servers (OPS) provide five operator positions in the UK and four in France. In normal operation these provide for a supervisory position and two or more operating positions. The UK control centre also has a major incident control centre (MICC) with a dedicated OPS.

EMS operations are possible simultaneously in both the UK and French control centres. However, only one control centre can have an active status which determines the nature of the possible operation.

21.5 SOFTWARE MANAGEMENT

PLCs, power distribution systems and SCADA systems all make use of software. In many cases, the software components can be seen as the main contributors to the systems' functionality. This use of software has many advantages but it also poses many problems which need to be addressed carefully if they are not to threaten project success.

21.5.1 Software – a special case

The use of software in control systems offers the engineer increased flexibility in the design and operation of systems. Often software allows system to provide functionality which could not otherwise be provided in a cost effective way. However, software development projects are renowned for being late, over budget and not meeting the requirements of the customer. The key to understanding why software development projects frequently possess these unfortunate characteristics is to look at how software development differs from other branches of engineering.

The problems presented by software are many and somewhat fundamental in their nature. The still maturing discipline of software engineering attempts to address these problems.

21.5.1.1 *Software is complex*

Software is a highly complex dynamic object, with even a simple programme having a large number of possible behaviour patterns. For most non-trivial software it is impossible to exhaustively test its behaviour[1] or prove that it will always behave as its specification requires. The difficulty of proving that a software system meets its specification is compounded by the lack of fundamental laws that can be applied to software. The mathematics underlying software engineering is still in its infancy compared with other branches of engineering.

21.5.1.2 *Software is discontinuous*

The discontinuous nature of software means that small changes in input values can result in large unexpected changes in the software and system behaviour. Small changes in the software itself can have similar results. As a result meaningful testing is much less straightforward than for analogue systems. Testing of a completed software system does have a place in providing confidence that it performs its functions correctly but more is required. Considerable effort needs to be expended on managing and assessing the software development process.

21.5.1.3 *Software changes present difficulties*

The range of functions a software system can perform and the apparent ease with which new software can be added makes software very attractive to engineers. However this is deceptive, once software is built it is difficult to change with confidence. Even minor changes can have dramatic and unforeseen effects on often unrelated parts of a system. Furthermore, as more changes are made the software architecture will tend to become increasingly complex and fragmented. Changes become increasingly difficult to implement satisfactorily. This fact should be borne in mind when requesting modifications to completed systems.

21.5.1.4 *Software is insubstantial*

The intangible nature of software means that you can not see, touch or feel software. As a result a software system is very difficult to appreciate until the very end of a development when the component parts are integrated. Unfortunately by this stage a high proportion of project resources will have been expended making any corrective actions expensive to say the least. Furthermore testing at this stage is only of limited use in providing confidence in the software.

21.5.1.5 *Software requirements are often unclear*

Software systems usually perform a very large number of diverse functions which can interact with each other in complex and subtle ways. It is very difficult for a customer to describe these functions precisely and this leads to unclear and changing requirements. This problem is made worse by the culture gap that frequently exists between customers and software developers. In other branches of engineering the specifier of a product will usually be experienced in the engineering discipline required to build that product. This situation rarely exists with software systems. As a result software systems are often specified in narrative English because the notations of software engineering are unfamiliar to the customer. The use of English (or other natural languages) can lead to ambiguities and inconsistencies in the specification which are then fed into the development process and only discovered late in the project when they are difficult and costly to correct.

21.5.2 Software life cycle

At the highest level a software development project should be managed in the same manner as any other engineering project. Thus a software development should follow a software project life cycle similar to that shown in Fig. 21.1.[2] Such a lifecycle has clearly defined phases, with each phase having defined inputs and outputs. The project should have recognized review points to aid control. Normally review points would occur at least at the end of each phase. The whole software and system development process should take place within a quality assurance system such as the ISO 9000 series. The lifecycle shown in Fig. 21.1 and described below is a lifecycle for software development, which should be integrated into the overall project lifecycle.

21.5.2.1 *Requirements specification phase*

The objective of the requirements specification phase is to produce a clear, complete, unambiguous, non-contradictory description of what a system is to do. The requirements specification should be fully understandable to both the customers and the developers. There may be a separate software requirements specification, but if not, the software requirements should be clearly separated and identifiable within the overall requirements specification. Errors at the requirements specification phase can have very serious consequences and therefore the developers should make a major effort to confirm its correctness.

When the requirements specification has been agreed a requirements test specification (often called an acceptance test specification) should be drawn up. This document should state those tests a system must pass for it to be acceptable to the customer. Should a system fail any of the acceptance tests the customer has the right for the problem to be fixed and re-tests performed. However, these tests cannot on their own ensure that the software is correct.

21.5.2.2 *Software design phase*

Using the requirements specification the developers will begin designing the software. As with any engineering discipline this is an essentially creative process which can be done in many different ways.

The objective of the software design phase is to decompose the software into a coherent set of self-contained modules which will each have their own specification and which can each be tested separately .The software design phase will often see the software development process disappear into a tunnel as far as the customer is concerned. Some time later a fully working system will emerge from the other side at the software validation phase. The work carried out within this tunnel is vitally important and it is well worth the customer understanding and monitoring what occurs.

A structured top down approach should be taken to this high level design of the software, producing a hierarchy of modules at different levels. A variety of techniques, often supported by automated tools, may be used during the design. Typical techniques include data flow diagrams, state transition diagrams, object-orientated notations and entity relationship diagrams. Any of these techniques should be supplemented by English language descriptions.

21.5.2.3 *Software module design phase*

The objective of the software module design phase is to perform the detailed design of exactly how each module will carry out its required task. The means by which this detailed design is expressed will vary depending on the type of system being developed and the tools used by the supplier. Typical approaches are to use logic diagrams, flow charts, pseudo code (programming language like statements), formal mathematical notation or decision tables. Alternatively the techniques used during the software design phase may continue to be used. Often a combination of such methods, supplemented by English language description is best. It is essential that the required inputs and outputs, their meaning and possible values are clearly identified for each module.

During the detailed design of a module the developer should produce a test specification detailing those tests that need to be carried out to confirm the correct functions of a module once coded.

21.5.2.4 *Code phase*

The objective of the code phase is to transform the software design specification and software module design specifications into a coherent computer programme or programmes.

It is important to ensure that the code produced is understandable to persons other than the author. In order to achieve this, project standards should be set up and adhered to code structures, format and commenting. The code produced should also be reviewed and changes to approved code are strictly controlled.

In principle the programming language or languages to be used could be selected at this stage. In practice it is likely that design constraints considered earlier will already have determined the language. Such considerations might be dictated by availability, experience or processor used in addition to the merits of a particular language. If possible a high level, structured language should always be preferred to using assembler.

21.5.2.5 *Software testing phase*

The software testing phase covers the testing of the software from individual modules to the complete software system. The phase therefore involves much more than testing against the acceptance test specification. The objective of the phase is to ensure that the software functions correctly, in so far as this can be achieved by testing. In order to satisfactorily test the individual modules it is likely to be necessary for much of this testing to take place in parallel with the coding phase, though conceptually it occurs afterwards.

Records should be kept of the testing of each individual module, of each group of modules as software integration proceeds and of the complete integrated software. These records should be considered part of the documentation of the software and should be retained either by the supplier or the customer. The customer should ensure that the testing process is monitored either directly or by a third party.

21.5.2.6 *Software/hardware integration phase*

The objective of the software/hardware integration phase is to combine the software and hardware into a coherent whole. The integration process involves further testing of the software and system, with further changes being made to the software to resolve any problems which arise. Frequently part or all of this phase must take place at the customer's site. It is essential that the activities of this phase particularly software changes and their testing are adequately controlled and recorded.

21.5.2.7 *Software validation phase*

The software validation phase occurs when the software is complete.

The objective of the phase is to ensure that the completed software complies with the software requirements specification. A variety of methods may be used including software and system testing and various levels of review of the software and system documentation.

The relationship between software validation and acceptance testing may vary depending on the type of project, the function of the software and the customer requirements. In some cases software validation is required as part of the acceptance testing before software installation on site. In other cases validation may be required after all commissioning adjustments have been implemented.

21.5.2.8 *Software maintenance phase*

Software maintenance differs from other maintenance activities in that is necessarily involves modifications to the software. These modifications may correct errors in the software, add facilities which should have been included originally or add new facilities. Often software maintenance involves upgrading to a new operating system and modifying existing software so that it works within the environment.

Because software maintenance always involves new changes to the software it requires careful control and regulation. For example the benefits or otherwise of each proposed change and any safety implications (e.g. IEC61508 compliance) should be carefully considered and analysed before the change is authorized.

21.5.3 Software implementation practice

The process of software development described in Section 21.5.2 provides a theoretical framework for the activities which an engineer can expect to see taking place. The concept of a software lifecycle and its associated documentation are well understood and accepted but interpretations vary. In particular, phase and document titles may not match those presented in Section 21.5.2. Nevertheless it should be possible to identify all the key features in any software development process (see Section 21.5.3.1).

In practice there are a variety of tools, methods and techniques which suppliers can and should use during the software lifecycle. There can be clear rules about which should be used and the choice may well affect the lifecycle and documentation set associated with the software. The important point is that none of the tools, methods or techniques is sufficient on its own. They should be used as part of an approach based on a coherent justifiable lifecycle and associated with comprehensive documentation and software project management techniques (see Section 21.5.4).

21.5.3.1 *Key lifecycle features*

The key features which should be evident in any software lifecycle are:

- a clear specification of the software identifies those requirements separately from the system requirements and separately from the software design;
- a software design which is recorded and goes through two or more stages of increasing detail before coding;
- software testing which is clearly specified, and which covers each stage of the code being developed, integrated and installed;
- final validation of the completed code against the requirements;
- control of changes to the complete software; and
- formal design and quality reviews at appropriate points in the life cycle.

21.5.3.2 *Software safety and reliability*

The safety aspects of systems containing software are often not appreciated by engineers. Software can often provide the potential for enhanced safety through enhanced functionality. However, the characteristics of software are such that special care is needed where it is to be used as part of a system which has the potential to harm or is otherwise required to be of high integrity. It is beyond the scope of this section to provide any detailed guidance on the issue, but Table 21.5 lists a few of the emerging draft and final standards and guidelines which

Table 21.5   Software safety standards

Standard	Title	Source	Date
HSE PES	Guidelines for programmable electronic systems	UK Health & Safety Executive	1987
MoD Interim Def Stan 00–55	The procurement of safety critical software in Defence equipment	UK Ministry of Defence	1991
MoD Draft Def Stan 00–56	Safety management requirements for defence systems containing programmable electronics	UK Ministry of Defence	1993
BSIEC 61508	Functional safety of electrical/electronic/programmable electronic systems	British Standards/IEC	2000

apply. The Institution of Electrical Engineers (now renamed The Institution of Engineering and Technology) also produce an excellent professional brief on the subject of safety-related systems.[3]

21.5.3.3 *Analysis and design methods and tools*

Various tools are available to assist with software specification, design, implementation and test. Different methods address different aspects of the software lifecycle and use different approaches. All provide at least some of the framework on which to base a software lifecycle.

Computer aided software engineering (CASE) tools and the methods on which they are based are frequently used as the foundation on which a software development project is planned. Such tools typically assist with specification, design and implementation and provide much of the necessary lifecycle documentation for those phases. The provision of such documentation by an automated tool helps ensure that it is consistent and follows a coherent format. Most importantly traceability of requirements, through to the final design and code, is also ensured. In many cases the methods and tools make extensive use of diagrams which helps make the designs understandable.

Formal methods provide a mathematically based approach to software specification and design. The principal attraction of such methods is that they allow a proof that the mathematical specification is internally consistent and that the completed code correctly implements the specification. At the time of writing these methods are not widely used and the necessary skills for their use are in short supply. In the future the use of such methods can be expected to increase, particularly for high integrity applications.

A proprietary code management tool, to control build configurations should be adopted by system developers. Such tools assist in providing librarian facilities in a multi-developer environment and ensure that all software modifications are recorded and incorporated into new system builds. The tools also provide configuration control facilities by version stamping individual files and enabling current and historic versions of a software system to be recovered and rebuilt.

Other methods and tools are available which are not so readily categorized. Static analysis can be used to analyse the software code and generate metrics which express various characteristics of the code as numbers. Combinations of these metrics can be used to help form a judgment about the quality of the code and its structure. Dynamic analysis can be used to exercise the code and collect data about its behaviour in use.

21.5.3.4 *Configuration management*

Configuration management of software systems should be applied during the development and operational life of the software in order to control any changes required and to maintain the software in a known state.

To achieve configuration management the components of a software system are partitioned to form configuration items. These encompass all design and test documentation as well as the constituent software components.

The concept of a baseline is applied to software once the build is in a known state, usually once the software integration phase in the development lifecycle is reached. Thereafter any changes required, resulting from anomalies or functional modifications, are controlled through a predefined change control process. The basic stages of the change control process are:

- identification of need for change;
- identify change implementation, assess impact and approve (or reject) implementation;
- audit change implementation;
- install modified software and update the baseline.

21.5.4 Software project management

This section sets of out areas in which the problems and techniques of software-based projects differ from those in more traditional manufacturing projects. None the less the basic issues of project management remain valid and to achieve success the following areas need to be addressed:

- definition of work scope;
- risk incurred;
- resources required and
- tasks and phases to be accomplished.

21.5.4.1 *Planning and estimating*

In common with any project, planning and estimation attempts to quantify what resource is required. Typically this is measured in man–months effort, the chronological duration, and task breakdown and other areas affecting cost. The complexity of software requirements and the difficulty of correctly defining them make resource requirements difficult to estimate. Over recent years a number of estimation techniques have evolved which attempt to quantify the likely costs and durations. The basis for the different techniques, in all cases, is based on past experiences and the function sizing of the whole computer-based system.

Each estimation technique has a number of common attributes:

- project scope;
- software metrics (measurements) forming the basis on which the estimates are to be made
- functional and task decomposition allowing estimation of individual items.

There are two basic categories of estimation techniques, size orientated and function oriented. An example of a size-orientated technique is the constructive cost model (COCOMO).[4] This computes development effort as a function of programme size and produces development effort (cost) and duration.

In contrast function-orientated techniques typically refer to a function point analysis[5] and considers effort associated with the number of user inputs and outputs, enquiries, files and interfaces. Once calculated function points are used to derive productivity, quality and cost measurements.

21.5.4.2 *Scheduling*

In a small software development a single software engineer may well analyse design, code, test and subsequently install a system. However, as project size and complexity increases more engineers must become involved. In a multi person project team there is a time overhead incurred in communication between team members. In addition when team members join project teams in an attempt to make up lost time, they need to learn the system, most likely from those already working on the project. In summary, as project size and complexity increase then the engineering effort required for implementation increases exponentially. If project development slips (or requires accelerating) adding new effort will typically increase the magnitude of any slippage (at least in the short term).

The basic issue to be considered is that people's working relationships and structures are essential for project success, but need careful structuring and management.

21.5.4.3 *Effort distribution*

All software estimation techniques lead to estimates of project duration in effort (typically man months). These assume an effort distribution across the development lifecycle of 40-20-40 (see Fig. 21.1). The 40-20-40 distribution puts the emphasis on the front-end analysis and design tasks and back-end testing.

21.5.4.4 *Progress monitoring*

The insubstantial nature of software makes progress very difficult to measure. A lot of resource is often required to complete a project which is reported as nearly complete. Typical figures quoted are 50% of resource to complete the last 10% of the project.[6]

By partitioning and reporting on software development activities down to a low level realistic measurement of progress becomes more practical. Because each basic task is small and self-contained it is relatively straightforward to identify whether it has been completed and thus estimate the progress which has been made.

REFERENCES

1. Leveson, N.6., Software safety: Why, what, and how, *Computing Surveys*, Vol. 18, No. 2, June 1986.
2. IEC, *Draft Software for Computers in the Application of Industrial Safety Related Systems*, IEC (Secretariat) 122, International Electrotechnical Commission, 1991.
3. IEE, *Safety-related System-Professional Brief*, Issue 1, January 1992, The Institution Electrical Engineers.
4. B. Boehm, *Software Engineering Economics*, Prentice-Hall, 1981.
5. A. J. Albrecht, 'Measuring Application Development Productivity', *Proc, IBM Applic. Dev, Symposium*, Monterey, CA, October 1979, pp. 83–92.
6. F. P. Brooks, *The Mythical Man–Month*, Addison-Wesley, 1975.

22 Project Management

22.1 INTRODUCTION

This Chapter describes the major project management techniques necessary for the evaluation, planning, monitoring and control of transmission and distribution projects. It looks at the situation primarily from the point of view of the client or distribution company management. In particular the chapter explains the importance of correct definition of the work to be performed together with the need for a mature approach to the client/consultant/contractor relationship. A project definition/questionnaire is included in Appendix A of this chapter.

Engineers may feel that high engineering standards and excellence in design should be the major factors in the award of contracts. It is a fact of life that in a highly competitive environment good design is only one part of the overall project process. Another way of looking at it is that good engineering design is that which provides best value for money. Indeed even compliance with the specification may be of secondary importance if the contractor is able to offer alternative equally viable schemes with substantial cost savings. Over-specification helps no-one. Low price, whole life cost, short and certain delivery and low cost financing are also major factors leading to the success of the project from both the client's and contractor's point of view.

22.2 PROJECT EVALUATION

22.2.1 Introduction

The electricity supply company and the network operating company (which may be a separate organization) must satisfy demand for power by the consumer and obtain sufficient revenue from sales to meet investor requirements and future expansion plans. In order to achieve these goals investment in generating, transmission and distribution plant is necessary. Money on capital projects is spent

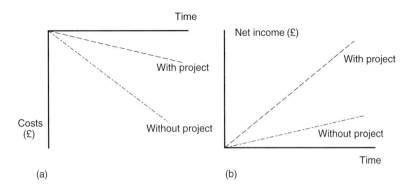

Figure 22.1   (a) Cost reduction project; (b) expansion project

now in the hope or expectation of sufficient returns or *profit* at a later date in the *future*. Investment may be for:

- Replacement of equipment possibly to reduce maintenance costs (cost reduction) – Fig. 22.1a.
- Expansion of transmission or distribution capability to reach more customers – Fig. 22.1b.
- Provision of new products.

The problem is to find 'good' projects by imagination, alertness and creativity (plus a degree of luck) in order to spot the investment opportunity. Imagine, for example, the problems that international aid agencies might have in trying to identify a 'good' project. They will be inundated with different schemes for projects but only a small number will be viable and be capable of bringing about the required benefits. It is necessary to look at both the *financial* and the *economic* costs and benefits of the project before a final investment decision may be made. Financial project investment assessments look at the project purely in monetary terms. Economic appraisals look beyond this and include such intangibles, converted to money terms, as general benefits to the community that the project will bring about. For example a distribution scheme might allow the community to stop chopping down trees for fuel (thereby saving for example the environment from soil erosion) and allow greater productivity in the community.

22.2.2 Financial assessment

22.2.2.1 *Annual rate of return*

This simple appraisal method looks at:

- Initial investment.
- Total cash inflows resulting from the project.
- Average annual profit.

The annual rate of return project appraisal method is simple, easy to understand and a good guide to the profitability of a capital project investment.
Consider the two projects X and Y below:

	Project X $(£ \times 10^6)$	**Project Y** $(£ \times 10^6)$
Initial investment (End of year 0)	-6	-12
Cash inflows from project (EOY 1)	$+3$	$+7$
Cash inflows from project (EOY 2)	$+4$	$+8$
Cash inflows from project (EOY 3)	$+8$	$+9$
(a) Total cash inflow	$+15$	$+24$
(b) Total net profit	$+9$	$+12$
(c) Average annual profit	$+3$	$+4$
(d) $\dfrac{\text{Average annual profit}}{\text{Initial investment}} \times 100\%$	$\% = 50\%$	$\frac{4}{12} = 33\%$

Project Y has higher total cash inflow (£24 × 10⁶ vs £15 × 10⁶)
Project Y has higher total net profit over 3 years (£12 × 10⁶ vs £9 × 10⁶)
Project Y has higher average annual rate of return or profit (£4 × 10⁶ vs £3 × 10⁶)
Project X has higher rate of return on investment (50% vs 33%)

The averaging process, however, eliminates possible important and very relevant information about the timing of the cash flows involved with the investment. For example consider two further projects V and W:

	Project V $(£ \times 10^6)$	**Project W** $(£ \times 10^6)$
Initial investment (End of year 0)	-6	-6
Cash inflows from project (EOY 1)	$+1$	$+6$
Cash inflows from project (EOY 2)	$+2$	$+2$
Cash inflows from project (EOY 3)	$+6$	$+1$
(a) Total cash inflow	$+9$	$+9$
(b) Total net profit	$+3$	$+3$
(c) Average annual profit	$+1$	$+1$
(d) $\dfrac{\text{Average annual profit}}{\text{Initial investment}}$	$\% = 17\%$	$\% = 17\%$

The cash inflows for the two projects are in reverse order with Project W recouping the initial investment sooner than Project V. The average annual rate of return on investment appraisal method for each project, however, gives the same assessment for both Project V and W. The useful earlier recouping of monies is not immediately apparent from this appraisal method. It is certainly feasible that the early cash generation from Project W could be put to good use elsewhere. Under Project W the extra £5 000 000 received in year 1 compared to Project V could be invested on the capital markets to yield further gains. Taking into account the 'opportunity cost' of capital Project W is superior. This is a typical example of the importance of cash generation for the contractor in construction projects. Generally construction yields quite low profit margins but the money sums involved are often large.

Comparatively little capital is tied up in plant in comparison to the manufacturing industry. The cash generated from the construction project is therefore reinvested by the construction company in other areas yielding higher returns.

22.2.2.2 Payback

To allow for the *timing* of returns from the project the payback method of assessment may be used. Again this is a simple project financial appraisal method which indicates how many years it will take before the original amount invested in the project is 'paidback', i.e. the time before cumulative returns exceed the initial investment with generally the shorter the period the better.

Consider two Projects S and T with the same initial investment as shown below:

	Project S ($£ \times 10^6$)	Project T ($£ \times 10^6$)
Initial investment (End of year 0)	−6	−6
Cash inflows from project (EOY 1)	+3	+8
Cash inflows from project (EOY 2)	+4	+5
Cash inflows from project (EOY 3)	+8	+2
(a) Total cash inflow	+15	+15
(b) Total net profit	+9	+9
(c) Payback period (years)	$1\frac{3}{4}$	$\frac{3}{4}$

Whilst the payback method of assessment has taken timing into account it still does not consider the maximum acceptable payback period. The method ignores cash receipts expected after payback but this, for the longer term thinker, could a be very important. For example although Project T above has a much quicker payback period the receipts from the project over the years seem to be diminishing whereas those from Project S are on the increase. The method is therefore a rough screening device and a measure of risk associated with the project. The method does not inform the investor about the overall profitability of the project.

22.2.2.3 Discounted cash flow

In order to better forecast and analyse a particular project investment more information about both the *amounts* of cash generated and the timing involved is required. Having said this no method of analysis gives a precise answer or avoids the risks involved. The timing assumed in the analysis may not be correct, the forecasted cash quantities may be inaccurate, the opportunity cost of capital may vary as interest rates and tax regimes change during the life of the project and non-financial aspects (cost benefit analysis, political upheaval, etc.) all need to be considered. A most important point is to remember that the cost of the analysis and the time to complete it must be kept in check and should not exceed the benefit obtained from it. In accounting terminology this is known as the 'materiality concept'.

The time value of money is nowadays generally understood since inflation is a topic much covered in the news. Money will depreciate in terms of its purchasing power over time if the inflation rate is higher than the returns received from investment. £100 invested @ 10% annual interest over a 3 year period will have a 'future' value of £133.10 at the end of 3 years. The 'present' value of this investment (without considering reductions in purchasing power due to inflation) is therefore £133.10 as calculated below:

EOY 0		£100.00		
EOY 1	£100.00 × 1.1 =		£110.00	
EOY 2	£110.00 × 1.1 =			£121.00
EOY 3	£121.00 × 1.1 =			£133.10

When comparing the cash returns over time from different project investment options at a given discount rate the higher the Net Present Value the better.

Consider £10 000 000 invested now in Project R to yield returns of:

+3 000 000 at EOY 1 (End of Year 1)
+4 000 000 at EOY 2
+5 000 000 at EOY 3

Should the distribution company on purely financial considerations make this investment or simply bank the money @ 10% interest per annum?

£10 000 000 invested for 3 years @ 10% p.a. will yield £13 310 000.

In comparison with this the individual End Of Year cash returns, without allowing for depreciation of money over time, total £12 000 000. These cash flows may be converted to an equivalent EOY 0 value totalling £9 786 000 by applying a 10% discount rate as shown.

From this analysis it can be seen that the returns over a 3 year period are not enough to justify the investment in Project R since with the 10% opportunity cost of capital it is better to more safely invest the monies elsewhere. This may not be the most important factor for a publicly owned utility, but it is a very relevant approach in the case of a privately funded or joint public/private finance project.

Table 22.1 shows the present value of £1.00 receivable at the end of n periods. Thus given an interest rate of 15% per annum, £1000.00 receivable at the end of 5 years (end of year 5 or EOY 5) may be calculated from the table by applying the appropriate discount factor to give a present value of £497.00.

Table 22.2 shows the present value of £1.00 per period receivable at the end of each of the next n periods. Therefore for an interest rate of 15% per annum,

Table 22.1   Present value of £1

Years hence	1%	2%	4%	6%	8%	10%	15%	20%	25%	30%	40%	50%
1	0.990	0.980	0.962	0.943	0.926	0.909	0.870	0.833	0.800	0.769	0.714	0.667
2	0.980	0.961	0.925	0.890	0.857	0.826	0.756	0.694	0.640	0.592	0.510	0.444
3	0.971	0.942	0.889	0.840	0.794	0.751	0.658	0.579	0.512	0.455	0.364	0.296
4	0.961	0.924	0.855	0.792	0.735	0.683	0.572	0.482	0.410	0.350	0.260	0.198
5	0.951	0.906	0.822	0.747	0.681	0.621	0.497	0.402	0.328	0.269	0.186	0.132
6	0.942	0.888	0.790	0.705	0.630	0.564	0.432	0.335	0.262	0.207	0.133	0.088
7	0.933	0.871	0.760	0.665	0.583	0.513	0.376	0.279	0.210	0.159	0.095	0.059
8	0.923	0.853	0.731	0.627	0.540	0.467	0.327	0.233	0.168	0.123	0.068	0.039
9	0.914	0.837	0.703	0.592	0.500	0.424	0.284	0.194	0.134	0.094	0.048	0.026
10	0.905	0.820	0.676	0.558	0.463	0.386	0.247	0.162	0.107	0.073	0.035	0.017
11	0.896	0.804	0.650	0.527	0.429	0.350	0.215	0.135	0.086	0.056	0.025	0.012
12	0.887	0.788	0.625	0.497	0.397	0.319	0.187	0.112	0.069	0.043	0.018	0.008
13	0.879	0.773	0.601	0.469	0.368	0.290	0.163	0.093	0.055	0.033	0.013	0.005
14	0.870	0.758	0.577	0.442	0.340	0.263	0.141	0.078	0.044	0.025	0.009	0.003
15	0.861	0.743	0.555	0.417	0.315	0.239	0.123	0.065	0.035	0.020	0.006	0.002
16	0.853	0.728	0.534	0.394	0.292	0.218	0.107	0.054	0.028	0.015	0.005	0.002
17	0.844	0.714	0.513	0.371	0.270	0.198	0.093	0.045	0.023	0.012	0.003	0.001
18	0.836	0.700	0.494	0.350	0.250	0.180	0.081	0.038	0.018	0.009	0.002	0.001
19	0.828	0.686	0.475	0.331	0.232	0.164	0.070	0.031	0.014	0.007	0.002	
20	0.820	0.673	0.456	0.312	0.215	0.149	0.061	0.026	0.012	0.005	0.001	
21	0.811	0.660	0.439	0.294	0.199	0.135	0.053	0.022	0.009	0.004	0.001	
22	0.803	0.647	0.422	0.278	0.184	0.123	0.046	0.018	0.007	0.003	0.001	
23	0.795	0.634	0.406	0.262	0.170	0.112	0.040	0.015	0.006	0.002		
24	0.788	0.622	0.390	0.247	0.158	0.102	0.035	0.013	0.005	0.002		
25	0.780	0.610	0.375	0.233	0.146	0.092	0.030	0.010	0.004	0.001		
30	0.742	0.552	0.308	0.174	0.099	0.057	0.015	0.004	0.001			
40	0.672	0.453	0.208	0.097	0.046	0.022	0.004	0.001				

Table 22.2   Present value of £1 received annually for n years

Years (n)	1%	2%	4%	6%	8%	10%	15%	20%	25%	30%	40%	50%
1	0.990	0.980	0.962	0.943	0.926	0.909	0.870	0.833	0.800	0.769	0.714	0.667
2	1.970	1.942	1.886	1.833	1.783	1.736	1.626	1.528	1.440	1.361	1.224	1.111
3	2.941	2.884	2.775	2.673	2.577	2.487	2.283	2.106	1.952	1.816	1.589	1.407
4	3.902	3.808	3.630	3.465	3.312	3.170	2.855	2.589	2.362	2.166	1.849	1.605
5	4.853	4.713	4.452	4.212	3.993	3.791	3.352	2.991	2.689	2.436	2.035	1.737
6	5.795	5.601	5.242	4.917	4.623	4.355	3.784	3.326	2.951	2.643	2.168	1.824
7	6.728	6.472	6.002	5.582	5.206	4.868	4.160	3.605	3.161	2.802	2.263	1.883
8	7.652	7.325	6.733	6.210	5.747	5.335	4.487	3.837	3.329	2.925	2.331	1.922
9	8.566	8.162	7.435	6.802	6.247	5.759	4.772	4.031	3.463	3.019	2.379	1.948
10	9.471	8.983	8.111	7.360	6.710	6.145	5.019	4.192	3.571	3.092	2.414	1.965
11	10.368	9.787	8.760	7.887	7.139	6.495	5.234	4.327	3.656	3.147	2.438	1.977
12	11.255	10.575	9.385	8.384	7.536	6.814	5.421	4.439	3.725	3.190	2.456	1.985
13	12.134	11.343	9.986	8.853	7.904	7.103	5.583	4.533	3.780	3.223	2.468	1.990
14	13.004	12.106	10.563	9.295	8.244	7.367	5.724	4.611	3.824	3.249	2.477	1.993
15	13.865	12.849	11.118	9.712	8.559	7.606	5.847	4.675	3.859	3.268	2.484	1.995
16	14.718	13.578	11.652	10.106	8.851	7.824	5.954	4.730	3.887	3.283	2.489	1.997
17	15.562	14.292	12.166	10.477	9.122	8.022	6.047	4.775	3.910	3.295	2.492	1.998
18	16.398	14.992	12.659	10.828	9.372	8.201	6.128	4.812	3.928	3.304	2.494	1.999
19	17.226	15.678	13.134	11.158	9.604	8.365	6.198	4.844	3.942	3.311	2.496	1.999
20	18.046	16.351	13.590	11.470	9.818	8.514	6.259	4.870	3.954	3.316	2.497	1.999
21	18.857	17.011	14.029	11.764	10.017	8.649	6.312	4.891	3.963	3.320	2.498	2.000
22	19.660	17.658	14.451	12.042	10.201	8.772	6.359	4.909	3.970	3.323	2.498	2.000
23	20.456	18.292	14.857	12.303	10.371	8.883	6.399	4.925	3.976	3.325	2.499	2.000
24	21.243	18.914	15.247	12.550	10.529	8.985	6.643	4.937	3.981	3.327	2.499	2.000
25	22.023	19.523	15.622	12.783	10.675	9.077	6.464	4.948	3.985	3.329	2.499	2.000
30	25.808	22.396	17.292	13.765	11.258	9.427	6.566	4.979	3.995	3.332	2.500	2.000
40	32.835	27.355	19.793	15.046	11.925	9.779	6.642	4.997	3.999	3.333	2.500	2.000

an 'annuity' of £1000.00 over 5 years has a total present value of £3352.00. This figure could also be derived from Table 22.1 by adding the individual discounted values of the £1000.00 received each year. For example:

EOY 1 present value = £870
EOY 2 present value = £756
EOY 3 present value = £658
EOY 4 present value = £572
EOY 5 present value = £497
————
£3353

An alternative measure of the financial acceptability of the project using discounted cash flow techniques is to assess the project's internal rate of return (IRR). This is the discount rate that exactly reduces the net present value to zero. The higher the IRR the better the return on the investment is considered to be. For project R above the IRR will be less than 10%. With a discount rate of approximately 8.9% the NPV is almost zero. Hence it can be immediately seen that investment in a bank with an interest rate of 10% is a preferable option. Again this analysis tells us nothing about cash flows beyond the 3 year period. It should be noted, however, that in the longer term (and over 25 years in the case of the 10% discount rate) cash flows far into the future have little present day value.

End of Year (EOY)	Cash Flows (£)	Discount Factor @ 10% (see Table 22.1)	Present EOY 0 Value (£)	Resultant NPV (£)
0	−10 000 000	1	−10 000 000	−10 000 000
1	+3 000 000	0.909	+2 727 000	
2	+4 000 000	0.826	+3 304 000	+9 786 000
3	+5 000 000	0.751	+3 755 000	

Net Present Value = −214 000

22.2.2.4 Sensitivity analysis

Obviously the end result of any such financial analysis can only be as good as the input data and original assumptions. Such items as interest rates, cost of materials, exchange rates and inflation may all change during the life of the project and have an effect on the viability of the project. For the more sophisticated analysis the sensitivity of the results to such changes are considered. With the use of spreadsheets on modern micro computers the cash flows are entered into a table and the NPV or IRR calculated. Variations in parameters can then be altered with the computer doing all the calculations to allow the

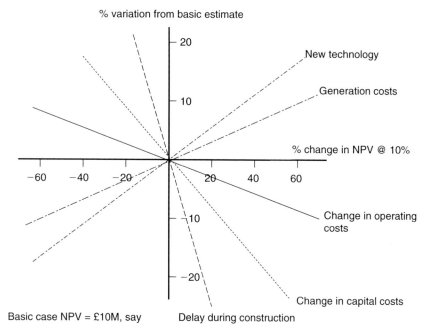

Figure 22.2   Graphical sensitivity analysis results for a typical transmission project

effect of such changes to be assessed. A graphical output such as that shown in Fig. 22.2 is typical of such a sensitivity analysis. The smaller the angle of the line to the *x* axis the greater the sensitivity of the estimate to this particular parameter change.

Whilst computers take much of the hard work out of discounted cash flow and associated sensitivity analysis it must be borne in mind that cash forecasts are undoubtedly open to wide margins of error. Care must therefore be taken, especially over the early years of the project, in estimating such cash flows as accurately as possible. In addition a whole host of economic and political factors must also be taken into account if the analysis is to be meaningful.

22.2.3 Economic assessment

22.2.3.1 *Principles*

The economic appraisal of electrical transmission and distribution schemes is generally a three stage process. Firstly the cost of the scheme has to be assessed – taking account not just of the initial capital cost but also of the costs likely to be incurred over the life of the project. Various technically viable schemes are therefore considered and costed. Secondly an estimation of the financial and economic benefits as revenues plus cost savings is made. Thirdly a comparison between the discounted benefits and costs is computed using discounted cash flow techniques.

The minimum amount of benefit relating to an electrification scheme can be measured by the amount of revenue collected from the estimated future potential consumers. In normal cases the benefits are in excess of this basic amount paid for the electricity because:

- electricity is cheaper than alternative sources of energy
- electricity is of superior quality to the alternatives
- electricity makes possible new and extra activities

In such cases the gross benefit equals the amount paid by the consumers for the electricity plus some 'surplus benefit'. Surplus benefit consists of one or more of the following:

- cost savings related to the alternative
- the value of difference in quality
- the value of any extra output or activity generated by lower costs and/or a change in quality.

In developing countries the willingness to pay for the incremental electricity sales may be estimated in terms of the costs which the consumers would have had to bear in order to meet their energy requirements in the absence of the proposed project. It may also be possible to quantify other benefits in money terms. The estimation of avoided costs yields a minimum measure of 'consumer surplus'. Often the benefits outweigh the price considerably. That is, 'consumer surplus' is high.

Actual and forecast consumption levels and consumer numbers are derived from field surveys and may be broken down into customer groups (low voltage domestic, high voltage domestic, low voltage commercial, high voltage commercial, low voltage industrial, high and extra high voltage industrial, institutional, rural, etc.) all having possible different tariff rates.

22.2.3.2 *Cost benefit analysis*

When consumers purchase electricity from the National Grid it is apparent that they do so because the benefits to them outweigh the price they have to pay for the service. The benefits can be any combination of the following:

- Resource saving – The price of the supply company's service is typically much cheaper than can be obtained from a substitute or from a private source of supply. Public electricity is cheaper than electricity from small individual diesel fuel powered generators (or from wind or solar, without a basic subsidy) for all but low levels of demand in remote areas.
- Superior quality energy supplies – Electric lighting is valued more highly by 'households' or consumers because it is of a higher quality than kerosene fuelled substitutes.
- Extra output – This may be produced on account of the reduced prices of the service (relative to substitutes), or by extra quality, or both. More lighting, more motive power and more business activity, may all be induced by

the cheaper or higher quality service that the transmission or distribution project offers.

The benefits estimation analysis methodology is concerned with the calculation of this incremental consumers' surplus.

Willingness to pay benefits can be estimated by placing forecast electricity consumption into two categories. Firstly, that electricity consumption which is considered to have been substitute for consumption of an alternative source of energy (a diverted market); secondly that consumption which would not have occurred in the absence of the project (a generated market).

Willingness to pay in the diverted market for each consumer group can be estimated with reference to the cost of the total displaced alternative source of energy. An estimate of the willingness to pay in the generated market is not so straightforward. Benefits are therefore defined as the avoided initial and recurring costs of the alternatives to incremental Grid Supply Company electricity. The method does not necessarily make specific allowance for incremental consumer surplus on the generated market as distinct from the diverted market. The avoided costs can be suitably adjusted to reflect the declining valuation of energy inputs with increasing consumption that would apply to the generated market.

Figure 22.3 outlines the methodology on a price/quantity demand graph. Line D_{Q0}–D_{P0} is to be interpreted as showing the maximum amount that a consumer would be willing to pay for each successive unit of electricity in a particular period. The total willingness to pay for the quantity Q_1 is the area under the D_{Q0}–D_{P0} line 0–Q_1–C–D_{Q0}. With the price P_1 the amount the consumer is required to pay is 0–Q_1–C–P_1. The difference between what the consumer is prepared to pay and that which is actually paid is often referred to as the 'consumer surplus', area P_1–C–D_{Q0}.

Electricity from the Grid as a result of a transmission and distribution project may be seen as a substitute for the use of, say, kerosene for lighting or for diesel, used for motive power directly or for diesel generation. In each case the 'consumer surplus' can be measured by the net difference in costs between using electricity from the Grid or a substitute. Again using Fig. 22.3 let P_1 be the unit price of electricity from a diesel generator and P_2 be the electricity supply company tariff for all the incremental electricity, quantity Q_2, assumed to be supplied as a result of the project under consideration. Given that the quantity Q_2 is supplied at the price P_2 the change in 'consumer surplus' as opposed to that existing if alternative energy supplies were to be used is measured by the area P_2–B–C–P_1. A minimum measure of this incremental 'surplus' is given by the area P_2–A–C–P_1 which represents the cost saving to those energy users in the diverted market. The cost savings can be used to obtain an approximate measure of any incremental 'surplus' for all these consumers for whom substitutes for the use of electricity from the Grid system are available.

The availability of lower cost electricity as a result of the project and the tariff structure may lead to an increase in planned output levels by new businesses which would have been unprofitable using the substitute forms of energy. A separate estimation of 'surplus' for such a generated market may also be attempted in a full economic study.

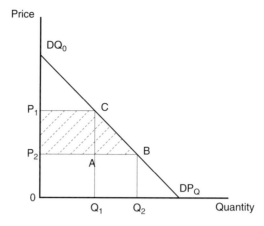

Figure 22.3   Demand curve for electrical power as an example of consumer surplus

22.2.3.3 *The difficulties*

The practical difficulty of converting perceived benefits into money terms that may then be used in a financial and economic analysis is open to priorities based upon political opinions. A famous early cost benefit analysis in the late 1960s for the assessment of the Victoria Underground Metro line in London used such factors as:

- waiting time for people held up in traffic jams in their cars if the line was not built
- pollution levels due to high traffic in London without the line
- stress levels on health of people being delayed in traffic jams.

Obviously such assessments will have a degree of subjectivity. The very important advantage of such cost benefit analysis techniques is to provide a level playing field for the comparative assessment of a variety of similar projects such that the 'best' may be selected. Such analysis has become an essential part of the project selection process demanded by the large aid agencies.

Consider as an example of the difficulties a rural electrification project in a developing country where the options include:

- a small hydro electrification scheme and associated distribution from the generators to the load centres
- a diesel generation scheme.

To an experienced engineer it is well known that the life of diesel generators in such applications is very short. Even if maintenance is good often foreign exchange constraints make the availability of spare parts very difficult. In comparison hydro schemes have extremely good records of reliability, long life and once built do not drain the foreign exchange reserves of the country on possible costly fuel imports. A hydro scheme is capital intensive and, since monies in the early stages of a project are emphasized in discounted cash flow analysis techniques, often small hydro schemes are rejected. In fact they are often able to provide a much better service to the community.

22.3 FINANCING

22.3.1 Responsibilities for funding

Borrowing money on the open market is expensive and so both client and contractor should attempt to keep these costs to a minimum. For example it may well be possible for a large and stable electricity supply company to arrange and pay for insurance costs associated with a particular construction contract because they are able to obtain preferential rates. If the client leaves all such costs to be borne by the contractor then they will only appear in the contractor's tender sum with a suitable mark-up. Therefore the client could well end up paying more for the project overall in the long run without this more mature approach to contract financing.

The principle to be adopted in a large construction contract between client and contractor is that the individual contract costs must be borne by the side best able to bear them if the costs are to be kept to a minimum. Such cost allocation must include risk costs; poor risk management is one of the major factors in total project cost, and a sensible contract will all ensure that risk responsibility lies where it is best managed.

22.3.2 Cash flow

Since a contractor starved of cash cannot function the following factors are important:

1. Sufficient advance payments to the contractor should be considered. The purpose of such payments is to cover contract 'front end' expenditure such as mobilization, early engineering work, payments to sub-contractors, commission and insurances. Typically such advances will be of the order of 10% of the contract value.
2. Progress payments to the main contractor on a regular basis related to the *value* of work actually achieved. These should not normally be related to payments made by the main contractor to the sub-contractors. An independent valuation by respectable, professional and independent quantity surveyors or consulting engineers on behalf of the client assists in reducing client/ contractor disputes.
3. Progress payments linked to the timely completion of specific defined parts of the works or milestones. These must be capable of easy definition and measurement.

4. Retentions kept by the client from interim progress payments during the course of the contract. The purpose of these is to act as an incentive for the contractor to finish the works.
5. Insurance bonds or bank guarantees to be provided by the contractor and held by an independent bank as surety that the contractor will complete the works. These are often capable of being redeemed by the client 'on first demand'. Following some abuse of this terminology in the 1970s such wording should be avoided if at all possible by the contractor before a contract is signed. A better wording might be, 'on first written demand by two authorized and agreed client signatories'. This makes snap decisions by the client, which may not be strictly in accordance with the contract, more difficult and gives a little more time for reflection. Such retentions are typically 5–10% of the contract value.
6. Payment documentation should be simplified as much as possible in order to avoid misunderstandings and non-payment by the client. Further the contractor will attempt to ensure that one payment is not conditional upon another unrelated event associated with the contract works.

22.3.3 Sources of finance

Sources of finance for a project may be internal to the client and taken out of investment capital provisions or reserves held in the accounts. Client support for the project may also arise from tax incentives or local currency loans.
    External sources of funds include:

- Export credits from the contractor's or major transmission and distribution plant manufacturer's countries.
- Commercial loans, often linked to export credits.
- Development aid in the case of developing countries (Asian Development Bank, European Development Bank, World Bank, etc.).
- Provision of future cash flows from future electricity sales associated with the project or counter-trade (barter) deals are also considered. Such cash flows may be used to justify capital raised specifically for the project (see also Section 22.3.6).

22.3.4 Export credit agencies

22.3.4.1 *Introduction*

Export credit agencies are generally set up or sponsored by their relevant government with the primary function of encouraging the export of manufactured goods. The export credit agencies offer guarantees for loan liabilities, preferential insurance rates to cover commercial and political risks, fixed interest rates for certain currencies below those available on the open market, inflation and exchange rate fluctuation protection.

The various criteria which need to be satisfied in order to obtain support from such agencies therefore include:

- majority of equipment and services to originate from the agency country
- financial soundness and reputation of contracting organization
- financial soundness and reputation of client organization
- the category status (based in part upon the Gross National Product, GNP, per capita, inflation, stability, etc.) of the client's country and classified relatively rich, middle income and relatively poor.

There are a number of alternative financing structures used by export credit agencies to support the payment terms associated with an export contract.

22.3.4.2 *Supplier credits*

The contractor or manufacturer takes the lead to arrange for payments, administration and funding to support costs until monies are received from the client or purchaser as the contract proceeds (see Fig. 22.4a).

22.3.4.3 *Buyer credits*

The client or purchaser arranges a loan agreement with a bank. The contractor then receives these funds as the contract proceeds. This is normally the most appropriate finance structure for major transmission and distribution lump sum turnkey projects involving interim or progress payments (see Fig. 22.4b). An advantage to the contractor over supplier credit is that the funding arrangements do not appear on the contractor's balance sheet accounts.

22.3.5 Funding risk reduction

Obviously overseas transmission and distribution projects involve additional financial risks compared to home based construction works. However, even home projects which involve a high level of imported materials will involve financial risk associated with foreign exchange fluctuations. These finance risks may be reduced under buyer or supplier credit schemes by placing the funding work in the hands of experts. A confirming house may, for example, act as a 'go between' client or contractor and the banks, credit agencies, insurance companies, etc. They are able to advise on the best funding arrangements to suit a particular project and are often able to speed up the financing process. However, all such precautions cost money to arrange and execute thereby reducing profit margins. The risks and costs must be carefully weighed up before commitment. Figure 22.5 shows the application of such a simple analysis which tends to highlight areas often overlooked when judging risk in purely money

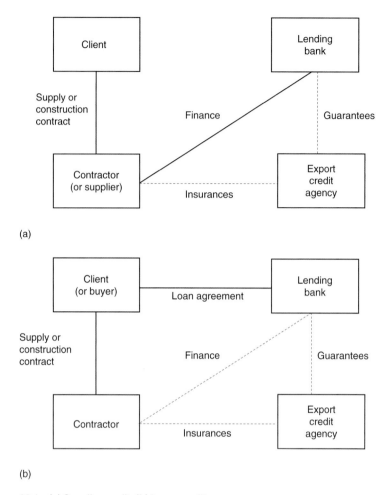

Figure 22.4   (a) Supplier credit; (b) buyer credit

terms. For example client/contractor organization structures, planning systems (PERT analysis – programme evaluation and review techniques using logical programme networks and critical path analysis), sophisticated insurance cover, etc. are also important.

22.3.6 Use of private finance

There is an increasing trend for public authorities around the world to make use of private capital to fund major utility or infrastructural projects. An entrepreneur or, more typically a consortium of firms (which may include contractors, consultants, manufacturers and sometimes banks) will raise the capital to undertake the project, will employ that money to execute the project and will be the managers of the assets resulting for an agreed period, known as the 'concession period'. The concession period may be as long as 30 years, during which time

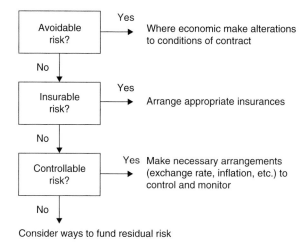

Figure 22.5   Financial risk minimization

they receive income, either in the form of 'rental' from the client body or by having rights (a 'concession') to receive payment direct from the end users (e.g. the electricity network operators using a new transmission system link). The consortium will normally be responsible for maintenance of the asset during this period, and will hand the asset over to the utility or government department concerned at the end of the concession.

The advantage of this approach to the client is that the money raised to buy and install the asset does not appear as a debt on the public authority's balance sheets, thus avoiding any effect on the financial ratios of the body concerned. The corresponding disadvantage may be that the cost of this capital, which is eventually paid for by the client through the rental or the transferred rights to income, is probably higher than the cost of capital raised by the public body itself. This will be the case if the financial markets see the risks of that public body reneging on its debts as being less than the risk of lending money to the concessionaire.

It will also be the case that the risks of construction cost over-run and higher than planned operating costs are transferred to the concessionaire. This should result in reduced costs overall if the bodies chosen by the concessionaire to construct and operate the asset are better able to manage these risks than the client.

For the consortium comprising the concessionaire this approach offers the opportunity to obtain improved return on their capital by building and operating the concession efficiently. There is a risk of a poor return, or even the loss of their capital, if the project costs more to build or to operate than estimated, and the original bid for the private finance project will have included some financial quantification of this risk.

A variant of the private finance project is a joint funding arrangement whereby the client provides just a part of the total capital, and the concessionaire provides the balance. The share of risks and responsibilities will then depend upon the negotiated arrangement in each case.

22.4 PROJECT PHASES

22.4.1 The project life cycle

A project is the process of creating a specific result. Infrastructure transmission and distribution development projects all tend to follow the same general phases as shown in Fig. 22.6.

The concept and planning stages will require the type of financial and economic evaluation together with finance resourcing as detailed in Sections 22.2 and 22.3. In a 'fast track' project, where the client requires very fast completion times, considerable overlapping of the different project phases occurs. This is not to be recommended unless the particular project work is well understood and has been completed before. Such 'fast track' project work has a history of escalating prices and in some cases much longer construction periods than originally envisaged. The most important point is to ensure clear *definition* of the exact project requirements. Although the different project phases cannot usually be compartmentalized into clearly defined boxes with rigid start and end dates, rules should be set up such that sufficient definition is available before commencing each project phase. For example so many percent of steel work design complete before ordering, so many percent orders placed before delivery to site, so much paving and access road laid before commencement of steelwork erection (to avoid the site becoming a quagmire in Winter), etc. Only in this way will rework costs be kept under control in a 'fast track' or indeed a more traditional project.

It is also important to understand the relative magnitudes of the financial commitments involved during the different life cycle project phases. Although planning applications and studies take time the actual expenditure during the concept and definition phases is relatively small. Expenditure will increase as the engineering design phase gets underway and continue to increase very steeply during the construction period. Such expenditure and progress through a project tends to follow an 'S-curve' shape as shown in Fig. 22.7. If a project has to be abandoned then it is obviously necessary to make such decisions as early as possible in order to avoid abortive expenditure.

Modern practice is to introduce a series of "gates" into the project life-cycle. These are formal interim project reviews and decision points to consider performance to date, to revalidate assumptions, and to consider whether to proceed with the project in the originally envisaged manner. The UK Office of Government Commerce (OGC) has issued excellent advice in this area and suggests "gate" reviews of the project strategy, the business case, the contract strategy, project initiation (go/no-go decision to let the major contract), handover, and feedback upon competition to maintain a "lessons learned" register. On large projects the development of the project strategy and the work to achieve the correct level of definition may involve considerable sums of money. As such this early work should be treated as a project in itself with all the rigours of standard project management discipline.

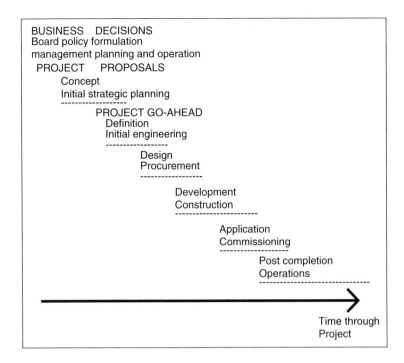

Figure 22.6   Project phases

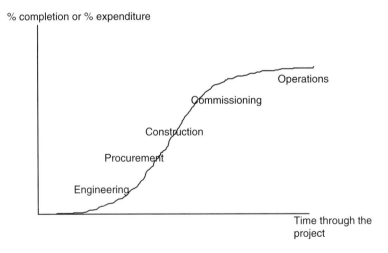

Figure 22.7   Typical project 'S-curve'

22.4.2 Cash flow

Typical project cash flows resulting from a substation construction project might be as shown in Fig. 22.8 where the revenue from the additional customers

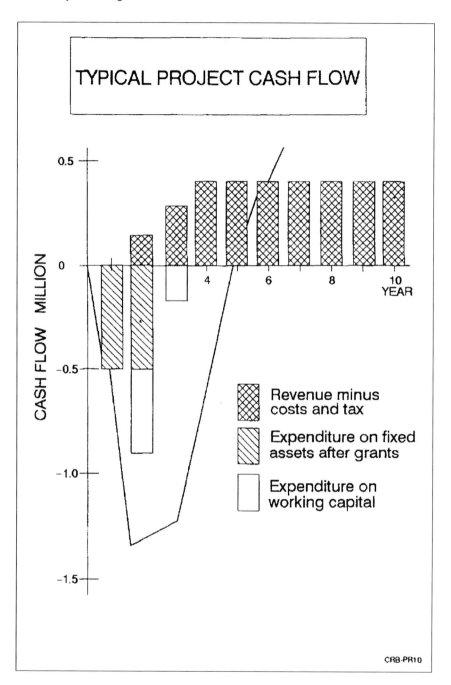

Figure 22.8   Typical project cash flow

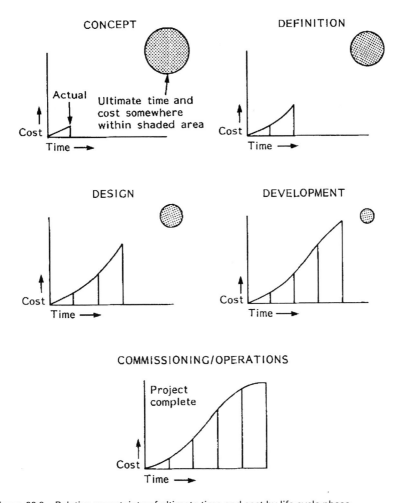

Figure 22.9   Relative uncertainty of ultimate time and cost by life cycle phase

is fully attributed to the project itself. Such projections are used in the initial financial assessment of the project and should be regularly updated as the project proceeds. There is always the necessity to appreciate the relative uncertainty of the ultimate time duration and cost during the different project life cycle phases (see Fig. 22.9).

22.4.3 Bonds

22.4.3.1 *General*

The contractor may be required to provide bank bonds as part of a construction contract. The bonds are paid for by the contractor as a small percentage of their full value. Such bonds are usually held by the banks and may be called upon by the client to be converted into cash payments if the contractor fails to perform in accordance with the contract. The bank and the contractor arrange a back-to-back agreement such that if such bonds are called for payment by the client then the contractor either directly, or using insurances, pays the bank the money owing. An explanation of the effect of bond guarantees on cash flows during a construction contract based upon the International FIDIC Terms and Conditions of Contract is given in Fig. 22.10.

22.4.3.2 *Tender bonds*

Initially a bond may be required from the contractor by the client as early as the tender stage before the final contractor has even been chosen. 'Tender Bonds' are introduced in order to discourage a large number of contractors from tendering for the works without any real intention of accepting the job. Such an occurrence would result in abortive tender evaluation by the client and possible costly retendering. Tender bonds may not be returnable to tenderers by the client after the tender process and they therefore act as an insurance to the client that only serious contenders apply for the work.

22.4.3.3 *Performance bonds*

During the course of the construction contract the client may wish the contractor to take out a 'Performance Bond'. This acts as a guarantee of due performance by the contractor and may be typically 10% of the total contract value.

22.4.3.4 *Maintenance bonds*

After construction the contractor usually has a continuing obligation to the client for the repair of faults in the works due to the contractor's bad workmanship or materials employed. This is known as the 'maintenance period' and is 1 or 2 years for mechanical and electrical projects. The client may withhold monies from the contractor to ensure that he will return to repair such defects during this maintenance period. The contractor may offer the client a 'maintenance period bond' in return for 100% cash payment upon handing over the completed works to the client. In this way the contractor obtains full monies but guarantees, against the possible encashment of the 'maintenance bond', a continuing obligation during the maintenance period.

22.4.4 Advance payments and retentions

The client may advance to the contractor say 10% of the contract value upon signing the agreement between client and contractor to carry out the works.

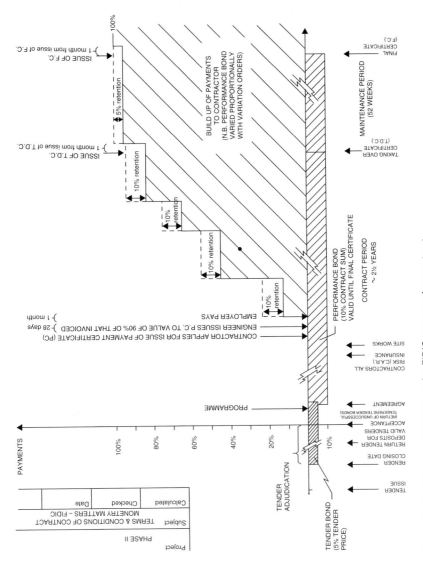

Figure 22.10   Finance payments under FIDIC construction contract

This may be tied to receipt of the contractors detailed programme for completion of the works or similar requirements. The idea is to assist the contractor in the early stages of the contract. Presumably the client has already budgeted for such provisions and is therefore best able to carry this burden which is intended to keep the contractor's prices down.

During the course of the contract retentions are made on progress payments to the contractor by the client. For example if a 30% progress payment is due to the contractor upon certified completion of 30% of the contract works then the client may deduct a 10% retention to recover in stages the originally made 10% advance payment. In addition retentions, say a further 5%, may be held by the client to cover the contractor's maintenance period obligations. An example of the effect of such advance payments and retentions upon the contractor's project cash flow is given in Fig. 22.10.

22.4.5 Insurances

A wide variety of insurances are required to be taken out by client and contractor both under national legal requirements (generally covering safety) and the specific requirements of the particular construction contract being employed. These include contractors all risks (CAR) insurance. In particular, insurances to cover damages to equipment from the time it is being manufactured, through transportation to site, and during erection and commissioning on site are necessary. The cost of such equipment, if damaged whilst in the care of the contractor, may be more than he could be capable of covering and lack of speedy replacements could put the project in jeopardy.

22.4.6 Project closeout

The possible long 'tail' on the project life cycle S-curve at the end of the project requires careful management. Often certain key items, which may not seem significant to the contractor, cause delay in receipt of final payments. For example such items as correctly delivered final commissioning test records, operations and maintenance manuals or as-built drawings are often all tied into release of payments even though the works themselves have been handed over to the client.

22.5 TERMS AND CONDITIONS OF CONTRACT

22.5.1 Time, cost and quality

Any project incorporates a degree of risk which once initiated may be countered by insurances, payment bonds, advance payments and retentions as described in Section 22.4. The type of contract employed to complete the project works is also important in order to match risks against either the client or contractor who

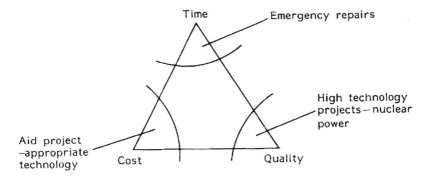

Figure 22.11   Relative risk to contractor and client for different types of construction

is best able to carry them. It is totally immature to expect a project to be other than a compromise between time, cost and quality (see Fig. 22.11). For example the priorities for different projects may be different and therefore demand different contractual treatment:

1. Construction of a nuclear power station will require the highest degree of quality at the expense of cost.
2. Repair of a damaged primary substation transformer might demand priority to be given to the time of the repair.
3. A Third World aid power distribution project on the other hand may involve very tight cost constraints leading to well tried and proven design of sufficient quality to give lasting service under a minimum of maintenance.

Different types of contract allow for better management of these different priorities and risks.

22.5.2 Basic types of contract

Figure 22.12 indicates the different relative levels of risk to client and contractor when using different forms of contract for the completion of the project works. Of absolute and key importance is the adequate *definition* of the works. A good guide to the preparation of specifications is given in BS 7373. The differences between these different forms of contract and typical uses are described below.

1. Cost reimbursible with % fee – The contractor agrees to carry out the work for whatever it actually costs him to complete it (as substantiated by receipts, time sheets, etc.) and then charges this amount plus a percentage fee based upon these costs. The disadvantage or risk to the client is that the contractor may not keep his costs under tight control. There may be no particular incentive for the contractor to keep his costs down since he will get a larger fee the longer and more expensive he makes the job. Such conditions of contract may

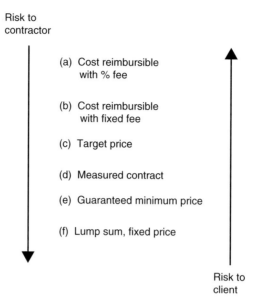

Figure 22.12   Relative risk to contractor and client for different types of construction contract

be necessary for research work where only a few contractors have the capability and the outcome may not be known for certain. Client and contractor must be in complete harmony, working for the same goal, for this type of contract to be considered.

2. Cost reimbursible with fixed fee – This form of contract puts a limit on the costs by imposing a fixed fee upon the contractor. Often this form of contract is used by engineering design consultants. Normally the reputation of the consultant is at stake and abuse of such conditions is unlikely; it still puts the client at risk if increased costs due to poor budgetary control by the contractor, but this may be a acceptable if it enables small specialist firms to be employed.

3. Target price – The contractor agrees to perform the works within a given cost ceiling and/or time frame. If the contractor manages to complete the works within budget or time frame then a bonus is paid. This type of contract has been very successful for such projects as motorway road repairs where rapid completion is required by the client and the incentive of a large bonus has driven such works to a successful conclusion by the contractor.

4. Measured contract – A Bill of Quantities is prepared to describe the works in great detail. Rates are attached to each item of work and the contractor is paid according to the amount of work performed. For example, a rate is applied to the supply and laying of cable in a trench per linear meter of cable laid.

Description	Unit	Rate/unit	Quantity	Total
185 mm^2 11 kV PILC cable laid in trench	linear metre	£100.00/m	1500 m	£150 000

There is a risk to the contractor if the bill of quantities does not define in sufficient detail the work involved. He may underprice the work at the tender stage and have no recourse if he did not fully understand the full scope of the work involved. It would be necessary to check in the example above if the rate should also include for the cutting of an asphalt surface, digging of the trench, the sand surround to the cables, inclusion of cable tiles, back filling the trench and possible reinstatement of the asphalt surface.

Interim payments may be made to the contractor on a regular basis based upon a measurement of the work completed. This type of contract is particularly common for building services work, cable laying and overhead line construction.

Variations to the estimated quantities in the original bill of quantities invariably occur in practice. As long as these increases or decreases do not materially affect the overall intent of the contract works (often judged by whether the overall contract value has changed by more than $\pm 15\%$) then the rates detailed in the contract remain valid. The risk to both contractor and client is therefore kept within manageable bounds.

5. Guaranteed minimum price – The client and contractor agree a guaranteed minimum price for the completion of the works. This may then be varied should the scope of the works change during the contract period. A guaranteed minimum price reduces the risk to the client but increases it for the contractor. This type of contract requires good definition and a minimum of interference and change requests by the client during the contract period.

6. Lump sum, fixed price – The client and contractor agree a fixed price for carrying out the work. The risk here is greatest to the contractor since unforeseen circumstances may alter the cost of the works considerably. The client has effectively placed the risks involved with unforeseen circumstances onto the contractor with this type of contract. Of course the contractor will price the works accordingly with a larger than normal contingency to cover any lack of definition. It is important with this type of contract that the client does not impose significant changes to the scope or definition of the work during the contract period. If the client does this then the contractor will be able to correctly claim for extra costs. A form of contract such as this is useful where the design, supply and installation of transmission and distribution equipment is required to be placed totally in the hands of a competent contractor.

22.5.3 Standard terms and conditions of contract

22.5.3.1 *Forms of contract*

A contract is an agreement between two or more parties such that if one party fails to do what he has promised another party will have legal remedy. The contract therefore embraces both statute and common law.

A variety of standard conditions of contract are available for transmission and distribution construction works. These may be broadly classified into whether the works consist of:

- Supply only of materials
- Supply of materials and supervision of erection/installation
- Supply and installation.

In addition a degree of design work may also be required from the contractor in all these variants.

Model forms of conditions of contract have been written for application within the engineering industry. Considerable thought has gone into these documents and many of the contract clauses are interrelated. Therefore any modification of one clause may have 'knock on' effects throughout the conditions of contract. Only after very careful consideration and certainly only after expert advice should any attempt be made to amend the standard model form conditions of contract.

IMechE/IEE Model Form 'A' and the more recent MF/1 Conditions are suitable for detailed design, supply and installation transmission and distribution plant construction contracts. Similar IMechE/IEE Model Forms are available for Supply Only (Model B2) and Supply and Supervision of Erection (Model B3) types of work. The ICE and FIDIC Model Forms are useful where a large element of civil engineering works is involved. IChemE forms of contract are especially suitable for large 'design and construct' projects. RIBA Model Forms are very different in concept and intended for building works contracts and not normally suitable for substation or overhead line construction.

Less adversarial terms and conditions are now gaining acceptance. The NEC, published by the Institution of Civil Engineers, is now well tried and tested and attempts to resolve 'claims' as they arise. Such contracts may also include associated client/contractor partnering arrangements.

Note:

IMechE	–	Institution of Mechanical Engineers
IEE	–	Institution of Electrical Engineers*
ICE	–	Institution of Civil Engineers
FIDIC	–	Federation International Des Ingéniurs - Conseils
RIBA	–	Royal Institute of British Architects
IChemE	–	Institution of Chemical Engineers
NEC/ECC	–	New Engineering Contract/Engineering Construction Contract

*The IEE became the IET (Institution of Engineering and Technology) in 2006, following changes and mergers agreed by the members in 2005.

22.5.3.2 *Role of the consulting engineer*

It is quite normal for the client to employ an 'Engineer' to act as an impartial and technically competent adjudicator between client and contractor for the

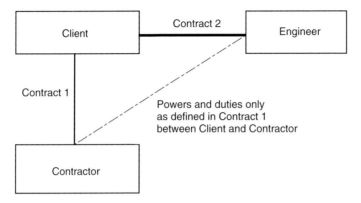

Figure 22.13   Privity of Contract – Typical Client/Contractor arrangement using a Consulting Engineer for transmission and distribution construction contracts

administration and perhaps the design of the contract works. The client sets up a contract between client and contractor and a separate contract between client and engineer. There is no privity of contract between contractor and engineer. However, the consulting engineer has specific duties and powers under the terms and conditions of the client/contractor contract and is able to instruct the contractor to perform during the course of the works. The arrangement is shown in Fig. 22.13.

The consulting engineer is often brought in by the client at an early stage in the project life cycle to carry out the technical system studies. The engineer may advise the client on the best form of contract for the particular work to be performed. The engineer may also assist or complete the financial and economic project evaluation and prepare the project outline design. This work is then converted into a tender document. The engineer will then supervise the issue of tenders and produce an independent tender adjudication from which the most appropriate contractor is selected to carry out the works. Typically, the engineer will check the contractors detailed designs during the contract period and also supervise the installation on site.

Often the client has considerable technical resource and may well have a contracts group which is capable of fulfilling all the roles of the consulting engineer listed above. In this case the client takes on the dual role of both client and engineer.

22.5.3.3 The design and construct contract

Recently there has been a trend away from the use of the consulting engineer in large mechanical and electrical (M&E) multi-discipline contracts. The client appoints a single 'management contractor' or 'main contractor' to complete the whole works on his behalf. This requires a mature approach to client/contractor relationships. The method is particularly suitable where the client does not have a large technical resource, is prepared to put faith into a contracting organization and will not interfere with the design process without understanding that this is likely to cause changes and will result in cost escalation. The advantage to the client is that the responsibility for interfaces between the different engineering disciplines or sub-contracts is all in the hands of a single main contractor.

If the work is well defined such an arrangement does not place undue risk on the contractor and, all things being equal, prices for the work will be competitive. However, more often than not the work will not be clearly defined (since at the time of tender it has not been fully designed) and the work will be described more in terms of a performance requirement. Such contracts may run into difficulties if risks have not been placed with those best able to handle them. For example it is difficult for a contractor to handle all aspects of planning permissions or to match the design of the plant to the client's particular operating procedures if these are not clearly specified at the time of tender. Since a contract should involve co-operation between client and contractor with the goals of getting the work completed on time, to a given cost and to a given standard such an equitable split of risks should be possible given a mature approach by both parties from the outset.

In order to try to reduce the antagonistic relationship between client and contractor much thought is currently being given to new engineering conditions of contract (see Section 22.5.3.1). Further partnering arrangements have been introduced whereby client and contractor meet on a regular basis at all levels within the organizations to resolve issues.

22.5.3.4 *Private finance contracts*

There is no standard form of private finance contract, although with experience some governments and public bodies are developing 'standards' to meet their particular requirements. The consequence of the lack of standardization, together with the large sums of money involved, the correspondingly significant risks and the special aspects of each major project, is that private finance projects usually involve a long period of contractual negotiation. Typically the client body will solicit interest by public notice, will invite tenders from four pre-qualified bidders, and on receipt of tenders will select one as 'preferred bidder'. Negotiations are then started with the preferred bidder, during which technical and financial options may be considered, risks re-apportioned and contract terms finalized, before the bidder puts in a 'best and final offer'. If this is unacceptable to the client, he may chose to re-start negotiations with one of the other bidders, but depending on the original terms of the bid will probably have to compensate the originally selected preferred bidder for the costs incurred during the negotiating period. The total period from bid invitation to award of the contract has been known to take more than 2 years, and legal and technical costs can be high.

The format of the contractual arrangements will vary according to the nature of the project and the finally agreed allocation of risks, but typically the successful firm or consortium of firms will establish a company for the purpose of holding and operating the concession. This is called a 'Special Purpose Vehicle' (SPV). The capital provided for the SPV will probably come only partly from risk capital put in by the owners; they will employ a financial consultant to advise on the optimum financial structure and to assemble a consortium of banks to lend the rest of the money needed for the project. This may involve a separate bidding process, to obtain the best offer.

In order to get the project built, the concession company will then let a contract to a construction company, or consortium of companies, specializing in the sector concerned. If one of the concessionaire partners is a contractor or a manufacturer they will be in receipt of at least part of the work. The form of this construction contract, which is in effect a sub-contract to the concession contract, will be more conventional – probably a design and construct contract with an element of partnership or risk-sharing (see Section 22.5.3.3). Often a further separate contract will be let for the operation of the asset. This enables the best specialists to be used in each area, and again one of the member companies, such as an electricity utility, may be the obvious choice. However if the operating contractor is different from the construction contractor it does leave a difficult contractual interface between the two when it comes to project handover, operational start-up problems, etc. One way of minimizing this is for the SPV to be more than a financial framework, and instead to take on staff seconded from partner companies and/or the client and to operate the concession itself.

It should be apparent from the foregoing that the costs of bidding and of establishing the SPV mean that this form of financing is only appropriate for large projects, and indeed it is only on large and expensive projects that public bodies will be concerned at the effect on their debt ratios.

22.5.4 Key clauses

22.5.4.1 *Introduction*

This section gives an explanation of the key clauses contained in model forms of contract applicable to transmission and distribution construction works. Figure 22.14 is a network showing the interrelationship of these different clauses and their relevance during the construction phase of the project life cycle.

22.5.4.2 *Programme*

The client may issue a *letter of intent* to the chosen contractor following the tender adjudication process. There is no assurance that this will result in a contract and such letters should be treated with caution.

The contract commencement and overall contract period is defined in the *letter of acceptance* unless stated elsewhere.

Where *time is of the essence* (a wording which introduces much risk to the contractor) the client has the right to terminate the contract (and payments to the

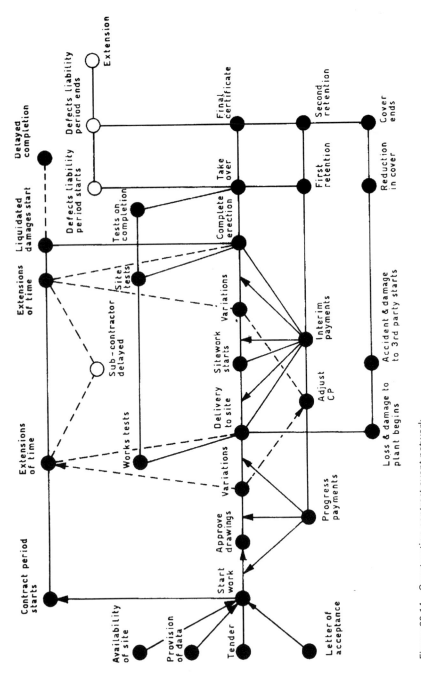

Figure 22.14   Construction contract event network

contractor) if the contractor fails to complete the work on time. An *extension of time* may be granted if the delay has been caused by client, contractor or circumstances beyond the control of either party (for example an industrial dispute in one of the equipment manufacturer's factories). It is in the client's interests to deal with extensions of time as they arise or on a regular basis. Extension of time claims should not be left until the end of a contract or the contractor may well decide to be late by a certain period and be the only party to have all the necessary paperwork to justify this.

Variation orders may be issued by the engineer to the contractor. Time related costs may be involved together with programme implications. The scope of the variations must not alter the original intent of the contract and are limited to perhaps $\pm 10\%$ of the contract value. As long as this limit is not exceeded the prices for the varied works must be based upon those prices included by the contractor in his original tender. It is in the contractor's interests if the client substantially alters the scope of the works since this will allow the contractor to fix prices for the excessive changes without recourse to open tender.

If the contractor fails to complete the work on time then he will be in breach of contract. In practice, things are not so clear cut on multi-discipline contracts. It is clear that if the work is delayed the client will probably suffer some loss of revenue which was originally intended to arise from the completion of the project. A *liquidated damages* clause is therefore placed in the contract as a genuine pre-estimate of the damage likely to flow from the breach. The damages, whether actually suffered or not by the client, are *not* a penalty. The liquidated damages clause tends to set down in advance a limitation on the contractor's liability. An example of the preparation of a liquidated damages claim is given in the case study included in Section 22.7.

During the course of the contract the engineer will witness or check *tests* carried out at the manufacturer's works, on site and upon completion of the works to show that all is in order.

Upon completion of the works the contractor receives a *Taking Over Certificate* (often tied to release of monies to the contractor). The contractor is then responsible for a *maintenance or defect liability period* when defects caused by bad workmanship or materials have to be rectified. At the end of the maintenance period the contractor receives a *Final Certificate* representing the end of the contractor's obligations under the contract. The contractor will, however, still remain liable under *tort*. If the contractor's conduct constitutes negligence then they should still have a liability in tort for a period of several years (6 years in UK) from the time the action was known or reasonably thought to be known.

22.5.4.3 *Payments*

Figure 22.10 shows the build up of monies flowing to the contractor during the course of the contract. The different payments, retentions and bonds are explained in Section 22.4. *Contract price adjustment* (CPA) clauses are often included in contracts involving the supply and installation of cables where the raw material and manufacturing labour costs are likely to fluctuate on the open market. The contractor should ensure that *Progress Payments* should include for variation orders and CPA if applicable. Plant for which progress payments have been made should be marked as the client's property. The client should also make sure that the contract allows temporary possession of dies, templates, etc. such that the work may be completed should the contractor become bankrupt during the course of the project.

22.5.4.4 *Insurances*

Civil engineering contracts usually have to include *contractors all risks* (*CAR*) insurance. Reasonable precautions must be taken to ensure that damages to both the works and plant are rectified. The insurance policies must be adequate since an indemnity does not alter the legal position and is only as good as the insurance cover or the financial status of the parties involved. Both client and contractor must therefore be aware of the financial standing of the other party before a contract agreement is signed. Indemnity clauses may be included in the contract to limit the contractor's liabilities under tort. In such circumstances the contractor remains legally liable but may have the client underwriting his liabilities.

22.6 TENDERING

22.6.1 Choosing the contractor

Contractor selection may be achieved by direct negotiation between client and perhaps only one well-known contractor or via the process of open competitive tendering. In order to avoid a large number of bids from, perhaps, unsuitable tenderers the client may restrict the number of prospective tenderers by an initial pre-qualification and use of a short list of suitable contractors. The advantages and disadvantages of these methods are described in Table 22.3. Note that the tendering process for large public works within the European Union (EU) is covered by legislation.

22.6.2 Estimating

The elements of the estimate for the works includes:

- cost
- contingency
- overheads and profits including a risk margin
- effect of the type of conditions of contract employed.

Table 22.3   Methods of tendering

Type of tender process	*Advantages*	*Disadvantages*
Open competitive tendering	(1) Allows many and perhaps new contractors to bid for the work. (2) Allows tender list to be without bias. (3) Ensures good competition. (4) Prevents contractors from fixing prices between themselves	(1) Long tender lists – wasteful of client and contractor time and monies. May discourage otherwise good contractors because of the remote chance of obtaining the contract. (2) Tender adjudication open to scrutiny and question if lowest bidder is not granted the contract.
Selective tendering	(1) Only competent contractors invited to tender for the works. (2) Cost of tender adjudication reduced. (3) Competing contractors may have freedom to allow adequate profit level and therefore flexibility and stability.	(1) Must avoid favouritism when drawing up list of tenderers. (2) Must continually review tender list in order to allow new contractors to compete. (3) Possible higher tender prices in comparison with open tendering. (4) Possible price fixing between small band of contractors. (5) May conflict with requirements of legislation or of funding authority.
Negotiated tenders	(1) Contractor may give advice to client during tender negotiations. (2) Work can commence at an early stage (materials may be ordered on a cost plus basis before overall contract arrangements have been agreed). (3) May be only way to encourage a realistic bid from a specialist contractor.	(1) Cost of the work may be higher than if completed on a Competitive Tender basis. (2) Only really applicable to specialist Contractors with good reputation. (3) May conflict with requirements of legislation or of funding authority.

The tenderer will cover risk by modifying the overall bid price and placing qualifications and clarifications in the offer in order to try and improve the definition. A bidding checklist is included in Appendix B of this chapter to assist in the process of appreciating some of the major factors involved. Contingency covers the probability of cost over/under-runs for specified items of cost estimate that may be estimated in money terms. Normally a proportion of total contingency is included in the final tender sum and not the total amount. Risk margin covers uncertainties which do not relate to specific items in the cost estimate but which are of a wider nature. This is often difficult to estimate in money terms and requires a considerable degree of expertise.

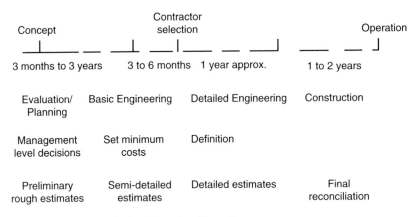

Figure 22.15   Estimates during the project life cycle

It is important for both client and contractor to have a clearly defined estimating policy. Without such it is possible to overestimate the cost of the works by double counting contingency allowances. Underestimating by not fully understanding the implications of the terms and conditions of contract is also an area of error.

The types of estimate required will vary from the rough estimate at the very early concept stages of the project to definitive estimates prior to the detailed engineering/construction phases.

22.6.3 Tender evaluation

Different countries and organizations have different methods of evaluating tenders. Usually very strict guidelines are enforced in order to avoid any favouritism for one contractor or another. It is not necessary to always accept the lowest bid price since the bid may not be technically compliant and may not have met the commercial conditions required. Therefore it must be made clear at the tender stage what tender evaluation method is to be employed so that there can be no complaints. Consider the following examples all of which are valid approaches:

1. The bid closest to the average of all bids might be accepted on the basis that the average bid price from a large range of tenderers might be considered the most correct estimate of the cost including a competitive profit.
2. The lowest bid is accepted provided that it is not less than, say, 80% of the client's original estimate for the work on the basis that this will give the most competitive, but still realistic, bid from a competent contractor.
3. The lowest bid provided it is not less than, say, 90% of the average of all bids on the basis that a very low bid from a tenderer might indicate that he does not have a full appreciation of the work involved.

For a typical transmission and distribution contract considerable emphasis will be placed upon technical compliance as well as cost. Therefore the technical

Table 22.4   Possible relative weightings for cost criteria

Technical	Management	Financial	Manufacturing	Quality control
40 Design approach	20 Plan	30 Price	20 Experience	20 Quality plan
25 Technical capability	26 Organization	15 Strength	15 Plan	25 Operational system
15 Development plan	14 Manpower	15 Accounting system	13 Facilities	25 Technical capability
8 Reliability	20 Controls	10 Capability	6 Skills	15 Reliability
7 Specification conformity	5 Experience	20 Cost control	10 Tooling	15 Record system
5 Release of software	10 Reliability	10 Estimating	16 Controls	
	5 Labour relations		12 Improvements plan	
			5 Training	
100 Total	100 Total	100 Total	100 Total	100 Total

merits of the different tenders have to be evaluated by a fair method of comparison. One such method is to apply a 'points system' score for each major part of the tender under such sections as:

- technical
- management
- financial
- manufacturing
- quality control

Each section is then broken down into criteria which are considered significant and a relative weighting as to the importance of each of these factors allocated. Such an approach might be as shown in Table 22.4.

The advantage of this approach is that it applies exactly the same criteria to each bid and allows examination of the fair approach adopted. The disadvantage is that critical factors may get overlooked. For example if the tenderer does not have the correct design approach then the overall project may fail to function correctly. A high point score has been given to reflect this in the above example. However in some cases such as for the development of a SCADA (supervisory control and data acquisition) system for distribution substations the correct design approach may outweigh all other factors and must be correct. For this reason substation and overhead line tenders are more normally evaluated on technical grounds item by item for compliance with the specifications. Engineering judgement is used to compare differences and their relevance without the 'straight jacket' of a points system.

22.7 MODEL FORMS OF CONTRACT – EXERCISE

The questions given in this section are designed to allow readers to familiarize themselves with standard Model Forms of Contract associated with design, supply and installation transmission and distribution project work. The questions are applicable to IMechE/IEE, MF/1, ICE and FIDIC Model Forms, copies of which will be needed to answer the questions. The relevant clauses should be quoted in the answers together with a summary and interpretation of these clauses.

1. The engineer (E) under a contract has appointed in writing an engineer's representative (R) with special powers. E is killed in an accident. R is aware of this but continues to issue instructions to the contractor.
   Should the contractor follow these instructions if he is aware of E's death?
   How should the situation be regularized?

2. A contractor issues a sub-contract for the supply of reinforcing rods for a substation concrete structure which houses plant which he is responsible for manufacturing and installing. Due to strike action in the sub-contractor's works, the main contractor is delayed for 28 days. He claims from the purchaser for a 28-day extension of time and associated costs. Advise the engineer on what action he should take.

3. A contractor who is in cash flow difficulties decides that he can ease his cash flow situation by delivering to site materials and plant well ahead of programme. His intention is to claim for the value of these materials in an interim payment certificate.
   Will this scheme work?
   What are the rights of the parties?
   What would be the situation if a receiving order (bankruptcy suit) were made against the contractor?

4. What rights has a purchaser against a contractor who persistently refuses to follow the engineer's reasonable orders?

5. What is the difference between a progress and an interim payment certificate?

6. How may a purchaser recover liquidated damages?

7. A contractor who is in financial difficulties asks the engineer to take over parts of the works as and when they are completed.
   Can the engineer do this?
   Can the contractor enforce it?
   What would be the situation if the engineer issues a qualified taking over certificate?
   Has the engineer the powers under the contract to do this?

8. A contract has been completed and after the issue of the final certificate the engineer discovers that the wrong grade of stainless steel has been used as specified in the contract technical specifications for the plant.
   What, if any, are the rights of the purchaser?

9. In what way do the rights of the contractor after suspension of the works by the engineer differ from those when delivery has been suspended?

10. After the works have been taken over, under what circumstances can the contractor have access to the works?

11. Under what circumstances can a purchaser refuse or restrict the payment of sums due to inflation of costs?

12. Under what circumstances can the engineer order overtime or other acceleration measures without extra cost to the purchaser?

13. In a £250 000 contract for the supply and erection of substation plant, the works are divided into three sections as follows:

Section A – Value £100 000   Completion date – 31.3.91
Section B – Value £100 000   Completion date – 14.4.91
Section C – Value £ 50 000   Completion date – 14.4.91

Due to the design of the works neither Section A or Section B can be used until Section C is completed.

Variations are ordered by the engineer with associated extensions of time as follows:

Section A – Value £2000   Extension of Time granted – 2 weeks
Section B – Value £5000   Extension of Time granted – 4 weeks
Section C – Value £500   Extension of Time granted – nil weeks

The actual completion dates are as follows:

Section A – 14.4.91
Section B – 26.5.91
Section C – 2.6.91

The percentages associated with liquidated damages included in the appendix to the conditions of contract are 1% of the contract value up to a maximum of 6%. As the client's advisor prepare a short paper to advise him what liquidated damages (if any) are due arising from the late completion of the works.

APPENDIX A

PROJECT DEFINITION/QUESTIONNAIRE
(for use by design consultants)

Schedule A – Client/consultant data

A1   Project and client
A2   Services required of the consultant
A3   Regulations and specifications

A4   Estimating data
A5   Documentation
A6   Progress inspection and shipping
A7   Site data and conditions
A8   Site construction and data
A9   Extent of work required by consultant
A10   Preferred suppliers and equipment

Schedule B – Local information

B1   Drawings, maps and clearances
B2   Details of existing network

Schedule C – Project data

C1   Substations
C2   Feeders
C3   Switchgear
C4   Transformers
C5   Busbars
C6   Isolation and earthing switches
C7   Protection
C8   Control and communications
C9   Overhead lines
C10   Underground cables
C11   Civil works

SCHEDULE A – CLIENT/CONSULTANT DATA

Schedule A1 – Project and client

Title of Project

Consultant's Job Number

Location of Project Works

General Description of Project Works

Client

Address

Name of Principal Contact(s) and Title

Telephone No.
Fax No.
Telex No.

Governing Body of Area
Address, Name of Principal Contact(s) and Title

Telephone No.	Fax No.	Telex No.

Local Agent
Address, Name of Principal Contact(s) and Title

Telephone No.	Fax No.	Telex No.

Schedule A2 – Services by consultant

Site Selection

Site Survey and Soil Mechanics

Development Studies

Economic and Financial Evaluation

Engineering Design (see note below)

Ordering

Expediting

Inspection at Works

Shipping and Insurance

Custom Duties and Clearance Through Port

Supervision of Construction and Erection

Supervision or Carrying Out Testing

Supervision or Carrying Out Commissioning

State Extent of Services for Initial Operation of Plant, if any

Selection of Operating Staff (Minimum Qualifications, etc.)

Other Services: General

At Site

Note: There are two main consultant engineering roles.

(a) The traditional British method in which the consultant produces the general design and specifications. The consultant controls the tender process and produces a tender evaluation report with recommendations. The appropriate vendor then produces the detailed design and materials lists. By this method major items of plant would be controlled by the contractor and ordered complete.
(b) The American method in which the consultant produces the detailed and complete design and materials lists. By this method all items of plant are generally subject to separate purchase orders.

The difference in the amount of work for the consultant entailed in these two methods is considerable. It is therefore essential at the outset that the method to be adopted is clearly understood by both client and consultant (see also the note to Schedule A3).

Schedule A3 – Regulations and specifications

Building Regulations or Acts

Electricity Regulations or Acts

Health and Safety Regulations or Acts, including medical requirements (vaccinations, etc.)*

Fire Regulations or Acts

Labour Laws

Import/Export Regulations

Company Law

Insurance Regulations

Currency Regulations

Other

Is there any preference for any standards which the works or plant is required to be built in compliance with, and what precedence is given:

British Standard Specifications (BS)
International Electrotechnical Commission (IEC)
Other (ANSI, DIN, VDE, etc.)

*In the UK the Construction (Design and Management) Regulations require the clear allocation of the duties of client, client's agent, planning supervisor, principle contractor and designer. It is the responsibility of the client to designate their planning supervisor, client's agent (if any) and principle contractor, although in practice a consultant or major contractor may provide advice to the client on such appointments.

Schedule A4 – Estimating data

State which of the following costs are applicable and who is to meet them.

Item	If Applicable		To be met by:	
		Client	Consultant	Vendor/ Contractor
Site Selection				
Purchase of Site				
Leases, Ownerships, Wayleaves, for the proposed site land				
Import duty				
Sales Tax				
Investment Tax				
Capital Levy				
Development Charge				
Shipping				
Freight Insurance (Marine – War Risks)				
Insurance of Plant at Site				
Special Dock Charges				
Handling Charges from Railhead or Port to Site				
Right of Way Charges				
Site Staff Salaries				
Site Staff Expenses				
Wayleaves for Overhead Lines, Cable Routes, etc.				
Site Storage				
Roads Outside Site Area				
Site services: Electricity Water Sewage				
Dredging or Land Reclamation				
Drainage and Sewage				

Are any of the above to be included in the consultant's financial capital estimates for the project?
Are there any local price controls?
Are import licences required?
Are export licences required?

Remarks:

Schedule A5 – Documentation

State client's and consultant's requirements for documents giving number of copies required and routing.

Item	No. Copies	Where Routed
Project Estimates		
Specifications (as issued)		
Completed Specifications (for bids, tenders, etc.)		
Tender Analysis and Recommendations		
Purchase Order Issued by: If issued by Client is Consultant to provide draft order?		
Progress Reports Frequency of Issue: Manufacture (Yes/No) Design (Yes/No) Inspection and Test (Yes/No) Construction (Yes/No)		
Progress Photographs		
Shipping Specifications		
Advice Notes		
Invoices		
Correspondence to Client		
Correspondence to Vendors		
Drawings for Approval: Building Detailed Plant Layout Cable and Overhead Line Routes Schematic and Wiring Diagrams Equipment		
Final Drawings on: Film Transparencies 35 mm Aperture Card Micro Film Computer Scan Disks Prints		
Operating and Maintenance Manuals		

State client's preferred size of drawings:

State dimensional units to be used on drawings:

List any special forms (progress, shipping, payment certification, etc.) which client prefers to use:

Schedule A6 – Progress, inspection and shipping

The following information should be submitted on separate sheets where consultant's services are employed for progress, inspection and shipping.

Note that where the consultant is also employed for design and ordering items 3, 4, 7 and 8 below will be required at an early stage for release of plant specifications and orders:

1. A copy of all orders giving full details and delivery requirements, etc.

2. Information regarding inspection as to what standards are applicable (IEC, etc.).

3. Details regarding shipping – shipping marks, marking of cases, packaging against special items such as corrosion, humidity, rain, high temperature, rough handling, etc.

4. Applicable shipping payment methodology – FOB, FAS, CIF, etc.

5. Import and export licenses, etc.

6. Insurance

7. Port of disembarkation and maximum port lifting capacity, capability to handle containers, etc.

8. Maximum weight and sizes for cases for road and rail transport to site.

Schedule A7 – Site data and conditions

1. LOCATION
State the location of the sites, if location plans are attached and if site layout plans are attached. If ordnance or geological maps are required state where available and attach if possible.

Is site suitable for future expansion?

2. SITE LEVELS
Datum level at site(s).

Datum reference.

Nearest bench mark (if any).

3. WEATHER CONDITIONS
Average rainfall per annum
Maximum rainfall per annum
Minimum rainfall per annum
Severity of rainstorms
Air temperature (max.)
Air temperature (min.)
Prevailing wind direction
Average wind velocity
Maximum wind velocity
Barometric pressure (max.)
Barometric pressure (min.)
Relative humidity (max.)
Relative humidity (min.)
Is condensation severe due to rapid temperature change with high humidity?
Maximum snowfall
Frequency of lightning
Sand and dust storms
Earthquakes
Frost penetration
Soil temperature at cable laying depth (max.)
Soil temperature at cable laying depth (min.)
Soil thermal resistivity
Ground resistivity (if known for different substation sites)

4. GEOLOGICAL DATA
State if any geological surveys are available and if possible give nature of ground strata.

State data regarding test holes made in recent times, if available. Give soil bearing pressure and characteristics, if known and if piling is likely.

Details of any foundation troubles experienced with existing buildings.

Detail any troubles with surplus surface water drainage.

Minimum depth of water table to be encountered.

5. SOIL CORROSION
What is experience regarding corrosion of buried pipework and cables? State if any known satisfactory practices.

6. SEWAGE AND DRAINAGE
What are the proposed arrangements for handling sewage and/or drainage?

7. LOCAL AUTHORITY WATER SUPPLY
What is the availability of water drawn from the local water authority?

8. ACCESS TO SITE
Describe the access to site(s):

(a) Personnel

(b) Light Goods

(c) Heavy Goods (transformers, etc.)

Nearest Airport	Distance from site(s)
Nearest Seaport	Distance from site(s)

Nearest railway station for:	Passengers	Distance from site(s)
	Goods	Distance from site(s)

9. HANDLING FACILITIES
Largest goods that can be handled by:
Airport
Docks
Rail
Road

10. REMARKS

Schedule A8 – Site construction and data

1. GENERAL CONSTRUCTION SERVICES
Describe generally the arrangements, and the extent of consultant's responsibilities for the following:

Supply point for temporary electrical power (state voltage, frequency and
phases plus kW or kVA capacity and tariff)

Site distribution of temporary supply (if any)

Temporary compressed air supplies including pressure

Water supplies – general, drinking, etc.

Water, gas, electricity, etc., for site personnel accommodation

Site telephones

Material testing

2. ACCOMMODATION
Working accommodation for consultant's staff

Working accommodation for contractor's staff

Living accommodation for consultant's staff

Living accommodation for contractor's staff

Medical facilities

Transport for labour to site (distances involved)

Roads linking site(s) with existing roads

Rail tracks linking site(s) with existing system

Storage facilities for plant on site

Unloading of plant on site

3. LABOUR
Engagement of labour – local laws applicable, working week, overtime limitations, etc.

4. EQUIPMENT
Instruments for testing and commissioning.

5. GENERAL
Will the consultant's resident engineer have authority to sanction local orders without reference to the client? If so to what value?

Schedule A9 – Extent of work by consultant

A – CIVIL WORKS	If in contract	New or extension	Client's preference or consultant's specification	Remarks and style
INITIAL WORKS				
Site clearance				
Demolition				
Site levelling				
Foundations:				
Piling				
Mass concrete				
Cellular concrete				
Other				
MAIN BUILDINGS				
Switch rooms:				
Main				
Substation				
Control rooms				
Battery rooms				
Guardhouse				
SWITCHYARDS				
Transformer bays				
Transformer plinths				
Line landing structures				
Support structures				
Earthing pits				
Surface dressing				
Tower foundations				
Fencing and gates				
TRANSPORT				
Roads				
Weigh bridges				
Jetties and wharves				
Harbour services				
SERVICES				
Incinerator				
Effluent disposal				
Surface water drains				
Water distribution				
ANCILLARY BUILDINGS, ETC.				
Offices				
Laboratory				
Workshops				
Stores				
Fire station				
Permanent cranes				
Canteen				
Recreation rooms				
Welfare and first aid centre				
ELECTRICITY				
WATER				
GAS				
LIQUID FUEL				
GAS				
TEMPORARY WORKS				

Schedule A9 – Continued:

A – CIVIL WORKS	If in contract	New or extension	Client's preference or consultant's specification	Remarks and style
SUBSTATIONS				
kV Outdoor switchgear				
kV Outdoor switchgear				
kV Outdoor switchgear				
kV Indoor switchgear				
kV Indoor switchgear				
kV Indoor switchgear				
Earthing switchgear				
Transformers kV/kV				
Transformers kV/kV				
Transformers kV/kV				
Reactors				
Liquid/dry earthing resistors				
Earthing arrangements				
Batteries				
Chargers				
DC switchgear				
Protection				
Control				
Alarms and annunciators				
Internal phone system				
Telemetry				
Metering				
Cabling kV				
Cabling kV				
Cabling kV				
Cabling Control				
LVAC supply and equipment				
Lighting				
HVAC				
Firefighting				
Testing equipment				
Erection				
Fitting				
Jointing				
Spares				
TRANSMISSION				
Overhead Lines and Underground Cables				

Schedule A9 – Continued:

A – CIVIL WORKS	If in contract	New or extension	Client's preference or consultant's specification	Remarks and style
Line supports				
Conductor				
Insulators				
Erection				
Cable laying				
Jointing				
Surveying				
Ground clearance				
Wayleaves				
DISTRIBUTION				
Overhead Lines and Underground Cables				
Line supports				
Conductor				
Insulators				
Erection				
Cable laying				
Jointing				
Surveying				
Ground clearance				
Wayleaves				

Are the following craftsmen and machines locally available?
Jointers
Wiremen
Electrical fitters
Overhead line erectors
Rough labourers

Form of contract required?

Notes:
1. FORM OF CONSTRUCTION CLASSIFICATION

(a) Brick

(b) Steel frame and brick

(c) Steel frames with asbestos covering

(d) Steel frames with sheet metal covering

(e) Wooden

(f) Reinforced concrete

(g) Concrete blocks

(h) Other forms of construction

(i) Consultant to recommend

1. Totally enclosed

2. Partially enclosed

3. Roof only

4. Outdoor

5. Consultant to recommend

Schedule A9 – Continued:

2. CONSTRUCTION
Is the Consultant to arrange for specialist erectors where necessary?
State arrangements.

Are the following craftsmen and machines available locally?

Excavators	Names of recommended local
Pile drivers	contractors and associated rates
Reinforcement benders and fixers	for craftsmen listed.
Carpenters and joiners	
Concreters	
Drain layers	If craftsmen are not available what
Brick layers	have been arrangements used in
Asphalters	the past?
Pairers and tilers	
Roofers	
Plumbers and hot water fitters	Are Quantity Surveyors to be
Welders	employed to prepare a Bill of
Glaziers	Quantities and should their fees
Painters	be incorporated in with those of
Rough labourers	the consultant?

3. FORM OF CONTRACT DESIRED
R.I.B.A (not generally recommended for substation, or overhead line and cable works)
I.C.E
F.I.D.I.C
I.Mech or Elec.E Model Forms
MF/1
NEC/EEC
Other
Partnering

Schedule A10 – Preferred suppliers and equipment

Insert names of suppliers, country of supply and/or local contractors (if any) preferred by the client.

1. CIVIL WORKS

2. ELECTRICAL (SUBSTATIONS)

3. ELECTRICAL (OVERHEAD LINES)

4. ELECTRICAL (CABLES)

5. CONTROL AND COMMUNICATIONS

SCHEDULE B – LOCAL INFORMATION

Schedule B1 – Drawings, maps and clearances

The following information should be supplied where required so that new plant may be designed to tie-in and operate satisfactorily with existing equipment and be arranged and installed to conform generally with existing installations and client requirements.

Single line diagrams

Ordnance maps of existing and proposed
overhead line and underground cable routes
(scales typically 1:200 000, 1:50 000, and 1:10 000)

Site layout drawings of existing substations and feeders

Power and control layout drawings

Layout and/or schematic of earthing system

Description of types of feeders and construction

Firm capacity of all substations and feeders

Maximum demands at all substations with duration of peak loads

General description of substations and existing equipment

Clearance requirements for all situations

Details of protection, control and telemetry equipment

Details and make of any compulsory equipment required for purposes of standardization by the client with the existing system

Fault levels at all existing switchboards

Details of any existing spare circuits

Note: Where possible items should be marked on the drawings to show details of the proposed new works.

Schedule B2 – Details of existing network

Design data from which existing equipment was supplied:

Switchgear

Transformers

Reactors

Batteries

Earthing equipment

Control and telemetry equipment

Underground cables

Overhead lines

Network details

Voltages

Phases

Frequency

Rate of rise of recovery voltage

Insulation levels

SCHEDULE C – PROJECT INFORMATION

Schedule C1 – Substations

1. Name of substation
2. Location of site
3. Drawing No. showing site layout
4. Drawing No. showing line diagrams
5. Existing, new or extensions to existing substation
6. Indoor or outdoor
7. Service conditions

Description	Existing	Required	Notes
8. Type of substation			
9. Substation attendance			
10. Remote control			
11. Central control			
12. Fire protection			
13. Heating/air conditioning (HVAC)			
14. LVAC loads			
15. Means of supplying services above			
16. DC supply			
17. DC voltages			
18. Firm capacity			
19. Earthing			

Schedule C2 – Feeders

Description	Incoming Existing/required	Outgoing Existing/required	Notes
1. Rated service voltage (kV)			
2. Highest service voltage (kV)			
3. Impulse withstand level (kV)			
4. Number of feeders			
5. Overhead or underground			
6. Feeder capacity (MVA)			
7. Conductor and size			
8. Metering			
9. Protection			
10. CTs			
11. VTs			
12. Are all feeders required covered in the Contract?			

Schedule C3 – Switchgear

Description	Existing	Required	Notes
1. Rated service voltage			
2. Type			
3. Manufacturer			
4. Indoor/Outdoor			
5. Busbars			
6. Busbar insulation			
7. Rated short circuit capacity			

8. Circuit current ratings:
   (a) busbars
   (b) feeder circuits
   (c) transformer circuits
   (d) bus sections
   (e) bus couplers			

9. Impulse withstand			

10. Minimum clearances:
   (a) phase to phase
   (b) phase to earth			

11. CTs and VTs			

Schedule C4 – Transformers

Description	Existing	Required	Notes
1. Number of units			
2. HV voltage			
3. LV voltage (state load or no-load)			
4. Step-up or step-down?			
5. Rating (ONAN/ONAF/OFAF)			
6. No. of phases			
7. Frequency			
8. Vector group reference			
9. Impedance (nom tap)			

10. Earthing:
   (a) HV
   (b) LV			

11. Connections:
   (a) HV
   (b) LV			

12. Impulse withstand:
   (a) HV
   (b) LV			

13. Tap changer:
   (a) auto or manual
   (b) local or remote
   (c) on load, off load, off circuit
   (d) max. current rating
   (e) tapping range ± %
   (f) tapping steps %			

14. Tertiary winding			
15. Tertiary impedance			
16. CTs			
17. Ancillary equipment			

Schedule C5 – Busbars

Description	Existing	Required	Notes
1. Rated service voltage			
2. Type of busbars			
3. Indoor or outdoor			
4. Single or duplicate			
5. No. of sections			
6. No. of bus couplers			
7. Busbar insulation			
8. Current rating			
9. Short circuit rating			
10. Minimum clearances: (a) phase to phase (b) phase to earth			

Schedule C6 – Isolators and earthing switches

Description	Existing	Required	Notes
1. Rated service voltage			
2. Type			Fault make/load break, off load, etc.
3. Manufacturer			
4. Type of operating mechanism			AC or DC motor drive
5. Current breaking capacity			
6. Indoor/outdoor			
7. Normal current rating			
8. Arc extinguishing device			
9. Locations: (a) transformer circuits (b) bus sections (c) feeders			

Schedule C7 – Protection

Description	Existing	Required	Notes
1. System voltage level			

2. Feeders:
    (a) Overhead lines
 	
    (b) Underground cables			

3. Transformer incomers			
4. Transformers			
5. Bus couplers			
6. Bus sections			
7. Bus zone			

Schedule C8 – Control and communications

Description	Existing	Required	Notes

1. Method of communication:
    (a) Speech
 	
    (b) Network information (system
         control and data acquisition)			

2. Form of control:			

3. Information transmitted:
    (a) No. of circuits
    (b) No. of alarms
    (c) No. of analogues
    (d) No. of annunciators
    (e) No. of circuit breaker controls			

4. Manufacturer of existing system			

Schedule C9 – Overhead lines

Description	Existing	Required	Notes
1. Route From: To:			
2. Route length			
3. Voltage			
4. Phases			
5. Frequency			
6. Power transmitted			
7. Number of circuits			
8. Type of support			
9. Single or double circuit			
10. Conductor protective coating			LV distribution systems
11. Type of crossarm			
12. Conductor configuration			
13. Type of conductor			
14. Size of conductor			
15. Conductors per phase			
16. Type of earth wire			
17. Size of earth wire			
18. Type of insulators			
19. Insulator profile			
20. Factors of safety used			
21. What is type of terrain: (e.g. wooded, built up area, hilly, flat, etc.)			
22. What is type of ground: (e.g. soil, sand, rock, etc.)			
23. Max. altitude along route			
24. Any other comments			

Schedule C10 – Underground cables

Description	Existing	Required	Notes
1. Route From: To:			
2. Route length			
3. Voltage			
4. Phases			
5. Frequency			
6. Power transmitted			
7. Number of circuits			
8. Type of cable			
9. Type of conductor			
10. Type of insulation			
11. Screening			
12. No. of cores			
13. Size of conductors			
14. Type of sheath			
15. Type of armouring			
16. Type of outer coating			Graphite for sheath integrity tests, etc.
17. Depth of cable laying			
18. Type of backfill			
19. No. of pilot cables laid with main cables			
20. Cable spacing			
21. Cable formation			
22. What is type of terrain: (e.g. wooded, built up area, hilly, flat, etc.)			
23. What is type of ground: (e.g. soil, sand, rock, etc.)			
24. Max. and min. altitude along route			
25. Any other comments			

Schedule C11 – Civil works

Description	Existing	Required	Notes
1. Substation			
2. Switch houses (kV):			
(a) No. of bays or switchgear units in panel			
(b) Separate control and relay rooms?			
(c) Control room			
(d) Battery room			
(e) Relay room			
(f) Guardhouse			
(g) Messing facilities			
(h) Toilet facilities			
(i) Method of construction (blockwork, brick, reinforced concrete, etc.)			
3. Switchyard			
4. Transformer plinths			
5. Transformer bays			
6. Switch bays (kV):			
(a)			
(b)			
(c)			
7. Earthing pits			
8. Fencing and gates			
9. Roads:			
(a) Internal to substation			
(b) External to substation (access roads)			
10. Plinths			
11. Trenches			
12. Surface water drainage			
13. Tower foundations			
14. Surface dressing			
15. Any other comments			

APPENDIX B

BIDDING CHECKLIST

This checklist is intended to assist clients and contractors when preparing or adjudicating tenders for transmission and distribution construction work. The preparation of tenders for large projects is a costly business. Working through the checklist will help to ensure that no major items have been overlooked.

A Initial contractor pre-selection
B The client/the contractor
C Competition
D Technical specifications
E Quality assurance
F Delivery
G Payments
H Estimates
I Documentation
J Terms and conditions of contract

A INITIAL CONTRACTOR PRE-SELECTION

A.1 Is this a serious bid from which a contract will follow or is the client looking for a measure of cost to gauge work in the distant future?
A.2 Is there a pre-qualification in order to gauge the capacity or capability of the contractor?
A.3 Are there at this stage any special terms and conditions of contract that would make the final contract too risky to consider further? For example consequential loss, extremely high liquidated damages, high performance bonds, ongoing defect rectification, etc., contract clauses.

B THE CLIENT/ THE CONTRACTOR

B.1 Is the client or contractor well known with the resources and past track record to consider further?
B.2 If this is an overseas job what experience does the client or contractor have of working in the country under consideration?
B.3 If supported by a international bank are client or contractor able to conform to the usually strict guidelines required?
B.4 Are special insurances required?

C COMPETITION

C.1 What other large contracts are currently on offer from this or alternative clients?

C.2 Does the contractor have other competitors and if so who are they? What are their strengths and weaknesses?

C.3 Are there any particular non-financial features that the client is looking for or which the contractor is able to offer which could differentiate the bid? (history of previous satisfactory work, training, spares availability, maintenance capability, etc.)

C.4 What technical, financial or marketing advantage would help to differentiate one contractor from another?

C.5 How many tenderers are being invited to bid? If more than four or five contractors are invited to bid for substation or transmission line work will this discourage high quality contractors from tendering?

D TECHNICAL SPECIFICATIONS

D.1 Has the contractor carried out this type of work before with a successful track record?

D.2 Have the standards to be adopted clearly specified and if so may they be altered?

D.3 Must the tender exactly conform to the specification or may fully compliant bids plus alternative bids be offered?

D.4 Are there any restrictions with regard to subletting any or all of the works by the contractor?

E QUALITY ASSURANCE

E.1 Are special QA conditions applicable?

E.2 Are special test procedures applicable?

F DELIVERY

F.1 If nominated suppliers are specified what is the client or contractor experience with this supplier in the past as regard to reliable and prompt delivery of equipment?

F.2 Are deliveries likely to be a critical factor in the project programme and if so is a critical path network programme analysis necessary?

F.3 Are delivery dates for manufactured goods (switchgear or transformers, etc.) negotiable?

F.4 Are penalties to be applied for late delivery of manufacturers goods, which may be outside the control of the main contractor, to be applied by the client?

F.5 What are the transportation arrangements? Is existing infrastructure adequate for transportation of materials to site?

F.6 If the site is unavailable on time or if materials require special warehousing is this covered under the contract?

F.7 Are procedures for monitoring the progress of the contract clearly defined?

G PAYMENTS

G.1 What contract bond arrangements are required? Who is best to carry these costs (client or contractor)?

G.2 Are progress payments involved and are the milestone events against which payment is to be released clearly defined?

G.3 Is the payment documentation clearly defined and kept simple in order to avoid client/contractor disputes?

G.4 Is foreign exchange required to fund the project? Are there any advantages in requesting payments to be made in multiple currencies? Are currency restrictions involved in the country where the construction work is to take place or where specific manufactured goods are to be purchased?

G.5 What insurances are required? Who is best to carry these costs (client or contractor)?

G.6 Are liquidated damages involved? Is this a contract where time is of the essence?

G.7 Does the contract allow for contract price adjustment formulae to cover, for example, changing prices for raw materials such as copper?

G.8 What is the inflation situation in countries of origin of particular manufactured goods required or in the country where construction is to take place?

H ESTIMATES

H.1 What is the cost of preparing this tender?

H.2 Does the work involve any special cost estimating procedures?

H.3 Have risk and contingency allowances been made in the cost estimates?

H.4 Is the technology well known or new and untried?

H.5 Are there any restrictive labour practices or labour laws? Are labour relations good in the area of construction?

I DOCUMENTATION

I.1 Are the requirements for documentation in terms of quality, quantity and types throughout the project life cycle clearly defined in the contract?

I.2 Are special storage conditions required for project lifetime records defined? What is the period for which such records must be maintained?

I.3 Are any special requirements covering vendor documentation, operation and maintenance manuals, as-built drawings, etc. defined?

J TERMS AND CONDITIONS OF CONTRACT

J.1 Has the legal department checked the contract document for key clauses?

J.2 What is the ruling law that governs the contract? Is it known and acceptable?

J.3 Does the contract allow for variations to the scope of the works? If so what is the maximum variation to the contract value allowable before rates may be varied?

J.4 Does the contract allow adequate cover for the related impact to time and cost associated with variations to the original scope of work? Are sufficient safeguards in place for the control and management of variations?

J.5 Does the contract recognize situations outside the control of client or contractor (*force majeure*)?

J.6 Have restrictions been placed regarding publicity or secrecy?

J.7 Is it clear who is responsible within the client, contractor and consultant organizations for the contract? What powers and authority do they have?

23 Distribution Planning

23.1 INTRODUCTION

This chapter describes the general distribution planning steps that may be taken in order to estimate the magnitude of the medium and low voltage distribution system loads to be supplied. It presents various load forecasting methods for estimating load development within the time period under review and within the specified geographic area under consideration. Such estimates and forecasts then allow the size of the necessary supply equipment and service overhead lines or cables to be calculated taking into account normal factors such as:

- continuous current rating
- line voltage regulation
- fault rating
- supply interference (motor starting, harmonic distortion, unbalance, etc.)
- supply security
- construction hazards and standards.

The overall efficiency of the distribution system is as important in load forecasting as energy consumption. Therefore load factor, maximum demand, diversity, losses and growth characteristics are particularly discussed. Modern distribution planning makes considerable use of computer modelling and equipment reliability statistics in order to assist with design optimization, and reference is made to such techniques.

23.2 DEFINITIONS

This section defines some load definitions and describes the terminology used in distribution planning.

23.2.1   Demand or average demand

'The demand of an installation or system is the load at the receiving terminals averaged over a specified interval of time'.

The load may be expressed as active power (kW) or reactive power (kVAr). The period over which the demand is averaged is known as the demand interval and may be governed by the thermal constant of the equipment or the duration of the load. Figure 23.1 illustrates average hourly loads (kW) over a

Average hourly loads (kW)
Example peak day

Morning			Afternoon/Evening		
Hour		Demand	Hour		Demand
From	*To*	*(kW)*	*From*	*To*	*(kW)*
Midnight	1 am	10	Midday	1 pm	13
1 am	2 am	8	1 pm	2 pm	15
2 am	3 am	6	2 pm	3 pm	16
3 am	4 am	7	3 pm	4 pm	19
4 am	5 am	8	4 pm	5 pm	21
5 am	6 am	9	5 pm	6 pm	24
6 am	7 am	10	6 pm	7 pm	27
7 am	8 am	12	7 pm	8 pm	30
8 am	9 am	15	8 pm	9 pm	28
9 am	10 am	14	9 pm	10 pm	23
10 am	11 am	13	10 pm	11 pm	19
11 am	Midday	11	11 pm	Midnight	13
		Σ = 123 kW			Σ = 248 kW

Total = 371 kWh

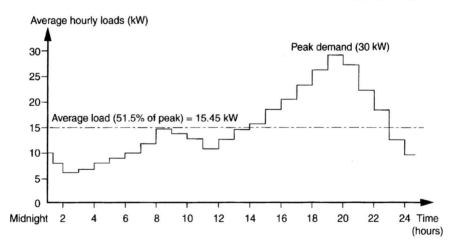

Figure 23.1   Average hourly loads for the example day

24-hour period. The demand interval must always be stated when describing average demand or the figure is meaningless:

$$\text{Average demand} = \frac{\text{Total energy (kWh)}}{\text{Total Period (hours)}}$$

From Figure 23.1:

$$\text{Average demand} = \frac{371\ \text{kWh}}{24\ \text{hours}} = 15.45\ \text{kW (based upon average hourly demands over a 24-hour period)}$$

23.2.2   Maximum demand (MD)

'The maximum demand of an installation or system is the greatest of all demands which have occurred during the specified period of time'.

The maximum demand may be expressed in kW, kVAr, etc. Both the demand interval (average hourly loads, etc.) and the time period (daily, weekly, etc.) must be defined for the expression to be meaningful. Figure 23.2 illustrates the variation in demand with demand interval. Loads normally alter through a

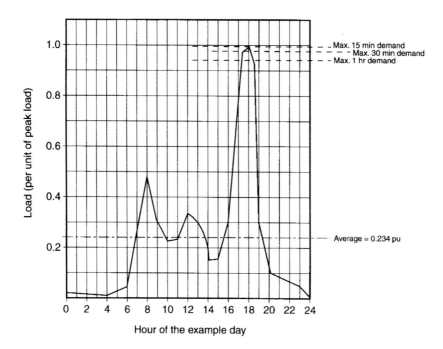

Figure 23.2   Variation in demand with demand interval (note that a lower maximum demand results from a larger demand interval because of such a smoothing effect)

24-hour period with clear peaks occurring. For example, the load increases in the morning as people get up to have breakfast and to go to work. Similarly with the advertisement intervals on the television in the evening, load peaks occur as viewers get up from watching a popular show to use electric kettles to boil water and make a cup of tea. A larger demand interval will have the effect of smoothing out such effects and will therefore normal result in a lower maximum demand.

23.2.3   Demand factor

'The demand factor is the ratio of the maximum demand of a system to the total connected load of the system'.

The total connected load of the system is defined as 'the sum of the continuous ratings of the load consuming equipment connected to the system'. Both the maximum demand and the total connected load should be expressed in the same units thus making the demand factor dimensionless. Again the demand interval and the period over which the maximum demand applies should be stated. The demand factor is most often used in association with a consumer's services rather than to a complete distribution system:

$$\text{Demand factor} = \frac{\text{Maximum demand of the system}}{\text{Total connected load}} \quad (\text{normally} \leq 1)$$

23.2.4   Utilization factor (UF)

'The utilization factor is the ratio of the maximum demand of a system to the rated capacity of the system'.

Both the maximum demand and the system rated capacity should both be expressed in the same units to make the utilization factor expression dimensionless. Again the demand interval and the period over which the maximum demand applies should be stated. The utilization factor indicates the degree to which the system is being loaded during peak load periods with respect to its capacity:

$$\text{Utilization factor, UF} = \frac{\text{Maximum demand of the system}}{\text{Rated capacity of the system}} \quad (\text{normally} \leq 1)$$

23.2.5   Load factor (LDF)

'The load factor is the ratio of the average load over a designated period of time to the peak load occurring in that period'.

To accurately define the load factor then the demand interval, the period to which the maximum demand and average load apply, the manner in which the maximum demand is measured and the load commodity involved should all be stated. The average and the peak demand loads should be expressed in the same units to make the expression dimensionless. Load factor is usually expressed as a percentage figure or a fraction. Fundamentally, the load factor indicates the degree to which the peak load is sustained during the period. In the United States the national average load factor is currently approximately 63%, whereas in a developing country it may be as low as 50%:

$$\text{Load factor, LDF} = \frac{\text{Average demand over designated period of time}}{\text{Peak load occurring in that period}}$$

$$(\text{normally} \leq 100\% \text{ or } \leq 1)$$

With reference to Figure 23.1, for the sample 24-hour period:

$$\text{Load factor, LDF} = \frac{\text{Average demand (kW)}}{\text{Peak demand (kW)}} \times 100\%$$

$$= \frac{15.45}{30.00} \times 100\%$$

$$= 51.5\%$$

23.2.6   Diversity factor (DF)

'The diversity factor is the ratio of the sum of the individual maximum demands of the various subdivisions of a system to the maximum demand of the whole system'.

Loads do not normally all peak at the same time. The sum of the individual peak loads will therefore inevitably be greater than the peak load of the composite system. The diversity factor normally has a value greater than unity and is only equal to unity if all the individual demands occur simultaneously. The coincident nature of load demands is of great importance to the distribution planning engineer as it is a key factor in the economic sizing of plant. Figure 23.3 shows the effects of coincidental and non-coincidental demands:

$$\text{Diversity factor, DF} = \frac{\sum(\text{individual maximum demands})}{\text{Maximum demand of the system}} \quad (\text{normally} \geq 1)$$

From Figure 23.3:

$$\text{DF} = \frac{10 + 9 + 10 + 14}{14}$$

$$= 3.07$$

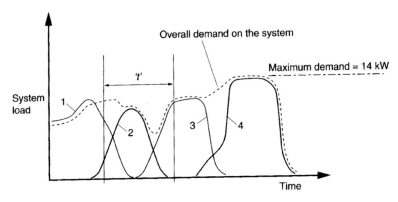

Diversified or coincidental demand = demand on the system during demand interval 't'

Non-coincidental demand   = sum of the demands on the system with no
                                     restriction to the demand interval
                        i.e. = 10 + 9 + 10 + 14
                             = 43 kW

(Usually non-coincidental demands are comprised of individual maximum demands.
Therefore the term is also referred to as the maximum non-coincident demand)

Figure 23.3   Coincidental and non-coincidental demands

23.2.7   Coincident factor (CF)

Some engineers prefer to have a factor which describes the characteristics of
loads that have a value equal to or less than unity. The reciprocal of the diversity factor is known as the coincident factor:

$$\text{Coincident factor, } CF = \frac{1}{DF} \quad (\text{normally} \leq 1)$$

The coincident factor is dependent upon the type of loads connected to the
system. Typically:

Loads	CF
Distribution transformers	0.74–0.83
Primary feeders	0.83–0.92
Substations	0.80–0.96

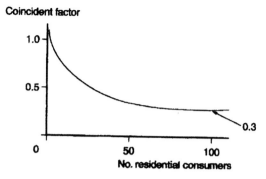

Figure 23.3(a)   Coincident factor

In general, and in the absence of other data:

$$CF = 0.5\left(1 + \frac{5}{2n + 3}\right)$$

where n is the number of loads connected to the system.

For residential areas in countries with developed economies the coincident factor tends to settle at approximately 0.5. However, caution must be applied and data should be collected to obtain meaningful information, as shown in Figure 23.3a, where CF settles at approximately 0.3.

23.2.8   Load diversity

'Load diversity is the difference between the sum of the peaks of two or more individual loads and the peak of the combined load'.

Since load diversity is the difference between two quantities of similar units (rather than a ratio) it is expressed in the units of the two demands being compared.

Referring to Figure 23.3:

$$\text{Load diversity} = \{\Sigma \text{ (individual maximum demands)}\} - (\text{maximum}$$
$$\text{demand of the system})$$
$$= (10 + 9 + 10 + 14) - 14$$
$$= 29\,\text{kW}$$

23.2.9   Loss factor (LSF)

'The loss factor is the ratio of the average power loss to the peak load loss, during a specified period of time'.

Since power losses are proportional to the square of the load current:

$$\text{Loss factor, LSF} = \frac{\text{Average (load)}^2}{\text{Maximum (load)}^2} \quad \text{or} \quad \frac{\text{Average loss}}{\text{Peak loss}}$$

From the simple average hourly load variation and square of the hourly demand patterns shown in Figures 23.1 and 23.2:

Average load $= 371\,\text{kWh}/24\text{ hours} = 15.45\,\text{kW}$ over a 24-hour period

Average (load)$^2 = 6849\,\text{kW}^2/24\text{ hours} = 285.37\,\text{kW}^2$ over a 24-hour period

Load factor $=$ Average load/Maximum load $= 15.45/30 = 0.52$ or 52%

Loss factor $=$ Average (load)^2/Maximum (load)^2
$$= 285.37/900 = 0.32 \text{ or } 32\%$$

Note: (Average load)$^2 \neq$ Average (load)^2 and for this example, of course, $238.7 \neq 285.37$.

In the United States the national average loss factor is currently approximately 45%.

Although loss factor cannot generally be expressed in terms of load factor, the limiting values of the relationship may be established and this is illustrated in Figures 23.4a and 23.4b. In more general terms if:

$x =$ peak load of duration, t

$y =$ minimum load of duration $(T - t)$

$$\text{Average load} = \frac{xt + y(T - t)}{T}$$

$$\text{Load factor, LDF} = \frac{xt + y(T - t)}{Tx} \qquad (23.1)$$

and if $y \rightarrow 0$, then:

$$\text{LDF} = \frac{xt}{Tx} = \frac{t}{T}$$

Peak loss $= x^2R$ for duration, t

Minimum loss $= y^2R$ duration $(T - t)$

$$\text{Average Loss} = \frac{x^2Rt + y^2R(T - t)}{T}$$

$$\text{Loss factor, LSF} = \frac{\text{Average loss}}{\text{Peak loss}} = \frac{\{x^2Rt + y^2R(T - t)\}}{T}/x^2Rt$$

$$= \frac{x^2t + y^2(T - t)}{Tx^2}$$

$$= \frac{t}{T} + (y/x)^2 \cdot (T - t) \qquad (23.2)$$

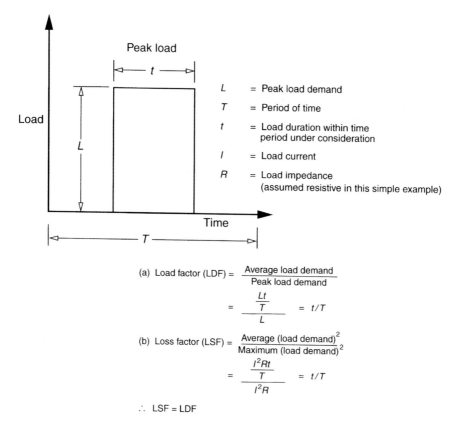

(a) Load factor (LDF) = $\dfrac{\text{Average load demand}}{\text{Peak load demand}}$

$$= \dfrac{\dfrac{Lt}{T}}{L} \qquad = t/T$$

(b) Loss factor (LSF) = $\dfrac{\text{Average (load demand)}^2}{\text{Maximum (load demand)}^2}$

$$= \dfrac{\dfrac{I^2Rt}{T}}{I^2R} \qquad = t/T$$

$$\therefore \ \text{LSF} = \text{LDF}$$

Figure 23.4(a)   Hypothetical load case where loss factor = load factor (LSF = LDF if the load remains at its peak value all the time that it is on, and zero for the remainder of the time period)

If $y \rightarrow 0$ and $x \rightarrow 0$, then:

$$\text{LSF} = \frac{t}{T}, \ \text{i.e. LSF} = \text{LDF}$$

Thus if the load remains at its peak value all the time that it is on, and zero for the remainder of the time period, then the loss factor (LSF) is equal to the load factor (LDF).

Further, if the following assumptions are considered:

$$\frac{T-t}{T} \rightarrow 1.0 \quad \text{and} \quad \frac{t}{T} \rightarrow 0 \quad \text{and} \quad \frac{y}{x} \text{ does not also approach zero}$$

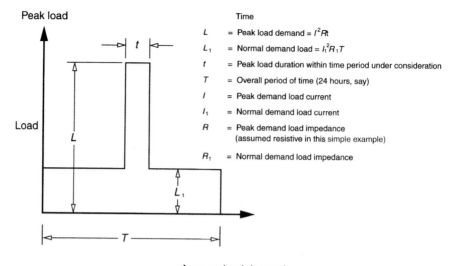

(a) Load factor (LDF) = $\dfrac{\text{Average load demand}}{\text{Peak load demand}}$

$$= \dfrac{\dfrac{L_1 T = (L - L_1)\, t}{T}}{L}$$

$\begin{aligned} t &\longrightarrow 0 \\ L_1 t &\longrightarrow 0 \\ L t &\longrightarrow 0 \end{aligned}$

$$\cong L_1/L$$

(b) Loss factor (LSF) = $\dfrac{\text{Average (load demand)}^2}{\text{Maximum (load demand)}^2}$

$$= \dfrac{I_1^2 R_1 T + I^2 R t - I_1^2 R_1 t}{T} \Big/ I^2 R$$

$\begin{aligned} t &\longrightarrow 0 \\ I_1^2 R_1 t &\longrightarrow 0 \\ I^2 R t &\longrightarrow 0 \end{aligned}$

$$\cong I_1^2 R_1 / I^2 R$$

$$\therefore\ \text{LSF} = (\text{LDF})^2$$

Figure 23.4(b)   Hypothetical load case where loss factor = load factor². (LSF = (LDF)² if the load has a sharp peak and then a fairly steady value for the remainder of the period under consideration.)

then rewriting equation (23.1) above:

$$\text{LDF} = \frac{t}{T} + \frac{(y)}{x} \cdot \frac{T - t}{T}$$

and applying these assumptions and comparing with equation (23.2) above:

$$\frac{t}{T} + \frac{(y)}{(x)} \cdot \frac{T - t}{T} = \frac{t}{T} + (y/x)^2 \cdot \frac{(T - t)}{T}$$

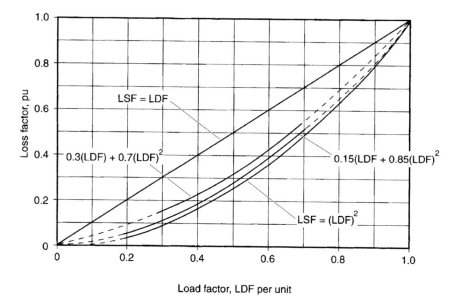

Typical distribution transformer loss factor $= 0.15(LDF) + 0.85(LDF)^2$

Typical feeder loss factor $= 0.3\ (LDF) + 0.7\ (LDF)^2$

Figure 23.4(c)   Curves of loss factor LSF as a function of load factor LDF

Thus if a load profile has a sharp peak and then a fairly steady load of a fixed value for the period under consideration then the loss factor is equal to the load factor squared, i.e. $LSF = LDF^2$.

The loss factor cannot be determined directly from the load factor because the loss factor is determined from the losses as a function of time, which in turn are proportional to the time function of the square of the load. However, a relationship has been calculated which gives a reasonable value of the 30-minute, monthly, kW loss factor in terms of the corresponding load as shown graphically in Figure 23.4c.

In general:

$$LSF = c(LDF) + (1 - c) \cdot (LDF)^2$$

where $c \cong 0.3$ for transmission systems
   and $c \cong 0.15$ for distribution systems.

Referring to Figure 23.1 for the sample day:

Loss factor,
LSF % $= \dfrac{\text{Square of all actual demands} \times 100\%}{\text{Square of peak demand} \times 100\% \text{ time of the demands}}$

$= \dfrac{(6849\,\text{kW})^2\ \text{hour}}{(30\,\text{kW})^2\ \times 24\ \text{hours}} \times\ 100\%$

$= 31.7\%$

23.2.10   Load duration

'Load duration is the relationship of demands and the duration of the demands over a specified time period.'

Referring to Figure 23.1 the hourly demands have been sorted in descending order and tabulated in Table 23.1 as shown below to give:

- Frequency   = Number of hours of occurrence for each demand
- Equal/Exceed   = Summation of frequencies

- Percent of peak $= \dfrac{\text{Demand (kW)}}{\text{Peak (kW)}} \times 100\%$

- Percent of duration $= \dfrac{\text{Equal/Exceed}}{\text{Specified time}} \times 100\%$

- Square of demands $= (\text{Demand})^2 \times \text{Frequency}$

The load duration parameters for example day have been plotted in Figure 23.5 (percent peak load vs percent duration). Technical losses are a function of the squares of the load current (amps) which is directly related to the squares of the demands. Figure 23.6 is a graph of the squares of the hourly demands for the examples day illustrated in Figure 23.1.

23.2.11   Loss equivalent hours

'Loss equivalent hours are the number of hours of peak loads which will produce the same total losses as is produced by the actual loads over a specified period of time'.

Both the actual and peak demand values must be chosen from the associated load duration:

$$\text{Loss equivalent hours} = \frac{\text{Square of all actual demands}}{\text{Square of peak demand}}$$

With reference to the load duration and loss table (Table 23.1):

$$\text{Loss equivalent hours} = \frac{6849\,\text{kW}^2}{900\,\text{kW}^2}$$
$$= 7.61 \text{ hours}$$

Table 23.1   Load duration and loss table (for the peak day described in Figure 23.1)

Demand (kW)	Frequency	Equal/exceed	Percent of peak (%)	Percent duration (%)	Square of demand
30	1	1	100.0	4.2	900
28	1	2	93.3	8.3	784
27	1	3	90.0	12.5	729
24	1	4	80.0	16.7	576
23	1	5	76.6	20.8	529
21	1	6	70.0	25.0	441
19	2	8	63.3	33.3	722
16	1	9	53.3	37.5	256
15	2	11	50.0	45.8	450
14	1	12	46.7	50.0	196
13	3	15	43.3	62.5	507
12	1	16	40.0	66.7	144
11	1	17	36.7	70.8	121
10	2	19	33.3	79.2	200
9	1	20	30.0	83.3	81
8	2	22	26.7	91.7	128
7	1	23	23.3	95.8	49
6	1	24	20.0	100.0	36
					$\Sigma = 6849$

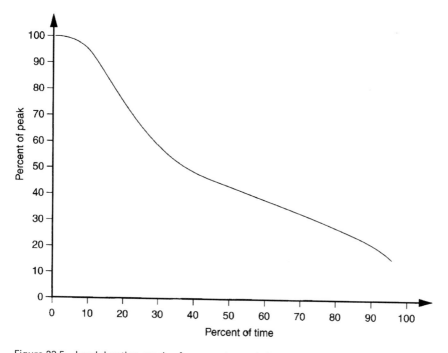

Figure 23.5   Load duration graph – for example, peak day

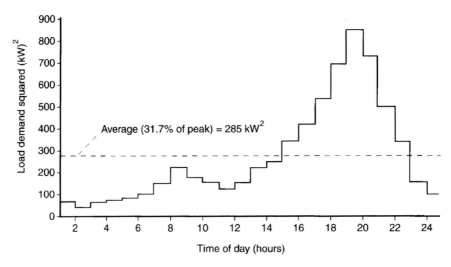

Figure 23.6   Graphical plot of squares of the hourly demands for the example day

The loss equivalent hours are also referred to as the 'Equivalent peak loss time' (EPLT). An alternative method of calculating this is:

$$\begin{array}{c}\text{Equivalent peak}\\\text{loss time, EPLT}\end{array} = \frac{\text{Average power loss} \times \text{hours in specified period}}{\text{Peak power loss}}$$

$$= \text{Loss factor} \times \text{hours in period}$$

Therefore for the example day in Figure 23.1:

$$\text{Loss equivalent hours} = \text{Equivalent peak loss time} = 31.7\% \times 24 \text{ hours}$$

$$= 7.61 \text{ hours}$$

23.2.12   Peak responsibility factor (PRF)

'The peak responsibility factor represents the contribution a component makes to the system demand losses at the time of system peak demand.'

Peak responsibility factor, $\text{PRF}_{(\text{distribution})}$:

$$= \frac{\text{Component load at time of referred component peak load}}{\text{Component peak load}}$$

$$(\text{normally} \leqslant 1)$$

Peak responsibility factor, $\text{PRF}_{(\text{system})}$:

$$= \frac{\text{Component load at time of system peak load}}{\text{System peak load}} \quad (\text{normally} \leqslant 1)$$

Typical $PRF_{(system)}$ values for different transformers in the system are:

Transformer type	$PRF_{(system)}$ ($\propto$ system loads)	$PRF^2_{(system)}$ ($\propto$ system losses)
General step-up transformer	1.0	1.0
Transmission substation transformer	0.9	0.81
Distribution substation transformer	0.8	0.64
Distribution feeder pillar transformer	(0.46–0.95), say 0.75	say 0.56

It should be noted that no-load losses are continuous and occur both during system peak demand and at other times. Generation therefore had to be designed to support these no-load losses.

Load losses vary with the load such that peak losses on a particular component of the overall distribution system occur at peak load on that component which may not be at the same as the overall system peak demand. Only a fraction of the individual component losses therefore contribute to the system peak demand.

23.3   LOAD FORECASTING

23.3.1   Users of load forecasts

Electricity supply authorities plan the capacity of their systems to meet the expected peak demand requirements. They maintain power (kW demand) and energy (kWh) forecasts as a basis for their physical network and financial planning.

In addition to demand forecasts the projected load curve, based upon hour-by-hour demand throughout the planning period, has an influence on the choice of generating capacity and the most economic order in which to bring different generating units onto the grid or distribution system. For example, fast run-up generating units (often with diesel or gas turbine prime movers) may be used to most economically satisfy short peak demands.

The combination of demand and energy forecasts form the basis for planning generating fuel requirements. They are the starting point for plant capacity and fuel strategies which are, in turn, translated into financial requirements. Fuel costs may vary substantially between different power stations or even generating plant within a particular station. It is therefore essential to normally employ the most cost effective and fuel efficient plant and only use more expensive plant for short periods (e.g. to deal with short peaks or short-term loss of wind or tidal power). Generating costs may be translated into fixed charges (generally associated with the capital plant and overheads required regardless of the actual power being generated, transmitted and distributed through the network) and variable costs directly associated with the energy demand (additional shift workers, additional fuel, etc.) Energy forecasts are the basis of revenue planning.

Forecasts also assist in the compilation of statistical data for the information of the public, government bodies, academic institutions and manufacturers. For example, manufacturers of electric supply equipment are able to gauge their future manufacturing output and marketing strategies from such data. In the short term the distribution planning process allows:

1. Relief of overloads in the distribution system.
2. Voltage control.
3. Reactive compensation (power factor correction).
4. Improvements in service quality for consumers.
5. Short-term system reinforcement and better provision of consumer connection requirements.

And in the longer term:

1. Pre-warning of changes in load and load usage.
2. Selection of the most appropriate primary distribution voltages.
3. Selection of substation capacity.
4. Determination of substation locations (at or near load centres).
5. Sub-transmission system requirements.
6. Long-term budgeting estimates.

23.3.2   The preparation of load forecasts

It is very important to estimate how the load will grow, the possible load growth rate, the load characteristics and magnitude together with the load location. Macro and micro load forecasting methods are therefore used and may be checked against one another.

In summary, distribution system planning:

1. Is a continuous process providing rapid evaluation and response to changes.
2. Has a planning period which reflects the lead times associated with project sanction (financial and economic appraisal and approval by the company and funding agencies), equipment procurement, installation and commissioning.
3. Should integrate into other areas of power system expansion as shown in Figure 23.7.
4. Provides a framework within which system efficiency (loss reduction campaigns, procurement policies, etc.) may be kept under review.

23.3.3   The micro load forecast

The *micro load forecast* is made up from small component parts and a separate forecast for each part is estimated. In microscopic estimations electricity demands

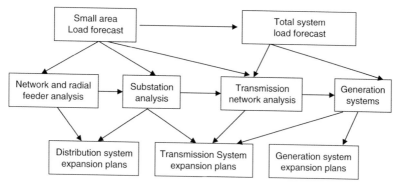

Figure 23.7   Distribution planning interfacing with other areas of the power system planning process

are estimated in terms of service classifications or consumer groups for pre-determined geographical areas and then integrated to produce peak power and energy demands for each such consumer group. The number of component parts used in the forecast is partly dependent on the value such complexity brings. A typical planning performance index (PPI) might be:

$$PPI = \frac{\text{Quality of the analysis}}{\text{Time to perform the analysis}}$$

Each part may be subdivided into a number of items and dimensions in which the forecast is made, e.g. maximum demand, energy consumption, numbers of consumers, population, load centres, tariff categories, diversity and losses, etc. Each item may, in turn, be further subdivided (e.g. the number of tariff categories) and inter-related with other items. Normally such micro load forecasts are based upon a combination of the following data:

- Extrapolations from historical data (sometimes using regression analysis).
- Data from power market surveys.
- Forecast of changes in population, housing, commercial, industrial, agricultural and other developments.
- Stated national, regional and local government policies, the electrical supply utility marketing plans and those of other relevant authorities.
- Expansion plans into currently non-electrified areas (often using results from past comparable experiences).
- The experience and judgement of the electrical supply utilities' forecasting department.

Since the micro load forecast requires the progressive amalgamation of data it would be ideal to commence with the smallest possible physical area and build such areas up to represent a district or region. In order to prepare the

Table 23.2   Micro forecast district/branch data collection proforma

District/Branch:	....
Prosperity	Very Prosperous, Prosperous, Average, Below Average, Poor (*delete inapplicable categories*)
Year Electrified:	....
Urban/Rural:	....

		Year after connection
		1 2 3 4 etc.
Domestic	– Number of households connected – Total number of households – % electrification – kWh/consumer – Total kWh	
Commercial	– Number of consumers – Total No. of possible consumers – kWh/consumer – Total kWh – List large unconnected consumers	
Industrial	– Number of consumers – Total No. of possible consumers – kWh/consumer – Total kWh – List large unconnected consumers	

economic input to the production of a micro forecast, it is necessary to assess how a district or branch load grows over a number of years. Table 23.2 is a useful data collection proforma.

The basic methods employed are:

(a) *Scratch pad methods*:
'Rules of thumb' are employed by experienced distribution system planners.

- kWh/month or year and load factor information gives a view on the kW demand.
- kW demand per substation or feeder and power factor at peak demand gives a view on the kVA demand per substation or feeder.
- Demand per substation or feeder and associated maps allow estimates of:
  - demand per km of overhead line or underground cables
  - demand per square km
  - demand per connected load (kW or kVA).
- Demand per square km coupled with information on the population and the customers per unit gives a view on the demand per customer.

Such methods are useful for well-understood areas and for relatively small expansion schemes. Such methods should not be used to support large investment proposals.

(b) *Trending*:
Regression curve fitting analysis is used on historical load growth information to estimate future load growth trends. This is an easy and simple load forecasting

method. However, it is not very accurate because it does not take into account new emerging or future dominant factors. The method may be enhanced by carrying out a market survey in order to allow the forecaster to identify and take into account likely changes from past rends and their causes. Linear programming methods may also be employed using multivariate analysis. However, it should be noted that a piecewise linear solution more closely approximates non-linear load growth.

(c) *End-use methods*:
Simulation land-based models are used to take into account such factors as:

- where people live
- where people work
- when people want power
- how people wish to use the power supplied.

This is a more advanced and accurate method of load forecasting and can be used to forecast the changing character of the load demand over long time periods. Such end-use methods may correlate land use with industrial/residential/commercial load demand and growth. The method necessitates time consuming data collection and computer analysis.

23.3.4   The macro load forecast

The *macro load forecast* is widely used by economic analysts. Such modelling focuses on the relationship between the growth of the national or regional economy and the total energy consumption required to achieve such growth. Energy consumption (as represented by electricity demand) may change linearly with the growth of the economy. In such cases it is valid to relate the growth of electricity consumption to the Gross National Product (GNP), population growth, power consumption in manufacturing, individual consumptive expense, etc.

Detailed information may be collected from a series of countries which are similar in their climate and level of economic development. Alternatively a standard regression model, which has already been tested, may be employed. The resultant growth trends predicted by these two macro load forecasting methods should be broadly similar. However, in developing countries things are seldom so simple. The load growth is often not governed by demand but by the ability of the electrical supply utility to build and finance the expansion of the network. Pre-investment studies normally take into account a range of confidence levels associated with the load demand forecast.

A typical macro load forecasting study for a developing country would follow the following process:

1. Collect all available data of historical population growth.
2. Analyse available population forecasts.
3. Collect historical Gross Domestic Product (GDP)/Gross Domestic Regional Product (GDRP) data.

4. Discuss and agree future GDP/GDRP forecasts with central planning authorities in country concerned, the World Bank, university economics specialists, etc.
5. Check future forecasts (e.g. elasticity analysis examines the relationship between Gross Nation Product per capita (GNP/capita and the electricity consumed per capita (kWh/capita).
6. Produce international model based upon countries in the same region.
7. Apply the GDP or GDRP/capita forecast to the international model.
8. Apply an existing regression analysis model and check against the international model.
9. Carry out sensitivity analysis based upon changes to the GDP or GDRP assumption.
10. Produce upper, medium and lower load forecasts.

A major disadvantage of the economics method of load forecasting is that it does not lend itself to forecasting the detailed geographical distribution of demand as required for the practical planning of transmission and distribution facilities. Such work is, however, carried out by economists as part of an overall power distribution project submission to funding agencies.

23.3.5   Nature of the load forecast

It is important to take into account the known constraints, such as technical or financial limitations, on supply expansion. The *attainable demand* is that portion of the demand for electricity that may be satisfied after taking into account such known constraints on supply. The potential load may be much greater than this attainable demand and there may be underlying demands which it may not be possible to supply because of physical or other constraints. Loads may be *suppressed* due to:

- bad voltage conditions at the consumers' terminals
- load shedding
- voluntary load shedding by selected co-operative consumers.

Potential customers may be placed on *waiting lists* because:

- Customers have been refused permission to connect loads to the network for technical or other supply constraint reasons.
- Lack of supply availability in the short term.

Captive plant (or *'distributed generation'*) may distort the load forecast because:

- Private electricity generating plant may already be available but not connected or synchronized onto the distribution network.

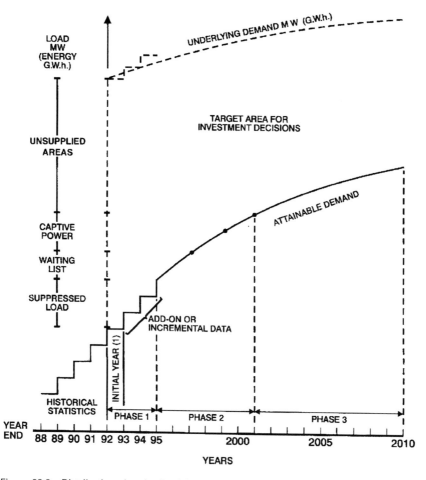

Figure 23.8 Distribution planning load forecast (The graph postulates a forecast prepared in 1994 with a forward projection to 2010.)

- Diesel engines may be currently used to drive pumps and other machinery which could be changed to electrically drive sources if available.
- Growth of domestic generation (solar power, fuel cells, etc.) may reduce effective demand growth.

Unsupplied areas:

- May have potential loads not included in the waiting lists.

The transition from underlying demand to satisfied demand may, therefore, have to take place over several years. Figure 23.8 shows how some of these factors are incorporated into a forecast prepared in 1994 for load growth projections to 2010.

23.4   SYSTEM PARAMETERS

23.4.1   Distribution feeder arrangements

Typical distribution feeder arrangements are shown schematically in Figure 23.9 and described below:

- *Simple ring* arrangements as used from primary or distribution switched substations offering a high level of supply security. Under fault conditions the faulty part of the ring may be isolated either manually or automatically and power delivered to the load via the healthy part of the circuit.
- *Interconnected or three legged ring* arrangements as used from primary or distribution switched substations. Used where the use of two simple rings is not possible geographically or where the load over the period under consideration does not warrant the security of supply offered by two separate ring feeder arrangements.
- *Radial tee* off a simple ring as used where the load is small and where it is not economic or practically feasible to include a ring feeder arrangement.
- *Radial feeders* from primary or distribution switched substations. May be used either where it can be forecast that the future load growth and associated extensions will warrant the eventual formation of a ring or where it is geographically impossible to provide a supply from a radial tee. An additional use of a radial feeder is in conjunction with an auto-recloser/sectionalizer scheme.
- *Express feeder* as used to establish a distribution switching substation as a sub-distribution point in the network. In such cases the distribution sub-station should employ a bus-section switch and have two or more incoming supply sources from the same primary substation such that a 'firm' supply is established.
- *Interconnected distributor* as used for essential (or very important person, VIP) services. In such cases further security of supply may be obtained by infeeds to the primary substation from different parts of the grid system.
- *Interconnector* as used to allow a partial alternative source to a distribution substation. In the same way as for the interconnected distributor, increased security of supply may be obtained by infeeds to the primary substation from different parts of the grid system.
- *Subring* as used to supply areas where a simple ring may be geographically hampered. By the careful positioning of isolators it is possible to isolate any fault within a small section of the circuit thereby enabling restoration of the supply to the remainder of the customers.

The designed rating of new circuits within any of these configurations must take account of:

- The maximum load to be expected under normal network conditions within the period of planning, based on the load forecast.
- The maximum loads to be expected under emergency conditions, given the chosen network reliability criteria.

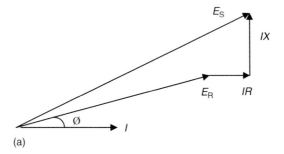

(a)

Figure 23.8(a)   Vector diagram

- The maximum and minimum voltage conditions allowable at all points on the network.

23.4.2   Voltage drop calculations

The vector diagram in Figure 23.8a applies to a three phase system where E_S and E_R are the phase to neutral sending and receiving voltages respectively. I is the line current of the three phase load. R and X are the line resistance and reactance with $\cos \emptyset$ being the system load power factor.

The voltage drop is given by the equation:

$$E_S - E_R = R \cdot I \cdot \cos \emptyset + X \cdot I \cdot \sin \emptyset$$
$$= I(R \cdot \cos \emptyset + X \cdot \sin \emptyset) \qquad (23.3)$$

For a 400 V phase to phase (nominal) distribution system:

$$E_S = \frac{400}{\sqrt{3}} = \frac{400}{1.732} = 230.94 \text{ V}$$

The line current for a balanced three phase load is obtained from the equation:

$$\text{kW} = \frac{\sqrt{3} \, EI \cos \emptyset}{10^3} \qquad \text{where } E \text{ is the phase to phase voltage (400 V)}$$

$$I = \frac{\text{kW} \cdot 10^3}{\sqrt{3} \, E \cos \emptyset} = \frac{\text{kW}}{0.6928 \cos \emptyset} \text{ amps}$$

Substituting for I in equation (23.3):

$$E_S - E_R = \frac{\text{kW}(R \cdot \cos \emptyset + X \cdot \sin \emptyset)}{0.6928 \cdot \cos \emptyset} \text{ volts per km} \qquad (23.4)$$

For a 1% volt drop design criteria:

$$E_S - E_R = \frac{E_S}{100} = \frac{230.94}{100} = 2.309 \text{ V}$$

Substituting for $(E_S - E_R)$ in equation (23.4):

$$2.309 = \frac{kW(R \cdot \cos\emptyset + X \cdot \sin\emptyset)}{0.6928 \cdot \cos\emptyset}$$

Assuming a unit length of 1 km of line:

$$kW \text{ km} = \frac{1.6 \cdot \cos\emptyset}{(R \cdot \cos\emptyset + X \cdot \sin\emptyset)} \qquad \text{for 1\% volt drop}$$

or

$$kW \text{ m} = \frac{1600 \cdot \cos\emptyset}{(R \cdot \cos\emptyset + X \cdot \sin\emptyset)} \qquad \text{for 1\% volt drop}$$

For distribution voltage systems at a power factor of 0.95 the volt drop factor for 1% volt drop is:

$$\frac{1520}{(0.95 \cdot R + 0.3122 \cdot X)} kW \text{ m} \qquad (\text{three phase}_{400 \text{ V, 0.95 pf, 1\% voltage drop}})$$

and for a power factor of 0.85:

$$\frac{1360}{(0.85 \cdot R + 0.5268 \cdot X)} kW \text{ m} \qquad (\text{three phase}_{400 \text{ V, 0.85 pf, 1\% voltage drop}})$$

Similarly for 11 kV phase to phase nominal distribution systems and 0.8 pf the volt drop factor for 1% volt drop is:

$$\frac{961}{(0.8 \cdot R + 0.6 \cdot X)} kW \text{ m} \qquad (\text{three phase}_{11 \text{ kV, 0.8 pf, 1\% voltage drop}})$$

Considering a 230 volt phase to neutral single phase distribution arrangement with identical phase and neutral conductors then for a 1% or 2.3 V voltage drop:

$$E_S - E_R = 2.3 = 2(R \cdot I \cdot \cos\emptyset + X \cdot I \cdot \sin\emptyset)$$
$$= 2I(R \cdot \cos\emptyset + X \cdot \sin\emptyset) \qquad (23.5)$$

$$I = \frac{kW \times 10^3}{230 \cdot \cos \emptyset} = \frac{kW}{0.23 \cdot \cos \emptyset}$$

Substituting for I in equation (23.5) above for the 1% voltage drop design criteria and assuming a unit route length of 1 km:

$$2.3 = \frac{2\,kW(R \cdot \cos \emptyset + X \cdot \sin \emptyset)}{0.23 \cdot \cos \emptyset} \times 1\,km$$

$$kW\,km = \frac{2.3 \times 0.23 \cdot \cos \emptyset}{2(R \cdot \cos \emptyset + X \cdot \sin \emptyset)}$$

$$= \frac{0.529 \cdot \cos \emptyset}{2(R \cdot \cos \emptyset + X \cdot \sin \emptyset)}$$

and

$$kW\,m = \frac{529 \cdot \cos \emptyset}{2(R \cdot \cos \emptyset + X \cdot \sin \emptyset)} \quad \text{for 1\% voltage drop}$$

Useful values of power factor for distribution systems that are often encountered are given in Table 23.3.

Table 23.3   Typical system power factors

System voltage (nominal)	Power factor (pf)	Voltage drop factor
11 kV (three phase)	0.8 lagging	$\dfrac{961}{(0.8 \cdot R + 0.6 \cdot R)}$ kW km
		(three phase 11 kV, 0.8 pf, 1% voltage drop)
400 V (three phase)	0.95 lagging	$\dfrac{1520}{(0.95 \cdot R + 0.3122 \cdot X)}$ kW m
		(three phase 400 V, 0.95 pf, 1% voltage drop)
230 V (single phase) – for predominantly heating and lighting loads	0.95 lagging	$\dfrac{502.55}{2(0.95 \cdot R + 0.3122 \cdot X)}$ kW m
		(single phase 400 V, 0.95 pf, 1% voltage drop)
400 V (three phase)	0.85 lagging	$\dfrac{1360}{(0.85 \cdot R + 0.5268 \cdot X)}$ kW m
		(three phase 400 V, 0.85 pf, 1% voltage drop)
230 V (single phase) – other loads	0.85 lagging	$\dfrac{449.65}{2(0.85 \cdot R + 0.5268 \cdot X)}$ kW m
		(single phase 400 V, 0.85 pf, 1% voltage drop)

23.4.3   Positive sequence resistance

The resistance of cables is dependant not only upon the physical make-up and conductor materials but also the operating temperature. Reference should be made to Chapters 12 (Cables) and 18 (Overhead Line Conductor and Technical Specifications). Based upon a 20°C everyday working temperature the AC resistance, R_{AC} ohms becomes:

$$\frac{R}{R_{AC}} = 0.01ab + 0.25b^2 + 4.7\frac{b^3}{a}$$

where R_{AC} = AC resistance in ohms

   R = DC resistance in ohms

   $a = \sqrt{(f/R)}$ f = system frequency (Hz)

   $b = (r_2^2 - r_1^2)/r_2^2$ r_1 = internal radius of conductor

                r_2 = external radius of conductor

                For solid conductors $b = 1$

For example, from cable manufacturers' information the DC resistance of DOG ACSR (assumed solid) conductor = 0.2733 Ω/km.

Then:

$$\frac{R}{R_{AC}} = 0.01 \cdot \sqrt{(50/0 \cdot 2733)} \cdot 1 + (0.25 \cdot 1^2) + \frac{4.7 \cdot 1^3}{\sqrt{(50/0.2733)}}$$

$$= (0.01 \cdot 13.526) + (0.25) + (4.7/13.526)$$

$$= 0.73276$$

$$\frac{R}{R_{AC}} = 1.36R$$

23.4.4   Inductive reactance

For a single circuit line the inductive reactance, $X_L = 2\pi f L = \omega L$ ohms:

$$X_L = \omega \frac{N_o}{2\pi} (\log_e \frac{d}{r} + \frac{1}{4n}) \; \Omega \text{ per km of conductor}$$

where L = inductance of the line (henrys)

   $\omega = 2\pi f$ and f = system frequency (Hz)

   N_o = space permeability = $4\pi 10^{-4}$ H/m

   d = GMD for conductors (m) – see Chapter 18, Section 18.7

   r = radius of conductor (m)

   n = number of strands in bundled conductor (assumed 1)

Again for DOG ACSR conductor with a GMD taken from manufacturer tables of 0.9 M:

$$X_{\mathrm{L}} = 2\pi 50 \times \frac{4\pi 10^{-4}}{2\pi} \left[\log_e \frac{0.9}{7.075 \times 10^{-3}} + \frac{1}{4.1}\right] \Omega \text{ per km of conductor}$$

$$= 0.06283 \, (\log_e 127.208 + 0.25)$$

$$= 0.06283 \, (\log_e 127.458)$$

$$= 0.3046 \, \Omega \text{ per km}$$

23.4.5   Economic loading of distribution feeders and transformers

Chapter 22 (Project Management) explains how investment decisions may be appraised using simple financial and economic analysis tools. The effect of inflation and the high cost of borrowing money to finance a project means that great importance must be paid to matching the initial capital expenditure for the distribution equipment against the future revenue stream resulting from the sales of electrical energy over the lifetime of the project. Therefore equipment must not be oversized or over specified. At the same time allowance must be made for the equipment to be capable of dealing with the factors listed in Section 23.4.1. Hence the importance of distribution planning and data collection to allow the utilization factor, load factor and other key parameters as described in Section 23.2 to be evaluated.

It is most important to be aware that discounted cash flow techniques in themselves do not give a 'correct' investment and distribution design answer. Such techniques should be used for comparative purposes with the goal of maximizing the returns in line with the electricity supply utility's performance measures and also minimizing depreciation, technical load and no-load losses, non-technical losses, maintenance costs, taxes, etc. Sensitivity analysis allows the supply utility to determine the minimum revenue requirements to support the proposed expansion project.

23.4.5.1   *Annual feeder costs*

The total annual cost of a feeder per phase, per unit length, C_{Feeder}, may be evaluated from an expression taking the form:

$$C_{\mathrm{Feeder}} = a + bL_{\mathrm{Feeder}}^{2} \qquad (23.6)$$

where a = function (annual fixed costs per unit length per phase)
   b = function (annual cost of load losses in £ per load² per phase per unit length) – no-load losses may be assumed negligible
   L_{Feeder} = total three phase feeder load (kVA)

23.4.5.2   *Annual transformer costs*

Annual transformer costs, $C_{\text{Transformer}}$, take a similar form to the annual feeder costs but also take no-load losses into account:

$$C_{\text{Transformer}} = d + e \cdot L_{\text{Transformer}}^2 \qquad\qquad (23.7)$$

where d = annual fixed costs
         e = annual operating costs taking no-load losses into account
$L_{\text{Transformer}}$ = transformer load (kVA)

23.4.6   System losses

System losses may be categorized as:

1. No-load power losses (transformer magnetizing currents, etc., see Chapter 14).
2. Load power losses ($I^2 R$ copper losses, eddy current losses, etc.).
3. Reactive losses (poor power factor, transformer losses, etc.).
4. Regulation losses (voltage drops).
5. Non-technical losses (illicit connections, poor tariff collection or metering).

The relative importance of these losses at different parts of the overall power system is illustrated below:

Part of reticulation	Overall losses (%)	Total losses (%)	Annual capital expenditure (%)
Generation	10	5–9.5	63
Transmission	30	1.5–3	12
Distribution	60	3–6	25

In order to help simplify load loss calculations the sum of the loads connected to an approximately uniformly loaded radial feeder may be lumped together at a set point along the feeder length. Consider the radial feeder below:

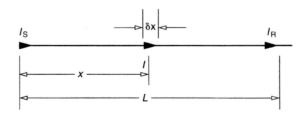

For a uniformly loaded distribution feeder the load losses (I^2r) are:

$$I = I_S + (I_R - I_S)\frac{x}{L}$$

and the three phase peak losses $= 3\int_0^L I^2 r\,dx$

where I_S = peak sending end current (A)
I_R = peak receiving end current (A)
r = resistance per unit length of feeder (Ω/km)
L = feeder section length (km)

$$\text{Therefore peak losses} = 3\int_0^L \left[I_S + (I_R - I_S)\frac{x}{L}\right]^2 dx$$
$$= rL(I_S^2 + I_S I_R + I_R^2)$$
$$= rLI_S^2\left(1 + \frac{I_R}{I_S} + \frac{I_R^2}{I_S^2}\right)$$

Let $\dfrac{I_R}{I_S} = b$, then peak three phase losses $= rLI_S^2(1 + b + b^2)$.

The expression $(1 + b + b^2)$ is known as the distribution factor where b is the ratio of the receiving end to the sending end current – see Figure 23.8b.

Average losses = Peak losses × Loss factor
Energy losses = Average losses × 8760
(over the year)

Demand losses = Peak losses × System PRF
Distribution feeder single phase peak losses $= \frac{2}{3}rLT_S^2(1 + b + b^2)$

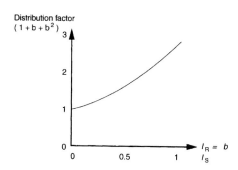

Figure 23.8(b)   Distribution factor

23.5   SYSTEM RELIABILITY

23.5.1   Introduction

Historically system availability has been assured by the use of heavy duty equipment and where necessary by the provision 'firm' supplies. A 'firm' supply point in a network is one where an outage due to a fault or during maintenance on one part of the system will not prevent a supply being available at that point. Duplication of equipment and alternative feed arrangements allow the supply to be restored either after a manual switching interval or automatically by the use of suitable switchgear, protection and control.

With advances in modern equipment manufacturing and reliability, coupled with less frequent maintenance requirements, it is possible to avoid full duplication of equipment and still effect an acceptable availability of supply.

When applying such principles consideration must be given to consumer satisfaction and the level of supply availability that consumers will still find acceptable.

Tables 23.4 and 23.5 detail typical input and output data upon which reliability studies may be based. Table 23.6 is an incident list for a feeder serving 200 customers (1000 kVA of connected load).

Table 23.4   Reliability studies – data collection

Parameter	Input data recorded
Outage date, time and duration	
Cause	
Feeder/other equipment (designation or type)	
Weather	
No. customers affected	
Division and district	
Comments/remarks	
Component that failed (identification number)	
Substation	
Voltage level	
Isolation component	
Pole/manhole, etc.	
Fuse/switchgear data	
Other location reference	
Overhead line – underground, etc.	
Action taken	
Multiple restoration	
Major electrical component	
Phasing	
Component that failed (data)	
Address of outage	
Amount of lost kW/kVA affected	
Manufacturer of failed component	
Type of damage	
Protective device failure details	

23.5.2   Reliability functions

23.5.2.1   *Introduction*

This section defines some terminology used by the American EPRI organization in distribution system reliability analysis. Examples, based upon the statistics shown in Table 23.6, are given to illustrate the usage of such data. The

Table 23.5   Reliability studies – output data

Parameter	Output data
Outage listing	
Subdivision by cause	
Performance indices (total system subdivisions, summaries and breakdown)	
Feeder – circuit trouble list	
Components that fail by cause	
Distribution of incidents, based on duration	
Voltage level	
Detailed location (feeder – circuit by cause)	
Performance in time periods by cause	
Protective component by subdivision of system	
Detailed location/feeder circuit by protective device	
Weather by subdivision of system	

Table 23.6   Incident listing for feeder serving some 200 customers (1000 kVA total connected load)

Incident	Customers interrupted	Customers affected by the restoration step(s)	Connected load (kVA) interrupted	Connected load affected by restoration step(s)	Duration (hours) of restoration step(s)	Total interruption duration (hours) for customers/ load affected by this restoration step(s)	Comments
1	40	40	150	150	0.5	0.5	Tree related
2	150	100	800	600	2.0	2.0	Transformer failure
		50		200	8.0	10.0	
3	0	0	0	0	5.0	5.0	Equipment outage which did not result in customer interruption
4	70	70	250	250	1.0	1.0	Scheduled maintenance
5	40	40	100	100	3.0	3.0	Fire under line

generalized formulae which describe each of the factors given below are included in order to allow the reader to program them into a desktop computer.

23.5.2.2   System Average Interruption Frequency Index (SAIFI)

This factor describes the historical interruptions performance of the system.

$$\text{SAIFI} = \frac{\text{Total customer interruptions}}{\text{Total customers served}} = \sum_{i=1}^{m} \frac{C_i}{C}$$

$$= \frac{40 + 150 + 70 + 40}{200} = 1.5 \text{ interruptions per year}$$

On average customers would expect to have between one and two interruptions during the year.

23.5.2.3   System Average Interruption Duration Index (SAIDI)

This factor describes the average duration of the interruptions or outages on the system:

$$\text{SAIDI} = \frac{\text{Total customer-hours interrupted}}{\text{Total customers served}} = \sum_{i=1}^{m} \sum_{j=1}^{ki} \frac{C_{ij} T_{ij}}{C}$$

$$= \frac{40(0.5) + 100(2.0) + 50(10.0) + 70(1.0) + 40(3.0)}{200}$$

$$= 4.55 \text{ hours per year}$$

Note that for the second incident described in Table 23.6 service restoration require two steps. One hundred customers were without service for 2 hours and 50 customers were without service for 10 hours. Each customer was without service for 4.55 hours during the year.

23.5.2.4   Customer Average Interruption Duration Index (CAIDI)

This factor describes the average customer outage duration:

$$\text{CAIDI} = \frac{\text{Total customer-hours interrupted}}{\text{Total customers interruptions}} = \sum_{i=1}^{m} \sum_{j=1}^{ki} \frac{C_{ij} T_{ij}}{\sum_{i=1}^{m} C_i}$$

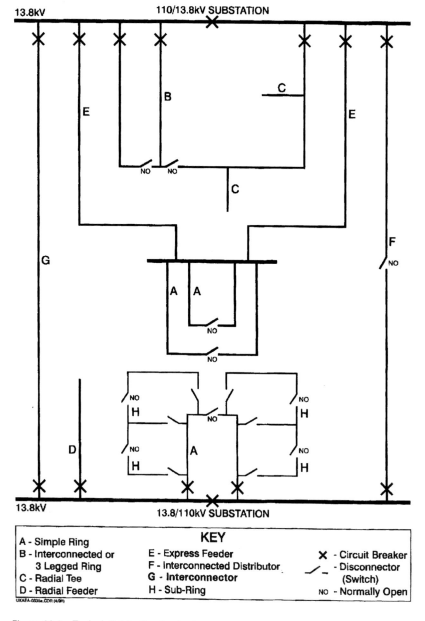

13.8kV

110/13.8kV SUBSTATION

13.8kV

13.8/110kV SUBSTATION

KEY

A - Simple Ring
B - Interconnected or
      3 Legged Ring
C - Radial Tee
D - Radial Feeder

E - Express Feeder
F - Interconnected Distributor
G - Interconnector
H - Sub-Ring

✗ - Circuit Breaker
／‒ - Disconnector
         (Switch)
NO - Normally Open

Figure 23.9   Typical distribution feeder types

$$= \frac{40(0.5) + 100(2.0) + 50(10.0) + 70(1.0) + 40(3.0)}{40 + 150 + 70 + 40}$$

$$= 3.03 \text{ hours per interruption}$$

Note that CAIDI = SAIDI/SAIFI.

23.5.2.5 *Average Service Availability Index (ASAI)*

This factor describes how closely the customer demand was met based upon a normally anticipated full 8760 hours of supply availability in the year:

$$
ASAI = \frac{\text{Customer-hours of service provided}}{\text{Customer-hours of service demanded (or anticipated)}}
$$

$$
= \frac{8760C - \displaystyle\sum_{i=1}^{m}\sum_{j=1}^{ki} c_{ij}T_{ij}}{8760C} = \frac{200(8760) - 910}{200(8760)} = 0.999481
$$

Therefore 99.95% of demand (customer-hours) was met. Alternatively the probability that the service was available at any time during the year was 0.9995.

23.5.2.6 *Average Load Interruption Frequency Index (ALIFI)*

This factor is analogous to the System Average Interruption Frequency Index (SAIFI) and describes the interruptions on the basis of connected load (kVA) served during the year by the distribution system:

$$
ALIFI = \frac{\text{Total load interruptions}}{\text{Total connected load}} = \sum_{i=1}^{m} \frac{L_i}{L}
$$

$$
= \frac{150 + 800 + 250 + 100}{1000} = 1.3
$$

Therefore there were 1.3 interruptions per kVA of connected load served during the year.

23.5.2.7 *Average Load Interruption Duration Index (ALIDI)*

This factor is analogous to the System Average Interruption Duration Index (SAIDI) and describes the number of hours on average that each kVA of connected load was without service:

$$
ALIDI = \frac{\text{Total kVA-hours interrupted}}{\text{Total connected kVA}} = \sum_{i=1}^{m}\sum_{j=1}^{ki} \frac{l_{ij}T_{ij}}{L}
$$

$$
= \frac{150(0.5) + 600(2.0) + 200(10.0) + 250(1.0) + 100(3.0)}{1000}
$$

$$
= 3.82 \text{ hours per year}
$$

Therefore each kVA of connected load was, on average, without power for 3.8 hours during the course of the year.

m = number of interruptions in a subdivision of the network (feeder, sub-station, operating district, etc.) for a given time period

ki = number of restoration steps associated with the ith interruption

C = total number of customers in the subdivision

L = total connected load (kVA) in subdivision

C_i = total customers interrupted by ith interruption $\quad C_i = \sum_{j=1}^{ki} C_{ij}$

L_i = total connected load (kVA) interrupted by ith interruption $\quad L_i = \sum_{j=1}^{ki} L_{ij}$

C_{ij} = number of customers restored during jth restoration step

l_{ij} = connected load restored during jth restoration step

T_{ij} = cumulative interruption duration (hours) for customers/load affected

By jth restoration step associated with ith interruption $\quad T_{ij} = \sum_{k=1}^{j} T_{ij}$

23.5.3 Predictability analysis

In order to have some knowledge about the reliability of a system component and when failures might occur it is first necessary to collect historical data. In this way, and for the particular application, the following questions may be addressed:

- How often does the system fail? (frequency)
- How long does it take to restore the system after failure? (duration)

And, of course, the distribution planning engineering needs to appreciate how much the system reliability is improved by a given action on a cost/benefit basis in order to aid investment decisions. Similarly, in contracts where system reliability is a contract condition, it is essential for the contractor to know the level of investment needed in order to meet the requirement, and for the client to satisfy himself that that the proposal offered will satisfy his needs.

If:

m = Mean Time To Failure = MTTF

r = Mean Time To Restore = MTTR

T = Mean Time Between Failures = MTBF

Then:

$$\lambda = \frac{1}{m} = \frac{1}{\text{MTTF}} \quad \text{(failure rate)}$$

$$\mu = \frac{1}{r} = \frac{1}{\text{MTTR}} \quad \text{(rate of transition form operating to failed state)}$$

$$f = \frac{1}{T} = \frac{1}{\text{MTBF}} \quad \text{(frequency of failure)}$$

Consider a simple distribution transformer with the failure pattern for the transformer as shown below:

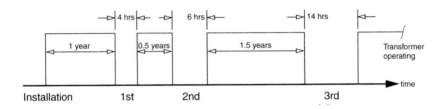

The distribution planning engineer needs to understand when such a transformer may fail in similar service conditions. This in turn might imply answering the following types of question:

- How often will it be damaged (by a falling tree, vandalism, etc.)?
- How often will power surges (lighting, etc.) occur that might cause it to fail?
- Are there any special conditions concerning this component?

In essence the time to failure for a particular installation is a random variable. However, practical precautions may be taken to increase component life since, as explained in Chapter 14, it is well known that insulation failure occurs more rapidly at higher operating (long overload period) temperatures.

From the data:

$$\text{Mean Time to Failure (MTTF)} = \frac{1 + 0.5 + 1.5}{3} = 1 \text{ year or 8760 hours}$$

$$\text{Mean Time Between Failures (MTBF)} = \frac{(1 \text{ yr} + 4 \text{ hrs}) + (0.5 \text{ yrs} + 6 \text{ hrs}) + (1.5 \text{ yrs} + 14 \text{ hrs})}{3 \text{ yrs}}$$
$$= 8768 \text{ hours}$$

$$\text{Mean Time To Restore (MTTR)} = (T - m) = 8 \text{ hours}$$

From the formulae above:

$$\lambda = \frac{1}{m} = \frac{1}{8670} = 0.0001142 \text{ failures/hour (1 failure/year)}$$

$$\mu = \frac{1}{r} = \frac{1}{8} = 0.125 \text{ restoration/hour}$$

$$f = \frac{1}{T} = \frac{1}{8768} \approx 0 \text{ failures/hour } (0.0001141 \text{ failures/hour or } 1.00091 \text{ failures/year})$$

A distribution system may be reduced, from its source to the load, to a single equivalent component with a composite failure rate (λ, failures per hour or per year) and a restoration time (r, meantime to restore supply in hours):

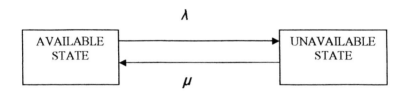

Note that:

$$\text{MTTF} + \text{MTTR} = \text{MTBF}$$

$$\therefore \frac{1}{\text{MTTF} + \text{MTTR}} = \frac{1}{\text{MTBF}}$$

If MTTF $\gg$ MTTR   then $\dfrac{1}{\text{MTBF}} \approx \dfrac{1}{\text{MTTF}}$ and $\lambda \approx f$

Long-term steady state availability $A = \dfrac{\mu}{\lambda + \mu} = \dfrac{m}{m + r} = \dfrac{\text{MTTF}}{\text{MTBF}}$

Unavailability $U = 1 - A = \dfrac{\lambda}{\lambda + \mu} = \dfrac{r}{m + r} = \dfrac{\text{MTTR}}{\text{MTBF}}$

Using logical functions where: logical   AND = ·
                                logical   OR = +

For a series system:

Source — [λ_1 r_1 / A_1] — [λ_2 r_2 / A_2] — Load

- Equivalent long-term steady state availability
$$A_{eq} = A_1 \cdot A_2$$

- Equivalent unavailability
$$U_{eq} = (U_1 + U_2) - (U_1 \cdot U_2)$$

- Equivalent failures
$$\lambda_{eq} = \lambda_1 + \lambda_2$$

- Equivalent mean time to restore
$$r_{eq} = \frac{\lambda_1 r_1 + \lambda_2 r_2 + \lambda_1 \lambda_2 r_1 r_2}{\lambda_1 + \lambda_2}$$
$$\approx \frac{\lambda_1 r_1 + \lambda_2 r_2}{\lambda_1 + \lambda_2}$$

For a parallel system:

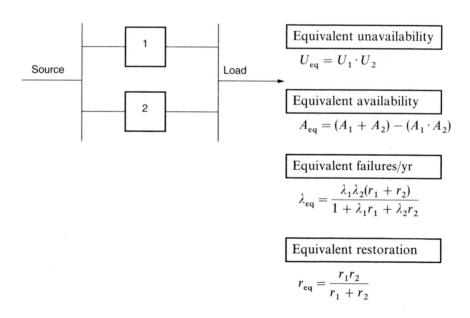

Equivalent unavailability
$$U_{eq} = U_1 \cdot U_2$$

Equivalent availability
$$A_{eq} = (A_1 + A_2) - (A_1 \cdot A_2)$$

Equivalent failures/yr
$$\lambda_{eq} = \frac{\lambda_1 \lambda_2 (r_1 + r_2)}{1 + \lambda_1 r_1 + \lambda_2 r_2}$$

Equivalent restoration
$$r_{eq} = \frac{r_1 r_2}{r_1 + r_2}$$

Consider the simple two transform 66/11 kV substation arrangement shown below:

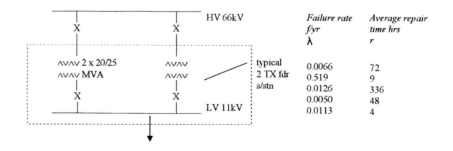

- For the one circuit case then:
  - Equivalent failure rate per source/load path to LV bus:
    λ_{eq} = 0.0066 + 0.519 + 0.0126 + 0.0050
    = 0.5432 failures/year
  - Average system down time per failure:

    $$r_{eq} = \frac{(0.0066 \times 72) + (0.519 \times 9) + (0.0126 \times 336) + (0.0050 \times 48)}{0.5432}$$
    = 17.72 hours/failure

  - Average annual outage time, $U'_{eq} = \lambda_{eq} \cdot r_{eq}$ = 9.62 hours/year.
- For the parallel supply feeder case:
  - Equivalent parallel feeder failure rate:

    $\lambda_{eq} = \lambda_1 \lambda_2 (r_1 + r_2)$

    $$= 0.5432^2 \times \frac{(17.72)}{8760} \times 2 = 0.00119 \text{ failures/year}$$

  - Equivalent parallel feeder system down time:

    $$r_{eq} = \frac{r_1 r_2}{r_1 + r_2} = 8.85 \text{ hours/failure}$$

- For the LV busbar which is in series with parallel supply paths:
  - Equivalent failure rate:

    λ_{eq} = 0.00119 + 0.0113 = 0.01249 failures/year

  - Mean time to restore:

    $$r_{eq} = \frac{\lambda_1 r_1 + r_2 r_2}{\lambda_1 + \lambda_2} = \frac{(0.00119 \times 8.86) + (0.0113 \times 4)}{0.01249} = 4.46 \text{ hours/failure}$$

  - Equivalent unavailability:
    U'_{eq} = 0.0557 hours/year

From this example it is seen that loss of LV supply is, of course, dominated by LV busbar failure since failures in other parts of the system are offset by the system having parallel supply paths. When studying such cases it is also necessary to consider the probability of both scheduled and a limited forced outage occurring simultaneously.

Scheduled outages are when equipment is deliberately taken out of service, for example during maintenance or testing operations. Forced outages are due to component failure or faults.

A range of proprietary 'System Reliability' software is available to enable computer analysis of more complex and extensive systems such as the total system illustrated in Figure 23.9. Some software uses the calculation method illustrated above, other packages use more sophisticated probability approaches. In a situation where reliability is a contract condition, care must be taken to ensure that the methodology used meets the required specification. Whatever the method, users must remember that the accuracy of such predictions depends very significantly on the accuracy of the data input, and that the use of equipment fault rates or fault restoration times obtained from another system or even from 'typical' statistics may be quite inappropriate in any one given power system.

23.6   DRAWINGS AND MATERIALS TAKE OFF

Drawings of the distribution network are normally maintained on a computer system using digitized maps and 'layered' data as described in Chapter 12. These maps show the routing of the distribution overhead lines, cables, substations and feeder pillars on the background of the normal street maps. Each and every part making up the system is given a unique identifier which links into the planned maintenance regime adopted by the utility. Such an approach also allows for the collection of statistical data for predictability analysis as described above.

A utility or consultancy will have developed standard ways of meeting its power distribution requirements over the years. Standard drawings linked to a computer database for such items as pole-mounted transformer arrangements, cable terminations, etc. and all the associated fittings which make up such an assembly will be recorded in this way. Maintenance or erection of new facilities then becomes a matter of planning the work (often using Programme Evaluation Review Techniques (PERT) with bar charts and critical path analysis), drawing out the required parts form the stores and programming the work into the overall plans.

The Westinghouse 'CADPAD' program is an advanced computer-aided planning system covering all the main areas required. Such a program allows for long-term planning of new feeders and may be used to maintain a distribution planning database giving feeder connections, capabilities, lengths, substation locations, etc. The program may be used to assist with the determination of alternative feeder arrangements taking into account switching, reinforcements and new feeders for minimum cost within existing transformer capacity, minimum cable lengths, minimum voltage drop, etc. It produces a substation summary with loadings and maximum voltage drop, etc. It produces a substation area so as to highlight possible future reinforcement requirements.

Load flow and fault level analysis is also part of the package with auxiliary programs covering regulation (reactive compensation), reliability analysis and protection co-ordination. Further the program may be used to hold system constraints and a log of equipment data for stores and ordering purposes. The program may be coupled with an interactive digitizing system to allow distribution planning drawing to link into the overall design.

24 Power Quality – Harmonics in Power Systems

24.1 INTRODUCTION

The term 'power quality' refers to the purity of the voltage and current waveform, and a power quality disturbance is a deviation from the pure sinusoidal form. Harmonics superimposed on the fundamental are one cause of such deviations, and this chapter describes the nature, generation and effects of harmonics on power supply systems, together with the limitation of such effects and harmonic studies.

The widespread and increasing use of solid state devices in power systems is leading to escalating ambient harmonic levels in public electricity supply systems.

These harmonic levels are subject to limitations in order to safeguard consumers' plant and installations against overheating and overvoltages. It is also incumbent upon individual consumers to ensure that their equipment does not produce harmonic levels that exceed such limits at the point of common coupling with other consumers. These are called 'emission limits'. Immunity standards set down the disturbance levels which equipment should be capable of tolerating without undue damage or loss of function. A third set of standards, for 'compatibility levels' has the function of enabling co-ordination and coherence of the emission and immunity standards (see Table 24.3).

In the UK limits at the point of common coupling are detailed in the Electricity Councils Engineering Recommendation, G5/4[1], although where necessary (see Section 24.7.1) a connection agreement may stipulate other limits for an individual consumer. Internationally the subject is covered by various parts of IEC 61000 (see Table 24.1) and acceptable levels of voltage disturbance to be established in power supply systems in Europe are provided in EN 50160 (but these are not formal compatibility levels). In the USA the appropriate standard is IEEE 519-1992.

Table 24.1   Useful selection of IEC standards (see also Chapter 20)

Document number	Document title
61000-1-1	Electromagnetic compatibility (EMC) – Part 1: General – Section 1: Application and interpretation of fundamental definitions and terms
61000-1-4	Electromagnetic compatibility (EMC) – Part 1–4: General – Historical rationale for the limitation of power-frequency conducted harmonic current emissions from equipment, in the frequency range up to 2 kHz
61000-2-2	Electromagnetic compatibility (EMC) – Part 2–2: Environment – Compatibility levels for low-frequency conducted disturbances and signalling in public low-voltage power supply systems
61000-2-4	Electromagnetic compatibility (EMC) – Part 2–4: Environment – Compatibility levels in industrial plants for low-frequency conducted disturbances
61000-2-6	Electromagnetic compatibility (EMC) – Part 2: Environment – Section 6: Assessment of the emission levels in the power supply of industrial plants as regards low-frequency conducted disturbances
61000-2-12	Electromagnetic compatibility (EMC) – Part 2–12: Environment – Compatibility levels for low-frequency conducted disturbances and signalling in public medium-voltage power supply systems
61000-3-2	Electromagnetic compatibility (EMC) – Part 3–2: Limits – Limits for harmonic current emissions (equipment input current $\leqslant$ 16 A per phase
61000-3-4	Electromagnetic compatibility (EMC) – Part 3–4: Limits – Limitation of emission of harmonic currents in low-voltage power supply systems for equipment with rated current greater than 16 A
61000-3-6	Electromagnetic compatibility (EMC) – Part 3: Limits – Section 6: Assessment of emission limits for distorting loads in MV and HV power systems – Basic EMC publication
61000-3-12	Electromagnetic compatibility (EMC) – Part 3–12: Limits – Limits for harmonic currents produced by equipment connected to public low-voltage systems with input current $>$ 16 A and $\leqslant$ 75 A per phase
61000-3-14	Electromagnetic compatibility (EMC) – Limits – Part 3–14: Assessment of emission limits for installations connected to LV power systems
61000-4-1	Electromagnetic compatibility (EMC) – Part 4–1: Testing and measurement techniques – Overview of IEC 61000-4 series (*Note*: There are at the time of publication 34 Sections to Part 4, covering different aspects of testing & measurement; many are relevant to the present topic.)

The major producers of harmonics are railway traction loads, large furnaces and large converter-controlled electric motor drives. Such harmonics are usually filtered on site so that they do not inject a harmful level of harmonic currents into the public electricity supply system. A further significant source of harmonics arises from the myriad of miscellaneous non-linear loads connected to the power system such as rectifiers, welders, discharge lamps, control systems, television sets, microwave ovens and computers, etc. It is fortunate that because of the arbitrary and independent nature of these loads a significant amount of harmonic cancellation occurs thus reducing the overall impact.

24.2 THE NATURE OF HARMONICS

24.2.1 Introduction

Power systems are generally linear and because of this each harmonic has an independent existence. For instance, there is no net power and energy generated between, say the fifth harmonic current and a seventh harmonic voltage, etc. This is very convenient since it greatly simplifies the treatment of harmonics and allows superposition techniques to be used in harmonic analysis. However, in the case of weak systems a detailed representation of any system non-linearity may be required as in such systems there can be interactions between harmonics of different order that are not predicted by linear time-invariant models.

24.2.2 Three phase harmonics

The general expression for harmonic currents in a three phase system is given by:

neutral current = red phase current + yellow phase current
+ blue phase current

For a balanced system:

$$I_{N} = I_{R} \sin n\omega t + I_{y} \sin n(\omega t + 4\pi/3) + I_{B} \sin n(\omega t + 2\pi/3) \qquad \text{(i)}$$

where n is the harmonic number.

From equation (i) it is clear why the third and all triplen harmonics are zero phase sequence in nature and must always have a neutral conductor to flow in or a delta connected winding in which to circulate. Furthermore, the fifth harmonic is seen to be backward rotating and therefore negative phase sequence in nature. The harmonic sequence is as follows:

Harmonic number	1	2	3	4	5	6	7	8	9	10	11	12	13
Harmonic sequence	+	−	0	+	−	0	+	−	0	+	−	0	+

For the general unbalanced case:

$$I_{N} = I_{R} \sin n\omega t + I_{Y} \sin n(\omega t + \phi_{Y}) + I_{B} \sin n(\omega t + \phi_{B}) \qquad \text{(ii)}$$

where $I_{R} \neq I_{Y} \neq I_{B}$ and $\phi_{Y} \neq 4\pi/3$ and $\phi_{B} \neq 2\pi/3$

In this case all harmonics will exhibit positive, negative and zero phase sequence components. That is, if the zero sequence currents have a neutral path in which to exist.

24.3 THE GENERATION OF HARMONICS

24.3.1 General

Harmonic distortion needs to be defined as either 'current distortion' or 'voltage distortion'. Non-linear loads, unlike linear loads, draw a non-sinusoidal current from a sinusoidal voltage supply. The distortion to the normal incoming sinusoidal current wave can be considered to result from the load emitting harmonic currents that distort the incoming current. These emitted harmonic currents, like any generated current, will circulate via available paths and return to the other pole of the non-linear load. In doing so, they cause harmonic voltage drops in all the impedances through which they pass which distort the normal supply sinusoidal voltage. The aim must therefore be to shunt the emitted harmonic currents into low impedance paths as close to the non-linear load as possible to minimise the resulting voltage distortion, as the voltage distortion will cause harmonic currents to flow in other linear and non-linear connected loads, such as motors, with deleterious effects. Zero sequence triplen harmonic currents present a further problem as they are constrained to zero sequence paths such as neutral conductors which can then become overloaded and present a serious risk as neutral conductors are not normally protected against overloading.

24.3.2 Transformers

Public electricity and industrial supplies are, to a first approximation, linear with the generated voltage being an almost pure sinusoidal wave. Virtually all harmonics are generated in non-linear loads and machine drives connected to the system. Exceptions to this are the magnetizing currents of transformers and the triplen currents that flow in generator neutral circuits. All other power system shunt equipment with non-linear characteristics, such as shunt reactors, static VAr compensators and static balancers, etc., can from the point of view of harmonic generation be regarded as non-linear loads.

Magnetic circuits in transformers and rotating machines operating under varying conditions of saturation have, since the earliest days, been known to produce power system harmonics. Typically a transformer magnetizing current (I_{mag}) will contain small third, fifth and seventh harmonic components as given in per cent by the following formula for the older stalloy-type transformer core steels:

$$I_{mag} = 100 \sin(\omega t - 78) - 39 \sin(3\omega t - 83) \\ + 18 \sin(5\omega t - 81) - 8 \sin(7\omega t - 80) \tag{iii}$$

Modern cold rolled grain oriented silicon steel, which has a squarer magnetizing characteristic produces significantly less third harmonic current.

Normally transformers are designed to operate up to the knee point of their magnetizing curve, but under conditions of magnetic saturation (caused by overvoltage or ferroresonance) the harmonic content of the magnetizing current can increase dramatically. Equipment containing saturable reactors, which deliberately exploit the magnetic saturation phenomenon, will therefore probably require harmonic filtering.

Remanent magnetism in transformers (caused for example by circuit breaker interruption of a fault when there is still a significant DC component) can persist for months or longer and, by displacing the B-H curve can result in magnetizing asymmetry and even harmonics in the magnetizing current.

24.3.3 Converters

Converter is the generic name given to rectifier and inverter systems. These systems range from simple rectifiers through to AC–DC–AC systems for the interconnection of major power networks such as the UK/France cross Channel power link and frequency changer systems for soft start and speed control of AC machine drives. In addition, particularly with the advent of the gate turn off thyristor (GTO), Flexible AC Transmission Systems (FACTS) of great technical and operational sophistication are gaining widespread use for the control and conditioning of power systems. All these systems can produce copious harmonics. The principal harmonic numbers may be filtered out on site but significant harmonic currents may still emanate from these systems onto the power network.

24.3.4 The thyristor bridge

Thyristor rectifier – inverter bridges are the basis of all the systems described in Section 24.3.3. Figure 24.1 shows a basic three-phase six-pulse thyristor-controlled bridge together with the idealized DC side voltage and AC current of one phase. Applying Fourier analysis to the square wave of phase current yields the following harmonic series:

$$I_{phase} = \frac{2\sqrt{3}}{\pi} I_d \left(\cos \omega t - \frac{1}{5} \cos 5\omega t + \frac{1}{7} \cos 7\omega t - \frac{1}{11} \cos 11\omega t \right.$$
$$\left. + \frac{1}{13} \cos 13\omega t - \right. \tag{iv}$$

This indicates that the fifth and seventh are the principal or characteristic harmonics of the six-pulse bridge. These harmonics would be present on the primary side of a star-star configured transformer feeding the bridge, However, if the bridge was fed from a star-delta vector group transformer then a

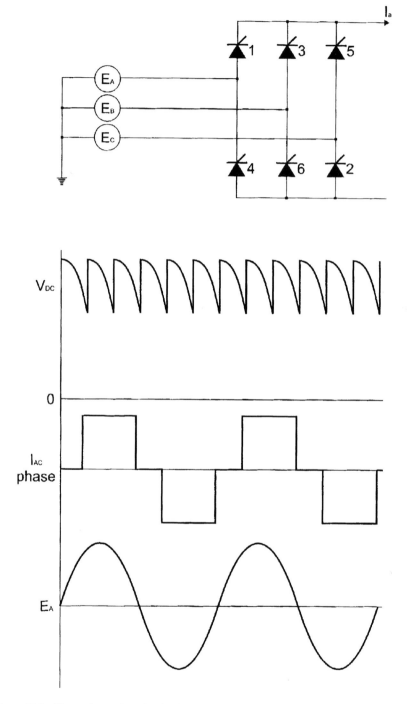

Figure 24.1   Three-phase, six-pulse thyristor-controlled bridge rectifier supplied from an infinite busbar

30 degree phase shift would have to be taken into account together with some adjustment for the transformer ratio. The harmonic series for this connection then becomes:

$$I_{phase} = \frac{2\sqrt{3}}{\pi} I_d (\cos \omega t + \frac{1}{5}\cos 5\omega t - \frac{1}{7}\cos 7\omega t - \frac{1}{11}\cos 11\omega t$$
$$+ \frac{1}{13}\cos 13\omega t + \qquad \text{(v)}$$

If these two converters are supplied from the same AC source and connected in series on the DC side we have a 12 pulse connection. Notice that in this case the fifth and seventh harmonics cancel out yielding the following harmonic series:

$$I_{phase} = \frac{4\sqrt{3}}{\pi} I_d (\cos \omega t - \frac{1}{11}\cos 11\omega t + \frac{1}{13}\cos 13\omega t$$
$$- \frac{1}{23}\cos 23\omega t + \frac{1}{25}\cos 25\omega t - \qquad \text{(vi)}$$

In equation (vi) the eleventh and thirteenth are the principal harmonics, equations (iv) and (vi) indicate that a polyphase bridge will produce harmonics of the order:

$$n = pk \pm 1 \qquad \text{(vii)}$$

Where p is the pulse order and k an integer. These harmonic series are idealized since in practice such converters will operate from supplies having a significant impedance. This will modify the converter response and waveforms. In addition, imperfections and unbalance in the power supply system and in the converter itself will increase the harmonic spectra produced. Table 24.2 shows the actual current spectra of a 70 kVA, six-pulse converter motor drive operating on an electrically weak system. Here phase unbalance has caused the individual phase harmonics to be dissimilar in magnitude and a third harmonic has appeared. Further, slight imperfections in the converter's firing angle control has given rise to small even harmonic terms in the spectra.

Therefore it should be appreciated that harmonic spectra encountered in practice may be significantly different from the idealized spectra anticipated from a particular installation.

24.3.5 Railway and tramway traction systems

24.3.5.1 *Introduction*

Rail traction locomotives can produce high power harmonics and it is normally impractical to completely filter these on the rolling stock. Filtering, if

Table 24.2   Practical harmonic current spectra produced by a 70 kVA, 415 V, six-pulse thyristor-controlled converter supplied from an electrically weak source

Harmonic number	Frequency (Hz)	G5/3 Limit (amps)	Red phase (amps)	Yellow phase (amps)	Blue phase (amps)
1	50		95.0	100.0	97.0
2	100	48	0.9	1.2	1.0
3	150	34	3.0	3.6	4.0
4	200	22	0.5	0.6	0.5
5	250	56	27.0	31.0	28.2
6	300	11	0.2	0.3	0.1
7	350	40	10.0	9.0	0.4
9	450	8	0.4	1.0	1.4
11	550	19	7.4	8.6	8.0
13	650	16	4.6	4.0	4.2
15	750	5	0.2	0.8	0.6
17	850	6	2.3	3.0	3.1
19	950	6	3.0	2.8	3.2
21	1050	4	0.2	0.6	0.4

required must therefore be carried out at the traction substations. The pragmatic approach is to supply the traction substations from a strong (high fault level) high voltage grid connection point if studies show that harmonics may be a problem at a lower voltage level on the network. Use of such a connection must be checked for economic viability since the higher the system voltage the higher the capital costs of the associated equipment.

24.3.5.2   AC traction

The motive power units of trains taking an AC supply comprise onboard single-phase transformers supplying the axle drive motors through one of a variety of converter systems. Older AC trains with diode/thyristor convertor systems produce lower range harmonics (100–750 Hz). More modern trains with GTO drives, pulse-width-modulated (PWM) systems and synthesized driving voltages produce harmonics in the higher ranges centred around, say, 1800 Hz, but of a lower pro rata magnitude. Although the harmonic spectra generated by modern rolling stock has improved, (reduced) significant third harmonics remain a feature of these systems.

Rail traction systems are rich in harmonics and the difficult assessment of the filtering requirements has to take into account that several trains of varying vintage and type operating at different duties will be supplied from any one traction substation at any given time. Further, since the traction load is essentially single phase it creates an unbalance on the three-phase supply source. The effect of phase unbalance is to impose both positive and negative fundamental harmonic phase-sequence currents on the supply system. For further discussion and an example of this see Sections 25.3 and 25.4, in Chapter 25. In practice this unbalance is partially reduced by connecting the different traction substations along the route of the railway line from different selected phase pairs of the three-phase supply system. However, this has only partial success because the loads on each substation traction transformer will be varying with time throughout the day. In addition, different substation transformers may be taken out of service at different times for maintenance and this may exaggerate the overall state of unbalance. Because of the phase pair connection no zero sequence components will be present but triplen harmonics with positive and negative sequence components will exist in the traction load current spectrum brought about by the phase unbalance.

24.3.5.3 *DC traction*

DC traction systems are normally supplied through three-phase rectifier banks, often combined in one piece of plant with a transformer, to comprise transformer-rectifiers which take supply at 11 kV or 20 kv and convert to a traction supply at 500 V, 650 V or even 1500 V DC. The problem of unbalance is avoided, because supply is taken equally from all three phases, but a harmonic distortion problem is created by the rectifiers. A six-pulse rectifier bank will generate 5th and 7th harmonics in the supplying system, while a 12-pulse bank will generate mainly 11th and 13th harmonics and a 24-pulse rectifier produces 23rd and 25th harmonics. Most public supply systems are already distorted with significant levels of the 5th harmonic, so any new DC traction rectifier system will probably be at least 12 pulse. The cost of a 24-pulse system may be difficult to justify; the cost of a 12-pulse system can be reduced in substations where there are pairs of transformer rectifiers. This is achieved by installing six-pulse units, but specifying the transformers of each pair to have a vector group with a 30° displacement between them, thus achieving an overall 12-pulse system while both transformer rectifiers are in operation and sharing load equally.

It should be noted that while a three-phase rectifier bank is a balanced load, where it is subject to unbalanced system voltages (e.g. unbalance already existing in the supplying power system) additional triplen harmonics are generated.

24.3.6 Static VAr compensators and balancers

These devices are discussed in more detail in Chapters 25 and 26. Their ability to control voltage by compensating for rapid changes in reactive power loading has resulted in their widespread use as elements in power transmission systems. They typically include in their assembly thyristor control equipment that inevitably generates its own harmonic currents which are very sensitive to the thyristor firing angle delay as shown in Fig. 24.2. They also contain capacitor banks, which are often split into sub units that double as the necessary harmonic filters (either shunt filters tuned to the characteristic frequencies of the converter or high pass filters) as shown in Fig. 24.6.

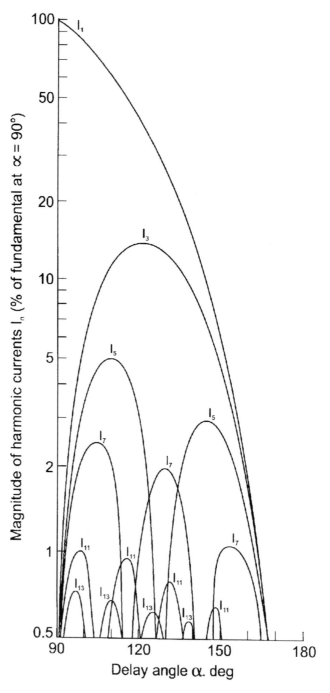

Figure 24.2   Harmonic current spectra, as a function of the firing angle of a six-pulse thyristor-controlled reactor

24.4 THE EFFECTS OF HARMONICS

24.4.1 Heating effects of harmonics

Harmonic currents flowing in rotating machines cause heating effects both in the conductors and in the iron circuit. In particular, eddy current losses are proportional to the square of the frequency. Further, some harmonics are negative phase sequence in nature and these give rise to additional losses by inducing higher frequency currents and negative torques in machine rotors.

Harmonic currents will tend to flow into the system capacitance and this can give rise to overloading of power factor correction capacitors and to the derating of cables.

The only meaningful way to sum harmonic currents is by their heating effect, that is, by their root mean square (r.m.s.) value. Thus the effectiveness of, say, a group of harmonic currents and the fundamental current is given in terms of their r.m.s. values as follows:

$$I_{rms} = \sqrt{\frac{I_1^2}{2} + \frac{I_2^2}{2} + \frac{I_3^2}{2} + \cdots + \frac{I_n^2}{2}} \qquad \text{(viii)}$$

For example, a 100% fundamental current with a 40% third, a 25% fifth and a 15% seventh harmonic will yield a total r.m.s. (thermal) current of 111% with a heating effect of $(111^2/100^2) \times 100\% = 123\%$ over that of the fundamental alone.

24.4.2 Harmonic overvoltages

Harmonic voltages, generated by harmonic currents flowing against impedance to the harmonic, can lead to significant overvoltages. Such effects are known to cause equipment failures, and capacitors are particularly susceptible. These overvoltages can be enhanced by system resonances whereby a given harmonic current may generate a disproportionately large harmonic voltage. Since, from the point of view of electric stress, the peak value of applied voltage is important, it is not appropriate in this case to take the r.m.s. value of a given harmonic voltage spectra. It is not possible to be certain of the changing phase relationship of the harmonics to the fundamental voltage. Therefore it is recommended that the arithmetic sum of the peaks of the fundamental and harmonic voltages are calculated when assessing the stresses placed on equipment due to harmonics. Such a pessimistic approach will ensure that the equipment, particularly capacitors, are generously rated and be less susceptible to overvoltage failure.

24.4.3 Resonances

Any inductive–capacitive–resistance (LCR) circuit, such as a power system, will exhibit a resonant response to one or more frequencies. Resonance is

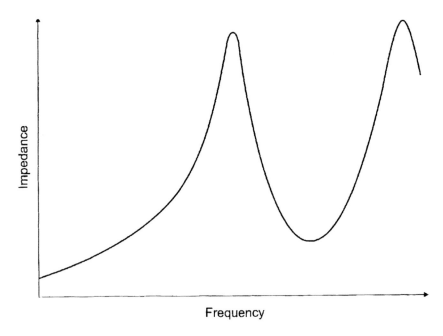

Figure 24.3   Typical power system harmonic impedance plot as a function of frequency

defined by the circuit becoming resistive with the reactive components can-
celling out. As a consequence the phase angle between the driving voltage and
the current becomes zero. Either side of the resonant frequency the circuit
becomes inductive or capacitive.

There are two types of resonant response. Series resonance is characterized
by the circuit impedance tending towards a small residual, largely resistive,
impedance. Consequently, in this response the circuit current will tend to be
high when the circuit is fed from a voltage source and large voltages will
appear across the reactive circuit components. Parallel resonance exhibits a
high impedance response which is still resistive. This response gives rise to
the generation of relatively high voltages across reactive components when
the circuit is fed from a constant current source. These characteristics are put
to good use in filter circuits.

An inspection of the frequency response (Figs 24.3 and 24.4) of a typical
power system impedance against increasing frequency shows a variable non-
linear response with peaks and troughs. These peaks and troughs are due to
resonances caused by the system capacitive and inductive reactance. The
peaks are parallel type resonance responses and the troughs are series resonance effects. The low impedance troughs will give rise to increased harmonic
currents of the appropriate frequency and these in turn can cause increased
harmonic voltages in other equipment.

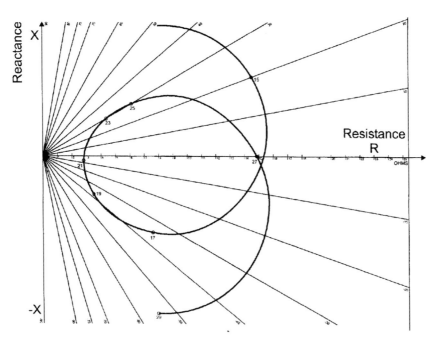

Figure 24.4   Typical power system harmonic impedance polar plot in the XR plane corresponding to Fig. 24.3

Such natural system resonances are not in themselves necessarily a cause for concern. It is only when such system responses, coupled with significant harmonic current inputs from non-linear loads, lead to excessive harmonic voltages that steps must be taken to limit the response. Nevertheless on some (high Q) systems with low damping, potentially huge harmonic voltages can be generated. Overvoltages as high as 120% or more have been encountered in studies and in actuality on some systems.

24.4.4 Interference

Power system harmonics may cause interference with communication, signalling, metering, control and protection systems either by electromagnetic induction or by the flow of ground currents. However, systems such as signalling circuits whose correct function is essential to safety, should have any sensitivity to harmonic interference designed out of them at the outset. Also, standby earth fault relays connected in the neutral of transformer circuits may employ third harmonic filters. These filters are designed to prevent anomalous relay operation from large discharge lighting loads which may generate triplen harmonics flowing in the neutral conductor. Incorrect earth fault residual current relay element operation may also be prevented by connecting the supply transformers to converter equipment in a delta configuration thus blocking the flow of zero sequence currents from converters to the power system. Other adverse effects of harmonics include:

- Overstressing and heating of insulation.
- Machine vibration.
- The destruction by overheating of small auxiliary components, e.g. small capacitors and motors.
- Malfunctioning of electronic devices.

24.5 THE LIMITATION OF HARMONICS

24.5.1 Harmonic filters

Harmonic filters are series or parallel resonant circuits designed to shunt or block harmonic currents. They reduce the harmonic currents flowing in the power system from the source and thereby reduce the harmonic voltage distortion in the system. Such devices are expensive and should only be used when other methods to limit harmonics have also been assessed. The application of filters in a given situation is not always straightforward. The filters themselves may interact with the system or with other filters to produce initially unsuspected resonances. Hence in all but the most simple cases harmonic studies should be used to assist with the determination of the type, distribution and rating of the filter group. Classical shunt filter circuits and their associated characteristics are shown in Fig. 24.5. Note that when the filter forms the capacitive section of an SVC, it is essential for it to be capacitive at fundamental frequency so it will produce the reactive power required.

The selectivity or tuning response of the simple single resonant frequency filter circuit is defined by its Q or quality factor:

$$Q = \omega L/R \tag{ix}$$

A high Q factor gives good selectivity (narrow frequency response) but the filter tuned circuit may be prone to drifting in its tuned frequency owing to changes in temperature or component ageing. Since slight changes in system frequency will cause detuning a less peaky filter response with a lower Q factor is more desirable to accommodate these changes. The tuned resonance frequency of a series LCR circuit is given by:

$$f = \frac{1}{2\pi}\sqrt{(1/LC)} \tag{x}$$

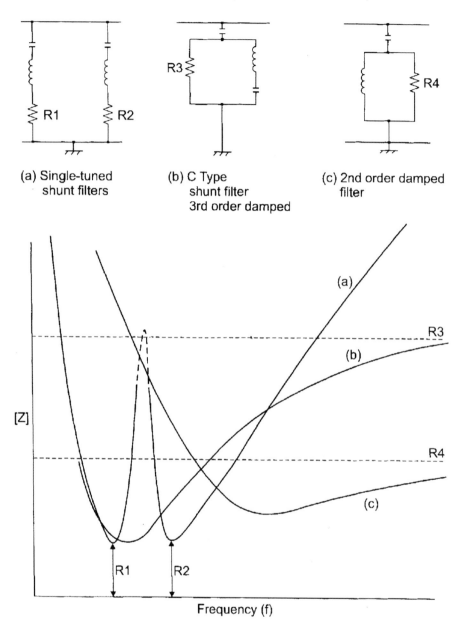

(a) Single-tuned
    shunt filters

(b) C Type
    shunt filter
    3rd order damped

(c) 2nd order damped
    filter

Figure 24.5   Typical harmonic filter characteristics

and the impedance at resonance is simply the residual reactor resistance, R. The detuning of filters for changes in harmonic frequency, can be expressed as:

$$\delta = \frac{\omega - \omega_n}{\omega_n} = \frac{\Delta f}{f_n}$$

(xi)

If changes in capacitance and inductance, due to temperature change and ageing are included, the detuning factor becomes:

$$\delta = \frac{\Delta f}{f_n} + \frac{1}{2}\left|\frac{\Delta L}{L_n} + \frac{\Delta C}{C_n}\right| \qquad\qquad \text{(xii)}$$

Active filters may be employed to overcome such effects such that the filter is constantly kept in tune by automatically varying the reactor by means of a control system to keep the inductor and capacitor voltages equal.

It is often the case that more than one harmonic is exceeding the harmonic limits set by the supply authority. Therefore more than one filter is necessary. However, as the number of shunt filters increases there is a tendency for these circuits to interact with the power system impedance to produce unwanted resonances involving other frequencies, if such harmonic frequencies exist on the system. A solution is to use a high pass shunt or C type filter arrangement whereby all frequencies above a certain harmonic are shunted to ground. A typical filter group is shown in Fig. 24.6.

The required optimum selectivity of the tuned filters depends upon the system impedance angle, Ø, at the point of filter connection, and the detuning factor, δ.

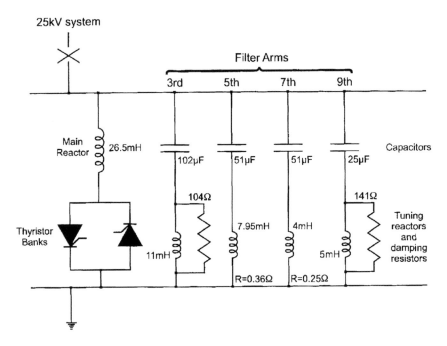

Figure 24.6   A balancer and filter group schematic

An approximation for a possible optimum Q value derived from a graphical construction by Arrillaga et al.[3,4] is given by the expression:

$$Q = \frac{1 + \cos \phi}{2\delta \sin \phi} \qquad \text{(xiii)}$$

Consider a converter connected to a 33 kV system with a 50 Hz + 1% frequency supply where studies have shown that there is a need for fifth and seventh harmonic filters. Suppose also that these studies show the need for 2 MVAr of reactive compensation for the converter. This could be conveniently split into two 1 MVAr units to form the filters; the 1 MVAr capacitors being more than adequate for the filter duty. Assume that the temperature variation for the inductors and capacitors is 0.01% per degree Celsius and 0.04% per degree Celsius respectively with a possible ambient temperature variation of 20°C above normal. Then from equation (xii):

$$\delta = \frac{\Delta f}{f_n} + \frac{1}{2}\left[\frac{\Delta L}{L_n} + \frac{\Delta C}{C_n}\right] = \frac{1}{100} + \frac{1}{2}(0.0001 \times 20 + 0.0004 + 20)$$
$$= 0.015$$

Now if the system impedance angle is 70 degrees, from equation (xiii):

$$Q = \frac{1 + \cos 70}{2(0.015 \sin 70)} = 47.6$$

then for each capacitor

$$\text{MVAr} = \frac{V^2}{X_c} \quad \text{and} \quad X_c = \frac{10^6 \ \Omega}{2\pi f C} \quad \text{(where } V \text{ is the line voltage in kV)}$$
$$C = \frac{\text{MVAr}}{2\pi f V^2} \times 10^6 = 10^6 \text{ microfarads (μF)}$$

Thus for the 5th harmonic:

$$C = \frac{10^6}{2\pi f V^2} = \frac{10^6}{2\pi 250(33)^2} = 0.584 \ \mu\text{F}$$

$$L = \frac{10^6}{(2\pi f)^2 C} = 0.694 \text{ H}$$

and hence

$$R = \frac{2\pi \times 250 \times 0.694}{47.6} = 22.9\,\Omega$$

and similarly for the 7th harmonic filter:

$C = 0.417\,\mu F$
$L = 0.496\,H$

And

$R = 22.9\,\Omega$

The more complex calculations associated with parallel high pass and C type filters are given in the references at the end of this chapter.

24.5.2 Capacitor detuning

It is possible for power factor correction capacitors, particularly on thyristor-controlled drives, to form a low impedance path or 'sink' for harmonics or to inadvertently resonate with one of the harmonics produced by the non-linear load. Symptoms are typically capacitor overheating, capacitor fuse protection operation or failure due to overstressing. A solution is to detune the capacitors from high harmonics by the insertion of a series reactor forming a tuned circuit with the resonant frequency typically around the fourth harmonic. The capacitor circuit then looks inductive to all harmonics above the fourth harmonic and resonance is quenched. This is a sufficiently common problem that power factor correction capacitor banks may be specified for installation with these detuning components at the outset. Examples of detuned power correction capacitor networks are shown in Fig. 24.7.

24.6 FERRORESONANCE AND SUBHARMONICS

24.6.1 Introduction

These phenomena involve real physical events and real practical problems on power systems. For example, when the British 275 kV and 400 KV transmission systems were installed, a number of ferroresonance events occurred leading to outages and damages to grid transformers. The failure of a large generator in Mohave power station in the USA, caused by subsynchronous resonance with the turbine natural frequency mode, is also well documented.

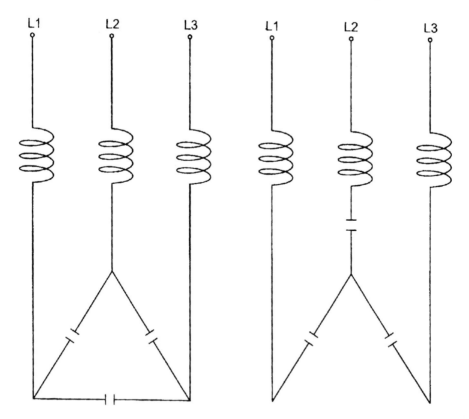

Figure 24.7   Detuned power factor correction capacitors

24.6.2 A physical description of ferroresonance

Ferroresonance is characterized in a circuit by the sudden departure from sinusoidal conditions and the emergence of current spikes reaching magnitudes of typically 2 to 5 per unit values. These current spikes arise from the magnetic cores of transformers or reactors going into brief saturation excursions. Such large current spikes give rise to system overvoltages reaching values in excess of 1.5 per unit as illustrated in Fig. 24.8c.

Ferroresonance and subsynchronous resonance can arise in power system circuits when capacitance is connected in series (and less commonly when connected in parallel) with non-linear inductive circuits such as transformers and reactors and when the voltage is sufficient to drive the non-linear inductance to near the knee point of the B-H curve. As the inductance falls at the knee point a stage may be reached where the residual inductance is in resonance with the capacitance at the driving voltage frequency. This causes a drop in the circuit impedance to the value of the residual resistance and a spike of current results that drives the inductive reactance well into saturation. The inductance

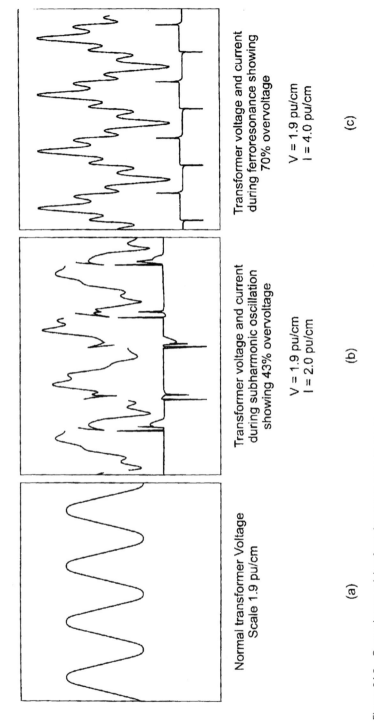

Normal transformer Voltage
Scale 1.9 pu/cm

(a)

Transformer voltage and current
during subharmonic oscillation
showing 43% overvoltage

V = 1.9 pu/cm
I = 2.0 pu/cm

(b)

Transformer voltage and current
during ferroresonance showing
70% overvoltage

V = 1.9 pu/cm
I = 4.0 pu/cm

(c)

Figure 24.8   Overvoltages arising from ferroresonance phenomena

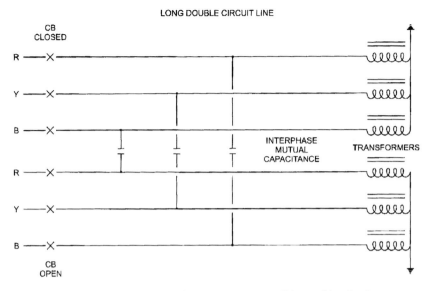

Figure 24.9   Double circuit line transformer feeders a possible condition for the energization of ferroresonance

then becomes very low, the resonance condition is destroyed, and since the voltage wave is now falling, the current rapidly falls to a low value. This whole process is repeated in the next half cycle yielding another current spike in the opposite direction. This is a simplistic explanation of a complex phenomenon since sometimes two spikes of current occur each half cycle. The potential for ferroresonance problems has ironically increased with the introduction of low-loss square law characteristic transformer and reactor steels. Such materials increase the inherent non-linearity of transformers and reduce system damping. Hence, ferroresonance is basically a fundamental system frequency event, but, because of the current spikes and voltage distortion a rich harmonic spectrum is generated.

A typical situation that arose on the British 275 kV grid system involved a double circuit line feeding two grid transformers as shown in Fig. 24.9. If one circuit was tripped out for whatever reason, that circuit should have been dead and was initially expected to be so. However, it transpired that if the double circuit line was long enough then there was sufficient intercircuit capacitance between the live and apparently dead circuits for a ferroresonance response to be excited in the transformer feeder circuit that had been switched out of service. The transformer was continuing to be fed by energy through the intercircuit mutual capacitance. The resulting spiky currents caused an alarming noise from the transformer core and some transformer failures resulted from overvoltage flashover effects. Such phenomena are now avoided by the use of operational rules that require the transformer to be initially switched off and isolated from its circuit and earthed, before the overhead line circuit is switched out.

Another well-documented case occurred in the USA on the Detroit-Edison Company electrical supply system. A 40 kV transformer had lost a phase and gone into ferroresonance with the system capacitance. The resulting overvoltages caused the failure of 39 surge arresters on the network.

In summary, the following parameters are important[12]:

- The characteristics of the iron of the transformer core.
- The designed flux level of the transformer.
- The level of the supply voltage compared to the nominal.
- The inductance of the relevant transformer winding when saturated.
- The load on any other windings on the transformer, and the coupling between the windings.
- The instantaneous flux in the transformer when the initiating incident occurs.
- The point on wave at which the switching event occurs.

Problems can be avoided by tackling these criteria, and by operational actions such as ensuring sufficient load remains connected to the system being switched and eliminating single-phase switching.

24.6.3 Subharmonics

Consider the same LCR circuit involving non-linear inductance as described above but energized by a voltage below the value sufficient to cause ferroresonance. As might be expected the circuit will behave in a linear manner. However, if this current is suddenly disturbed by, say, a switching event or a transient voltage fluctuation, the circuit may jump into a subharmonic response characterized again by spiky currents and overvoltages but at a frequency that is a sub-multiple of the fundamental frequency. The subsynchronous frequency may be typically one third (16 2/3 Hz for a 50 Hz fundamental power frequency system) or less likely one fifth of the fundamental frequency. A transformer undergoing this subharmonic response will exhibit a waveform typically as shown in Fig. 24.8b and generate loud audible vibrations.

24.6.4 Interharmonics

Between the harmonics of the power frequency voltage and current, further frequencies can be found which are not an integral multiple of the fundamental. These can appear as discrete frequencies or as a wide-band spectrum. Summation effects are not likely and need not be considered (IEC 61000-2-1). The main sources of interharmonics are static frequency converters, cycloconverters, induction motors, welding machines and arc furnaces. One effect of interharmonics is the disturbance of ripple control systems, another is a flicker effect. The latter occurs with discrete frequencies close to the fundamental.

24.7 HARMONIC STUDIES

24.7.1 The requirement

The basis under which a utility will consider acceptance of a load which may further add to the harmonic distortion on its supply network is twofold:

a) The level of distortion already existing – to which its customers are already subject (this may be expressed as the effect of other distorting loads).
b) The increase in disturbance which the subject load will cause.

This approach is inherent in the IEC 61000 standards, and in the UK's G5/4 guide. The utility will assess both these aspects in order to assess the likely harmonic content in its mains after connection of the load. A comparison is then made with the "planning levels" (see Table 24.3) of the affected harmonics in the standard chosen, and if the G5/4 approach is used, acceptance normally given if the new level is lower than the planning level. If it is not, usually some mitigating action will be required before the load is accepted; if existing distortion levels are high this may result in very significant and expensive mitigation investment for the new consumer. The IEC approach sets an emission limit on the new connected load which, if achieved will ensure the planning level is maintained provided other consumers are treated in the same way. It acknowledges that if existing network distortion levels are low, a utility may exceptionally accept individual loads which exceed the emission limit. Guidance is also given on calculating the combined effect of other specific existing loads, and the effect of existing distortion introduced from higher voltage levels must be considered. (IEC 61000-3-6).

Table 24.3   Terms relating to power system EMC (from IEC 61000-3-6)

Term	Explanation
Planning level	Level of harmonic distortion that can be used for planning purposes in evaluating the impact on the supply system of all consumer loads. They are set by the utility and can be considered as internal quality objectives. They are equal to or lower that compatibility levels.
Compatibility level	The reference value for co-ordinating the emission and immunity of equipment which is part of, or supplied by, a supply network to ensure the EMC in the whole network. Compatibility levels are generally based upon the 95% probability levels of entire systems using distributions which represent both time and space variations of disturbances.
Immunity level	The maximum level of a given electromagnetic disturbance incident on a particular device, equipment or system for which it remains capable of operating at the required degree of performance.
Emission level	At each (inter)harmonic frequency, the emission level from a distorting load is the (inter)harmonic voltage or current which would be caused by the load into the power system if no other distorting load were present.

In order to determine the level of distortion existing, it is necessary to take accurate measurements at key points in the system. To determine the effect of the additional load it is necessary to undertake system studies.

24.7.2 The studies

To calculate the effect of a potentially distorting load it is necessary both to determine the characteristics of the load, and of the network to which it is connected. In a relatively simple case where one continuous industrial process is supplied through a rectifier, a computer model of the load and rectifier installation can determine the maximum level and spectrum of harmonics generated into a given impedance. A much more complex task is to set up a model of the harmonic impedance of the network to which the load will be connected, and into which its harmonics will be generated. It is not just that in old networks data is often unreliable; account must also be taken of all the different network configurations that might realistically be encountered, so that potentially a great many study scenarios have to be considered. To reduce the extent of the work, it is sometimes realistic to treat parts of a network very remote from the subject investigation as 'lumped' impedances. Guidance is given in Ref. (14).

Once a network model, or suite of models, has been created, and the harmonic generating potential of the load has been determined, any one of a number of proprietary harmonic system study software packages can be used to find out the effect of connecting the load, taking account of the existing voltage and current distortion in the network. One methodology is broadly as follows[11]:

$$Vhp = Vhr + Vhi$$
$$Vhr = Vhm \times (Zhf/Zhe)$$
$$Vhs = Vhl - Vhr$$

Where Vhp = predicted harmonic voltage distortion at harmonic order h
Vhr = calculated harmonic distortion caused by network reconfiguration
Vhi = Calculated harmonic distortion due to new or incremental current injection
Vhm = measured existing harmonic distortion
Vhs = calculated spare margin for additional load
Vhl = the limit under the standard applied (eg G5/4)
Zhf = the future harmonic impedance
Zhe = the existing harmonic impedance

Greater complexity is encountered when the load is of a widely varying nature, because (for example) the level of harmonics generated in a rectifier depends on the load and electrical environment in a manner that is almost random[10] and even more care has to be taken when the varying load appears over a number of different supply points, such as in a traction system, since the net effect on the networks and its connected customers will depend upon the combined effect of the different injections at each instant.

24.7.3 Measurements

Bearing in mind that the effect of the many distorting sources feeding into a supply network is to cause a rapidly fluctuating mix of harmonic spectra, it is necessary to undertake accurate measurements to determine the true maximum levels of harmonic pollution present. The technology of harmonic measuring equipment is undergoing rapid development, and it is wise to check the market and compare with the requirements of the project, before choosing instrumentation. In a recent project in London[13] the selected equipment simultaneously read phase currents and voltages, and digitized its readings into 256 segments of every fundamental cycle. It used a fast fourier transform to calculate the harmonic currents and voltages. Analysis of readings over a significant period enabled the worst levels of pollution to be selected as the basis for system studies. Further testing was undertaken to check performance after connection of the new loads, both to ensure limits were maintained and to check the accuracy of the computer predictions of the design studies. In order to be sure that system performance could be correctly related to load behaviour, the use of GPS techniques was necessary to ensure highly accurate timing of simultaneous readings.

Guidance on measuring techniques is provided in IEC 61000-4 and IEEE 1159.

24.8 CASE STUDIES

There are now many published case studies, examples of which are given in Refs (13) and (15). A whole series of worked examples are provided in IEC 61000-3-6.

REFERENCES

1. Engineering Recommendation G5/4, *Limits for harmonics in the United Kingdom Electricity Supply System.* UK Electricity Council [Now Energy Networks Association (ENA)].
2. G. Say, *The performance and design of alternating current machines*, 3rd Edition, Pitman & Sons, London, 1965.
3. Arrillaga, Smith, Watson and Wood, *Power system harmonic analysis*, John Wiley & Sons, 1997.
4. Arrillaga, Bradley and Boddger, *Power system harmonics*, John Wiley & Sons, 1985.
5. Gaibrois and Bacvarov, Open phase conductor and load transfer combine to produce ferroresonance. *McGraw-Edison Company Power System Division Quarterly*, Vol. 8, 1978, pp.193–198.
6. Yacamini, Power system harmonics, *IEE Power Engineering Journal*, August 1994.

7. Howroyd, Public supply system distortion and unbalance from single phase AC traction. *IEE Proceedings*, Vol. 124, October 1997.

8. Barnes, Harmonics in power systems. *IEEE Power Engineering Journal*, Vol.3, No.1, January 1989.

9. Ainsworth, *Filters, Damping Circuits and Reactive Voltages in HV DC Current Converters and Systems*. MacDonald, London, 1965.

10. Wang, Pierrat , Feuillet and Shi, *Simulation study of random properties of harmonic currents generated by static power converters*. UPEC 1991.

11. Jutty and Gasrcha, *Southern New Trains Programme Power Quality Studies*, IEE Conference publication Power it's a quality thing, 16/2/2005.

12. Steed, Ferroresonance – mystery, mythology and magic! *IEE Power Engineering Journal*, June 1996.

13. Ponsonby and Hardy, *Power Quality on London Underground's Power system – a survey of recent work*. Recent Developments in Railway Electrification IEE Conference Publication 2004/10458, p. 58.

14. A. Robert and T. Deflandre, *Guide for assessing the network harmonic impedance*, WGCC02, Electra, August 1996.

15. Bletterie and Brunner, Solar Shadows, *IEE Power Engineer*, February/March 2006, p. 27.

25 Power Quality – Voltage Disturbances

25.1 INTRODUCTION

This chapter considers those aspects of power quality related to disturbance of the voltage level. This includes short-term interruptions and long-term deviations from the declared voltage level. With the increasing dependence of industry and commerce on electronic equipment and processes with sophisticated controls, this is an area of increasing concern to power supply utilities and to their contractors and consultants. Some standards relating to this subject are given in Table 25.1. The IEC approach is as explained in Chapter 24 (Section 24.1) – i.e. to set limits for equipment connected to utility systems – but the thresholds of irritability provided are equally valid for users already connected to an existing supply system, and indicative 'planning limits' provided are quality targets for a supplying utility. Standards of supply system quality to be expected in Europe (for systems up to 35 kV) are set out in EN 50160.

While there are major efforts being made by various standards authorities to arrive at common criteria, the environment, priorities and requirements of different nations and their power systems mean that variations in approach are inevitable. It is absolutely essential that any consultant engineer or contractor with design or procurement responsibility checks the particular requirements of the country in which a project is to be executed, and indeed of the power network involved.

25.2 THE NATURE AND CAUSE OF VOLTAGE DISTURBANCES IN POWER SYSTEMS

25.2.1 Short-term interruptions and voltage dips and peaks

Short interruptions (termed 'discontinuities' in some standards) to a consumer's supply, lasting not more than 1 minute (which for the purposes of power quality include short time voltage reductions to less than 0.1 pu) are typically caused by

Table 25.1 Standards relating to voltage disturbances (see also Chapter 20)

Document number	Document title
IEC 61000-1-1	Electromagnetic compatibility (EMC) – Part 1: General – Section 1: Application and interpretation of fundamental definitions and terms
IEC 61000-2-8	Electromagnetic compatibility (EMC) – Part 2–8: Environment – Voltage dips and short interruptions on public electric power supply systems with statistical measurement results
IEC 61000-3-3*	Electromagnetic compatibility (EMC) – Part 3–3: Limitation of voltage fluctuations and flicker in low-voltage supply systems for equipment with rated current ≤16 A
IEC 61000-3-5	Electromagnetic compatibility (EMC) – Part 3: Limits – Section 5: Limitation of voltage fluctuations and flicker in low-voltage power supply systems for equipment with rated current greater than 16 A
IEC 61000-3-6	Electromagnetic compatibility (EMC) – Part 3: Limits – Section 6: Assessment of emission limits for distorting loads in MV and HV power systems – Basic EMC publication
IEC 61000-3-7	Electromagnetic compatibility (EMC) – Part 3: Limits – Section 7: Assessment of emission limits for fluctuating loads in MV and HV power systems – Basic EMC publication
IEC 61000-3-11*	Electromagnetic compatibility (EMC) – Part 3–11: Limits – Limitation of voltage changes, voltage fluctuations and flicker in public low-voltage supply systems – Equipment with rated current ≤75 A and subject to conditional connection
IEC 61000-3-14	Electromagnetic compatibility (EMC) – Limits – Part 3–14: Assessment of emission limits for installations connected to LV power systems
IEC 61000-4-1*	Electromagnetic compatibility (EMC) – Part 4–1: Testing and measurement techniques – Overview of IEC 61000-4 series. (Note there are at the time of publication 34 sections to Part 4, covering different aspects of testing and measurement; many are relevant to the present topic)
IEEE1159-1995	IEEE Recommended Practice on Monitoring Electrical Power Quality
EN 50160	Voltage characteristics of electricity supplied by public distribution systems
ER P28	Engineering Recommendation on planning limits for voltage fluctuations caused by industrial, commercial and domestic equipment in the UK – Engineering Management Conference, Utilisation Consultancy group
ER P29	Engineering Recommendation on planning limits for voltage unbalance in the UK

* For these standards there is also an identical standard (with the same number) in the BSEN series.

the operation of automatic reclosing systems. Voltage dips can be caused by any of a number of events including the starting current of large motors, inrush currents occurring on connection of certain reactive loads and system faults. Faults on the transmission system result in dips of relatively high retained voltage over a large area, while distribution system faults tend to cause dips of lower retained voltage and longer duration but effective over a smaller area.

The term 'sag' is also used, particularly in American usage; it is often a synonym for 'dip', but can sometimes be intended as a discriminatory term designating a longer-term reduction in voltage. A dip is defined in terms of duration and 'retained voltage', the latter usually expressed as the percentage of nominal rms voltage remaining at the lowest point of the dip. Attempts have

Table 25.2 Some definitions of power quality disturbances based on IEEE Standard 1159-1995 (after Putrus[3])

Disturbance type		Typical duration		Typical voltage magnitude
Short Duration Variation				
	Sag	Instantaneous	0.5–30 cycl	0.1–0.9 pu
		Momentary	30 cycl–3s	0.1–0.9 pu
		Temporary	3 s–1 min	0.1–0.9 pu
	Swell	Instantaneous	0.5–30 cycl	1.1–1.8 pu
		Momentary	30 cycl–3s	1.1–1.4 pu
		Temporary	3 s–1 min	1.1–1.2 pu
	Interruption	Momentary	0.5 cycl–3 s	<0.1 pu
		Temporary	3 s–1 min	<0.1 pu
Long Duration Variation				
	Sustained interruption		>1 min	0.0 pu
	Under voltages		>1 min	0.8–0.9 pu
	Over voltages		>1 min	1.1–1.2 pu
Voltage imbalance				
	Magnitude imbalance		Steady state	
	Phase imbalance		Steady state	
Waveform distortions				
	DC offset		Steady state	0–0.1%
	Harmonics		Steady state	0–20%
	Interharmonics		Steady state	0–2%
	Noise		Steady state	0–1%
Voltage flicker			Intermittent	0.1–7%
Power frequency variations			<10 s	0.95–1.05 pu

been made to standardise all these terms – see Tables 25.2 and 25.3 – but they are not universally applicable. For the IEC a voltage dip has an amplitude greater than 10% of the nominal voltage, and a short supply interruption is simply a dip with 100% amplitude (see Fig. 25.1).

Motor drives, including older variable speed drives (VSDs), are particularly susceptible to dips and interruptions because the load still requires energy and takes it from the inertia of the drive. Different machines will slow at different rates, potentially resulting in chaos in a system requiring overall process control. (Most modern electronic VSDs can tolerate a significant variation in supply voltage without changing motor speed from the set value. The VSD maintains constant input power by increasing current as the voltage falls, the motor speed being set by the frequency of the VSD inverter).

Another area of sensitivity is IT equipment, which can suffer from shutdown or a 'crash'. In this context the term 'crash' is used for loss of a computer involving data loss. 'Shutdown' is usually used for a computer system where a voltage threshold has been set at sufficiently high level that there is sufficient residual energy in the power supply for the computer system to transfer data to non-volatile memory in an orderly manner before the processor actually stops.

Table 25.3 Some voltage disturbance terms defined (based on IEC standards; compare Table 25.2)

Term	Definition
Voltage variation	An increase or decrease of voltage normally due to variation of the total load of a distribution system or part of it.
Rapid voltage change	A single rapid variation of the rms value of a voltage between two consecutive levels which are sustained for definite but unspecified durations.
Voltage fluctuation	A series of voltage changes or a cyclic variation of the voltage envelope.
Flicker*	Impression of unsteadiness of visual sensation induced by a light stimulus whose luminance or spectral distribution fluctuates with time.
Supply voltage dip	A sudden reduction of the supply voltage to a value between 90% and 1% of the declared voltage, followed by a voltage recovery after a short period of time. Conventionally the duration of a voltage dip is between 10 ms and 1 min.
Supply interruption	A condition in which the voltage at the supply terminals is lower than 1% of the declared voltage.
Temporary power frequency overvoltage	An overvoltage, at a given location, of relatively long duration.
Transient overvoltage	A short duration oscillatory or non-oscillatory overvoltage usually highly damped and with a duration of a few milliseconds or less.
Voltage unbalance	In a three-phase system, a condition in which the rms values of the phase voltages of the phase angles between consecutive phases are unequal.

*:Voltage fluctuation can cause changes in the luminance of lamps which can create the visual phenomenon called flicker. Above a certain threshold flicker becomes annoying. The annoyance grows very rapidly with the amplitude of the fluctuation but at certain repetition rates even very small amplitudes can be very annoying (see Figure 25.2).

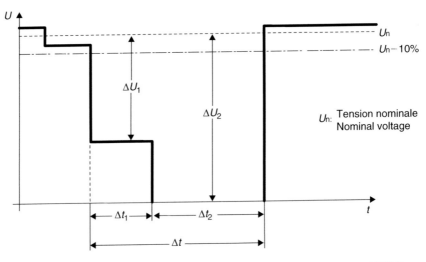

Figure 25.1   Illustration of a voltage dip (ΔU_1) or a voltage interruption ($\Delta U_2 = 100\%$) (from IEC 61000-2-1)

The high voltage threshold needed to do this often means that a relatively small voltage dip will initiate a computer shutdown denying the computer any chance of riding through the dip.

Discharge lamps will extinguish and may not restrike, contactors trip and thyristor bridges in inverter mode suffer commutation failure.

Short-term peaks, surges or swells (the term 'swell' for a voltage excursion above nominal is more common in American usage) can be caused by the sudden disconnection of load; by some types of system fault or by the maloperation of regulating equipment; peaks of a transient nature, generally measured in milliseconds for switching surges or microseconds for impulse spikes, can also occur.

25.2.2 Voltage fluctuations

Voltage fluctuations are described by the IEC as cyclical variations of the voltage envelope or a series of random voltage changes of up to $\pm10\%$ of the nominal. In low-voltage networks domestic appliances are significant sources, but each appliance will affect only a limited number of consumers.

Fluctuations resulting from industrial loads can affect many consumers, the main effect usually being flicker (see Section 25.2.3). Also step voltage changes due to the connection or disconnection of large loads or capacitor banks can fall into the category of voltage fluctuations (see also Section 25.2.6).

Equipment with significant time constants, such as heating elements, is largely unaffected, but some electronic and IT equipment can be seriously affected.

25.2.3 Voltage flicker

Flicker is fundamentally a term used to describe the behaviour of lighting, and its perceived effect on the human observer. Historically various experiments have been undertaken to determine the relationship between perception or irritation on the one hand and both frequency and magnitude of lighting level disturbance on the other. The concern of the power system engineer is that the most usual cause of flicker is a distortion of the voltage waveform of the electric supply to a lamp, and flicker is the most likely cause of complaint arising from voltage fluctuations. Curves exist, derived from lighting perception results, relating percentage voltage dip, frequency of disturbance and acceptability. An example is given in Fig. 25.2, but such curves must be treated with caution because they make assumptions as to lamp characteristics.

The IEC approach is more sophisticated, and takes account of both short-term sensitivity (a value calculated every 10 minutes) and long-term sensitivity (a combination of 12 short-term values). Intensity of flicker annoyance can be measured with a flickermeter (IEC 60868). IEC 61000-3-7 provides indicative levels for acceptable annoyance on a network, but recognises that the absolute limits will vary between utilities depending on specifics of the loads served and the supply network. In the UK, for example, utilities apply Engineering

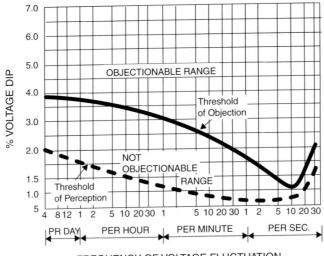

Figure 25.2   Curve illustrating the principle of the flicker problem (derived from IEEE 141)

Recommendation P28 to determine acceptability. In Europe EN 50160 gives power system flicker levels at consumers' terminals.

Typical causes of annoying levels of flicker are arc furnaces, welding loads, rolling mills and processes involving frequent but random motor starts.

It is interesting to consider that since flicker irritation to humans is a function of the perceptible change in brightness of the lighting (even though the exact relationship is difficult to formalise, as stated above), and incandescent lighting is more affected by supply voltage fluctuation than fluorescent lighting due to the latter's phosphor coating. As low-energy compact fluorescent lamps become more widely used, it could be that the problem of flicker may consequently reduce.

25.2.4 Slow-voltage fluctuations

So-called 'slow'-voltage variations, which fall within the limit of $\pm 10\%$ for voltage fluctuations (see Section 25.2.2), but are caused over a long period by the effect of load changes in the network, are not generally considered as voltage disturbances. They will be managed by conventional network development planning, and the use of automatic tap-changing on transformers, perhaps with the addition of line-drop compensation.

25.2.5 Voltage unbalance

Voltage unbalance is a condition in which the three-phase voltages differ in amplitude or are displaced from their normal 120° phase relationship, or both. The degree of unbalance is usually defined by the ratio of the negative sequence voltage component (see Chapter 26) to the positive sequence component.

Negative-phase sequence (NPS) unbalance appears in networks supplying such single-phase loads as modern 25 kV traction systems. NPS components influence other equipment such as rotating machines and rectifiers connected to the network. NPS voltages produce NPS currents that increase the stator and rotor losses and reduces equipment life because of the associated temperature rises involved. (A continuous operation at $10°K$ above the normal recommended operating temperature can reduce rotating machine life by a factor of two). IEC 60034-1 imposes a 1% NPS voltage limit on the supply feeding machines. However EN 50160 points out that in some areas unbalances up to 3% can be expected, and indicates that an acceptable supply system standard is that 'under normal operating conditions, during each period of one week, 95% of the 10-minute mean rms values of the NPS component of the supply voltage shall be within the range 0% to 2% of the positive-phase sequence component'.

Polyphase converters are also affected by an unbalanced supply, which causes an undesirable ripple component on the DC side and non-characteristic harmonics on the AC side (see Section 24.3.3).

A significant cause of voltage unbalance at medium and high voltage is single-phase railway or tramway traction load. At transmission voltages (which can then be reflected into lower-voltage levels) it can also result from unequal mutual inductance between untransposed conductors on long transmission lines. At low-voltage (distribution) level it is usually the result of unbalanced connection of single-phase loads on a three-phase system.

25.2.6 Step-change events

Step-changes in voltage are caused by events such as capacitor switching or transformer tap-changing. Provided such events are planned, infrequent and of a controlled magnitude they will not normally constitute a nuisance and they are rarely considered when assessing a system's power quality. However, the existence of such events in a given system does have to be considered when designing compensation equipment for other voltage disturbances. For example, 'hunting' between tap-changers and static VAr compensators (see Section 25.3.2) has to be avoided.

25.3 SOLUTIONS

25.3.1 Energy Storage

The ideal solution to the problem of variations in supply voltage is an instantly accessible store of electrical energy which can supply energy to 'top up' voltage below the nominal and can absorb electrical power to reduce voltage above the nominal. Historically non-electrical energy storage has seemed more practical, bearing in mind the magnitude of the required store, particularly if short interruptions are to be covered. Very fast-acting systems based on flywheels have been successfully developed[1] but have been difficult to apply economically except in special circumstances. Hydraulic systems have not been developed on a scale sufficient for power system compensation. Chemical energy storage includes batteries, and these were widely used in the early days of direct current electricity distribution. For today, they may be applicable to small individual installations (such as UPS (Uninterruptible Power Supply) systems for sensitive IT units, where they are certainly valuable), but their main drawback is low power and energy density, resulting in space and/or cost penalties[2]. Large-scale chemical storage techniques have been developed (e.g. the Regensys™ system) but at the time of publication have not been widely applied – physical space requirement is one negative factor.

Capacitors are an option for power quality applications where the energy requirement is not large, and the capability of commercially available equipment is increasing as time passes. The Dynamic Voltage Restorer (DVR™) is an example, with module ratings of 2 MVA or 5 MVA. It is designed to deal with voltage dips of up to 50%.

25.3.2 Balancing

The optimum solution of out-of-balance problems (see Section 25.2.5) depends significantly on the nature and magnitude of the problem. At low voltage it can be remedied by re-connection of unbalanced loads, or by the installation of a static balancer – a three-phase transformer core carrying two windings per phase in zig-zag connection (Fig. 25.3). This is a robust and simple application particularly common in rural areas. The Steinmetz balancing system may be used in urban areas or at higher voltages (see Fig. 25.4).

In the case of single-phase AC railway systems, again the simplest solution is balancing the single-phase load across the three phases. Multiple infeeds to the traction overhead catenary system are made along the length of the railway line. The single-phase demand from any one source at any one time is then usually sufficiently small to avoid significant out-of-balance problems. Where such multiple infeeds are not possible inductive and capacitive reactive components have to be added across the unused phases to 'balance' out the single-phase traction load. A practical example of such a balancer arrangement, which is of the largest of its kind to be installed in the world to date, has been used for the Channel Tunnel (again, Fig. 25.4 explains the compensation arrangements).

25.3.3 Static VAr compensators

These devices have their origins in efforts to control the 'flicker' produced by arc furnaces, but their ability to rapidly respond to changes in reactive power loading has resulted in their widespread use as elements in power transmission systems. They have been used to deal with voltage dips, fluctuations, flicker and unbalance. Such compensators are formed from a parallel connection of capacitors and thyristor-controlled reactors. The thyristor control varies the lagging reactive current so that the compensator can either generate capacitive VArs to

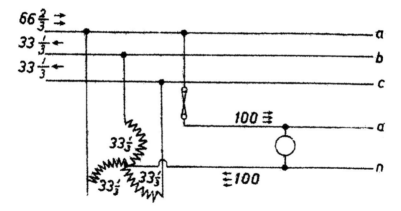

Figure 25.3   Application of an AC three-phase static balancer

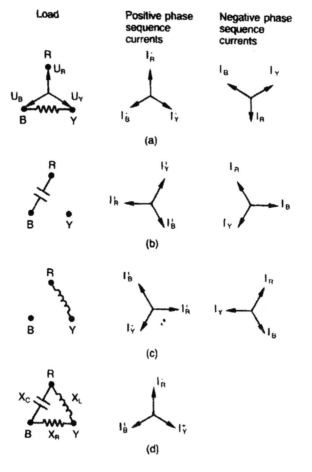

Figure 25.4   Balancing of single-phase load. $X_C = X_L = 3R$

support the voltage or generate lagging vars in order to reduce the voltage (see Section 26.8.5). Thyristor control equipment inevitably generates its own harmonics which are very sensitive to the thyristor firing angle delay as shown in Fig. 24.2. The equipment capacitor arms are often split into sub units to act as the necessary harmonic filters as shown in Fig. 24.6 and sometimes the capacitors are also thyristor switched.

Further explanation of the principle of reactive compensation is given in Chapter 26 (Section 26.8.5).

25.3.4 The STATCOM

A development of the well-proven SVC, the STATCOM is a voltage sourced converter which uses power electronic switches (Gate Turn Off Thyristors (GTOs)) to derive an approximately sinusoidal output voltage from a DC source (typically a capacitor). It is coupled to the system being compensated via an inductive impedance of low per unit value, and acts very much as a synchronous compensator, but with a vastly faster response. It has a natural tendency to compensate for changes in system voltage, and can do so very quickly. Unlike a constant impedance device such as a capacitor or inductor whose output current will decrease with voltage, it will continue to generate its maximum output current even at low system voltages.

It is an option for compensating voltage dips, surges, unbalance and flicker and generally takes less space than an SVC, but as a more sophisticated device it is more expensive, power for power and does not yet have the proven reliability of an SVC. Nevertheless many have already been installed in transmission schemes to provide reactive compensation. An example is given in Ref. (4).

In principle by replacing the storage capacitor with a chemical storage system or even a super-conducting storage device, the STATCOM offers the possibility of becoming a fast response system energy store.

25.4 CASE STUDY

Tenders have been received for a large 25 kV, 80 MVAr static balancer to help balance single-phase traction loads across a three-phase incoming supply. The system is fed from a single relatively weak 132 kV, 800 MVA low fault level supply connected in turn from a 400 kV, 6 GVA minimum fault level primary substation source. The maximum level of allowable unbalance has been set by the requirements of IEC 60034-1 for consumers and by the Electrical Supply Utility at the 400 kV point of common coupling as a 0.25% NPS restriction. A simple relationship exists between the maximum allowable unbalance load, S_{LOAD} (MVA), the system fault level, F_{SOURCE} (MVA) and the percentage NPS:

$$\text{Maximum NPS restriction, NPS (\%)} = [S_{LOAD} \text{ (MVA)/System fault level,} F_{SOURCE} \text{ (MVA)]} \cdot 100$$

Hence the maximum level of unbalance at the 400 kV source is 0.25% on 6 GVA = 15 MVA. This is equivalent to the power demand of a single high power Channel Tunnel 'Le Shuttle' train.

The original intention has been to supply the single-phase traction loads which are spread across the three phases by three single-phase transformers connected in delta on the primary side and in star on the secondary side with a common neutral connection for the traction return current. The single line diagram is shown in Fig. 25.5. Specialist firms have mentioned in their tenders that whilst their balancer will reduce NPS components in accordance with the Steinmetz balancing principle shown in Fig. 25.4 the effective 132/25 kV single-phase transformer connections will have a high impedance to zero-phase sequence (ZPS) components (see Chapter 14). Such components are, of course, inherent as the return current in a traction system design and will introduce 25 kV traction supply voltage regulation difficulties.

The Project Manager calls into his office the Project Technical Manager and the Associate Director-Systems Studies to have explained to him the situation and asks what can be done to resolve the transformer problem as tenders for the equipment are to be released shortly.

1. Could the situation be improved by increasing the fault level at the primary source substation? Is this a practical proposition in the short term?

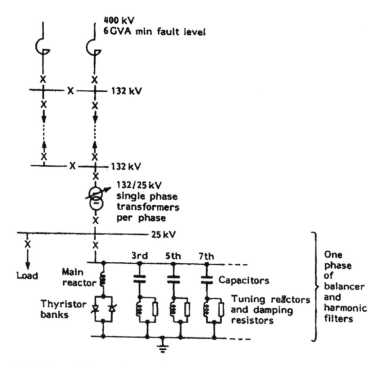

Figure 25.5   Case study – traction supply system: initial design

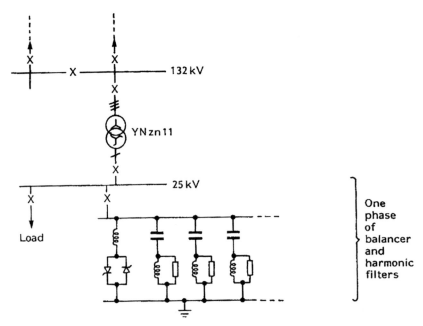

Figure 25.6   Case study – traction supply system: final transformer connection design

2. Could the balancer be introduced at the 132 kV level to improve the situation? What would be the relative cost implications of static var compensation at 132 kV compared to 25 kV? If a thyristor-controlled balancer generates harmonics will the necessary filters also cost more at 132 kV? Will such filters 'suck in' harmonics from the supply source as well as from the balancer itself?

3. Could an alternative transformer arrangement help to reduce the voltage regulation problem caused by ZPS losses? What alternative transformer connection would you recommend?

In practice the transformer arrangement was changed from single-phase transformers to three-phase star – zig-zag star connections. The single line diagram of the final arrangement is shown in Fig. 25.6.

REFERENCES

1. C. Tarrant, Revolutionary flywheel energy storage system for quality power, *IEE Power Engineering Journal*, June 1999, p. 159.
2. J. Baker and A. Collinson, Electrical energy storage at the turn of the millennium, *Power Engineering Journal*, June 1999, p. 107.
3. G. Putrus, Disturbance extraction for power quality monitoring and mitigation, *IEE Seminar 'Power it's a quality thing'*, February 2005.
4. D.J. Hanson et al., STATCOM: a new era of reactive compensation, *IEE Power Engineering Journal*, June 2002, p. 151.

26 Fundamentals

26.1 INTRODUCTION

This chapter summarizes some of the fundamental concepts necessary to appreciate the workings of the component parts of a transmission and distribution power system. It is assumed that the reader will already have been introduced to these topics in a theoretical manner during his or her student days. It is therefore a recap of the main points and the practical application of mathematical methods to the solution of real engineering problems. Examples are, therefore, included throughout this chapter.

26.2 SYMBOLS AND NOMENCLATURE

26.2.1 Symbols

The following symbols are used throughout this book:

C	Capacitance	(farads, μF, etc.)
E	Electromotive force – emf.	(volts, kV, etc.)
f	Frequency	(Hz)
H	Inertia constant of a machine	(MJ/MVA rating)
h	h operator or 120° operator	(h $= -0.5 + $ j0.866)
I, I_r, I_{r1}	Current, red phase current, red phase positive sequence current	(amps, kA, etc.)
J	Moment of inertia or work	(kgm^2) (joules, Nm, etc.)
j	j operator or 90° operator	$(\text{j}^2 = -1)$
L	Inductance	(henrys, mH, etc.)

n	Speed	(rev/s, etc., $n = \omega/2\pi$)
P	Active power $= UI\cos\phi$	(watts, kW, MW –
		1 horse power $= 746$ watts)
pf	Power factor $-\cos\phi$	
Q	Reactive power $= UI\sin\phi$	(VAr, kVAr, MVAr, etc.)
R	Resistance	(ohms, %, pu, etc.)
σ	Resistivity	(ohm metres)
S or VA	Apparent power $= UI$	(Voltamps, kVA,
		etc. $= \sqrt{[P^2 + Q^2]}$)
T	Torque	(N m)
t	Time	(second)
U or V	Voltage	(volts, kV, etc.)
X	Reactance	(ohms, %, pu, etc.
		$-jX$ capacitive and
		jX inductive)
x'_d	Transient reactance	
x''_d	Subtransient reactance	
Z	Complex impedance	(ohms, %, pu, etc.)
		$Z = \sqrt{[R^2 + \{jX\}^2]}$
Z_1	Positive sequence impedance parameter	
Z_2	Negative sequence impedance parameter	
Z_0	Zero sequence impedance parameter	
η	Efficiency	(% or pu)
θ	Load angle	(degrees)
ϕ	Phase angle difference	(degrees)
ω	Angular velocity	(radians/s)

26.2.2 Units and conversion tables

Introduction

The International metric system of units (SI) is used for the basic quantities of mass, length, time, electric current and temperature. Imperial and US units are still referred to for older and North American equipment. Conversion tables are included for the quantities commonly encountered in transmission and distribution engineering.

For example, it may be necessary to joint an existing imperial standard cable, say 0.3 in., with a modern XLPE 185 mm^2 cable, or understand US cable manufacturers' conductor sizes which may be quoted in circular mils.

Table 26.1   Basic SI units

Quantity	Unit	Symbol
Mass	Kilogram	kg
	gram	$g = 10^{-3}\,kg$
	tonne	$t = 10^{3}\,kg$
Length	Metre	m
	millimetre	$mm = 10^{-3}\,m$
	centimetre	$cm = 10^{-2}\,m$
	kilometre	$km = 10^{3}\,m$
Time	Second	s
Electric current	Ampere	A
Temperature	Kelvin	K

Table 26.2   Multiplication factors

Decimal power	Nomenclature	Symbol
10^{12}	Tera	T
10^{9}	Giga	G
10^{6}	Mega	M
10^{3}	Kilo	k
10^{2}	Hecto	h
10^{1}	Deca	da
10^{-1}	Deci	d
10^{-2}	Centi	c
10^{-3}	Milli	m
10^{-6}	Micro	μ
10^{-9}	Nano	n
10^{-12}	Pico	p

Units and conversion tables

Mass   kilogram (kg)

1 oz	= 0.028 349 kg	
1 lb	= 0.453 5924 kg	
1 stone	= 6.350 29 kg	
1 slug	= 14.5939 kg	
1 cwt	= 50.8023 kg	
1 ton (UK)	= 1.0160 ton	= 1016.05 kg

Length   metre (m)

1 mile	= 1609 m
1 yard	= 0.9144 m
1 ft	= 0.3048 m
1 in.	$= 2.54 \times 10^{-2}\,m$
1 mil (0.001 in.)	$= 2.54 \times 10^{-5}\,m$

Area   square metre (m^2)
$$1\,yd^2\ =\ 0.8361\,m^2$$
$$1\,ft^2\ =\ 0.0929\,m^2$$
$$1\,in.^2\ =\ 6.4516 \times 10^{-4}\,m^2$$
$$1\,acre = 4046.86\,m^2 = 0.404686\,ha$$
$$1\,ha\ \ =\ 1.0 \times 10^4\,m^2$$

Volume cubic metre (m^3)
$$1\,yd^3\ \ \ \ \ \ \ = 0.7646\,m^3$$
$$1\,ft^3\ \ \ \ \ \ \ \ = 2.8317 \times 10^{-2}\,m^3$$
$$1\,in.^3\ \ \ \ \ \ = 1.6387 \times 10^{-5}\,m^3$$
$$1\,gal\,(UK) = 4.5461 \times 10^{-3}\,m^3 = 1.2009\,gal\,(US)$$
$$1\,pt\,(UK)\ \ = 5.6826 \times 10^{-4}\,m^3$$
$$1\,gal\,(US)\ = 3.785 \times 10^{-3}\,m^3 = 0.8327\,gal\,(UK)$$
$$1\,pt\,(US)\ \ = 4.7318 \times 10^{-4}\,m^3$$
$$1\,litre,\,l\ \ \ \ = 1.0 \times 10^{-3}\,m^3$$
$$1\,ml\ \ \ \ \ \ \ \ = 1.0 \times 10^{-6}\,m^3 = 1.0\,cm^3$$

Density kilogram per cubic metre (kg/m^3)
$$1\,ton/yd^3 = 1328.94\,kg/m^3$$
$$1\,lb/ft^3\ \ \ = 1.6019 \times 10^1\,kg/m^3$$
$$1\,lb/in.^3\ \ = 2.7680 \times 10^4\,kg/m^3$$
$$1\,slug/ft^3 = 515.379\,kg/m^3$$

Time   second (s)
$$1\,minute,\,min = 60\,s$$
$$1\,hour,\,h\ \ \ \ \ = 60\,min = 3600\,s$$
$$1\,day\ \ \ \ \ \ \ \ \ = 24\,h = 1440\,min = 86400\,s$$
$$1\,year\ \ \ \ \ \ \ = 365\,days = 8760\,h$$

Frequency Hertz (Hz)
$$1\,Hz\ \ \ \ \ \ \ \ \ \ \ \ \ = 1\,periodic\,event\,per\,second = 1/s$$

Velocity metre per second (m/s)
$$1\,mph\ \ \ \ \ \ = 1.609\,km/h = 0.4470\,m/s$$
$$1\,ft/s\ \ \ \ \ \ \ = 0.3048\,m/s$$
$$1\,knot\,(UK) = 0.5148\,m/s = 1.85318\,km/h$$
$$1\,km/h\ \ \ \ \ = 0.2778\,m/s$$

Angle   radian (rad)
$$2\pi\,rad\ \ \ \ \ \ = 360°$$
$$\pi/180\,rad\ \ = 1°$$
$$1\,minute,\,' \ = 1°/60$$
$$1\,second,\,'' = 1°/360 = 1'/60$$

(Note: solid angle, steradian, sr, $1\,sr = 1\,m^2/m^2$)

Angular velocity   radian per second (rad/s)
Acceleration   metre per second squared (m/s^2)
$$1 \, \text{ft/s}^2 = 0.3048 \, \text{m/s}^2$$

Angular acceleration   radian per second squared (rad/s^2)

Volume flow rate   cubic metre per second (m^3/s)
$$1 \, \text{gal/s} = 4.54609 \times 10^{-3} \, \text{m}^3/\text{s} = 4.54609 \, \text{l/s}$$

Temperature Kelvin (K) and thermal quantities
$$1°F = (9/5°C) + 32$$
$$1°C = (°F - 32) \times 5/9$$
$$K = °C + 273.15$$

Thermal conductivity   watt per Kelvin metre (W/Km)
Heat transfer coefficient   watt per Kelvin square metre (W/Km^2)
Thermal capacity   Joule per Kelvin (J/K)

Moment of inertia   kilogram square metre (kg m^2)
$$1 \, \text{lb ft}^2 \ = 4.2140 \times 10^{-2} \, \text{kg m}^2$$
$$1 \, \text{slug ft}^2 = 1.3558 \, \text{kg m}^2$$
$$1 \, \text{lb in.}^2 \ = 2.9264 \times 10^{-4} \, \text{kg m}^2$$

Momentum   newton-second (Ns)
$$1 \, \text{lb ft/s} = 0.1383 \, \text{kg m/s}$$
$$1 \, \text{Ns} = 1.0 \, \text{kg m/s}$$

Force   newton (N)
$$1 \, \text{tonf} = 9.9640 \, \text{kN}$$
$$1 \, \text{lbf} = 4.4482 \times 10^{-3} \, \text{kN}$$
$$1 \, \text{pdl} = 0.1383 \, \text{N}$$

$$1 \, \text{kgf} = 9.8066 \, \text{N}$$
$$1 \, \text{dyn} = 1.0 \times 10^{-5} \, \text{N}$$
$$1 \, \text{N} = 1 \, \text{kg m/s}^2$$
(Note: 1 kgf refers to 1 kilogram force)

Torque   newton-metre (Nm)
$$1 \, \text{tonf ft} = 3.0370 \, \text{kN m}$$
$$1 \, \text{lbf ft} = 1.3558 \, \text{N m}$$
$$1 \, \text{pdl ft} = 0.0421 \, \text{N m}$$
$$1 \, \text{lbf in.} = 0.1130 \, \text{N m}$$

$$1 \, \text{kgf m} = 9.8066 \, \text{N m}$$
$$1 \, \text{N m} = 1.0 \, \text{joule, J} = 1 \, \text{watt second, Ws}$$

Pressure   pascal (Pa)
$$1 \, \text{tonf/ft}^2 = 107.252 \, \text{kPa}$$
$$1 \, \text{Tonf/in.}^2 = 15.4443 \, \text{MPa}$$
$$1 \, \text{lbf/ft}^2 = 47.8803 \, \text{Pa}$$
$$1 \, \text{lbf/in.}^2 \, (\text{psi}) = 6.8948 \, \text{kPa} = 0.0703 \, \text{kgf/cm}^2$$

$$1\,\text{Pa} = 1.0\,\text{N/m}^2 = 1.0 \times 10^{-6}\,\text{N/mm}^2$$
$$1\,\text{bar} = 1.0 \times 10^5\,\text{Pa} = 10^5\,\text{N/m}^2 = 0.1\,\text{MPa}$$
$$= 1.0 \times 10^{-1}\,\text{N/mm}^2 = 10\,\text{N/cm}^2$$
$$1\,\text{atm} = 1.0132\,\text{bar} = 0.1013\,\text{MPa}$$
$$1\,\text{torr} = 133.322\,\text{Pa} = 1\,\text{mm Hg}$$
$$1\,\text{mm H}_2\text{O} = 9.8067\,\text{Pa} = 0.0001\,\text{kgf/cm}^2$$

Energy – work, heat   joule (J)
$$1\,\text{therm} = 1.05506 \times 10^8\,\text{J} = 105.506\,\text{MJ}$$
$$1\,\text{BThU} = 1.05506 \times 10^3\,\text{J}$$
$$1\,\text{hp/h} = 2.68452 \times 10^6\,\text{J}$$

$$1\,\text{J} = 1.0\,\text{Nm} = 1\,\text{Ws} = 1\,\text{kg m}^2/\text{s}^2$$
$$1\,\text{kilowatt hour, kWh} = 3.6\,\text{MJ}$$
$$1\,\text{electron volt, eV} = 1.60219 \times 10^{-19}\,\text{J}$$

Power   watt (W)
$$1\,\text{hp} = 745.700\,\text{W}$$

Heat flow rate
$$1\,\text{ton of refrigeration} = 3516.85\,\text{W}$$
$$1\,\text{BThU/h} = 0.293071\,\text{W}$$

$$1\,\text{W} = 1.0\,\text{J/s} = 1\,\text{Nm/s} = 1\,\text{VA}$$
$$(\text{DC or AC @ unity pf})$$

Electrical quantities
Potential difference   volt   (V)
Current   ampere   (A)   (Often abbreviated to simply amp or amps)
Resistance   ohms   (Ω)   (Conductance measured in siemens, $S = 1/\Omega$)
Charge   coulomb   (C)   (Quantity of electricity, $C = 1\,\text{As}$)
Capacitance   farad   (F)   (Practical components expressed in μF,
                                            10^{-6}F or in terms of kVAr. $1\,\text{F} = 1\,\text{C/V}$)
Electric flux density   coulomb per square metre (C/m^2)
Field strength   volt per metre   (V/m)
Inductance   henry   (H) $(1\,\text{H} = 1\,\text{Wb/A})$
Magnetic flux   weber   (Wb) $(1\,\text{Wb} = 1\,\text{Vs})$
Magnetic flux density   tesla   (T) $(1\,\text{T} = 1\,\text{Wb/m}^2)$
Magnetic field strength ampere per metre (A/m)

Lighting quantities
$$1\,\text{lm/ft}^2 = 10.7639\,\text{lx}$$
$$1\,\text{cd/ft}^2 = 10.7639\,\text{cd/m}^2$$
$$1\,\text{cd/in.}^2 = 1550.0\,\text{cd/m}^2$$

$$1 \text{ foot lambert} = 3.42626 \text{ cd/m}^2$$

Luminous flux	lumen	(lm)	(1 lm = 1 cd sr)
Illuminance	Lux	(lx)	(1 lx = 1 lm/m²)
Luminous intensity	candela	(cd)	
Luminance	candela per square metre (cd/m²)		

26.3 ALTERNATING QUANTITIES

Sinusoidal alternating quantities

Instantaneous values may be expressed in terms of peak value and angular position. Consider the alternating voltage with an instantaneous value, u, peak value, U_{max}, and angular position, ϕ degrees, then:

$$u = U_{max} \sin \phi$$

- The waveform repeats itself every cycle or 360°.
- The duration of each cycle is the periodic time, t seconds.
- The frequency, f Hz, is the number of cycles per second so $f = 1/t$ Hz.
- For an angular velocity of ω radians per second and at an instant t seconds:

$$u = U_{max} \sin \omega t$$

- The angular velocity, ω radians per second = angular movement/time

$$= \frac{2\pi}{1/f} = 2\pi f$$

$$u = U_{max} \sin 2\pi ft$$

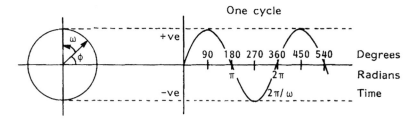

Figure 26.1　Sinusoidal waveform

Root mean square (rms) values

The alternating value may be replaced by an effective steady state value with an equivalent heating effect known as the rms value. For a sinusoidal current with a maximum value, I_{max}, then:

$$I_{rms} = I_{max}/\sqrt{2} = 0.707 I_{max}$$

or in general for a quantity, Y, and period, T:

$$Y_{rms} = \sqrt{1/T} \int_0^t y^2(t)\, dt = \sqrt{1/2\pi} \int_0^{2\pi} y^2(t)\, d\omega t$$

Average values

The average value of the sinusoidal wave:

$$I_{av} = 2/\pi \cdot I_{max} = 0.637\, I_{max}$$

or in general for a quantity, Y, and period, T:

$$Y_{av} = 1/T \int_0^t y(t)\, dt$$

The form factor is the ratio of the rms to mean value for the particular waveform. For the generalized case form factor $= Y_{rms}/Y_{av}$ and for a sinusoidal waveform, $y = a \sin \omega t$ with an integration interval from 0 to $T/2$, the form factor $= 0.707/0.637 = 1.1$. The more peaky the wave shape the greater will be the form factor (see Fig. 26.1).

Integration period for tariff calculations

Power levels are measured at regular intervals and mean powers calculated over an integration period depending upon the tariff policy, typically 10 or 15 minutes. The mean power for the waveform shown in Fig. 26.2 takes the form:

$$\bar{P}_{\Delta T} = 1/\Delta T \left(\sum_{i=1}^{n-1} \frac{[P_i + P_{i+1}]}{2} \cdot \Delta t_i \right)$$

Power definitions encountered in equipment sizing problems

Conductors, transformers, motors, etc. need to be designed according to their thermal behaviour when submitted to cyclic loading. The equivalent sizing current, I_{eq} rms, is defined as the current producing the same power losses over

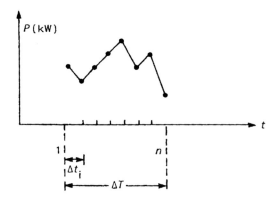

Figure 26.2   Example of variation of power with time

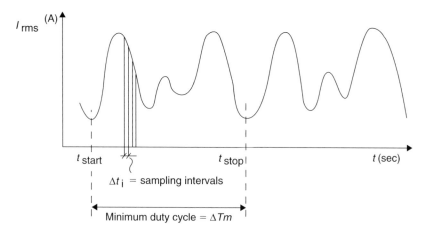

Figure 26.3   Loading cycle waveform

the minimum identified duty cycle. Consider the loading cycle given in Fig. 26.3 then:

$$I_{eq}\mathrm{rms}\big|_{\Delta Tm} = \sqrt{\sum_{i=1}^{n-1} \left[\frac{I_{i+1} + I_i}{2}\right]^2 \Delta t_i / \Delta Tm}$$

where $\displaystyle \Delta Tm = \sum_{i=1}^{n-1} \Delta t_i$

A typical value for the sampling interval, Δt_i, is about 10 seconds.

   Values of rms currents related to the thermal constants of the equipment may be calculated by adapting the above formula; modifying ΔTm to typically 10 minutes or less, and producing 'sliding' results from multiple calculations changing t_{start}. It is important to note that the thermal behaviour of the equipment is

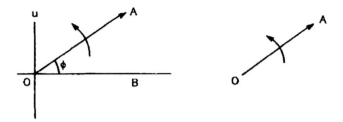

Figure 26.4   Vector representation

related to the apparent passing through currents. When loads are given in terms of kW all necessary corrective factors need to be introduced (efficiency, load power factor) prior to the equivalent current calculation.

26.4 VECTOR REPRESENTATION

An alternating quantity may be represented as a vector having both direction and magnitude. The vector representation of the sinusoidal waveform takes the form of a vector line OA rotating at a constant angular velocity ω radians per second. At any instant of time, t, OA $=$ OB $\sin \phi =$ OB $\sin \omega t$. The length of the vector is conventionally made proportional to the rms value of the quantity being represented rather than the more mathematically correct peak or maximum value. The vector arrow head represents the polarity of the quantity and conventionally vector quantities are assumed to have positive polarity with the arrow head pointing away from the source (see Figure 26.4). This applies equally to sequence currents.

Figure 26.5a shows a generator feeding an unbalanced three phase load. The three phases in phase sequence order of rotation – red, yellow, blue (r, y, b) – rotate in an anticlockwise direction with a neutral connection, n. Suffix labels may be added for clear identification of polarity.

Figure 26.5b shows the phase-to-neutral generator voltage vectors U_{rn}, U_{yn}, U_{bn}. U_{rn} is the positive direction of the source, r, phase voltage vector, U, relative to the neutral. Interphase voltage vector representation, U_{rb}, etc. is shown in Fig. 26.5c. Current vectors are labelled with a suffix according to phase. For example, Fig. 26.5d shows current vectors for an inductive load with the phase currents lagging the phase voltages by the different phase angles $\phi_{1,2,3}$. I_b is the blue phase current. When the current divides into more than one circuit a double suffix label nomenclature may be used. I_{rn} would represent the component of r phase current acting in a circuit between the r phase and the neutral. Since the convention is for all current vectors, including the neutral current, I_n, to act away from the source then $I_n = -(I_r + I_y + I_b)$ as shown in Fig. 26.5g.

Under balanced, pure sinusoidal (no harmonics), three phase conditions voltage and current vectors would vector sum to zero. Residual currents occur under unbalanced load, asymmetrical fault or distorted waveform conditions.

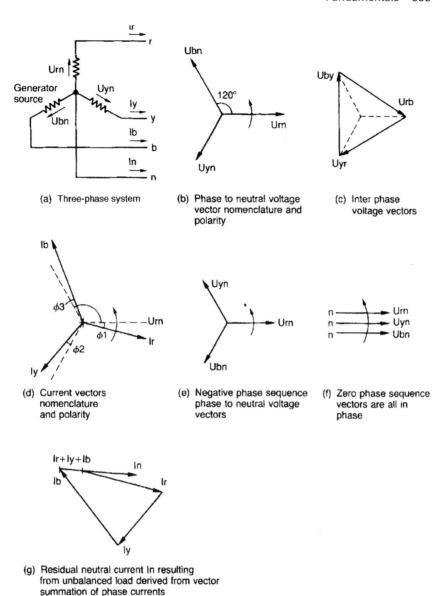

(a) Three-phase system

(b) Phase to neutral voltage
vector nomenclature and
polarity

(c) Inter phase
voltage vectors

(d) Current vectors
nomenclature
and polarity

(e) Negative phase sequence
phase to neutral voltage
vectors

(f) Zero phase sequence
vectors are all in
phase

(g) Residual neutral current In resulting
from unbalanced load derived from vector
summation of phase currents

Figure 26.5 Vector representation for an unbalanced three phase inductive load

Such residual currents may be detected by the parallel connection of current transformers as shown in Fig. 26.6a – $I_{\text{residual}} = I_r + I_y + I_b$. In a similar way residual voltages, such as may occur during power system fault conditions, may be detected by an open delta voltage transformer secondary winding. $E_{\text{residual}} = E_{\text{re}} + E_{\text{ye}} + E_{\text{be}}$ as shown in Fig. 26.6b.

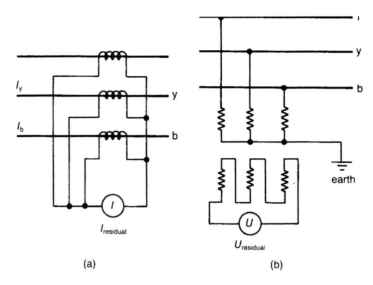

(a)   (b)

Figure 26.6   Measurement of residual currents and voltages

In order to further explain this nomenclature the vector representations of a symmetrical three phase fault, two phase-to-earth fault, phase-to-phase fault and a single phase-to-earth fault are given in Fig. 26.7a–d (single point earthing and neglecting capacitive currents).

An understanding of vectors allows the engineer to obtain a pictorial representation of system conditions. With modern desktop computing techniques being available actual problems are not solved by graphical methods. Instead sketches of vector relationships allow a rough check on calculated results and the conditions which such results represent. Consider this case study example.

During distribution transformer commissioning in the Middle East the inspector is requested by the consulting engineer to carry out quick phasing checks on the transformers before final terminations and energization. The 13.8/0.380 kV full load, 1000 kVA, delta/star transformers have come from a new supplier complete with supposedly satisfactory manufacturer's routine factory test certificates. The connections for this test are sketched out by the engineer and passed to the inspector who goes away grumbling, saying that they had never bothered with this in the past. For an explanation of transformer vector grouping and voltage ratio see Chapter 14.

Let R, Y, B represent the primary red, yellow and blue phase connections and r, y, b, the supposedly correct secondary phase connections. The following voltages were measured after connecting together the primary and secondary red phase transformer terminals and using a three phase variac on the HV winding set to about 100 V:

Rr potential difference $= 0\,\text{V}$
Ry potential difference $= 2.8\,\text{V}$ (corresponding to a secondary 390 V
                    phase to phase voltage)

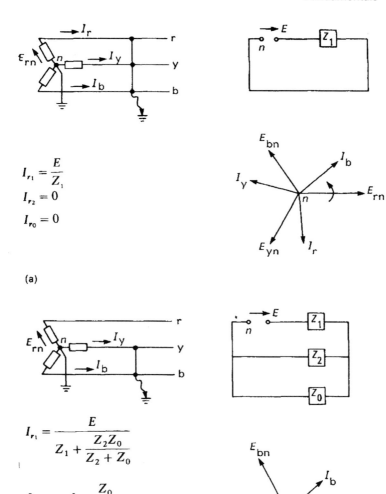

(a)

$$I_{r_1} = \frac{E}{Z_1}$$

$$I_{r_2} = 0$$

$$I_{r_0} = 0$$

(b)

$$I_{r_1} = \frac{E}{Z_1 + \dfrac{Z_2 Z_0}{Z_2 + Z_0}}$$

$$I_{r_2} = -I_{r_1} \frac{Z_0}{Z_2 + Z_0}$$

$$I_{r_0} = -I_{r_1} \frac{Z_2}{Z_2 + Z_0}$$

Figure 26.7 (a) Three phase fault; (b) two phase to earth fault (y–b–earth); (c) phase to phase fault (y–b); (d) single phase to earth fault (r–earth)
Note: E = line to neutral voltage

Rb potential difference = 2.8 V
Yr potential difference = 100 V (corresponding to a primary
                     voltage of 13.8 kV)
Yy potential difference = 98 V
Yb potential difference = 98 V

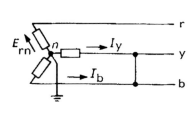

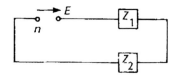

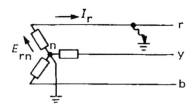

$$I_{r_1} = \frac{E}{Z_1 + Z_2}$$

$$I_{r_2} = -I_{r_1}$$

$$I_{r_0} = 0$$

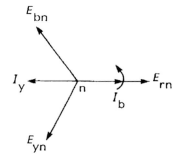

(c)

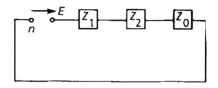

$$I_{r_0} = I_{r_1} = I_{r_2}$$

$$= \frac{E}{Z_0 + Z_1 + Z_2}$$

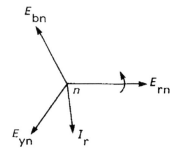

(d)

Figure 26.7   Continued

Br potential difference = 100V
By potential difference = 98 V (if the phasing was correct you would
                                  expect the measured By voltage to be greater
                                  than the measured Bb voltage)
Bb potential difference = 98 V

The inspector considers something is adrift but cannot figure out the cause of
these odd results. By using a vector representation of the situation the inspector

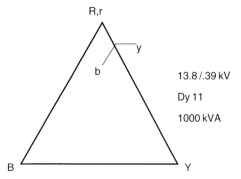

Figure 26.8   Transformer vector diagram

and engineer sit down together and come up with the opinion that the red and blue secondary phase connections have been incorrectly labelled and in fact have been reversed. This is confirmed with a phase rotation meter. By the application of a vector representation of the problem a better understanding is obtained. By checking first, abortive and expensive transformer cable termination work is avoided. Try plotting out the vector diagram for this test (see Fig. 26.8).

26.5 VECTOR ALGEBRA

26.5.1 The j operator

The j operator swings vector quantities through 90° in an anticlockwise direction.
The vector operator, j, has a numerical value $\sqrt{-1}$ such that $j^2 = -1$. The angle between the vector and the reference is known as the argument; $\tan \phi = BA/OB = X/R$ so $\phi = \tan^{-1} X/R$. The argument may be positive or negative depending upon rotation from the reference line. Inductive reactance is positive $(+jX)$ and capacitive reactance negative $(-jX)$ (see Fig. 26.9).

The series circuit

An equivalent single phase series circuit fed by an alternating voltage, V volts, containing resistance, R ohms, inductive reactance, X_L ohms, and capacitive reactance, X_C ohms, is shown in Fig. 26.10 together with the associated vector diagram.

$$\text{Total reactance, } X = X_L - X_C$$

$$\text{Total impedance, } Z = \sqrt{(R^2 + X^2)} = \sqrt{[R^2 + (X_L - X_C)^2]}$$

$$\text{Power factor, } \cos \phi = R/Z = \frac{R}{\sqrt{[R^2 + (X_L - X_C)^2]}}$$

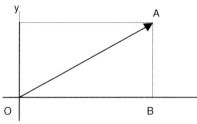

The vector OA is made up from the vector addition of components OB and BA.

OA = OB + jBA

tan O = BA/OB

In electrical circuits OB could represent a resistance of value R Ohms and BA an inductive reactance component of X ohms. Then the impedance Z ohms = $R + jX$.

Figure 26.9   Vector representation

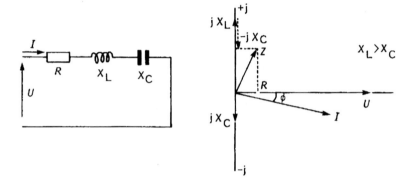

Figure 26.10   The series circuit

There are three cases to consider:

1. $X_L > X_C$ such that the inductive reactance predominates. The current will lag the applied voltage by a phase angle $\phi = \tan^{-1} X_L/R$ (neglecting capacitive reactance).
2. $X_L = X_C$ such that the total reactance is zero. The effective impedance is purely resistive and the current is in phase with the applied voltage.
3. $X_L < X_C$ such that the capacitive reactance predominates. The current will lead the applied voltage.

The parallel circuit

An equivalent parallel circuit together with its associated vector diagram is shown in Fig. 26.11.

26.5.2 Exponential vector format

The magnitude of the vector impedance $|Z| = \sqrt{R^2 + X^2}$
    Therefore $R = |Z|\cos\phi$ and $X = |Z|\sin\phi$.
    $Z = |Z|\cos\phi + j|Z|\sin\phi = |Z|(\cos\phi + j\sin\phi) = |Z|e^j$ where ϕ is in radians.

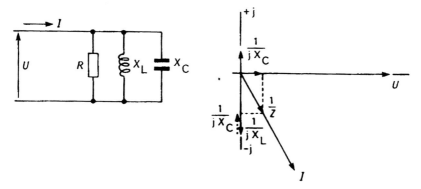

Figure 26.11 The parallel circuit

26.5.3 Polar co-ordinate vector format

$Z = |Z| \angle \phi$ where $|Z| = \sqrt{R^2 + X^2}$ and $\phi = \tan^{-1} X/R$.

26.5.4 Algebraic operations on vectors

1. Addition

   Consider the quantities $Z_A = R_A + jX_A = 2 + j2$ and $Z_B = R_B + jX_B$
   $= 4 + j1$

   $Z = Z_A + Z_B = 2 + j2 + 4 + j1 = (4 + 2) + j(2 + 1) = 6 + j3$.

   In general $Z = Z_A + Z_B + \cdots = (R_A + R_B + \cdots) \pm j(X_A + X_B + \cdots)$

2. Subtraction

   $Z = Z_A - Z_B = (R_A - R_B) \pm j(X_A - X_B)$

3. Multiplication

   $Z = Z_A \times Z_B = R_A R_B + R_A jX_B + jX_A R_B + jX_A jX_B$
   $= R_A R_B + j^2 X_A X_B + j(R_A X_B + X_A R_B)$
   $= R_A R_B - X_A X_B + j(R_A X_B + X_A R_B)$

For hand calculations it is easier to use the exponential or polar formats for multiplication and division:

$$Z = Z_A \times Z_B = |Z_A||Z_B|e^{j(\phi A + \phi B)} \quad \text{or} \quad |Z_A||Z_B| \angle \phi_A + \phi_B$$

4. Division

$$Z = Z_A \div Z_B = |Z_A| \div |Z_B| e^{j(\phi A - \phi B)} \quad \text{or} \quad |Z_A| \div |Z_B| \angle \phi_A - \phi_B$$

For example:

$Z_A = R_A + jX_A = 2 + j2 = 2.83 \angle 45°$
and $Z_B = R_B + jX_B = 4 + j1 = 4.12 \angle 14°$

So

$$Z_A \times Z_B = 11.66 \angle 59° \quad \text{and} \quad Z_A \div Z_B = 0.69 \angle 31°$$

26.5.5 The h operator

The h operator swings vector quantities through 120° in an anticlockwise direction.

The vector operator, h, has a numerical value

$$1 \cdot e^{j120} = (-0.5 + \sqrt{3}/2)$$
$$= -0.5 + j0.866$$

such that $h^2 = 1 \cdot e^{j240} = -0.5 - j0.866$. The operator is useful when dealing with three phase symmetrical systems which have 120° phase separation.

An example of electrical power system network reduction using these mathematical techniques is given later in this chapter.

26.6 SEQUENCE COMPONENTS

26.6.1 Theoretical background

Sequence components are used in power system analysis as an artificial and theoretical mathematical tool. They allow a three phase unbalanced system to be represented as three separate balanced vector systems with equal 120° phase relationships. This greatly simplifies calculations since by using symmetrical components the sums need only be made for one phase just as with balanced loads or three phase symmetrical faults. The phase relationships and the magnitudes of the sequence components can be calculated and the unbalanced phase conditions determined by simple vector addition. Despite any degree of load unbalance in the three phase vector system it may be represented by its three positive, negative and zero sequence balanced vector systems. So:

$$I_r = I_{r1} + I_{r2} + I_{r0}$$
$$I_y = I_{y1} + I_{y2} + I_{y0} = h^2 I_{r1} + h I_{r2} + I_{r0}$$
$$I_b = I_{b1} + I_{b2} + I_{b0} = h I_{r1} + h^2 I_{r2} + I_{r0}$$

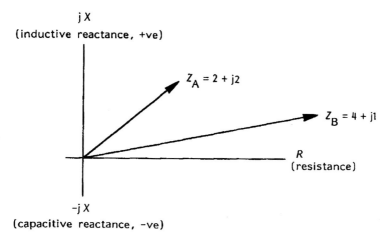

Figure 26.12   Vectors considered in 26.5.4 (addition, subtraction, multiplication and division of vectors)

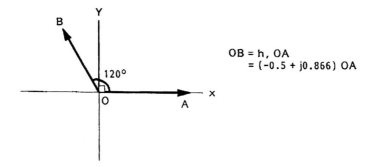

Figure 26.13   The h operator

and the sequence components in terms of the phase components are:

$$I_{r0} = (I_r + I_y + I_b)/3$$
$$I_{r1} = (I_r + hI_y + h^2I_b)/3$$
$$I_{r2} = (I_r + h^2I_y + hI_b)/3$$

where the suffix numbers 1, 2 and 0 represent the positive, negative and zero phase sequence components of the red (r), yellow (y) and blue (b) phases. Negative phase sequence vectors would reach maximum values in the order r, b, y as shown in Fig. 26.5e. Zero sequence vectors are all in phase rotating at the same angular velocity as shown in Fig. 26.5f. Because of the power and user-friendly nature of modern computing systems it is unlikely that the engineer will have to do hand calculations involving sequence components except in very simple cases. The point is that engineers will discuss effects in terms of sequence components and therefore an understanding of the concepts is essential in advanced electrical engineering.

26.6.2 Calculation methodology and approximations

The examples given in Fig. 26.7a–d cover the main power system fault cases to be encountered in practice. In all cases E represents the phase-to-neutral voltage (line voltage $\div \sqrt{3}$). Where the fault is associated with different phases from those given in Fig. 26.7a–d then it is merely a question of altering the nomenclature whilst maintaining the correct phase rotation and relationships. The following table describes these substitutions with suffix references, e.g. y', being the actual phase(s) involved and those without a suffix being the substituted phase(s) for calculations in accordance with Fig. 25.7a–d.

Fault conditions		*Substitutions* (as per Fig. 26.7a–d)	
1 phase to earth:	r'–earth	r–earth:	$r' = r, y' = y, b' = b$
	y'–earth	r–earth:	$y' = r, b' = y, r' = b$
	b'–earth	r–earth:	$b' = r, r' = y, y' = b$
phase to phase:	y'–b'–earth	y–b–earth:	$r' = r, y' = y, b' = b$
(and to earth)	r'–b'–earth	y–b–earth:	$r' = b, y' = r, b' = y$
	r'–y'–earth	y–b–earth:	$r' = y, y' = b, b' = r$

The principles to follow for hand calculations are as follows:

1. Determine the appropriate sequence network for the type of fault involved.
2. Collect necessary data on sequence impedances in the network. Note:

   - Zero sequence impedance data is only required for faults involving earth.
   - Z_1 normally equals Z_2 for static plant such as feeders and transformers but not for generators.
   - Z_0 may be very different from Z_1 or Z_2 and will require data collection. See Fig. 14.7 for the effect of transformer connections on Z_0.
   - For cables
     three core $Z_0 \approx 3Z_1$ to $5Z_1$
     single core $Z_0 \approx 1.25Z_1$
   - For overhead lines, single circuit
     no ground wire $Z_0 \approx 3.5Z_1$
     steel ground wire $Z_0 \approx 3.5Z_1$
     non-magnetic ground wire $Z_0 \approx 2Z_1$
   - For overhead lines, double circuit
     no ground wire $Z_0 \approx 5.5Z_1$
     steel ground wire $Z_0 \approx 5Z_1$
     non-magnetic ground wire $Z_0 \approx 3Z_1$.

3. Calculate an equivalent single impedance, the total positive, negative and zero sequence impedances from source to fault.
4. Connect the equivalent sequence impedances in the correct manner for the type of fault involved.

5. Calculate the phase currents involved from the equations in Fig. 26.7a–d.
6. Note that the driving source emf, E, is the phase-to-neutral voltage.
7. Neutral earthing resistances should only be included in the zero sequence network at three times their value. All three components of the zero phase sequence currents (but none of the positive or negative phase sequence currents) flow in the earth resistance or impedance.

A typical hand calculation is given later in Section 26.7.

26.6.3 Interpretation

26.6.3.1 *Zero phase sequence*

Zero phase sequence components involve the neutral and arise from asymmetrical earth fault conditions and unbalanced loads. 3rd, 6th, 9th, ..., etc. triplen harmonics also form a zero sequence set. The three zero sequence components are equal in magnitude and phase. They can therefore only flow where a path exists for their return to the neutral. The zero sequence impedances of power system components (generators, overhead lines, cables, transformers, etc.) are considerably different from the positive and negative impedance values and data should be obtained from the manufacturers. For example, the negative sequence reactance, X_0, for generators may be some 20% less than the positive sequence X_1 reactance but depends on winding pitch and exact machine construction. The passage of zero sequence components through three phase dual winding transformers depends upon the system earthing and winding configuration. A delta winding forms a closed circuit to triplen harmonics although the inclusion of a delta tertiary winding on star/star transformers is not essential and more detail is given in Chapter 14. For three core cables the zero sequence resistance, R_0, is the core resistance, R_C, plus three times the sheath resistance, R_S, with ratios of X_0/X_1 in the range 3–5. For single core cables $R_0 = R_C + R_S$ where R_S is the sheath resistance of each cable and typical X_0/X_1 ratios in the range 1 to 1.3. With resistance earthed cable networks the ohmic value of the earthing resistor may 'swamp' the cable impedances involved and this may help simplify hand calculations.

26.6.3.2 *Negative phase sequence*

Negative phase sequence components also arise from asymmetrical faults but not necessarily involving earth. In addition, 2nd, 5th, 8th, 11th, ..., etc. harmonics have a 120° negative phase relation. Positive and negative sequence impedances are the same for normal three phase power transformers, overhead lines and cables. Negative phase sequence (NPS) unbalance is discussed in Chapter 25 (Section 25.2.5).

26.7 NETWORK FAULT ANALYSIS

26.7.1 Introduction

Chapter 1 describes the analysis of power system networks using computing techniques. Sometimes it is necessary to carry out simple hand calculations in order to get a feel for the correctness of such an analysis or to quickly determine the magnitudes involved. Where the networks have multiple power infeeds and parallel paths such hand calculations rapidly become complex and very time consuming. Some of the simplifications described in Chapter 1 are necessary in order to reduce the algebra but still give meaningful results. This section describes the use of per unit, percentage and ohmic representation of network components. It describes network reduction and representation for the hand calculation of three phase symmetrical and asymmetrical fault levels in a power system network.

Generally cable and overhead line impedances are quoted in ohmic format whereas transformers and generators in percentage reactance terms. It is therefore necessary to be able to rapidly convert from one format to another so that calculations may be completed all in one system.

The fault value, F_{MVA}, is the three phase fault value expressed in MVA at the system voltage at the fault point in the system. The symmetrical fault current, I_F, in kA is obtained by dividing the fault value in MVA by $\sqrt{3} \times$ the line or phase-to-phase voltage, U, in kV. The term 'short circuit power' is no longer used in IEC recommendations since the short circuit current, I_F, provides more theoretically correct and useful information.

$Z_\%$ = % impedance expressed as a percentage at a stated MVA rating	S (%)
F_{MVA} = three phase fault value in MVA at the fault point in the system	(MVA)
Z = ohmic impedance of the circuit or equipment = R + jX	(ohms)
I = specified current	(kA)
S = MVA rating of the equipment in MVA, or the nominal MVA base on which the calculations are being made	(MVA)
E = phase-to-neutral voltage	(kV)
U = phase-to-phase voltage	(kV)
I_F = symmetrical fault current in kA $= F_{MVA} \div \sqrt{3} \cdot U$	(kA)
A and B = known quantities	

26.7.2 Fundamental formulae

26.7.2.1 *Percentage impedance notation*

1. $Z_\% = Z \times I/E \times 100$ (MVA is usually used instead of current when dealing with % impedances since MVA is proportional to current at a given voltage)

2. $Z_\% = S/F_{MVA} \times 100$
3. $Z_{\%@AMVA} = A (Z_{\%@BMVA})/B$
4. $Z_{\%@100MVA} = 100 (Z_{\%@BMVA})/B$
5. $F_{MVA} = 100 \cdot S/Z_\%$

Note that before applying the percentage impedance formulae all impedances must be expressed at the same MVA rating.

% impedance example

- A 500 kVA distribution transformer has a percentage reactance of 5%. What is its percentage reactance on a 100 MVA base?

  From formula 4. above:

$$X_{\%@100\ MVA} = \frac{100(X_{\%@0.5\ MVA})}{0.5} = \frac{100 \cdot 5}{0.5} = 1000\%$$

- What is the symmetrical three phase fault level on the secondary side of a 35 MVA, 132/11 kV nominal, 20% reactance transformer assuming zero source impedance? (see Figure 26.17)

  From formula 5. above:

$$F_{MVA} = \frac{100 \cdot S}{X_\%} = \frac{100 \cdot 35}{20} = 175\ MVA$$

- What is the symmetrical three phase fault level on the secondary side of a 35 MVA, 132/11 kV nominal, 20% reactance transformer assuming a primary side fault level of 1000 MVA?

  From formula 2. above the source reactance,

$$X_\% \text{ (100 MVA base)} = \frac{S}{F_{MVA}} \times 100$$
$$= \frac{100 \cdot 100}{1000} = 10\%$$

  From formula 4. above the transformer reactance on a 100 MVA base:

$$X_{\%@100\ MVA} = 100(X_{\%@B\ MVA})/B = \frac{(100 \cdot 20)}{35} = 57.14\%$$

Impedance from source to fault location on secondary side of transformer = $10 + 57.14 = 67.14\%$

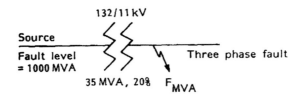

Figure 26.17   Percentage impedence example

Fault level corresponding to this impedance from formula 5. above:

$$F_{\text{MVA}} = \frac{100 \cdot S}{X_{\%}} = \frac{100 \cdot 100}{67.14} = 148.9, \quad \text{say} = 150 \text{ MVA}$$

Note that we have only considered reactance in these examples in order to simplify matters and because the transformer reactance will in practice swamp resistive effects at these sort of voltage levels. Note also that the inclusion of the source impedance in this example has not appreciably changed the magnitude of the result. The transformer itself acts as the main fault limiting reactance in the circuit from the source to the fault.

26.7.2.2 Per unit impedance notation

The per unit impedance of a circuit or a piece of equipment is the impedance voltage drop in the circuit or in the equipment when it is carrying a specified current and is expressed and as a decimal fraction of the line-to-neutral voltage. The only difference between the percentage impedance, $Z_{\%}$, and per unit impedance, Z_{PU}, notation is that $Z_{\text{PU}} = Z_{\%} \div 100$. The 20% reactance of the 35 MVA transformer in the example above is equivalent to 0.2 per unit reactance on a 35 MVA base and 0.57 per unit on a 100 MVA base.

26.7.2.3 Ohmic impedance notation

6. $Z_{\text{o}} = U^2 / F_{\text{MVA}}$

7. $Z_{\text{o}@ A \text{ kV}} = (Z_{\text{o}@ B \text{ kV}}) \cdot A^2 / B^2$

8. $F_{\text{MVA}} = U^2 / Z_{\text{o}}$

Note before applying these formulae all impedances must be expressed at the same selected voltage level.

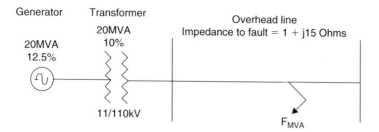

Figure 26.18   Impedence conversion example (see 26.7.2.4)

26.7.2.4 *Percentage and ohmic impedance conversions*

Calculations are most conveniently carried out using either the percentage or ohmic impedance methods depending upon the number of circuit components already expressed in either format. The idea is to reduce the number of conversions from one format to another.

9. $Z_{\%} = \dfrac{100 \cdot Z_{o} \cdot S}{U^2}$

10. $Z_{o} = \dfrac{Z_{\%} \cdot U^2}{100 \cdot S}$

- A 20 MVA, 11 kV generator has a percentage transient reactance of 12.5%. It is connected to a 110 kV overhead line with an overall impedance of $1 + j15$ ohms via a 11/110 kV, 20 MVA, 10% reactance transformer. Taking only reactive effects into account what is the three phase fault level at the end of the overhead line?

Working in percentage reactance values on a 100 MVA base:

$$\text{Generator } X_{\%@100 \text{ MVA}} = \frac{100 \cdot 12.5}{20} = \text{j}62.5\%$$

$$\text{Transformer } X_{\%@100 \text{ MVA}} = \frac{100 \cdot 10}{20} = \text{j}50\%$$

$$\text{Overhead line } X_{o@110 \text{ KV}} = \text{j}15 \text{ ohms}$$

$$X_{\%@100 \text{ MVA}} = \frac{100 \cdot Z_{o} \cdot S}{U^2} = \frac{100 \cdot \text{j}15 \cdot 100}{110^2} = \text{j}12.4$$

Total percentage reactance from source to fault $= \text{j}124.9\%$

Three phase fault level at end of overhead line $F_{\text{MVA}} = \dfrac{100 \cdot S}{Z_{\%}} = \dfrac{100 \cdot 100}{\text{j}124.9}$

$$= 80 \text{ MVA}$$

and the corresponding fault current $I_F = 80/\sqrt{3} \cdot 110 = 0.42 \text{ kA } (\approx 90° \text{ lagging})$

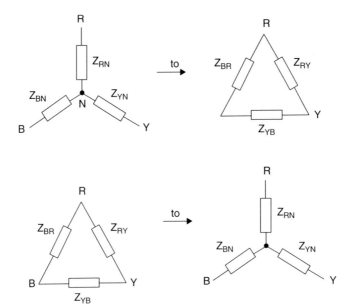

Figure 26.19   Star/delta and delta/star transformations

26.7.2.5 *Star/delta conversions*

Equivalent delta impedances:

$$Z_{RY} = Z_{RN} + Z_{YN} + \frac{Z_{RN} \cdot Z_{YN}}{Z_{BN}}$$

$$Z_{YB} = Z_{YN} + Z_{BN} + \frac{Z_{YN} \cdot Z_{BN}}{Z_{RN}}$$

$$Z_{BR} = Z_{BN} + Z_{RN} + \frac{Z_{BN} \cdot Z_{RN}}{Z_{YN}}$$

(see Fig. 26.19)

26.7.2.6 *Equivalent star impedances*

$$Z_{RN} = \frac{Z_{RY} \cdot Z_{BR}}{Z_{RY} + Z_{YB} + Z_{BR}}$$

$$Z_{YN} = \frac{Z_{YB} \cdot Z_{RY}}{Z_{RY} + Z_{YB} + Z_{BR}}$$

$$Z_{BN} = \frac{Z_{BR} \cdot Z_{YB}}{Z_{RY} + Z_{YB} + Z_{BR}}$$

(See Fig. 26.19.)

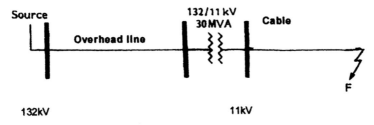

Figure 26.20 Simple radial network (see 26.7.3)

26.7.3 Simplified network reduction example

Consider the simple radial network single line diagram shown in Fig. 26.20. This example shows how to reduce the network into its associated positive, negative and zero sequence components and then how to arrange these components into the appropriate sequence networks for the solution of a three phase fault, a single phase earth fault and a phase-to-phase fault. Such hand calculations quickly become complex if resistive and reactive components are used. Therefore this example has been simplified by considering only the inductive reactance of the source, overhead line and cable involved. The example goes on to calculate the voltages and currents at the different points in the network. It then introduces a suitable protection scheme and illustrates the relay settings or discrimination using computer-generated relay protection curves.

1. Network parameters
   (a) Source fault level   = 3500 MVA
   (b) 132 kV overhead line:
       Positive sequence reactance, $jX_{1(OHL)}$ = 12.42 ohms
       Zero sequence reactance, $jX_{0(OHL)}$ = 33.2 ohms
   (c) 30 MVA, 132/11 kV transformer:
       Positive sequence reactance, $jX_{1(TX)}$ = 10% (on 30 MVA rating base)
   (d) Transformer earthing resistance   = 5 ohms
   (e) 11 kV cable reactance to fault:
       Positive sequence reactance, $jX_{1(C)}$ = 5 ohms
       Zero sequence reactance, $jX_{0(C)}$ = 5 ohms (assumed single core
                                                          cable $X_0/X_1 = 1$)

2. Per unit values
Use a 100 MVA base and 132 or 11 kV as the voltage base for the calculations. Therefore the base reactance $X_{132\,kV\ BASE} = kV$ base^2/MVA base = $132^2/100 = 174.24$ ohms. Next convert all components to per unit (pu) ohmic values.

The 132 kV system source may be represented by an equivalent star-connected solidly earthed generator. Then the source generator positive sequence unit reactance,

$jX_{1(SOURCE,pu)} = kV\ pu^2/MVA\ pu = (132/132)^2/(3500/100) = 0.02857\ pu$

Let $jX_{0(SOURCE,pu)} = 2.5jX_{1(SOURCE,pu)} = 2.5 \times 0.02857 = 0.0714\ pu$

$jX_{1(OHL,pu)} = 12.42/174.24 = 0.0713\ pu$

$jX_{0(OHL,pu)} = 33.2/174.24 = 0.19\ pu$

$jX_{1(Tx,pu)} = 10\% \times 100/30 = 0.1 \times 3.34 = 0.334\ pu$

Let $jX_{1(TX,pu)} = jX_{0(TX,pu)} = 0.334\ pu$

The base reactance on the 11 kV side of the transformer,

$$X_{11\,kV\ BASE} = kV\ base^2/MVA\ base = 11^2/100 = 1.21\ ohms$$

The per unit value of the earthing resistor, $R(pu) = 5/1.21 = 4.132\ pu$

$jX_{1(C,pu)} = 5/1.21 = 4.132\ pu$

$jX_{0(C,pu)} = 4.132\ pu$

3. Sequence networks

The positive, negative and zero sequence networks from the source to the fault may now be drawn (Fig. 26.21). The connections for three phase and single phase-to-phase faults are as shown in Section 26.6.

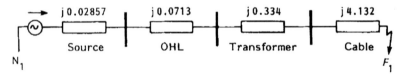

POSITIVE SEQUENCE NETWORK

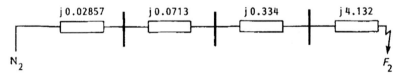

NEGATIVE SEQUENCE NETWORK

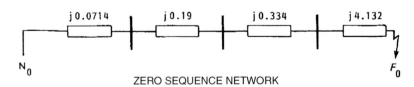

ZERO SEQUENCE NETWORK

Figure 26.21   Positive, negative and zero phase sequence networks. Note that the zero phase sequence network is the phase to earth zero phase sequence impedance plus three time the earth return impedance for earth faults

4. Fault conditions

Base current

The per unit base current at 11 kV

$$I_{BASE(pu)} = \text{MVA base}/(\sqrt{3} \times 11 \times 10^3)$$
$$= 100/(\sqrt{3} \times 11 \times 10^3)$$
$$= 5.248\text{k A}$$

The per unit base current at 132 kv,

$$I_{BASE(pu)} = \text{MVA base}/(\sqrt{3} \times 132 \times 10^3)$$
$$= 100/(\sqrt{3} \times 132 \times 10^3)$$
$$= 0.4374\text{ kA}$$

Three phase fault

The conditions at the fault are:

$$|I_r| = |I_y| = |I_b| = |I_F|$$
$$I_r = I_F \angle 0°$$
$$I_y = I_F \angle -120°$$
$$I_b = I_F \angle 120°$$
$$I_0 = I_r + I_y + I_b = I_F(1 \angle 0° + 1\angle -120° + 1 \angle 120°) = 0$$
$$I_1 = 1/3(I_r + hI_y + h^2 I_b) = 1/3I_F(1 + [1 \angle 120° \cdot 1\angle -120°]$$
$$+ [1 \angle -120° \cdot 1 \angle 120°]) = 1/3I_F(3) = I_F$$
$$I_2 = 1/3(I_r + h^2 I_y + hI_b) = 1/3I_F(1 + [1 \angle -120° \cdot 1 \angle -120°]$$
$$+ [1 \angle 120° \cdot 1\angle 120°]) = 1/3I_F(0)$$
$$= 0$$

Only the positive sequence current flows for the three phase fault conditions. The three phase fault current

$$I_{F(pu)} = U_{(pu)}/X_{1(pu)} = \frac{1}{j4.5658} = -j0.219 \text{ pu} = 0.21.19 \angle -90°$$
$$|I_F| = 5.248 \cdot 0.219 = \underline{1.15\text{ kA}}$$
$$I_r = 1150 \angle -90° \text{ A}$$
$$I_y = 1150 \angle 150° \text{ A}$$
$$I_b = 1150 \angle 30° \text{ A}$$
$$V_r = V_y = V_b = 0$$

The resulting fault currents on the high voltage side of the 132/11 kV power transformer may also be calculated taking into account any phase shift due to the transformer vector group.

Single phase-to-earth fault
For a single red phase-to-earth fault:

$$I_y = I_b = 0, \quad I_0 = I_1 = I_2, \quad V_r = 0, \quad X_{0(pu)} = (j4.466 + 3 \times 4.132)$$

$$\begin{aligned}
I_{0(pu)} &= 1_{(pu)}\text{voltage}/(X_{1(pu)} + X_{2(pu)} + X_{0(pu)}) \\
&= 1/(j4.5658 + j4.5658 + j4.466 + 12.396) \\
&= 1/18.4 \angle 47.6° \\
&= 0.054 \text{ pu} \angle -47.6°
\end{aligned}$$

The phase fault currents may be calculated from equation (26.1) (see below). Then:

$$\begin{aligned}
I_{r0} &= 1/3(I_r + I_y + I_b) \\
I_{r1} &= 1/3(I_r + hI_y + h^2I_b) \\
I_{r2} &= 1/3(I_r + h^2I_y + hI_b)I_{r0}
\end{aligned}$$

Substituting $I_y = I_b = 0$ in these equations gives $I_{r0} = I_{r1} = I_{r2} = I_r/3$

$$\begin{aligned}
I_{r1} &= E/(Z_1 + Z_2 + Z_0) \\
&= 1/(j4.5658 + j4.5658 + j4.466 + 12.396 \\
&= 1/(18.4 \angle 47.6°) \\
&= 0.05435 \angle -47.6° \\
I_{r(pu)} &= 3 \times 0.05435 \angle -47.6° \\
&= 0.1631 \angle -47.6° \text{ pu} \\
I_r &= 0.1631 \times 5.248 = 0.85 \angle -47.6° \text{ kA}
\end{aligned}$$

The sequence voltages at the fault are given by equation (26.2) below.

Phase-to-phase fault
The symmetrical components of current for a y–b phase-to-phase fault are again given by equation (26.1) below, where:

$$I_r = 0, \quad I_y = -I_b, \quad I_{r2} = -I_{r1} \quad \text{and} \quad I_{r0} = 0$$

$$\begin{aligned}
I_{r1(pu)} &= V/(Z_1 + Z_2) = 1/(j4.5658 + j4.5658) = -j0.109 \text{ pu} \\
I_{r2(pu)} &= j0.109 \text{ pu} \\
I_{r0(pu)} &= 0 \\
I_{r(pu)} &= I_{r1} + I_{r2} + I_{r0} = -j0.109 + j0.109 + 0 = 0 \\
&\quad \text{(red phase not involved)}
\end{aligned}$$

$$\begin{aligned}
I_{y(pu)} &= h^2I_{r1} + h^2I_{r2} + I_{r0} = -j0.109(-0.5 - j0.866) \\
&\quad + j0.109(-0.5 + j0.866) + 0 \\
&= -0.189 \text{ pu} \\
I_Y &= -0.189 \cdot 5.248 = -0.996 \angle 180° \text{ kA} \\
I_b &= -I_y = 0.996 \angle 0° \text{ kA}
\end{aligned}$$

Since there is no flow of current to ground for the phase-to-phase fault at the fault point, the presence or absence of a grounded neutral does not affect the fault current. With the connection from the transformer neutral to earth Z_0 is infinite and $V_{r0} = 0$. With $V_Y = V_b$ the symmetrical voltage components are given by the matrix equations (26.3) and (26.2) below, i.e.:

$$\begin{pmatrix} V_{r0} \\ V_{r1} \\ V_{r2} \end{pmatrix} = \frac{1}{3} \begin{pmatrix} 1 & 1 & 1 \\ 1 & h & h^2 \\ 1 & h^2 & h \end{pmatrix} \begin{pmatrix} V_r \\ V_y \\ V_b \end{pmatrix} \quad \text{and} \quad \begin{pmatrix} 0 \\ V_{r1} \\ V_{r2} \end{pmatrix} = \begin{pmatrix} 0 \\ V \\ 0 \end{pmatrix} - \begin{pmatrix} Z_0 & 0 & 0 \\ 0 & Z_1 & 0 \\ 0 & 0 & Z_2 \end{pmatrix} \begin{pmatrix} 0 \\ I_{r1} \\ -I_{r2} \end{pmatrix}$$

So $V_{r1} = V_{r2} = V - I_{r1} \cdot Z_1$

$= 1 - (-j0.109) \cdot (j4.5658)$

$= 0.5 \, \text{pu}$

$V_r = V_{r1} + V_{r2} + V_{r0} = 0.5 + 0.5 + 0 = 1 \, \text{pu or } 1.11/\sqrt{3} = 6.35 \angle 0°\text{kV}$
$V_y = h^2 V_{r1} + h V_{r2} + V_{r0} = 0.5(-0.5 - j0.866) + 0.5(-0.5 + j0.866)$
$\quad = -0.5 \, \text{pu or } -0.5 \cdot 11/\sqrt{3} = 3.17 \angle 180° \, \text{kV}$

$V_b = V_y = -0.5 \, \text{pu or } 3.17 \angle 180° \, \text{kV}$

$\quad V_{ry} = V_r - V_b = 1 + 0.5 = 1.5 \, \text{pu or } 1.5 \cdot 11/\sqrt{3} = 9.53 \angle 0° \, \text{kV}$

$\quad V_{yb} = 0$

$\quad V_{br} = 1.5 \, \text{pu or } 1.5 \cdot 11/\sqrt{3} = 9.53 \angle 180° \, \text{kV}$

Phase-to-phase to ground fault
The following conditions apply at the fault point:

$V_y = V_b = 0$
$I_r = 0$

The symmetrical voltage components are given by matrix equation (26.3) below.

So $V_{r0} = V_{r1} = V_{r2} = V_r/3$

$\quad I_{r1} = V/(Z_1 + \{Z_2 \cdot Z_0/Z_2 + Z_0\})$

$\quad = 1/j4.5658 + \{j4.5658(12.396 + j4.466)/j4.5658$
$\quad\quad + j4.466 + 12.396\}$

$\quad = 0.119 \angle -82.5° \, \text{pu}$

$\quad V_{r1} = V - I_1 \cdot Z_1$

$\quad = 1 - 0.119 \angle -82.5° \cdot 4.5658 \angle 90°$

$\quad = 0.4667 \angle -8.75°$

$\quad I_{r2} = -Vr_2/Z_2$

$\quad = -0.4667 \angle - 8.75°/4.5658 \angle 90°$

$\quad = 0.102 \angle 81.25° \, \text{pu}$

$$I_{r0} = -V_{r0}/Z_0$$
$$= -0.4667 \angle -8.75°/(12.396 + j4.466)$$
$$= 0.0354 \angle 151.42° \, \text{pu}$$

$$I_{r(pu)} = I_{r1} + I_{r2} + I_{r0} = 0.119 \angle -82.5° + 0.102 \angle 81.25°$$
$$+ 0.0354 \angle 151.42° \, \text{pu}$$
$$= 0 \, \text{pu}$$

$$I_{y(pu)} = h^2 I_{r1} + h I_{r2} + I_{r0} = (-0.5 - j0.866) \cdot (0.119 \angle -82.5°)$$
$$+ (-0.5 + j0.866) \cdot (0.102 \angle 81.25°)$$
$$+ 0.0354 \angle 151.42°$$
$$= 0.238 \angle 173.85° \, \text{pu}$$
$$\text{or } 0.238 \angle 173.85 \cdot 5.248 = 1.25 \angle 174° \, \text{kA}$$

$$I_{b(pu)} = h I_{r1} + h^2 I_{r2} + I_{r0} = (-0.5 + j0.866) \cdot (0.119 \angle -82.5°)$$
$$+ (-0.5 + j0.866) \cdot (0.102 \angle 81.25°)$$
$$+ 0.0354 \angle 151.42°$$
$$= 0.145 \angle 10° \, \text{pu}$$
$$\text{or } 0.145 \angle 10° \cdot 5.248 = 0.76 \angle 10° \, \text{kA}$$

$$V_r = V_{r1} + V_{r1} + V_{r2}$$
$$= 1.4 \angle -8.75° \, \text{pu}$$
$$V_y = V_b = 0$$
$$V_{ry} = V_r - V_y = 1.4 \angle -8.75° \, \text{pu or } 1.4 \angle -8.75° \cdot 11/\sqrt{3}$$
$$= 8.89 \angle -8.75° \, \text{kV}$$
$$V_{yb} = 0$$
$$V_{br} = V_b - V_r = 1.4 \angle 171.25° \, \text{pu or } 1.4 \angle 171.25° \cdot 11/\sqrt{3}$$
$$= 8.89 \angle 171.25° \, \text{kV}$$

Thus, by using the equations expressed below in matrix format the currents and voltages for the different types of fault can be calculated:

$$\begin{pmatrix} I_{r0} \\ I_{r1} \\ I_{r2} \end{pmatrix} = \frac{1}{3} \begin{pmatrix} 1 & 1 & 1 \\ 1 & h & h^2 \\ 1 & h^2 & h \end{pmatrix} \begin{pmatrix} I_r \\ I_y \\ I_b \end{pmatrix} \tag{26.1}$$

$$\begin{pmatrix} V_{r0} \\ V_{r1} \\ V_{r2} \end{pmatrix} = \begin{pmatrix} 0 \\ V \\ 0 \end{pmatrix} = \begin{pmatrix} Z_0 & 0 & 0 \\ 0 & Z_1 & 0 \\ 0 & 0 & Z_2 \end{pmatrix} \begin{pmatrix} I_{r0} \\ I_{r1} \\ I_{r2} \end{pmatrix} \tag{26.2}$$

$$\begin{pmatrix} V_{r0} \\ V_{r1} \\ V_{r2} \end{pmatrix} = \frac{1}{3} \begin{pmatrix} 1 & 1 & 1 \\ 1 & h & h^2 \\ 1 & h^2 & h \end{pmatrix} \begin{pmatrix} V_r \\ V_y \\ V_b \end{pmatrix} \tag{26.3}$$

$$\begin{pmatrix} I_r \\ I_y \\ I_b \end{pmatrix} = \begin{pmatrix} 1 & 1 & 1 \\ 1 & h^2 & h \\ 1 & h & h^2 \end{pmatrix} \begin{pmatrix} I_{r0} \\ I_{r1} \\ I_{r2} \end{pmatrix} \quad (26.4)$$

$$\begin{pmatrix} V_r \\ V_y \\ V_b \end{pmatrix} = \begin{pmatrix} 1 & 1 & 1 \\ 1 & h^2 & h \\ 1 & h & h^2 \end{pmatrix} \begin{pmatrix} V_{r0} \\ V_{r1} \\ V_{r2} \end{pmatrix} \quad (26.5)$$

The arrangement of the different sequence impedances in networks to describe the different fault conditions are shown in Fig. 26.7a–d.

26.8 DESIGN OPTIMIZATION

26.8.1 Introduction

Transmission and distribution networks represent a huge capital investment with the purpose of delivering generated power to the consumer. For example, distribution networks represent some 40% of the total generation, transmission and distribution plant costs. The main factors involved in the optimization of the engineering associated with transmission and distribution networks are described in this section.

It is necessary at the project planning and investment stages to optimize the kWh cost of the network by adopting efficient transmission and distribution systems, low loss design and high reliability equipment. The interactive design processes involved for new networks and network extensions are shown in Fig. 26.22. Some of the major considerations are listed below:

- Power density.
- Load diversity.
- Type of load (single, two, three phase, power factor, harmonics, etc.).
- LV single or three phase distribution, use of special distribution schemes (single wire earth return, single phase two wire networks, etc.).
- Total load.
- Load growth rate.
- Climatic and environmental conditions.
- Interconnection possibilities with adjacent networks.
- Consumer requirements (continuity of supply, voltage/frequency fluctuations).
- Electrical supply utility and national standards/regulations.
- Fault and reliability statistics.
- Necessity or otherwise for an intermediate MV network upstream of the LV distribution.
- Earthing arrangements.
- Radial or meshed networks to match system reliability requirements.

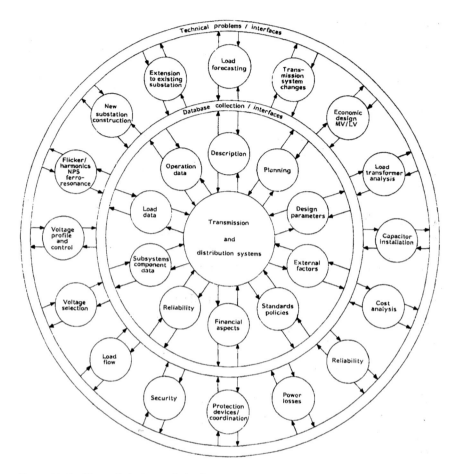

Figure 26.22   Network design optimization processes

26.8.2 Technical problems

26.8.2.1 *Voltage selection*

The choice of operating voltage is made according to:

1. Economics   –   Related to the power delivery requirements and distances involved.

2. Existing Standards and Policies   –   IEC, BS, DIN, etc.

3. Safety   –   Reduced voltages may be imposed for special applications such as portable hand tool power.

Table 26.3   Insulation class temperature limits

Insulation class	Y	A	E	B	F	H	C
Temperature (°C)	90	105	120	130	155	180	>180

The voltage delivered to the consumer must be maintained between specified limits. Special attention must be paid at the design stage to:

- heavy load conditions (thermal limits);
- light or no-load conditions (possible overvoltages);
- transmission of reactive power flows (reactive compensation);
- transformer tap range and voltage regulation.

Possible transient phenomena that could affect the consumers supply must also be investigated and suitable compensation equipment included for in the project budget proposals (flicker, harmonics, negative sequence unbalance, ferro resonance, etc.).

Insulating materials have specific operating temperature limits (Insulation Class) in accordance with IEC 60085 based upon acceptable life under normal operating conditions. Because of the possibility of operation under cyclic loading or for short durations at higher than normal temperatures the manufacturers' experience should be sought to ensure correct insulation temperature classification for specific applications (see Table 26.3).

26.8.2.2 *Conductor selection*

Copper or aluminium alloy conductor selection based upon existing local preferences and market prices. Detailed choice takes into account:

1. Climatic and environmental conditions and basic current carrying capability.
2. Permissible voltage drops including:
   - ohmic voltage drops corresponding to a decrease in transmission efficiency,
   - reactive voltage drops related to the network voltage profile.
3. Short circuit withstand capability taking into account relay protection settings and avoiding possible annealing.
4. Safety conditions and loop impedance applying to step, touch and mesh voltages (see Chapters 12 and 18).
5. Possible security considerations (likelihood of copper theft).

26.8.2.3 *Equipment selection*

The following factors are involved:

1. Financial and economic considerations and limitation of capital outlay (see Chapter 22).

Table 26.4a   IP coding-first numeral

First characteristic numeral	Degree of protection against contact with live parts and the ingress of foreign bodies
0	No protection of persons against contact with live or moving parts inside the enclosure. No protection of equipment against ingress of solid foreign bodies.
1	Protection against accidental or inadvertent contact with live or moving parts inside the enclosure by a large surface of the human body as, for example, a hand, but no protection against deliberate access to such parts. Protection against ingress of large solid foreign bodies of diameters greater than 50 mm.
2	Protection against contact with live or moving parts inside the enclosure by fingers. Protection against ingress of medium sized solid foreign bodies of diameters greater than 12 mm.
3	Protection against contact with live or moving parts inside the enclosure by tools, wires or such objects of thickness greater than 2.5 mm. Protection against ingress of small solid foreign bodies of diameters greater than 2.5 mm.
4	Protection against contact with live or moving parts inside the enclosure by tools, wires or such objects of thickness greater than 1 mm. Protection against ingress of small solid foreign bodies of diameters greater than 1 mm.
5	Complete protection against contact with live or moving parts inside the enclosure. Protection against harmful deposits of dust. The ingress of dust is not totally prevented, but dust cannot enter in an amount sufficient to interfere with the satisfactory operation of the equipment enclosed.
6	Complete protection against contact with live or moving parts inside the enclosure. Protection against ingress of dust.

2. Operational aspects related to client's requirements.
3. Standardization in order to limit operation, maintenance and replacement costs.
4. Technology evolution in order to avoid rapid obsolescence and sooner than anticipated new equipment investment costs.
5. Network evolution with particular attention to short circuit power levels and equipment that allows easy extensions at competitive cost (see Chapters 1 and 3).
6. Environmental and climatic aspects – tropicalization, ratings at elevated temperatures, enclosure protection. Degrees of protection are defined by International Protection (IP) codings. The coding gives the level of protection against contact with live parts and the ingress of foreign bodies (exposure of live parts to touch, ingress of dust, etc.), degree of protection against ingress of liquids (water jets, condensate, etc.) as shown in Tables 26.4a and 26.4b.

A third numeral designates protection against mechanical damage, from 0 (no protection) to 9 (20 joules).

Table 26.4b   IP coding-second numeral

Second characteristic numeral	Degree of protection against ingress of a liquid
0	No protection.
1	Protection against drops of condensate: drops of condensate falling vertically on the enclosure shall have no harmful effect.
2	Protection against drops of other liquids: drops of falling liquid shall have no harmful effect when the enclosure is tilted at any angle upto 15° from the vertical.
3	Protection against rain: water falling as rain at an angle equal to or less than 60° with respect to the vertical shall have no harmful effect.
4	Protection against splashing liquid: liquid splashed from any direction shall have no harmful effect.
5	Protection against water jets: water projected by a nozzle from any direction under stated conditions shall have no harmful effects.
6	Protection against conditions on ship's decks (deck watertight equipment): water due to heavy seas shall not enter the enclosures under prescribed conditions.
7	Protection against immersion in water: it must not be possible for water to enter the enclosure under stated conditions of pressure and time.
8	Protection against indefinite immersion in water under specified pressure: it must not be possible for water to enter the enclosure.

The degrees of equipment protection available as manufacturers standard products for indoor substation designs are as follows:

Metal enclosed switchgear to 36 kV	up to IP34	typically IP31
Dry type transformers	up to IP31	typically IP23
Earthing resistors	up to IP31	typically IP23
400 V distribution boards	up to IP41	typically IP31
Battery chargers	up to IP31	typically IP21
Marshalling cubicles	up to IP43	typically IP31
Lighting fittings	up to IP54	IP20 internal fluorescent/diffuser IP54 corrosion resistant/ weatherproof IP44 industrial
Lighting switches	up to IP54	typically IP31–IP44
Emergency lighting fittings	up to IP54	typically IP31
Socket outlets (exterior use)	up to IP54	typically IP44
Motors (transformer cooling fans)		typically IP44

26.8.2.4 *Protection device selection*

The following factors are to be considered:

1. Safety and reliability in order to limit the extent of perturbations on the network and power supply interruption in case of a fault.

2. Finance and economics.
3. Network operating principles, radial or meshed circuits, automatic reclosure, etc.
4. Standardization to limit operation, maintenance and replacement costs. Note some electrical supply utilities use different main and back-up distance protection to avoid risk of maloperation due to an inherent fault in one model.
5. Technology evolution. Whilst solid state relays have considerable flexibility in setting characteristics some electrical supply utilities in developing countries request electromechanical relays. They are able to repair these in house without foreign exchange constraints on ordering sophisticated spare parts.

26.8.2.5 *Safety device selection*

The following problems are to be considered:

1. Transient overvoltage phenomena, use of surge arresters.
2. Neutral earthing (grounding) policy.
3. Earthing (grounding) and bonding policy.
4. Safety devices – key interlocks, indicators.
   (See Chapters 8–10.)

26.8.3 Loss reduction

26.8.3.1 *Introduction*

Losses are divided into technical and non-technical categories.
  Technical losses arise from:

- Load losses (I^2R) in overhead lines, cables and transformers.
- Reactive losses generally caused by low power factor at the consumers' terminals.
- Iron losses due to magnetization of transformer cores.
- Auxiliary loads at generating stations.
- Corona losses due to high voltage stresses on overhead lines.
- Joint losses due to poor clamping or finishing.
- Cable sheath losses.
- Leakage current losses. (insulator pollution, etc.)

Non-technical losses are reduced by better management and administration aided by suitable metering, meter calibration and computerization of billing in order to ensure efficient bill collection.

26.8.3.2 *General principles*

Unacceptable distribution feeder (overhead line or cable) losses are brought about by low power factor, undersized and/or high impedance conductors and connections. Precautions to be taken against loss and overloaded transformers are described in Chapter 14. In economic terms loss reduction measures should be continued up to the point where a marginal increment in capital cost will be exactly counterbalanced by the consequent decrease in the value of the losses. Refer to Chapter 22 for an explanation of some of the economic and financial terms used in this section.

Generation and transmission circuits are normally monitored by energy meters and special attention is paid to loss reduction in the design process. Independent economic studies have shown that the additional capital expenditure in reducing losses on distribution systems that have become inefficient can be more cost effective than installing additional generation. It is therefore fully justifiable also to show an interest in loss reduction on distribution systems. The cost of energy has increased several fold in real terms over the last 20 years whereas the cost of basic raw materials (copper and aluminium) has increased with inflation thus remaining essentially constant in real terms. Distribution design has traditionally been based on thermal and acceptable voltage drop parameters.

Distribution system technical losses cannot be so readily quantified as at transmission voltage levels because energy metering on a per feeder basis is often not installed.

26.8.3.3 *System load factor*

Certainly at the higher voltage levels the loading of feeders is normally less than the maximum current carrying capacity. This is because, for system security reasons, the lines are duplicated and therefore normally carry less than half the line thermal rating. The load factor at the generation level taking into account the various system diversity factors, gives load factors throughout the network under consideration at the different voltage levels. The loss load factors may then be calculated from equations of the form:

$$\text{Loss load factor} = (a \cdot \text{LF}) + (b \cdot \text{LF}^2)$$

where   LF = annual load factor
   a and b = constants

An example of results for a densely populated city 132/33/11/0.4 kV reticulation system in a developing country is shown in Table 26.5.

Table 26.5   Urban system load factor for a developing country

System voltage level (kV)	Annual load factor	Annual loss load factor (constants: a = 0.3 and b = 0.7)
132		
132/33	0.53	0.36
33		
33/11	0.47	0.30
11		
11/0.4	0.376	0.21
0.4		

26.8.3.4 System power factor

Each part of the network will have an inherent power factor arising from the line, cable and transformer impedance together with the major contribution from the load itself. The current flowing in the network is inversely proportional to the power factor. Since the major feeder losses are proportional to the square of the current, the system losses will also be proportional to the inverse of the square of the power factor. For example, if the load power factor is 0.5, the associated losses will be $(1/0.5)^2$ or 4 times the loss that would arise with a unity power factor. Therefore a low power factor will not only be a cause of poor system voltage regulation but will also significantly contribute to distribution system losses and associated cost.

26.8.3.5 Power factor correction methodology

To correct a given power factor from $\cos \varphi_1$ to an improved power factor $\cos \varphi_2$, at an active power, P (kW), requires a capacitor of reactive power rating:

$$Q_C = P(\tan \varphi_1 - \tan \varphi_2) \, \text{kVAr}$$

(See Fig. 26.23.)
Low power factor may be dealt with by:

- correction by the manufacturers of electrical apparatus,
- correction by the supply utility,
- correction by the consumer,

Correction by the consumer has the great advantage of correcting the power factor nearest to the source which enables the widest possible benefit to be achieved. In addition, it puts the onus on the 'bad' consumer to pay for the correction of his own loads. This is achieved by a combination of legal conditions of supply and by tariff penalties. In practice, the cost of installing special

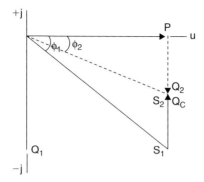

Figure 26.23   Power factor improvement from $\cos\phi_1$ to $\cos\phi_2$

metering to measure power factor plus the cost of reading such meters and computing the penalty usually means that this approach is limited to the larger commercial or industrial consumers. Generally, figures in the range of 0.85 to 0.90 pf are taken as acceptable power factor levels before penalties are applied. It is important not to overcompensate when applying power factor correction capacitors since this could lead to an increase in receiving end voltage outside desirable limits. Typically, a power factor of approximately 0.95 is an acceptable limit to aim for.

Correction by the manufacturers and importers of electrical apparatus works well in countries with an adequate system of legal controls and customs regulations. This form of control is most effective with domestic appliances (lights, fans, air-conditioning, refrigerators, etc.) where legal requirements for correction to, say, 0.9 pf can be enforced. The process is less effective for large motor drives where the running power factor depends as much on the application of the machine as the machine itself.

In countries with long transmission and distribution lines and difficulties in the application of legal controls the electrical supply utility often has to take the appropriate power factor correction measures. Where energy metering is installed the capacitive correction can be derived in accordance with the formula:

$$Q_C = (\text{kVArh} - \text{kWh} \times \tan\phi_2)/t$$

where t is the integrated operating time and kVArh and kWh are the reactive and active metering information.

Figure 26.24 shows the basic principles involved for an economic assessment of capacitive correction. As power factor is improved by the addition of capacitor kVAr, the net load power factor improves and the current drawn by the load decreases which, in turn, reduces system losses. The cost of the capacitors is proportional to the kVAr installed. This analysis is a simplification (which allows direct use of the tariff rates in force by the utility) since it assumes that the cost of energy can be expressed as a single figure of

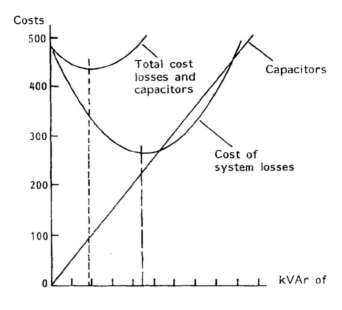

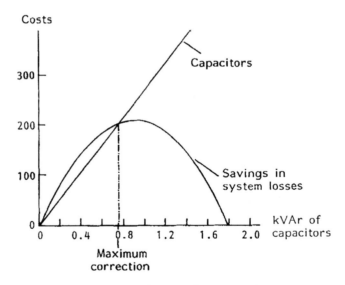

Figure 26.24   Application of capacitors for power factor correction

price per kWh. In practical terms the cost of losses is made up from two components:

1. Demand power. The sum of the incremental discounted long range marginal costs over the lifetimes of capital generation, transmission and distribution plant (£/kW/year). Economic lifetimes for plant are open for debate but

as long as the same times are used for a variety of scenarios a good comparison is possible. Typical orders could be:
- generation plant – 25 years
- transmission plant – 40 years
- distribution plant – 30 years.

2. Energy costs. The unit costs weighted to reflect the fact that maximum losses occur only during peak demand and taken as average marginal energy costs (£/kWh) over a year. Also losses are quoted both in energy and power terms; the former being the most common since it can be readily calculated by deduction of sales from energy generated. Losses not associated with the load losses should be deducted (e.g. power station auxiliary energy loss, non-technical losses, no-load transformer and iron losses, etc.) in the analysis. The results of such a simplified analysis will give indicative rather than definitive guidance taking into account the economics involved.

Section 26.8.6 gives more details of economic power factor correction calculation procedures.

26.8.3.6 *Loss evaluation*

In practice, the supply network will consist of a mixture of plain feeders, spurs and regular load tappings. It will have provision for spare capacity, include duplication for security reasons, and will have been designed for voltage regulation as well as to thermal ratings. Experience and judgement are necessary to leave margins for planned and ultimate loadings, especially in developing countries. Therefore universal guidelines for the selection of feeders on a minimum economic cost basis are difficult to establish but the following procedure using spreadsheet software is useful:

- Select the conductor.
- Establish the economic loading.
- Take into account regular load distribution along feeders to give a revised economic loading ($x \approx 1.7$).
- Check voltage regulation at new economic loading.
- If the voltage regulation is within limits and the planned load is less than the economic loading, the selected conductor is satisfactory subject to a thermal rating check.
- Multiply planned loading by an appropriate factor:
  $x \approx 2$ for duplicated feeders or open rings
  $x \approx 1.5$ to 1.7 for networks with additional interconnections to alternative sources of supply.
  Check that this loading is below the thermal rating.
- Select next larger conductor size and repeat process if initial selection fails this procedure.

Table 26.6   Cable capital costs and ratings

6.35/11 kV cable type	Capital cost	Annual cost/km discounted at 15% for 20 years
35 mm^2, 3 core, Al conductor	£6096/km	£976
95 mm^2, 3 core, Al conductor	£12800/km	£2045
185 mm^2, 3 core, Al conductor	£16634/km	£2657
185 mm^2, 3 core, Cu conductor	£24855/km	£3971

6.35/11 kV cable type	Current rating (@ 30°C)	MVA rating (@ 11 kV)	Conductor resistance (Ω/km)	Conductor reactance (Ω/km)
35 mm^2, 3 core, Al conductor	110	2.096	1.12	0.114
95 mm^2, 3 core, Al conductor	200	3.811	0.411	0.106
185 mm^2, 3 core, Al conductor	280	5.335	0.211	0.0977
185 mm^2, 3 core, Cu conductor	350	4.668	0.128	0.0977

Table 26.7   Cable losses

6.35/11 kV cable type	Cable losses (kW/km) 0.5 MVA through load transfer	Cable losses (kW/km) 1.0 MVA through load transfer	Cable losses (kW/km) 2.0 MVA through load transfer	Cable losses (kW/km) 3.0 MVA through load transfer	Cable losses (kW/km) 4.0 MVA through load transfer
35 mm^2, 3 core, Al conductor	2.3	9.3	37.0	83.3	148.1
95 mm^2, 3 core, Al conductor	0.8	3.4	13.7	30.6	54.8
185 mm^2, 3 core, Al conductor	0.4	1.7	7.0	15.7	27.9
185 mm^2, 3 core, Cu conductor	0.3	1.1	4.2	9.5	16.9

26.8.3.7 MV underground cable loss optimization example

1. Select cables taking into account costs and ratings (see Table 26.6).
2. Calculate the losses ($3 \times I^2 R$) arising on each cable size per km for fixed increments of full load transfer (see Table 26.7).
3. Compute the cost of cable capital and losses to transfer increments of through load power per km per annum (see Table 26.8).

    A range of costs associated with the cable losses have been shown in the tables derived from:

    - The discounted incremental capital costs plus the economic marginal unit cost of loss energy.
    - Actual tariff information. In some developing countries the cost the consumer is asked to pay for electricity may not be based on purely financial considerations.

Table 26.8   Total annual cable costs

6.35/11 kV cable type and 0.5 MVA load transfer	Cable losses (kW/km)	Cost of losses (£/kW/km/annum) High	Low	Cost of capital (£/annum)	Total cost (£/annum) High	Low
35 mm², 3 core, Al conductor	2.3	388	247	976	1344	1223
95 mm², 3 core, Al conductor	0.8	135	91	2045	2180	2136
185 mm², 3 core, Al conductor	0.4	69	47	2657	2726	2704
185 mm², 3 core Cu conductor	0.3	42	28	3971	4013	3999

6.35/11 kV cable type and 1.0 MVA load transfer	Cable losses (kW/km)	Cost of losses (£/kW/km/annum) High	Low	Cost of capital (£/annum)	Total cost (£/annum) High	Low
35 mm², 3 core, Al conductor	9.3	1474	988	976	2450	1964
95 mm², 3 core, Al conductor	3.4	541	363	2045	2586	2408
185 mm², 3 core, Al conductor	1.7	278	186	2657	2935	2843
185 mm², 3 core Cu conductor	1.1	168	113	3971	4139	4084

6.35/11 kV cable type and 1.0 MVA load transfer	Cable losses (kW/km)	Cost of losses (£/kW/km/annum) High	Low	Cost of capital (£/annum)	Total cost (£/annum) High	Low
35 mm², 3 core, Al conductor	37.0	5895	3952	976	6871	4928
95 mm², 3 core, Al conductor	13.7	2180	1450	2045	4225	3495
185 mm², 3 core, Al conductor	7.0	1111	744	2657	3768	3401
185 mm², 3 core, Cu conductor	4.2	674	453	3971	4645	4423

6.35/11 kV cable type and 3.0 MVA load transfer	Cable losses (kW/km)	Cost of losses (£/kW/km/annum) High	Low	Cost of capital (£/annum)	Total cost (£/annum) High	Low
35 mm², 3 core, Al conductor	–	–	–	976	–	
95 mm², 3 core, Al conductor	30.6	4868	3263	2045	6913	5308
185 mm², 3 core, Al conductor	15.7	2499	1675	2657	5156	4332
185 mm², 3 core, Cu conductor	9.5	1516	1016	3971	5487	4987

6.35/11 kV cable type and 3.0 MVA load transfer	Cable losses (kW/km)	Cost of losses (£/kW/km/annum) High	Low	Cost of capital (£/annum)	Total cost (£/annum) High	Low
35 mm², 3 core, Al conductor	–	–	–	976	–	
95 mm², 3 core, Al conductor	54.8	8721	5801	2045	10 766	7846
185 mm², 3 core, Al conductor	27.9	4443	2978	2657	7100	5635
185 mm², 3 core, Cu conductor	16.9	2695	1807	3971	6666	5778

4. The results from this particular study, for the given tariff rates, discount factors and capital costs, indicate that the I^2R cable losses are relatively insignificant for practical cable loadings compared to the capital costs of the cable itself. It should be noted that an overhead line distribution system would

Table 26.9 Comparison of cable ratings

6.35/11 kV cable	Cable thermal M VA rating (@ 11 kV)	Economic optimum rating (MVA)		Economic rating % of thermal rating	
		High	Low	High	Low
35 mm², 3 core, Al conductor	2.1	0.9	0.9	42	42
95 mm², 3 core, Al conductor	3.8	1.3	1.4	35	41
185 mm², 3 core, Al conductor	5.3	2.5	2.6	47	49
185 mm², 3 core, Cu conductor	4.7	4.4	4.6	66	69

involve capital costs approximately a third of the equivalent underground cable costs and consequently the losses would have more significance. The economic optimum loadings generated by such techniques may be compared with the cable thermal ratings as shown in Table 26.9.

5. The conclusions from the study are shown in graphical format in Fig. 26.25 giving the total costs of transmitting fixed levels of load through the different cable sizes considered over the high and low cost range. Because of the high capital costs of cables it is economic to load cables close to their thermal ratings.

26.8.4 Communication link gain or attenuation

The power gain or attenuation of a communication link or sound levels is often measured in terms of decibels (dB), being one-tenth of a Bel:

$$\text{Gain or attenuation measured in dB} = \pm 10 \log_{10}(P_2/P_1)$$

The reference power unit has to be specified. For example, in a fibre optic link this may use a reference relative to 1 mW or 1 μW and the overall system gain or attenuation would then be referred to in terms of dBm or dBμ. The advantage of using logarithmic scales is that the gain or attenuation of the individual components of an overall communication link may be numerically added to derive the overall system gain or attenuation effect.

Since power, P, is proportional to the square of the voltage, U^2 ($P = U^2/R$) then for cases where a common resistive impedance ($R_1 = R_2$) is used:

$$\text{Gain or attenuation measured in dB} = \pm 10 \log_{10}(U_2^2/U_1^2)$$
$$= \pm 20 \log_{10}(U_2/U_1)$$

Table 26.10 gives conversions between power or voltage gain and dB.

If the impedance matching values are different ($R_1 \neq R_2$) then:

$$\text{Gain measured in dB} = 20 \log_{10}(U_2/U_1) - 10 \log_{10}(R_2/R_1)$$

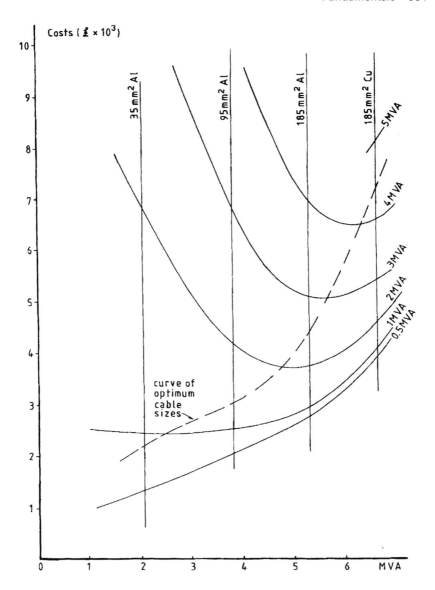

Figure 26.25 Total costs of transmitting fixed levels of load through different sizes of 11 kV cable (higher energy costs)

26.8.5 Reactive compensation

The apparent power $S(VA)$ consumed by a load or in a network is made up of real power P (W) and reactive power Q (VAr) such that $S = P + jQ$. The convention is for the reactive power, Q to be positive for an inductive load or lagging power factor condition and negative for a capacitive load or leading power

Table 26.10   Relative gain in dB for different power and voltage ratios

Power ratio	Voltage ratio	Gain (dB)
1.00	1.00	0.00
1.50	1.22	1.76
2.00	1.41	3.01
2.50	1.58	3.98
3.00	1.73	4.77
3.50	1.87	5.44
4.00	2.00	6.02
4.50	2.12	6.53
5.00	2.24	6.99
5.50	2.35	7.40
6.00	2.45	7.78
6.50	2.55	8.13
7.00	2.65	8.45
7.50	2.74	8.75
8.00	2.83	9.03
8.50	2.92	9.29
9.00	3.00	9.54
9.50	3.08	9.78
10.00	3.16	10.00
50.00	7.07	16.99
100.00	10.00	20.00
10^3	31.62	30.00
10^4	100.00	40.00
10^5	316.23	50.00
10^6	1000.00	60.00

factor condition. Thus an inductive reactor is said to absorb VArs and a capacitor to generate VArs. Chapters 1 and 12 cover examples of the addition of reactive components into networks to improve system stability, voltage regulation or power factor. The addition of such components allows the economic transfer of power, optimizes plant utilization, improves transient stability and assists in system voltage control.

Shunt capacitor banks may be added to networks to compensate for the mainly inductive voltage drop over long heavily loaded lines. Some switching arrangements are usually required since such capacitor banks may result in higher than required receiving end voltages under lightly loaded conditions. The special considerations to be taken into account when switching capacitor banks are covered in Chapter 13.

Shunt reactors may also be used for voltage regulation purposes. For example, the load on a network in the Middle East or the USA, say, may considerably increase during the Summer months because of an increase in the usage of air conditioners which could well form a major part of the total connected system load. During the lightly loaded Winter months the reactors may be used to maintain voltage levels within specified limits.

Under normal conditions in a well-designed network the VAr compensation requirements will change relatively slowly and will not be sensitive to small load changes. Reactive compensation may be achieved in such cases with switched

shunt reactor or capacitor banks. In addition to the mechanical switching of reactive components variable control of shunt reactors and stepped control of capacitors are now feasible using modern thyristor power electronic devices (static VAr
compensation, SVC). Such systems improve the VAr control and reduce unnecessary system losses. The capacitors generate VArs and the reactors absorb VArs
and to operate in both generation and absorbtion modes at least one of banks
must be variably controlled. This is normally achieved by variable reactor thyristor control and mechanical- or thyristor-switched capacitor control.

Thyristor-switched capacitors are characterized by:

- stepwise control;
- short switching delay (often within 1/2 to 1 cycle);
- low (controlled) inrush transients;
- low losses at low VAr output;
- minimum harmonic generation.

Thyristor controlled reactors are characterized by:

- continuous control;
- short operating delay (less than 1/2 cycle);
- low transients;
- harmonic generation (due to fast switching) which requires the addition of
  filters. Note that the capacitor arms of a combined reactor and capacitor
  bank may be tuned with series inductive reactance to act in a secondary role
  as filter components.

Figure 26.26a shows one phase arm of a rather specialized application of these
principles as used for the balancing of a single phase traction load onto a three
phase supply. The current in the reactor arm may be phase angle controlled by
back-to-back thyristors thus effectively altering the total reactance in circuit and
the absorbed VAr. The generated VAr contribution is achieved from the capacitor arms which are stepped controlled by thyristors switching in or out the fixed
capacitors. The combination of stepped VAr generation and continuously variable VAr absorption may be arranged to give an overall continuously variable
static VAr compensation over the range required. Figure 26.26b shows the output from such a Static VAr Compensator (SVC) arrangement which is characterized by:

- continuous control;
- practically no transients;
- low generation of harmonics (using the capacitor arms as filter elements);
- low losses;
- flexible control and operation.

Separately from reactive compensation series reactors may also be added into
networks to reduce the fault levels downstream of the reactors. The introduction of such series reactors into the network may be cheaper than having to

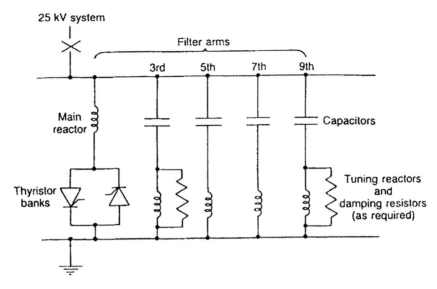

Figure 26.26(a) Diagram of one phase of a balancer

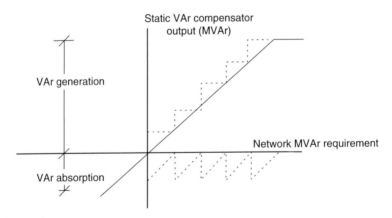

Figure 26.26(b) Explanation of the principle of continuously variable static VAr compensation (SVC) as seen in one phase. The dotted steps correspond to the thyristor-switched capacitor VAr generation which is not continuously controlled but switched in stages either in or out of circuit. Since the reactor arm is continuously controlled the resultant SVC output is continuous.

replace switchgear with equipment of higher fault ratings. The sizing of such reactors may be derived using the formulae given in Section 26.7.

26.8.6 Power factor correction calculation procedures

The calculation of the optimum technical and economic power factor correction capacitor size by a distribution utility to improve power factor, assist voltage regulation and reduce technical losses as outlined in Section 26.8.3.4 may be based on the following formulae and methodology:

1. Calculate the following variables:

$$V = L \times E \times F \times 8760 \times [1 - (1 + I)^{-Y}]/I$$
$$K = 2 \times \text{PF} \times (1 - \text{PF}^2)^{1/2}$$

where

Variable	Units	Description
PF	Power factor (cos ϕ)	Average load power factor
L	Per unit (pu)	Net losses on the system (load losses to the point of connection of correction capacitors)
F	–	Load loss factor (at the capacitor connection point)
I	pu	Interest (or discount) rate
Y	Years	Estimated life of capacitor installation
E	£/kWh	Energy cost at capacitor connection point
C	£/kVAr	Installed cost of capacitors
V	£/kW (load)	Capitalized value of losses on the existing system

2. Calculate the quantity of capacitive compensation:
   (a) Optimum correction:

Optimum capacitive kVAr per kW of load $= V \times K - C/(2 \times V \times \text{PF}^2)$

New power factor $= \dfrac{1}{[(C/(2 \times V \times \text{PF}^2))^2 + 1]^{1/2}}$

   (b) Maximum correction:

Evaluate $V \times K/2$ and compare with C

If $C > V \times K/2$ then calculate the maximum correction as given below
If $C < V \times K/2$ then correct to near unity power factor

Maximum capacitive kVAr per kW of load $= V \times K - C/(V \times \text{PF}^2)$

New power factor $= \dfrac{1}{[(C/(V \times \text{PF}^2 - K/(2 \times \text{PF}^2))^2 + 1]^{1/2}}$

Notes:
(a) In some cases the price of capacitors may be too high in relation to the value of loss reduction to make it worthwhile to use this method of loss reduction. This will occur if the installed cost of capacitors (C) exceeds the product $V \times K$.
(b) In a system with high losses and with cheap capacitors, near 100% power factor correction is possible with a possible positive net return on investment. The capacitor cost (C) will then be less than the product $(V \times K)/2$. However, the optimum correction formula will give realistic results and assist in the determination of the desired level of correction.
(c) If the price of capacitors falls within the range $(V \times K)$ and $(V \times K)/2$ then more capacitive correction could be applied above that given by the optimum formula in 2(a) (up to the limit given by the maximum formula 2(b)) and still give a net saving in costs. This could be desirable if the capacitors also give other benefits such as assisting voltage regulation.

3. Some typical examples:
A spreadsheet (Table 26.11) may be used to set up different values for the variables described in 1. above and then used to plot the graphs given in Fig. 26.24.

Table 26.11

Variable	Calculation 1	Calculation 2	Calculation 3	Units
PF	0.75	0.85	0.85	Power factor ($\cos \phi$)
L	0.09	0.09	0.12	Per unit (pu)
F	0.4	0.4	0.4	–
I	0.12	0.12	0.12	pu
Y	10	10	10	Years
E	0.027	0.027	0.027	£/kWh
C	27	27	33	£/kVAr
V^*	47.5	47.5	63.4	£/kW (load)
K^*	0.9922	0.8955	0.8955	$2 \times PF \times (1 - PF^2)^{1/2}$
$Q_c\%$ (optimum)	0.38	0.23	0.26	IVAr per kW load-optimum correction
pf* (resultant new optimum)	0.9	0.93	0.94	Power factor ($\cos \varphi$) (optimum conditions)
Q_c^* (maximum)[a]	0.77 $Q_c = V \times K - C/(V \times PF^2)$	0.46 $Q_c = V \times K - C/ (V \times PF^2)$	0.51 $Q_c = V \times K - C/ (V \times PF^2)$	kVAr per kW load-maximum correction
pf* (resultant new maximum)	1.0	0.99	0.99	Power factor ($\cos \phi$) (maximum conditions)

Notes: *calculated values
[a] For maximum correction evaluate $V \times K/2$ – if $C > V \times K/2$ then calculate result using maximum correction formula – if $C < V \times K/2$ then calculate to near unity power factor.

4. Capacitor values:
Capacitors may be used on single or three phase circuits where they may be connected in either star or delta configurations with the following capacitor values (see Fig. 26.27):

(a) Single phase:

$$Q_C = U_C^2 \times 2\pi f \times C \times 10^{-3}$$
$$I_C = Q_C/U_C$$
$$X_C = U_C \times 10^3/I_C$$

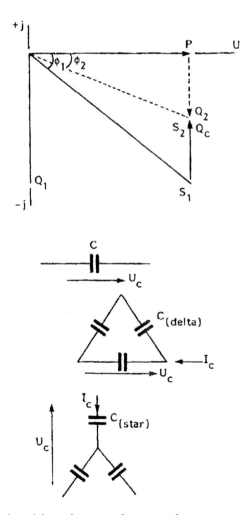

Figure 26.27   Single and three phase capacitor connections

(b) Three phase delta connection

$$Q_C = 3 \times U_C^2 \times 2\pi f \times C_{(delta)} \times 10^{-3}$$
$$I_C = Q_C/\sqrt{3} \times U_C$$
$$X_C = \sqrt{3} \times U_C \times 10^3/I_C$$

(c) Three phase star connection

$$Q_C = U_C^2 \times 2\pi f \times C_{(star)} \times 10^{-3}$$
$$I_C = Q_C/\sqrt{3} \times U_C$$
$$X_C = U_C \times 10^3/\sqrt{3} \times I_C$$

where Q_C = capacitor rating (kVAr)
U_C = voltage across capacitor (V)
I_C = current through capacitor (A)
f = operating frequency (Hz)
X_C = capacitive reactance (Ω)
C = capacitance (μF)

For a given reactive capacitance, Q_C (kVAr), the three phase delta connection is advantageous since $C = C_{(star)} = 3 \times C_{(delta)}$. Also, operation at supply frequency or voltage different from those quoted as the rated capacitor values will require a correction to the capacitor ratings.

REFERENCES

General

1. V. Rojansky, *Electromagnetic Fields and Waves*, Dover Publications Inc., New York.
2. *Electrical Engineering Handbook*, Siemens Publications.
3. M. Laughton and M. Say, *Electrical Engineer's Reference Book*, Butterworth Heinemann.
4. D. Fink and H. Wayne Beaty, *Standard Handbook for Electrical Engineers*, McGraw Hill.
5. J. Edminister, *Electrical Circuits*, Schaum's Outline Series, McGraw Hill.
6. Dr Ing Richard Ernst, *Comprehensive Dictionary of Engineering and Technology*, Oscar Brandstetter Verlag, Berlin.
7. *IEC Multilingual Dictionary of Electricity, Electronics and Telecommunications*, Sales Dept., P.O. Box 131, 3 rue de Varembe, 1211 Geneva 20, Switzerland.
8. *Catalogue* of *I EC Publications*, Sales Dept., P.O. Box 131, 3 rue de Varembe, 1211 Geneva 20, Switzerland.
9. O. Guthmann et al., *Switchgear Manual*, Brown Boveri & Cie, Aktiengesellschaft, Mannheim, Germany.

10. S. Austen Stigant and A. C. Franklin, *The J&P Transformer Book*, Newnes-Butterworths.
11. R. T. Lythall, *The J&P Transformer Book*, Newnes-Butterworths.
12. E. W. G. Bungay and D. McAllister, *BICC Electric Cables Handbook*, BSP Professional Books.
13. M. O'C. Horgan, *Competitive Tendering for Engineering Contracts*, E&F.N. Spon Ltd.
14. Russell D. Archibald, *Managing High-Technology Programs and Projects*, John Wiley & Sons.

Transmission networks

1. C. A. Gross, *Power System Analysis*, John Wiley & Sons.
2. Turan Gonen, *Electrical Power Distribution System Engineering*, McGraw Hill.
3. *Electrical Transmission and Distribution Reference Book*, Westinghouse Publications.
4. *Electrical Utility Engineering Reference Book – Distribution Systems*, Westinghouse Publications.
5. *Protective Relays Application Guide*, GEC Measurements, St. Leonard's Works, Stafford, England.
6. *IEEE Guide for Safety in AC Substation Grounding*, IEEE Power Engineering Society approved by the American National Standards Institute.
7. H. W. Denny, *Grounding for the Control of EMI*, Don White Consultants Inc., State Route 625, P.O. Box D, Gainesville, Virginia 22065, USA.
8. A. A. Smith Jr., *Coupling of External Electromagnetic Fields to Transmission Lines*, Interference Control Technologies, Inc.
9. Reinhold Rudenberg, *Transient Performances of Electrical Power Systems*, MIT Press.

Consumer networks (industrial, domestic and general purpose)

1. I. Lazar, *Electrical Systems Analysis and Design*, McGraw Hill.
2. G. G. Seip, *Electrical Installations Handbook – Parts* 1, 2 & 3, Siemens Aktiengesellschaft, John Wiley & Sons.

Applied mathematics for electrical networks

1. K. Arbenz and A. Wohlhauser, *Advanced Mathematics for Practising Engineers*, Artech House Inc.
2. K. Arbenz and J. C. Martin, *Mathematical Methods of Information Transmission*, Artech House Inc.

3. O. C. Zienkiewicz and K. Morgan, *Finite Elements and Approximation*, John Wiley & Sons.

Economics

1. T. W. Berrie, *Power System Economics*, IEE Power Engineering Series 5.
2. L. A. Williams, *Microcomputers and Marketing Decisions*, IEE Management and Technology Series 5.

Computing methods related to electrical engineering

1. Glenn W. Stagg and Ahmed H. El Abiad, *Computer Methods in Power Analysis*, McGraw Hill International Editions.
2. Herman Kremer, *Numerical Analysis of Linear Networks and Systems*, Artech House Inc.
3. J. Arrillaga, C. P. Arnold and B. J. Harker, *Computer Modelling of Electrical Power Systems*, John Wiley & Sons.
4. Ruckdeschel, *Basic Scientific Subroutines*, Vols 1 & 2, McGraw Hill Book Company.
5. D. Ellison and J. C. T. Wilson, *How to Write Simulations Using Microcomputers*, McGraw Hill Book Company.
6. F. Neelamkavil, *Computer Simulation and Modelling*, John Wiley & Sons.
7. J. R. Leigh, *Modelling and Simulation*, IEE Topics in Control Series 1.
8. W. R. Benett Jr., *Scientific and Engineering Problems Solving with the Computer*, Prentice-Hall.

Index